The Correspondence of MICHAEL FARADAY

Volume 4

Plate 1. Matthew Noble's 1854 bust of Faraday, presented to the Royal Institution by the Executors of James Walker. See letter 2964.

The Correspondence of MICHAEL FARADAY

Volume 4

January 1849-October 1855

Letters 2146-3032

Edited by
Frank A J L James

Published by the Institution of Electrical Engineers

Published by: The Institution of Electrical Engineers, London, United Kingdom

© 1999: Selection and editorial material, The Institution of Electrical Engineers

This publication is copyright under the Berne Convention and the Universal Copyright Convention. All rights reserved. Apart from any fair dealing for the purposes of research or private study, or criticism or review, as permitted under the Copyright, Designs and Patents Act, 1988, this publication may be reproduced, stored or transmitted, in any forms or by any means, only with the prior permission in writing of the publishers, or in the case of reprographic reproduction in accordance with the terms of licences issued by the Copyright Licensing Agency. Inquiries concerning reproduction outside those terms should be sent to the publishers at the undermentioned address:

The Institution of Electrical Engineers,
Michael Faraday House,
Six Hills Way, Stevenage,
Herts. SG1 2AY, United Kingdom

While the editor and the publishers believe that the information and guidance given in this work are correct, all parties must rely upon their own skill and judgment when making use of them. Neither the editor nor the publishers assume any liability to anyone for any loss or damage caused by any error or omission in the work, whether such error or omission is the result of negligence or any other cause. Any and all such liability is disclaimed.

The moral right of the authors to be identified as authors of this work has been asserted by them in accordance with the Copyright, Designs and Patents Act 1988.

British Library Cataloguing in Publication Data

A CIP catalogue record for this book
is available from the British Library

ISBN 0 86341 251 3

Printed in England by Short Run Press Ltd., Exeter

Contents

Plates	vii
Acknowledgements	ix
Editorial Procedure and Abbreviations	xv
Note on Sources	xxv
Introduction	xxvii
Biographical Register	li
The Correspondence	1
Previous Publication of Letters	909
Bibliography	917
Index	947

To Jozefa Hermaszewska, 'Babcia', for being such a source of support and help over the years

Plates

1. Matthew Noble's 1854 bust of Faraday. Dust jacket and frontispiece
2. Emil Heinrich du Bois-Reymond. 78
3. Harriet Jane Moore. 101
4. Friedrich Wilhelm Heinrich Alexander von Humboldt. 172
5. Lambert-Adolphe-Jacques Quetelet by Jean-Baptiste Madou. 182
6. Faraday by Thomas Herbert Maguire. 307
7. Prince Albert. 341
8. Faraday in his laboratory by Harriet Jane Moore, 1852. 356
9. Faraday by George Richmond, 1852. 373
10. Dominique François Jean Arago. 576
11. Macedonio Melloni. 597
12. Faraday giving his card to Father Thames. 883

Acknowledgements

I acknowledge the gracious permission of Her Majesty the Queen to publish the letter in the Royal Archives, Windsor Castle, and to consult the diary of Queen Victoria.

It is with great pleasure and gratitude that I again thank the Institution of Electrical Engineers for the financial support without which the project to locate, copy and edit all extant letters to and from Faraday would not have been possible. Furthermore, I thank them for the support which made possible the publication of this volume. I am grateful to the Royal Institution for the provision of all the essential support they have given for this work and to my friends and colleagues there for their unceasing support and interest. It is also a pleasure to acknowledge the support of the British Academy for a grant which supported Ms Marysia Hermaszewska to make the initial transcriptions of the letters in the Guildhall Library and also those for 1855.

I thank the following institutions and individuals for permission to publish the letters to and from Faraday which are in their possession: The Director of the Royal Institution (and for plates 1-11); the Institution of Electrical Engineers; the Elder Brethren of Trinity House for the letters in the Guildhall Library; the Syndics of Cambridge University Library and, for the letters in the archives of the Royal Greenwich Observatory, the Director of the Royal Greenwich Observatory; the President and Council of the Royal Society; the Oeffentliche Bibliothek der Universität Basle; the Master and Fellows of Trinity College Cambridge; Smithsonian Institution Libraries, Washington; the British Library Manuscript Department; Joan Ferguson; the Bibliothèque Publique et Universitaire, Geneva; the Bodleian Library, Oxford, the British Association for the letter deposited in its collection and the Earl of Lytton for the letter deposited in the Lovelace-Byron collection; Mrs Elizabeth M. Milton; the Archives of the Science Museum Library, London; National Research Council Canada; the American Philosophical Society Library, Philadelphia; the Trustees of the British Museum; Mr Dennis Embleton; the Trustees of the National Library of Scotland; the Berkshire Record Office; the Archives de l'Académie des Sciences de Paris; Professor Kathleen Tillotson and Mr Graham Story for the permission of Mr Christopher Dickens to publish the letters of Charles Dickens; Mr W.A.F. Burdett-Coutts; the Trustees of the Wellcome Trust; University College London Library; the Handschriftenabteilung, Staatsbibliothek Preussischer Kulturbesitz, Berlin; St Andrews University Library; the Department of Rare Books and Special Collections, McGill University Libraries and the Chemistry Department of McGill University, Montreal; the John Rylands University Library of Manchester; the College Archives, Imperial College of Science, Technology and Medicine, London; the Houghton Library, Harvard University; the Francis A. Countway Library of Medicine, Boston; the Librarian of the University of Bristol; Académie Royale des Sciences, des Lettres et des Beaux-Arts de Belgique; Rosalind Brennand; Drs Günther and Rosemarie Gerisch; Professor George W. Platzman; Department of Special Collections, University Research Library, University of California, Los Angeles; the Sidney M. Edelstein Library, the Hebrew University, Jerusalem; Department of Geology, National Museum of Wales; Hollandsche Maatschappij der Wetenschappen; Haverford College Library; Special Collections Library, Duke University; the Archives, California Institute of Technology; the Bibliothèque Nationale, Paris; the Royal Society for the encouragement of Arts, Manufactures and Commerce; Bayerische Staatsbibliothek; Biblioteca dell'Archiginnasio, Comune di Bologna; Birmingham Public Library; the University of Birmingham; the Trustees of the Boston Public Library; the Mugar Memorial Library, Boston University; British Geological Survey Archives; the Burndy Library, Massachusetts Institute of Technology; C.J. Kershaw; the Rare Book and Manuscript Library, Columbia University Library; Cumbria Record Office (Carlisle); Miss Lois Hodgkin for the letter in Durham County Record Office; Mrs Elizabeth Faraday Baird; Håndskriftafdelingen, Det Kongelige Bibliothek, Copenhagen; Herbert T. Pratt; the Historical Society of Pennsylvania; the Hydrographic Office, Taunton; Professor Jorge C.G. Calado; Herr K.W. Vincentz; Special Collections and Archives, Knox College Library, Galesburg; Kunstsammlungen der Veste Coburg; Lacock Abbey Collection, National Trust Fox Talbot Museum; Liebig Museum, Giessen; the Maddison Collection, Templeman Library, University of Kent at Canterbury; the President and Fellows of Magdalen College, Oxford; the City of Manchester Arts and Leisure Committee; Massachusetts Historical Society; the late Mr and Mrs S. Aida; Mrs Raven Frankland; Museo di Storia della Fotografia Fratelli Alinari; Museum of History of Science, Oxford; the National Library of New Zealand; Heinz Archive and Library, National Portrait Gallery; the British Museum (Natural History); the Rare Books and Manuscripts Division, New York Public Library; University of Newcastle upon Tyne Library; Niedersächsische Staat- und Universitätsbibliothek Göttingen; Northeastern Science Foundation, Brooklyn College of the City University of New York; the Peirpont Morgan Library, New York; the Queen's University of Belfast; Rijksarchief in Noord-Holland; Mr Roy

Deeley; Dr Roy G. Neville of the Roy G. Neville Historical Chemical Library, California; the Director of the Royal Botanic Gardens, Kew; Manuscripts and Archive Collection, British Architectural Library, Royal Institute of British Architects; Library and Information Centre, Royal Society of Chemistry; the Royal Society of South Africa; the Somerset Record Office; Southwark Local Studies Library; St Bride Printing Library; Department of Special Collections, Stanford University Libraries; the Royal Swedish Academy of Sciences for the letter in the Stockholms Universitetsbibliotek; the Sutro Library, San Francisco; The Birr Scientific and Heritage Foundation, courtesy of the Earl of Rosse; Torquay Natural History Society; the University Museum, Oxford; the Harry Ransom Humanities Research Center, University of Texas at Austin; the Handskriftsavdelningen, Uppsala Universitetsbibliotek; Special Collections, Vassar College Libraries, Poughkeepsie, New York; Wayne Lee Radziminski; Whitby Literary and Philosophical Society; Dr Y. Watanabe; the Manuscripts and Archives, Yale University Library. All Crown copyright material in the Public Record Office and elsewhere is reproduced by permission of the Controller of Her Majesty's Stationery Office.

I wish to thank the staff of all the institutions listed above for helping me locate the letters to and from Faraday in their possession and in most cases providing me with photocopies and answering follow up questions. Particular thanks should go to Mrs Irena McCabe and latterly my staff at the Royal Institution, Mrs E.D.P. Symons and her staff at the Institution of Electrical Engineers, Ms Mary Nixon and her staff at the Royal Society and Mr Adam Perkins of Cambridge University Library.

Although the following institutions do not have any Faraday letters in their archives which are published in this volume, I thank them for answering queries concerning letters in this volume: the Athenaeum Club, Surrey Record Office, Friends' House, the Archives of the Institution of Civil Engineers, Scottish Record Office, the Principal Registry of the Family Division of the High Court (in High Holborn), the General Register Office (in the Family Records Centre), the Royal Astronomical Society, the London Library, Westminster Record Office, Doctor Williams's Library, the Library of the Society of Antiquaries, Kingston Record Office and last, but not least, the National Register of Archives whose resources once again directed my attention to the location of many letters.

Many friends and colleagues have helped in locating letters and dealing with queries and I wish to thank the following particularly: Dr Mari E.W. Williams, Professor John Krige, Professor Raymond Klibansky, Dr Anita McConnell, Ms Anne Secord and Dr Shigeo Sugiyama for doing the initial ground work of locating Faraday letters in Paris, Geneva, Montreal, Bologna, Cambridge and Japan respectively. I also thank Professor Colin Russell for facilitating access to the Frankland archive. I thank Dr J.V. Field (for help with the Greek and Latin and for answering various art historical and other queries), Mr J.B. Morrell (for help relating to the British Association and John Phillips), Dr M.B. Hall (for advice and information on the Royal Society), Professor A.R. Hall (for references to various pieces of Newtoniana), Mr Geoffrey King (who placed the almost the entire London Sandemanian community on his genealogy programme thereby saving me an immense amount of tedious work in sorting out manually the relations between individuals), Dr Willem Hackmann (for discussions on nineteenth century electricity), Dr Simon Schaffer and Dr Robert Iliffe (for information on eighteenth century natural philosophy), Dr Gloria Clifton (for information on the scientific instrument trade), Herr Michael Barth (for translating the letters from German), Dr Larry J. Schaaf (for discussions on the history of photography), Professor W.H. Brock (for help with Liebig related queries), Dr Allan Chapman (for discussions on nineteenth century astronomy), Commander A.E. Fanning (for help with compass technology), Mr James Hamilton (for discussions on nineteenth century science and art), Dr Peter Nolte and Dr Ulf Bossel (for information about Schoenbein and his family), Dr G.M. Prescott Nuding (for discussions of Faraday's iconography), Professor Ryan Tweney (for many discussions on Faraday's work) and Professor David Knight (who acted as a court of final appeal for many obscurities). Finally, in this paragraph, I should like to acknowledge the immense generosity over the years of Mr John Thackray in providing information on the history of nineteenth century geology; he tragically died earlier this year at far too early an age.

Furthermore, it is with great pleasure that I acknowledge the help and hospitality I have received from many members of Faraday's extended family. In particular, Mr Michael A. Faraday (for providing me with additional information to that contained in his and the late Dr Joseph E. Faraday's Faraday genealogy) and Miss Mary Barnard (who provided me with a detailed genealogy of the Barnard family whose traditional business of gold and silversmithing she continues). I am grateful to Mr Gerard Sandeman (the last remaining Elder of the Glasite / Sandemanian Church) for allowing me access to the nineteenth century records of the Sandemanian Church in London and Old Buckenham and who also provided much valuable genealogical information. I also thank Mrs Isobel Blaikley and Mrs Molly Spiro for continuing to ferret out material from the rest of their family and for introducing me to various members. These are too numerous to mention fully here, but in this volume I should particularly like to note the generosity of Mr Martin Conybeare in placing his album of Faraday letters on permanent loan in the Royal Institution.

I thank Professor Geoffrey Cantor, Dr Sophie Forgan and Professor David Gooding for their valuable advice and comments on the introduction and for many stimulating discussions on Faraday. Professor Cantor also informed me of many additional places where I could locate Faraday letters and generously shared much useful information about members of the Sandemanian community in the nineteenth century which he gathered in the course of writing his book on Faraday's religion. Professor Gooding also transferred the large number of large databases relating to Faraday's correspondence from my old machine onto a new Mac.

Last, but not least, I thank my wife, Joasia, who was again able to translate the letters from the French and the Italian.

Editorial Procedure and Abbreviations

All letters to and from Faraday which have been located in either manuscript or in printed form have been included in chronological order of writing. The term letter has been broadly construed to include not only extracts from letters where only these have survived, but also reports on various matters which Faraday submitted to institutions or individuals. What has not been included are scientific papers written in the form of a letter, although letters which were deemed worthy of publication, *subsequent* to their writing, are included as are letters to journals, newspapers etc. Letters which exist only in printed paraphrase form have not been included. Letters between members of Faraday's family, of which there are relatively few, are included as a matter of course. Of letters between other third parties only those which had a direct effect on Faraday's career or life are included; the large number of letters which simply say what an excellent lecturer, chemist, philosopher, man etc Faraday was, or the letters (rather more in this period than previously because of his views on table turning) that are critical of him, are not included.

The aim has been to reproduce, as accurately as the conventions of typesetting will allow, the text of the letters as they were written. The only exceptions are that continuation words from one page to the next have not been transcribed and, as it proved impossible to render into consistent typeset form the various contractions with which Faraday and his correspondents tended to terminate their letters, all the endings of letters are spelt out in full irrespective of whether they were contracted or not. Crossings out have not been transcribed, although major alterations are given in the notes.

It should be stressed that the reliability of the texts of letters found only in printed form leaves a great deal to be desired as a comparison of any letter in Bence Jones (1870a, b) with the original manuscript, where it has been found, will reveal. The punctuation and spelling of letters derived from printed sources has been retained.

This volume contains a number of letters between Faraday and John Tyndall. Most of these letters now only exist in the form of typescripts prepared by Tyndall's widow, Louise (1845-1940), as part of her project to write a life of her husband which was never completed. Only a very few original manuscripts of these letters have been found. Obvious minor typographical errors (for instance "thw" for "the") have been silently corrected. Otherwise the same editorial policy has been adopted for these typescripts as for the rest of the correspondence.

Members of the Society of Friends and those closely associated with them had a strong aversion to using the names of the months (which they regarded as pagan) in dating letters. Instead they numbered the months: thus, for example, 6 6mo 1851 should be read as 6 June 1851.

Each letter commences with a heading which gives the letter number, followed by the name of the writer and recipient, the date of the letter and its source. There is, occasionally, a fifth line in the heading in which is given the number that Faraday allotted to a letter. He only numbered letters to provide himself with a reminder of the order of a particular series of letters which invariably referred to some matter of controversy; in this volume letters relating to the circumstances surrounding Brande's retirement from the Royal Institution, Faraday's argument with Becquerel on the magnetic properties of oxygen and the problems involved with Brodie's request for more resources in the Royal Institution are so numbered and all of these issues are discussed in the introduction. Following the main text of a letter, endorsements and the address are always given. The postmark is only given when it is used to date a letter or to establish that the location of the writer was different from that of the letter head.

The following symbols are used in the text of the letters:

[some text]	indicates that text has been interpolated.
[word illegible]	indicates that it has not been possible to read a particular word (or words where indicated).
[MS torn]	indicates where part of the manuscript no longer exists (usually due to the seal of the letter being placed there) and that it has not been possible to reconstruct the text.
<some text>	reconstructs the text where the manuscript has been torn.
[sic]	indicates that the peculiar spelling or grammar in the text has been transcribed as it is in the manuscript. The use of this has been restricted as much as possible to rare cases. Hence, for example, Faraday's frequent spelling of "Herschell" is not followed by [sic].

[blank in MS]　　　　　indicates where part of the text was deliberately left blank.

The following abbreviations are used in the texts of the letters:

CB	Companion of the Order of the Bath
CE	Civil Engineers
DCL	Doctor of Civil Law [also given as LLD occasionally]
DV	Deo volente [God willing]
FE	Friday Evening
FGS	Fellow of the Geological Society
FRS	Fellow of the Royal Society
GCB	Grand Cross of the Bath
HRH	His/Her Royal Highness
MA	Master of Arts
MD	Doctor of Medicine
MP	Member of Parliament
MRI	Member of the Royal Institution
NB	North Britain (i.e. Scotland)
PGS	President of the Geological Society
PM	*Philosophical Magazine*
QED	quod erat demonstrandum [which was to be proved]
RA	Royal Artillery or Royal Academy
RE	Royal Engineers
RI	Royal Institution
RM	Royal Military [Academy]
RN	Royal Navy
RS	Royal Society
US	United Services [Club]
US(A)	United States (of America)
VP	Vice President

Britain did not decimalise its currency until 1971 and is still half-heartedly trying to metricate its weights and measures, although in scientific and technical writings this latter has been largely completed. During the nineteenth century the main unit of currency was the pound (£) which was divided into twenty shillings (s) of twelve pennies (d) each. The penny was further sub-divided into a half and a quarter (called a farthing). A sum such as, for example, one pound, three shillings and sixpence could be written as 1-3-6 with or without the symbols for the currency values. Likewise two shillings and six pence could be written as 2/6; this particular coin could be called half a crown. There was one additional unit of currency, the guinea, which was normally defined as twenty one shillings. There is no agreed figure by which the value of money in the nineteenth century can be multiplied to provide an indication of what its value would be now and, as this is one of the more contentious areas of economic history, no attempt will be made here to provide such a figure.

The following give conversion values for the units used in the correspondence as well as their value in modern units. For mass only the Avoirdupois system is given as that was most commonly used. But it is important to remember that both the Apothecaries' and Troy systems were also used to measure mass and that units in all these systems shared some of the same names, but different values. For conversion figures for these latter (and also for other units not given here) see Connor (1987), 358-60.

Temperature
> To convert degrees Fahrenheit (F) to degrees Centigrade (C), subtract 32 and then multiply by 5/9.

Length
> 1 inch (in or ")   =   2.54 cm
> 1 foot (ft or ')   =   12 inches = 30.48 cm
> 1 yard (yd)   =   3 feet   = 91.44 cm
> 1 mile   = 1760 yards   = 1.6 km

Volume
> 1 cubic inch (ci)   = 16.38 cc
> 1 pint   =   4 gills   = .568 litres
> 1 gallon   =   8 pints   = 4.54 litres
> 1 bushel   =   8 gallons = 36.3 litres

Mass
> 1 grain (gr)   = .065 gms
> 1 ounce (oz)   = 28.3 gms
> 1 pound (lb)   = 7000 grains
>   =   16 ounces = .453 kg
> 1 stone   =   14 pounds = 6.3 kg
> 1 hundredweight (cwt) = 112 pounds = 50.8 kg
> 1 ton   =   20 cwt   = 1.02 tonne

The Notes

The notes aim to identify, as far as has been possible, individuals, papers and books which are mentioned in the letters, and to explicate events to which reference is made (where this is not evident from the letters). In correspondence writers when discussing individuals with titles used those titles, but as British biographical dictionaries use the family name this is given, where necessary, in the notes.

The biographical register identifies all those individuals who are mentioned in three or more letters (in either text or notes). The register provides a brief biographical description of these individuals and an indication of where further information may be found. No further identification of these individuals is given in the notes. Those who are mentioned in one or two letters are identified in the notes. While information contained in the genealogies of various Sandemanian families has been invaluable, this information has been checked against that available in the General Register Office (GRO). In these cases, and others, where the GRO is cited, the year of death is given followed by the age at death. If this agrees with information derived from other sources, then the year of birth in given in preference to the age.

References in the notes refer mainly to the bibliography. However, the following abbreviations are used to cite sources of information in the notes:

AC	Alumni Cantabrigienses
AD	Alumni Dublinenses
ADB	Allgemeine Deutsche Biographie
AG	Almanach de Gotha
AuDB	Australian Dictionary of Biography
Bx	Boase Modern English Biography, volume x
BDPSDG	Bénézit Dictionnaire Peintres, Sculpteurs, Dessinateurs et Graveurs
BNB	Biographie Nationale de Belgique
CCD	Crockford Clerical Directory
CP	Complete Peerage
DAB	Dictionary of American Biography
DBF	Dictionnaire de Biographie Française
DBI	Dizionario Biografico degli Italiani
DBL	Dansk Biografisk Leksikon
DHBS	Dictionnaire Historique et Biographique de la Suisse
DMB	Dictionary of Mauritian Biography
DNB	Dictionary of National Biography (if followed by a number then it refers to that supplement; mp = missing person supplement)
DNZB	Dictionary of New Zealand Biography
DQB	Dictionary of Quaker Biography (typescript in Friends' House London and Haverford College Pennsylvania)
DSB	Dictionary of Scientific Biography
EI	Enciclopedia Italiana
GDMM	Grove Dictionary of Music and Musicians
LUI	Lessico Universale Italiano
NBU	Nouvelle Biographie Universelle
NDB	Neue Deutsche Biographie

NNBW   Nieuw Nederlandsch Biografisch Woordenboek
OBL   Oesterreichisches Biographisches Lexikon
Px   Poggendorff Biographisch-Literarisches Handwörterbuch, volume x
POD   Post Office Directory (see below)
RI MM   Greenaway et al (1971-6). This is followed by date of meeting, volume and page number
TBKL   Thieme-Becker Künstler-Lexikon
WWW   Who Was Who

Reports of lectures in the newspaper press and references to plays, poems and pieces of music are given only in the notes. From 1851 the Royal Institution published accounts of the Friday Evening Discourses in its *Proceedings*. These reports are listed in the bibliography and when cited in notes, indication is made that these were Discourses. References to the *Gentlemens' Magazine* and the *Annual Register*, as well the daily and weekly press are likewise given only in the notes. The following directories are cited in the notes:

Imperial Calendar
Post Office Directory (POD)

Citations to these directories, unless otherwise indicated, refer to the edition of the year of the letter where the note occurs. Both these directories universally make the claim to contain up to date and complete information. This was frequently far from the case and this explains apparent discrepancies which occur.

Faraday's Diary. Being the various philosophical notes of experimental investigation made by Michael Faraday, DCL, FRS, during the years 1820-1862 and bequeathed by him to the Royal Institution of Great Britain. Now, by order of the Managers, printed and published for the first time, under the editorial supervision of Thomas Martin, 7 volumes and index, London, 1932-6 is cited as Faraday, *Diary*, date of entry, volume number and paragraph numbers, unless otherwise indicated.

Faraday's "Experimental Researches in Electricity" are cited in the normal way to the bibliography, but in this case the reference is followed by "ERE" and the series and paragraph numbers, unless otherwise indicated. Two of Faraday's papers (1852d, 1855b) were published in the *Phil.Mag.*, as he regarded them as "of a speculative and hypothetical nature" (1852d, p.401), but in them he continued the paragraph numbering of his "Experimental Researches" published in the *Phil.Trans*. To help locate references within these papers, I have allocated them the series numbers 29a and 29b respectively and they are thus cited in the notes in square brackets.

John Tyndall's manuscript diary in the Archives of the Royal Institution is cited as Tyndall, *Diary*, followed by date, volume and page numbers.

Manuscript abbreviations

The following are used to cite manuscript sources where the primary abbreviation is used twice or more. (NB Reference to material in private possession is always spelt out in full). These abbreviations are used in both the letter headings and the notes:

AC MS	Athenaeum Club Manuscript
APS	American Philosophical Society
AS MS	Académie des Sciences Manuscript
BeRO	Berkshire Record Office
BL add MS	British Library additional Manuscript
BM	British Museum
CA	Central Archives
DWAA MS	Department of Western Asiatic Antiquities Manuscript
BN	Bibliothèque Nationale
Bod MS	Bodleian Library Manuscript
BPUG MS	Bibliothèque Publique et Universitaire de Genève Manuscript
BRAI ARB	Bibliothèque royale Albert Ier, Académie royale de Belgique
BrUL MS	Bristol University Library Manuscript
CITA	California Institute of Technology Archives
DUL	Duke University Library
FACLM	Francis A. Countway Library of Medicine
GL MS	Guildhall Library Manuscript
GRO	General Register Office (see above)
HCL	Haverford College Library
HLHU	Houghton Library, Harvard University
HMW	Hollandsche Maatschappij der Wetenschappen
IC MS	Imperial College Manuscripts
LP	Lyon Playfair papers
IEE MS	Institution of Electrical Engineers Manuscript
SC	Special Collection
2	David James Blaikley Collection
3	S.P. Thompson Collection
22	W.H. Preece Collection
71	W.E. Staite Collection
JRULM	John Rylands University Library of Manchester

MU		McGill University
	MS 173	Fredrick Hendrick Papers. Letters bound in Babbage (1864)
NLS MS		National Library of Scotland Manuscript
NMW		National Museum of Wales
NRCC ISTI		National Research Council Canada Institute for Scientific and Technology Information
PRO		Public Record Office, Kew
	ADM1	Admiralty Secretariat Papers
	ADM12	Admiralty Digest
	ADM88	Surveyor of the Navy Digest
	BJ1	Kew Observatory Papers
	HO107	Returns of the 1851 census
	LC6	Records of Levées
	WO44	Ordnance Office in letters
RAW		Royal Archives, Windsor
RGO		Royal Greenwich Observatory
	6	Airy papers
RI MS		Royal Institution Manuscript
	F1	Faraday Collection
	A-G	Letters from Faraday
	H-K	Faraday's portrait albums
	L, N	Miscellaneous letters to and from Faraday
	F2	Faraday's experimental notebooks
	F3	Bound offprints of Faraday's papers
	F4	Faraday's notes of lectures
	G	Papers of W.R. Grove
	GB	Guard Book
	GM	Minutes of General Meeting
	HD	Papers of Humphry Davy
	JB	Papers of John Barlow
	T	Papers of John Tyndall
	TS	Typescript volumes
RS MS		Royal Society Manuscript
	241	Faraday's Diploma book
	AP	Archived Papers
	Bu	Buckland Papers
	Cert	Certificate of Fellow Elected
	CM	Council Minutes (printed)
	CMB	Committee Minute Books
	67	Excise Committee Minutes
	90C	Minutes of Committee of papers, 1828-1852
	HS	Herschel Papers

MC	Miscellaneous Correspondence
MM	Miscellaneous Manuscripts
PT	Manuscript of *Phil.Trans.* papers
RR	Referees Reports
RSA MS	Royal Society of Arts Manuscript
SAU MS	St Andrews University Manuscript
JDF	Papers of J.D. Forbes
SELJ MS	Sidney M. Edelstein Library Jerusalem Manuscript
SI	Smithsonian Institution Library
A	Archives
JHC	Joseph Henry Collection
D MS	Dibner Collection
SM MS	Science Museum Library Manuscript
SPK	Staatsbibliothek Preussischer Kulturbesitz
DD	Dokumentensammlung Darmstaedter F 1 e 1831 (2). Autograph I/1478/23
SuRO	Surrey Record Office
TCC MS	Trinity College Cambridge Manuscript
UB MS NS	Universität Basle Manuscript Nachlass Schoenbein
UCL	University College London
UCLA UL	University of California, Los Angeles, University Library
ULC	University Library Cambridge
Add MS 7342	Thomson Papers
Add MS 7656	Stokes Papers
Add MS 8177	Henslow Papers
ULL	University of London Library
WIHM MS	Wellcome Institute for the History of Medicine Manuscript
FALF	Faraday autograph letter file

Note on Sources

In the British Library there are two letters (BL add MS 46404, f.10-11) written by Angela Georgina Burdett Coutts in 1849 and 1852 which are catalogued, albeit with a query, as being to Faraday. These letters were clearly written in a Royal Institution context. However, there is no evidence from any of the letters published in this volume that these British Library letters were to Faraday and they have thus been omitted.

Introduction

The 887 letters in this volume, of which nearly two thirds are published for the first time, deal with many of the far reaching changes that occurred in Faraday's life during the first half of the 1850s, that is between the ages of 58 and 64. Some of the associations that he had developed during the 1820s came to an end in the 1850s. For example, he resigned from membership of the Athenaeum Club and from the Professorship of Chemistry at the Royal Military Academy, Woolwich. He came perilously close to a second (and thus final) exclusion from the Sandemanian Church and played a role in the schism that saw the exclusion of the Edinburgh meeting house from all the other Sandemanian and Glasite communities in England and Scotland. In his scientific research Faraday undertook major pieces of work on what he called atmospheric magnetism and on lines of forces as well as a highly significant, but shorter, piece of work on telegraphic retardation. He continued to play a prominent role in the Royal Institution both in its running and in delivering lectures. In the broader world Faraday publicly attacked spiritualist phenomena, especially table turning, complained about the polluted state of the Thames, continued to provide advice to Trinity House, served on a Jury of the Great Exhibition and undertook work for Admiralty in the urgent context of the Anglo-French war against Russia. All these activities were undertaken against a background of continuing ill-health with the added worry that his wife, Sarah, nine years his junior, was also beginning to suffer from serious illnesses.

For the six full years covered by this volume the average number of letters written per year is 130 compared with 101 in volume three. This increase is due to the steady year by year rise in the number of letters from 102 in 1849 to a peak of 161 in 1853 followed by a slight tailing off. In general Faraday's circle of correspondents during the first half of the 1850s remained remarkably similar to that of the 1840s[1]. He did lose through their deaths a number of correspondents such as his old friend Richard Phillips[2] and Macedonio Melloni[3]. The death of the former is not mentioned in the letters, but the latter is[4]. Melloni, as usual, complained about his position in Naples[5] and asked Faraday, among others, not only to help him specifically[6], but also to help rally support generally for the cause of Italian independence and unity. This latter included gaining the support of Abraham Lincoln[7] and Faraday helped to pass on a message to him from Melloni[8].

The only major additions, all in a Royal Institution context, to Faraday's circle of correspondents were Harriet Jane Moore, John Tyndall and Henry Bence Jones. There were, of course, many additional new men and women with whom Faraday corresponded for a short period on specific topics. Examples of these would include Paul Frederick Henry Baddeley, Benjamin Collins Brodie, Josiah Latimer Clark, Charles John Huffam Dickens, Harriet Martineau, George Ransome, William Stevenson, George Gabriel Stokes and Joseph John William Watson.

A number of members of Faraday's extended family died during this period. His father in law, Edward Barnard, died in early 1855. This event is not mentioned in the extant correspondence, though it may have prompted Faraday to leave directions as to the disposal of some of his scientific manuscripts a few days later[9]. Faraday's brother in law Adam Greenlaw Gray[10] died at the end of 1849 while Sarah's brother in law George Buchanan died in 1852. Neither of these events were recorded in the extant correspondence. However, the deaths of two of Sarah's nieces, Margery Ann Faraday and Mary Boyd, were noted in letters written in 1850 and 1853 respectively[11].

Most of these relations had some form of connection with the Sandemanian Church. The overall silence on family matters observed in the extant correspondence, when one would expect, especially at the time of death, considerable discussion, is mirrored by the overall silence about church affairs. There are only a couple of major episodes relating to Faraday's involvement in the church dealt with by letters in this volume. Furthermore, these letters are drawn from the papers of the Edinburgh Church rather than from the London Church or from what has survived of Faraday's papers. But we must always remember that Faraday, when in London, would have attended the Sandemanian Meeting House each Sunday as well as Wednesday evenings; he seems to have once used the latter meeting as the reason for not accepting an invitation to an event organised by the Lord Mayor[12]. Furthermore, during the period covered by this volume Faraday twice visited the small Sandemanian community in Old Buckenham in 1850[13] and 1852[14]. Yet even in the letters discussing these visits there is only one that deals with the theological state of that small Norfolk community[15]. One is forced to conclude that Faraday, and his executors following him, deliberately sought to keep his religious life private; Faraday for instance, explicitly declined to comment on his religion when more or less invited to do so by Barlow[16].

The two events preserved in the papers of the Edinburgh Meeting House, do give an indication of the sort of turmoils that the London Church went through. The first was specific to Faraday and the other a general schism from the Edinburgh Church. In late 1850 Faraday came to have doubts about the Biblical basis of "the view the Church takes of not receiving an excommunicant more than once" as a clearly distraught Sarah Faraday wrote to the Edinburgh Elder William Buchanan at three o'clock one morning[17]. Faraday had been excluded from the Church in 1844[18] and it would have been ironic in the extreme if his second, and thus final exclusion, had come about because he disputed the Biblical basis on which he would have been finally excluded. Within a space of about a week the matter had been sorted out and Faraday had submitted to the will of the church[19].

The second event, which started in late 1854, dealt with the Biblical injunction to "abstain from blood & from things strangled"[20]. In the view of the Glasite Church in Dundee this injunction included game which, it would appear, was eaten by members of the church; indeed it is clear that Faraday enjoyed pheasant[21]. However, Dundee wrote to the London Church pointing out the error of their ways which London accepted[22] as did the congregations in Old Buckenham, Newcastle and Chesterfield[23]. The congregation at Edinburgh, did not accept the Dundee interpretation and refused to hear the Elders from Dundee and London who were sent to remonstrate with them. The result was a schism in early 1855 between Edinburgh and the other Glasite and Sandemanian communities in the rest of the country[24]. Thus Faraday had to sever, for the time being, relations with the Buchanan family which had been close and which still included his widowed sister in law Charlotte Buchanan[25]. This episode is but one demonstration of the way in which the church dominated Faraday's life. A small, but telling example, of the influence of Sandemanianism on Faraday is his refusing an invitation to attend the funeral of the Duke of Wellington held at St Paul's Cathedral on 18 November 1852. As places for this funeral were in great demand, it is hard to imagine that many other people joined Faraday in not accepting the invitation[26]. Although he simply declined, not giving his reasons, his grounds would have been that a Sandemanian should not attend an Anglican service[27].

The connection between church and family expressed itself in the relations of Faraday and Sarah with some of their nieces. During the period covered by this volume Jane Barnard came to live with them in the Royal Institution, but her presence is not felt strongly in these letters. Of great concern to the Faradays throughout this period, however, were the problems of Caroline Deacon. In 1846 she had married Thomas John Fuller Deacon, at the rather late age of twenty nine, and the following year she had their only child, Constance, of whom Faraday was very fond[28]. However, Thomas Deacon appeared to have serious problems, the precise nature of which are not clear. In May 1850 he was committed to a mental asylum[29]. Within a year he was out and the family had moved to Newcastle where Faraday had given him some translation work to do[30] which suggests that Deacon had not found employment. The situation had not improved by September 1851 when Faraday wrote a long comforting theological letter to Caroline[31] and by March 1852, Faraday was actively lobbying, without success, for Deacon to be appointed Assistant Secretary of the Royal Society of Literature[32]. In July that year Faraday exhorted Caroline to "cheer up"[33], but her problems were still a subject for discussion for the Faradays when they were on holiday in Wales in 1853. By this time the Deacons seemed to have acquired a house and the Faradays sent Caroline five pounds help them settle in[34]. The acquisition of a house does suggest that Deacon had found employment and that the problems had abated; indeed two years later Faraday referred to Deacon's "chemical pursuits"[35].

Faraday's theological outlook played a major role in his approach to the table moving craze that swept England in 1853. It was inevitable that Faraday, as one of the leading scientific figures of the day, would be asked for his opinion of the subject and indeed he described the applications to him for such an opinion as "numerous"[36]. However, the number of letters which deal with table moving that have survived, and are thus published here, is not large. The attrition rate for this material can no doubt be explained by the note on a wrapper which once contained some of this material: "The greater part destroyed by D.J.B. 19/4/14"[37].

A primary reason for Faraday's involvement was the widespread belief that the tables moved due to some form of electric or magnetic force. The earliest reference in the correspondence is in mid May 1853 when George Wingrove Cooke asked Faraday for his opinion on the subject[38]. This was then followed by two letters from John Allen and William Edward Hickson containing detailed descriptions of table turning events[39]. Faraday evidently responded to Hickson's letter negatively and was rewarded with an invitation to visit him in Kent to witness the phenomenon[40] which Faraday declined[41]. The next letter to have survived was anonymous[42] and that, together with its silliness, clearly provoked a very sharp response from Faraday which elicited an appropriate apology[43]. As a consequence of these few letters which have survived and the many that were destroyed, Faraday, at the end of June, turned his experimental attention to table turning. He attended two seances at John Barlow's house[44] where he was able to devise an experimental procedure which showed that the phenomenon was due to "a *quasi* involuntary muscular action"[45] on the part of those moving the tables and not some known or unknown force.

Unlike mesmerism, on which Faraday had kept his scepticism private in the 1840s when inundated with letters on the subject[46], with table turning he decided to make his position public. The same day as the second seance at Barlow's house, Faraday wrote an article for the *Athenaeum*, outlining the results of his experiments which was published on 2 July[47] and in which he made it clear that he could not "undertake to answer such objections as may be made"[48]. Clearly Faraday suffered no illusion that his was going to be the last word on table moving. The day after he wrote this article, Faraday wrote a shorter account of his experiments in a letter to the *Times* which was published on the last day of June[49]. At the end of this letter he blamed the widespread public belief in table turning on the educational system: "I think the system of education that could leave the mental condition of the public body in the state in which this subject has found it must have been greatly deficient in some very important principle". This letter was quickly reprinted in the *Illustrated London News*[50] which two weeks later published an illustration of the apparatus Faraday had used in his experiments[51].

Very quickly Faraday received letters of support from the scientific and education communities. Herschel, Alexander Bath Power, Andrews and Tyndall all wrote to support Faraday's stance[52]. However, Faraday's fear that he would meet with criticism was well founded. Some was private criticism such as that by Elizabeth Browning[53], who in several letters to her friends expressed her outrage at Faraday's views accusing him of "arrogance & insolence"[54]. Some wrote to Faraday directly; Robert Espie, for instance, wrote a fairly abusive letter[55]. However, this is one of the few letters to have survived out of the many that Faraday must have received during July. As with letters written before his *Times* letter, the attrition rate of letters arguing against Faraday was high and few have survived[56]. The number of letters he did receive prompted him to write to Caroline Deacon on 23 July that "I have been shocked at the flood of impious & irrational matter which has rolled before me in one form or another since I wrote my times letter"[57]. He went on to tell Deacon: "I cannot help thinking that these delusions of mind & the credulity which makes many think that supernatural works are wrought where all is either fancy or knavery are related to that which is foretold of the latter days & the prevalence of unclean spirits"[58]. By quoting from The Revelation of St John the Divine, Faraday indicated that not only had his sense of scientific propriety been transgressed, but so had his Sandemanian beliefs. It was probably this combination that led Faraday to feel that it was incumbent on him to publicly denounce table moving rather than let it rest as he had done with mesmerism a few years before. Because of the affair over Andrew Crosse's acarii in the late 1830s[59] (an issue which reemerged briefly in 1850[60]), Faraday was well aware of the pitfalls involved in entering the fray in such areas. In the end Faraday came to have contempt for table turners, especially as they attempted to circulate rumours that he had changed his mind[61]. It is difficult to gauge what effect Faraday's efforts had on the table turning fashion, other than it was open to wildly differing interpretations. In later years Faraday's name was invoked both by supporters of research into spiritualist phenomena (such as Crookes[62]) on the grounds that Faraday had taken the subject sufficiently seriously to experiment on it, and by opponents (such as Tyndall[63]) on the grounds that Faraday had provided an entirely satisfactory explanation of such occurrences.

Faraday's criticism of the system of education, which permitted, in his view, table turning to achieve such wide spread popularity, led the Royal Institution to arrange a course of Lectures on Education. A committee (comprising Barlow, Bence Jones[64] and William Frederick Pollock) was formed in February 1854 for the purpose of organising these lectures[65]. Unfortunately, as with so many Royal Institution committees in the nineteenth century, its minutes have not survived. Unusually for the Royal Institution, where lecture courses were normally delivered by one individual, this course was made up of a series of seven single lectures by some very eminent men of science[66] ("great guns" as Bence Jones put it[67]) given on Saturday afternoons between the end of April and the beginning of June. At the end of February, and again in March, Faraday wrote to Whewell to add his weight to the invitation for him to deliver a lecture in the series[68]. Whewell agreed, following Faraday's second letter, and indeed delivered the opening lecture of the series before Prince Albert on 29 April 1854[69]. Faraday sought to avoid providing a lecture for the series, but the Managers and the Committee insisted and Faraday delivered the second lecture in the series, also before Prince Albert, on 6 May 1854[70] (after the lecture Faraday spent half an hour alone with the Prince in the Library[71] where they seemed to have talked about propulsion methods for ships[72]). Faraday's reluctance to deliver a lecture in the series seems to have been genuine. He wrote some of it over Easter while staying at Redhill. From there Sarah Faraday told Caroline Deacon that Faraday was not enjoying writing it since, unusually, possibly uniquely since the late 1810s, he had to write the lecture out entirely for reading without any experimental demonstrations. He was also, according to Sarah, concerned that the lecture would be "keenly criticised"[73]. In the lecture, Faraday returned to the attack on table turning and gave an autobiographically based account of how the judgement should be educated[74].

Not only did the Royal Institution pay each of the lecturers in the series ten pounds[75], it also, and again most unusually, arranged to have the lectures printed and published at its own cost[76]. From 1851 the Royal Institution had again taken up publishing some of its work in the form of *Proceedings* which contained published versions of Friday Evening Discourses, but it had never previously published courses of afternoon lectures. Each of the Education Lectures was separately printed as a pamphlet, but they were continuously paginated, so that they could be collected together in a single volume[77]. By the end of May, Faraday had corrected the proofs of his lecture[78]; on 10 June he presented a copy to Prince Albert[79], the same day as it was advertised[80]; and on 1 July it received a reasonable, though by no means brilliant, review in the *Literary Gazette*[81].

The role that Faraday played in the Lectures on Education was consistent with the role he played generally in managing the affairs of the Royal Institution during this period, which in turn was a continuation of what he had done during the latter half of the 1840s. He delivered eighteen Friday Evening Discourses between 1849 and 1855 which continued to be immensely popular. As George Eliot[82] put it after attending to Faraday's Discourse on the magnetic properties of gases on 24 January 1851[83], "Faraday's lectures are as fashionable an amusement as the Opera"[84]. Faraday also delivered a course of afternoon lectures after Easter each year from 1849 and 1853 and four courses of Christmas lectures for children. He was asked, by Barlow, to secure prominent men of science to provide Friday Evening Discourses[85] and lecture courses in much the same way as he persuaded Whewell to lecture in the series on Education. Thus, for example, he was successful in persuading Airy to deliver Discourses[86], but not Whewell[87] or Sabine[88].

Faraday continued to oversee the physical fabric of the building in his role as Superintendent of the House[89] and was involved in smoothing over problems as they arose in the Royal Institution. For example he dealt with the issue of the admission of the guests of lady members to Discourses[90] and was instrumental in ensuring, in the Spring of 1852, that what could have been an ugly row over the retirement of William Thomas Brande from the Professorship of Chemistry was avoided. Following the reform of the Royal Mint, it had been decided that its employees, of whom Brande was one, could not hold other paid appointments and he thus chose to retire from the Royal Institution. The announcement of this to the members of the Royal Institution was badly handled and unfortunate rumours began to circulate which annoyed Brande intensely. Faraday had to spend some time in persuading him that they were without foundation[91].

Brande's retirement bought to the fore the question of who should constitute the audience for lectures at the Royal Institution. This issue had been addressed in late 1846 when Brande's commitment to deliver laboratory lectures to students from St George's Hospital (and elsewhere) had been reduced with help from Brodie[92]. By 1849 Brodie was seeking to secure his position in the Royal Institution and this led to a fairly acrimonious correspondence with Faraday. Brodie wanted to make the lectures more academic and educational, but the Managers and Faraday were not convinced that this was the best course for the Royal Institution to take. The Managers thus declined Brodie's request for extra laboratory facilities and for an assistant[93]. This decision Brodie called "parsimonious and inconsiderate"[94] which prompted Faraday to provide a robust defence of both the Managers and Brande[95]. The result of this was that Brodie resigned and Brande was asked to take on the lectures again[96] which he did until his retirement with help from other chemists including Bence Jones and August Wilhelm Hofmann[97]. With the retirement of Brande the question of who should succeed him and what role his successor should play became a matter of some urgency. Very quickly Alexander William Williamson offered to deliver a course of lectures on chemistry which the Managers accepted[98].

However, due to the active lobbying of Faraday[99] and Bence Jones, Tyndall succeeded Brande with the title of Professor of Natural Philosophy in June 1853[100] and the laboratory lectures were discontinued thus ending the Royal Institution's long connection with the London medical institutions. Tyndall had met Faraday in the summer of 1850[101] and in the ensuing years had struck up a friendship with him. Faraday agreed to provide Tyndall with a reference for a position at Toronto University should he be asked for one[102]. It was not, however, until after Tyndall had delivered his first Discourse in February 1853 that he came onto "Dear Tyndall" terms with Faraday[103]. The relationship between Faraday and Tyndall is a highly curious one. Although Tyndall courted Faraday's good opinion of him (including, on occasion, quoting from the Bible[104]), he was quite capable of disagreeing publicly and strongly with the views of some of Faraday's other friends, particularly Plücker, Thomson and Whewell. In one instance, at the 1855 Glasgow meeting of the British Association, when Tyndall had sent Faraday a toned down account of what had been a sharp exchange with Thomson, Faraday felt compelled to chide Tyndall, of whom he certainly had a high opinion as a lecturer[105], and give him some advice on how to deal with such matters[106].

Faraday himself was no stranger to controversy during the first half of the 1850s. In addition to the arguments over table turning and Brande's retirement he also got involved in three other disputes. The first began at the end of 1850 when Antoine-César Becquerel wrote to lobby for Faraday's support for the election of his son, Alexandre-Edmond Becquerel, to the Académie des Sciences. He told Faraday that his son had discovered that oxygen was a magnetic substance and asked if Faraday had done any work on this[107]. In his reply[108] Faraday said that he had published on the magnetism of oxygen in 1847[109] and sent Becquerel a marked copy of his paper. Becquerel responded by seeking to maximise the importance of his son's work and consequently to play down the significance of Faraday's results[110]. In his reply[111], which terminated the correspondence, Faraday disputed Becquerel's claims and also some of his son's results.

The second dispute involved Faraday allocating credit for the discovery of a link between the sunspot cycle and the variations in terrestrial magnetism to Johann Rudolf Wolf. This he had done in a letter to Wolf in August 1852[112] which Wolf had published in a pamphlet[113].

Sabine, with whom Faraday normally enjoyed good relations, claimed he had made the discovery earlier; he was annoyed and wrote to Faraday to tell him so in early 1853[114]. Faraday's response was a letter mostly concerned with atmospheric magnetism, with a couple of lines at the end referring to Wolf, but not answering Sabine's criticism directly[115]. Sabine's reply made no mention of the matter[116].

The third dispute, also in early 1853, began when Bence Jones dedicated a book to Faraday which discussed the work of Emil Du Bois-Reymond on animal electricity[117], some of which had been carried out and demonstrated at the Royal Institution[118]. The book contained criticisms of the work of Carlo Matteucci and drew from him a short, but very hostile letter, to Faraday[119]. Faraday responded at length saying that he was "startled" by the receipt of Matteucci's letter. He went on to try and smooth things over, by discussing his own experiences, and by explaining his attitude towards dedications[120] (which was the same as the view he later expressed to Benjamin Humphrey Smart[121]).

The disputes with Becquerel and with Sabine arose while Faraday was undertaking his major piece of scientific work during the first half of the 1850s on what he termed atmospheric magnetism. Following his discovery of the magnetic nature of oxygen, Faraday sought to explore the implications of this discovery. In this work he moved away from the exclusive consideration of his own experimental results towards attempting to understand, using mainly graphical methods, the large quantity of data that had become available following publication of the results of the magnetic observatories established throughout the world from the late 1830s onwards[122]. The work of Wolf and Sabine are other instances of attempts to make some sort of sense of the overwhelming quantity of information on the earth's magnetic behaviour that was now available. It would also seem possible that Faraday maintaining, for a time, his correspondence with Stevenson and with Baddeley was aimed at helping his understanding of atmospheric phenomena.

The earliest reference to Faraday's interest in this subject occurs in his diary on 23 July 1850[123]. Here he stated the basic idea that he was to explore during the rest of the year. Since the magnetic strength of oxygen weakened as the temperature increased[124], perhaps the source of variation of terrestrial magnetism was linked to the warming and cooling of the atmosphere due to night and day and the change of the seasons: "Perhaps the cause here of the daily variation and even of the larger annual variations"[125]; he mentioned this idea to Whewell a month later, though he asked him to *"keep this to yourself"*[126]. Since, as Faraday pointed out, two ninths of the atmosphere by weight was made up of highly magnetic oxygen, he found it *"an impossible thing to perceive"*[127] that its changes in temperature did not affect its magnetic condition and thus what was measured as magnetic variation. Throughout the summer, Faraday developed this idea, which "works out beautifully" he told Whewell[128]. He experimented on his hypothesis and also analysed the data from a number of magnetic observatories, including unpublished data which Sabine let him have[129]. He showed in two long papers in the *Philosophical Transactions* that there existed a correlation between the temperature of the atmosphere and magnetic variation[130]. The reception of this work, for instance, by Herschel and by Humboldt was equivocal[131], although Harris was more enthusiastic[132].

In turn Faraday's work on atmospheric magnetism was part of a broader effort to elaborate his theory of the field and establish the reality of lines of force. This was a continuation of work started in the 1840s on diamagnetism on which Faraday published new results in 1850 and 1851[133]. In the course of this work, Faraday adopted the term "paramagnetic" following correspondence with Whewell[134]. To reinforce his idea of the reality of lines of force, Faraday fixed iron filing diagrams in waxed paper and in late 1851 sent them to his friends[135]. It was this reality he argued for, presenting the iron filing diagrams as evidence, in two papers in 1852[136].

It is well known that Faraday's field theory was not accepted by men of science in Britain for a few years and was not taken up in Germany for several decades[137]. What contributed greatly to its being developed in Britain was the need for a theory of long distance telegraph signalling. Such a theory was developed from Faraday's ideas, when it was found that the apparently more precise action at a distance theories could not produce an adequate theory. Faraday's role in this process resulted from an invitation to witness, with Airy and others, some observations made by Josiah Latimer Clark of telegraphic retardation. This phenomenon occurred when Clark passed signals from the Electric Telegraph Company's station at Lothbury Wharf to Manchester and back several times. Faraday saw these experiments on 4 and 15 October 1853[138]; on the 15th he watched the retardation through 1600 miles of cable. Faraday then began to develop a theory of the phenomenon in correspondence with both Airy[139] and Clark[140] and by the end of October was referring it to his theory of induction proposed in the 1830s[141]. In his Friday Evening Discourse of 20 January 1854[142] Faraday discussed the theory of the phenomenon in which he argued that the wire and its gutta percha insulation formed a leyden jar and was therefore referable to his theory of induction. Quickly thereafter he wrote about this work to several of his friends[143]. When the lecture was published in early February, it provoked another set of letters between Airy and Faraday on the subject[144] and later in the year formed one of the subjects discussed in a series of long letters between Melloni and Faraday[145] which was ended prematurely with the death of Melloni from cholera[146]. With the development of Faraday's theory, long distance telegraph signalling became a possibility and thus drew strong attention, in Britain at least, to the explanatory power of field theory. The consequences of this would become apparent during the following two decades.

In early 1855 Faraday collected most of the papers he had published since 1846 into the third (and final) volume of his *Experimental Researches in Electricity*[147]. The volume included series 19 to 29 of "Experimental Researches in Electricity" published in the *Philosophical Transactions*, the two papers published in the *Philosophical Magazine* where Faraday continued the paragraph numbering from series 29[148] and a number of other shorter papers. Although Faraday dated his preface to volume three as "January, 1855", one paper[149] included was published after this date and the volume was not published until April[150].

This work on telegraphic retardation was only a small part of Faraday's continuing involvement with various practical projects, some with a public dimension, some conducted with great secrecy, during this period. Projects with a public element with which Faraday was involved included the Great Exhibition, advising the Ipswich Museum on their exhibits and providing advice and evidence on conservation matters to various public bodies. Faraday's ability to carry out such a diverse range of tasks is perhaps best captured by Airy's comment to him in a letter of January 1850: "You have so much of the civil-engineer's talent of applying new means to new purposes"[151].

Faraday was concerned, to a limited extent, with the planning for the Great Exhibition in 1851[152]. He was a member of the Jury for Class I, "Mining, Quarrying, Metallurgical Operations, and Mineral Products"[153] which kept him busy during the early summer[154] and for which he received a medal from Prince Albert[155]. It is clear from the correspondence for 1851, that Faraday received a large number of visitors from overseas who had come to London for the Exhibition. In this context Faraday continued his friendship with one of the engineers who worked on the Exhibition, Charles Fox, and his family[156].

The Great Exhibition seems to have acted as a spur for improving displays in provincial museums. Certainly, while the Great Exhibition was under construction, Henslow was busy initiating improvements to the Ipswich Museum. Furthermore, Ipswich was to be the location for the meeting of the British Association that year at which Prince Albert would be present. Henslow knew Faraday from their work for the University of London and therefore had no hesitation to "appealing to the highest authority" when asking Faraday for advice on the display of chemical science[157]. Faraday was more than happy to provide both advice and exhibits for the museum which he did in 1851 and again in 1853[158]. He was elected an Honorary Member of the Museum in May 1851[159].

The state of the pictures in the National Gallery led to the establishment of a Select Committee in 1853 to investigate the problem.

Faraday was sent a sample of an old oil painting to see what he could do about cleaning it[160] and he reported his findings to the Committee on 10 June as part of his evidence which amounted to ten printed pages. Faraday was also asked about conservation issues by the British Museum, for instance on the best way of cleaning metal objects[161], on copying Assyrian sculpture[162] and on conserving glass[163]. Faraday, who had given up professional work, did not charge for any of this advice, but in 1853 he did ask to borrow a large silica crystal from the British Museum for his magnecrystallic work. This request was initially turned down by one of the Museum's staff[164], whereupon Faraday wrote a formal letter to the Trustees asking to borrow the crystal and reminding them of all the advice he had given them[165]. In a covering letter to Henry Ellis, Faraday wrote: "I think it can hardly be needful that I should make such application, or that I should move such bodies as the Royal Society or the British Association to make such application, to a Secretary of State for a purpose so simple"[166]. The implied threat was effective and Faraday was allowed to borrow the crystal[167].

Less publicly, but nevertheless effectively, Faraday continued to advise Trinity House on lighthouse matters. Somewhat mundanely Faraday analysed chemically oils[168], waters[169] and paints[170] used by the Corporation in their lighthouses and also continued to give advice on his ventilating chimney. This last piece of work entailed inspecting the equipment at lighthouses in Norfolk (twice)[171], the Isle of Wight (thrice)[172], Portland Bill[173] and elsewhere.

Faraday also oversaw the first serious trials of electric light as a means of lighthouse illumination which had been proposed by Watson in November 1852[174]. This proposal basically involved passing an electric current from a battery across a carbon arc[175]. Faraday was quickly asked by Trinity House to test Watson's light and for Watson to communicate directly with Faraday about the practical arrangements[176]. Within the month it was agreed that Watson's light would be first tested for eight hours and if that was successful then the light would be tested for at least eight hours each day for a fortnight (apart from Sundays). Faraday concluded his letter confirming these arrangements with Watson with a plea that "above all I most earnestly beg of you to produce no imperfectly prepared apparatus or any thing that requires excuse"[177]. Watson seems to have taken Faraday's advice on board since he agreed not to commence the trials until after the middle of January[178]. However, no evidence has been found which suggests that these trials took place, and the next mention of Watson's light is in a letter of June 1853, five months later, where Faraday was again asked by Trinity House to test it. By this time Watson had transferred his patents for electric light to the Electric Power and Colour Company[179]. Yet again it would appear that the trials did not take place then, and it was not till more than a year later in July 1854, that Watson again proposed the trials[180].

These trials were undertaken[181] and Faraday submitted a 4200 word report on Watson's electric light[182]. Faraday started with the positive aspects of the light. He noted that it would shine for more than eight hours a day and had worked for five days; that the light was brighter than current sources of illumination; and that the carbon rods could be changed in less than a second. The problem was, Faraday pointed out, that in order to keep the cost of the light to a minimum (always a major concern with Trinity House), the chemical products produced by the battery would have to be collected and sold. In turn this would mean building a large battery room at each lighthouse which would have to be some distance from the lighthouse to avoid the fumes produced by nitric acid (this problem had caused one of the men at the trial to spit blood and for Faraday to suspend the trial). Furthermore, accommodation would have to be provided for the three battery men required. Faraday pointed out the problems that would arise by having two sets of employees working at a lighthouse; otherwise Trinity House would need to become a manufacturing and commercial body. He also, more technically, noted that the light flickered too much and that to maintain it would require skills and aptitude on the part of lighthouse keepers which they did not possess. He concluded: "Much, therefore, as I desire to see the Electric light made available in lighthouses, I cannot recommend its adoption under present circumstances. There is no human arrangement that requires more regularity and certainty of service than a lighthouse. It is trusted by the Mariner as if it were a law of nature; and as the Sun sets so he expects that, with the same certainty, the lights will appear". Aside from the poetic imagery of this passage, the striking comparison Faraday makes between the laws of nature (which God had written into nature at the Creation[183]) and a human technology, emphasises the moral seriousness with which Faraday regarded his lighthouse work. For Faraday, within this theological imperative, the overall system of lighthouse illumination, was far more important than installing a particular new technology – a point he made clearly in his report: "For much as I may desire that the Electric light, with its special advantages, may find ultimately its full application in light houses, I am bound before all other things to consider the security and the constancy of the service required".

Watson, who was clearly irritated with Trinity House's decision not to proceed with his electric light, wrote to Faraday asking for the reasons he had given to Trinity House that had led to the rejection of his light which he regarded as "of great national importance"[184]. Faraday responded by saying that he could not reveal the content of his report as that belonged to Trinity House, and that in effect Watson was not an expert on lighthouses and could not see the whole picture. He added that had he seen a promising way towards a practical electrical light for lighthouses he would have pursued it himself. He concluded by saying that it was of great national importance that the lighthouse service was not jeopardised by installing inappropriate technology[185]. Thus ended the first attempt to use electric power for lighthouse illumination.

Faraday also undertook some secret work for the government. For instance in 1852 the Ordnance Office sent Faraday a new type of French shell which had been captured off Gibraltar and Faraday was very quickly able to provide the Office with a chemical analysis of the shell[186]. However, it was the major political and military crisis of the 1850s, the Anglo-French war against Russia fought between 1854 and 1856, that led to Faraday's most important piece of secret government work. The aim of the war, which saw the largest deployment of British forces in Europe between 1815 and 1914, was to prevent the expansion of the Russian Empire westward into Europe, particularly at the expense of the Turkish Empire. To achieve this aim it was decided to attack and capture Russia's two main naval bases, Cronstadt in the Baltic Sea and Sebastopol in the Black Sea and to impose a blockade on Russian trade. Cronstadt was to be blockaded and attacked by the Royal Navy. Sebastopol was to be besieged both by sea and by an Anglo-French Army (with a large component of mercenaries from other countries) landed on the Crimean peninsular. Throughout this volume of correspondence there are references in the letters of Faraday and others, to the impending crisis (generally called 'The Eastern Question') and then to the war itself. So deep was the impact of the war, that Faraday used the siege of Sebastopol as a metaphor for his approach to studying magnetism: "The secret of magnetic action is like a Sebastopol at least in this point that we have to attack it in every possible direction and make our approaches closer and closer on all the sides by which we can force access"[187].

Faraday's attitude towards the war was ambivalent. A month before its declaration, on 22 March 1854, Tyndall recorded that Faraday "seems to think that the government has done every thing possible to secure peace, but sooner than retreat in dishonour he, for his part, would lose half his income"[188]. Slightly more than a year later, Faraday wrote to De La Rive criticising the way the war had gone in terms of honour: "How the whole surface of the earth seems about to be covered with the results of evil passions"[189]. Whatever Faraday's views on the war were, he helped the government in its prosecution as much as he could.

In the Summer of 1854 Admiral Thomas Cochrane, 10th Earl of Dundonald, proposed to attack Cronstadt using fireships filled with burning sulphur, the fumes from which would overcome the defenders and allow the capture of the fortress. This proposal by Dundonald, who as a senior naval officer, had been considered for the command of the Baltic Fleet[190] could not be ignored. Thus a highly secret committee was established to consider the proposal. Its members were the Commander in Chief Devonport William Parker, a Lord of the Admiralty Maurice Frederick Fitzhardinge Berkeley, the Inspector General of Fortifications John Fox Burgoyne and the Admiral of the Fleet Thomas Byam Martin.

Martin, who had had previous dealings with Faraday[191], wrote in early August 1854 to Faraday with a list of questions to which the committee required answers so that it could assess the feasibility of Dundonald's proposal. Most of these questions were of a strictly naval character and Faraday declined to answer them on the grounds that he lacked sufficient data or knowledge. On the general scheme he said that it was doubtful if it could be made to work as it was too dependent on variables such as wind and anyway the defenders would probably have time to take defensive precautions[192]; he later, as usual, declined payment for this advice[193]. Faraday's report, apart from the first sentence where he denied any especial knowledge of the matter, was quoted in full in the report[194] which the committee submitted, on 9 August 1854, to the First Lord of the Admiralty, James Robert George Graham. On this basis of this report, and thus of Faraday's comments, Dundonald's proposal was rejected. It was floated again later when the war appeared to be going badly, but was again rejected[195]. The general outlines of the plan were not published until the early twentieth century[196] and it was not until 1946 that the Admiralty file containing the details was declassified and released.

The other area where Faraday may have given advice was on the construction of prefabricated hospital wards. In response to criticisms of the medical facilities in the Black Sea, the minister of war, Benjamin Hawes, commissioned his brother in law, Isambard Kingdom Brunel, to build a prefabricated hospital that could be transported to Turkey. One of the wards was erected at Paddington Station to be inspected by politicians and men of science. These included Faraday who Brunel asked particularly to examine the ventilation[197]; unfortunately, Faraday's comments have not been found.

Although Faraday was kept busy with his various public commitments, he also resigned some of them during this period. For example he retired from being Professor of Chemistry at the Royal Military Academy, Woolwich, in February 1852[198], but he seems to have been thinking about this since the previous December when he had written a long letter to Joseph Ellison Portlock about some proposals by the Academy to modify his course of instruction[199]. His resignation from this position, which he had held since 1829[200] meant that he would lose an annual income of £200. No doubt this reduction explains why at exactly the same time Faraday resigned from membership of the Athenaeum Club, of which he had been the founding Secretary in 1824, pleading diminution of income as the reason[201].

All this work in science and for the state and its agencies was conducted against a background of Faraday's ill health. A glance at the index under "Faraday, health of" reveals the large range of symptoms from which he suffered during this period. Most of the serious problems, that is the giddiness, lack of memory, headaches etc, were continuations of the problems that he had suffered from throughout the 1840s. In addition he had serious dental problems in the summer of 1850 when eight teeth were extracted and others filled[202] and he suffered from an attack of deafness in late 1853[203]. In August and September 1852 Faraday suffered from what he called "one of the old time depressions"[204] and Sarah reported that he had "been so depressed & unable to bear a part, even in casual conversation"[205]. It is not clear from the context exactly what the problem was; no document has been found which uses the term 'depression' earlier in reference to Faraday, but, as Cantor has noted, accounts of Faraday's health were policed carefully by his early biographers[206]. It is thus difficult to determine how much significance should be attached to these comments which were given as the reason for not attending the Belfast meeting of the British Association.

In addition to Faraday's continual health problems, Sarah, nine years his junior, started also to suffer from bad health. At the same time as Faraday was suffering from depression, in August and September 1852, she was having difficulty in walking any significant distance[207]. Thereafter she is occasionally referred to as infirm[208]. For much of 1854, starting in February[209], she experienced deafness[210]. In May she went to see the ear surgeon Joseph Toynbee[211], but the following month Faraday wrote from the Isle of Wight to Bence Jones asking for advice[212] and she was still deaf at the end of August[213]. Thereafter there are no further references to this problem and presumably it cleared itself up. Bence Jones only gradually became physician to the Faradays. He first treated Sarah in 1851[214], but there is no indication that he dealt with Faraday until 1854 when Benjamin Collins Brodie also gave advice[215]. Brodie, in the early 1850s, was the doctor of choice for Faraday and he also treated Sarah[216].

Part of the Faradays' strategy for dealing with their illnesses was to go away from London[217]. Usually they visited the seaside at Brighton (for winter breaks), Hastings and the Isle of Wight or the Sandemanian communities at Old Buckenham and at Newcastle. They took a holiday in Scotland in the Summer of 1849 (possibly also to visit the Glasite communities there) and one in North Wales in July 1853. During the first half of the 1850s there were no visits to the Continent as there had been in the 1840s. Another part of their strategy was to rent houses on the outskirts of London and for Faraday to commute to the Royal Institution. They had done this a couple of times during the 1840s, but during this period it became an annual occurrence: Wimbledon (Spring 1849), Norwood (Summer 1850), Blackheath (Autumn 1851), Hampstead

(Autumn 1852), Hammersmith (Autumn 1853), Redhill (Spring 1854), Surbiton (Autumn 1854), Sydenham (Autumn 1855). However, such stays away from the Royal Institution did not entirely free Faraday from the pressures of fame and the bustle of Albemarle Street. His identity would become quickly known in the locality[218] and when staying in Surbiton he was invited by the Mayor of Kingston to the Venison Feast[219]. Although these stays away from London seem to have been partially intended to allow Faraday to do some work, they are also an indication that he was beginning to adopt a more relaxed approach to life. This is also evident from some of his letters. For example he attended the dinner of the Royal Geographical Society in 1853, the first time, as he noted, that he had been to such an event for many years[220]. He also visited the opera on several occasions[221], although the extent of our knowledge of these visits may be only a function of the letters that have survived.

Faraday's fame continued to grow during the first half of the 1850s. In the scientific community, Faraday was compared by three of his correspondents with Isaac Newton[222], something that had not been done before; Faraday's responses, if any, to this comparison, have, unfortunately, not been found. Faraday continued to be visited by members of the scientific community from overseas who recorded their impressions of him. For example Hermann Helmholtz, who visited in 1853[223], wrote: "[Faraday] is as simple, charming, and unaffected as a child; I have never seen a man with such winning ways. He was, moreover, extremely kind, and showed me all there was to see. That, indeed, was little enough, for a few wires and some old bits of wood and iron seem to serve him for the greatest discoveries"[224]. Bence Jones commissioned George Richmond to paint him[225]; James Walker commissioned Matthew Noble to sculpt him[226]. In 1852 Harriet Moore (with whom, as their letters show, the Faradays had formed quite a close friendship) executed a series of watercolours of Faraday's basement laboratory in the Royal Institution including one showing Faraday working in his laboratory[227]. But perhaps the most public visual image of Faraday was his appearance in a full page *Punch* cartoon handing his card to Father Thames. This was published shortly after his letter to the *Times* in July 1855 criticising the state of the river[228]. Faraday reappeared in *Punch* the following month when, with the war against Russia apparently still going badly, the magazine called for Faraday to be brought in to suggest how chemistry should be used to defeat the Russians[229]. Aside from confirming that Faraday's consideration of Dundonald's plan had indeed remained secret, this suggestion by *Punch* also provides evidence that the public at large viewed Faraday as the person to turn to for advice on using science in practical contexts.

It was doubtless the popular fame enjoyed by Faraday which prompted the editor of *Household Words*, Charles Dickens, early in the magazine's history to ask Faraday for permission to produce an abstract of two of his lecture series. Faraday agreed to this and lent Dickens his lecture notes[230]. In return Dickens may have given Faraday a copy of *David Copperfield*[231]. Dickens had already mentioned Faraday in the first volume of *Household Words* and it is perhaps this that summed up Faraday's fame in the early 1850s: "On its [the gymnotus] first arrival in England, the proprietors offered Professor Faraday (to whom this country may possibly discover, within the next five hundred years, that it owes something) the privilege of experimenting upon him for scientific purposes ..."[232].

1. For instance Faraday continued corresponding with George Biddell Airy, Thomas Andrews, John Barlow, Arthur-Auguste De La Rive, Jean-Baptiste-André Dumas, William Robert Grove, Jacob Herbert, John Frederick William Herschel, Justus Liebig, Edward Magrath, Thomas Phillipps, Joseph Antione Ferdinand Plateau, Lyon Playfair, Julius Plücker, Lambert-Adolphe-Jacques Quetelet, Edward Sabine, Christian Friedrich Schoenbein, William Thomson, Benjamin Vincent, Charles Richard Weld, William Whewell and a few others to a lesser extent.
2. Richard Phillips (1788-1851, DNB). Chemist and curator of the Museum of Practical Geology, 1839-1851. (Those discussed only in the introduction are identified in the notes here. All other individuals are identified in the appropriate place in the Correspondence).
3. Other friends and correspondents of Faraday's who died during this period include Arthur Aikin, Dominique François Jean Arago, William Brockedon, Edward Cowper, Andrew Crosse, Edmund Robert Daniell, Henry Thomas De La Beche, Edward Forbes, Ada Lovelace, Gideon Algernon Mantell, John Martin, Hans Christian Oersted and Nathaniel Wallich.
4. Letters 2875, 2881, 2890, 2902.
5. Letters 2255 and 2762.
6. Letter 2255.
7. Abraham Lincoln (1809-1865, DAB). Lawyer and politician in Illinois.
8. Lincoln to Melloni, 1853, Hertz (1931), **2**: 623-5.
9. Letter 2932.
10. Adam Greenlaw Gray (1778-1849, GRO). Retired saddler.
11. For Faraday see letters 2283 and 2284 and for Boyd see letter 2703.
12. Letter 2676.
13. Letters 2297, 2298, 2299 and 2320.
14. Letters 2548, 2549, 2551.
15. Letter 2551.
16. Letters 2889 and 2890. For his executors see below.
17. Letter 2335.
18. See Cantor (1989) and Cantor (1991a), 61-3, 275.
19. Letter 2340. On this episode see Cantor (1991a), 275-7, 345.
20. Letter 2918 quoting from Acts 15: 29.
21. Faraday to Barlow, 2 January 1845, letter 1668, volume 3.
22. Letter 2918.
23. Letter 2931.
24. Letter 2935.
25. On this episode and Faraday's role in it see Cantor (1991a), 68-70.
26. Letters 2583 and 2584.
27. See Cantor (1991a), 89.
28. See, for example, letter 2531.
29. Letter 2284.
30. Letter 2407.
31. Letter 2462.
32. Letter 2503.
33. Letter 2551.
34. Letter 2703.

35. Letter 3021.
36. Letter 2691.
37. IEE MS SC 2. DJB was David James Blaikley (1846-1936) who oversaw the ultimate disposal of Faraday's archive. See volume 1, p.xxiv.
38. Letter 2674.
39. Letters 2675 and 2677.
40. Letter 2678.
41. Letter 2679.
42. Letter 2686.
43. Letter 2688.
44. Faraday's notes are in IEE MS SC 2. The chemist and scientific journalist William Crookes (1832-1919, DSB) later claimed in Crookes (1871), 11 that he was present on one of these occasions. Although Crookes is not listed by Faraday in the manuscript as being present, there were a number of people there whom Faraday did not identify.
45. Letter 2691.
46. See volume 3, p.xxxviii.
47. "Professor Faraday on Table-Moving", *Athenaeum*, 2 July 1853, pp.801-3.
48. *Ibid.*, , p.802.
49. Letter 2691.
50. *Ill.Lond.News*, 2 July 1853, **22**: 530.
51. *Ill.Lond.News*, 16 July 1853, **23**: 26.
52. Letters 2693, 2965, 2697 and 2698.
53. Elizabeth Browning, née Barrett (1806-1861, DNB). Poet.
54. Browning to Ogilvy, 21 July 1853, Heydon and Kelley (1974), 100-4. See also Browning to Mitford, 15 July 1853, Raymond and Sullivan (1983), 388-91 and Browning to Chorley, 10 August 1853, Kenyon (1897), **2**: 127-31.
55. Letter 2701.
56. Letters 2712, 2724, 2748 and 2752.
57. Letter 2703. See also Faraday's letter to Schoenbein two days later, letter 2705.
58. Letter 2703.
59. Secord (1989).
60. Letters 2277, 2279 and 2280.
61. Letters 2769 and 2770.
62. Crookes (1871).
63. Tyndall (1879), 496. This article, first written in 1864, did not originally mention Faraday by name, but referred to him as "a celebrated philosopher". Tyndall (1871), 427.
64. Who seems to have been one of the leading lights behind the idea. See Tyndall, *Diary*, 4 February 1854, **5**: 316 and Forgan (1977), 219-20.
65. RI MM, 6 February 1854, **11**: 44.
66. For the details of the lectures see letter 2804.
67. Tyndall, *Diary*, 4 February 1854, **5**: 316.
68. Letters 2797 and 2804.
69. Letter 2806. The Committee explicitly thanked Faraday for using his influence with Whewell. RI MM, 20 March 1854, **11**: 52.
70. Letter 2808.
71. Tyndall, *Diary*, 7 May 1854, **5**: 343.
72. See letters 2838 and 2839.
73. Letter 2819.
74. Faraday (1854f). For discussions of the lecture see Crawford, E. (1985), Gooding (1991) and Cantor (1991b).
75. RI MM, 20 March 1854, **11**: 52.
76. RI MM, 8 May 1854, **11**: 60 and 15 May 1854, **11**: 62.
77. Anon (1855).
78. Letter 2842.
79. Letter 2853.
80. *Lit.Gaz.*, 10 June 1854, p.532.

81. *Lit.Gaz.*, 1 July 1854, pp.612-3.
82. George Eliot (Mary Ann Evans, 1819-1880, DNB under Cross). Journalist and writer.
83. Faraday (1851a). Friday Evening Discourse of 24 January 1851.
84. George Eliot to Charles and Caroline Bray, 28 January 1851, Haight (1954), 341-4.
85. See, for example, letters 2710 and 2717.
86. Letters 2223 and 2917.
87. Letter 2225.
88. Letter 2596.
89. See letters 2215, 2298, 2524 and 2710. There are many more examples of this aspect of Faraday's work in RI MM.
90. Letters 2154, 2155, 2371 and 2380.
91. Letters 2516, 2524, 2525, 2527 and 2529.
92. Forgan (1977), 141-2.
93. Letters 2206 and 2207.
94. Letter 2209.
95. Letter 2210.
96. RI MM, 3 December 1849, **10**: 226 and 11 December 1849, **10**: 229. See also letters 2242 and 2243.
97. See Forgan (1977), 142-8 for a detailed discussion of this episode.
98. Letter 2532.
99. RI MM, 23 May 1853, **11**: 13-14.
100. RI MM, 6 June 1853, **11**: 19. On his appointment see Forgan (1977), 151-4.
101. Letter 2308.
102. Letter 2452.
103. Letter 2639.
104. Letter 2861.
105. See letters 2463 and 2636.
106. Letter 3027.
107. Letter 2363.
108. Letter 2364.
109. Faraday (1847a).
110. Letter 2372.
111. Letter 2373.
112. Letter 2560.
113. Wolf (1852c).
114. Letter 2613.
115. Letter 2615.
116. Letter 2618.
117. Bence Jones (1852).
118. Letters 2520, 2521 and 2522.
119. Letter 2640.
120. Letter 2647.
121. Letters 3016 and 3017.
122. Cawood (1977, 1979).
123. Faraday, *Diary*, 23 July 1850, **5**: 10954-63.
124. A result he had established in Faraday (1847a), 417.
125. Faraday, *Diary*, 23 July 1850, **5**: 10958.
126. Letter 2317.
127. Faraday (1851d), ERE26, 2847.
128. Letter 2322.
129. Letter 2325.
130. Faraday (1851d, e), ERE26 and 27 dated 14 September 1850 and 16 November 1850 respectively.
131. Letters 2394 and 2427.
132. Letter 2415.
133. Faraday (1850, 1851c), ERE23 and 25.

134. Letters 2310, 2311, 2317, 2320 and 2322. See Faraday (1851c), ERE25, 2790 for the first use of the term.
135. Letters 2475, 2485 and 2494. This was repeating the epistemological strategy that Faraday had pursued in 1822 to demonstrate the reality of electro-magnetic rotations (Gooding (1985), 120-2) and in 1846 in distributing pieces of heavy glass to allow colleagues to see the magneto-optical effect (Introduction to volume 3, p.xxxvii). He also seems to have done this in 1850 with some of his work on atmospheric magnetism. See letter 2341.
136. Faraday (1852c, d), ERE29 and [29a].
137. Hunt (1991), 1.
138. Faraday, *Diary*, 15 October 1853, **5**: pp.393-408.
139. Letters 2742, 2743, 2749, 2746, 2767.
140. Letter 2751.
141. Letter 2749.
142. Faraday (1854a).
143. Letters 2780, 2781 and 2782.
144. Letters 2791, 2792, 2793 and 2794.
145. Letters 2813, 2834, 2846, 2862, 2865 and 2870.
146. Letter 2875.
147. Faraday (1855d).
148. Faraday (1852d, 1855b), [ERE29a and 29b].
149. Faraday (1855c).
150. Letter 2974.
151. Letter 2257.
152. Letter 2261.
153. Letter 2405.
154. Letter 2432.
155. Letter 2472.
156. Fox, F. (1904), 7-8. See Joseph Henry's Diary for 31 July 1837, Reingold and Rothenberg (1972-99), **3**: 426-7.
157. Letter 2362.
158. Letters 2366, 2367, 2754, 2756, 2757, 2758 and 2783.
159. Letter 2424.
160. Letter 2684.
161. Letter 2446.
162. Letter 2466.
163. Letter 2744.
164. Letter 2721.
165. Letter 2722.
166. Letter 2723.
167. Letter 2728.
168. Letters 2528, 2539 and 2543.
169. Letters 2230, 2231, 2385, 2386, 2533, 2822, 2825, 2975, 2976, 2997 and 2998.
170. Letters 2288, 2319, 2507, 2513, 2518, 2533, 2655, 2661, 2692, 2810, 2940, 2956 and 2958.
171. Letters 2726 and 2907.
172. Letters 2204, 2238 and 2692.
173. Letter 2238.
174. Letter 2599.
175. For the details see Watson, (1853).
176. Letter 2606.
177. Letter 2608.
178. Letters 2609 and 2614.

179. Letter 2687.
180. Letter 2868.
181. Detailed notes of the trials, though not in Faraday's hand, are in an untitled notebook in IEE MS SC 2.
182. Letter 2878.
183. Cantor (1991a), 201-5.
184. Letter 2894.
185. Letter 2895.
186. Letter 2500 and 2501.
187. Letter 3027.
188. Tyndall, *Diary*, 22 February 1854, **5**: 322.
189. Letter 2965.
190. Lambert, A.D. (1990), 34.
191. Faraday to Martin, 12 June 1844, letter 1591, volume 3.
192. Letter 2871.
193. Letter 2887.
194. PRO ADM1 / 5632.
195. Lloyd, C. (1946).
196. Douglas and Ramsay (1908), **2**: 340-2.
197. Letter 2947.
198. Letter 2495.
199. Letter 2478.
200. Drummond to Faraday, 16 December 1829, letter 416, volume 1.
201. Letters 2479 and 2480. The annual subscription was six guineas.
202. Letters 2309 and 2315.
203. Letters 2761 and 2764.
204. Letter 2563.
205. Letter 2569.
206. Cantor (1991a), 261-2.
207. Letters 2556 and 2569.
208. Letters 2799, 2819, 2890 and 2965.
209. Letter 2787.
210. Letter 2819.
211. Letter 2833.
212. Letters 2860 and 2863.
213. Letter 2890.
214. Letter 2471.
215. Letter 2873.
216. Letter 2581.
217. See, for example, letters 2412, 2604, 2605 and 2869 which make this point explicitly.
218. See, for example, letter 2171.
219. Letter 2890.
220. Letter 2680.
221. See letters 2835, 2991 and 3009.
222. Letters 2415, 2861 and 3014.
223. Letter 2718.
224. Koenigsberger (1906), 110-11.
225. Plate 9. See letters 2504 and 2523.
226. Plate 1. See letter 2964.
227. Plate 8. Curiously none of the letters deal with Moore's paintings.
228. Letter 3003; plate 12.
229. "The Complaint of Chemistry", *Punch*, 4 August 1855, **29**: 44.
230. Letters 2291, 2292 and 2295.
231. Letter 2355.

232. "A Shilling's Worth of Science", *Household Words*, 1850, 1: 507-10, p.509. For his work on the gymnotus see Faraday, *Diary*, 26 November 1838 and 3 December 1838, 3: 5064-9 and Faraday (1839a), ERE15.

Biographical Register

This provides information on those individuals who are mentioned in three or more letters in this volume.

ABEL, Frederick Augustus (1827-1902, DNB2): Professor of Chemistry at Royal Military Academy, Woolwich, 1852-1888.

ACLAND, Henry Wentworth (1815-1900, DNB1): Dr Lee's Reader of Anatomy at Christ Church, Oxford, 1845-1858.

AIKIN, Arthur (1773-1854, DNB): Chemist who was formerly Secretary of the Society of Arts.

AIRY, Elizabeth (d.1879, age 76, GRO): Sister of George Biddell Airy. She lived at the Royal Greenwich Observatory.

AIRY, George Biddell (1801-1892, DSB): Astronomer Royal at the Royal Greenwich Observatory, 1835-1881.

AIRY, Richarda, née Smith (d.1875, age 70, GRO): Wife of George Biddell Airy whom she married in 1830. See his entry in DNB.

ALBERT FRANCIS CHARLES AUGUSTUS EMANUEL, Prince (1819-1861, DNB): Husband of Queen Victoria.

AMPERE, André-Marie (1775-1836, DSB): French physicist. Inspector General of the French university system. Taught philosophy (from 1819) and astronomy (from 1820) at University of Paris.

ANDERSON, Charles (d 1866, age 75, GRO): Originally a Sergeant in the Royal Artillery. Joined Faraday on 3 December 1827 as assistant on the project to improve optical glass. Bence Jones (1870a), 1: 398. After the end of the glass project, in 1830, Faraday paid him out of his own pocket (Faraday to South, 3 February 1865, IEE MS SC 3) before he was appointed Royal Institution Laboratory Assistant in 1832. RI MM, 5 November 1832, 8: 28.

ANDREWS, Jane Hardie, née Walker: Married Thomas Andrews in 1842. See his DNB entry.

ANDREWS, Thomas (1813-1885, DSB): Professor of Chemistry at Queen's College, Belfast, 1845-1879.

ARAGO, Dominique François Jean (1786-1853, DSB): French physicist and astronomer. Professor of Descriptive Geometry at Ecole Polytechnique. Director of the Paris Observatory. Co-editor of *Ann.Chim.* Permanent Secretary of the Académie des Sciences from 1830.

BABBAGE, Charles (1791-1871, DSB): Mainly worked on inventing mechanical calculating machines.

BADDELEY, Paul Frederick Henry (1806-1882, Crawford, D.G. (1930), 99): Surgeon in the Bengal Army, 1829-1855.

BARKER, Thomas (1825-1866, Cantor (1991a), 299): Merchant's clerk and a Deacon in the London Sandemanian Church.

BARLOW, Cecilia Anne, née Law (d.1868, age 72, GRO): Married John Barlow in 1824. See his entry in AC.

BARLOW, John (1798-1869, AC): Secretary of the Lecture Committee at the Royal Institution from 1841 and Secretary of the Royal Institution, 1843-1860.

BARNARD, Edward (1767-1855, GRO): Father of Sarah Faraday. Silversmith. See Grimwade (1982), 430-1.

BARNARD, Frank (1828-1895, GRO): A nephew of Sarah and Michael Faraday.

BARNARD, Jane (1832-1911, GRO): A niece of Michael and Sarah Faraday. Lived with them at the Royal Institution from at least 1851. Census returns for 1851, PRO HO107/1476, f.64.

BATE, Bartholomew (1806-1895, GRO): Son of Robert Brettel Bate.

BECQUEREL, Alexandre-Edmond (1820-1891, DSB): French physicist.

BECQUEREL, Antoine-César (1788-1878, DSB): French chemist who supported the contact theory of the Voltaic cell.

BELL, Thomas (1792-1880, DNB): Dental surgeon at Guy's Hospital, 1817-1861. Secretary of the Royal Society, 1848-1853.

BENCE JONES, Henry (1814-1873, DNB under Jones): Physician at St George's Hospital, 1846-1862. A Visitor, 1851-1853, and, from 1853, a Manager of the Royal Institution.

BERZELIUS, Jöns Jacob (1779-1848, DSB): Swedish chemist. Professor of Medicine and Pharmacy at Karolinska Institutet, Stockholm.

BIOT, Jean-Baptiste (1774-1862, DSB): French physicist.

BOIS-REYMOND, Emil Heinrich du (1818-1896, DSB): Electro-physiologist. Instructor in anatomy at Berlin Academy of Art, 1848-1853. Lecturer (1854) and then Associate Professor at University of Berlin, 1855-1858.

BOOSEY, Thomas (1795-1871, B1): Publisher and an Elder of the London Sandemanian Church, 1843-1860.

BRANDE, William Thomas (1788-1866, DSB): Professor of Chemistry at the Royal Institution, 1812-1852. Superintendent of Machinery, 1825-1852, and then of the Coining and Die Department, 1852-1866, at the Royal Mint.

BRAYLEY, Edward William (1802-1870, DNB): Writer on science.

BREDA, Jacob Gisbert Samuel van (1788-1867, NNBW): Secretary of the Hollandsche Maatschapij der Wetenschappen from 1839 and Director of the Teyler Museum.

BREWSTER, David (1781-1868, DSB): Man of science. Worked chiefly on optics. Principal of St Andrews University, 1838-1859.

BRODIE, Benjamin Collins (1783-1862, DSB): Surgeon at St George's Hospital.

BRODIE, Benjamin Collins (1817-1880, DSB): Secretary of the Chemical Society, 1850-1856.

BRUNEL, Isambard Kingdom (1806-1859, DNB): Civil engineer. Worked on Great Western Railway and many other projects.

BUCHANAN, George (c1790-1852, DNB): Scottish civil engineer and Edinburgh Sandemanian.

BUCHANAN, William (1781-1863, DNB): Lawyer and Elder of the Edinburgh Sandemanian church.

BUCKLAND, Mary, née Morland (1797-1857, Burgess (1967), 5-6, 69): Married William Buckland in 1825. See his DNB entry.

BUCKLAND, William (1784-1856, DSB): Geologist. Dean of Westminster, 1845-1856.

BURDETT COUTTS, Angela Georgina (1814-1906, DNB2): Philanthropist and heiress who lived at 1 Stratton Street (near Albemarle Street).

CHANCE, James Timmins (1814-1902, DNB2): Birmingham glass manufacturer.

CHRISTIE, Samuel Hunter (1784-1865, DSB): Professor of Mathematics at the Royal Military Academy, Woolwich from 1838. Secretary of the Royal Society, 1837-1854.

CLARK, James (1788-1870, DNB): Court physician. Member of the Senate of the University of London, 1838-1865. Influential in establishing the Royal College of Chemistry.

CLARK, Josiah Latimer (1822-1898, DSB): Assistant electrical engineer to the Electric and International Telegraph Company, 1850-1860.

CLARKE, Charles Mansfield (1782-1857, DNB): Physician.

CONOLLY, John (1794-1866, DNB): Physician specialising in mental illness who owned a private asylum in Hanwell.

COULOMB, Charles Augustin (1736-1806, DSB): French physicist.

COWPER, Edward (1790-1852, DNB): Professor of Manufacturing Art at King's College, London, 1840-1852.

COX, William (d.1864, age 66, GRO): A hotelier in Jermyn Street and table turner.

CROSSE, Andrew (1784-1855, DNB): Amateur man of science.

DANIELL, John Frederic (1790-1845, DSB): Professor of Chemistry at King's College, London, 1831-1845. Foreign Secretary of the Royal Society, 1839-1845.

DAVY, Humphry (1778-1829, DSB): Professor of Chemistry at the Royal Institution, 1802-1812. Knighted 1812, created Baronet, 1818. Secretary of the Royal Society, 1807-1812, President, 1820-1827.

DEACON, Caroline, née Reid (1816-1890, Reid, C.L. (1914)): Niece of Sarah Faraday. Married Thomas John Fuller Deacon on 18 June 1846, GRO.

DEACON, Constance (1847-1924, GRO): Daughter of Caroline and Thomas John Fuller Deacon.

DEACON, Henry (1822-1876, DNBmp): Industrial chemist.

DEACON, Thomas John Fuller (d.1901, age 78, GRO): Nephew in law of Sarah Faraday.

DE LA BECHE, Henry Thomas (1796-1855, DSB): Director of the Geological Survey from 1835.

DE LA RIVE, Arthur-Auguste (1801-1873, DSB): Professor of General Physics at Geneva from 1823 and, from 1825, of Experimental Physics. Swiss ambassador to London in 1850.

DE LA RIVE, Jeanne-Mathilde, née Duppa (1808-1850, DHBS): Writer and historian. Married De La Rive on 18 August 1826. Died 18 August 1850.

DE LA RIVE, Louise-Victoire-Marie, olim Fatio née Fatio (1808-1874, Choisy (1947), 50): Married Arthur-Auguste De La Rive on 24 May 1855.

DE LA RIVE, William (1827-1900, Choisy (1947), 51): Swiss politician and writer. Son of Arthur-Auguste and Jeanne-Mathilde De La Rive.

DE LA RUE, Warren (1815-1889, DSB): Printer and amateur astronomer.

DICKENS, Charles John Huffam (1812-1870, DNB): Journalist and novelist.

DRAPER, John William (1811-1882, DSB): Professor of Chemistry at New York University, 1839-1882.

DREW, John (1809-1857, B1): Astronomer and meteorologist in Southampton.

DUMAS, Ernest-Charles-Jean-Baptiste (1827-1890, DBF): Metallurgist and Director of the Mint at Rouen from 1852.

DUMAS, Hermenie, née Brongniart: Married Dumas in 1826. See Crosland (1992), 184.

DUMAS, Jean-Baptiste-André (1800-1884, DSB): Professor of Chemistry at the Sorbonne, 1841-1868. French Minister of Agriculture, October 1850 to January 1851. Co-editor of *Ann.Chim.*

EICHTAL, Adolphe Seligman d' (1805-1895, DBF): French banker and politician.

ELLIS, Henry (1777-1869, DNB): Librarian of the British Museum, 1827-1856.

EULER, Leonhard (1707-1783, DSB): Swiss mathematician and natural philosopher.

FARADAY, James (1817-1875, GRO): Gas engineer. Nephew of Faraday.

FARADAY, Sarah, née Barnard (1800-1879, GRO): Daughter of Edward and Mary Barnard. Married Faraday on 12 June 1821.

FEILITZSCH, Fabian Carl Ottokar von (1817-1885, P1, 3): Professor of Physics at University of Greifswald.

FLAUTI, Vincenzo (1782-1863, LUI): Mathematician and Secretary of the Royal Academy of Sciences, Naples.

FORBES, James David (1809-1868, DSB): Professor Natural Philosophy at Edinburgh University, 1833-1860. Secretary of the Royal Society of Edinburgh, 1840-1860.

FOX, Charles (1810-1874, DNB): Civil engineer who worked on the Great Exhibition and its move to Sydenham. Knighted 1851.

FOX, Mary, née Brookhouse: Married Charles Fox in 1830. See DNB under his entry.

FRANCIS, William (1817-1904, Brock and Meadows (1984), 97-128): Publisher.

FRANKLAND, Edward (1825-1899, DSB): Professor of Chemistry at Civil Engineering College, Putney, 1850-1851 and then at Owen's College, Manchester, 1851-1857.

FREDERICK WILLIAM IV (1795-1861, NDB): King of Prussia, 1840-1861.

GASSIOT, John Peter (1797-1877, DSB): Wine merchant and electrician.

GAUSS, Carl Friedrich (1777-1855, DSB): Director of the Göttingen Observatory, 1807-1855.

GORDON, Alexander (1802-1868, B1): Lighthouse engineer.

GRAHAM, Thomas (1805-1869, DSB): Professor of Chemistry at University College London, 1837-1854.

GRANT, Miss: Friend of Barlow.

GREY, Charles (1804-1870, B1): Private Secretary to Prince Albert, 1849-1861.

GROVE, William Robert (1811-1896, DSB): Lawyer and man of science.

GYE, Frederick (1810-1878, DNB): Manager of Covent Garden Opera, 1848-1878.

HARRIS, William Snow (1791-1867, DNB): Plymouth man of science who worked on electricity, particularly lightning conductors.

HENRY, Joseph (1797-1878, DSB): Secretary of the Smithsonian Institution, 1846-1878.

HENSLOW, John Stevens (1796-1861, DSB): Rector of Hitcham, 1837-1861. Professor of Botany at Cambridge University, 1825-1861. President of the Ipswich Museum, 1850-1861.

HERBERT, Jacob (d.1867, age 79, Gent.Mag., 1867, **3**: 262): Secretary of Trinity House, 1824-1856. Chaplin [1950], 183.

HERSCHEL, John Frederick William (1792-1871, DSB): Man of science who worked on astronomy, chemistry and physics. Lived and worked at Collingwood near Hawkhurst from 1840. Master of the Mint, 1850-1855.

HICKSON, William Edward (1803-1870, B1): Owner and editor of *Westminster Review*, 1840-1852.

HOFMANN, August Wilhelm (1818-1892, DSB): Professor of Chemistry at the Royal College of Chemistry, 1845-1865.

HOLLAND, Henry (1788-1873, DNB): Fashionable physician. Physician Extraordinary to Queen Victoria from 1837 and Physician in Ordinary to Prince Albert from 1840.

HUMBOLDT, Friedrich Wilhelm Heinrich Alexander von (1769-1859, DSB): German man of science and traveller.

HUNT, Robert (1807-1887, DNB): Chemist who worked particularly on photography. From 1845, Keeper of the Mining Record Office. Lecturer on mechanical science at the Royal School on Mines from 1851 and on experimental physics from 1853.

JOULE, James Prescott (1818-1889, DSB): Manchester physicist.

LACEY, Henry (d.1862, age 44, GRO): Porter at the Royal Institution, 1844-1862. RI MM, 5 February 1844, **9**: 289-90 and 5 May 1862, **11**: 423.

LAMONT, Johann von (1805-1879, DSB): Director of Bogenhausen Observatory from 1835 and Professor of Astronomy at Munich from 1852.

LATHAM, Peter Mere (1789-1875, DNB): Physician in London.

LEIGHTON, John (c1777-1857, Cantor (1991a), 301): Printer and Deacon of the Sandemanian Church from 1813.

LEIGHTON, Stephen (d.1881, age 83, GRO): Printer and an Elder of the London Sandemanian Church.

LIEBIG, Justus von (1803-1873, DSB): Professor of Chemistry at Giessen University, 1825-1851. Moved to Munich in 1852. Worked especially on organic and agricultural chemistry.

LLOYD, Humphrey (1800-1881, DSB): Irish physicist.

LOGEMAN, Wilhelm Martin (1821-1894, NNBW): Scientific instrument maker in Haarlem.

LOVELACE, Augusta Ada King, née Byron, Countess (1815-1852, DNBmp): Daughter of Lord Byron. Friend of Charles Babbage whom she helped with his calculating machines.

LYELL, Charles (1797-1875, DSB): Geologist.

MACOMIE, Alexander: A Deacon in the London Sandemanian Church. Otherwise unidentified.

MADAN, Frederick (d.1863, age 66, *Gent.Mag.*, 1863, **15**: 810): An Elder Brother of Trinity House, 1837-1863. Chaplin [1950], 86, 93.

MAGNUS, Heinrich Gustav (1802-1870, DSB): Professor of Technology and Physics at Berlin University, 1845-1870.

MAGRATH, Edward (d.1861, age 70, GRO): Secretary of the Athenaeum Club, 1824-1855. Waugh [1894].

MANBY, Charles (1804-1884, DNB): Civil engineer and Secretary of the Institution of Civil Engineers, 1839-1856.

MARCET, François (1803-1883, *Ann.Reg.*, 1883: 142): Anglo-Swiss man of science.

MARCET, Jane, née Haldimand (1769-1858, DNB): Popular scientific writer.

MARTIN, David W. (1798-1884, Cantor (1991a), 301): A member of the London Sandemanian Church.

MARTIN, Thomas Byam (1773-1854, DNB): Admiral of the Fleet, 1849-1854.

MARTINEAU, Harriet (1802-1876, DNB): Historian and writer.

MATTEUCCI, Carlo (1811-1868, DSB): Italian physiologist and physicist.

MELLONI, Macedonio (1798-1854, DSB): Director of the Physics Conservatory in Naples.

MELSENS, Louis Henri Frédéric (1814-1886, P2, 3): Professor in the veterinary school in Brussels.

MERRYWEATHER, George (1793-1870, Browne (1946), 132-4): Physician and curator of the Whitby Museum, 1847-1861.

MEYERBEER, Giacomo (1791-1864, GDMM): German composers of French operas.

MILNE-EDWARDS, Henri (1800-1885, DSB): French zoologist.

MOORE, Harriet Jane (d.1884, age 82, GRO): Painter and member of the Royal Institution, 1852-1881.

MURE, William (1799-1860, DNB): Conservative MP for Renfrewshire, 1846-1855. Chairman of the Select Committee on the National Gallery.

NEWMAN, John (d.1860, age 77, GRO): Scientific instrument maker.

NEWTON, Isaac (1642-1727, DSB): Natural philosopher.

NOBERT, Friedrich Adolph (1806-1881, DSB): German scientific instrument maker.

NORTHUMBERLAND, Algernon Percy, 4th Duke of (1792-1865, DNB): President of the Royal Institution, 1842-1865.

OERSTED, Hans Christian (1777-1851, DSB): Danish natural philosopher and Director of the Polytechnic Institute in Copenhagen, 1829-1851.

OWEN, Richard (1804-1892, DSB): Hunterian Professor of Comparative Anatomy at the Royal College of Surgeons, 1836-1856.

PAKINGTON, John Somerset (1799-1880, DNB): Conservative MP for Driotwich, 1837-1874.

PALAGI, Alessandro: Physicist in Bologna.

PARADISE, William (d.1866, age 78, GRO): A member of the Newcastle Sandemanian Church. Cantor (1991a), 68.

PELLY, John Henry (1777-1852, DNB): Deputy Master of Trinity House, 1834-1852. Arrow (1868), 44.

PELTIER, Jean Charles Athanase (1785-1845, DSB): French physicist.

PERCY, Grace, née Piercy (d.1880, age 64, GRO): Married John Percy in 1839. See under his entry in DNB.

PERCY, John (1817-1889, DSB): Metallurgist in Birmingham and the taught at Royal School of Mines, 1851-1879.

PHILLIPPS, Thomas (1792-1872, DNB): Antiquary and bibliophile.

PHILLIPS, John (1800-1874, DSB): One of the founders of the British Association and its Assistant General Secretary until 1859.

PLATEAU, Joseph Antione Ferdinand (1801-1883, DSB): Professor of Physics at the University of Ghent, 1835-1872. Totally blind from 1843.

PLAYFAIR, Lyon (1818-1898, DSB): Special Commissioner and member of the Executive Committee of the Great Exhibition. Secretary for Science at the Department of Science and Art, 1853-1855.

PLUCKER, Julius (1801-1868, DSB): Professor of Physics, 1847-1868, at the University of Bonn.

POGGENDORFF, Johann Christian (1796-1877, DSB): Editor of *Annalen der Physik und Chemie*, 1824-1877.

POISSON, Siméon-Denis (1781-1840, DSB): French mathematical physicist.

POLLOCK, Juliet, née Creed (d.1899, age 80, GRO): Writer. Married William Frederick Pollock in 1844.

POLLOCK, William Frederick (1815-1888, DNB): A Master of the Court of Exchequer, 1846-1886.

QUETELET, Lambert-Adolphe-Jacques (1796-1874, DSB): Astronomer at the Brussels Observatory from 1828 and Permanent Secretary of the Brussels Academy from 1834.

RANSOME, George (1811-1876, Markham (1990), 25): Secretary of the Ipswich Museum, 1846-1852.

REES, Richard van (1797-1875, BNB): Professor of Mathematics and Physics at the University of Utrecht, 1838-1867.

REGNAULT, Henri Victor (1810-1878, DSB): Professor of Physics at the Collège de France, 1841-1854.

REICH, Ferdinand (1799-1883, ADB): Professor of Physics at Freiberg, 1827-1866.

REID, David (1792-1868, Reid, C.L. (1914)): A brother in law of Sarah Faraday. Reid, C.L. (1914) gives him as a silversmith, but White (1847), 114 also lists him as a chronometer and watchmaker in Newcastle.

REID, Edward Ker (1821-1885, Fallon (1992), 244-7): London silversmith and a Deacon of the London Sandemanian Church. Cantor (1991a), 301.

REID, Margery Ann (1815-1888, Reid, C.L. (1914)): Niece of Sarah Faraday. Lived with the Faradays at the Royal Institution between about 1826 and 1840.

REID, Elizabeth (b.1830, Reid, C.L. (1914)): A niece of Sarah Faraday. Made Confession of Faith in London Sandemanian Church on 4 July 1847. Excluded 24 August 1851 but Restored; London Meeting House Records.

RIESS, Peter Theophilus (1804-1883, ADB): Professor of Physics at the Academie der Wissenschaften in Berlin from 1842

ROSE, Heinrich (1795-1864, DSB): Professor of Chemistry at Berlin University from 1835.

ROSSE, William Parsons, 3rd Earl of (1800-1867, DSB): Irish astronomer. President of the Royal Society, 1848-1854.

RUHMKORFF, Heinrich Daniel (1803-1877, DSB): German scientific instrument maker who worked in Paris.

SABINE, Edward (1788-1883, DSB): Colonel in Royal Artillery. Worked on terrestrial magnetism. Foreign Secretary of the Royal Society, 1845-1850 and Treasurer, 1850-1861.

SCHLAGINTWEIT, Adolph (1829-1857, ADB): Geologist at the University of Munich.

SCHLAGINTWEIT, Hermann Rudolph Alfred (1826-1882, ADB): German geologist and explorer.

SCHOENBEIN, Berta (1846-1927, private communication from Peter Nolte and Ulf Bossel): Daughter of Christian Friedrich Schoenbein and Emilie Wilhelmine Luise Schoenbein.

SCHOENBEIN, Christian Friedrich (1799-1868, DSB): Professor of Physics and Chemistry at University of Basle, 1835-1852 and the Professor of Chemistry.

SCHOENBEIN, Emilie (1836-1859, GRO): Daughter of Christian Friedrich Schoenbein and Emilie Wilhelmine Luise Schoenbein.

SCHOENBEIN, Emilie Wilhelmine Luise, née Benz (1807-1871, Bidlingmaier (1989), 35): Married Christian Friedrich Schoenbein in 1835.

SCHOENBEIN, Fanny Anna Franziska (1840-1921, private communication from Peter Nolte and Ulf Bossel): Daughter of Christian Friedrich Schoenbein and Emilie Wilhelmine Luise Schoenbein.

SCHOENBEIN, Wilhelmine Sophie (1838-1914, private communication from Peter Nolte and Ulf Bossel): Daughter of Christian Friedrich Schoenbein and Emilie Wilhelmine Luise Schoenbein.

SCHWABE, Samuel Heinrich (1789-1875, DSB): German amateur astronomer.

SHEPHERD, John (d.1859, age 63, GRO, B3): Deputy Master of Trinity House, 1852-1859.

STEVENSON, William (1820-1883, Duns (1883)): Amateur meteorologist in Duns.

STOKES, George Gabriel (1819-1903, DSB): Fellow of Pembroke College, Cambridge, 1841-1857. Lucasian Professor of Mathematics, University of Cambridge, 1849-1903. Taught at the Royal School of Mines, 1849-1856. Secretary of the Royal Society, 1854-1885.

SYKES, William Henry (1790-1872, DNB): Member of the Board of Directors of the East India Company.

TALBOT, William Henry Fox (1800-1877, DSB): One of the inventors of photography.

TAYLOR, Alfred Swaine (1806-1880, DNB): Medical jurist.

TAYLOR, Richard (1781-1858, DNB): Publisher and one of the editors of the *Philosophical Magazine*.

THOMSON, William (1824-1907, DSB): Fellow of Peterhouse, Cambridge, 1845-1852. Professor of Natural Philosophy, University of Glasgow, 1846-1899.

TOWLER, George: Pharmacist in Norwich. Otherwise unidentified.

TWINING, Thomas (1806-1895, DNB): Writer on technical education.

TYNDALL, John (1820-1893, DSB): Taught science at Queenwood College, 1847-1848 and again 1851-1853. Studied at University of Marburg, 1848-1850. Professor of Natural Philosophy at the Royal Institution, 1853-1887. An editor of the *Phil.Mag.*, 1854-1863.

VICTORIA, Queen (1819-1901, DNB): Queen of England, 1837-1901.

VINCENT, Benjamin (1818-1899, B3): Elder of the London Sandemanian Church, 1849-1864. Assistant Secretary of the Royal Institution from 1848 and Librarian, 1849-1889.

VINCENT, William R. (1798-1867, Cantor (1991a), 302): Member of the London Sandemanian Church.

VOLPICELLI, Paolo (1804-1879, P3): Secretary of the Pontifical Academy of Sciences, 1847-1877.

VOLTA, Alessandro Giuseppe Antonio Anastasio (1745-1827, DSB): Italian natural philosopher.

VULLIAMY, Lewis (1791-1871, DNB): Architect to the Royal Institution.

WALKER, James (1781-1862, Smith, D. (1998)): Civil and marine engineer.

WARREN, Samuel (1807-1877, DNB): Writer and lawyer.

WATERHOUSE, George Robert (1810-1888, DNB): Keeper of the Mineralogical and Geological Collections at the Royal Institution, 1851-1857.

WATSON, Joseph John William: Worked on electrical light. Possessed a PhD from an unidentified university and is listed as a member of the Geological Society between 1852 and 1888. See James (1997), 294.

WEBER, Wilhelm Eduard (1804-1891, DSB): German physicist.

WELD, Charles Richard (1813-1869, DNB): Assistant Secretary of the Royal Society, 1843-1861.

WHEATSTONE, Charles (1802-1875, DSB): Professor of Experimental Philosophy at King's College, London, 1834-1875. Worked on sound and electricity.

WHEWELL, William (1794-1866, DSB): Master of Trinity College, Cambridge, 1841-1866. Primarily an historian and philosopher of science.

WHITELAW, George (d.1872, age 68, GRO): Publisher's manager and an Elder of the London Sandemanian Church. Cantor (1991a), 302.

WIEDEMANN, Gustav Heinrich (1826-1899, DSB): Professor of Physics at University of Basle, 1854-1863.

WILKINS, William Crane: Lighthouse and patent lamp manufacturer of 24 and 25 Longacre. POD.

WILLIAMSON, Alexander William (1824-1904, DSB): Professor of Practical Chemistry at University College London, 1849-1887 and from 1855-1887 also Professor of Chemistry.

WILSON, George (1818-1859, DNB): Chemical lecturer and writer in Edinburgh.

WOHLER, Friedrich (1800-1882, DSB): Professor of Chemistry at Göttingen, 1836-1882.

WOLF, Johann Rudolf (1816-1893, DSB): Professor of Astronomy at the University of Bern, 1844-1855.

WROTTESLEY, John (1798-1867, DNB): Astronomer. President of the Royal Society, 1854-1858.

The Correspondence

Letter 2146
Faraday to Charles Richard Weld
8 January 1849[1]
From the original in APS Misc MS Collection

R Institution | 8 Jany 1848 [sic]

My dear Sir
    I am happy to say that in England I am not a Sir and do not intend (if it depends upon me) to become one. The Prussian Knighthood[2] I am in hopes will appear in the list[3] in its due form for in that I do feel honored, in the other I should not.
    Ever Truly Yours | M. Faraday
C.R. Weld Esq | &c &c &c

Endorsement: This letter was written in answer to one from Mr Weld desiring to know whether there was truth in a statement made by a writer in the Edinburgh Review (No *Jan* 1848 [sic]) that Faraday was now *Sir Michael*[4].

1. Dated on the basis of the references contained in this letter.
2. See Humboldt to Faraday, 18 August 1842, letter 1420, volume 3 and Bunsen to Faraday, 21 August 1842, letter 1421, volume 3.
3. This was noted in the List of Fellows of the Royal Society in November 1848, but not in that of November 1847.
4. This was asserted in [Joyce] (1849), 48.

Letter 2147[1]
Faraday to William Buckland
12 January 1849[2]
From the original in RS MS Bu 154

Royal Institution | 12 Jany. 1848 [sic]

My dear Dean
    I shall be hard at work at the Trinity House tomorrow evening or I should have made an exertion to reach the Deanery, as it is I shall not be able.

My Soap bubbles were all very good but my Carbonic acid was too recently prepared, indeed only the moment before[.] I had learnt the lesson before but in the hurry of the moment forgot it again.

I wish I could come tomorrow night that we might blow Soap bubbles against each other. What a beautiful & wonderful thing a soap bubble is?

My sincere respects to Mrs. Buckland[.]

Ever Truly Yours | M. Faraday

Revd. Dr. Buckland | &c &c &c

1. This letter is black-edged, due to the death of one of Sarah Faraday's brothers, the silversmith William Barnard (1801–1848, GRO, Grimwade (1982), 431) on 20 October 1848.
2. Dated on the basis that this letter refers to Faraday's fourth Christmas lecture delivered on 6 January 1849 "On the Chemical History of a Candle", RI MS GB 2: 50. For the reference to soap bubbles see Faraday (1861), 108.

Letter 2148[1]
Faraday to Henry Holland
17 January 1849
From the original in WIHM MS FALF

R Institution | 17 Jany 1849

My dear Sir

I shall be at the Great Magnet tomorrow from 11 o clk until 1 or 2 o'clk and if any later hour is more suitable to you I can hold matters on until 3 or 4 oclk.

Ever Yours | M. Faraday

Dr. H. Holland | &c &c &c

1. This letter is black-edged. See note 1, letter 2147.

Letter 2149[1]
Faraday to Robert Hunt
22 January 1849
From the original in RI MS F1 C27

R Institution | 22 Jany 1849

My dear Sir

Very large magnets have no particular temptation for me nevertheless I should like to see them if time place & opportunity should serve but of that your note gives me no idea.

Ever Truly Yours | M. Faraday
Robt Hunt Esq | &c &c &c

1. This letter is black-edged. See note 1, letter 2147.

Letter 2150[1]
Faraday to Charles Barry[2]
24 January 1849
From the original in SI D MS 554A

Royal Institution | 24 Jany 1849

My dear Sir
  I am very sorry to say that I have strained my power so far as to be almost unable to give. I send you one ticket[3] and cannot do more[.] I have asked Mr. Barlow but he with great regret says he cannot help me. It is very possible you may fall in with some M.R.I who has not used his tickets if you do secure them & in such case if you can cancel mine do: because it will so far ease the number of mine which will appear at the examination[.]
  Ever Truly Yours | M. Faraday
C. Barry Jur Esq | &c &c

1. This letter is black-edged. See note 1, letter 2147.
2. Charles Barry (1823–1900, B4). Architect.
3. That is for Faraday's Friday Evening Discourse of 26 January 1849, "On the Crystalline Polarity of Bismuth and other bodies, and its relation to the magnetic force", *Lit.Gaz.*, 10 February 1849, pp.96–7.

Letter 2151
Faraday to Edward Magrath
1 February 1849
From the original in WIHM MS FALF

Brighton | 1 Feby 1849

My dear Magrath
  I cannot be in town on Monday[1] at Mr Tuppers[2] ballot[3]. I hope there is nothing left undone by me that I can do here or ought to do.
  Ever Truly Yours | M. Faraday

1. That is 5 February 1849.

2. Arthur Chilver Tupper (d.1876, age 60, GRO). Brother of the popular writer Martin Farquhar Tupper (1810 - 1889, DNB). See Hudson (1949), 5.
3. Tupper was proposed by Brande and seconded by Faraday for membership of the Athenaeum Club. AC MS Certificate Book, 1849–50, number 1952. He was elected in 1849, Waugh [1894], 147.

Letter 2152
Julius Plücker to Faraday
7 February 1849
From the original in IEE MS SC 2
Dear Sir!

I thank you very much for your kind letter of the 14th of Dec.[1] by which you give me a short notice of your newest discoveries. You may think that I was a[n]xious to repeat them instantly, but being a poor german professor, obliged to give 3 lectures a day & charged too with different functions at the University, I found time to work only a few days ago. As my results are in some respect different from yours, I take [the] liberty to comunicate the following ones.

1. I first tried antimony and was extremely surprised to find that its magnecrystallic axis (perpendicular to the chief cleavage planes) points not axially but *equatorially*. It has been confirmed by all different modes of suspension. I could not attribute the different effects to the state of impurity of my specimen of Antimony, nevertheless I tried pure antimony taken from the laboratory of Professor Bischof[2]: but allways the same result.

2. Secondly I tried very fine and pure bismuth, and was satisfied to find its magnecrystallic power *very strong*, the magnecrystallic axis pointing axially quite so as you describe it. Then I experimented with the less perfect cleavage planes and I found that there is another corresponding magnecrystallic axis, that axis belonging to the chief cleavage planes being more affected by the magnet than the other ones. The following experiment may prove it.

A plate of bismuth bounded by chief cleavage planes, showed in these planes several parallel lines, indicating a second less perfect cleavage. If suspended horizontally in the level of the poles (the principal magnecrystallic axis being vertically and the two poles as near as possible) it pointed like a comon diamagnetic body, but lifted up one centimeter or nearly so, it turned instantly taking such a position, that the parallel lines mentioned above pointed equatorially and continued to do so even at a height of 10–15 centimeter[s] above the level of the poles. It proves that there is another magnecrystallic axis contained in a plane perpendicular to these lines and perpendicular itself to the less perfect cleavage planes.

If you think my observations wrong, I'll send you in a letter some of the pieces of antimony and bismuth used in my experiments.

My head turns round by the multitude of new facts. I am about to get new ones.

I hope you will realize Your imagination to see Germany, and then I hope too your first stay will be at Bonn. In this case I shall be happy to give you any information you like, and if you should desire to work then I'll be your assistant (My Electromagnet will become more powerfull than it is now, I think it proper to put on it 100 pounds more of copper wire). You have many things better in England than we have in Germany, but our climate, namely in the spring, is better than yours, and I am certain, that even a short stay in our country wou⟨ld⟩ do good to your health. No doubt our political s⟨tate⟩ is a disturbed one[3], but travelling or residing in our country you will scarcely perceive it; none of the English families, living here, left Bonn for that reason.

I did not yet see any of my papers in the English dress[4], though Mr Frances [sic] promised me some copies of them. I join to this letter a copy of my last papers[5] and will be very glad to see yours about the magnecrystallic axes[6].

I am Sir | very truly yours | Plücker
Bonn | 7 Febr. 1849
PS. Excuse Sir my bad English.

Address: A | Monsieur Faraday | &c &c &c | Royal Institution | London

1. Faraday to Plücker, 14 December 1848, letter 2136, volume 3.
2. Carl Gustav Christoph Bischof (1792–1870, DSB). Professor of Chemistry and Technology at the University of Bonn from 1819.
3. A reference to the instability of the governments of Prussia and of the other German states (including the overthrow of some) during 1848. See *Ann.Reg.*, 1848, 355–71, 375–400.
4. Plücker (1849e, f).
5. Probably Plücker (1848a, b, c).
6. Faraday (1849a, b), ERE22.

Letter 2153[1]
Faraday to William Buchanan
15 February 1849
From the original in the possession of Joan Ferguson

Royal Institution | 15 Feby 1849

My very dear & kind brother

Your letter arrived yesterday and was a great solace to us both comforting us with that comfort with which I trust you are yourself sustained and as you most truly say what have we to expect but frequently recurring blanks among our friends in their places as to this life and the speedy termination of our own[.] This morning we received accounts from Dundee that our dear friend Mr. John Duff was departed and at this moment the body of our late beloved brother Mr. Stopard[2] who died last Sabbath morning[3] lies for interment here. There is great happiness in the midst of the sorrow wherewith we think of the departure of such as these & our close friend your brother[4] for they were witnesses to the end & evidences that God both can & will keep his people who put their trust in him[.]

We have had several continual communications with George [Buchanan] about his family but yours were for the time the latest words and made us very thankful in the assurance they gave us that Charlotte[5] & all the others were improving steadily[.]

I am very glad to have the least hopes of your coming here this season & the same expectation will rejoice many others as they come to know it. If not inconvenient to your labours at the Houses[6] I hope you will come to us[.] We shall be ready & most happy to see you i.e. if it be the will of God to keep us in life until then for not only do the cases of those around but my own increasing hints of weakness remind me that we should say if the Lord will we shall do this or that. But if in the course of time you come up and we are here come to us i.e. if in all things it is suitable to you and others do not appear to you to have a higher claim.

I have almost forgotten the first matter which you mentioned, but I understand: who indeed has not felt the pressure of events[.] Say no more about it until it is entirely convenient to you[.]

My wife is out but still I say give our most affectionate remembrance to Mrs. Buchanan[7] and — I have no means at hand of recalling the name:- your daughter our friend at the Ferry and believe me
    Ever Your Affectionate & Grateful Friend | M. Faraday
William Buchanan Esq | &c &c &c

1. This letter is black-edged.
2. William Stoppard (d.1849, age 76, GRO). Farmer and Sandemanian.
3. That is 11 February 1849.
4. David Buchanan (1779–1848, DNB). Edinburgh journalist and Glasite.
5. Charlotte Buchanan, née Barnard (1805–1866, GRO). Sister of Sarah Faraday and wife of George Buchanan.
6. A reference to some piece of business in Parliament.
7. Elizabeth Buchanan, née Gregory. Wife of William Buchanan. See DNB under his entry.

Letter 2154[1]
Faraday to Angela Georgina Burdett-Coutts
23 February 1849
From the original in the possession of W.A.F. Burdett-Coutts
    Royal Institution | 23 Feby 1849.
My dear Miss Coutts
    The moment our Managers attention was called to the subject last Monday they resolved that our Lady Members should have the right of bringing their two friends to the reserved seats at *any hour* on the Friday Evening[2]. I am anxious to say this to you for I made one mistake to a Lady and I called upon you about another or what I feared was another and am glad to have this mode of clearing them all away[.]
    Ever Very Truly Yours | M. Faraday

1. This letter is black-edged. See letter 2153.
2. This resolution was not minuted in RI MM, 19 February 1849, **10**: 161–4.

Letter 2155[1]
Faraday to Louisa Maria Kerr[2]
23 February 1849
From the original in Boston University Library, Newell Collection

Royal Institution | 23 Feby 1849

My dear Madam

It gives me great pleasure to say that the Managers have ordered that *Lady Members* have the privilege of bringing their friend or two friends with them into the seats reserved for Members on Friday Evenings[3][.]

Ever Truly | Your Obedient Servant | M. Faraday

Mrs. Kerr

1. This letter is black-edged. See letter 2153.
2. Louisa Maria Kerr, née Hay (d.1890, age 70, GRO). Composer (see Ebel (1913), 76). Member of the Royal Institution, 1848–1890.
3. See letter 2154 and note 2.

Letter 2156
James Clark to Faraday
24 February 1849[1]
From the original in IEE MS SC 2

My dear Dr. Faraday,

I may tell *you* that the Prince does not care how much time you take in your lecture[2]. If it is an hour and a half I know he will not think it too much. I mention this in case you should think it necessary on his account to limit yourself to the hour. He will most likely be at the Institution a little before the 3 o'clock lecture. If you think I can give you any further information on the matter pray let me know.

You will be glad to hear that the College of Chemistry is flourishing - we have the rent of the big house in Hanover Square & now confine members to the laboratories. The College will soon be a very valuable institution, and I can assure you that your countenance has done it good service.

Very truly yours | Ja Clark

Brook Street | Saturday evening

1. Dated on the basis of the reference to Faraday's lecture.
2. Faraday gave a special Discourse "On Magnetic and Diamagnetic Bodies" in the presence of Prince Albert and 330 others on Monday 26 February 1849. See RI MM, 19 February 1849, **10**: 162–4 for the administrative arrangements and for accounts of the Discourse see Bence Jones (1870a), **2**: 243–4 and Curwen (1940), 233 . Faraday's notes are in RI MS F4 G31. Albert's attendance at the lecture was noted by Queen Victoria in her Diary (RAW MS).

Letter 2157
James Clark to Faraday
26 February 1849[1]
From the original in IEE MS SC 2

Brook Street | Monday evening

My Dear Dr. Faraday,

Although I have no doubt the Prince expressed to yourself his great satisfaction with your lecture[2], I cannot resist telling you how much he expressed himself to me pleased with the manner as well as the substance of your lecture. Indeed so much was he pleased with it, I expect he will come occasionally to the lectures at the Institution.

You managed the whole thing admirably. Your peroration I thought particularly happy.

Very truly yours | Ja. Clark

1. Dated on the basis of the reference to Faraday's lecture.
2. See note 2, letter 2156.

Letter 2158[1]
Faraday to Edmund Belfour[2]
3 March 1849
From the original in BL add MS 42240, f.20

Royal Institution | 3 Mar 1849

Sir

I beg to return you my sincere thanks for the favour of an admission to the Lectures at the College of Surgeons[3][.] I hope to profit by some of them though the appointed days & hours are in very frequent collision with my duties here & at Woolwich[4][.]

I am Sir | Your Very Obliged Servant | M. Faraday
Edd. Belfour Esq | &c &c &c

1. This letter is black-edged. See letter 2153.
2. Edmund Belfour (d.1865, age 75, B1). Secretary of the Royal College of Surgeons, 1814–1865.
3. This was a set of twenty four lectures by Richard Owen on the "Generation and Development of Animals" which commenced on Tuesday 13 March 1849 and continued on succeeding Thursdays, Saturdays and Tuesdays at 4pm. See *Lancet*, 1849, 1: 247.
4. That is Faraday's lectures at the Royal Military Academy.

Letter 2159
Faraday to Arthur Aikin
8 March 1849
From the original in Stanford University Library MS M121/4/1

R Institution | 8 Mar 1849

My dear Aikin

Can you inform me on the following points? Was not the lime light used very early in the Society of Arts room? at what date? is it recorded in the Transactions or in print? Was it or not the first proposal or application of the lime light?

I have a vague notion that the lime light was shewn first there & if so I have no doubt you can tell me at once[.]

Ever Yours | M. Faraday
A. Aikin Esq | &c &c &c

Letter 2160
Arthur Aikin to Faraday
9 March 1849[1]
From the original in IEE MS SC 2

7 Bloomsbury Square | 9 March

Dear Faraday

I have this morning looked over all the reports of the Committee of Chemistry during the time that I was Secretary to the Socy of Arts, but without finding any communication on the Lime-light. I am perfectly certain that Mr Gurney[2] exhibited this light one evening in the Society's great room shewing among other things the brilliancy & precision of the prismatic colours produced by means of it: but as no record exists of this I presume it was a mere exhibition unaccompanied by any written communication.

In April 1823 an improvement to the apparatus for the safe use of the oxyhydrogen blowpipe, by Mr Gurney[3], was favourably reported on by the Committee of Chemistry[4] & probably his exhibition of the lime-light took place about the same time.

Yours very truly | A. Aikin
M. Faraday Esq

1. Dated on the basis that this is the reply to letter 2159.
2. Goldsworthy Gurney (1793–1875, DNB). Inventor.
3. Gurney (1823). His blowpipe won the Gold Isis Medal of the Society of Arts.
4. RSA MS Minutes of Committees, 1822–1823, Chemistry Committee 11 April 1823, pp.30-2 and 26 April 1823, pp.40–1 discussed only this blowpipe. Faraday was not present at either of these meetings.

Letter 2161
John Conolly to Faraday[1]
19 March 1849
From the original in RI MS F1 I93a

The Lawn House | Hanwell | March 19, 1849

Sir,

I was very glad to hear yesterday that you were expected to pay a visit to Mrs. Giles[2]. It will give her great pleasure, and I do not think that it will produce any unfavourable effect.

I enclose you a paper of the trains (printed by the patients in the Hanwell Asylum). Each train from town reaches Southall a few minutes after the time of arrival at Hanwell; and from the Southall Station a conveyance may be readily had to Miss Dence's[3] house at Hayes. The journey from Paddington to Hanwell occupies about 20 Minutes.

I do not know what days or hours will suit you best; and perhaps you will prefer choosing your own:- but if your arrangements would permit you to leave town on *Friday* morning by train at $\frac{1}{2}$ past 8 & to breakfast here, I could have the pleasure of taking you to Hayes, & of coming back with you in time for the train to town at 2.

Or, on that, or any other day this week, except Thursday[4], I shall be at home to dinner at six. I need not add what pleasure it would give me & my family to see you.

Hayes is about 3 miles & a half beyond Hanwell.

Believe me dear Sir, | With great respect | Yours very Sincerely | J. Conolly

1. Recipient identified on the basis of provenance of manuscript and its location.
2. Possibly a relation by marriage of Faraday's.
3. Charlotte Dence (age 39 in 1851 Census returns, PRO HO107 / 1697, f.481). Part proprietor of a private lunatic asylum for gentlewomen at Grove House, West End Green, Hayes.
4. That is 22 March 1849.

Letter 2162
Faraday to T.L. Shuckard[1]
20 March 1849
From the original in MU Chemistry Department

R Institution | 20 Mar 1849

Sir
   The effect is electric and well known. Schoenbein who discovered the substance in Gun cotton & Collodion as it is usually called pointed it out.
   I am Sir | Very Truly Yours | M. Faraday
T.L. Shuckard

1. Unidentified.

Letter 2163
Jacob Herbert to Faraday
22 March 1849
From the original in GL MS 30108/1/49

Trinity House, London, | 22nd March 1849

Sir,
   The attention of the Elder Brethren having been sometime since drawn to the great breakage of Cylinders which takes place at the Dungeness Light House, and which, they are of opinion, arises from the inoperative State of the Ventilating Apparatus thereat, - they have accordingly directed me to transmit to you the accompanying copy of a Letter from Mr. Arthur Watson[1], the Principal Keeper of that Light House[2], which relates to this Subject, And they will be obliged by your communicating to me any opinion you may be enabled to form in respect of this Ventilating Apparatus not acting properly. -
   I beg to observe in reference to this Subject that the Lantern of the Dungeness Light House is 18. feet diameter, but only $4\frac{1}{2}$ feet of height in Glass.-
   I am, | Sir, | Your most humble Servant | J. Herbert
M. Faraday Esqre | &c &c &c

Endorsed by Faraday: 28 Mar 1849. | Called at Trinity house & talked with *Mr. Herbert* & *Captn Madden*. Found that the chief point was a reverse or wrong action of the pipes but whether occasional or constant much or little could not find - must write to the keeper - | Wrote to Dungeness & Freshwater Light Keepers | 3 April 1849[3]

1. Otherwise unidentified.
2. Watson to Herbert, 10 February 1849, GL MS 30108/1/49.
3. Letters 2166 and 2167.

Letter 2164
Joseph Antione Ferdinand Plateau to Faraday
25 March 1849
From the original in RI MS F2G, tipped in between pp. 2054–5 and 2058–9

Gand, 25 Mars 1849

Mon cher Monsieur Faraday.

Permettez-moi de vous offrir un exemplaire du mémoire que je viens de publier[1]. Ce travail constitue la suite de celui que j'ai eu l'honneur de vous envoyer il y a quelques années[2], et au sujet duquel vous avez bien voulu m'écrire une lettre flatteuse que je conserve comme un témoignage de vos bons sentiments pour moi. Dans ce premier mémoire, je n'ai guères eu recours qu'à l'expérience; aussi renferme-t-il plus d'un chose hasardée, et même de petites erreurs théoriques. Dans le mémoire actuel, au contraire, la theorie et l'expérience marchent de front, et se prêtent un mutuel appui. Vous y verrez se produire sur une grande échelle, des phénomènes de l'ordre de ceux auxquels on a donné l'épithète de *Capillaires* à cause de leur exiguité; vous y trouverez en même temps une suite de confirmations inattendues de l'admirable théorie sur laquelle repose l'explication des phénomènes capillaires; enfin, vous arriverez à une application qui consiste dans la théorie complète d'une phénomène dont l'étude expérimentale a formé la matière de l'un des plus beaux mémoires[3] de Savart[4].

J'ai reçu l'exemplaire que vous avez bien voulu m'envoyer de votre mémoire sur les nouveaux phénomènes dont vous avez enrichi la science[5]; Je vous en remercie, et je saisis cette occasion un peu tardive de vous exprimer toute mon admiration pour ces brillantes découvertes. On pouvait penser que c'était assez pour votre gloire d'avoir ajouté à vos travaux antérieurs la découverte de l'induction électro-dynamique avec toutes ses conséquences si extraordinaires; eh bien non! voilà que vous constatez d'une manière inespérée l'influence des courants électriques sur la lumière, puis l'universalité de l'action du magnétisme. En vérité, créer une nouvelle branche de la physique, ce n'est qu'un jeu pour vous.

A propos de magnétisme, causons un peu, si vous le voulez bien. Vos belles expériences sur les gaz vous ont conduit à établir l'état magnétique ou dia-magnétique d'un gaz donné, par rapport à un autre gaz donné; mais, ainsi que vous l'avez fait remarquer, elles ne permettent pas de constater si tel gaz est, par lui-même, magnétique ou dia-magnétique. Or, il m'a semblé qu'il serait possible d'arriver à cette connaissance absolue, et cela au moyen d'un procédé que je vais avoir l'honneur de vous soumettre.

Supposons l'électro-aimant renfermé dans une cage transparente remplie d'un gaz donné. Si ce gaz est magnétique, ses molécules seront attirées pas les pôles de l'électro-aimant, et elles seront, au contraire, repoussées si le gaz est dia-magnétique. Or, dans le premier cas, l'attraction des pôles aura nécessairement pour effet d'augmenter la densité du gaz autour de ces mêmes pôles, et, dans le second cas, la répulsion devra, au contraire, diminuer cette densité. Si donc le gaz est magnétique, la densité de la couche qui environne les pôles ira en croissant rapidement depuis une petite distance de la surface du métal jusqu'à cette même surface, et, si le gaz est dia-magnétique, ce sera un décroissement rapide de densité qui aura lieu. Par conséquent, lorsque un rayon lumineux traversera très obliquement la couche dont il s'agit, il sera quelque peu dévié dans un sens ou dans l'autre, suivant que la couche sera condensée ou dilatée.

Cela étant, et l'électro-aimant étant supposé vertical, placez verticalement derrière lui, à la distance d'une dixaine de pieds, par exemple, une feuille de papier blanc sur laquelle vous aurez marqué un point noir, et faites en sorte que ce point soit à la hauteur des pôles; puis, avant de faire agir le courant, placez-vous du côté opposé, à une distance au moins aussi grande de l'électro-aimant, et de manière que la droite qui va du point noir à votre oeil rase la surface supérieure de l'un des pôles; enfin, faites agir le courant. Alors, si le gaz est magnétique, le point noir devra paraître s'élever un peu au-dessus de la surface du pôle, et, si le gaz est dia-magnétique, le point devra disparaître derrière ce même pôle.

Il est inutile de vous faire remarquer que, dans cette expérience, l'oeil devra être bien immobile, et que, parconséquent, il faudra regarder à travers un petit trou percé dans une plaque portée par un support fixe. Il me semble, en outre, que les pôles ne devront pas être munis des armatures coniques dont vous vous êtes servi pour vos expériences: car les pointes de ces armatures étant très voisines, le magnétisme de chacun des pôles doit être en partie dissimulé par celui de l'autre; je crois que les pôles devraient être terminés par des surfaces horizontales très légerement convexes. Le plus ou moins d'effet dépendra surtout de la force de l'électro-aimant; mais je pense qu'on pourrait augmenter ce même effet, en plaçant les deux pôles suivant la ligne qui va du point noir à l'oeil: car alors le rayon, après avoir traversé la couche qui environne l'un de pôles, traverserait ensuite celle qui environne l'autre, et sa déviation serait doublée. Peut-être aussi serait-il bon de remplacer le point noir par une ligne noire horizontale et suffisamment longue: cette ligne devrait paraître brisée. Si l'action était trop faible, vous pourriez regarder à travers une lunette munie d'un fil horizontal. Enfin, il est possible que l'action de la pile échauffe notablement les fils de l'électro-aimant, d'où résulterait un courant d'air ascendant et dilaté, qui pourrait devenir une cause de déviation du rayon lumineux; dans ce cas, il faudrait garantir les pôles de ce courant d'air chaud, au moyen d'écrans convenables.

Je vous expose ces idées telles qu'elles me sont venues à l'espirit, et vous en ferez l'usage qu'il vous plaira; seulement, si vous les mettez en pratique, j'attends de votre bonte que vous me fassiez part de résultats positifs ou négatifs auxquels vous serez arrivé[6].

Puis-je espérer que vous aurez l'obligeance de faire remettre les exemplaires ci-joints à Sir J. Herschel et à Messieurs Wheatstone et Grove dont j'ignore les adresses, ainsi qu'à Société Royale?

Tout à vous | Jh. Plateau | professeur à l'Université, place du Casino 18.

P.S. Je m'aperçois qu'au commencement de ma lettre, en parlant de mon premier mémoire sur les masses liquides, je semble faire le procès aux méthodes expérimentales. Telle n'a pas été mon intention; j'ai trop souvent moi-même employé ces méthodes pour ne pas en reconnaître toute l'importance; j'ai voulu dire uniquement, que comme le sujet de ce premier mémoire pouvair être abordé à la fois par la théorie et par l'expérience, j'ai eu tort, dans ce cas, de m'en tenir à l'expérience seule.

Endorsed by Faraday: See M.S. notes of Expts 10 Octr. 1849 &c (10277 &c to 10301)

TRANSLATION

Ghent, 25 March 1849

My dear Mr Faraday,

Permit me to offer you a copy of a paper that I have just published[1]. This work follows on from that which I had the honour of sending you some years ago[2], about which you very kindly wrote me a flattering letter, which I have kept as proof of your benevolent feelings towards me. In the earlier paper, I hardly referred to experimentation; it contains more than one guess and even contains some small errors of theory. The current paper, on the contrary, uses theory and experiment equally and both lend support to each other. You will see produced on a grand scale, the kind of phenomena which has been called *Capillary* because of their exiguity; you will find at the same time a whole series of unexpected confirmations of the admirable theory on which rests the explanation of capillary phenomena; finally, you will come to an application which lies in the complete theory of a phenomenon of which the experimental study has formed the subject of one of the most beautiful papers[3] by Savart[4].

I received the copy that you very kindly sent me of your paper on the new phenomena with which you have enriched science[5]. I thank you, and take this rather belated opportunity to express to you all my admiration for these brilliant discoveries. One might have thought that it was enough for your glory to have added to your earlier works the discovery of electro-dynamic induction with all its extraordinary consequences; but no! here you are discovering in an unexpected way, the influence of electric currents on light, and the universality of magnetic action. In truth, creating a new branch of physics, is but a game for you.

Concerning magnetism, let us talk a little, if you do not mind. Your beautiful experiments on gas have led you to establish the magnetic or diamagnetic state of a given gas, in relation to another given gas; but, just as you have pointed out, they do not allow you to determine whether a particular gas is, of itself, magnetic or diamagnetic. Now it seems to me that it would be possible to arrive at this absolute knowledge, and that by means of a method that I shall have the honour of submitting to you.

Let us suppose that an electro-magnet is enclosed in a transparent cage filled with a given gas. If this gas is magnetic, its molecules will be attracted by the poles of the electro-magnet, and, on the other hand, they will be repelled if the gas is diamagnetic. Now, in the first case, the attraction of the poles will necessarily have the effect of increasing the density of the gas around these same poles, and in the second case, the repulsion will, on the other hand, diminish this density. Thus if the gas is magnetic, the density of the layer which surrounds the poles will grow rapidly from a little distance away from the metal right up to its surface, and, if the gas is diamagnetic, there will be a rapid decrease in the density which will take place. Consequently, since a ray of light will cross the layer in question very obliquely, it will be deviated a little in one direction or another, depending on if the layer is condensed or dilated.

This being the case, supposing the electro-magnet is vertical, place vertically behind it, at a distance of ten or so feet, for example, a sheet of white paper on which you will have marked a black point, and do it in such a way that this point is at the height of the poles; then, before you switch the current on, put yourself on the opposite side, at a distance at least as far from the electro-magnet and in such a way that the straight line which goes from the black point to your eye skims the top surface of one of the poles; finally, switch the current on. Now if the gas is magnetic, the black point will appear to rise slightly above the surface of the pole, and if the gas is diamagnetic, the point will disappear behind this same pole.

It is unnecessary to tell you that for this experiment, the eye must not move, and that consequently one must look through a little hole pierced in a sheet, held by a fixed support. It seems to me, moreover, that the poles must not be fitted with the conical armatures that you used in your experiments: for the points of these armatures being very close, the magnetism of each of the poles must in part be dissimulated by that of the other; I believe that the poles should end in horizontal surfaces which are very slightly convex. The greater or lesser effect will depend above all on the strength of the electro-magnet; but I think the effect could be increased by placing the two poles following the line which goes from the black point to the eye: for then the ray, having crossed the layer which surrounds one of the poles, would then cross the one which surrounds the other, and its deviation would be doubled. Perhaps it would also be a good idea to replace the black point by a sufficiently long horizontal black line: this line would appear to be broken. If the action were too weak, you could look through a magnifying lens fitted with a horizontal wire. Finally, it is possible that the action of the pile could significantly heat the wires of the electro-magnet, from which would result a current of rising and dilated air, which could become a reason for the deviation of a ray of light; in this case, one must protect the poles from this current of warm air by means of suitable screens.

I expound these ideas just as they have come into my mind, and you can make whatever use you wish of them; only, if you put them into practice, I would be very grateful if you could convey to me the positive or negative conclusions that you reach[6].

May I hope that you will have the kindness to convey the enclosed copies to Sir J. Herschel and to Messrs Wheatstone and Grove, whose addresses I do not possess, as well as to the Royal Society?

All yours | Jh. Plateau | Professor at the University, place du Casino 18

P.S. I have realised that at the beginning of my letter, speaking of my first paper on liquid masses, I seem to have put experimental methods in the dock. This was not my intention; I have all too often myself employed these methods not to recognise their importance; I wanted only to say, that the subject of this first paper ought to have been tackled both by theory and by experimentation. I was wrong, in that case, to rely to experimentation only.

1. Plateau (1849).
2. Plateau (1843). See Plateau to Faraday, 15 May 1844, letter 1586, volume 3.
3. Savart (1833).
4. Félix Savart (1791–1841, DSB). French physicist.
5. Faraday (1847b).
6. Faraday did not receive this letter until 29 September 1849 (Faraday, *Diary*, 10 October 1849, 5: 10277). The following month he conducted experiments stemming from the letter. See Faraday, *Diary*, 10, 12 and 15 October 1849, 5: 10277–10301.

Letter 2165
Faraday to Apsley Pellatt[1]
29 March 1849
From the original formerly in the possession of the late Mr and Mrs Aida

Royal Institution | 29 Mar 1849

My dear Pellatt
I have received & rejoice in your book[2] & am very much obliged to you indeed that you have remembered me amongst those worth remembering on this occasion. As yet I have only peeped into it but set it aside for a good examination[.]
Ever Truly Yours | M. Faraday
Apsley Pellatt Esq | &c &c &c

1. Apsley Pellatt (1791–1863, DNB). Glass manufacturer.
2. Pellatt (1849).

Letter 2166
Faraday to Arthur Watson[1]
3 April 1849
From the original copy in GL MS 30108/1/49
1 Is there any return of air down the ventilating tubes and does it affect the light and how?
2 If any return is it constant or only now & then?
3 If only now & then what appears to be the cause of it?
6 How is the wind & weather at those times?
4 Are both the upper & lower tier of lamps affected - or one more than the other?
5 Are all the lamps on one tier affected alike?
8 How does air enter into the lanthorn to ventilate it?
7 Is there a stove *in* the lanthorn and where does its chimney go; through the roof or into the cowl.
9 Is there any door between the lanthorn & watch room?
10 Is there any door between the watch room & the tower?
3 April 1849

Royal Institution | 3 April 1849
Sir
In reference to a communication of yours to the authorities of the Trinity house[2] I have, by their authority[3], to request that you will have the

Letter 2168

goodness to send answers to the enclosed questions + to my address at the Royal Institution Albemarle Street London, with any other information that you think will illustrate the case[4][.]
    I am | Yours &c | M. Faraday
Mr Watson | Lighthouse keeper | Dungeness Light Establishment | Kent.
+ as above

1. Lighthouse keeper at Dungeness. Otherwise unidentified.
2. Watson to Herbert, 10 February 1849, GL MS 30108/1/49.
3. See letter 2163.
4. Faraday's abstract of Watson's reply, dated 7 April 1849, is in GL MS 30108/1/49.

Letter 2167
Faraday to William Barchard[1]
3 April 1849
From the original copy in GL MS 30108/1/49
                                          Royal Institution | 3 April 1849
Sir
    By desire of the Trinity house authorities[2] I have to request you will be good enough to send answers to the enclosed questions + addressed to me Royal Institution Albemarle St. with any other information relating to the action of the ventilating pipes
    I am Sir | Your &c | M. Faraday
Mr Wm Barchard | Needles lighthouse | near Fresh Water | Isle of Wight
+ as above[3]

1. Keeper of the Needles Lighthouse. Otherwise unidentified.
2. See letter 2163.
3. That is in letter 2166.

Letter 2168
William Barchard[1] to Faraday
11 April 1849[2]
From the original copy in GL MS 30108/1/49
Sir
    In answer to your letter[3] I have sent you these few observations as follows

1. There is constant air down from the top which returns from the ventilator but do not affect the lights.
2. The lights are most affected by foggy weather also by a S.W. wind.
3. The lights on both tiers has the same effect[.]
4. There is a stove in the lanthorn, the chimney is through the roof not into the cowl[.]

There is a door between the watch room & the tower also a door in the lanthorn to walk on the outside on the Gallery.

I remain | Yours &c | William Barchard

Mr Faraday

1. Keeper of the Needles Lighthouse. Otherwise unidentified.
2. Date given by Faraday.
3. Letter 2167.

Letter 2169
William Thomson to Faraday
21 April 1849
From the original in IEE MS SC 2

(No 2) College, Glasgow | April 21, 1849

My dear Sir

I have just received your kind letter and I hasten to let you know that I am not now at Cambridge. I expect however to arrive there on Friday week (May 4) and if that is not too late, I shall then most gladly undertake your commission and send the copies of your paper[1] to the gentlemen you mention - and to the Society[2]. I shall be extremely obliged to you for the copy which you design for me. I am particularly anxious to become acquainted with the subject of your recent researches of which I have heard and read some very imperfect accounts, as it is a subject in which I am greatly interested. I hope to have an opportunity in London in May or June of hearing one of your lectures as I have done before with so much pleasure.

I am not sure whether I have ever mentioned to you that I have been made Professor of Natural Philosophy in this University[3], and that I am therefore at home here during at least the winter six months of the year. My father[4] until his death[5] was Professor of Mathematics here, and thus my early associations are connected with the place where it is likely I shall remain for life. For the present however I retain my fellowship in St. Peter's College, and therefore I usually spend much of the summer six months at Cambridge.

I remain, My dear Sir, | Your's very truly | William Thomson
Prof. Faraday

1. Faraday (1849a, b), ERE22.
2. Presumably the Cambridge Philosophical Society.
3. Thomson had already informed Faraday of this. See Thomson to Faraday, 11 June 1847, letter 1998, volume 3.
4. James Thomson (1786–1849, DNB). Professor of Mathematics at the University of Glasgow, 1832–1849.
5. On 12 January 1849.

Letter 2170
Faraday to William Robert Grove
26 April 1849
From the original in RI MS G F25

R Institution | 26 April 1849

My dear Grove
    I forget many things but am quite ashamed I should have forgotten you[.]
    Ever Yours | M. Faraday

Letter 2171
Arthur Aikin to Faraday
26 April 1849[1]
From the original in RI MS Conybeare Album, f.48

7 Bloomsbury Square | 26 April

Dear Faraday
    My niece Mrs Le Breton[2] with her husband[3] & family, as well as my sister[4] who is an inmate of theirs, hearing that you are become a temporary resident in Wimbledon are very desirous of showing you & Mrs Faraday all neighbourly attention. With this view she will take the liberty of calling on Mrs F & has desired me to send you this notice in order that when she calls she may not seem to be altogether a stranger[.]
    Yours very truly | Ar. Aikin
M. Faraday Esq

1. Dated on the basis that Faraday was at Wimbledon at this time. See letters 2176, 2178 and 2182. See also Margery Ann Reid's recollection of the visit in her diary of April 1853, RI MS F13B, pp. 11–12.
2. Anna Letitia Le Breton, née Aikin (1808–1885, DNB). Writer.

3. Philip Hemery Le Breton (d.1884, age 77, GRO). Barrister who married Anna Letitia Aikin in 1833. See DNB under her entry.
4. Lucy Aikin (1781–1864, DNB). Writer.

Letter 2172
Faraday to Jacob Herbert
27 April 1849
From the original copy in GL MS 30108/1/49

27 April 1849 | Royal Institution

My dear Sir

In reference to the ventilation of the Dungeness & Needles lighthouses I drew up ten questions (of which I send you copies) and sent them to the keepers of these lighthouses respectively[1]. I send you the answers[2] by which you will see how impossible it is for me to form any judgement on such vague information. That from the needles is as scant as it is possible to be and that from Dungeness is in direct contradiction to Captn Madan & Mr. Wilkins in respect of the air tubes in the watch room. I think it would be better that I should go to one or both but that I could not do just now because of my lectures here. After the 10th of June I shall be free[3]. I believe the *lanthorn* itself requires ventilation & air[.]

I am Dear Sir | Very Truly Your | M. Faraday
Jacob Herbert Esq | &c &c &c

1. Letters 2166 and 2167.
2. Letter 2168.
3. Faraday delivered a course of eight lectures on static electricity on successive Saturdays after Easter which were due to end on 9 June 1849; however, see note 1, letter 2178. His notes are in RI MS F4 J10.

Letter 2173
Faraday to Miss Miles[1]
28 April 1849
From the original in the possession of Dennis Embleton

Royal Institution | 28 April 1849

My dear Miss Miles

I am promised for June to Mr Clowes[2] for the name of *Prichard*[3] & also for December to Dr Todd[4] for another candidate. But if for any reason *Prichard* should be removed from the list then I shall have great pleasure in sending you my Proxy[.]

Pray give my respects to Mr and Mrs. Miles[5][.]

Yours Very faithfully | M. Faraday

1. Unidentified. This letter refers to the provision a place for an orphan in the London Orphan Asylum in Clapton, founded in 1813, of which Faraday became a subscriber in 1831. See Faraday to Roberts, 24 January 1832, volume 2, letter 533. Each subscriber had one or more votes (depending on the size of their subscription) for suitable candidates for a place in the orphanage. Votes could be transferred by proxy between subscribers. For accounts of the London Orphan Asylum see Alvey (1990) and Bache (1839), 58–65. Miss Miles is not listed as a subscriber, but there were a large number of anonymous and pseudonymous subscribers.
2. William Clowes (1807–1883, DNB). Printer. He had been a subscriber since 1846. See the 1850 *Report of the London Orphan Asylum* SuRO 3719/1/13, p.76.
3. James Henry Prichard appears in the list of orphans in the 1850 *Report of the London Orphan Asylum* SuRO 3719/1/13, p.34. His father had been a compositor in Stamford Street.
4. Possibly Robert Bentley Todd (1809–1860, DNB). Physician at King's College Hospital, 1840–1860. He is not listed as a subscriber to the London Orphan Asylum.
5. Both unidentified.

Letter 2174
Julius Plücker to Faraday
30 April 1849
From the original in IEE MS SC 2
Sir!
I received your kind letter[1], by which you announce to me your last paper[2]; I am very anxious to get it and will read it, with the greatest attention.
The new facts I alluded to in my last letter[3] are the following ones. Since my first experiments I was convinced, that there ought to be an influence of Magnetism on crystallisation and I expected, that any salt crystallising *slowly* between the poles of a strong magnet, would have its optical axes perpendicular to the line joining the two poles. Having tried in vain to prove it experimentally I did not speak about it. But when I repeated your last experiments with bismuth, I thought this metal exce[e]dingly proper, to be subjected to the former experiments, and this time I had a full success. Melted bismuth, crystallising slowly between the poles of a strong Magnet gets such a crystalline structure that the chief cleavage plan[e] becomes perpendicular to the line joining the two poles. I prove the same, even without cleaving the crystallized bismuth, by the following experiment. A piece of bismuth, crystallised between the two poles, takes, whatever may be its shape, when suspended in such a way, that it may turn freely round a vertical axis allways and exactly (according to the magnecrystallic action) the same position it had during the crystallisation.

This result proves strikingly that the force acting on the magnecrystallic axis (and also I think on the optic axes) is a *molecular* one. That has been allways my opinion.

All my experiments confirm that this force produces, as you call it an effect of position only and not an effect of place[.] When I say "repulsion of the optical axes" as I may say "attraction of the magnecrystallic one" I meaned only to explain the facts, without anticipating any conception about the nature of the acting forces.

About the 20th of March I gave my paper to Poggendorff[4], as soon as I get a copy of it, I'll send it to you by post.

Since I have been at Paris. Tired by lectures & a "changement d'air" was necessary for my health. Being restored by travelling, I'll find time to go on in my researches, even, I may say, in spite of the government, which does not at all favour them.

Yours very sincerely | Plücker
Bonn 30th of April | 1849

Address: Professor Faraday | &c &c &c | Royal Institution | London

1. Not found.
2. Faraday (1849a, b), ERE22.
3. Letter 2152.
3. Plücker (1849a).

Letter 2175
Faraday to J. Miller[1]
c May 1849[2]
From the original in HCL, Quaker Collection

Norfolk St | Saturday [sic][3] Morning

My dear Sir

It is not wonderful that we should miss one another in London - especially as on my part I am living by order of my medical friend[4] out of town at Wimbledon. We have an evening meeting of the members of the Institution to night at $\frac{1}{2}$ past 8 if it would give you pleasure to be there I should be most happy to see you[.] Tomorrow I must lecture[5] & until after the lecture must keep my thoughts together & see no one[.] The lecture begins at 3 & is over at 4 o'clk. After lecture I shall of course be there[.] if you are passing either at 3 or 4 o'clk come in if you feel inclined[.]

On Monday again I shall be at home *I am almost sure* all the morning until 12 or 1 o'clk[.]

Ever My dear Sir | Very Truly Yours | M. Faraday
Revd. J. Miller | &c &c &c

1. Unidentified.
2. Dated on the basis that Faraday was at Wimbledon at this time. See letters 2176, 2178 and 2182.
3. Friday must have been intended.
4. That is Peter Mere Latham.
5. Faraday delivered a course of eight lectures on static electricity on successive Saturdays after Easter which were due to end on 9 June 1849; however, see note 1, letter 2178. His notes are in RI MS F4 J10.

Letter 2176
Faraday to John Barlow
7 May 1849
From the original in RI MS F1 C29

Wimbledon[1] | 7 May 1849.

My dear Barlow

I write a brief note to say that I am improving & hope to be ready for next Saturday[2] but this easterly wind is not good. I intend to remain here nursing until Thursday or Friday and expect a good result. There is a meeting of the University of London Senate & Graduates on Wednesday Evening[3]. I do not suppose you care about going - & I dare say if you do you have an invitation. Still if you have not & *should* like to go you will find a couple of invitation cards on my mantle piece which I have a right to use. Of course I shall not be there under present circumstances[.]

Ever Truly Yours | M. Faraday

1. "R Institution" is crossed out underneath Wimbledon.
2. That is 12 May 1849 when Faraday was due to deliver one of his lectures in his course of eight on static electricity which he gave on successive Saturdays from 21 April to 9 June 1849; however, see note 1, letter 2178. His notes are in RI MS F4 J10.
3. There was no meeting of the Senate that day.

Letter 2177
Peter Mere Latham to Faraday
7 May 1849
From the original in RI MS Conybeare Album, f.31

36 Grosvenor St. May 7, 1849

My dear Friend

I wish you were getting well a little faster than you are. Either your Doctor wants skill or the elements are against you. We will suppose the latter. For the sudden change from warm to cold in the last two days has stopt the convalescence of many an invalid. Mrs Faraday asks permission to give you a little wine. It is an experiment; but one which might fairly be tried. Take then half a glass of sherry after dinner tomorrow; and if it do not heat you or make the throat more husky & you are manifestly refreshed by it, take as much or even a little more on the following day.

My kind remembrance to Mrs Faraday.

Always, my dear friend, | Yours most truly | P.M. Latham

Letter 2178
Faraday to John Barlow
8 May 1849
From the original in RI MS F1 C30

Wimbledon | 8 May 1849

My dear Barlow

Your letter & Mr. Brandes offer[1] are very kind indeed but I am happy to say I feel no doubt about next Saturday[2]. My present purpose is to be in town in the course of Friday to prepare &c.

Perhaps a short advertisement on Saturday Morning[3] saying Mr Faradays 2nd lecture will be delivered this day &c &c might be advisable[.] If you think so will you tell *Lacy*[.]

Ever Yours | M. Faraday

I intend to write to Mr. Brande to thank him. MF

1. Because of his illness, Faraday had not been able to deliver two of his course of eight post Easter Saturday lectures on static electricity which he had commenced on 21 April; he had not been able to lecture on 28 April and 5 May 1849. RI MM, 7 May 1849, **10**: 184 records that Brande offered to undertake Faraday's lectures for him. Barlow was asked to inform Faraday of this offer. Faraday made up the missed lectures on 12 and 16 June 1849 (RI Lectures Index). Faraday's lecture notes are in RI MS F4 J10.
2. That is 12 May 1849.
3. Such an advert was placed in the *Morning Post*, 12 May 1849, p.1, col. e.

Letter 2179
John Martin[1] to Faraday
15 May 1849
From the original in IEE MS SC 2

Lindsey House, Chelsea | May 15th 1849

My dear Faraday,

Will you oblige me by presenting to the library and reading room of the Royal Institution[2] the accompanying copies of my recent Thames and Metropolis Improvement Plan 1849[3] Plan of London Connecting Railway

1845[4], Plan for ventilating coal mines 1849[5] - and reprint of report upon the Thames Improvement Plan 1836[6].

I remain, faithfully yours | John Martin
Dr. Faraday

Address: Dr. Faraday

1. John Martin (1789–1854, DNB). Historical and landscape painter. He also worked on the disposal of London's sewage.
2. These gifts were noted in RI MM, 4 June 1849, **10**: 193.
3. This is a map of London showing Martin's plans.
4. Martin (1845).
5. Martin (1849b).
6. Anon (1849).

Letter 2180
Willem Vrolik[1] to Faraday
16 May 1849
From the original in RS MS 241, f.116

Amsterdam ce 16 Mai 1849

Monsieur,

J'ai l'honneur de Vous informer que Sa Majesté le Roi des Pays-Bas[2] a approuvé Votre nomination comme Membre Associé de la première classe de l'institut royal des Pays-Bas.

La classe espère que Vous voudrez bien l'appuyer par Vos lumineuses talents dans le but scientifique, qu'elle s'est proposé et que Vous connaitrez par les mémoires qu'elle publie. Par conséquent elle se flatte que Vous voudrez bien accepter le titre qu'elle Vous a décerné, et qui vient de recevoir la sanction royale.

Je Vous prie de me croire avec les sentiments les plus distingués, | Monsieur | Votre devoué Serviteur. | W. Vrolik | Secrétaire perpetuel de la | première classe de l'institut Royal des | Pays-Bas.
A Monsieur | M.Faraday | à Londres.

TRANSLATION

Amsterdam, this 16 May 1849

Sir,

I have the honour of informing you that His Majesty the King of the Netherlands[2] has approved your nomination as Associate Member of the first class of the Institut Royal des Pays-Bas.

The class hopes that through your illustrious talents you will support it in the scientific goals it has set itself which you will know from the papers it publishes. Consequently it flatters itself that you will wish to accept the title it has awarded and which has just received royal sanction.

I ask you to believe me with the most distinguished sentiments | Sir | Your Devoted Servant | W. Vrolik | Permanent Secretary of the first class of the Institut Royal des | Pays-Bas.

To Mr | M. Faraday | in London

1. Willem Vrolik (1801–1863, NNBW). Dutch physiologist.
2. William III (1817–1890, NNBW). King of the Netherlands, 1849–1890.

Letter 2181
Isambard Kingdom Brunel to Faraday
16 May 1849[1]
From the original in RI MS F1 N3/22

May 16

My dear Sir

A very intelligent american Mr Francis[2] is sent over to England at the expence of a company or association to enquire thoroughly into the use of corrosive sublimate for preserving timber and particularly to investigate the causes of its present disuse. I have given him all the information in my power but as he says he dare not go back to America without seeing you. Knowing your engagements I have not complied with his request to give him a note to you without first ascertaining if it suits you or whether it is too great an inconvenience.

Trusting you are in good health

Believe me dear sir | Your faithfull | I.K. Brunel

1. Recipient and date established on the basis that letter 2182 is the reply.
2. James Bicheno Francis (1815–1892, DAB). American civil engineer.

Letter 2182
Faraday to Isambard Kingdom Brunel
17 May 1849
From the original in BrUL MS

Wimbledon | 17 May 1849.

My dear Brunel

I am here under Doctors orders and though recovering am not allowed to be much in town. However I shall be in town tomorrow afternoon (Friday) and also part of Saturday but as that is my lecture day I cannot see any one until after 4 oclk on that day[1][.] Will any of the spare pieces of time suit your friend[2]. Perhaps he would like to be with us on the Friday Evening[3]. If so you can give him a ticket[.] I shall be most happy personally to see him & am only sorry I am of necessity thus restrained[.]

Bad memory. I cannot remember the name of your Street, or where to direct to you. Your letter does not help me so that I must send this to the Institution first & am afraid I shall lose 2 or 3 deliveries by so doing. Pray forgive me.

Ever Truly Yours | M. Faraday

1. Faraday delivered a course of eight lectures on static electricity on successive Saturdays after Easter which were due to end on 9 June 1849; however, see note 1, letter 2178. His notes are in RI MS F4 J10.
2. James Bicheno Francis (1815–1892, DAB). American civil engineer mentioned in letter 2181.
3. The Friday Evening Discourse on 18 May 1849 was delivered by Edwin Sidney (d.1872, age 74, B6), Rector of Little Cornard, 1847–1872 and lecturer at the Royal Institution and elsewhere. His topic was "On the Geographical Distribution of Corn Plants". For an account see *Athenaeum*, 26 May 1849, pp.546–7.

Letter 2183
Julius Plücker to Faraday
20 May 1849
From the original in IEE MS SC 2

Allow me, Sir, to communicate to you several new facts, which, I hope, will spread some light over the action of the Magnet upon the optic and magnecrystallic axes.

I. The first and general law, I deduced from my last experiments is the following one

"There will be *either repulsion or attraction* of the optic axes by the poles of a Magnet, according to the crystalline structure of the crystal. If the crystal is a *negative* one, there will be *repulsion*, if it is a *positive* one there will be *attraction*"

The crystals most fitted to give the evidence of this law are *diopside* (a positive crystal) *cyanite, topaze* (both negatives) and other ones, crystallising in a similar way. In these crystals the line (A) bisecting the acute angles, made by the two optic axes, is neither perpendicular nor parallel to the axis (B) of the prism. Such a crystal, suspended horizontally like a prism of turmaline, staurotite or "cyanure rouge de fer & potasse" in my former experiments, will point neither axially nor equatorially, but will take allways a fixed intermediate direction. This direction will continually change if the prisme will be turned round its own axis B. It may be proved by a simple geometrical construction, which shows, that during one revolution of the prism round its axis (B), this axis without passing out of the two fixed limits C & D, will go through all intermediate positions. The directions C & D, where the crystall returns, make *either* whith [sic] the line joining the two poles, *or* with the line perpendicular to it, on both sides of these lines, angles equal to the angle included by A and B: the first being the case, if the crystal is a *positive* one, the last if a *negative* one. There it follows that if the crystal by any kind of horizontal suspension may point to the poles of the Magnet, it is a *positive* one; if it may point equatorially it is a *negative* one. This last reasoning conducted me at first to the law mentioned above.

The magnecrystallic axis, I think, is, optically speaking, the line bisecting the (acute) angles made by the two optic axes, or in the case of one single axis, the axis itself. The crystals of bismuth and arsenic are positive crystals, Antimony, according to my experiments, is a negative one. All are uniaxal.

II The cyanite is by far the most interesting crystal, I examined till now. If suspended horizontally it points very nicely, *only by the magnetic power of the Earth*, to the north. It is a true compass needle, and more than that, you may comand over its declination. If for instance, you suspend it in such a way that the line A bisecting the two optic axes of the crystal, be in the vertical plane passing through the axis B of the prism, the crystal will point exactly as a compass needle does. By turning the crystal round the line B you may make it point exactly to the north of the Earth &c[.] The crystal does non [sic] point according to the Magnetisme of its substance, *but only by following the magnetic action upon its optical axes*. This is in full concordance to the different law of diminution by distance of the pure magnetic and the optomagnetic action. If you approach to the north end of the suspended crystal the south pole of a permanent magnetic bar, strong enough to overpowering the magnetism of the Earth, the axis B of the prism will make with the axis of the bar (this bar having any direction whatever in the horizontal plane) an angle exactly *the same* it made before with the meridian plane: the crystal being directed either more towards the East or more towards the West.

The crystal showed, resembling for that also to a magnetic needle, strong polarity: the same end being allways directed to the north. I dare say, if it may be a *polarity of the optomagnetic power*. Two questions too may easily be answered. 1° Is the north pole indicated by the form of crystallisation 2° did the crystal get, when formed, its polarity by the magnetism of the Earth. Between the poles of the strong Electromagnet the permanent polarity disappeared as long as the Magnetism was excited.

I am obliged by the new facts, mentioned above, to take up my former memoir[1], I must reproduce it under a quite new shape. I'll examine again the rock crystal, which being acted upon weakly by a magnet induced me to deny in that memoir, what I ascertain now and what I thought most probably, as soon as I got the first notice of your recent researches[2]. (That you will find in the Memoir given to Mr Poggendorf[f] 2 or 3 months ago[3].) Perhaps the exceptional molecular condition of rock crystal, as indicated by the passage of light t[h]rough it, will produce a particular magnetic action.

I should be very obliged to you, if you would give notice of the contents of my present letter to Mr. de la Rive, when he calls on you, as he intended to do. I showed him several of my experiments when he passed through Bonn the 12th of Mai. The following day I got the different results, mentioned above.
My best whishes [sic] for your health!
Very truly yours | Plücker
Bonn, the 20th of May | 1849.

Endorsed by Faraday: On the Magnetic relations of the Positive & Negative optic axes of crystals by Professor Plucker of Bonn in a letter to, and communicated by Dr. Faraday[4].
Address: Professor M. Faraday | &c &c | Royal Institution | London

1. Plücker (1847).
2. See Faraday to Plücker, 14 December 1848, letter 2136, volume 3.
3. Plücker (1849a).
4. The title of Plücker (1849b) which published this letter.

Letter 2184
Jean-Baptiste-André Dumas to Faraday
c21 May 1849
From the original in IEE MS SC 2
Mon Cher Confrère et ami
Cette lettre vous sera remise par M. D'Eichtal l'un des plus honorables et des plus distingués parmi les Membres de la chambre des députés de l'ancien tems. Il veut bien se charger de vous remettre en même occasion, des échantillons qui vous intéresseront je l'espère. Le premier consiste en une reproduction obtenue sur un enduit d'amidon, au moyen de l'Yode. Le procédé de M. Niepce[1] vous est bien connu, mais personne n'est en état de le pratiquer aussi habilement qui lui et son

oeuvre peut vous le prouver. La gravure très imparfaite qui se trouve reportée sur la glace amidonnée y a produit une image plus agréable à l'oeil qu'elle ne l'étoit elle même[2]. Vous remarquerez que j'ai fait encadrer la gravure dans un Cadre fait avec du bois injecté par le procédé[3] de M. Boucherie[4]. il y en a divers échantillons.

Vous trouverez en second lieu, une série d'épreuves obtenues soit sur pierre, soit sur acier ou cuivre, par des gravures en relief, tirées avec de l'encre ordinaire. C'est un problème très heureusement résolu par nos artistes et très proprement fournit des papiers qui résistent aux tentatives de faux, soit pour les effets de commerce, soit pour les actes publics. examinez ces papiers avec une bonne loupe et vous serez frappé de la pureté des lignes et de la perfection du tirage, qui est plus facile avec l'encre aqueuse qu'avec l'encre grasse d'Imprimerie. C'est une nouvelle ressource pour l'Industrie et pour les arts.

M. D'Eichtal vous entretiendra lui même d'un objet nouveau plein d'intérêt. Il s'agit de l'oxide de zinc et de fer dérivée qu'il substitue aux matières à bas de plomb employées en peinture. J'ai employé ses couleurs pour des Laboratoires et des amphithéatres avec le plus complet succès. Nous avons à paris déjà une grande expérience de ce procédé qui me semble destiné à jouer un rôle très considérable dans le monde industriel. Je suis persuadé que vous en tirerez à Londres le parti le plus avantageux, tant à cause de l'Innocuité des produits que de leur résistance à l'action du gaz hydrogène sulfuré.

Enfin, M. D'Eichtal vous dira quelques mots d'une découverte faite dans mon laboratoire par un de mes éleves qui avoit bien voulu m'accompagner à Londres et qui a eu l'honneur de vous voir, M. Melsens. Il s'agit d'un procédé merveilleux qui permet d'extaire de la canne à sucre, *tout le sucre* qu'elle contient[5]. L'expérience avoit si bien réussi avec les Betteraves que j'ai voulu la vérifier sur la canne. J'ai fait venir une Centaine de livres de canne à sucre fraiches d'Andalousie et elles ont été traitées sous mes yeux avec le résultat le plus satisfaisant et le plus décisif.

M. Melsens vouloit que je vous fisse connaitre Son procédé, je m'y suis refusé. Je sais par ma propre expérience combien un secret embarasse son dépositaire. Si, cependant, vous permettiez qu'on vous le fit savoir et si vôtre gouvernement y mettait le moidre intérêt, vous consentiez à examiner la question, sans lui donner vôtre opinion, Je Vous transmettrais tout ce que Je sais à ce sujet.

M. Melsens lui même iroit au besoin répéter sous vos yeux toutes ses expériences. Ce seroit un immense service rendu à M. Melsens, qui en est bien digne par son dévouement à la Science et à qui Je porte un intérêt de père.

Je suis très à court de nouvelles purement Scientifiques. Vous savez déja par nos comptes rendus que M. Boutigny[6] a trouvé le moyen de plonger la main dans la fonte en fusion, sans le moindre accident⁷. L'Epreuve du feu est expliquée désormais. Il étoit bien dû à la France, *ou l'on joue avec le feu*, tous les jours, de voir cette découverte faite par un des Savans qu'elle compte dans son Sein. La recette de M. Boutigny pourra être utile à nos hommes politiques.

Je me recommande à toute votre amitié. Vous verrez que le Dept. du Nord vient de m'envoyer à l'assemblée législative, plaignez moi et soyez bien assuré que s'il avoit été possible de refuser ce dangereux honneur, je me serois empressé de la faire, pour me consacrer tout entier à mes travaux. heureusement, qu'il reste encore au monde un coin paisible où la philosphie conserve un asile. Vous étiez digne plus que personne qu'il vous fut réservé, Jouissez de vôtre bonheur et qu'il vous soit longtems conservé.

Mille amitiés | Dumas

Made Dumas se rappele au bon souvenir de Madame Faraday à qui Je vous prie de présenter mes respectueux hommages.

Endorsed by Faraday: Received 23 May 1849

TRANSLATION
My Dear Colleague and Friend

This letter will be given to you by M. D'Eichtal, one of the most honourable and most distinguished amongst the Members of the chamber of deputies of former times. He has kindly agreed at the same time to pass on some samples which I hope will interest you. The first consists of a reproduction obtained on a coating of starch, by using iodine. The procedure used by M. Niepce[1] is well known to you, but no one is able to put it into practice as ably as he himself can and his work can prove it to you. The very imperfect engraving which has been reproduced on a starched sheet of glass has produced an image more pleasing to the eye than it was in the original[2]. You will notice that I have put the engraving in a box made with wood injected by the method[3] of M. Boucherie[4]. There are various samples.

Secondly, you will find a series of proofs obtained on rock, steel and copper, for relief engravings, drawn with ordinary ink. This is a problem very happily resolved by our artists and very properly supplied with papers which resist any attempt at forgery, be it for commercial reasons or for public acts. Examine these papers with a good magnifying glass and you will be struck by the purity of the lines and the perfection of the print, which is much easier with waterbased ink than with oily printing ink. This is a new resource for industry and the arts.

M. Eichtal will tell you himself of a new object which is full of interest. It concerns the oxide of zinc and derived iron which he substitutes for lead based materials used in painting. I have used his colours in laboratories and in lecture theatres with total success. In Paris we already have a great deal of experience of this procedure which seems to me to be destined to play a most important role in the industrial world. I am sure that you, in London, will put it to the best use, both because of the innocuousness of the products and because of their resistance to the action of hydrogen sulphide gas.

Finally, M. D'Eichtal will say a few words on a discovery made in my laboratory by one of my pupils, who was kind enough to accompany me to London and who had the honour of seeing you, M. Melsens. It concerns a marvellous method which enables *all the sugar* contained in sugar cane to be extracted[5]. The experiment was so successful with beet that I wanted to verify it on cane. I imported about a hundred pounds of fresh sugar cane from Andalusia, and it was treated under my eyes, with the most satisfying and definitive results.

M. Melsens wanted me to make known his method to you, but I refused. I know from my own experience how awkward it is to be the guardian of a secret. If, however, you allowed us to make it known to you and if your government expressed the slightest interest in it, you consented to examine the question, without giving him your opinion, I would transmit to you all I know of the subject.

M. Melsens himself would, if necessary, repeat before your eyes all his experiments. It would be an enormous service given to M. Melsens, who is worthy of it because of his devotion to Science and in whom I take a fatherly interest.

I am very short on purely scientific news. You already know from our *Comptes Rendus* that M. Boutigny[6] has found a method of plunging his hand into molten iron, without the slightest accident[7]. The proof of fire is explained from now on. It fell to France *where one plays with fire* every day, to see this discovery made by one of the savants that it holds in its bosom. M. Boutigny's recipe could be useful for our politicians.

I recommend myself to all your friendship. You will see that the Department du Nord has just sent me to the legislative assembly. Pity me and rest assured that had it been possible to refuse this dangerous honour, I would have been anxious to do so, in order to concentrate entirely on my work. Happily, there is still a peaceful corner of the world where philosophy provides a sanctuary. You, more than any other person, were worthy that it was reserved for you. Rejoice in your good fortune and may it remain like that for a long time.

My best regards | Dumas

Madam Dumas asks to be remembered to Mrs Faraday, to whom I beg you to pay my respectful homage.

1. Claude Félix Abel Niepce de Saint-Victor (1805–1870, NBU). French photographer.
2. On Niepce's photographic methods see Gernsheim and Gernsheim (1955), 148–9.
3. See Boucherie (1840).
4. Auguste Boucherie (1801–1871, Vapereau (1880), 272). French physician and chemist.
5. Melsens (1849).
6. Pierre Hippolyte Boutigny (1798–1884, Oursel (1886), 1: 127). French pharmacist.
7. Boutigny (1849).

Letter 2185
Faraday to Julius Plücker
23 May 1849
From the original in NRCC ISTI

R Institution | 23 May 1849

My dear Plucker

Not 10 minutes ago I received your letter of the 20th instant[1] & write at once to congratulate you on the beautiful facts you describe. How wonderfully this branch of Science is progressing! - I saw De la Rive two days ago & he gave me the pieces of bismuth from you about which I say nothing because I have not experimented with them & I conclude you have not yet received my paper[2]. I shall see de la Rive in a day or two & will show him yours.

I am not quite sure what you would like but I think (if I can get the letter back from De la Rive in time) I shall send it to Mr Taylor to print in the next Number of the Phil Mag[3][.]

In haste but | Most truly Yours | M. Faraday
Professor Plucker | &c &c &c

Address: Professor Plücker | &c &c &c | University | Bonne | on the Rhine

1. Letter 2183.
2. Faraday (1849a, b), ERE22.
3. Plücker (1849b).

Letter 2186
Faraday to Willem Vrolik[1]
25 May 1849
From the original in Rijksarchief in Noord-Holland, archief Koninklijk Instituut van Wetenschappen, inv. nr. 5

Royal Institution - London | 25, May 1849.

Sir

I have the honor to acknowledge the receipt of your letter of the 16th instant[2] and beg you will do me the favour to express the high sense I entertain of the approbation of the Royal Institute and the mark of honor it has conferred upon me for which I am most grateful and beg to return for it my sincere thanks. I can only say, under a consciousness of failing powers with increasing years, that if any thing could make me increase my exertions in the cause of science it would be such approval as that you give me[.]

Permit me to offer for yourself personally the expression of my highest esteem & respect and believe me to be

Sir | Your Most faithful & Obedient Servant | M. Faraday
M. W. Wolik [sic] | &c &c &c &c

1. Willem Vrolik (1801–1863, NNBW). Dutch physiologist.
2. Letter 2180.

Letter 2187
Faraday to Edward Magrath
25 May 1849
From the original in RI MS F1 C28

R Institution | 25 May 1849

Dear Magrath

I send the tickets[.]

Some person has sent me the enclosed. They have put your name instead of Mr. Dances[1] I suppose[2]:-

Do you know what Arnett's[3] books of the Ancients is?[4] Can you tell me when you & I were first acquaint? and whether & when we were members of Dorset Street Society?[5]

Ever Truly Yours | M. Faraday

Endorsed by Magrath: Answered the same day see over leaf

1. William Dance (1755–1840, DNB). One of the founders of the Royal Philharmonic Society.

2. Arnett (1837), 203–5 gave a short account of Faraday's life (derived largely from Anon (1835)). On p.204 Magrath, rather than Dance, is stated to have given Faraday the tickets to attend Davy's lectures at the Royal Institution in 1812.
3. John Andrews Arnett. Pseudonym of John Hannett. Otherwise unidentified.
4. Arnett (1837).
5. That is the City Philosophical Society on which see James (1992b).

Letter 2188
Edward Magrath to Faraday
25 May 1849
From the original copy in RI MS F1 C28

Athenaeum | 25 May 1849.

Dear Faraday

Until I received your note[1] and the Extract I had never heard of Arnetts[2] 'Books of the Ancients'[3][.] Some Weeks since my Brother Henry[4] told me had seen in a fugitive publication a paragraph which may probably be the same & which I have in my hand for the first time. I had forgotten the matter altogether[.]

I think you and I were first acquainted in the same year you were appointed to the Royal Institution[5]. I know this much that it was during the following year you went with Sir Humphrey Davy to Paris and I heard you say that you were there when Napoleon[6] returned from the Campaign in Russia (1812)[7].

I have never kept a dairy and must therefore in all matters of my past Life refer to memory alone. I have never paid the slightest attention to the many publications in which your name and mine have appeared in connection - because there has been in them for the most part more of error than truth, and also because I have always systematically abstained from all connection or communication with the press whether they chose to treat me roughly or smoothly. One is sure to get rough treatment if you notice mistatements or shew the slightest sensitiveness at personal attacks[.]

Whenever I have been asked proper questions about you I have answered frankly and openly and in some instances have corrected misinformation which (if true) might have gratified my vanity, but I did not choose to indulge it at the expense of truth. For instance I have been often asked if I knew you in early youth. My answer has been invariably No! Not until you became a member like myself, of the Dorset St. Society to which we were introduced at different periods by John Tatum[8][.]

E. Magrath

Athenaeum | Friday 25 May 1849.

1. Letter 2187.
2. John Andrews Arnett. Pseudonym of John Hannett. Otherwise unidentified.
3. Arnett (1837).
4. Otherwise unidentified.
5. That is in 1813.
6. Napoleon Bonaparte (1769–1821, NBU). Emperor of France, 1804–6 April 1814. Exiled on Elba until 1 March 1815. Lost battle of Waterloo and finally abdicated on 22 June 1815.
7. The chronology here is incorrect. Napoleon had returned from Russia in December 1812. Faraday saw him in Paris a year later on 19 December 1813. Bowers and Symons (1991), 33.
8. John Tatum (d.1858, age 86, GRO). Silversmith of 53 Dorset Street. Appears in London directories until 1827. Probably the same as John junior noted in Grimwade (1982), 677. Founder of the City Philosophical Society on which see James (1992b).

Letter 2189
Faraday to Julius Plücker
31 May 1849
From the original in NRCC ISTI
My dear Plucker

The inclosed will appear in the Phil Magazine tomorrow morning[1]. I have altered a word or two here & there. I hope that what I have done meets your approbation & is agreeable to you. I could not resist my desire to make it known[.]

Ever Truly Yours | M. Faraday
31 May 1849

Address: Professor Plucker | &c &c &c | University | Bonn | on the Rhine

1. Plücker (1849b).

Letter 2190
Julius Plücker to Faraday
2 June 1849
From the original in IEE MS SC 2
Sir, I thank you very much for the unexpected kindness, with which you received my last comunication[1]. A letter not being a memoir, I am very satisfied you sent mine to be printed by Mr. Taylor[2]. My bad English, I know very well, is to be changed before going to press; therefore I am much obliged to you, for having altered some expressions. There is only one word, introduced by you, which I do not understand. You say "It is a

true compass-needle and more than that, *you may obtain its declination*"³[.]
My meaning was "you may dispose on the direction, it shall take" or "you may give to it any declination you like, from about 25° to the East to 65° to the West".

I'll go again to work as soon as I get crystals, I am expecting from Berlin. Since my last letter I tried only a few crystals, ascertaining all the general law. I found that angite is acted upon by the magnetic power of the Earth, quite in the same way as cyanite.

I examined also rock crystal. If you suspend a prism of it horizontally it will, according to the *smal[l]er* or *greater* distance from the poles, point either *equatorially* by the diamagnetism of its substance or *axially, its optic axis being attracted*. Why did I not try that before? To excuse my stupidity I might write a long psychological memoir.

I did not yet receive your paper⁴.

Most sincerely | Yours | Plücker
Bonn 2th of June 1849

Address: Professor M. Faraday | &c &c | London | Royal Institution

1. Letter 2183.
2. Plücker (1849b) of which Faraday had sent an offprint with letter 2189.
3. Plücker (1849b), 451.
4. Faraday (1849a, b), ERE22.

Letter 2191
Faraday to Jean-Baptiste-André Dumas
5 June 1849
From the original in AS MS

Royal Institution | 5 June 1849

My very dear friend

I expect to see M. de la Rive in half an hour who afterwards will leave London for Paris and I rejoice in the opportunity (though a hasty one) of acknowledging both your letters & the presents which you sent me by M Eichtal¹. I grieved very much for a long time thinking of the unapt and adverse circumstances under which you must as a man of peace order & science have felt yourself oppressed and I thought I knew how far the fine natural tone of your mind would make these things distasteful to you. But now I hope things are better not externally merely but as respects the feelings of mens mind[s] and to hear that *you* are nominated a Deputy and intend to act in that position makes me think that matters must be righting fast.

Mr. Niepce's[2] result[3] which you sent me is beautiful and has excited the admiration of many. It keeps very well for the present but I conclude will not be permanent. The frame also of dyed wood is very beautiful. Surely that must be a valuable result.

M. D'Eichtal is busy in some communications which if I understand him rightly may concern both you and me about Sugar[4]. From all he says I shall be most happy to be joined in the matter i.e. provided it is necessary here for I am never willing to be the depository of a *secret* unless there be a necessity. However as he understands the matter thoroughly and will take the trouble of guarding it aright I will leave it all to him to explain. All I can say is that I feel it a great pleasure & a great honor to be joined with you in any thing - or in the smallest matter.

I am greatly behind in Scientific reading & hardly know what is doing - and my encouragement to read is sadly diminished by the daily consciousness that I cannot keep what I read. Still as I know you feel an interest in me let me say that I am pretty well & cheerful & happy in mind.

My wife desires her kindest remembrances to you and Madame Dumas. We often speak of the kindness we received together[5] & wonder you could so consent to loan your time and powers idling with us. And now my dear friend with the most earnest wishes for your health & happiness and that of those around you who make your happiness

Believe me to be | Ever | Most truly Yours | M. Faraday
A Monsieur | Monsieur Dumas | &c &c &c

Address: A Monsieur | Monsieur Dumas | Professor | &c &c &c &c | à Paris

1. See letter 2184.
2. Claude Félix Abel Niepce de Saint-Victor (1805–1870, NBU). French photographer.
3. See Gernsheim and Gernsheim (1955), 148–9.
4. Melsens (1849).
5. See Sarah Faraday and Faraday to Reid, 28 and 29 July 1845, letter 1762, volume 3.

Letter 2192
James Prescott Joule to Faraday
5 June 1849
From the original in RI MS Conybeare Album, f.36

New Bailey St, Salford, Manchester | June 5th 1849

My dear Sir,

I beg to enclose a paper on the Mechanical Equivalent of Heat which I should feel obliged by your communicating to the Royal Society[1]. I hope it is in time to be read this session as I understand there are yet two meetings to be held[2]. If there be anything which requires addition or alteration I should be glad to supply it, particularly with regard to the sketch of the history of the mechanical Doctrine[3]. I can only say that I have endeavoured to make the paper as perfect as possible, and that it is the result of nearly a year's labour. Trusting it will interest you and that it will be treated with favour by the Society

I remain | Dear Sir | Ever Yours Respectfully & truly | James P. Joule
Prof Faraday DCL, FRS. | &c &c &c

1. Which Faraday did (see letter 2193). The paper was published as Joule (1850).
2. The paper was read on 21 June 1849.
3. Joule (1850), 61–4 gave this historical account.

Letter 2193
Faraday to Charles Richard Weld
6 June 1849
From the original in the Royal Society of South Africa

R Institution | 6 June 1849

My dear Sir

I beg to communicate the accompanying paper to the R.S[1].
Ever Truly Yours | M. Faraday
R. Weld Esq | &c &c &c

1. Joule (1850); see letter 2192. For a discussion of this letter see Spargo (1992).

Letter 2194
Faraday to Charles Caleb Atkinson[1]
6 June 1849
From the original in UCL, College Correspondence 1849

Royal Institution | 6 June 1849

Sir

I grieve to perceive that there is some misunderstanding at University College; with which I am, to my great surprize, in part mixed up, much against my inclination. I cannot but suppose that the whole is due to some explainable mistake. I would not intrude into the matter, except under the necessity of vindicating my own consistency. I have uniformly for many years past refused to give certificates of eligibility; and I have, in accordance with that resolution, given none for any of the candidates for the vacant chair at your college. I have been spoken to by several respecting the persons who are candidates[2]. I have refused to compare them, or give an opinion on their eligibility to the particular position now to be filled: but have answered enquiries as to general scientific rank as accurately as I could. I had been given to understand that I had been quoted for an opinion of Dr. Percy, very unlike that which I was known to entertain; (but there must, I think, here be some mistake.) I repeated my opinion in conversation to Dr Grant[3], and he has very faithfully reported it, in his letter of 18th May[4]. What I (for my own consistency) wish you and all concerned to understand, is, that neither in that conversation (or letter) or at any other time have I departed from my rule of not giving a certificate of fitness for a particular office[.]

Allow me to hope that you will make this letter known to the Council, or to any other body in the College, which may otherwise think I have falsified my rule of conduct in these matters:- and allow me, further, to hope that all uneasy feeling in regard to this affair will in the end, and shortly be entirely removed[5][.]

I have the honor to be | Sir | Your Very Obedient Servant | M. Faraday

C.C. Atkinson Esq | &c &c &c | Secretary

1. Charles Caleb Atkinson (1793–1869, B1). Secretary of University College London, 1835–1867.
2. That is for the Professorship of Practical Chemistry at University College to which Alexander William Williamson was appointed. UCL MS Council Minutes, **4**, 16 June 1849.
3. Robert Edmond Grant (1793–1874, DNB). Professor of Comparative Anatomy and Zoology at University College London, 1827–1874.
4. Although the appointment was made by the Council of University College, the Senate of the College had an advisory role. Grant, as Dean of the Medical Faculty, chaired the committee which the Senate established to consider the applicants for the position. Though not supported either by the committee or by the Senate, Grant championed Percy's appointment and was strongly opposed to Williamson's. This position he made plain in a minute in UCL MS Senate Minutes, **2**, 18 May 1849 where he praised Percy by writing that his abilities as an analytic chemist had been "attested by the highest authorities". This view is presumably a reference to the content of Grant's letter of the same date (and may even be the same document) discussed here by Faraday. The immediate cause of Faraday's annoyance must have lain with the Council discussion of the matter on 2 June 1849, but the minutes (UCL MS Council Minutes, **4**, 2 June 1849) do not reveal the nature of the debate.
5. UCL MS Committee of Management Minutes, 13 June 1849, noted that this letter, together with letter 2196, was read to the Committee of Management who forwarded them to Senate and its committee. UCL MS Senate Minutes, **2**, 15 June 1849, noted the letters and overruled further delaying tactics by Grant so that Williamson was appointed the following day (see note 2).

Letter 2195
Faraday to John Leighton[1]
7 June 1849
From the original in RI MS F1 C31

Royal Institution | 7 June 1849.

Sir

I received your parcel and letter and am very much obliged by your kindness. I had seen several of your works before and now Mrs. Faraday has had the opportunity of looking at them. I hope you find Art as productive as business. It is very evident from your works that it is a far more agreeable pursuit and it is a great happiness when the means by which we get our bread are in their nature pleasant to us. Wishing you every success in life that is good for you

I am | Dear Sir | Very Truly Yours | M. Faraday
John Leighton Jur | &c &c &c

1. John Leighton (1803–1868, Cantor (1991a), 301). Sandemanian and artist.

Letter 2196
Faraday to Charles Caleb Atkinson[1]
8 June 1849
From the original in UCL, College Correspondence 1849

R Institution | 8 June 1849

Dear Sir

I have received your note. I am so anxious not to be misunderstood that I venture further to say I trust no one can interpret my letter[2] as meaning any unfitness on the part of Dr. Percy for the *particular* office. I want to stand perfectly free of any opinion in that respect as regards any & all the candidates[3][.]

Ever Truly Yours | M. Faraday
Charles Atkinson Esq | &c &c &c

1. Charles Caleb Atkinson (1793–1869, B1). Secretary of University College London, 1835–1867.
2. Letter 2194.
3. See the notes to letter 2194.

Letter 2197
Faraday to L. Thompson[1]
9 June 1849
From the original in SI D MS 554A

R Institution | 9 June 1849

My dear Sir

Olifiant gas requires a pressure of $42\frac{1}{2}$ atmospheres to liquefy it at 30°F. I have not succeeded in liquefying coal gas or light hydro-carbon but it requires a much higher pressure than that. Again Babbage I think bored a hole in a very compact limestone rock, poured in acid & closed it up hoping to confine the gas & blow up the rock but the gas gradually found its way through the body of the rock[2][.]

Ever Truly Yours | M. Faraday

L. Thompson

1. Unidentified.
2. For Babbage's interest in the liquefaction of gases see Babbage (1832), 234–6.

Letter 2198
Faraday to Benjamin Hawes[1]
13 June 1849
From the original in DUL, H.A.J. Wilder Scrapbook, L–5868, p.100

Royal Institution | 13 June 1849.

My dear Sir

I received your letter of the 9th instant; and I have also received a letter from the discoverer of the process of extracting sugar[2], and another from M. Dumas stating to me the results which he witnessed[3]. I am willing to answer your enquiry as far as lies in my power: provided I can guard myself from the possible charge, hereafter, of having given a hasty and ill considered opinion. Let me remind you, therefore, that circumstances connected with the season & the time of the year[4] forbid that I should have the opportunity at present of seeing the operation of extraction performed: and that I can only know the process by description; can only judge of it by a consideration of the principles of chemical action which it involves, and can only be aware of the effects by the testimony of M. Dumas and the discoverer.

Having, then, considered the communications carefully, I see nothing to make me doubt that the facts are as stated:- namely that a very large proportion, approaching towards the whole, of the sugar in the Sugar cane may be extracted from it in the form of unchanged white crystalline sugar:

and at an expence not greater and probably very much less than that of the present process.

I am My dear Sir | Ever Most Truly Yours | M. Faraday
Benjn Hawes Esq. M.P. | &c &c &c

1. Benjamin Hawes (1797–1862, DNB). Whig politician. Under Secretary for the Colonies, 1846–1851.
2. Louis Henri Fréderic Melsens. See Melsens (1849).
3. Letter 2184.
4. Faraday delivered a course of eight Saturday lectures on static electricity after Easter. For his notes see RI MS F4 J10.

Letter 2199
Arthur-Auguste De La Rive to Faraday
14 June 1849
From the original in IEE MS SC 2
Mon cher Monsieur,

Me voici à Genève depuis deux jours; je ne suis resté à Paris que trois à quatre jours. J'ai trouvé nos amis si préoccupé de la politique & du choléra que j'ai cru qu'il n'y avait grand chose à espérer d'eux au point de vue de la Science. Arago avec qui j'ai passé une heure, ne m'en a presque pas parlé. Regnault a été cependant plus scientifique; mais il est découragé & peu entrain de continuer ses travaux. Le jeune Becquerel (le père etait absent) m'a entretenu de ses dernières expériences qui me paraissent assez curieuses mais qui ont encore besoin de vérification.

J'ai été bien heureux de me retrouver chez moi au milieu de tous les miens que j'ai trouvés en très bonne santé; ma femme a été très sensible à votre bon souvenir & à celui de Made Faraday & elle me charge bien de la rappeler a son tour à votre bonne amitié. Je n'ai point vu Mr. Dumas; c'est à Made. Dumas qui j'ai remis votre lettre[1]; elle avait l'air bien & elle m'a beaucoup demandé de vos nouvelles & de celles de Made Faraday.

J'ai vu à Paris de tristes effets du cholera, des quartiers dans lesquels une charette receuillait les morts; & cependant il régnait au milieu de ce fléau envoyé par la Providence une apparence d'indifférence qui faisait mal. On dansait & on chantait à l'un des bouts de Paris pendant qu'on mourrait à l'autre. Ce peuple est courageux, mais il n'est pas *sérieux*.

Le but de ma lettre n'est pas de vous parler uniquement de mon voyage & de mon arrivée. J'en ai un autre plus important, c'est de vous demander d'avoir la bonté de m'écrire le plus tot que vous pourrez, *un seul mot* pour me dire si vous avez effectivement fait mon expérience, comment elle a réussi & les particularités qu'elle vous a présentées. Vous m'obligerez infiniment en me faisant cette communication; j'espère que vous ne me trouverez pas trop indiscret. Si en même temps vous avez quelque nouvelle scientifique à m'apprendre, vous savez le plaisir que vous me ferez en me la donnant. - Peut-être avez-vous à l'occasion de l'expérience dont je vous ai parlé, fait quelques observations qui m'auraient échappé.-

Mille remerciements des jolis moments que j'ai passés avec vous a Londres, mes compliments bien respectueux à Made Faraday

Votre tout dévoué & affectionné | Auguste de la Rive
Genève | le 14 juin 1849

Address: Prof Faraday | Royal Institution | Albemarle Str | Londres

TRANSLATION
My dear Sir,

I have been back here in Geneva for two days now; I stayed in Paris only three or four days. I found our friends so preoccupied with politics and cholera, that I realised there was nothing great to hope for from the point of view of Science. Arago, with whom I spent one hour, hardly spoke of it to me. Regnault was however a little more scientific; but he is discouraged and very little inclined to continue his work. Young Becquerel (the father was away) talked to me of his latest experiments which seem interesting, but which still need to be verified.

I was very happy to find myself back at home, amongst my own, whom I found to be in good health. My wife was very touched by your good wishes and those of Mrs Faraday and she has asked me to remember her to your kind friendship. I did not see M. Dumas; it was to Madam Dumas that I gave your letter[1]. She seemed well and asked me a great deal about your news and that of Mrs Faraday.

I saw in Paris the sad effects of cholera; districts in which a cart collected the dead; and yet there reigned in the middle of this scourge sent by Providence an apparent indifference which made one uneasy. There was dancing and singing at one end of Paris whilst people were dying at the other. This is a courageous people, but they are not *serious*.

The reason for my letter is not to tell you only of my travels and my return. I have another more important reason, namely to ask you to have the kindness to write as soon as you can *just one word* to tell me if you have effectively done my experiment, how it succeeded and the peculiarities it presented to you. You would oblige me infinitely in communicating this to me; I hope you do not find me too indiscreet. If at the same time you have any scientific news to tell me, you know the pleasure that you will give me. Perhaps when you did the experiment about which I spoke, you might have noted some observations I missed.

A thousand thanks for the fine moments that I spent with you in London. My most respectful compliments to Mrs Faraday.
Your most devoted and affectionate | Auguste de la Rive
Geneva | 14 June 1849

1. Letter 2191.

Letter 2200
Faraday to Jean-Baptiste-André Dumas
18 June 1849
From the original in AS MS

Royal Institution | 18 June 1849

My dear friend

My first thought in writing or thinking of you is are you happy - for so much turns up near & about you that seems to me incompatible with your habit of mind & occupation that I mourn a little at times. What comes to pass direct from the hand of God, as the serious illness of your Son[1], we must indeed strive to receive with patience but that which is evolved through the tumults & passions of man does not bring with it that chastening & in some degree alleviating thought. I rejoice however to hear that your Son is better and that in respect of him you and Madame Dumas are relieved from all present anxiety. I think I remember him well as he went with us through the Jardin des Plantes[2].

In reference to M. Melsens matter[3] I have written[4] to our Colonial Secretary Mr Hawes[5] and could not say other than what you have said but M. D'Eichthal will inform you of that matter. Surely it must become very important & I hope will on one way or another produce its fitting return to M Melsens.

I think I wrote a short time ago by De la Rive[6] and fear that you or still more Madame Dumas will have reason to be weary of my letters but I trust in the kindness of both for forgiveness. Ever my dear friend Yours most faithfully | M. Faraday

M. Dumas | &c &c&c

1. Ernest-Charles-Jean-Baptiste Dumas.
2. In July 1845. See Bence Jones (1870a), **2**: 222.
3. Melsens (1849).
4. Letter 2198.
5. Benjamin Hawes (1797–1862, DNB). Whig politician. Under Secretary for the Colonies, 1846–1851.
6. Letter 2191. See letter 2199.

Letter 2201
Faraday to Thomas Twining
19 June 1849
From a photocopy in RI MS

Royal Institution | 19 June 1849

My dear Sir

I am ashamed that I have not acknowledged your kindness before but the occupation caused by many little matters at this time of the year & the manner in which through an infirm memory they push each other out of my mind must be my excuse. The flowers which you have been so kind as to send us are very pleasant & very acceptable & the kindness of the act far more acceptable. My wife joins with me in offering sincerest thanks to you for them.

I am My dear Sir | Very faithfully Yours | M. Faraday
Thos. Twining Esq | &c &c &c

Letter 2202
Faraday to Henry Thomas De La Beche
29 June 1849
From the original in NMW

R Institution | 29 June 1849

Dear De la Beche

Lyell said he would move you to put the R Institution down in a list of places to which the Government wanted to give 40 copies of a work[1] (Forbes[2] I think). We are making exertions in our Library & if you can help us in this case do for I say it *modestly* we desire your help[.]

Ever Yours | M. Faraday

1. Forbes, E. (1849) was presented by the Geological Survey. This gift was noted in RI MM, 2 July 1849, **10**: 205.
2. Edward Forbes (1815–1854, DSB). Palaeontologist at the Geological Survey.

Letter 2203
Faraday to Henry Thomas De La Beche
29 June 1849[1]
From the original in NMW

R Institution | 29 June

Thanks My dear De la Beche[.] You are ever very kind to me & to us[.]

Most truly Yours | M. Faraday

1. Dated on the basis that this is the letter of thanks for De La Beche's agreement to present a copy of Forbes, E. (1849) to the Royal Institution, referred to in letter 2202.

Letter 2204
Faraday to Jacob Herbert
30 June 1849
From the original copy in GL MS 30108/1/49

Royal Institution | 30 June 1849.

My dear Sir

I started on Monday morning (25th instant) going to the Dungeness and Needles lighthouses and returned on Wednesday night; and now proceed to give you the results of my visit, which I hope was not in vain[1].

Arriving at Dungeness lighthouse about two hours before sunset, I found, upon enquiry, that the evil which had presented itself was not the deposition of moisture upon the glass or any part within the lanthorn, for in that respect the keeper reported that the ventilating apparatus answered perfectly: - but, that when the lamps were all alight then, at certain of them, from two to four in number according to the wind &c there was a return of air down the ventilating pipes which would occasionally disturb the flames and sometimes extinguish them. After carefully examining the draught at each separate pipe and also in & about the cowl, I came to the conclusion that this was due, not to deficiency of *general ventilation* but to some obstruction of the *lamp ventilation*, of such a nature as to make it insufficient for the whole number of lamps (eighteen); so that when all were alight a part of the burnt air ascending from the flames, being unable to pass away above, returned down those tubes which had the weakest ascensive power and thus established as it were for the time a returning syphon action.

Upon this view I proceeded to examine in detail the upper part of the apparatus, and found that one of the deflector plates had been set so low as to approximate over much to the aperture beneath it, and was likely enough to cause obstruction sufficient to account for the effect observed.

Thinking there was time before sunset, by vigorous exertion, to correct this state of things; and being well aided by the keeper & assistant keeper; we dismounted the upper part of the apparatus, and by means of a carpenters chisel & a pair of pincers, cut away three pieces from the copper tube so as to make the aperture beneath the deflector plate double what it had been before, and then restored all to its place. This was done before sunset; and upon the lamps being lighted I had the satisfaction of finding that every one drew well; that there was an *indraught* at the top of each glass; and that all parts seemed to work harmoniously together instead of being in opposition to each other. Such was the case all night long. The keepers, also, were conscious of the difference of action from the first moment of lighting and expected good results; the wind changed in the night, and rose at one time considerably without producing any ill effect. I believe that the evil is entirely remedied; nevertheless I shall be glad to hear the results in a week or two, and accordingly left instructions with the keeper to report the state of things to you shortly.

The next morning I started for the Needles lighthouse but could not reach it before 9 o clk at night. I found this to be a very different case from the former; for the keepers here made no complaint respecting the burning of the lamps: but on the other hand stated that the lanthorn itself was close and that moisture appeared on the windows. On examining each of the lamps in succession (thirteen in number), I found the draught right being up every tube. At the openings above the ball and between the pan & the roof into the cowl. the draught was upwards & outwards but feeble only & at times it was *reversed*, so that air entered from the cowl into the lanthorn which ought not to happen: - the keeper said it was often down there especially with the wind at SSW. These indications all agreed in shewing that the fault was not in the lamp ventilation tubes as at Dungeness but in the *general ventilation* of the lanthorn.

Then as respects the lanthorn itself:- First I found it very hot & close with a little dampness on two of the windows away from the lamps and also on the inside of the roof in the same part. I do not doubt that in cold weather the windows are often moist. I found scarcely any current at the stair case up into the lanthorn (the doors below being exactly as they are obliged to have them when the lamps are burning), and I do not think any air enters the lanthorn that way. There are six ventilators consisting of tubes about $1\frac{1}{2}$ inches in diameter which are fixed in six alternate sides of the polygon (of sixteen sides) and which in the figure I have numbered from 1 to 6. they rise up just under the glass in these respective places.

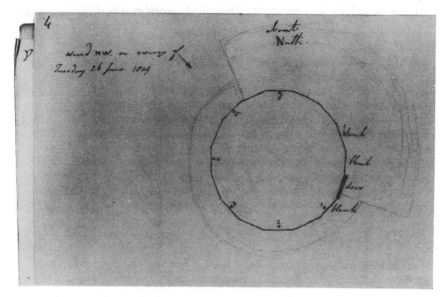

The wind was moderate at NW. & at the ventilator No. 5 (directly opposed to it) there was a strong amount of air *into* the lanthorn. At No. 6. the current wavered being mostly up into the lanthorn but also often down & out of it. At No. 4 the current was weakly up into the lanthorn. At No. 3. there was no current or else it was out of the lanthorn[.] At No. 2 it was almost always out with considerable strength and at No. 1 or the leeward ventilator it was a constant strong current outwards able to draw the flame of a lamp down with such force as almost & even quite to blow it out. Such was the state of things at this evening time with the lamps burning & the wind at N.W.; and it was not surprizing that the glass should be dim from dew and the air damp for instead of the bad air leaving the lanthorn well & freely at the upper part by the cowl, it was occasionally returning downwards at that point and on the other hand ventilators Nos. 1 & 2 which should have admitted air into the lanthorn were permitting it to issue outwards with considerable power. The *general* ventilation of the lanthorn was evidently all wrong.

The next morning when the lamps were out and the place at common temperatures, I caused the doors & passages to be arranged as when the lamps are burning, and then carefully examined every point of the lanthorn, to ascertain what were the *natural draughts* produced in it by its position upon a high & sudden bluff & by the form of the ground & buildings. The building as is well known is a low broad tower & the cottages & outhouses around it, the gallery, & the conical roof all tend to give it the form of a pyramid or cone. up to the very cowl itself[.]

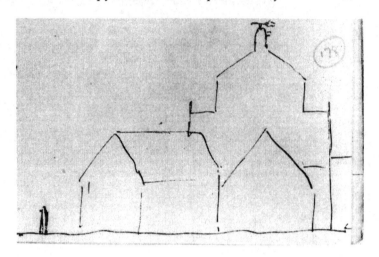

The wind was now about W. and I found the draught at ventilator 1 almost always *outwards* & that strongly. At No 2 it was the *same*. At No. 3. inwards almost constantly. At No 4. In & *out*, at No. 5 *out* chiefly & at No. 6 in & *out* - & evidently much affected by the eddies. So here again as last night what air was drawn into the lanthorn at the windward side - tended to go out at the leeward side of the place.

In the entrance from the stairs there was *no* draught either up or down; just as last night.

At the upper part of the lanthorn was the entrance into the cowl the air *entered* continually from without inwards i.e. between the roof & the edges of the catch pans; *contrary to every intention & desire*[.]

There is a stove in the middle of the lanthorn & its chimney goes through the roof of the building about 18 inches from the cowl & rises above the roof about 3 feet, the draught was almost constantly up this cold stove chimney and very good[.]

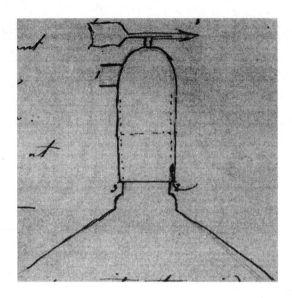

Believing that I should find the chief source of all the entering & adverse currents at the cowl to be due to *its form & position* I went outside & examined the effect produced there by the wind. I found a strong current *entering* the cowl at 2. so as to draw a piece of tape up into the inside a weak current out at 1. and also at 3.

Thus it was manifest that in consequence of the shape of the roof and the closeness of the cowl upon it, the wind was drawn in with much force at 2; and that though part of the air escaped at 1 and also at 3. that a large part *entered* into the lanthorn and actually inverted the use of the cowl making it in very blowing weather worse than nothing: for instead of the fowl air escaping there it will be kept down & even driven inwards and of course deposits the moisture from the burning lamps upon the glass. This effect which would probably occur more or less with every roof & cowl so circumstanced in shape & position as these is very likely greatly increased by the form of the surrounding buildings & land and this consists with the effect being worst when the wind is at S.S.W.

When the circumstances that conduce to the total result are so many and so difficult to estimate as to their separate effect, it is impossible to say before hand that any particular or proposed change would be certain to ease the evil, and therefore in offering *one* or *more* of the five following propositions I do it with a reservation of this kind[.]

1. The aperture round the cowl at the outside is large enough to allow of the introduction of the little finger, and only 3 inches above the projecting run, & 7 inches above the top of the conical roof; hence the wind which is driven over & up the roof enters immediately into this opening

(placed as it were to receive it & carry it with the lanthorn) & hence I believe a great part & *perhaps* the whole of the bad effect. This I expect would be almost entirely obviated by raising the cowl so that this aperture should be a couple of feet (at least) above the top of the roof, and by putting a deflector ledge about an inch & $\frac{1}{2}$ in width beneath it. I should not be surprized if this change alone were able to remedy the greater part if not the whole of the evil.

2. If from any circumstance of construction the cowl could not be allowed to be any higher than it is then I think if the part from a to 2 were made shorter so as to increase the distance between the aperture 2.3 & the roof from 7 to 20 or more inches and the deflector ring plate applied beneath it, much good might be expected[.]

3. The exit aperture *e* is only $5\frac{1}{2}$ inches in diameter. This I think is too small under the circumstances and is in fact the reason why air issues out of the lanthorn by the leeward ventilators. If made 7 or even 8 inches in diameter I should expect improvement[.]

4. If the ventilators were larger and as freely open as they now are and the one or two to windward were open & the others *shut* the effect would be to drive the air out of the lanthorn by the cowl, instead of at the leeward ventilators. In that case a dispensing plate which could be easily arranged & shifted from one to another would require to be hung over the inner entrance of the acting ventilators. As it is, the ventilators to leeward or the outgoing ventilators should never be left open.

5. From the present action of the stove chimney at common temperatures there is every reason to believe that if the upper tube from the lamps went through the roof in like manner & were like it raised three feet above the surface of the roof and terminated at the top by a like ventilator then the draught of the lamp air outwards would be nearly perfect & quite sufficient:- but the tube would probably need covering with a second jacket to keep it warm in cold weather or else the moisture from the lamps might condense & run back or even freeze on the inside & so cause some inconvenience[2][.]

I am | My dear Sir | Your Very faithful Servant | M. Faraday
Jacob Herbert Esq | Secretary &c &c &c | &c &c &c

1. See letter 2172.

2. This letter was read to the Trinity House Court, 3 July 1849, GL MS 30004/24, p.161 and was referred to the Deputy Master (John Henry Pelly) and Wardens. The Trinity House Wardens Committee, 9 October 1849, GL MS 30025/19, p.296 approved Faraday's suggestions and ordered them to be put into effect.

**Letter 2205
Warren De La Rue to Faraday
6 July 1849
From the original in IEE MS SC 2**

No. 7 St. Mary's Road Canonbury | July 6th 1849

My dear Sir

This is a begging letter - if not convenient to you to comply with my request lay it on one side and do not trouble yourself any more about it. I apply to you on behalf of a worthy man judging by his works and acts - a Mr. Nobert of Greifswald in Pomerania - a maker of microscopes and other philosophical instruments - he says "I have already asked so much & yet, cannot refrain from enquiring whether you could procure from Mr. Faraday a piece of his heavy glass"[.]

The applicant has accomplished, at my suggestion, some dividing on glass of such extreme fineness that it is a question whether the physical properties of light do not prevent our resolving it - the divisions being about the one hundred & ten thousandths of an inch apart from centre to centre and therefore considerably less than the wave-length of even the violet ray.

I have but just received the specimen and have not yet examined it but it will I have little doubt serve to elucidate some of the properties of light - it is a mechanical wonder at any rate and reflects the highest credit on the patience and skill of Mr Nobert. This is the man who wishes for a bit of your glass. I do not think that it will be lost on him if you have it to spare - but I am quite aware that these applications have been repeatedly made since you terminated your experiments[1] and shall therefore not be at all surprised to learn that this comes too late.

Yours Very truly | Warren de la Rue

Michael Faraday Esq | &c &c &c

1. On Faraday's glass work see James (1991).

Letter 2206
Benjamin Collins Brodie to Faraday
7 July 1849
From the original in IEE MS SC 2
Faraday number 1

13 Albert Road | Regent's Park | 7th July | 1849.

My dear Sir,

In compliance with your request, I send you a memorandum of some points of our late conversation with reference to the affairs of the Institution, and more especially as to the means of rendering useful and efficient the Laboratory lectures, as a course of scientific instruction[1].

In the present more advanced and exact state of Chemical Science, the means to this end are of a very different character to what they were twenty years ago. A course of scientific lectures to be truly useful, should afford the opportunity of seeing the careful and refined processes by which the latest results of chemical science have been arrived at. These, as far as possible, should be conducted in the lecture-room, and where, from the length or the delicacy of the experiments this is not practicable, it is very desirable that the chemical preparations, which are their result, should be exhibited, and a demonstration given of their general properties and reactions. If such lectures do not afford these opportunities, they lose their experimental character and much of their interest and utility. For these are precisely the advantages of the instruction of a lecture, over the instruction of a Book.

The arrangement of apparatus for these purposes, and the making of preparations, can only be done by one, who has gone through a course of chemical Instruction. In the lectures last year I greatly experienced the want of assistance for these purposes, not from any absence of care or intelligence on the part of the assistant to the Institution[2], but from the want of that preliminary knowledge which alone could enable him to fulfil the duties, I have mentioned. I have a private chemical assistant, who often went to the Institution to arrange and to assist at the lectures, and who under my direction made many chemical preparations for the purposes of the lecture, both last year and the year before when I assisted Mr Brande. For these lectures, however, the Institution afforded many advantages, which we should not have for a course on Organic Chemistry. Both from the greater familiarity of Anderson, with the class of experiments in the lectures, and from the nature of the preparations required. The preparations of organic chemistry cannot be bought, and require to make them, much trouble and skill. This is, of course, by no means peculiar to this class of preparations, but, I mention it, as in them the collection of the Institution is singularly deficient.

From these considerations it appears to me that a regular chemist, as assistant to the laboratory lectures is almost essential to their efficient working. And I will here answer your question as to the expences this would bring on the Institution, and the most practicable means of securing the ends I have proposed. To my own assistant I gave at first £100 and now give £110 a year. He has considerable chemical experience and is constantly in my laboratory and I think it would hardly be necessary to have as additional assistant at first, one whose services were quite so valuable. But I do not think less could reasonably be given than £80 a year with the prospect of advancement to £100. The other class of expences, attendant on Chemical Apparatus and preparations are of a more uncertain character, and would greatly depend on the quantity of work which was done. Such operations carried on for 10 months would cost more than for six. The principal expenses being the purchase of material and the breakage of glass and apparatus. A sufficient stock to start with the Institution already possesses. As however it would be essential to my superintendence of the work of such an assistant, that the greater part should be carried on in my own laboratory, which I cannot leave, and where I have the necessary means for the purpose. If the Institution were to think the plan generally desirable, it would be far better for them to find the Assistant, and enter into some arrangement with me as to the whole other expenses of the lectures, to which a certain sum might be devoted, I will say for example £50, for the current expenses of the year. In this case any further outlay I should incur on my own responsibility. The time of the assistant not occupied for the special purposes I have mentioned, I would endeavour to employ for worthy scientific ends.

As to the general question of how far it is desirable for the Institution to support this class of lectures I shall not offer an opinion. This is connected with considerations as to the purposes, for which the Institution exists, and the ends, which we may hope for it to fulfill, of which you and others who have been long engaged in its management are far better judges than I can be. In judging however of the success of such lectures, it is to be borne in mind that the Institution has not, like a College or University a certain class of its members who are, I may say, compelled to attend such lectures and the return to any outlay on their account is, by no means, to be looked for in money. The scientific labour of scientific men is like a gift from them to the world and, I think, we must not be disappointed if what our Institution does in this direction has also to be viewed in this light.

I am, my dear Sir, | very faithfully yours | B.C. Brodie
Professor Faraday | Royal Institution.

Endorsed by Faraday: Private to be read to Managers but not communicated to them[3]

1. Brodie and Brande the previous year had jointly delivered a course of lectures on chemistry in the Royal Institution to the medical students of St George's Hospital (see RI MM, 1 May 1847, **10**: 1). On this letter and the episode generally see Forgan (1977), 141–8.
2. Charles Anderson.
3. There is no indication from the minutes that this letter was read to the Managers; however, see letter 2207. At RI MM, 9 July 1849, **10**: 208 Faraday presented a short note on the role of chemistry in the Royal Institution together with its cost. It was then resolved to invite Brodie to deliver another course of lectures, *ibid.*, 209.

Letter 2207
Faraday to Benjamin Collins Brodie
9 July 1849
From the original copy in IEE MS SC 3
Faraday number 2
Private

Royal Institution | 9 July 1849

My dear Sir

I received your letter[1] & to day raised the whole question of the Laboratory before the Managers[2][.] I read them your letter to me & also placed before them the present disposition of the means directed to the chemical department[.] They find these to amount at present to £635 per annum and would not feel themselves justified in going to any further expence in that direction. I called their attention to the sums received by Mr. Brande & myself & Anderson which amounting to as much as £460 they did not seem to wish to alter *at present*[3].

Under these circumstances they resolved to ask whether you were willing to give another course of what we called Laboratory Lectures & you will hear accordingly from the Secretary[4].

Ever My dear Sir | Very Truly Yours | M. Faraday
B.C. Brodie Esq | &c &c &c

Endorsed by Faraday: Copy

1. Letter 2206.
2. See notes 1 and 3, letter 2206.
3. RI MM, 9 July 1849, **10**: 208.
4. *Ibid.*, 209, that is John Barlow.

Letter 2208
Faraday to Arthur-Auguste De La Rive
9 July 1849
From the original in BPUG MS 2316, f.65-6

Royal Institution | 9 July 1849.

My dear De la Rive

Though I have delayed writing this letter until the last minute still I have nothing satisfactory to tell you for as yet I have not made your experiment[1]. I gave orders to Newman at once for an apparatus but illness rendered me unable to follow him up or even to go on regularly with my lectures[2] and when at last he produced an apparatus it would not do. I have waited till today for a perfect one but have not yet received it & as we leave London directly for 6 or 7 weeks in the North I must defer the result until I come back. I have no doubt of a repetition in every point of the results you have obtained and I hope you will before then have given them to the world.

My thoughts are sluggish & heavy or I would say fifty things to you for though I have little to *tell* there is much I could *ask* of you. But head ach[e] & weariness make me quiet[.] I am afraid that the condition of Italy[3] sadly affects her scientific men for I had a letter from Majocchi[4] the other day seeming to say that he was driven from Turin & knew not what to do. For me, who never meddle with politics and who think very little of them as one of the games of life, it seems sad that Scientific men should be so disturbed by them and so the progress of pure undeviating unbiassed philosophy be so much & so often disturbed by the passions of men.

Ever my dear De la Rive | Yours Most Truly | M. Faraday

Address: A Monsieur | Professor A. de la Rive | &c &c &c | Geneva | Switzerland

1. Referred to in letter 2199.
2. See note 1, letter 2178.
3. A reference to the Austrian invasion of Piedmont in 1849. On this see *Ann.Reg.*, 1849, **91**: 278–90.
4. Giovanni Alessandro Majocchi (d.1854, P2). Professor of Physics and Mathematics at Milan.

Letter 2209
Benjamin Collins Brodie to Faraday
10 July 1849
From the original in IEE MS SC 2
Faraday number 3

13 Albert Road | July 10

My dear Sir,
So far as the decision of the Managers rests in grounds of oeconomy it appears to be both parsimonious and inconsiderate[1]. The Institution has a large surplus income, as much as the whole Chemical expenses, and it is of very little moment whether £100 or £200 be devoted to a good scientific purpose. If the end be not a worthy or suitable end, by all means let it be abandoned. This however is not the question. In proposing to me to take the lectures, it is acknowledged that they are desirable. The statements I have made as to the inefficient means of the Institution for carrying them out, are neither denied or discussed. But I am practically told that the changes, I desire, if made at all, must be made by me without their aid. This would truly result in the fact that I should devote the time of my private assistant, whose time is of value to me for my own experiments, to the service of the Institution or else have an additional assistant. While the question at issue is, whether the Institution is to find £100 more or less. This is very disheartening and certainly my first impulses would be to throw the thing entirely up. I have however written to Barlow to say that I will send my answer in a few days time, when I will write again[2].

Very faithfully yours | B.C. Brodie
Professor Faraday

Endorsed by Faraday: 1849

1. See letters 2206 and 2207 and notes therein.
2. Brodie did not subsequently deliver a course of lectures at the Royal Institution.

Letter 2210
Faraday to Benjamin Collins Brodie
16 July 1849
From the original copy in RI MS F1 C32
Faraday number 4[1]

Filey | Yorkshire[2] | 16 July 1849

My dear Sir
I do not at all share in your feeling as regards the Managers & their judgment[3][.] If their Professor of Chemistry[4] neither asks for nor suggests

any change either in the general system or the assistants it is hardly possible that they can imagine or have any reason for making a change without also having a strong judgment *against him*; and that I suppose they have not[.] Whatever prosperity the Institution *now has* has been acquired during his Professorship and he would have a right to say so supposing that any difference of opinion or discussion arose. I conclude that such thoughts as these must be on the mind of several of the Managers & I do not wonder at it, and though I may wish to very earnestly to infuse young blood & new Science into the chemical part of the establishment; I cannot overrule what others may think the common sense view of the case. To those who see the surface or even who may see some depth beneath the surface, it may appear quite unnecessary to make a change whilst the men who have thus far brought the Institution on its way are in their places: and who shall say they are wrong.

As regards the proposed lectures you write that your first impulse was to throw them entirely up, and I would say now *do not give them* unless you in your mind leave the Institution authorities as free to close them at the end of the season as you feel yourself to be either now or at any other time. There are probably *three* views of these lectures:- your own which is to make them strict logical expositions of chemical science & by that useful & very important:- then mine which is that though strict logical accurate & excellent, they are of little practical or influential character except in association with a practical school; *which you admit cannot exist in the Institution* - & thirdly the possible view of the Managers who have I think wished only for such simple elementary developments of chemical principles as should, like Mr Brandes former lectures, give a correct but popular idea of chemistry to their members who not being students are amateurs only & hardly that[.]

Remember also as respects your judgment of the Managers that your letter[5] was a private letter to me & that though I read it to the Managers they have it not in their possession and could not even lay it before their Professor of Chemistry if they desired[.] You can of course make it a letter to them by referring to it & expressing your wish; if you saw any end to be gained by that: but of that I must leave you to be the judge.

Ever Very Truly Yours | M. Faraday
B.C. Brodie Esq | &c &c &c

Endorsed by Barlow: Withdrawn See Man Min Feb: 4 1850 par J.B. Sec. RI[6]

1. On the copy in IEE MS SC 3.
2. "R Institution" crossed out and "Filey | Yorkshire" substituted.
3. Expressed in letter 2209. See also notes 1 and 3, letter 2206.
4. William Thomas Brande.

5. Letter 2206.
6. This letter was read to RI MM, 11 December 1849, **10**: 229 and entered in the letter book. However, at the following meeting Faraday conveyed the request that this letter be withdrawn from the Managers' records. This request was granted, RI MM, 4 February, **10**: 237.

Letter 2211
William Thomson to Faraday
24 July 1849
From the original in IEE MS SC 2

9 Barton Street, Westminster | July 24, 1849

My dear Sir

In the conversation which we had, about the beginning of this month, I mentioned several objects of experimental research which occurred to me as of much importance with reference to a theory of Diamagnetic, and, still more, of Magnecrystallic, action. I now take the liberty of addressing to you a few memoranda on the subject.

1. If a ball, cut out of a crystal of bismuth, be placed so as to be repelled by a magnet, will the repulsion not be stronger when the magnecrystallic axis is held perpendicular to the lines of force than when it is held in the direction of these lines? (Reference to § 2552 of your Researches[1]).

2. It would be a valuable acquisition to our experimental elements if a ball cut from a crystal of bismuth were suspended in the manner described by you in § 2551[2], and experiments were made by varying the length of the lever, and altering the general dispositions, so as to perceive cases in wh the tendency to move, due to the repulsive action, might be exactly balanced by a tendency to move in the contrary direction arising from the magnecrystallic action. A sketch, with dimensions, of the arrangements in any such case of equilibrium, would be most valuable.

3. In such a case as the preceding, if the strength of the magnet, (a pure electro magnet, without soft iron, would be the most satisfactory kind for such an experiment) be increased or diminished, will the equilibrium remain undisturbed?

4. Is the repulsion on a non-crystalline or crystalline diamagnetic ball or the attraction on a ferro-magnetic ball, exactly proportional to the square of the strength of the magnet? Thus in any case of pure repulsion, or of pure attraction, if the strength of the magnet be doubled, would the force be quadrupled; if the strength of the magnet be increased threefold, would the force be increased ninefold? In this investigation, as in the preceding, a pure electro-magnet would be the best, since in such a magnet the strength may be altered in any ratio, which ratio may be

measured with much precision by a torsion galvanometer, while the character and form of the lines remains absolutely invariable.

5. How are crystals of magnetic iron ore related to other crystals in their magnetic properties? Are they intrinsically polar, or are they merely axial? For example, if, supposing that to be possible, a crystal of magnetic iron ore have its polarity reversed, will it remain permanently magnetized in this reverse way? or, if a crystal of magnetic iron be demagnetized, will it remain non-magnetic? Will it not, in virtue of an intrinsic tendency to magnetization, gradually become magnetized in its original way?

I have a small ball of loadstone from the Island of Elba, which I have employed in place of a needle in a torsion galvanometer, & which appears to be susceptible of inductive action like soft iron (returning apparently to its primitive magnetic state when the inducing magnet is removed); and to be susceptible of this action to a greater degree when its axis is along the lines of inducing force than when it is perpendicular to them. My means of experimenting are however so very limited that I cannot be confident with reference to any such conclusions.

My intended departure for Norway, of wh I spoke to you, has been necessarily delayed for a fortnight by important & unexpected business. I hope tomorrow however to be on my way to Copenhagen by steamer.

I hope you are at present enjoying a pleasant and refreshing tour, as I heard today at the Royal Institution that you are travelling.

Believe me, my dear Sir, Your's very truly | William Thomson

1. Faraday (1849b), ERE22, 2552.
2. Ibid., 2551.

Letter 2212
Faraday to T.E. Venables[1]
4 August 1849
From the original in SELJ MS 6/1-4

Glasgow | 4 August 1849

Sir

Want of health has obliged me for a long time to lay bye all practical chemistry & I could now say very little to you about the bleaching question that could be of use[.] I suppose the bagging you complain of is of fibre that has undergone little or no preparatory process but *in that respect* is as nearly raw as may be & may require the alternate action of lye & chlorine to bleach it which is practised (or was a few years ago [)] by the linen weavers. But I really know nothing of the present state of the practical working of the subject, & I am sorry for it[.]

I am Sir | Your Very Obedient Servant | M. Faraday
T.E. Venables Esq | &c &c &c

1. Unidentified.

Letter 2213
Franz Adam Petřina[1] to Faraday
5 August 1849
From the original in RS MS 241, f.117
Perillustris ac Doctissime | Domine!

Facultas philosophica almae caes. reg. Universitatis Pragenae exactis centum lustris natalium[2] suorum memoriam celebrans, Te, vir, perillustris, ab Tua praeclara de scientiis praesertim de Physica merita, suis membris adnumerasse, sibi gratulatur.

Gratissime itaque officio meo satisfaciens, hac diploma Tibi, vir doctissime, tradere posse, maximo mihi duis honori.

Praeterea Te sincerrime rogo, ut, quae addita sunt diplomata, indiatis viris permittere velis.

Vale et fave Tuo amicissimo | sodali | Dr Francesco Petrina | caes. reg. prof. Physices | et Decano facult. philos.
Dabam Pragae 5 to Augusti 1849.

Endorsed by Faraday: See the Diploma in case. *Date 28 August 1848.*

TRANSLATION
Very illustrious and most learned | Sir!

The Faculty of Philosophy of the Imperial Royal University of Prague, celebrating the memory of an exact number of centuries from its foundation[2], is pleased to invite you, most illustrious Sir, on account of your famous eminence in the sciences, particularly physics, to number yourself among its members.

So I am most happy in fulfilling my official duty in being able to send you, most learned Sir, this diploma, which you would do me great honour [to accept].

Further I request you most sincerely to be so kind as to forward the additional diplomas to the persons mentioned.

Farewell and look favourably your most loving | colleague | Dr Francesco Petrina | Imperial Royal Professor of Physics | and Dean of the Faculty of Philosophy
Written at Prague, 5th August 1849

1. Franz Adam Petřina (1799–1855, P2). Professor of Physics at the University of Prague, 1844–1855.
2. That is the fifth centenary of the founding of the Charles University, Prague, in 1348.

Letter 2214
Julius Plücker to Faraday
10 August 1849
From the original in IEE MS SC 2

Bonn, | Aug. 10th, 1849

Dear Sir!

I feel myself very much obliged to you for having proposed me a member of the Royal Institution[1], and find no words to express my thanks in a proper way; but believe me Sir the kindness you showed to me on several occassions gave to me the greatest satisfaction I ever felt in my scientific career.

Permitt me to offer to you a "Resumé" of all my researches on Magnetism til[l] July 1849. Since my last letter I had scarcely any time to continue them. I found only crystals of Oxide of Tin showing a *very* strong magnetic polarity in the direction of their single axis; they were directed very well by the Earth. Then I examined most attentively the sulfate of iron. The line attracted by the poles of the Magnet is not perpendicular to the cleavage planes but makes with them an angle of 75°. According to that you will observe that a piece of such a crystal, bounded by cleavage planes, points differently, when turned round its magnecrystallic axis, this axis being allways horizontal. The difference is measured by an angle of 15° on both sides. The line attracted by the poles is one of the mid[d]le lines between the optic axes (which include an angle of 90°), the other one being not at all affected by the Magnet. In this (exceptional) case the resulting effect cant be deduced from the attraction of both the optic axes. Therefore I inquired, if the action may directly depend on the distribution of the Ether within the crystals, all the lines of less elasticity being attracted, the lines of greater elasticity repelled by the poles. But this law does not hold. Therefore new investigations only may give the true and complete law of nature.

I cut out of a very nice crystal of sulphate of iron a cube, two surfaces of which were perpendicular to the mid[d]le line attracted by the poles. By a sensible balance I found *no* difference in the magnetic attraction whatever a surface might be put on the approached poles of the Magnet. This result, fully according to your experiments, appears to me very strange: the directing power of the mid[d]le line being in this case so very strong.

These last days I tried again to prove that there is a diamagnetic polarity. The mutual action between magnetised iron being many thousand times stronger than that of magnetised iron on diamagnetic bismuth, you may never expect to see any mutual action between two pieces of diamagnetised bismuth. Such an action must be many - many *million* times weaker. But if you give to a piece of bismuth being acted upon by a magnet and suspended within a copper wire, by means of a current sent t[h]rough this wire alternately in opposite direction[s], a new diamagnetic polarity, the repulsion may be altered in the ratio of the intensity of the diamagnetic polarity, given to the bismuth by the Magnet, to to [sic] the intensity of that polarity, altered by the current. I[n] this way I succe[e]ded to show by means of the balance that a cylindre of bismuth obtained by the wire a magnetic polarity opposite to that which a magnetic body would obtain under the same conditions. But the action is very weak and I must before I may pronounce on this important point, repeat the same experiment in a varied way.

Being elected by the University of Bonn a deputy to the deliberations on the Universities's reform, which will take place at Berlin by ordre of the Prussian government in the month of September, I am not able to accept the kind invitation from Birmingham[2]. But I hope tis not the last time I crossed the Channel.

I had the pleasure to see Prof. Wheatstone here at Bonn, and was exce[e]dingly glad to learn from him, you were now of very good health.

Most truly | Yours | Plücker

1. See RI MM, 4 June 1849, **10**: 192 for this nomination.
2. For the meeting of the British Association.

Letter 2215
Faraday to John Barlow
18 August 1849
From the original in RI MS F1 C33

Royal Institution | 18, Aug 1849

My dear Barlow

A short letter is better than none and so I write for as Mr. Vincent was not here today when your letter came & I was I ventured to open it to see if there was any thing I could do at once for you. Accordingly I have sent to Jones[1] & Yarrel[2] to instruct them to send you the *Observers*[3] &c as you desire[.]

We reached home yesterday & as far as I can see all is going well here but the Painters are yet in the house & much work is going on[.] As to what has been passing I know little or nothing as yet & shall let Mr. Vincent tell you all that. Mr. Mason[4] is still alive[5] but I am to hear about him & his wife[6] on Tuesday or Wednesday[7][.]

We came home yesterday & are both better for our trip. We had much bad weather in the North & there have been heavy storms here also. On Monday[8] Anderson begins his holiday as I am now here to take charge[.]

I hope you & Mrs. Barlow have had good health no accidents and much enjoyment. I cannot remember whether we were at Avranches[9] or not so cannot guess how you may be enjoying yourselves but I hope it is heartily & with all the success you thought of. I have no news for you because I have heard none myself but when you come back I will shew you a letter I had from Brodie[10] immediately after mine to him[11] upon the last decision of the Managers[12] & mine in answer to his last which I had to write with some particularity & care[13]. You need know nothing of either until you see them, and the matter I conclude does not press[.]

With kindest remembrances to Mrs Barlow in which my wife would certainly join me if she knew I was writing I am
  My dear Barlow | Ever Yours | M. Faraday
Rev John Barlow M.A. | &c &c &c

Address: Revd John Barlow MA | &c &c &c | Avranches | Department de la Manche | en France

1. Edward Jones (d.1850, age 67, *Gent.Mag.*, 1851, **35**: 107). Newsagent of 34 Bury Street.
2. William Yarrell (1784–1856, DNB). Zoologist and Newsagent of 34 Bury Street.
3. That is the newspaper.
4. William Mason (d.1849, age 66, GRO). Librarian of the Royal Institution.
5. Mason died on 20 August 1849.
6. Margaret Mason. RI MM, 5 November 1849, **10**: 212 declined to give her an allowance.
7. That is 21 or 22 August 1849.
8. That is 20 August 1849.
9. Faraday did not get that far in 1845. Bence Jones (1870a), **2**: 214–21.
10. Letter 2209.
11. Letter 2207.
12. RI MM, 9 July 1849, **10**: 208–9.
13. Letter 2210.

Letter 2216
Faraday to George Macilwain[1]
20 August 1849
From the original in RI MS F1 C34

R Institution | 20 Aug 1849

Thanks my dear Sir for your kindness in sending me your lectures on *Fever*[2] which I received on my return from Scotland & shall endeavour to read with understanding & profit[.]

Ever My dear Sir | Yours faithfully | M. Faraday
Geo Macilwain Esq | &c &c &c

1. George Macilwain (1797–1882, DNB). Surgeon to the Finsbury Dispensary.
2. Presumably a collection of Macilwain's "Lectures on Fever" which were published in the *Med.Times*, 1849, **19** and **20**.

Letter 2217
Faraday to Charles Richard Weld
21 August 1849
From the original in RS MS MM 14.63

Royal Institution, | 21 Aug 1849

My dear Sir

I returned to town only a few days ago & now send you back Mr. Wards[1] paper[2] with the following remarks.

The paper does not carry conviction to my mind and as yet I retain my own view of *revulsion* &c.

The point whether a substance like *lead* or *zinc*, can, in a *perfectly pure state* & free from any ordinary magnetic substance assume either the magnetic or the diamagnetic condition according to the degree of magnetic force to which it is subjected - though assumed in the paper is not I think yet settled by the experiments in the paper: and would require far more care about even than appears from the paper to have been taken[.] Yet upon that point depends all the rest - the Sluggishness & contrary revulsions &c. I do not say it may not be so but as yet I have seen no results either in my own experiments or those of others that prove it to be so. Plucker himself I believe doubts whether a *a perfectly pure substance* can become both Magnetic & diamagnetic.

The hypothesis at the end is all dependant on this questionable point.

At pp.32–33 of the MS. the writer appears quite unaware of my suppositions in Nos. 2429, 2430, 2431 of my old Exp. Researches[3] or of the

Experiments & investigations made by Weber[4] Plücker[5] Reiss[6] in support of that view[.]

Still I have no right to decide in a case of difference of conclusions where I am one of the parties concerned & I am anxious never to stand in the way of the publication of opinions which are contrary to my own - & therefore must leave it to others to judge whether the communication is proper for insertion in the Phil Transactions[7].

Ever My dear Sir | Very Truly Yours | M. Faraday
R. Weld Esq | &c &c

---

1. William Sykes Ward (1813–1885, B6). Lawyer and Secretary of the Leeds Philosophical and Literary Society, 1840–1869.
2. W.S. Ward, "On some phenomena and motions of metals under the influence of Magnetic Force", RS MS AP 32.22. This paper had been read to the Royal Society on 21 June 1849 and summarised in *Proc.Roy.Soc.*, 1849, **5**: 855–6.
3. Faraday (1846c), ERE21, 2429–31.
4. Weber (1848, 1849).
5. Plücker (1849e).
6. This would appear to be slip of the pen for Reich. See Reich (1849). Faraday links this set of references in Faraday (1850), ERE23, 2640.
7. The discussion of this paper was postponed on 18 October 1849 and again on 22 November 1849, but the paper was archived on 10 January 1850. RS MS CMB 90c.

Letter 2218
Faraday to John Percy
7 September 1849
From the original in APS Misc MS Collection

Royal Institution | 7 Septr 1849

My dear Sir

Dr. Gustav Magnus Professor of Natural Philosophy in the University of Berlin has just left me. He intends being at the Meeting at Birmingham[1] and I write to ask you to say so at the Reception rooms that they may instruct him what to do when he presents himself on Wednesday next[2]. Phillips told me that they were always glad to know before hand of any eminent philosopher. He is a very pleasant man & talks good english[.]

Ever Yours | M. Faraday
Dr Percy

---

1. Of the British Association.
2. That is 12 September 1849.

Letter 2219
Faraday to Sarah Faraday
13 September 1849
From Bence Jones (1870a), 2: 249-50

Birmingham[1], Dr Percy's: | Thursday evening, September 13, 1849. My dearest Wife, - I have just left Dr. Percy's hospitable table to write to you, my beloved, telling you how I have been getting on. I am very well, excepting a little faceache; and very kindly treated here. They all long most earnestly for your presence, for both Mrs. and Dr. Percy are anxious you should come; and this I know, that the things we have seen would delight you, but then I doubt your powers of running about as we do; and though I know that if time were given you could enjoy them, yet to press the matter into a day or two would be a failure. Besides this, after all, there is no pleasure like the tranquil pleasures of home, and here - even here - the moment I leave the table, I wish I were with you IN QUIET. Oh! what happiness is ours! My runs into the world in this way only serve to make me esteem that happiness the more. I mean to be at home on Saturday night[2], but it may be late first, so do not be surprised at that; for if I can, I should like to go on an excursion to the Dudley caverns[3], and that would take the day.

Mr. Daniel[4] called on me to-day with a pressing invitation for you and me to his house, for which I thanked him sincerely, as he deserved to be thanked, but I could give no hopes of that.

Write to me, dearest. I shall get your letter on Saturday morning, or perhaps before.

Love to father, Margery, and Jenny, and a thousand loves to yourself, dearest,

From your affectionate husband, | M. Faraday

1. Faraday was in Birmingham to attend the meeting of the British Association of which he was Vice President that year. On the Thursday evening, Faraday explained electric light at the soirée held at the Town Hall (See *Athenaeum*, 22 September 1849, p.958 and *Birmingham J.*, 15 September 1849, p.7, col. d). Faraday also spoke at the Chemical Section on the Friday (*Athenaeum*, 22 September 1849, p.963).
2. That is 15 September 1849.
3. See *Athenaeum*, 22 September 1849, p.958-9 for an account of the visit, though it is not clear if Faraday attended.
4. Edmund Robert Daniell (d.1854, age 61, GRO, B1). Secretary of the Royal Institution, 1826-1843 and one of the Bankruptcy Commissioners for Birmingham.

Letter 2220
Jacob Herbert to Faraday
11 October 1849
From the original in GL MS 30108/1/49

Trinity House London.- | 11th October 1849.-
Sir,
The Board has had under consideration your report upon the defective ventilation of the Lanterns of the Dungeness and Needles Light Houses, and your proposals for remedying the same[1], - and the Board approving your suggestions for effecting this object I am directed to acquaint you therewith, - and to inform you that Mr. Wilkins has received instructions to carry into effect your recommendations as respects the Lantern at the Needles, to put himself in communication with you thereon, and execute what you may require. - In regard to the Lantern at Dungeness your suggestions have been already effectually carried out with reference to the intended advantage. -

I am | Sir, | Your most humble Servant | J. Herbert
M. Faraday Esq

1. See letter 2204.

Letter 2221
Faraday to John Percy
17 October 1849
From Bence Jones (1870a), 2: 250

Royal Institution: October 17, 1849.
My dear Percy, - I cannot be on the Committee; I avoid everything of that kind, that I may keep my stupid mind a little clear. As to being on a Committee and not working, that is worse still.

I wish we could get to Birmingham, and use your kindness, but that may not be. My working time is from October to December, and I am fully in it: I am sorry to say, as yet, with negative results. Still I must work[1].

Ever yours and Mrs. Percy's, | M. Faraday

1. See Faraday, *Diary*, 10–15 October 1849, 5: 10277–301.

Letter 2222
Faraday to James Morris[1]
26 October 1849
From the original in UCL, tipped in Tyndall (1870), History of Science Sources C

Royal Institution | 26 October 1849

Sir

I hasten to acknowledge the gift of the Transactions of the Royal Society of Mauritius and to offer my sincere and grateful thanks for them and I desire also to express to yourself the feelings which your very kind words raise up within me. As long as I have health & ability I will endeavour to deserve such good opinion[.]

I am Sir | Your Very Obliged Servant | M. Faraday
Jas. Morris Esq | &c &c &c

1. James Morris (1810–1869, DMB). Professor of Classics at the Royal College, Mauritius, 1845–1849 and Secretary, Royal Society of Arts and Sciences of Mauritius. At this time a Mauritian representative in London.

Letter 2223
Faraday to George Biddell Airy
29 October 1849
From the original in RGO6 / 403, f.113

Royal Institution | 29 Octr. 1849

My dear Airy

Am I to hope that Barlow is not mistaken and that there is a possibility of our having you at one of our Friday Evenings next season?[1] The thought is so pleasant to me that however small the chance may be I cannot resist helping to nurse it a little for I do not know any thing that would give me greater pleasure or do more honor to the Season here[.]

With kindest remembrances to Mrs Airy I am My dear Sir | Ever Yours faithfully | M. Faraday
G.B. Airy Esq | &c &c &c

1. See *Athenaeum*, 23 March 1850, pp.315–7 for an account of Airy's Friday Evening Discourse of 15 March 1850, "On the present State and Prospects of the Science of Terrestrial Magnetism".

Letter 2224
Faraday to Thomas Twining
29 October 1849
From a photocopy in RI MS

R Institution | 29 Octr. 1849

My dear Sir
    After much consideration & desire to accept your kind offer my wife feels obliged to decline it. Many things tend in that direction so as to constrain her to follow it and amongst them the daily expected death of a dear Cousin[1].
    Ever my dear Sir | Very Truly Yours | M. Faraday
T. Twining Esq | &c &c &c

1. Unidentified.

Letter 2225
Faraday to William Whewell
29 October 1849
From the original in TCC MS O.15.49, f.25

Royal Institution | 29 Oct. 1849

My dear Sir
    Mr. Barlow asks me to support his application to you for a Friday Evg in the coming season[1]. I would most willingly do so for our sakes and my own gratification but considering to whom I write I feel I ought not to say more than if such a thing comes within the scope of your convenience & willingness it will be the source of very high delight and of something more to me[.]
    Ever My dear Sir | Your Obliged Servant | M. Faraday
Revd. Dr. Whewell | &c &c &c

1. Whewell did not give a Friday Evening Discourse during 1850.

Letter 2226
William Whewell to Faraday
31 October 1849
From the original in TCC MS O.15.49, f.61

Trin. Lodge, Cambridge | Oct. 31, 1849

My dear Dr Faraday

I had not answered Mr Barlow's application sooner because I was not certain that I could comply with it to any good purpose. Your joining in it adds much to my wish to assent[1], but still I do not see the possibility of doing so. I could have wished to say something about the constitution of matter, but I dare not do so till I see my way better, or at least till I can bring out the difficulties more clearly. Your paper about axiality[2] is highly important in its bearing upon this point, but I cannot yet bring it into any definite relation with other things. I am at present in far too puzzled a condition to pretend to teach any one. If my ideas should grow any clearer I may perhaps beg to express them if you are willing to hear them; but I fear that will not be this year[3].

Always my dear Dr Faraday | Yours most truly | W. Whewell
Dr Faraday

1. In letter 2225.
2. Faraday (1849a b), ERE22.
3. Whewell did not give a Friday Evening Discourse during 1850.

Letter 2227
George Biddell Airy to Faraday
31 October 1849
From the original press copy in RGO6 / 403, f.114

Royal Observatory Greenwich | 1849 October 31

My dear Sir

Certainly I will give you a lecture as soon as I can find or think of a subject[1]. Mere skits of information on current changes in astronomical instruments &c which are very proper for the R. Aston. Society would not be proper for you.

But I do hereby promise to rack my brain (when it is not overreached by other things) to find a subject[2],-

and am | Yours most truly | G.B. Airy
Michael Faraday Esq | &c &c &c

1. See letter 2223.
2. See *Athenaeum*, 23 March 1850, pp.315–7 for an account of Airy's Friday Evening Discourse of 15 March 1850, "On the present State and Prospects of the Science of Terrestrial Magnetism".

Letter 2228
Jacob Herbert to Faraday
31 October 1849
From the original in GL MS 30108/1/49

Trinity House London | 31 October 1849.

My dear Sir,
 I have been instructed to transmit to you the accompanying Copy of a Letter from the Principal Keeper of the High Light House at Portland[1], having reference to the ventilation of the Lantern thereof, in connection with the Hot Air Stove, - and to signify the request of the Board, to be favor'd with your observations thereon. -
 I remain, | My dear Sir, | Very faithfully your's | J. Herbert
M. Faraday Esq

1. This letter, dated 16 October 1849, is in GL MS 30108/1/49.

Letter 2229
Faraday to Carlo Matteucci
5 November 1849
From Bence Jones (1870a), 2: 250-1

Royal Institution: November 5, 1849.

My dear Matteucci,-.... I have lately been working for six full weeks trying to procure results, and have indeed procured them, but they are all negative[1]. But the worse of it is, I find on looking back to my notes, that I ascertained all the same results experimentally eight or nine months ago[2], and had entirely forgotten them. This in some degree annoys me. I do not mean the labour, but the forgetfulness, for, in fact, the labour without memory is of no use.
 Still I have a thousand causes of thankfulness, and am not repining, only explaining. If I could have my own way, I would never write you a letter without some scientific point in it. As it is, the chances are they will be as barren as this one.
 Yours most truly, | M. Faraday

1. Faraday, *Diary*, 25 September to 6 November 1849, **5**: 10239–10412 record experiments mostly on diamagnetism.
2. There are no notes in Faraday, *Diary*, **5** covering this period.

Letter 2230
Jacob Herbert to Faraday
9 November 1849
From the original in GL MS 30108/2/88

Trinity House, London, | 9th November 1849.

Sir,

I am instructed to forward to you the accompanying Bottle of Water which has been received from the Trevose Head Light Establishment, together with an Extract of a Letter from the Light Keeper thereat and a copy of my Letter to him upon the Subject thereof[1]. And I am to signify the Request of the Elder Brethren that you will favor them by examining the said water and furnishing them with your opinion as to it's wholesomeness for Drinking or Culinary purposes.

I am, | Sir, | Your most humble Servant | J. Herbert

M. Faraday Esqre | &c &c &c

1. Watson to Herbert, 26 September 1849 and Herbert to Watson, 10 October 1849, GL MS 30108/2/88.

Letter 2231
Faraday to Jacob Herbert
15 November 1849
From the original copy in GL MS 30108/2/88

Royal Institution | 15 Novr. 1849

My Dear Sir

The water which I received from you from the Trevose Head lights[1], contains a portion of lead in solution which though small undoubtedly renders the water unfit for cooking or use in articles of food. There is every reason to believe that this lead is derived from the roof, & partly by the action of the Salt water spray for part the lead is in solution as a chloride and there is also common salt present. Except indeed for these substances the water is remarkably free from salt & would without them be very soft & pure[2][.]

I am My dear Sir | Very Truly Yours | M. Faraday

1. See letter 2230.
2. This letter was read to the Trinity House By Board, 20 November 1849, GL MS 30010/36, pp.464–5 with a note that the matter had been referred to James Walker who had proposed a solution to the problem which the Board agreed to pursue.

Letter 2232
Emil Heinrich du Bois-Reymond to Faraday
15 November 1849
From the original in IEE MS SC 2

Sir,

According to your own statement, natural p⟨hilo⟩sophy is indebted for the most important s⟨teps⟩ it has made at your hands to the strong conviction you always felt, that the various forms of force have one common origin, any form of force admitting of being converted into another under appropriate circumstances[1]. Nor have you, on several occasions, refrained from extending this view even to the mysterious agent of the nerves, and, in your paper on the Gymnotus[2], you have suggested some experiments for the purpose of discovering some new relation between nervous power and electricity.

I therefore venture to hope that you wil⟨l⟩ look with some interest on the results of an experimental inquiry, in which I have been engaged for these last eight years. This inquiry has led me to the most striking facts bearing upon the long-suspected identity of the nervous and muscular power and the electro-chemical form of force. The greater part of my investigations are detailed at length in the two accompanying volumes[3], which I beg you to accept as a proof of my deepest veneration.

I am very sorry to find, from your 'Experimental Researches' you do not read German[4]. Unfortunately, the subject I have treated is such a complicated one, and the various series of experiments are so extensive, that the shortest extract, to be intelligible at all, would far exceed the limits of a letter and probably of your patience also. You will, however, perhaps be kind enough to prevail upon some friend of yours to bring you and the scientific public of England acquainted with some parts, if not the whole, of my work; and, at least, you will see by a mere inspection of the plates, that it is not without having previously laid a new and somewhat large experimental groundwork, that I am so bold as, once more and that so positively, to bring forward theoretical views like those of old Priestley[5] and Galvani[6].

The second copy you would greatly oblige me by presenting, as a token of respect, to the Royal Society.

I am, Sir, yours most respectfully | Dr. E. du Bois-Reymond
Berlin, 21. Carlstr. | November 15, 1849.

Address: To | Mr. Faraday | Professor of Chemistry in the | Royal Institution, London.

1. Faraday (1846a), ERE19, 2146.
2. Faraday (1839a), ERE15.
3. Bois-Reymond (1848–9).

Plate 2. Emil Heinrich du Bois-Reymond.

4. Faraday (1838c), ERE13, note to 1635.
5. Joseph Priestley (1733–1804, DSB). Natural philosopher.
6. Luigi Galvani (1737–1798, DSB). Italian anatomist and physiologist.

Letter 2233
Faraday to James David Forbes
20 November 1849
From the original in SAU MS JDF 1849/121

R Institution | 20 Novr. 1849

My dear Sir
    The best pieces of heavy glass are gone but I send you one of the remaining pieces & hope it will answer your purpose[1][.] It contains some bubbles & I am not quite sure about the annealing. I will not however detain it for examination as I have not a better[.]
    Ever Truly Yours | M. Faraday
Professor Forbes | &c &c &c

1. Forbes may have wanted this glass for some work on radiant heat referred to in Shairp et al. (1873), 198 but which did not produce any results.

Letter 2234
Faraday to Jacob Herbert
21 November 1849
From the original copy in GL MS 30108/1/48

Royal Institution | 21 November 1849

My dear Sir,
    My call yesterday was only to say that as Mr Wilkins was about to send to the Needles[1] to day or tomorrow & would be a fortnight engaged in the business I prepared to leave town on the 3rd or 4th of Decr. & might be at the Portland Lighthouses on the 5th or 6th[2]. You wished me to let you know & said you said you would either give me a note or write to the Principal keeper.
    Meeting Captain Madan yesterday I endeavoured to refer him to some observations which I thought I had made in reply to a letter of yours of the 9th Decr. 1847[3] requesting me to examine Mr A. Gordons Catadioptic arrangement at the Trinity House on the Tuesday following. The observations were to the following purport. Mr Gordons object is, to use an Argand lamp of the same size & power as that in use by the Trinity house but by means of a different reflector & a ring lens, to send far more

light forwards towards the sea horizon than is so sent by the Trinity reflector. The Trinity reflector gathers & disposes of the light of 235° (measured on a great circle passing through the lamp flame & the axis of the reflector) & allows 125° to pass unused. Mr. Gordons reflector & lens together gather up & dispose of 320° & allow only 40° to escape[.] This I conclude is the chief foundation of the hope which is entertained by the inventor that it will *practically* surpass the Trinity reflector but there is at the same time another ground of expectation built upon the fact that in Mr Gordons reflector most of the impinging rays are reflected at a much smaller angle with the reflecting surface than in the Trinity house reflector & will therefore (according to a supposed law of reflexion which holds good indeed with such bodies as glass water marble &c) be more abundantly reflected.

With respect to this second point of a more abundant reflexion as the incident ray forms a smaller angle with the reflecting surface Newton[4] believed that to be the case with metals as it is with ordinary bodies and most writers on optical science including Herschell[5] & others have since his time repeated[6] the statement probably on his authority but Mr Potter[7] a few years ago showed that it was *not true* for metals for that they reflected at least as much if not more light when the ray was incident perpendicularly upon their surface as when incident at very small angles. Upon the occasion of your letter to which I have already referred I had two silvered reflecting surfaces prepared & observed & confirmed the quantity of light reflected at the same time at very high & very small angles[8] & could find no difference: in that respect therefore I am satisfied that any point of a Trinity House reflector acts upon the light as perfectly as any point of Mr. Gordons reflector.

With respect to the *first* point:- Mr. Gordons arrangement takes up a far greater number of rays of the sphere of light from the lamp than the Trinity reflector calling the whole quantity of light is 100 the Trinity reflector leaves about a fourth or rather more untouched whilst Mr. Gordons only allows about a thirty second part to escape and there can be no doubt that if all the rest were sent forward in the required beam of 15° of divergence the effect ought to surpass that of the simple Trinity arrangement[.] It is the province of trial & experiment only to decide whether these are so sent forward, but even theoretically it does not follow from the construction that they should go in the desired or expected direction & for the following reasons. The centre of the Argand lamp flame is in the Trinity arrangement $3\frac{1}{2}$ inches distant from the nearest point of the reflector that being the focal distance. Now the divergence of the *whole* beam of light depends upon *this distance* & *the size* of the flame & they are such as to give the required 15° of divergence or thereabouts. In Mr Gordons arrangement the flame has the same size but is only $1\frac{1}{2}$ inches distant from the reflector which gives a divergence so great that if the reflector alone were used the light would be spread over a space of 5 or 6 times as large as that covered by the Trinity reflector & would be proportionately feeble. It only requires the removal of the ring lens from Mr. Gordons arrangement (which I had done in my trial examination), and then the inferiority of Mr Gordon's reflector to the Trinity House reflector from this cause is even in a room at once manifest though it receives from the lamp a much greater number of rays. It is true that the ring lens gathers up much of this outstanding light, but it cannot correct the large dispersion and also in its action of bringing these rays nearer to parrallism [sic] with the axial ray it cannot be correct at one & the same time for the light from the reflector & for the light which it receives *directly* from the lamp. Trial & observation therefore only can decide how far this gathering up has been done and an observation *near at hand* cannot shew by its effect what would be the result *at a distance*. A very powerful light might be produced at 100 or any other number of feet or yards by the collecting & crossing of the rays at that spot and yet at a point further out as 2, 3, or 4 miles a comparatively poor effect be obtained. Trial only & at different distances can give a true comparison with the simpler & so far more correct apparatus of the Trinity house[9][.]

I am | My dear Sir | Ever truly Your | M. Faraday
Jacob Herbert Esq | &c &c &c

1. See letters 2204 and 2220.
2. See letter 2228.
3. Herbert to Faraday, 9 December 1847, letter 2032, volume 3. Faraday's reply has not been found.
4. See Newton's letter of 4 May 1672, *Phil.Trans.*, 1672, 7: 4057–9.
5. Frederick William Herschel (1738–1822, DSB). Astronomer and natural philosopher.
6. Herschel, F.W. (1800), 64–5.
7. Richard Potter (1799–1886, DNB). Professor of Natural Philosophy at University College London, 1841–1843 and again 1844–1865.
8. Potter (1830) which cited both Newton and Herschel.
9. This letter was read at Trinity House By Board, 27 November 1849, GL MS 30010/36, p.477.

Letter 2235
Jean-Baptiste-André Dumas to Faraday
December 1849[1]
From the original in IEE MS SC 2
Mon cher ami

M Milne Edwards membre de l'Institut et mon meilleur ami croit avoir besoin d'une lettre d'Introduction auprès de Vous. Je ne la croyais pas nécessaire, mais j'obéis à son désir. Il auroit besoin dans un interérêt d'enseignement, puisqu'il me remplace comme doyen de la faculté des sciences, de bien connaitre premièrement l'organisation de Vos sociétés polytechniques et adélaide, secondement celle de l'école de Wolwich; Il ne peut mieux s'adresser qu'à vous et je serais très reconnaissant des moyens que Vous lui donnerez pour prendre une idée juste de ces deux genres d'établissements.

Mille amitiés nouvelles | A. Dumas

TRANSLATION

My dear friend,

Mr Milne Edwards, member of the Institut and my best friend, believes he needs a letter of introduction to you. I did not think it necessary, but I am complying with his wishes. In the interests of education, as he is replacing me as dean of the faculty of sciences, he would like to get to know firstly the organisation of your polytechnic institutes and the Adelaide [Gallery], secondly, that of the Woolwich Academy; he could not do better than to address himself to you and I would be most grateful for any means you can give him to get a proper idea of these two types of establishments.

A thousand renewed best wishes | A. Dumas

1. Dated on the basis that letter 2246 is the reply.

Letter 2236
Thomas Graham to Faraday
3 December 1849
From the original in IEE MS SC 2

4 Gordon Square | December 3, 1849.

Dear Faraday

Your understanding of the passage in my paper[1] to which you refer in your note of Dec. 1[2] is perfectly right. So also is the illustration you give afterwards speaking of *weight* instead of volume. The numbers, however, which you give in illustration of the last, 2172 of time & 5292 of time are, I find, from a table of results (page 384) which illustrate a *divergence* from the law, owing to the capillary being too short to give resistance enough. The one number should be double of the other, according to the law.

Air or coal gas of 4 atmos. density should pass through a long tube, such as Perkins[3] 1 inch drawn-iron tubes[4], with 4 times the velocity of the same gas of 1 atmos. density, for equal volumes; & consequently with 16 times the velocity for equal "weights", the propulsive force being the same in both experiments. Hence a great facility in conveying coal gas in a *dense* state, by these pipes for long distances - such as many miles.

In the table head of page 385, would you be so good as to alter with your pen the first row of figures under 1 atmosphere from

$$1095 \;.\; 1096 \;.\; 1095.5 \;.\; 1095.5$$
$$\text{into}$$
$$1105 \;.\; 1107 \;.\; 1106 \;\;\;.\; 1106.$$

Your calling my attention to the point has led me to discover that I had calculated the last column of that table from a set of observations on air of 1 atmos. made last, & which I have omitted to give in the paper, & not from those observations for 1 atmos which are actually given.

It has been probably this which put you out. With the change, the numerical relations of the calculated times appear at once - 553 being $\frac{1}{2}$ of 1106, &c.

Very faithfully yours | Tho. Graham
M. Faraday Esq.

1. Graham (1849).
2. Not found.
3. Jacob Perkins (1766–1849, DAB). American inventor.
4. That is the form of tube which Perkins had made for his steam engine.

Letter 2237
Julius Plücker to Faraday
4 December 1849
From the original in IEE MS SC 2
Dear Sir!

A few days ago I received your long expected paper[1]. Belonging to sulfate of iron I found my observations mentioned in the latin Memoir[2] in complete accordance with yours. Your arsenic is diamagnetic, mine magnetic; nevertheless in both cases the axis is attracted by the poles. Respecting Antimony our observations dont agree. I examined a great number of crystals and allways the optic axis, which is perpendicular to the cleavage plane, was repulsed like in calcareous Spar. A piece of crystallised Antimony I brought from Paris (containing there is no doubt a small quantity of iron), showed very complicated phenomena. There was diamagnetic magnetic and crystallamagnetic action: all these three actions varying differently with distance.

Being returned from Berlin, - where I lost a good time in fruitless deliberations on University reform - took up again my magnetic researches. During the last two years I have been much tantalized, not being able to imagine any force whatever producing the paradox phenomena presented by crystals. Now I may reduce all to the comon law of magnetic attraction and repulsion and confirm my theoretical views as well by calcul[ati]ons as by direct experiment. I gave an incomplete notice of it to Mr Poggendorff[3] and I am now writing a more elaborate Memoir[4]. I dare not believe you will adopt my views, but nevertheless let me beg to give in a few words an idea of it.

I got by experiment new proofs of diamagnetic polarity, induced by electric currents: Ampère would say there are in diamagnetic bodies induced currents going round the molecules in a direction opposite to that in magnetic bodies. (There is some analogy in this explication with that given by Fresnel[5] of the phenomena of light passing along the axis t[h]rough rock crystal[6] and in any direction whatever t[h]rough the well known fluids, there being circular rotations of the ether in opposite direction.) That is also I think the first explication you gave of diamagnetic action, which I adopted myself since I found, that there were no necessity to adopt for the diamagnetic and magnetic force a different law of diminution by distance, and since I thought it without doubt that Bismuth gets polarity by becoming diamagnetic. Join to this hypothesis the new one, that in magnetic crystals the polarity of the molecules be induced in different directions with a different facility, energy or stability; that the same take place belonging to the diamagnetic polarity in diamagnetic crystals - and you may explain all phenomena. The meaning of the new hypothesis will be completely understood, when for instance I say that a cylindrical bar of soft iron gets its polarity more easily along the axis of the bar, and with more stability and power, than in any other direction; quite the contrary takes place, when the bar will be reduced to a plate.

By putting thin iron bars into a piece of wood or copper, you may, by choosing properly their directions, imitate all experiments with magnetic crystals.

Allow me to explain a single case, for instance that of a prism of turmaline [sic]. By considerations taken from the elasticity of the ether, in accordance with its conducting power for Electricity, you may deduce, that the magnetic polarity of such a bar is most easily and strongly excited perpendicularly to its axis. Suppose it be so.

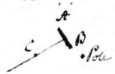

Now take a bar C of any indifferent substance, suspended horizontally between the two poles, to which ends a small iron bar AB is attached.

Such a bar will point equatorially with great energy: *the iron going away from the poles*. In that way you may, in the case of turmaline reproduce the magnecrystallic action by small iron bars, perpendicular to its axis. You may too reproduce the magnetic action of its mass by a small iron wire passing through it in the direction of its greatest dimension i.e. its axis. Thus you will have two groups of forces, acting one oppositely to the other. You may take only two forces, one driving A towards the nearer pole, the other one driving

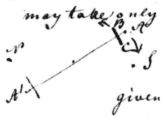

away from it the bar BC. By giving to the two bars BC and AA' convenient relative thickness, you will, for a given distance of the poles, have a state of equilibrium; and, the poles being far enough, the equilibrium will subsist in every position of the system of the two bars. Then you may mathematically prove that the system will point either *axially* or *equatorially* if you go either *farther* or *nearer* with the poles. &c. &c. An iron helix

will point equatorially &c &c.

Tis a consequence of my theoretic views, confirmed by experiment, that the attraction of a given crystal by the poles of the Magnet be only dependant of its exterior form, but independant of the direction of its optical or crystallographical axis with regard to the poles of the Magnet.

Some weeks ago, in repeating Boutigny's[7] experiments[8], I put my hand into melted iron at 1500 C, without feeling within any heat whatever. Tis curious.

My best thanks for your paper. At the same time I got Mr. Grove's[9]. When I write again I will have repeated his interesting experiments as well as Zantedchi's[10] very strange ones[11]. Unhappily for myself nearly all my time is absorbed by public lectures at the University.

Very truly | Yours | Plücker
Bonn | 4th of December | 1849.

Address: Professor Faraday | &c &c &c | Royal Institution | London

1. Faraday (1849a, b), ERE22.
2. Plücker (1849c).
3. Plücker (1849d), 427–31.
4. This paper was sent to the Haarlem Society of Sciences in December 1849 where it remained unpublished until it was published as Plücker (1852a). See p.1 of paper.
5. Augustin Jean Fresnel (1788–1827, DSB). French physicist.
6. Fresnel (1827).
7. Pierre Hippolyte Boutigny (1798–1884, Oursel (1886), 1: 127). French pharmacist.
8. Boutigny (1849). See also Plücker (1849d), 425–7.
9. Probably Grove (1849).
10. Francesco Zantedeschi (1797–1873, P2, 3). Professor of Physics in Padua.
11. See Plücker (1851a).

Letter 2238
Faraday to Jacob Herbert
8 December 1849
From the original copy in GL MS 30108/1/49

Royal Institution | 8 Decr. 1849

My dear Sir

I returned home last night and proceed to give you in brief the result of my visit to the Needles & Portland lights. I found the work at the Needles light drawing to a conclusion and so far advanced that I could decide upon the effect to be produced:- for the ventilators were in place & the cowl had been lengthened and guarded according to my instructions. I am happy to say that the result is perfect and the state of things utterly unlike what it was at the time of my visit in June[1]. I fortunately had the best opportunities of examining the effect produced for on Tuesday Night[2] (when I was there) there was a very hard frost (which in London made the streets very dangerous), and on Wednesday morning about 2 o'clk this changed (with rain) to violent wind at S & by E which is the wind the keeper complains of as the very worst for them. I could hardly keep my feet on the downs on my way from the lighthouse to the Alum bay hotel. Yet both in the time of the frost & of the wind, the glass was perfectly clear & dry and also the inside of the metal roof. There were no signs of condensation or of moisture in the lanthorn. All the lights burnt well and clear. The ventilation was right at every lamp, & a strong current existed *out* of the lanthorn into & through the cowl the whole time. Even when the lamps were extinguished on the Wednesday morning this current continued well & powerful:- so that there is no chance of any such entrance of air *into* the lanthorn by the cowl as occurred when I was here before. I believe that the evil which existed is perfectly remedied. The keepers & Mr Wilkins workman testified of the clear state of the glass ever since the cowl had been raised & guarded.

I then proceeded to the Portland lights[3] and passed part of Thursday with the night there. As to the lower light I found it in perfect order relative to the ventilation & no trace of dimness or moisture on the glass. The keeper said he had had no sweating since the ventilation pipes had been put up.

At the upper light, the lamps & their ventilation was quite right and though we had during the night & morning a strong gale of wind with rain from the S and SE yet all acted well & the glass remained clear. The keeper reports that nothing is wanting in this respect.

But there is here a stove at the bottom of the staircase tower leading up into the lanthorn, which they have to use in cold weather. But when at such times the wind is between NNE and E the smoke regurgitates into the tower issuing in a dense volume from the ash pit & door of the stove; & it ascends into the lanthorn in such quantities as actually to obscure the lights[.] The keeper spoke of one time when there were two East Indiamen lying to the West (as I understood him) of the light to which he thought of going & stating that as the wind then was they must be prepared for changes in the appearance of the light[.]

After much & careful examination I think I have come to a right conclusion as to the chief & almost sole source of this effect. The stove is surrounded by a *jacket* which is intended to produce warm air & from this a pipe 12 feet long & 4 inches in diameter passes upwards through the floor of the lanthorn terminating on the upper surface of the floor by a ventilation plate not more than 2 inches in diameter, though the air therefore round the stove is made very hot only a little of it passes by this aperture & the chief effect of the stove in warming the lanthorn is produced by its heating the air in the staircase which then enters the lanthorn by the door. In cold nights the keeper says he has to make the funnel pipe of the stove red hot that he may have sufficient warmth in the lanthorn. It is this funnel pipe which is chiefly in fault. It is 6 inches in diameter and after leaving the stove in a horizontal direction it rises vertically 5 feet 3 inches then passes horizontally through the wall of the tower to the outside being 7 feet in length in that point, (of which 2 feet are inside, 2 feet in the thickness of the wall & about 3 feet outside) and then it rises nearly vertically for about 25 feet more:- being for nearly half that distance against the outside of the small projecting staircase & for the rest of the distance 12 or 13 feet in the open air. The position of this metal pipe is nearly NE of the tower[.]

Now this pipe is exposed to the cooling influence of the wind in all that part outside the tower;- and out of a length of 37 feet. 28 feet are on the outside:- of the remaining 9 feet four are horizontal & there is only 5 feet 3 inches of vertical smoke pipe in the tower which being kept hot can tend to create draught. As to the outside part whenever the wind blows from NNE to E the whole 28 feet are exposed to its chilling influence; and as the stove is only used on cold nights it may be imagined what this influence is. If the wind blows from the opposite side the chimney is protected in part of its course & the effect is not so bad. I believe that when the wind is NNE or to E on a frosty night this pipe is cooled below the freezing point that the smoke within is cooled & being charged with sooty particles & carbonic acid gas is too heavy for the ascensive power of the hot smoke in the 5 feet 3 inches of vertical pipe in the tower & therefore descends & enters the tower through the fire door & ash pit & continues to descend until the tower doors are opened & a counter effect from the wind obtained.

The wind was at S.W when I was there in the evening & therefore I could not expect to find the stove in its particular bad state. Nevertheless I had a fire lighted in it that I might examine the condition of things. It drew very well having the ascensive power of the 5 feet 3 inches of funnel pipe in the tower & also the effect of the sheltered part outside the tower but when I applied my hand to it at the part where it rose above the staircase & left of its shelter I had full proof of the effect of the wind; for though it was not cold at the time the wind was getting up:- & though the pipe was warm or even hot a foot or two below the top of the shelter & the hot smoke continually moving upwards through it as soon as it had risen less than a foot above the shelter it was quite cold.

The keepers tell me the funnel pipe has been lengthened two or three times but this has done no good. Really such additions have done more harm than good for they have only increased the length of the heavy columns of cold smoke within which a wind at NNE or E produces.

The remedy I propose is this:- to dismiss the warm air tube altogether and in its place to take the funnel pipe of the stove up the middle of the stair tower through the stone floor & up the middle of the lanthorn until it rises to the level of the iron cross & then to take it obliquely to & through the roof & raise it about 3 feet above the roof where it shall terminate in a head of which I will immediately speak. In such an arrangement the flue would be kept hot in the building; would warm the air both of the staircase & lanthorn and would not (I believe) consume one third of the coals which are burnt in a cold night under the present arrangement and which amounts at times to a bushel per night[.]

Before I left London I had a certain form of head called a smoke disperser & six feet of a funnel pipe made. These I took with me to experiment with. Removing one of the four old ventilating covers on the roof of the Needles light house I projected this head through about 3 feet. The wind as I have said was a gale but in whatever position I placed the head, turning it in all directions, still its effect (due to the construction) was to create a powerful draught *up* the funnel pipe from below. I am satisfied this head is an excellent thing. It has no cowl or swinging or moving parts but is a fixture & can be made perfectly firm & stable. When the model comes up from the needles I will send it to the Trinity House & in the meantime recommend it for the head of the stove chimney in the present case.

Should the Board consider my proposition favourably & decide upon making the alteration I propose I have taken all the measurements & positions of the parts and could I think instruct the workmen & Mr. Wilkins perfectly in the arrangement of the pipe &c[4].

I am | My dear Sir | Ever Truly Yours | M. Faraday
Jacob Herbert Esq | &c &c &c

1. See letter 2204.
2. That is 4 December 1849.
3. See letters 2228 and 2234.
4. This letter was read to the Trinity House By Board, 11 December 1849, GL MS 30010/36, pp.485–6. It was agreed to make the alteration.

Letter 2239
Faraday to Julius Plücker
11 December 1849
From the original in NRCC ISTI

Royal Institution | 11 Decr. 1849.

My dear Plucker

I received your last letter[1] a day or two ago and I think I have one before that to acknowledge dated August[2]. I am very thankful to you for them & they are always a great pleasure. Your last views are very interesting but my head aches too much just now for me to say I have considered them and indeed they are fitted to remain in the mind for meditation again & again until by the growing up of facts they are developed confirmed & extended as the future progression of discovery may be. One part of your letter I do not quite understand where you say that it is a consequence of your theoretical views confirmed by experiments that the attraction of a given crystal by the poles of a magnet be *only* dependant of its exterior form but *independant* of the direction of its optical or crystallographical axes with regard to the poles of the magnet[.] For my own part I believe at present that the subjections of any crystal to the magnetic force depends upon its internal structure - or rather *on the forces* which give it its particular structure and that in any such crystal that line which coincides with the Magnetic axis may conveniently be called the Magnecrystallic axis. I do not suppose it necessary that the Magencrystallic axis should coincide either with the Crystallographical axis or with the Optic axis but I conclude that a very definite relation of these axis will in every case be found and that though they be convenient forms of expression in reference to three sets of phenomena that the ruling power is *one* and that when we properly understand it we shall see that *one law* will include all these phenomena and so all the forms of expression as *axes* &c by which we for the time represent them.

I have been at work endeavoring to establish *experimentally* any character of polarity in bismuth &c when in the magnetic field. I am sorry to say that I can get no stronger facts than those in my original paper and no stronger persuasion than that I gave in Par 2429, 2430[3]. Weber did me the honor to work upon this thought[4]. I believe I have obtained the effects he obtained in a far higher degree up to deflexions of 40°, 50°, or 60° but if they are the same they are not effects of polarity. I am just writing the paper[5].

I shall see Mr. Grove in an hour or too [sic] and shall tell him you have his paper[6].

Ever Yours Very Truly | M. Faraday
Profr. Plücker | &c &c &c

Address: Professor Plücker | &c &c &c | University | Bonn | on the Rhine

1. Letter 2237.
2. Letter 2214.
3. Faraday (1846c), ERE21, 2429–30.
4. Weber (1848, 1849).
5. Faraday (1850), ERE23.
6. Probably Grove (1849).

Letter 2240
Faraday to Ernest-Charles-Jean-Baptiste Dumas
11 December 1849
From the original in BN N.A. Fr. 2480, f.227

Royal Institution | 11 Decr. 1849
My Dear Sir

I was grieved to see a card with your name on it & the name of my friend Dr. Milne Edwards but without an address lying on my table when I returned home from the Sea side and I only this morning learned that you were somewhere in Leicester Square. I have just returned from the Sabloniere where I was told that you were gone out & Dr. M. Edwards gone to Paris. I have a meeting at 4 o'clock with our managers but shall endeavour between this & then to call again that I may see the son of my kind friend. Besides I think that you & I are personally acquainted for surely I had a most pleasant afternoon & evening with you & your father in the Jardin des Plants and also with Mama[1]. If I should be so unfortunate as to miss you do not fail to give my kindest & affectionate remembrances to M. Dumas and as far as it is proper to Madame[.] There are many on the continent whom I respect & esteem very highly but there are none other to whom my feelings are of the same kind as to them.

To Dr. M. Edwards too do not forget me and though I have a bad memory ask him whether he thinks I can have forgotten his kindness to me when I was at your house that he asks for a letter of introduction? - not that I at all object to it since it brought me the handwriting of M. Dumas[2][.]

I suppose I should hardly know you again if I saw you. Yet believe me to be Most Truly Yours | M. Faraday
M. Ernest Dumas | &c &c &c

1. In July 1845. See Bence Jones (1870a), 2: 222.
2. Letter 2235.

Letter 2241
Joseph Antione Ferdinand Plateau to Faraday
14 December 1849
From the original in IEE MS SC 2

Gand, | 14 décembre 1849.

Mon Cher Monsieur Faraday

J'ai reçu votre excellente lettre, et vous pouvez vous figurer tout le plaisir qu'elle m'a fait éprouver. De pareils témoignages d'approbation de la part d'un homme tel que vous sont l'une des plus douces compensations que je puisse recevoir aux peines et aux fatigues que m'a causées mon travail; ils sont, en outre, un bien vif stimulant pour m'engager à poursuivre activement mes recherches. Une chose cependant m'a beaucoup contrarié: C'est la nullité absolu d'effet dans l'expérience relative à la couche gazeuse en contact avec les poles d'un aimant. Mais vous ne me dites point si vous avez opéré sur un gaz unique, sur l'oxygène pur, par exemple, or, si votre électro-aimant était simplement plongé dans l'air, il serait possible que les pouvoirs magnétiques des deux gaz dont l'air se compose fussent de nature contraire, et qu'ainsi leurs effets se détruisissent mutuellement, du moins en très-grande partie. Si votre expérience a été faite avec un gaz unique, je ne conçois rien à l'absence totale d'effet, et ce résultat négatif me semble inexplicable: Car vos autres expériences montrent que l'action exercée par les aimants sur les gaz n'est pas si petite, et qu'elle se manifeste à des distances très-sensibles; Comment se ferait-il donc que ces attractions ou répulsions si notables ne produisissent, dans la couche gazeuse environnante, aucune condensation ou expansion appréciable?

J'ai maintenant à vous demander un léger service; j'ai adressé à Mr. Taylor, l'éditeur du philosophical magazine, un exemplaire de mon mémoire[1]; je lui ai écrit, en même temps, que j'espérais qu'il accorderait à cette *deuxième série* les honneurs de la traduction et de l'insertion dans les *Scientific Memoirs*, comme il les avait accordés à la *première*[2]; j'ai ajouté qu'il pouvait, à ce sujet, vous demander votre avis. Lorsque j'ai écrit cela à Mr Taylor, je ne connaissais pas encore votre opinion sur mon travail; maintenant que vous m'avez exprimé un jugement si favorable, j'espère que vous voudrez bien appuyer ma demande auprès de Mr Taylor, et, s'il y accède, l'engager à m'envoyer un exemplaire de la traduction[3].

La prière que je viens de vous adresser est peut-être indiscrète; mais les sentiments d'affection que vous voulez bien me témoigner, me donnent l'espoir que vous ne la considérerez pas comme telle

Agréez, Mon cher Monsieur Faraday, l'assurance de mes sentiments de respectueuse amitié | Jh Plateau
Professeur à l'Université, place | du Casino, 18.

Address: Monsieur | le professeur Faraday à l'Institution Royale, | Albemarle Street, | Londres.

TRANSLATION

Ghent, 14 December 1849

My Dear Mr Faraday,

I received your excellent letter, and you can imagine the pleasure it gave me. Similar expressions of approval from a man such as yourself are the sweetest recompense that I can receive for the painful and tiring nature of my work. They are, moreover, a very good incentive for me actively to pursue my research. One thing, however, has bothered me greatly: the complete lack of any effect in the experiment concerning the layer of gas in contact with the poles of a magnet. But you do not tell me whether you did the experiment on a elementary gas, on pure oxygen, for example, or if your electro-magnet was simply placed in air. It could be that the magnetic properties of the two gasses of which air is composed were of a contrary nature, and that thus their results cancelled each other out, at least for the most part. If your experiment was conducted on a elementary gas, I cannot understand the complete absence of any effect and this negative result seems to me to be inexplicable: for your other experiments show that the action exerted by magnets on gasses is not all that small, and that it is shown at very perceptible distances. How is it possible that such notable attractions or repulsions do not produce, in the surrounding layer of gas, any appreciable condensation or expansion?

I would now like to ask you a small favour. I sent to Mr Taylor, editor of the *Philosophical Magazine*, a copy of my paper[1]; I wrote to him, at the same time, that I hoped that he would accord to this *second series* the honour of a translation and that he would insert it in the *Scientific Memoirs*, as he had done with the *first*[2]; I added that he could, on this subject, ask for your opinion. When I wrote to Mr Taylor, I was not yet aware of what you thought of my work; now that you have expounded such a favourable judgement, I hope that you will wish to support my request to Mr Taylor, and if he agrees, to ask him to send me a copy of the translation[3].

The request I have just made is perhaps a little indiscreet; but the sentiments of affection that you kindly show me, give me the hope that you will not consider it as such.

Please accept, My Dear Mr Faraday, the assurance of my sentiments of respectful friendship | Jh Plateau
Professor at the University, Place du Casino, 18.

1. Plateau (1849).
2. Plateau (1843, 1844).
3. Plateau (1849) was translated into English as Plateau (1852).

Letter 2242
Benjamin Collins Brodie to Faraday
15 December 1849[1]
From the original in IEE MS SC 2
Faraday number 5

13 Albert Road | Regents Park | Decr. 15.

My dear Sir,

I received yesterday a note from Mr Brande, informing me that the Managers had requested him to undertake the course of Lectures[2], which I had resigned[3]. This is the only intimation I have received that my letter to Mr Barlow had been read at the Board[4].

In compliance, with what I understood to be your wish, I have entirely abstained from asking your advice or opinion on this matter, which otherwise I might have sought. Nor do I now write again to trouble you with it; But only to express to you my most grateful recollection of the kind interest you have taken in these lectures and in my other scientific occupations; to have enjoyed which, even for a time, I shall ever esteem one of the truest pleasures and advantages of my life.

Very sincerely yours | B.C. Brodie
Professor Faraday | Royal Institution

1. Dated on the basis that letter 2243 is the reply.
2. See RI MM, 11 December 1849, **10**: 229.
3. See letters 2206, 2207, 2209, 2210 and 2215.
4. RI MM, 3 December 1849, **10**: 226 notes the reading of Brodie's letter.

Letter 2243
Faraday to Benjamin Collins Brodie
17 December 1849
From the original copy in IEE MS SC 3
Faraday number 6

Royal Institution | 17 Decr. 1849

My dear Brodie

I owe you many & sincere thanks for your kind note[1]. As to your letter to the Secretary[2] which was of course read to the Managers[3], it contained so absolute a negative on your part to their request[4] that every body felt there was no more to say upon the matter. The Secretary might & very probably by this time has acknowledged the receipt of it[.]

And now my dear Sir though it was this affair that chiefly made you & me known to each other and though it has ended otherwise than I hoped still I shall not as regards ourselves let matters return to their former state. I hope much from you and shall as long as I remain in life look with expectation & I trust rejoicing to your course[.] If any word from me is of the least value as a word of encouragement or exhortation; I say *proceed, advance*.

Here things have reverted very much to their former state & I rather think perhaps fitly. The time was probably too soon for any change. But when such an one as myself gets out of the way then new conditions, new men, new views, and new opportunities; may allow of the development of other lines of active operation than those heretofore in service; and then perhaps will be the time for change.

Ever My dear Sir | Very Truly Yours | M. Faraday
B.C. Brodie Esq | &c &c &c

1. Letter 2242.
2. John Barlow.
3. RI MM, 3 December 1849, **10**: 226.
4. That Brodie continue to deliver the chemical lectures to the medical students of St George's Hospital.

Letter 2244
Faraday to John Lindley[1]
27 December 1849
From the original in Royal Botanic Gardens, Kew, MS Letters to Lindley, A-K

Royal Institution | 27 Decr. 1849.

My dear Lindley

Mr. Chater[2] has placed in my hands certain documents from which I make a selection imagining that you will think them quite enough. Take care of them as I have made myself responsible for their return. I will call on you on Saturday[3] about $\frac{1}{2}$ past 11 o'clk as I understand that time is convenient to you.

No. 1. Mr. Chaters letter to me.

No. 2. Act Geo III, 1794, page 431[4] *"blown plate or cast plate glass"* shewing that *both* were *plate*. Cast plate was not made here until about 1770 hence blown plate which was made long before that is named first[.]

No. 3. Act Geo III 1805, page 1481[5], *"rough plate glass and ground or polished plate or crown glass"* - shewing the meaning of *rough* in contradistinction to *ground* or *polished*.

No. 4. Act Geo III, 1809 p733[6]. - As No. 3.

No. 5 and *6* Two bills of the London Plate Glass company dated 1809 with numerous items of *rough plate and plates* - they making only blown plate glass.

No. 7. Tariff of the London Plate Glass company 1812 pp. 20, 21. *"rough Plates"* they still making only *blown plate glass*. In 1814 they began to make *cast plate glass* and accordingly issue a Supplement[.]

No. 8. of the larger sizes which they could then produce.

No. 9 Mr. Cookson's[7] letter and testimony that there was no *cast* plate glass before 1773 but blown plate 100 years previous:- *blown rough plates* a common article in this country before *cast* plate glass made.

No. 10. Cooksons Tariff for 1812[8]. At the end *"thick rough plates for skylights"*. They did not cast any plate glass at that time or earlier than 1817.

No. 11. Mr. Chance's testimony on a piece of glass:- that *rough blown plate* was known and in common use in this country before cast plate was made here[.]

No. 12. Letter. Peter Howard[9] - a glass maker for 50 years to the same purport[.]

No. 13. Letter Mr. Sims[10] a glass dealer for 46 years to the same purport.

On Saturday I will bring with me a very scarce book called the *Plate Glass book* printed & published in *1760*[11] i.e. about *13 years before Plate glass was cast in England* in which the term *rough plate* is familiarly used over &

over again, in the Title page, in the tables, and in above ten different parts of the work.

Ever Truly Yours | M. Faraday
Dr. Lindley | &c &c &c

1. John Lindley (1799–1865, DSB). Professor of Botany at University College London, 1828–1860.
2. Presumably of Chater and Hayward. Glass, lead, oil and colour merchants. POD.
3. That is 30 December 1849.
4. 34 Geo III, Cap. 27 "An Act for Granting to His Majesty certain additional Duties on Glass imported into, or made in, Great Britain", p.431.
5. 45 Geo III, Cap. 122 "An Act for charging additional Duties on the Importation of Foreign Plate Glass into Great Britain", p.1481.
6. 49 Geo III, Cap. 98 "An Act for repealing the several Duties of Customs chargeable in Great Britain, and for granting other Duties in lieu thereof", p.733.
7. Isaac Cookson (1776–1851, B1). Newcastle glass manufacturer.
8. For the interest of the Cookson family in glass manufacturing see Hedley and Hudleston [1964], 23–4.
9. Unidentified.
10. Robert Sims (d.1864, age 81, GRO). Sandemanian and plate glass merchant.
11. A later edition of Anon (1757) which went through many editions.

Letter 2245
Hans Christian Oersted to Faraday
27 December 1849
From the original in IEE MS SC 2

Copenhagen | the 27 Dec 1849

My dear Sir
Permit me to recall myself in your remembrance in introducing to the honour of your acquaintance the bearer of this letter, Mr. Colding[1], Inspector of the Waterworks of Copenhagen and at the same time a successful experimental and mathematical investigator of several philosophical questions. He is a former pupil of our Danish polytechnical school and since many years well known to me as a man of the most respectable character. He visits England particularly in the view to be acquainted with the great improvements, which are made in your Country in regards to the waterworks, sewers and gas-pipes; but he will, as much as his limited time permits him, take an interest in all other scientific objects. I beg that you will be so kind as to favour him with your good counsels and to recommend him to such gentleman, who can facilitate his pursuits.

I shall avail myself of this opportunity to give you a short notice of my continued diamagnetical researches which still I have now been obliged to interrupt through the now ending year.

In the meetings of the Royal Society of Copenhagen the 5th and 19th January this year I have communicated some researches upon diamagnetism[2]. Most of them are already contained in a printed French notice of which I sent a copy to you in the October 1848[3]; but since the publication of this notice I have been able to show some of the phenomena in a clearer light, and to state some others more correctly. I find that the positions taken by the suspended diamagnetical needles are determined by the borders of the magnetical body. In order to give a clearer view of the phenomena, I shall first suppose that one of the poles of the electromagnet bears a rectangular parallelopipedic [sic] polar piece of soft iron, and that two of its sides are horizontal, the four others perpendicular. If an attractive diamagnetical needle (for instance brass) is suspended vis-à-vis one of the perpendicular sides, the action of the magnetisme will make it point at this surface; but if the needle is suspended above one of the superior borders or below one of the inferior borders the magnetisme will make it parallel to the border. A repulsive diamagnetic body takes in all cases the opposite positions and is thus parallel to the perpendicular side vis-à-vis of which it is suspended, but perpendicular to it and to one of the borders, when it is placed above a superior or below an inferior border.

The same principle holds good in regard to all other forms of the poles of the magnet, when each of the electro-magnetical poles is provided with its polar pieces they will act together after the same principles but of course with much more effect.

My new & often repeated experiments have given me a more correct view of the distribution of magnetisme in the diamagnetic body.

The diagrams here joined will illustrate this distribution.

N and S represent the ends of two parallelopipedic polar pieces of an electromagnet nsns the transversal section of an attractive diamagnetical needle which suspended above the polar piece will take a position parallel to the two neighbouring borders. In this case the distribution of the magnetisme in the needle is such as the small letters s and n indicate; but if the diamagnetic needle in the contrary is a repulsive one, it takes the position perpendicular to the two neighbouring borders.

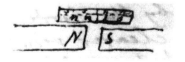

The distribution of its magnetisme is likewise here indicated by the small letters *n* and *s*.

When an attractive diamagnetic needle of inferior power is suspended in a powerfully attractive fluid and exposed to the action of the electromagnet, it will obtain the same magnetical distribution as a repulsive diamagnet, which is quite in conformity with your discoveries. In order to discover the distribution of magnetisme in the diamagnetical needle I make use of very thin iron wires bend in different manner, mostly as

or

You have in earlier years had the kindness to send me your series of researches, of which I possess the first 17[4]. Through I have the whole series in the Transactions, I should be glad to be in possession of the continuation of these immortal papers. I consider already those, which I have, as one of the most distinguished ornaments of my library.

I am, dear Sir, with sincere admiration | most faithfully Yours | H.C. Orsted
To | Doctor Michael Faraday Esq

Address: To | Doctor Michael Faraday Esq | 21 Albemarle-Street

1. Ludvig August Colding (1815–1888, DSB). Danish physicist and municipal engineer.
2. *Oversigt Kongelige danske Videnskabernes Selskabs Forhandlinger*, 1849, pp.2–9.
3. Oersted (1848).
4. See Oersted to Faraday, 5 September 1843, letter 1518, volume 3 and Faraday to Oersted, 31 October 1843, letter 1535, volume 3.

Letter 2246
Faraday to Jean-Baptiste-André Dumas
28 December 1849
From the original in BN N.A. Fr. 2480, f.229

Royal Institution | 28, Decr. 1849

My dear friend

I feel as if I were writing a last letter to you; for I can hardly think your duty & high occupation, apart as it is from every link that can recall or relate to a remembrance of me, can ever leave you a moment for imagination to travel hitherward: but, whatever our different destinies in, and paths through, life; that you may be prosperous in your proceedings & happy in your heart and home is the earnest hope & wish of one who will never forget you. It was a great grief to me to miss the meeting with your Son[1]. They told me at the Hotel he would leave England in the evening & I found afterwards he was a day or two longer in London. I think I must have made some mistake between him & the son whom I met with at the Jardin des Plants[2]; however, I trust he will be as kind as his father, and excuse any written mistakes in my note to him[.]

Though I was very glad to see your handwriting[3], yet I must confess, that, in the matter of M. Milne Edwards, I felt somewhat humbled in the thought that he would not come to me without a letter:- does he think that I have forgotten the Evening when he & I passed over from your house to his?[4] he was too kind, for me to lose the remembrance of it. If you see him again tell him I am almost inclined to reproach him.

But I must consider your time and thoughts. I do not know where I ought to direct this letter, but, for the old associations sake, shall send it to the Institute: and as I have a paper in hand[5], when it is printed I shall still send you a copy, and so on indeed till I hear I ought not. With our kindest remembrances to Madame Dumas and to both your sons I remain Ever My dear friend
    Yours | M. Faraday
Mons | M. Dumas | &c &c &c

1. Ernest-Charles-Jean-Baptiste Dumas. See letter 2240.
2. In July 1845. See Bence Jones (1870a), **2**: 222.
3. Letter 2235.
4. On 24 July 1845. Bence Jones (1870a), **2**: 222.
5. Faraday (1850), ERE23.

Letter 2247
Faraday to William Robert Grove
28 December 1849
From the original in RI MS G F26
Private

                                        Royal Institution | 28 Decr. 1849.
My dear Grove
    Brodie tells me that you must have mistaken him in supposing that he could obtain no opinion from me regarding his lectures[1]. What he

intended to convey (I cannot answer for his words) was that he could obtain no advice or opinion from me *at the last* i.e. at the time of your dinner conversations as I had declined having any thing more to say. This clears up what seemed to be a contradiction & I promised to speak to you about it[.]

Ever Yours | M. Faraday

1. See letter 2242.

Letter 2248
Faraday to Harriet Jane Moore[1]
1850[2]
From Bence Jones (1870a), 2: 256–7
Your note is a very kind one, and very gratefully received; I wish on some accounts that nature had given me habits more fitted to thank you properly for it by acceptance than those which really belong to me. In the present case, however, you will perceive that our being here[3] supplies an answer (something like a lawyer's objection) without referring to the greater point of principle. I should have been very sorry in return for your kindness to say *no* to you on the other ground, and yet I fear I should have been constrained to do so.

1. The recipient identified by Bence Jones (1870a), **2**: 256 as the same as that of letter 2318.
2. Bence Jones (1870a), **2**: 256 says this letter was written before letter 2318 and was a reply to a dinner invitation.
3. That is Brighton. Bence Jones (1870a), **2**: 256.

Letter 2249
Julius Plücker to Faraday
4 January 1850
From the original in IEE MS SC 2
Dear Sir!

My best thanks for your kind letter of the 11th of December last since[1]. Since - I think so at least - I got a satisfactory explication of all the questioned phenomena, discovered by you and by myself. This explication is founded on the principles exposed in my last letter[2]. The indications already given by me, are written down under the preoccupation of my mind, that there ought to be in crystals, brought between the two poles, a *conflict* of an attractif and a repulsif power, as it is the case with charcoal.

Plate 3. Harriet Jane Moore.

But we dont want such a coexistence of two opposite actions. In magnetic crystals tis *allways* a comon magnetic attraction, producing the known effects; in diamagnetic crystals tis *allways* a diamagnetic repulsion. The *freely* suspended magnetic crystal will allways go to the pole, the diamagnetic away from it: but, when the crystal *is obliged to turn round a vertical line*, the forces producing the rotatory motion of the crystal, deduced by mechanic law from the original ones, will, by different distance from the pole, act in an *opposite* way.

I adopt the views you explained 2439, 2440³. There is no difference at all between Magnetism and Diamagnetism, only the kind of inducing them is the opposite one in both cases. In diamagnetic bodies the "coercitif force" is greater than in magnetic ones. From that I deduced by new experiments, the known observations made by charcoal. Belonging to crystals the only hypothesis I adopt is, that such a crystal take *magnetic* polarity *with different facility in different directions* if it is a *magnetic* crystal, and *diamagnetic* polarity if it is a diamagnetic one. The molecules of positif crystals, when magnetic or diamagnetic by induction, will have the line joining their poles (completely or at least by preference) directed parallel to the axis; in negatif crystals this line is perpendicular to the optic axis. The turmaline for instance may be represented by lines of molecules, parallel to its axis, having their poles perpendicular to these lines, and may be imitated by a row

(a line) of small pieces of iron wire, arranged in this way

If such a line is brought, like the crystal between the two poles, you will find by a comon calculus, *that the molecules are, according to their distance from the centrum, either repelled or attracted by the poles*. Instead of the analytical results I have joined to my letter a figure, representing them.

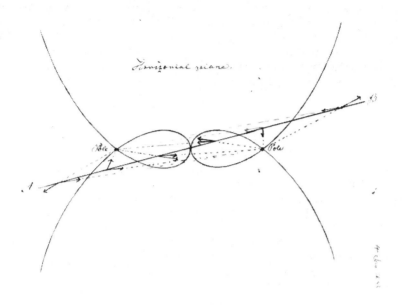

The two curves are the geometrical locus for the neutral points, where is respecting to each of both poles neither attraction nor repulsion. The kind of action on the molecules of the line AB is indicated by arrows. From this figure you may deduce the phenomena exhibited by turmaline.

The figure remains the same when there is a diamagnetic repulsion of the crystallic molecules: the terms attraction and repulsion are only to be commutated.

Every magnetic molecule, having two poles, is attracted by the pole of the Electro-magnet but the calculus proves that, the distance of the two poles of the molecule being *exceedingly small* with relation to the distance from the pole of the Magnet, this attraction is to be neglected with relation (par rapport) to the power which gives to the molecule a rotatory motion, whatever may be the direction of this motion. Tis a case quite analogous to that of a compass needle, acted upon by the Earth.

That is the explication you asked from me in your last letter. Hence it follows, according to my experiments made by the balance, that even when the attraction of the mass of a crystal be not absolutely independent of the position of its axes, the difference *of action* cant produce the questioned effect.

But I fear, Sir, you will not return to you own first conception, which includes diamagnetic polarity, and therefore you will be against my theoretical views. Nevertheless you will think it right from my part, to defend them as long as I am not convinced of the contrary. Therefore let me conclude by in[di]cating a very easy experiment, which, according to me proves, against Weber's theory, that the questioned diamagnetic polarity may be a *permanent* one, like in steel, even, for a short time, strong enough to be demonstrated by experiment[.]

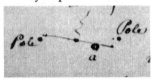

Let a bar of bismuth be suspended in the common way between the two poles and by the torsion of the silk thread put against a piece of glass or copper (a). It will be, with greater force, retained in the same position, when repelled by the magnet, but, if you change by means of a gynotrope, the polarity of the magnet, the bar, in the first moment, will be decidedly *attracted* by the poles.

Yours | very truly | Plücker
Bonn the 4th of January

Address: A | Monsieur Faraday | &c &c &c | Royal Institution | London
Postmark: 1850

1. Letter 2239.
2. Letter 2237.
3. Faraday (1846c), ERE21, 2439-40.

Letter 2250
Faraday to Julius Plücker
8 January 1850
From the original in NRCC ISTI

Royal Institution | 8 January 1850

My dear Plücker

I mean to write you a short letter though I only received yours yesterday[1] and I have not much to say[.] But I want to establish on your mind very clearly that you must not think I deny all that I do not admit. On the contrary I think there are many things which may be true and which I shall receive as such hereafter though I do not as yet receive them but that is not because there is any proof to the contrary but that the proof in the affirmative is not yet sufficient for me. It is so as regards the diamagnetic polarity of bismuth phosphorous &c my view is just the same as it was when I wrote paragraphs 2429–2430[2]. I think that may be the true view but I do not think that such a state is as yet proved.

Now I rejoice to see all your active reasonings experiments and consideration and am sure that your exertions are & will be successful in clearing up the present mystery of the subject. I am not quite sure I understand all the meaning of your last letter but that is because of my own ignorance[.] You also do me the honor to refer to paragraphs 2439, 2440 and I am not quite sure whether you mean them or the paragraphs 2429, 2430 and that makes a little confusion in my mind. Besides which, my memory is so bad that I cannot recall the purport of the enquiry I may have made of you in my last letter[3] & so am not sure of the bearings of the answer you have sent me. How continually I grieve that I cannot read German for if I could I suppose I could follow you in Poggendorf[f] very quickly. But I have no right to complain for Providence has been very kind to me.

I finished a paper at the close of the last year and sent it on the first of January to the Royal Society[4]. It was devoted to a rigid examination of the assumed polarity of bismuth phosphorous &c[.] I think my apparatus must at the last have been very much like Webers but I really regret to say that my conclusions were against the polarity. I obtained effects with bismuth of the kind Weber describes but I obtained far greater effects of the same kind with Gold silver copper tin & lead. On examining them I found they were all effects of the currents induced in the mass of metal employed as a whole and not of any polarity. When I divided the metal so as to interfere with these currents as by making up a cylinder from wires then the action was entirely stopped, whereas when I divided it so as to have the currents uninterfered with as by making the cylinder of discs of metal then the action as good as ever[.] The *reverse* of this was the case when polarity was concerned as in iron, for then division into wires did not interfere with the final effect but division into discs did.

I have also tested the question of polarity or of induced currents in the mass by reference to the *time* during which the moving metal was in motion i.e to the velocity of the motion, and which in the form of apparatus is in striking contrast, as presented by the two view[s]. Velocity does nothing in the case of polarity as with iron but the result depends entirely upon the whole amount of journey or motion; in the case of the induced currents it does every thing.

I have also constructed a communtator which because of the existence of a difference in the place of maximum action of polarities & place of maximum action of induced currents can separate or oppose or combine the results of these two kinds of action in any degree. All these means & investigations shew that as yet the effects obtained with bismuth have been the result of induced currents in the mass & not of any polarity, and that if there be such a polarity we have no proofs of it as yet beyond the first simple fact of a repulsion from the poles of the magnet. I have endeavoured in every way to repeat Reich's experiment as described by Weber & can not obtain the result[5].

All this does not make me deny the polarity, but only the sufficiency of the proofs thus far advanced. My opinion is exactly the same as when I wrote 2429, 2430. You will find in my paper when it appears what I think the explication of your experiment with bismuth, it is simply a case of revulsion (2310, 2315, 2338)[6], due to currents in the mass: this revulsion I have of late raised to a very high degree by division into discs & destroyed altogether even in Gold silver & copper by division into wires. Yet if the effects were due to polarity the reverse of these results ought to take place[.]

Ever My dear Plucker | Most Truly Yours | M. Faraday
Professor Plücker | &c &c &c

Address: Professor Plücker | &c &c &c | University | Bonn | on the Rhine

1. Letter 2249.
2. Faraday (1846c), ERE21, 2439–40.
3. Letter 2239.
4. Faraday (1850), ERE23.
5. See Weber (1849), 478–80 for a discussion of Reich (1849).
6. Faraday (1846b), ERE20, 2310, 2315, 2338.

Letter 2251
William Thomson to Faraday
10 January 1850
From the original in IEE MS SC 2

2 College, Glasgow | Jan 10, 1850

My dear Sir

By the same post I send two papers, one by myself, and the other by a brother of mine[1], which were communicated to the Royal Society of Edinburgh last spring[2].

In the paper by my brother some very remarkable reasoning founded on the principles of "Carnot's[3] Theory of the Motive Power of Heat", is adduced to show that the freezing point of water must be lower under a high pressure than it is when only subjected to the pressure of the atmosphere.

I have today succeeded in verifying by experiment this conclusion; and I think the fact that an increase of pressure lowers the freezing point of water may be considered as satisfactorily established.

My brother gives in his paper an estimate of the amount by which an additional atmosphere of pressure will lower the freezing point of water, wh[ich] he deduced, by Carnot's method of reasoning, from Regnault's experiments on the pressure and latent heat of vapour of water near the freezing point. The number which he gives is .0075 of a degree centigrade or .0135 of a degree Fahrenheit, of lowering due to 1 additional atmosphere of pressure.

After several unsuccessful attempts to test this result by experiment, I got a thermometer made with *ether* as the liquid, wh[ich] is so sensitive that the column of ether in the tube rises by more than two inches for 1° Fahr. of elevation of temperature. After graduating the tube very rudely, I had the whole thermometer sealed up in a glass tube to prevent the bulb from being exposed to the pressure to be applied to the water in the experiment. I then took an Oersted's apparatus for compressing water and filled it with lumps of clean ice and water; and I put it in a guage [sic] for the pressure, and my ether thermometer in its glass envelope.

Although the thermometer is excessively sensitive, it remained absolutely steady in the mixture of ice & water, until a pressure of about 9 atmospheres was applied. The column of ether then ran rapidly down, & nearly settled about $7\frac{1}{2}$ divisions of my scale lower than its primitive position. I increased the pressure up to about 19 atmospheres above the original atmospheric pressure and the column of ether descended farther, very rapidly. After giving it ample time to settle, during which the pressure was kept constant, I found that it stood $17\frac{1}{2}$ divisions below its primitive position. When the pressure was suddenly removed, the column of ether instantly began to run up, and it ran above its primitive position on account of a good deal of ice having been melted, and a lead ring I employed to keep a space clear for reading the thermometer having descended below the top of the bulb.

I have found by a very rough comparison with a common mercurial thermometer, that a degree Fahr. corresponds to about 70 divisions of my scale. Hence $17\frac{1}{2}$ of my divisions are equivalent to .25 of a Fahrenheit degree or a quarter ———. According to my brother's prediction, the lowering of temperature would have been .256 of a degree, if the pressure added had been exactly 19 atmospheres. This is only six thousandths of a degree more than what I found by experiment.

When I found how very close the agreement of my results with Theory was, I was very much surprised; and, when I consider the roughness of my experiments to find the value of a division of my scale;

and the uncertainty of my pressure guage, I cannot but attribute it partly to chance that I have fallen upon a result agreeing so very closely with the indication of Theory.

I hope you will excuse my having troubled you with all this. I have done so because I thought that, the lowering of the freezing point of water being a very remarkable fact apparently well established now, you might possibly feel some interest in having it brought under your attention[.]

I remain, My dear Sir, | Your's most truly | William Thomson
Prof. Faraday

1. James Thomson (1822–1892, DNB). Civil engineer.
2. Thomson, W. (1849). Thomson, J. (1849).
3. Nicholas Léonard Sadi Carnot (1796–1832, DSB). French physicist.

Letter 2252
Faraday to William Thomson
12 January 1850
From the original in ULC Add MS 7342, F34

R Institution | 12 Jany 1850

My dear Sir

I have received your papers[1] & note[2] & thank you heartily for them[.] The experiment & whole course of reasoning on the freezing point of water is most interesting. I remember in old times Mr. Perkins[3] had an idea that he could freeze water Acetic acid &c at a higher temperature under pressure than at common pressures[4] it is therefore the more striking on my mind to find the contrary to be the real truth[.] I have not read the papers as yet & therefore do not know whether the reasoning is general or not for all bodies, so I may be asking a foolish question in asking whether according to the theory the same results ought to occur with all bodies - as for instance with Glacial Acetic acid Spermaceti &c &c[.]

Ice is a body larger in volume than water so it might seem that if we could compress ice it might tend to squeeze it into water or else require a lower temperature to keep it as ice[.] What would be the effect with a body which instead of expanding should contract in solidifying?

Ever My dear Sir | Very Truly Yours | M. Faraday
W. Thomson Esq | &c &c &c

1. Thomson, W. (1849). Thomson, J. (1849).
2. Letter 2251.
3. Jacob Perkins (1766–1849, DAB). American inventor.
4. Perkins (1826).

Letter 2253
Faraday to Richard Owen
14 January 1850[1]
From the original in BL add MS 39954, f.138

Royal Institution | 14 Jany 1849 [sic]

My dear Owen

I have had your letter and am very sorry for the cause; and far more for the Dean of Westminsters[2] sake than our own. I had before to give an Evening for him on account of illness and from what I knew then from Mrs. Buckland should never have thought of asking him for one. But when he advertised in the newspapers that he would give a lecture on Artesian wells in this house, I felt that not to accept it would be to insult him; & told Barlow so[3]. Barlow is out about the matter. Our only plea for the change must be the indisposition of the Dean for otherwise certain wrong interpretations which the Dean himself hinted at to me & which he seemed to fear; are sure to be made by some who think his views not quite correct[.]

What would have happened if we had gone on with another offer which he voluntarily made of giving *Six* lectures here for some Charity!!![4]

Ever yours | M. Faraday
Richard Owen Esq | &c &c &c

1. Dated on the basis of the references given in note 3.
2. William Buckland.
3. The probable arrangements for Friday Evening Discourses RI MS GB **2**: 55 noted that Buckland was to give a Discourse on artesian wells on 18 January 1850. Instead William Robert Grove gave a Discourse "On some recent Researches of Foreign Philosophers". See *Athenaeum*, 26 January 1850, p.106 for an account.
4. For this see RI MM, 3 December 1849, **10**: 226 and 11 December 1849, **10**: 228.

Letter 2254
William Thomson to Faraday
14 January 1850
From the original in IEE MS SC 2

2 College, Glasgow, | Jan 14, 1850

My dear Sir

I am extremely glad that you have been so much interested by the subject of my last letter[1].

The conjecture which you make regarding the effects of pressure in altering the temperature at wh[ich] any liquid becomes solid, is entirely in accordance with our views. My brother's[2] reasoning would lead to the conclusions that the melting point of any solid wh[ich] contracts on becoming fluid would be lowered by pressure, & that the melting point of any solid wh[ich] expands on becoming fluid would be *raised* by pressure. His investigation would be applicable to all such cases, with but slight modification of the forms of expression in the cases when there is expansion on melting, and, in every case in wh[ich] the latent heat necessary to melt a cubic foot of the solid, and the volume of the liquid into wh[ich] it is melted, are known, his investigation would lead to a knowledge of the actual amount of alteration in the temperature of melting to be expected from any given, not excessively great pressure.

It would be very interesting to experiment on various solids, especially the metals, both simple metals, & alloys such as Rose's fusible metal[3].

It would also be very interesting to experiment upon the effects of pressure on liquid water, above & below the temperature 39.1, of greatest density, in altering its temperature. Regnault finds that, at the temperatures at wh[ich] he experimented, a sudden pressure of 10 atmospheres did not make as much as $\frac{1}{50}$ of a centigrade degree of elevation in the temperature[4]. Hence a very delicate thermometer would be necessary, but I do not think it w[oul]d be impossible to actually exhibit the *lowering* of temperature wh[ich] a sudden pressure applied to water below 39 ought (according to Carnot's[5] Theory) to produce. If a very delicate minimum thermometer of any kind could be constructed, it might be put into a vessel much stronger than any transparent vessel could be made, and two hundred, or more, atmospheres of pressure might readily be applied to the water.

I have today made very careful experiments to ascertain the value of a division of my ether thermometer. By taking a range from about 31° to about 34° Farh. and using an old mercurial thermometer of Crichton's[6] with an ivory scale, not at all delicate, as the standard, I find that 71 of my divisions correspond to a degree Fahr. Without a better standard thermometer I cannot get a more accurate estimate; but I think what I mention may be relied on within $\frac{1}{20}$ of the whole amount (I am sure 74 must be greater, and 67 less than the number of my divisions which correspond to a true Fahr. degree[)].

I think on the whole the number .25 which I mentioned in my last letter cannot differ by excess or defect by more than $\frac{1}{10}$ of the whole amount, or $\frac{1}{40}$ of a degree, from the lowering of temperature actually produced by 19 atmospheres.

I remain, Dear Sir, | Your's most truly | William Thomson
Prof. Faraday

1. Letter 2251. See also letter 2252.
2. James Thomson (1822–1892, DNB). Civil engineer.
3. See Rose to Faraday, 27 October 1846, letter 1923, volume 3.

4. Derived from data in Regnault (1847).
5. Nicholas Léonard Sadi Carnot (1796–1832, DSB). French physicist.
6. Unidentified, but see Clifton (1995), 70–1.

Letter 2255
Macedonio Melloni to Faraday
14 January 1850
From the original in IEE MS SC 2

Naples le 14 du 1850

Cher et illustre ami!

Vous recevrez par la voie de M. H. Bossange[1] commissionnaire à Paris plusieurs exemplaires de mon premier volume sur la *thermochrôse*[2]: ayez la bonté de les faire parvenir à leurs addresses et d'accueillir avec bienveillance celui qui vous est destiné. La préface vous dira le but de l'ouvrage; *le resumé* vous presentera le tableau des principaux résultats qu'il contient; et la *table raisonnée* vous apprendra l'ordre et la distribution des matières.

Aussi n'aurais-je rien à ajouter sur ce travail s'il ne s'agissait que des seuls intérêts scientifiques. Mais, comme d'autres intérêts pourraient s'y rallier et réagir, en quelque sorte, sur les progrès de la branche de physique dont il est question dans mon livre, permettez-moi, de grace, quelques remarques.

D'abord, ce n'est pas une compilation de mes précédents mémoires sur la chaleur rayonnante; mais un choix de certaines parties de ces mémoires modifiées par plusieurs séries d'expériences inédites et de nombreuses additions de faits entièrement nouveaux: Ce tout réuni en un seul corps de doctrine fondé sur des preuves expérimentales et des argumentations que je crois tout-à-fait rigoureuses. J'ajouterai que quelquesunes des nouvelles expériences décrites dans mon ouvrage me paraissent d'une importance capitale; et peut-être votre opinion ne s'écartera-t-elle par beaucoup de la mienne, lorsque vous aurez pris connaissance de la méthode indiqueé page 129 et suivantes pour démontrer que le thermomulplicateur donne les véritables rapports d'energie entre les rayons calorifiques, et que vous aurez examiné l'artifice employé pour prouver que toute sorte de chaleur rayonnante subit à la surface des appareils thermoscopiques noircis le même degré d'absorption (104).

Vous trouverez de nombreuses innovations dans le 5me § du Chap IV qui traite de l'action des couleurs proprement dites sur la transmission rayonnante de la chaleur. Le 2d et 3me § du même chapitre vous offriront des preuves irréfragables de l'existence de la thermochrôse dans les rayons calorifiques et les milieux susceptibles de la transmettre en conservant leur forme rayonnante. Je signalerai enfin à votre attention les théorèmes de l'hétérogénéité des flux de chaleur obscure qui forment la totalité des radiations des corps chauffés à de basses températures, et la *grande majorité* du rayonnement des flammes et des corps incandescents (pag. 290, 306 et suiv). Ces flux hétérogènes de chaleur obscure constituent, comme vous verrez, (§ 4 du Chap IV) la base fondamentale sur laquelle je m'appuie pour montrer que le rayonnement lumineux et le rayonnement calorifique possèdent la même constitution, dérivent d'un agent unique, et forment une suele série de radiations, dont une partie opère sur l'organe de la vue, et l'autre ne se devoile à nos sens que par les phénomènes qui accompagnent l'echauffement des corps - Je ne sais si je me trompe - Mais une théorie diametralement opposée à celle que m'avaient suggerée les résultats de mes premières recherches sur la chaleur rayonnante, une théorie qui convertit en autant d'arguments favorables les singulières oppositions observées entre la transmission calorifique et la transmission lumineuse justifie, mieux que toute autre chose, la nouveauté des éléments scientifiques contenus dans mon livre: car tous les auteurs de physique que j'ai pu consulter gardent encore la silence sur la portée des différences présentées par les deux transmissions, ou leur donnent une interpretation contraire à ma manière de voir.

Maintenant, savez-vous pourqoi j'insiste tant sur tout cela? ... Parceque, après avoir acquis, par un travail assidu, une position sociale tolerable, j'en ai été injustement privé (1) et que vous pouvez contribuer indirectement à diminuer, et peut-être même à reparer tout à fait les caprices du sort, en me proposant de nouveau pour la médaille de Rumford[3]. Cette médaille a été la première cause du succès qu'ont obtenu mes études sur la chaleur rayonnante, et vous verrez dans mon introduction que je ne le cache pas au public. Je ne sais quel secret pressentiment me dit qu'en m'honorant une seconde fois de ce haut temoignage de son estime la Société Royale de Londres arrêtera les persécutions intentées contre moi, m'ouvrira l'accès à quelque place nouvelle, et me donnera ainsi la tranquillité d'esprit et *les moyens indispensables* pour continuer mes recherches sur l'*Optique calorifique* - Ces dernières expressions vous disent assez clairement, mon illustre confrère et ami, que tout en ayant besoin de votre assistance je ne pretends pas invoquer les seuls secours de l'amitié - les titres scientifiques d'abord - les sentiments après - La recompense à laquelle j'aspire ne me porterait point bonheur, si elle ne venait qu'à la suite d'une operation de camaraderie - Je désire bien ardemment que vous trouviez le tems et la volonté de lire mon petit volume, pour vous assurer s'il satisfait, comme je le crois, aux conditions exigées par le programme de Rumford ... Mais dans le cas où cela ne pût avoir lieu, je vous prierais de charger de cette mission l'un de vos plus savants collegues, tels que M. Herschel par exemple, et de n'avancer votre proposition au *Comité* qu'après vous être intimement

convaincu de la verité des faits énoncés - Ici se termine mon rôle d'avocat
... il ne me reste plus qu'à vous souhaiter toute sorte de bonheure, à renouveller les voeux tant de fois émis pour le retablissement complet de votre santé: Si éminemment utile au véritable progrès de la science, et à me [MS torn] avec toute l'effusion du coeur votre très affec. et très reconn serviteur et ami Macédoine Melloni

P.S. En présentant l'exemplaire destiné à l'Institution Royale veuillez avoir la bonté de l'accompagner avec mes plus vifs remercîments pour la haute marque d'estime qu'elle a bien voulu m'accorder. La lettre de Mr J. Barlow qui m'annonçait ma nomination d'associé étranger d'une Corporation, rendue si celebre par vos immortelles découvertes, ne m'a été remise que fort tard[4], à cause des troubles politiques et des nombreux changements qui sont succedés en un très court espace de tems dans le personnel de notre Ministère des affaires étrangères - Je ne voudrais pas que mon silence m'eût fait passer aux yeux de vos heureux auditeurs comme sottement orgueilleux ou malhonnête ... Dites-leur, je le repète, que j'ai été on ne peut plus sensible à l'honneur reçue, et que je leur en serai toujours reconnaissant. Parmi les papiers que M. Barlow a eu la complaisance de m'envoyer j'y trouve un catalogue qui contient (en 1849) les noms de deux napolitains morts depuis quelques années - ce sont les cav. *Monticelli*[5] et *Sementini*[6]. Veuillez en avertir l'Administration de l'Institution afin qu'on puisse les effacer du catalogue de l'année courante.

(1) Pour vous prouver en quelques mots l'injustice de ma destitution et *de l'ordre d'exil porté contre moi et suspendu par l'intercession de l'Ambassadeur de Prusse*, il suffira de dire, que pendant le Gouvernement des deux années qui viennent de s'écouler, j'ai refusé, sous trois Ministères les honneurs de la Vice-présidence de l'Instruction publique, et repoussé les offres de naturalisation napolitaine, afin d'éviter la probabilité de ma nomination de pair ou de député. Je voulais eviter jusqu'au soupçon d'aspirer à sortir de la position scientifique que je devais aux bons offices de Mes. Humboldt et Arago près le Roi de Naples[7] ... Vous voyez que ma delicatesse a reçu une fort belle recompense! Ce Pays fourmille de gens intrigants, ignorants et bassement envieux: l'un d'eux, tout puissant aujourd'hui dans la Camarilla, aspirait à une place d'Académicien qu'il ne pût obtenir, et s'imagina avoir manqué le coup à cause de mon vote contraire. Voilà, dit-on, l'origine de ma disgrace.

Address: A Monsieur | Monsieur Michel Faraday | des *Sociétés Royales* de Londres et d'Edimbourg. | de *l'Institut* de France, des Académies R. I. | de Berlin, Turin, Stokholm, St Petersbourg && | A *l'Institution Royale* | *Londres*

TRANSLATION

Naples 14th day of 1850

Dear and illustrious friend,

You will receive via M. H. Bossange[1], [my] agent in Paris, several copies of my first volume on the *thermocouple*[2]: please be good enough to send them on to their addressees and to receive kindly the one which I intended for you. The *preface* will tell you what the work aims to achieve; the *summary* will present a table of the principal results contained in the work; and the *contents* will tell you the order and distribution of the subjects.

I would also have nothing to add to this work if it were solely a question of scientific interest. But, since other interests could be won over and react, so to speak, to the progress of the branch of physics which is dealt with in my book, permit me, I beg you, a few comments.

Firstly, this is not a compilation of my previous papers on radiant heat; but a selection of various parts of these papers, modified by several series of unpublished experiments and numerous additions of entirely new facts: all this is brought together in one sole body of doctrine based on experimental proofs and arguments that I believe to be of the utmost rigour. I would add that some of the new experiments described in my work seem to me to be of fundamental importance; and perhaps your opinion will not be very different from mine, when you have become familiar with the method, indicated on page 129 et seq, to show that the thermomultiplier gives accurate measures of energy from radiant heat, and when you have examined the ingenious device used to prove that all types of radiant heat subject the surface of blackened thermoscopic instruments to the same degree of absorption (104).

You will find numerous innovations in the 5th section of Chapter IV which deals with the action of actual colours on the transmission of radiant heat. The 2nd and 3rd sections of the same chapter will offer you unassailable proofs of the existence of a thermal link between heat rays and the media likely to transmit them in their radiant form. Finally I would like to draw to your attention the theorems on the heterogeneity of the fluxes of dark heat, which include all the radiations of materials heated to low temperatures, and the *greatest majority* of the radiation of flames and incandescent materials (pages 290, 306 et seq). These heterogeneous fluxes of dark light constitute, as you will see, (section 4 of Chapter IV) the fundamental base on which I rely to show that light rays and heat rays possess the same constitution, derive from a single agent, and form a single series of radiations, of which one part operates on the eye and the other becomes apparent to our senses only by the phenomena that accompany the heating of materials - I do not know if I am wrong - but a theory diametrically opposed to that suggested by the results of my first researches into radiant heat, a theory which converts into so many favourable arguments the singular oppositions observed between the transmission of heat and the transmission of light, justifies more than any other thing, the newness of the scientific elements contained in my book: for all the authors on physics that I have been able to consult are still silent on the significance of the differences presented by the two transmissions, or give them an interpretation contrary to my view.

Now, do you know why I insist so much on all that? ... Because, after having acquired, through hard work, a tolerable social position, I had it unjustly taken away from me (1) and that you can contribute indirectly to diminish and perhaps even to repair completely the whims of fate, by proposing me once again for the Rumford medal[3]. This medal was the primary cause of the success obtained by my studies on radiant heat and you will see from my introduction that I do not hide this fact from the public. I do not know what secret presentiment tells me that in honouring me a second time with this high testimony of its esteem, the Royal Society of London will stop the persecution instituted against me, will open up access to some new place and will thus give me the peace of mind and the *indispensable means* to continue my research on *heat optics*. These last expressions tell you clearly enough, my illustrious colleague and friend, that although I am in need of your help, I do not pretend to invoke merely the assistance of friendship - scientific titles first - feelings after. The recompense to which I aspire would not bring me happiness if it was merely an expression of friendship. I ardently desire that you find the time and the will to read my little volume, to assure yourself that it satisfies, as I believe it does, the conditions required for the Rumford medal. But if that is not possible, I would ask you to put in charge of this mission one of your wisest colleagues, such as Mr. Herschel, and not to advance your proposition to the *Committee* until after you had intimately convinced yourself of the truth of the stated facts - Here I end my role as lawyer ... I have but to wish you every sort of happiness, to renew the wishes that I have so many times expressed for the complete restoration of your health, which is so eminently necessary for the true progress of science and to [MS torn] with all the affection of my heart your very affectionate and very grateful Servant and friend Macédoine Melloni

P.S. When you present the copy destined for the Royal Institution, please have the kindness to accompany it with my most sincere thanks for the high mark of esteem that it has wished to accord me. Mr Barlow's letter which announced my nomination as Foreign Associate of an institution rendered so famous by your immortal discoveries, was given to me but recently[4], due to political troubles and numerous changes that occurred in a very short space of time in the personnel of our Ministry of Foreign Affairs - I would not like my silence to be interpreted by your happy listeners as foolish pride or dishonesty ... Tell them, I repeat, that I could not have been more touched by the honour I received and that I shall always be grateful for it. Amongst the papers that M. Barlow very kindly sent me, is a list which contains (in 1849) the names of two Neapolitans who died several years ago - they are Messrs *Monticelli*[5] and *Sementini*[6]. Please bring this to the attention of the Administration of the Institution so that they can be eliminated from this year's list.

(1) To prove to you in a few words the injustice of my destitution and of the *exile order made out against me and suspended through the intercession of the Ambassador of Prussia*, it will be enough to say that during the Government of the past two years, I refused, under three Ministers, the honour of the Vice Presidency of Public Education, and I rejected offers of Neapolitan naturalisation, in order to avoid the probability of being nominated a peer or a deputy. I wanted to avoid the tiniest suspicion of aspiring to leave the scientific position which I owed to the good offices of Messrs Humboldt and Arago with the King of Naples[7] ... You see that my care has received a very beautiful recompense! This country swarms with plotters, ignorant people and those who are basely envious: one of them, all powerful today in the Camarilla, aspired to a place in the Academy which he could not obtain and he has imagined that he was not nominated because I voted against him. There, they say, is the origin of my disgrace.

1. Unidentified.
2. Melloni (1850).
3. Melloni had won the Rumford medal in 1835. See Melloni to Faraday, 27 December 1834, letter 750, volume 2. He was not awarded it again, and indeed was not considered. See RS CM 14 November 1850, **2**: 165–6.
4. Melloni was elected an Honorary Member of the Royal Institution on 5 June 1849, RI MS GM **5**: 482. Faraday was one of his proposers.
5. Teodoro Monticelli (1759–1845, LUI). Chemist and geologist in Naples.
6. Luigi Sementini (1777–1847, P2). Professor of Chemistry at Naples.
7. Ferdinand II (1810–1859, DBI). King of Naples, 1830–1859.

Letter 2256
Faraday to Emil Heinrich du Bois-Reymond
15 January 1850
From the original in SPK DD

Royal Institution | 15 Jany 1850
Dear Sir

I this day received your kind present of books (your great work[1]) and also the letter[2]. I regret that I have no better thanks to offer you than those of a man who cannot estimate the work properly. I look with regret at the pages which are to me a sealed book and but that increasing infirmities too often warn me off I would even now attack the language of science & knowledge for such the German language is.

M. Magnus whom I rejoice to call a friend told me of your great experiment in which from the muscular excitement of the living human being you obtained a current of Electricity. I endeavoured a few months ago to procure the result but did not succeed no doubt being unacquainted with all the precautions needful & the exact manner of proceeding[3]. I was at fault - and now I am so engaged by the duties of my station & the Season that I have no time for anything else. During the season I trust to pick up the information that will give me success the next time that I try[.]

The second copy of your work is already on the road to the Royal Society and I shall do all I can to direct the attention of the men of Science & others to the copy you have sent me by placing it before them on the table of this Institution[.]

I am Sir | Your Very obliged & grateful Servant | M. Faraday
Dr. E. du Bois Reymond | &c &c &c

1. Bois-Reymond (1848–9).
2. Letter 2232.
3. Faraday, *Diary*, 1 October 1849, 5: 10270–6.

Letter 2257
George Biddell Airy to Faraday
28 January 1850
From the original press copy in RGO6 / 403, f.128

Royal Observatory Greenwich | 1850 Jan 28

My dear Sir

I send you copies of two or three small papers. You have so much of the civil-engineer's talent of applying new means to new purposes that I would almost hope that my lecture to the Astronomical Society, of December 14, might interest you[1]. Two of the papers contain statements relating to Lord Rosse's telescope[2], the rest I fear are worth little.

I have scarcely had a moment yet to think of my intended Lecture at the Royal Institution[3]. But I perceive that I shall want the following apparatus amongst others:

A helix of wire producing with a Galvanic current the effect of a magnet.

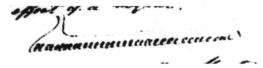

Something to show the effect of thermo-galvanic currents[.]

Most likely you have such things producible without the trouble of making or even looking for them; so, it would save some expence and trouble[.]

I am, my dear Sir, | Yours very truly G.B. Airy

Michael Faraday Esq DCL | &c &c &

1. Airy, G.B. (1849–50). This paper dealt with the possibility of using electric signals to determine the difference in longitude between different places and for aiding the making of astronomical observations generally.
2. One of these would have been a report of a verbal account by Airy of the progress made by Lord Rosse in mounting his mirror, *Month.Not.Roy.Ast.Soc.*, 1849–50, **10**: 20–1. The other might possibly have been Airy, G.B. (1848–9).
3. See *Athenaeum*, 23 March 1850, pp.315–7 for an account of Airy's Friday Evening Discourse of 15 March 1850, "On the present State and Prospects of the Science of Terrestrial Magnetism".

Letter 2258
George Towler to Faraday
30 January 1850
From the original in IEE MS SC 2

Fye Bridge, Norwich, Jan 30, 1850.

Sir,

It is under the impression that the following communication is a matter of great scientific importance that induces me once again[1] to address you upon a subject with which you are so intimately identified, and in which I take so deep an interest.

I have upon a former occasion taken the liberty of imparting my magnetic views to you, seeking the favour of your esteemed opinion thereon. In these views you must have at once discovered my incredulity in the hypothesis of the 'two fluids,' and consequently my dissent from the notions entertained regarding 'terrestrial magnetism'.

In the course of investigation which I have pursued in reference to these phenomena, I met with every encouragement to continue to maintain these views, & at the same time to convince me of their mechanical origin - that they were due to motion, which motion was carried as per se, by that class of bodies called 'magnetic' which of course if not to be called perpetual, would be of great duration, terminating only with the existence of the metal supporting it.

To prove to you that magnetic bodies carry on a perpetual motion per se, is the object of the present paper.

That magnetic phenomena are due to matter, and that matter in motion, I conceive there can be no necessity for me to contend. In this I believe you readily acquiesce, and it is simply to the agency by which a permanent motion is sustained by bars of iron &c, that there is any occasion for me to explain.

In the first place I must draw your attention to the great fact, indeed I may call it the magnetic basis, it is this.

Magnetic bodies as far as the fluid particles in their interstices are concerned *are non resisting mediums,* that is fluid particles whilst in the interstices of such bodes lose by virtue of their locality, a large amount of their *gravity,* and at the same time a corresponding amount of their *inertia.* Which being the case such internal medium is readily acted on by minute external forces, such as those generated by slight disturbances of the surrounding atmosphere.

To prove this, it is simply necessary to remark what you have frequently observed, the causes and consequences of which, have as frequently escaped your well known penetration.

When a magnet attracts a distinct unmagnetized suspended needle, the needle as you are well aware, at the moment it is so attracted, has become a magnet, were this not the case the attraction would not have ensued. The pole of the needle remote from the magnet has magnetism induced in it, as well as that in conjunction, *and the magnet has consequently generated a motion of the fluid particles through the interstices of the needle, for once there no motion there could be no magnetism* there.

For the magnet to do this *the resistance of the fluid in the interstices of the needle can be no greater than the pressure brought to bear upon it, consequent upon the disturbance of fluid about the magnet.*

The pressure of the fluid in its circulation round the magnet is so minute, that it requires a very delicate indicator to appreciate it, from which some approximate idea may be formed of the amount of resistance which the particles taken as a medium pervading the metal can offer to the pressure positive, or negative, emanating from the magnet.

Incredible as this may at first sight appear, it is an indubitable fact supported on irrefragable testimony and is but the extension of a well known principle, steel and iron are well known conductors, but the extent to which they possess this power has never been conceived.

Magnetic motion is primarily derived from the impulse communicated in induction.

The continuation or durability of the motion is the direct consequence of the first impulse in connexion with the principle of non resistance.

When a short bar of iron is made to conduct it does two things. It augments the preexisting quantity of fluid in the space into which the fluid enters, on departing from the bar, and at the same time & in the same ratio decreases the preexisting quantity in a given space at the opposite extremity, from which the fluid it conducts is taken up.

The increment of fluid on the one extremity, and the decrement on the other, immediately open curves of communication on all sides between the poles of the bar, and not only between the poles but every point on the surface on each side of the neutral point, excepting those immediately abutting on that point in this communication kept up, for all points of the magnet either take in, or give out fluid, which as it issues from one side, is deflected over to a corresponding point on the other, where the magnet is taking in fluid.

In these curves lie the constant impulse. It may be urged that the pressure of such a motion in the surrounding atmosphere must be very small - undoubtedly it is, but at the same time, the resistance of the internal fluid of the bar is less. The absence of resistance is equivalent to a large amount of momentum.

The external pressure produces the internal fluxion & the internal fluxion reproduces the elements of the external pressure, each reciprocating the other. The internal motion causes the external disturbance, which generates a pressure, which pressure overcomes the inertia of the internal fluid, and thereby drives it out of the bar occasioning the external disturbance &c.

This hasty sketch is by far to[o] brief to do justice to so important a question, but it is amply sufficient to convince you, that this is a new reading of magnetic origin.

The evidences contributing to the support of this are universal, and require only to be entered upon to be confirmed.

I have allowed this matter to remain in abeyance for some length of time, which reflexion convinces me is unjustifiable, and I trust by a lucid explanation of this truly important end, to the promoters of physical research that their well known zeal in the persuit [sic] of science, will be the means of placing it in a sphere of utility, to which end I beg most respectfully to commend it to your valuable consideration.

Your obedient Servant | G. Towler

Magnetic motion may very aptly be described as that of a wheel or vortex of fluid, the segment of which b.c.d.

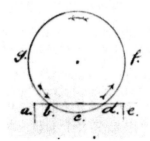

in each revolution it makes, passes through a vacuum a.e. by which it derives the impulse, by which its constant rotation is performed.

The magnet fulfills the office of an indestructible vacuum, in as far as it can never be destroyed by the ordinary operation of fluid particles, nor can a permanent equilibrium be formed between the external & internal medias, as the particles of the internal, are differently conditioned, being exempted from the gravity and resistance to motion which the external encounter.

When a motion of fluid is effected in the bar of iron a.e. and the fluid b.c.d is forced out in the direction a.e., the fluid b.c.d is forced out in the direction a.e., the fluid b.c.d is impelled to f., at the same time the fluid between g & a. enter into the bar, and take the place vacated by b.c.d. and the particles which were between e.f are now located between f & g.

In this motion the external presence upon b.c.d is a maximum at a. & a minimum at e. That there is no pressure upon it at e. and all the pressure that is found upon it is at a. acting in the direction a.e. For as the fluid at e. is progressing to f. it cannot press upon d. and the fluid g.a being in the act of entering into a, all its pressure must be upon b.c.d in the line a.e[.]

By this change of place of the media surrounding the magnet, no change is made as to the relative amount or direction of the pressure at a. for if this is once made to exceed that at e. there must be a constant motion through b.d. in the direction a.e provided the internal resistance is less than the external pressure.

As the motion at a & e takes place simultaneously the fluid g.a is disturbed at the same time, that the fluid e is, and consequently the fluid g-f at the same time as b.c.d, so that if the internal equilibrium is destroyed in the line or direction a.e the curve e.f.g.a is the direction of the external fluid resulting therefrom.

Repulsion is the consequence of the opposition of two or more rotating halos or atmospheres if I may be allowed the phrase, the currents of which are flowing in opposite directions. Fig 1. Attraction that of the conjunction of two or more rotating atmospheres, the currents of which flow in the same direction & the fluid which flows from one magnet, flows through the opposite one, and the two atmospheres more or less approximate & form themselves into one enlarged atmosphere regulated by time & distance - fig 2[.]

Polarity and its concomitants, are due to the operation of foreign or contingent forces operating upon these intrinsic forces of the magnet, these contingent forces having no connexion with or origin in magnetism.

The reason why light bodies are not moved by the action of fluid encompassing the magnet, is that the fluid which can either issue from or enter into feriginuous bodies, can pass through the interstices of bodies less dense without hindrance, and consequently they oppose no resistance to the fluid nor the fluid to them.

It is not the simple passage of the fluid from, or to a magnet which passes through a bar of iron, which produces magnetism; but the formation by a bar of iron of an atmosphere or envelope of media, differing in density from the media extending beyond it, and corresponding to that of the magnet which induces it.

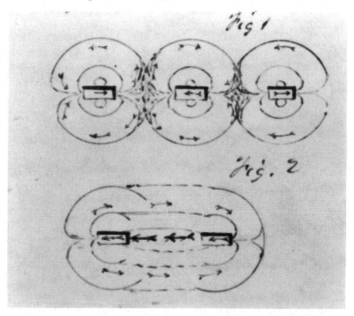

The following propositions I believe to be fully established.

---

1. When an unmagnetized suspended needle is disturbed & oscillates on the approach of a magnet, the magnet is the cause of the disturbance & oscillation.
2. Matter is a component part of motion that is, without matter, motion is a nonentity.
3. There is no state in which a body can be in, between rest and motion. That is if it ceases to move, it must come to rest, and if it be at rest, it cannot at the same time be in motion.
4. That the sensible motion of the suspended bar, is derived from the invisible, and insensible motion pertaining to, and existing about the magnet externally - which is a motion of particles or molecules of matter.

5. The external motion of fluid about the magnet is possessed of a force or pressure which generates motion through bars of iron & which are placed in connexion with them.
6. The resistance of a bar of iron placed in connexion with a magnet is less than the force of the magnet, that is the impulse of the magnet on the bar is greater than the resistance of the fluid in its interior.
7. And the external impulse of the magnet is greater than its own internal resistance.
8. There is no permanent magnetism where there are not opposite poles, That is when there is no positive there is no negative magnetism, & vice versa.
9. That these poles must essentially be within a given distance of each other. And that the intensity of the magnetism is in a ratio with their distance.
10. That they reciprocate and carry on a mutual action between each other. Each pole of the magnet is constantly magnetizing the other, in every respect the same as both magnetize all iron which is placed near them.
11. That the surrounding media enters in on all sides of one half and issues from all sides of the opposite half, except those points immediately surrounding the neutral line fig 3

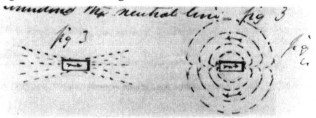

It then describes curves round the body on all sides. fig 4.
13. These curves are the direct consequence of the mutual action of one pole on the other, that is of the increment of fluid at one, and the decrement at the other.
14. That the external circulation is productive of a maximum & minimum pressure upon the internal fluid.
15. That the maximum is different over one half and the minimum over the other half.
16. That these act conjointly and uniformly upon the internal fluid and thereby destroy the internal equilibrium.
17. When the internal equilibrium is destroyed the equilibrium of the external media is at the same time destroyed also. And an increment and decrement of the external media in relative proportion and opposite positions is effected which form the elements of the external pressure.

1. See Towler to Faraday, 22 April 1841, letter 1347, volume 3.

Letter 2259
Faraday to George Biddell Airy
31 January 1850
From the original in RGO6 / 403, f.132

Royal Institution | 31 Jany 1850

My dear Sir

I received your letter[1] two or three days ago but waited until I obtained your papers which have just arrived[.] I beg to thank you for them and rejoice to find that Electricity may probably be honored by being admitted into the service of Astronomers[2]. I had heard of its successful illumination of the Oxford Heliometer[3] before and if it does as well in the observation of transits then I hope before very long to pay a visit to Greenwich & see it performing duty[.]

As to the apparatus I have no doubt we have such spirals & thermo-electric apparatus as you speak of & I trust all else you will need[.] When you let me know I will do every thing I can to make the Evening as agreeable to you as I am sure it will be to every body else[4][.]

With kindest respects to Mrs. Airy

I am | My dear Sir | Most faithfully Yours | M. Faraday
The Astronomer Royal | &c &c &c

1. Letter 2257.
2. See notes 1 and 2, letter 2257.
3. See *Month.Not.Roy.Ast.Soc.*, 1849–50, **10**: 21–2 which noted that the scale of the heliometer was illuminated by a wire heated by powerful magnets. However, p.25 noted that this statement was incorrect and that a Grove cell was used to supply the light.
4. See *Athenaeum*, 23 March 1850, pp.315–7 for an account of Airy's Friday Evening Discourse of 15 March 1850, "On the present State and Prospects of the Science of Terrestrial Magnetism".

Letter 2260
Faraday to Royal Society
9 February 1850
From the original in RS MS RR 2.155

I consider Matteucci a much fitter judge of the matters treated of in his paper than I am and for that reason alone should think it fit to be printed[1]. The reading of the paper confirms that conclusion.

M. Faraday
9 Feby. 1850

1. Matteucci (1850a).

Letter 2261
Faraday to John Scott Russell[1]
12 February 1850
From the original in RSA MS John Scott Russell Collection, volume 3, f.196

Brighton | 12 Feby 1850

My dear Russel

I intended to be with you on tomorrow but ill health will prevent me from leaving this place so that I shall not be able to be with you at any of the hours specified[2].

Ever Truly Yours | M. Faraday

J. Scott Russel Esq | &c &c &c

1. John Scott Russell (1808–1882, DNB). Secretary of the Society of Arts, 1845–1850.
2. This was a meeting at Buckingham Palace to discuss the lists for the Great Exhibition. For the minutes see RSA MS John Scott Russell Collection, volume 3, f.94–6. At this point (or at least by 21 February) Faraday was a member of the sub-section committee dealing with exhibits of the "Vegetable Kingdom", *ibid.*, f.72.

Letter 2262
Thomas Phillipps to Faraday
22 February 1850
From the original copy in Bod MS Phillipps-Robinson c.507, f.186–7

Middle Hill | Broadway | 22 F 50

My dear Sir

A lady has asked me for a Gentleman one of her friends, my permission to let him use my Ticket for the Evening Meetings on Friday at your Institution.

As I am not aware that I can transfer my Ticket, I write to you to ask the *real state of the Case*.

If there is one thing more than another why I regret my distance from London it is my inability to attend your beautiful Lectures on Electricity. For as it is the Science which governs the World, He who develops it the most, approaches the nearest to the knowledge of His Maker.

I hope you are in the enjoyment of the best Health & with my very best regards believe me always

My dear Sir | Most truly Yours | Thos Phillipps

Michl Faraday Esq

Letter 2263
Faraday to Lambert-Adolphe-Jacques Quetelet
25 February 1850
From the original in BRAI ARB Archives No 17986 / 989

Royal Institution | 25 Feby 1850

My dear friend

I must write you a letter that I may say in it how great pleasure I have had in reading and studying the third part of your Essays on the climate of Belgium; i.e. your results in atmospheric electricity[1]. They are, I think, very admirable; and I admire the truly philosophic spirit in which you have been content to give them, without any addition of imagination or hypothesis. They are *facts* and ought not too hastily to be confounded with opinion; for the facts are for all time whilst opinion may change as a cloud in the air. I think, you know, that I cannot adopt Peltiers views of the relation of the Earth & space[2]; and I was encouraged, therefore, to hold more confidently to my own conclusions in that respect, when I saw how carefully you abstained from any phrase that might commit you to the expression of such an opinion. I took the liberty of giving our Members here an account of your results, and they appeared to be most highly interested in them[3]. In doing so I pointed out your philosophic caution, and expressed my opinion that such was the true method by which advances in science in this very difficult part could be really made.

I have just received from you a few leaves in which I find a letter to you from young M. Peltier[4]. It is quite natural that he should hold to his fathers views, but he must remove the fundamental objection before he can make any impression, at least on my mind. That objection is, that it is absolutely impossible to charge any body with one electricity independent of direct relationship with the other electricity. Or in other words that it is absolutely impossible that the earth as a whole, or any other single body, as a globe, should have negative electricity appearing on its surface or be *driven into its interior* merely by variation in the electric intensity of the whole surrounding space. If an insulated ball of metal or earth be suspended within a much larger sphere of metal, or wire gauze or any thing else (to represent the space action) which can be charged simultaneously in all parts, no amount of charge which can be given to the sphere representing space, can induce any charge on the ball; nor would the discharge of that space electricity induce any charge on the ball:- and further; that representative of space could not exert any inducing action inwards;- nor could it receive charge, unless it could induce equivalently to something external & outside of itself;- and even in so doing would shew no sign of action inwards.

I have carefully considered all the reasonings and views which Peltier has put forth that seem to bear upon or touch this point; but with the best judgment I could exercise have come to the conclusion that none of them do really touch it.

Ever My dear Sir | Your Very obliged & faithful | M. Faraday
A Monsieur | Monsieur Quetelet | &c &c &c

Address: Professor Quetelet | &c &c &c | Royal Observatory | Bruxelles

1. Quetelet (1849).
2. See also Faraday to Schoenbein, 18 February 1843, letter 1471, volume 3.
3. See *Athenaeum*, 9 February 1850, pp.161–2 for an account of Faraday's Friday Evening Discourse of 1 February 1850 "On the Electricity of the Air".
4. This is in Quetelet (1850), 5–13. Ferdinand Athanase Peltier, who seems to have been a physician, wrote Peltier, F.A. (1847), an account of his father.

Letter 2264
Lambert-Adolphe-Jacques Quetelet to Faraday
5 March 1850
From the original in IEE MS SC 2
Académie Royale | *DES SCIENCES, DES LETTRES ET DES* | *BEAUX ARTS* | *DE BELGIQUE.* | *Bruxelles, le* 5 mars 1850.
Mon cher et illustre confrère,

Je vous suis bien reconnaissant pour la lettre amicale que vous m'avez fait l'honneur de m'envoyer[1]. Je ne pouvais recevoir aucun encouragement plus flatteur, que celui du savant qui s'est placé, par ses travaux, au plus haut point de la science que je cultive. Je suis heureux surtout d'avoir obtenu votre asssetiment pour la réserve que j'ai mise dans mon travail[2], réserve qu'on pouvait taxer d'extrême timidité. Quand un homme tel que Vous, conserve encore des doutes, je suis très excusable en effet de garder le silence; c'est même une loi à mes yeux!

Je me suis toujours attaché avec un soin spécial de l'étude de l'électricité de l'air, parcequ'il me semble qu'il n'y a pas de progrès possible dans la météologie, sans une parfaite connaissance de l'électricité atmosphérique. Aussi tous les amis des sciences doivent se féliciter de Vous voir tourner les yeux vers cet élément; c'est de Vous qu'ils attendent surtout l'explication du rôle qui joue l'électricité dans les grands phénomènes du globe.

Nous nous occupons beaucoup dans ce moment d'un projet d'établissement de télégraphes électriques dans notre Royaume[3]: j'en ai déjà écrit plusieurs fois à M. Wheatstone qui est le représentant naturel de cette branche importante de physique. J'irai probablement en Angleterre pour cet objet; et je serai charmé dans cette circonstance de pouvoir revenir à ses lumières et aux vôtres.

Je continue avec activité l'histoire du climat de la Belgique; j'espère pouvoir bientot vous en addresser un nouveau chapitre.

Agreez, mon cher et illustre confrère, les nouvelles expressions de mes sentiments les plus distingués et les plus affectueux. | Tout à vous | Quetelet

TRANSLATION
Académie Royale | *DES SCIENCES, DES LETTRES ET DES* | *BEAUX ARTS* | *DE BELGIQUE.* | *Brussels*, 5 March 1850.
My dear and illustrious colleague,

I am most grateful to you for the friendly letter that you honoured me by writing[1]. I can receive no more flattering encouragement than that from a savant who has placed himself, by his works, at the highest point of the science that I cultivate. I am happy above all to have gained your approval for the reserve with which I presented my work[2], reserve which one could say was extreme timidity. When a man such as you still harbours doubts, I can be easily excused for keeping silent; I would even say it was the law from my point of view.

I have always attached special attention to the study of the electricity of the air, since it seems to me that there cannot be any possible progress in meteorology without a perfect understanding of atmospheric electricity. Also all the friends of science must be congratulating themselves to see you turn your eyes to this subject; it is from you that they above all expect an explanation of the role that electricity plays in the great phenomena of the globe.

We are currently very occupied with a project to establish an electric telegraph in our Kingdom[3]. I have written of it several times to Mr Wheatstone who is the natural representative of this important branch of physics. I shall probably come to England because of this; and I shall be delighted in these circumstances to be able to return to his wisdom and yours.

I am busy continuing the history of the Belgian climate; I hope soon to send you a new chapter.

Please accept, my dear and illustrious colleague, new expressions of my most distinguished and most affectionate sentiments | All yours | Quetelet

1. Letter 2263.
2. Quetelet (1849).
3. See Quetelet to Faraday, 8 May 1846, letter 1877, volume 3.

Letter 2265
Faraday to Henry Deacon
7 March 1850
From a typescript in RI MS
My dear Sir,
I have just put up your name as a Candidate at the Chemical Society. When you are elected the Secretary Mr. Warrington[1] will write to you. He spoke to me of a kind of ornamental glass having the appearance of *red porphyry* of *Jasper* containing crystallized Protoxide of copper in a colourless glass. I told him I was sure you would be glad to have any enquiries made of you and would be not merely a nominal but a bonafide member of the Society.
Ever Very Truly Yours, | M. Faraday

1. Robert Warington (1807–1867, DNB). Chemist to the Society of Apothecaries, 1842–1867, and Secretary of the Chemical Society, 1841–1851.

Letter 2266
Faraday to Edward Magrath
11 March 1850
From the original in Cumbria Record Office MS DHC/1/16

R Institution | 11 Mar 1850

Dear Magrath
If you have an opportunity help us[1][.]
Ever Yours | M. Faraday

Address: Edward Magrath Esq | &c &c &c | Athenaeum | Pall Mall

1. With this letter there is a London Orphan Asylum (on which see note 1, letter 2173) card for Francis Thomas Farnes (b.1843, GRO) which stated that his father had been employed as a brass founder by James Faraday. London Orphan Asylum records in SuRO indicate that Farnes was not admitted. Magrath had been a subscriber to the asylum since 1831, see the 1850 *Report of the London Orphan Asylum* SuRO 3719/1/13, p.147.

Letter 2267
Thomas Phillipps to Faraday
13 March 1850
From the original copy in Bod MS Phillipps-Robinson e.385, f.65r–66v[1]
Michael Faraday Esq | Royal Institution London

MH 13 Mh 50

My dr Sir,
Will you allow me to introduce my Son in law to you the Revd Jno Fenwick[2] & his wife[3] my daughter, & to request yr good offices in obtaining for them Tickets for the Friday Evening Lectures while they stay in London. I take this opportunity of asking you whether I can transfer my right of Entrée to any other Gentleman during my absence from Town? I have been told it is sometimes done, but I had rather apply to you as the Fountain head.
With my best wishes for yr good health believe me always my dr Sir
Most sincerely yours TP

1. The original of this letter was enclosed in Phillipps to K.S.W. Fenwick, 13 March 1850, Bod MS Phillipps-Robinson e.385, f.65v.
2. John Edward Addison Fenwick (d.1903, age 78, GRO, CCD). Perpetual Curate of Needwood, 1853–1893.
3. Katherine Somerset Wyttenbach Fenwick, née Phillipps (d.1913, age 90, GRO). Phillipps's third daughter; see his DNB entry.

Letter 2268
Faraday to Hans Christian Oersted
15 March 1850
From the original in Håndskriftafdelingen, Det Kongelige Bibliothek, Copenhagen, MS Ørsted 1–2

Royal Institution | London | 15 March 1850

My dear Sir
I received your very kind letter 2 or 3 weeks ago[1] and was very greatly gratified that you should remember me. Since then I have waited in hopes I should see Mr. Colding[2] your friend. But as I have heard not of him and leave town in a few day[s] I thought I would not longer delay writing a word or two in acknowledgement of yours. This is a time of the year in which formal matters occupy me so much that (together with a system soon wearied) they prevent me from working to any good purpose so that I have little or nothing to say. I have it is true sent a paper to the Royal Society two or three months ago which was read lately[3] and in it I describe my failure to produce the results of Weber Reich & some others or (of such as were produced) my reference of them to other principles of action than those they had adopted. This branch of science is at present in

a very active & promising state[.] Many men (and amongst them yourself) are working at it and it is not wonderful that views differ at first. Time will gradually sift & shape them & I believe that we have little idea at present of the importance they may have 10 or 20 years hence[.]

As soon as my paper is printed I shall send it to you and I hope with copies of those you have not received. I thought I had sent you all in order, for it was to me a delight to think I might do so. I do not know what can have come in the way of them, but if I have copies left you shall have them with the next paper[.]

I am constrained to make this letter a short one as much through the paucity of matter as the want of time. Hoping it will find you in excellent health

I am My dear Sir | Your very Obliged & faithful | Servant - | M. Faraday
Professor Oersted | &c &c &c

1. Letter 2245.
2. Ludvig August Colding (1815–1888, DSB). Danish physicist and municipal engineer.
3. Faraday (1850), ERE23. Read on 7 and 14 March 1850.

Letter 2269
Warren De La Rue to Faraday
15 March 1850[1]
From the original in IEE MS SC 2

Royal Institution | Friday 15th 1850

My dear Sir

I have bought up the slide Mr. Nobert sends for your acceptance. You will, with the unassisted eye, perceive only two lines one broad, the width of which is $\frac{1}{2}$ a Paris line or the $\frac{1}{24}$th of a Paris inch (the 12th part of the old pied du roi) divided into 500 lines - there are actually 500 lines there I can vouch for, having counted them. The narrow line contains 15 series of these wonderful lines of Noberts, getting gradually finer - the enclosed piece of paper contains a table of their values the first two columns are the numbers given by Nobert the last two are the values in relation to an English inch calculated by me. I believe they are correct but I intend calculating them again and will then print off a few tables for such of my friends as possess these slides. Until I send you one you must not rely absolutely on the English values given. I have made measurements of some of the lines and find them very nearly concordant with the assigned values by Nobert.

I remain | My dear Sir | Very Truly Yours | Warren de la Rue
Michael Faraday Esq | &c &c &c

Nobert's Lines

Series	Lines contained in each series	Value in the old French line	Value in English inches being the preceding multiplied by 0.0888	Number of lines in an English inch
		‴		
1	7	0.001 000	.0000 8880	11261
2	8	0.000 850	.0000 7659	13056
3	9	0.000 730	.0000 6482	15426
4	10	0.000 620	.0000 5506	18163
5	11	0.000 550	.0000 4884	20475
6	13	0.000 480	.0000 4262	23461
7	15	0.000 400	.0000 3552	28153
8	17	0.000 350	.0000 3108	32175
9	19	0.000 300	.0000 2664	37537
10	21	0.000 275	.0000 2442	40950
11	23	0.000 250	.0000 2220	45045
12	24	0.000 238	.0000 2134	47327
13	26	0.000 225	.0000 1998	50050
14	27	0.000 213	.0000 1891	52870
15	29	0.000 200	.0000 1776	56306

Endorsed by Faraday: Noberts line | &c &c

1. Dated on the basis that this letter comes before letter 2273.

Letter 2270
Faraday to Thomas Phillipps
16 March 1850
From the original in Bod MS Phillipps-Robinson c.507, f.188–9

Royal Institution | 16 March 1850

My dear Sir

I received your note[1] - but have not seen Mr. Fenwick[2] and do not know where to send tickets. In answer to your enquiry I may say that either of two things may be done[.] You may have many tickets at once and having signed them can give them in charge to a friend to use gradually. Or sometimes a relation or friend has signed for the party (writing for instance your name as a signing clerk might do)[.] The latter plan we avoid as much as possible because as you will see if it were common persons not authorized might venture to fill up a card[.]

I think I will send you some tickets and as there are ten more nights you have a right in the whole to 20 tickets. On your request the rest shall be sent[.]

Ever My dear Sir | Very Truly Yours | M. Faraday
Sir Thos. Phillip[p]s Bart | &c &c &c
I send 13 tickets herewith | MF

1. Letter 2267.
2. John Edward Addison Fenwick (d.1903, age 78, GRO, CCD). Perpetual Curate of Needwood, 1853–1893.

Letter 2271
Faraday to Charles Hyde[1]
19 March 1850
From the original in SI D MS 554A

Royal Institution | 19 Mar 1850

Sir

I regret I cannot comply with your request but it is against my long established & invariable rule[.]

I am Sir | Your Very Obedient Servant | M. Faraday
Chas. Hyde Esq | &c &c &c

1. Unidentified.

Letter 2272
Faraday to Thomas Twining
19 March 1850
From the original in the possession of Günther Gerisch

R Institution | 19 Mar 1850

My dear Sir

Finding a copy of Sir W. Snow Harris' paper containing a description of his Electrometer I have added it to this note. You will see the description page 214 Par. 7[1]. I perceive the date is 1834. I could not at all remember it.

Ever Your Obliged | M. Faraday
- Twining Esq | &c &c &c

1. Harris (1834a), 214.

Letter 2273
Warren De La Rue to Faraday
23 March 1850
From the original in IEE MS SC 2

110 Bunhill Row | March 23rd 1850

My dear Sir

I enclose a printed table of Nobert's lines arranged in series. The single band on your slide[1] is the $\frac{1}{24}$th of a Parisian inch divided into 500 lines.

On the envelope enclosing your note for him you have directed for Paris, but M. Nobert is a german residing at Griefswald in Pomerania; would be troubling you too much, as you have been kind enough to acknowledge his little present, to write another envelope directed as above; as you have left the present one unsealed I will transfer the enclosure?

I was not aware last Friday[2] week that my name had been proposed for the Royal Society and that you had been good enough to sign the nomination paper[3], otherwise I would have taken that opportunity of thanking you; I beg that you will now accept my best thanks for your great kindness; I will endeavour by my work to prove myself worthy of it when my leisure permits me to devote myself to pure science.

I remain | Yours Very truly | Warren De la Rue
Michael Faraday Esq | &c &c &c

1. Referred to in letter 2269.
2. That is 15 March 1850.
3. RS MS Cert 9.252. Faraday signed from personal knowledge. De La Rue was elected a Fellow on 6 June 1850.

Letter 2274
Christian Friedrich Schoenbein to Faraday
27 March 1850
From the original in UB MS NS 389

My dear Faraday

Our Chief Magistrate Burgomaster Sarasin[1] friend to your friend and a liberal patron to science taking a trip to England will be kind enough to deliver these lines and the papers laid by into your hands and I am sure you will be glad to make the acquaintance of the highly worthy gentleman.

The paper in octavo deals with the voltaic pile[2] and that in quarto contains an account of my recent researches on ozône[3] of which I talked in my last letter to you[4]. To give you a substantial proof of the correctness of my statements I send you a little bit of peroxide of silver and nitrate of potash both the substances having been prepared by the means of ozône.

Little being known of ozône in England don't you think the subject fit for being once treated before one of the Friday meetings of the Royal Institution[5][.] It allows of a great number of striking experiments to be made. Should you like the Idea I would give you a list of those I think to be the most interesting and instructive ones.

As you know no doubt Mr. Henry[6], the Chymist who is Headbrewer in some great brewery of the City, pray let him have the enclosed.

My best compliments to Mrs. Faraday and my kindest regards to yourself

Your's | most truly | C.F. Schoenbein
Bâle March 27, 1850.

1. Felix Sarasin (1797–1862, DHBS). Cloth manufacturer and politician.
2. Schoenbein (1849a).
3. Schoenbein (1849b).
4. The last located letter from Schoenbein to Faraday is c. October 1848 (letter 2109, volume 3). This passage suggests that there may have been intervening letters which have not survived.
5. Faraday (1851g), Friday Evening Discourse of 13 June 1851.
6. Unidentified.

Letter 2275
Faraday to Charles Wentworth Dilke[1]
29 March 1850
From the original in RI MS F1 N1/20

Brighton | Friday 29 Mar.

My dear Sir

Your letter finds me here & gives me hardly time to say that I cannot be in town next Wednesday[2] or I would have been at the Adelphi.[1]

Ever Truly Yours | M. Faraday

C.W. Dilke Esq | &c &c &c

Endorsed: 1850

1. Charles Wentworth Dilke (1810–1869, DNB). Active member of the Society of Arts and a member of the Executive Committee of the 1851 Great Exhibition.
2. That is 3 April 1850. This was the day of election for members of the Council of the Society of Arts. See RSA MS, Council Minutes, 27 March 1850, **2**: 211 which instructed that the nominations be sent out to members instantly.

Letter 2276
George Wilson to Faraday
29 March 1850
From the original in IEE MS SC 2

March 29, 1850 | 24 Brown Square Edinb

Michael Faraday D.C.L. &c &c

Sir

I trust you will forgive the liberty I take in addressing you, although personally unknown to you. Mr George Buchanan kindly offered me an introduction two years ago, but I have always felt so reluctant to occupy your valuable time that I have never availed myself of his kindness. The visit, however, of my friend Dr Stenhouse[1] to London induces me to trouble you with the copy of a paper[2], with a view to ask your answer to this question "Is it possible to render a gas absolutely anhydrous?" I do not use the word *absolutely* in an unqualified sense as excluding the possibility of moisture being present through the practical difficulties attending the realisation of the drying process; but as referring to the sufficiency of the process could it be realised in practice. In other words are our present methods of drying gases, theoretically perfect, & fitted to secure the deprivation of moisture, or are they essentially imperfect and from their nature certain to leave a certain amount of water in every gas?

My object in asking the question, is to request your advice as to the best method of drying gases, with a view to its application to researches resembling those recorded in the accompanying paper. If you would kindly glance at your leisure at Section V, page 489[3], entitled *"On the methods applicable to the drying of gases"* you would see in the compass of some two pages, the difficulty which has arrested my researches, and which your great experience in the liquefaction of gases, probably enables you to remove.

I know too well your many occupations to wonder if it is out of your power, to write a reply. But, perhaps, you will find time to send a verbal message through Dr Stenhouse, if unable to write. I have enclosed two other papers and remain

Your Obedient Servant | George Wilson

1. John Stenhouse (1809–1880, DNB). Scottish chemist.
2. Wilson (1848).
3. *Ibid.*, 489–92.

Letter 2277
Edward Cowper to Faraday
1 April 1850
From the original in IEE MS SC 2

6 Campden Hill Villas | Kensington | Apr. 1, 1850

My dear Sir

I well remember when Mr Crosse's Acarus, & his letter were exhibited in the Library of the Royal Institution[1], how careful you were to explain to the Audience in the Theatre that you placed them on the table, without giving the slightest opinion either one way or the other, respecting the production of the said Acarus, & you particularly desired that no person would consider any thing that might be said or published on this subject or on any other, as your opinion unless published by yourself.

Now in Miss Martineaus history of England during the 30 year Peace, she has given some account of Mr Crosse's experiments & of their being repeated by Mr. Weekes[2] & then she refers to you in the following words,
(Vol 2 page 451).

"At a Lecture at the Royal Institution in 1837 Mr Faraday avowed his full belief of the facts stated by Mr Crosse, similar appearances having presented themselves to him in the course of his electrical experiments:

but he left it doubtful whether it was a case of production or revivification" Annual Register 1837 Chron. 21³[.]
(The words Annual Register &c in the Margin)
I was about writing to Miss Martineau to set her right on this subject, by giving her my own remembrance of what you *did* say, but as she might think it possible that you had said "that similar appearances had presented themselves to you in the course of your electrical experiments," on some occasion when I was not present, it appeared to me better to write to you, that I might be quite correct in correcting her - and again you might prefer that I should take no notice of the mistake, if so I hope you will excuse me troubling you about the matter.

Yours sincerely | Edw Cowper
M. Faraday Esq FRS &c &c

1. On 17 February 1837. See RI MS F4E, p.5.
2. William Henry Weekes (1790–1850, Twyman (1988)). Physician and electrical researcher in Sandwich.
3. Martineau (1849–50), **2**: 451. Despite Faraday's several rebuttals that he had not produced living creatures in his laboratory similar to Crosse's acarii (Faraday to Jerdan, 2 March 1837, letter 977, volume 2; Faraday to the Editor of the *Times*, 3 March 1837, letter 978, volume 2; Faraday to Schoenbein, 21 September 1837, letter 1030, volume 2), *Ann.Reg.*, 1837, **79**: 21 reported Faraday as having done so. On this episode see Stallybrass (1967) and Secord (1989).

Letter 2278
Wilhelm Martin Logeman to Faraday
5 April 1850
From the original in RI MS F1 L4/2a
Mr. M. Faraday Esq. | London.
Sir!

I have the honour to send you by this a magnet of 0,98 English Pounds or 0,449 kilogrammes weight, which, when loaded with convenient precautions such as have been indicated among others by Haecker[1], in Poggendorff's Annales der Physik und chemie, vol 57 page 335[2], can support a load of more as 27 English pounds or 12,30 kilogrammes. This power is constant and does not diminish even when the anchor is forced abruptly from the poles, several times in succession.

The numerous experiments of Haecker have enabled him to fix the power of a magnet of n kilogrammes weight at $10{,}33 n^{\frac{2}{3}}$ [3], and examining those produced by the best Workshops in Europe, they are found not to attain or at least not to surpass considerably the power, indicated by this formula. The magnet you receive here, has twice this power and, with a piece of post paper, interposed between the poles and the soft iron armature, he will support stil[l] a load equal at least to that supported directly by the best magnets of the same weight hitherto produced.

This magnet is constructed after a new method, the fruit of the investigations made by Mr. Elias[4] of this town. I am able to procure magnets of the same quality supporting 400 and even 600 English Pounds at very moderate prices.

I hope Sir, you may agree this magnet as a mark of my regard for you. If after examining it, you find it worth your approbation perhaps you may have the kindness to honour me with some lines and to communicate it among the scientific public in England in such a manner as you might think most convenient.

I am, Sir with much respect | Your humble servant | W.M. Logeman.
Haarlem 5 April | 1850.
Adres./ W.M. Logeman, | Optician | in Haarlem (Holland).
Our small horseshoe magnet weighs 7lbs. $14\frac{1}{2}$ oz. is nearly as the
                                        lifts 40lbs or 41 lbs law $10.33.W^{\frac{2}{3}}$
Logeman magnet weighs 0,98lbs.
                        lifts 26lbs.

Endorsed: No 4489.
Address: Mr. Faraday Esq | London

1. Paul Wolfgang Haecker. Jungnickel and McCormmach (1986), 123 identify him as a scientific instrument maker of Nuremberg.
2. Haecker (1842), 335.
3. *Ibid.*, 326.
4. Pieter Elias (1804–1878, NNBW). Scientific instrument maker in Haarlem.

Letter 2279
Faraday to Harriet Martineau
11 April 1850
From the original copy in IEE MS SC 3

Royal Institution | 11 April 1850

My dear Madam

I am sorry to find that in your great work you have been led at page 451, vol II[1] into an error respecting me by an authority which you might well think sufficient but which is inaccurate[2]. I cannot understand how the error arose at first but it appeared in the papers and I found it necessary in a letter to the Editor of the Literary Gazette (4 March 1837, page 147[3]) to correct it. The error probably passed from the papers into the Annual Register[4] & from that into the far more important position it holds in your History[.]

I send you a letter from a friend of mine[5] with whom you probably are acquainted that you may see from his testimony what really passed; it agrees with that of all those I have spoken to who were then present. Perhaps you would not mind taking the trouble of returning me his letter[.]

I hope you will forgive me for writing to you about this matter[.] I feel it a great honor to be borne in your remembrance but I would not willingly be there in an erroneous point of view[.]

I have the honor to be | My dear Madam | with every respect | Your faithful humble Servant | M. Faraday
Miss Martineau

1. Martineau (1849–50), **2**: 451.
2. See note 3, letter 2277.
3. Faraday to Jerdan, 2 March 1837, letter 977, volume 2.
4. *Ann.Reg.*, 1837, **79**: 21.
5. Letter 2277 from Edward Cowper.

Letter 2280
Harriet Martineau to Faraday
13 April 1850
From the original in IEE MS SC 2

Ambleside | April 13th

My dear Sir

I am greatly obliged to you, & to your correspondent[1] (who is, however, a stranger to me) for correcting the mistake in my history regarding your countenance of the Acarus Crossi[2]. I am anxious to be informed of every error of statement in my history, as there will be a new edition next year[3], & it is of some importance that I should be set right in time. It never occurred to me to doubt the authority of the Annual Register in a matter of such straight-forward contemporary statement & it is really difficult to see how one can make sure of one's material. However, I will take care that this mistake is rectified, & will explain the case whenever I have opportunity.

Believe me, dear Sir, with | the highest respect, your obliged | Harriet Martineau
I return Mr Cowper's note[4].

1. Edward Cowper.
2. Martineau (1849–50), **2**: 451. See note 3, letter 2277.
3. Martineau (1858), 592 did introduce a degree of doubt into Faraday's statements.
4. Letter 2277.

Letter 2281
Paul Frederick Henry Baddeley to Faraday
18 April 1850
From the original in IEE MS SC 2
Professor Farriday London.

Lahore 18h Apr 1850.

My dear Sir,
I have only an hour or two to spare, before the Indian mail leaves this, to give you a few notes regarding Dust storms, which are very prevalent in this part of India during the dry months of April, May & June - that is before the setting in of the rainy season.

My observations on this subject have extended as far back as the hot weather of 1847 when I first came to Lahore and the result is as follows. Dust storms are caused by spiral columns of the Electric fluid passing from the atmosphere to the earth - they have an onward motion - a revolving motion, like revolving storms at sea - and a peculiar spiral motion from above downwards - like a corkscrew. It seems probable that in an extensive dust Storm there are many of these Columns, moving on together, in the same direction - and during the continuance of the storm, many sudden gusts take place at intervals during which time, the *Electric* tension is at its maximum. These storms hereabouts, mostly commence from the N.W. or W. and in the course of an hour more or less they have nearly completed the circle, & have passed onwards.

Precisely the same phenomena, in kind is observable in all kinds of D.St. from the sturdy one of a few inches in diameter to some that extend for 50 miles & upwards; the phenomena is identical.

And it is a curious fact that some of the Smaller Dust storms, occasionally seen in extensive & arid plains - both in the Country & in Affghanistan above the Bolon Pass - called in familiar language "Devils" - are either stationary, for a long time, that is upwards of an hour - or nearly so - & during the whole of this time the dust & minute bodies on the ground are kept whirling about into the air. In other cases these small dust storms - are seen slowly advancing & when numerous usually proceed in the same direction - Birds - Kites & Vultures are often seen soaring first high up above these spots; apparently following the direction of the column, as if enjoying it.

My idea is that the Phenomena connected with Dust storms are identical with those present in water spouts & white Squalls at sea - and revolving storms - & tornadoes of all kinds - and that they originate from the same Cause - viz moving columns of electricity.

In 1847 when at Lahore - being desirous of ascertaining [word illegible] the nature of Dust Storms I projected into the air an insulated copper wire on a Bamboo - on the top of my house - & brought the wire into my room & connected it with a gold leaf electrometer and a detached wire communicating with the Earth - a day or two after, during the passage of a small dust storm, I had the pleasure of observing the electric fluid passing in vivid sparks from [one] wire to another - and of course strongly affecting the electrometer.

The thing was now explained - and since then I have by the same means - obsd at least 60 dust storms of various sizes - all presenting the same phenomena in kind.

I have commonly obsd that towards the close of a storm of this kind, a fall of rain suddenly takes place and instantly the stream of electricity ceases or is much diminished - and when it continues, it seems only on occasions when the storm is severe - and continues for some time after it. The Barometer steadily rises - thro' out - in this pt of the world the fluctuations of the Barometer column is very slight - seldom more than 2 or 3 tenths of an inch at a time.

The average height at Lahore is 1.180 corrected for temperature - indicating I suppose above 1150 feet above the level of the sea taking 30 Inch as the standard.

A large Dust storm is usually preceded by certain peculiarities in the dew point and the manner in which the particles of Dew is deposited on the Bulb of a thermometer. My mode of taking the Dew point is to plunge a Common thermometer in a little Ice, let it run down 20 or 30° take it out shake dry - hold it up to the light and observe the bright spot, & continue to shake off the dew so long as it is deposited and dulls the bulb at the instant it clears off mark the temp. This I have compared frequently with Daniell's Hygm cooled by means of Chloroform - & find them both correspond with the greatest accuracy.

This is a digression - but I have no time to arrange & must therefore put down my remarks as they occur to me.

The dew point varies very much but is usually many degrees below the temp of air 20 to 50° or more.

It also varies according to the time of year during November last mean te[m]p of D.P. was about 47°. That of the air about 71°.

In January 1850 D.P. 43° in the air 61° & the mean temp of self Reg then 45.4.

In February 1850 mean of D.P. 48° and air 64°.5.

April 1850 mean temp of D.P. so far is about 60° & the air 84°.

The sparks or the stream of electricity as it is seen passing from one wire to the other, is in some cases, and during high tension doubled or trebled thus and is never straight but invariably more or less crooked.

Various kinds of sparks as seen - at times - one end of the wire has a star and from the wire when held just beyond striking distance, a brush is seen - curved which when viewed th[r]o' a lens seems composed of a stream or curved brush of bright globules like a shower of mercury.

The manner in which the Electricity acts upon the dust and light bodies it meets with in its passage, is simple enough. I suppose the particles similarly electrified and mutually repulsive - and then together with it whirling motion communicated to them are whisked into the air, - the same takes place, when the electricity moves over water - the surface of the water becomes exposed to the Elec. agency and its particles rendered mutually repulsive, are in the same way whirled into the air.

At sea the water spout is thus formed - first of all is seen the cloud descending - & beneath may be obsd the water in a cone misty and aggitated - soon the cloud is seen to approach & join the latter, involving both extremities in one column having a spiral motion & on it moves or continues stationary. The power of Electy in raising bodies, when combined with this peculiar whirling motion, will account for fish &c being carried up in its vortex and afterwards discharged to a distance on the earth. The motion of the dust storm may be described thus

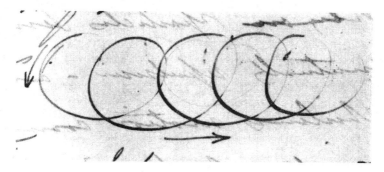

Or by spinning a tee-totum, on a drop of Ink and the way in which bodies are projected may be in like manner described by letting fall a drop of ink on the centre of a teetotum while spinning - in this case the particles of ink are thrown off at tangents ever varying, as the centre moves - and perhaps it will be found that when these kind of storms pass thro forrests - trees uprooted are distributed something in this manner.

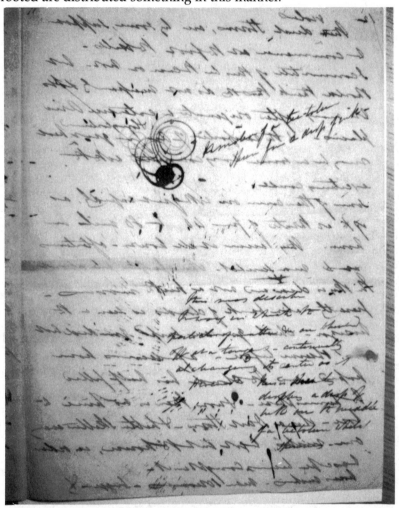

The violent dust storms, are by some, supposed to commence at the foot [of] the hills. I cannot tell if this be the case or not - but should think that they do not necessarily do so since many often originate in extensive

Arid plains - & the rarefaction of air from great & long continued heat may be in some way connected with the exciting cause.

Some of them come on with great rapidity, as if at the rate of from 40 to 80 miles an hour. They occur at all hours - oftentimes near sunset.

The sky is clear and not a breath moving - presently a low bank of clouds is seen on the horizon - which you are surp[rise]d you did not observe before - a few seconds have passed and the cloud has half filled the hemisphere - & now there is no time to lose - it is a Dust Storm - & helter skelter every one rushes, to get into the house, in order to escape being caught in it - horses rushing - men screaming - buggies & tearing allowing [sic] like mad - on rolls the Cloud - majestic - cloud involving cloud - infolding itself - and like a mass of dark rocky mountain - imperilling men - & ready to fall.

Now comes the rush of wind and darkness black as night, follows, shut your mouth & eyes - don't stir or you'll hit a wall, or get a kick - then wait, till it has blown over - may be 2 hours or more or only $\frac{1}{2}$ an hour if more lucky - and then you must wait patiently, and think of the disagreeable position you suffered & then a wash you must have, to get rid of the dirt & dust.

The Electric fluid continues to stream down the conducting wire unremittingly during the continuance of the storm. It sparks oftentimes upwards of an inch in length - and emitting a crackling sound - its intensity varying upon the force of the storm - and as before said, more intense during the gusts.

Many dust storms occur at Lahore and in the Punjaub generally during the hot and dry months - as many as 7 and 9 in one month.

One that occurred last year in the month of August, seemed to have come from the direction of Lica, on the Indus to the West & by S. of Lahore - and seemed to have a N-Ely direction. An officer travelling and at the distance of 20 miles or so from Lica, was suddenly caught in it - his tent was blown away & he himself knocked down - & nearly suffocated by the sand. He stated to me that he was informed by one resident at Lica, that so great was its force at the latter place, as to crack the walls of a substantial brick dwelling in which the above Officer had lately resided & uprooted some trees about.

The instant the insulated wire is involved in the electric current marked by the column of dust - down streams the electricity my wire is thus placed.

I have sometimes attempted to test the kind of Elect. & find that it is not invariably in the same state - sometimes appears + on other times - & changing during the storm.

One day I caused the current to pass thro' a solution of Cyanide of Silver - so as to affect a small piece of copper which was rapidly covered with a coating of Silver - which upon drying peeled off - in this case the Cyanide of Silver was pure - without any salt - but in subsequent attempts to silver a wire in this way, I have not succeeded only a very slight deposit taking place which was not increased by long exposure to the influence but in all the cases I tried subsequently to the one first alluded to, the oxide of silver was diss[olve]d in Cyanide of Potassium.

In the course of time bright minute crystals were formed - transparent & colourless - on a copper coin.

Not knowing anything about the science of electricity - I am unable to institute experiments, such as might suggest itself who was well

acquainted with the subject but it is possible you may see in the few crude observations I have now made something on which some useful fact may be elicited & if you think it worth while to inform me what kind of experiments to institute, I shall be happy to do so for you if I can & let you know the result.

A very dry climate this - not more than 12 or 14 Inch of rain falls in the year.

Soil a light clay soil - productive - a deposit from the river composed of fine [word illegible] buff or light slate coloured sand much peroxide of Iron & [word illegible] lime, beneath this stratum is often Quartz sand with mica - and magnetic Iron stone often in this sand or the dep[osi]t - from 12 to 60 feet - bed of potters clay - here & there.

Rain water very pure - almost as much so as distilled water - River [word illegible] stream - from 300 to 1800grs of fine mud in one cubic foot of water. The larger amount during floods or when the [four words illegible] of river is rapid 7 or 8 miles an hour.

Winds are very variable all the year round but chiefly from the NW.

I have not had time to read over what I have written, so excuse mistakes and style.

Believe me my dear Sir | Yours Faithfully | P. Baddeley | Surgeon Ast Dr | Lahore

These obsns have never been published nor are they at all known save that in my last Annual report to the Medl. Board at Calcutta, I have made a few allusions to the subject - but no one there is likely to take any interest in the matter, nor consider it so much [word illegible][1].

Address: Professor Farriday, F.R.S. | London | For[warde]d by John Fleming[2] Esq | Bootle, Liverpool.

1. Large portions of this letter (with no figures) were published in the form of a letter to the editors of the *Phil.Mag.* as Baddeley (1850a) and this was reprinted in Baddeley (1850b).
2. Liverpool POD gives John Fleming in charge of a boarding school in Bootle. See letter 2399 which identifies him as Baddeley's brother in law.

Letter 2282
Nathaniel Wallich[1] to Faraday
24 April 1850
From the original in RI MS Conybeare Album, f.37

Athenaeum 24 April | 1850

My dear Sir

Permit me to have the honor of introducing the bearer Dr. Beron[2] a meteorologist of distinction who has been strongly recommended to me by Profr von Martius[3] the celebrated botanist of Munich. Dr Beron has expressly directed his attention to the subject of terrestrial magnetism upon which, as well as on other similar subject[s] he is engaged in publishing a work[4]. He is most anxious to submit his views to You, and relying on your kind indulgence I venture to furnish him with these lines, remaining with the highest respect,

Dear Sir | Yours very sincerely | N. Wallich

Profr M. Faraday | &c &c &c F.R.S.

1. Nathaniel Wallich (1786–1854, DNB). Superintendent of the Calcutta Botanic Gardens, 1815–1850.
2. Peter Beron (1798–1871, Schischkoff (1971)). Bulgarian philosopher.
3. Karl Friedrich Philipp von Martius (1794–1868, DSB). Professor of Botany in Munich.
4. See Beron (1850).

Letter 2283
Faraday to James Faraday
c29 April 1850
From Bence Jones (1870a), 2: 233

If the loss be sudden and grievous to us[1], how much more so must it be to you; and indeed we feel *deeply* for you. Let us hope and think that strength will be given you to bear it as a man; patiently, as one to whom grief and adversity does not come bringing only distress and unavailing sorrow, but deeper thoughts and instruction which afterwards produces good fruit. There are none of us who do not need such teaching, but it is hard to bear; and indeed, my dear J., it is *very hard* when it comes with such an overwhelming flood as that which has just reached you. I know that no words of mine are fitted to comfort you, but I seek only to sympathise, and you may believe how earnestly I do so when I at the same moment think what my state would have been had your loss been mine.

Give a kiss to the children from me. Remember me to your mother[2], and think of me as your very affectionate uncle, M. Faraday

1. This was the death on 27 April 1850 of Margery Ann Faraday, née Reid (1820–1850, Reid, C.L. (1914)), a niece of Sarah Faraday's who married James Faraday on 30 April 1846. She died following the birth of Lucy Reid Faraday (1850–1900, GRO under Boyd) on 22 April 1850.
2. Margaret Faraday, née Leighton (d.1868, age 78, GRO). Widow of Faraday's brother, the gas engineer, Robert Faraday (1788–1846, GRO) whom she married in 1815 (see letter 53, volume 1).

Letter 2284
Faraday to William Buchanan
2 May 1850
From the original in the possession of Joan Ferguson

Royal Institution | 2 May 1850

My dear Friend

I received your letter & the check yesterday. Mr. David[1] has been a very pleasant inmate in our house and gained much upon us by his kind manner good intentions & earnest desire as I firmly believe to do something for himself. All that class of anxieties which you may have seen that I had at first have been entirely removed and my only sorrow is that having so strong & proper a desire to take any employment that might offer he still cannot obtain it. Poor fellow I pity him very much for I see that his present position is a serious drawback to his happiness[.] He does not appear to know any thing of book keeping or of any other branch of knowledge which one might use as a means of adding weight to ones enquiries but he seems determined to acquire such knowledge. He is still to look about him but we have encouraged him to stay with us a little longer if he sees any hope of its being useful (and we are very glad to have him in such case) so that perhaps he may not leave next Saturday[2] but remain with us until the middle of next week[.]

Our family troubles are progressing[.] Mrs. James Faraday[3] died last Saturday[4] I think & I attended her funeral to day[.] My nephew is in deep grief. Poor Thomas Deacon is worse in mind - knows no one about him - & is at times violent. We are constrained to transfer him to an Asylum to which he will be taken this morning. We have also reason now to fear for his bodily strength which at present is failing[.] His wife[5] is a very great example of Patient suffering & yet of active help & action where needed. She is with her Grandfather[6] & the Aged & the Young, mutually & beautifully support & help each other having that hope to speak of which can cheer in every trouble that comes over us in this life[.]

Ever My dear friend | Yours Affectionately | M. Faraday
Wm. Buchanan Esq | &c &c &c

1. David Buchanan (c.1823–1890, AuDB). Fifth son of William Buchanan. Emigrated to Australia in 1852.
2. That is 4 May 1850.
3. Margery Ann Faraday, née Reid (1820–1850, Reid, C.L. (1914)), a niece of Sarah Faraday's who married James Faraday on 30 April 1846. She died following the birth of Lucy Reid Faraday (1850–1900, GRO under Boyd) on 22 April 1850.
4. That is 27 April 1850.
5. Caroline Deacon.
6. Edward Barnard.

Letter 2285
Faraday to Henry Bence Jones
3 May 1850
From the original in RI MS F1 C35

R Institution | 3 May 1850

My dear Sir

I hasten to thank you for the copy of Fownes'[1] Manual[2] which was very unexpected[.] It is a most interesting remembrance of that excellent Man & Philosopher and a very pleasant token of your own kind feelings to me[.]

Ever My dear Sir | Yours Truly | M. Faraday
Dr. Bence Jones | &c &c &c:

1. George Fownes (1815–1849, DSB). Professor of Chemistry to the Pharmaceutical Society, 1842–1846.
2. Fownes (1850).

Letter 2286
Faraday to William Buchanan
8 May 1850
From the original in the possession of Joan Ferguson

Royal Institution | 8 May 1850

My dear friend & Brother

I can neither write to you nor hold my peace. I cannot let you think that the sudden calamity to our dear & widowed friend[1] & to you all did not strike us with sad force and yet I fear to say much for words seem trivial in the presence of such a grief. May you be comforted from God by the words wherewith you have comforted us and may we be all resting on him through that hope which is given to the guilty & to the weakest by our Lord Jesus Christ. In the other matter all we can do shall be done[2][.]

Believe me | My dear friend | Very affectionately Yours | M. Faraday
W. Buchanan Esq | &c &c

1. Unidentified.
2. Presumably a reference to David Buchanan (c.1823–1890, AuDB). Fifth son of William Buchanan. Emigrated to Australia in 1852. See letter 2284.

Letter 2287
Faraday to Christian Friedrich Schoenbein
11 May 1850
From the original in UB MS NS 390

Royal Institution | 11 May 1850

My dear Schoenbein

I have seen Burgomaster Sarasin[1] who has very kindly brought me your papers & letter[2]. I wish I could shew him any useful attention but you know what an out-of-the-world man I am. Your German papers[3] are very tantalizing I know the good there must be within & yet I cannot get at it. But now my thoughts are on Ozone[.] I like your idea of an Evening here but it cannot be this season for the arrangements are full[4]. Yet that in some degree suits me better for though I should like to give it, I am a slow man (through want of memory) and therefore require preparation. Now I shall lock up your letters & reread them & also the papers but let me pray you to send me a list of the experiments which you know to suit a large audience also if you can the references to the best French (or English) papers giving an account of its development & progress. Also your present view - also the best and quickest mode of making Ozonized air & such other information as I shall need. Probably other matter will arise before 1851 and I will get possession of it as we go along. If you come over here you shall give the subject yourself i.e if you can arrange & keep to time &c if not I must do my best[.] But every year I need more cramming even for my own particular subjects. Now do not delay to send me the list of experiments because you suppose there is plenty of time &c &c but let me have them that I may think over them during the vacation. I should like to do the matter to my own satisfaction: there are however very few things in which I satisfy myself now. I hoped to have had a paper[5] to send you ere this but Taylor is slow in the printing. Give our kindest remembrances to Madame Schoenbein

Ever My dear friend | Yours truly | M. Faraday

Address: Dr. Schoenbein | &c &c &c | University | Basle, | on the Rhine

1. Felix Sarasin (1797–1862, DHBS). Cloth manufacturer and politician.
2. Letter 2274.
3. Schoenbein (1849a, b).
4. Faraday (1851g), Friday Evening Discourse of 13 June 1851.
5. Almost certainly Faraday (1850), ERE23.

Letter 2288
Jacob Herbert to Faraday
16 May 1850
From the original in GL MS 30108/1/50&51

Trinity House, | 16th May 1850.
My dear Sir,
The Committee for Lights having reason to suppose that the White Lead supplied for the Corporation's Service of which the accompanying quantity is a Sample is not of a pure quality - They request you will favor them by testing the same, and reporting the result, together with your opinion of the nature of the Ingredients, if any, which have been mix'd therewith, and the relative proportions thereof, and of the pure Lead[1].-
I remain, | My dear Sir, | your's very faithfully, | J. Herbert
M. Faraday Esqre | &c &c &c

1. Faraday's analysis is in GL MS 30108/1/50&51.

Letter 2289
Faraday to Thomas Romney Robinson[1]
20 May 1850
From the original in ULC Add MS 7656, TR57

Royal Institution | 20 May 1850
My dear Robinson
All the best glass is gone long ago but I send you a piece of such as I have and if that will do any good I can spare you a piece or two more like it.
I shall send you a paper soon (nothing particular) which I sent in to the Royal Society at the end of last year & of which I have only now received copies[2]. Is not that keeping up with the Rail road & the Electric telegraph?
I do not know any body who could prepare the report you speak of. My memory is now so bad that I not only forget the matter itself but I forget also the men who have it. I wish very much such a summary existed[.]
Ever My dear friend Very | Truly Yours | M. Faraday
Rev. Dr. Robinson | &c &c &c

1. Thomas Romney Robinson (1792–1882, DNB). Director of the Armagh Observatory, 1823–1882.
2. Faraday (1850), ERE23.

Letter 2290
Faraday to James Walker
20 May 1850
From the original in SI D MS 554A

Royal Institution | 20 May 1850

My dear Sir
I was so hurried at the receipt of your last two notes that I had not time to answer them. Let me now thank you for all your kindness in your earnest endeavours to supply me with the models[1]. The three of the Edystone are very interesting but it was out of my power to go into their merits[.]

Ever Yours truly | M. Faraday
Jas Walker Esq | &c &c &c

1. In his third lecture, "A Lamp" of his course of six lectures "Upon some points of domestic chemical philosophy" delivered on 11 May 1850 Faraday showed a model of a lighthouse. See Faraday's notes in RI MS F4 J19, p.11.

Letter 2291
Charles John Huffam Dickens to Faraday
28 May 1850
From the original in IEE MS SC 2

Devonshire Terrace | Twenty Eight May, 1850.

Dear Sir
I take the liberty of addressing you as if I knew you personally; trusting that I may venture to assume that you will excuse that freedom.

It has occurred to me that it would be extremely beneficial to a large class of the public, to have some account of your late lectures on the breakfast-table[1], and of those you addressed, last year, to children[2]. I should be exceedingly glad to have some papers in reference to them, published in my new enterprize "Household Words". May I ask whether it would be agreeable to you, and, if so, whether you would favor me with the loan of your notes of those Lectures for perusal?

I am sensible that you may have reasons of your own, for reserving the subject to yourself. In that case, I beg to assure you that I would on no account approach it.

With great respect and esteem, I | remain Dear Sir | Your faithful Servant | Charles Dickens
Michael Faraday Esquire

1. Faraday's course of six lectures "Upon some points of domestic chemical philosophy" were delivered on 27 April, 4, 11, 18, 25 May and 1 June 1850. Faraday's notes are in RI MS F4 J19. These were condensed to "The Mysteries of a Tea-kettle", *Household Words*, 1850, **2**: 176–81.
2. These were Faraday's Christmas Lectures "On the Chemical History of a Candle". For the prospectus see RI MS GB 2: 50. These lectures were condensed as "The Chemistry of a Candle", *Household Words*, 1850, **1**: 439–44.

Letter 2292
Charles John Huffam Dickens to Faraday
31 May 1850
From the original in IEE MS SC 2

Devonshire Terrace | Thirty First May, 1850.
My dear Sir
    I really cannot tell you how very sensible I am of your great kindness, or what an honor I feel it to be to have interested you in my books[1].
    I think I may be able to do something with the Candle[2]; but I will not touch it, or have it touched, unless it can be re-lighted with something of the beautiful simplicity and clearness of which I see the traces in your notes.
    Since you are so generous as to offer me the notes of the lectures on the breakfast table[3], I will borrow them when you have done with them, if it be only for my own interest and gratification. I deeply regret now, not having heard the lectures to children, as it would have been a perfect delight to me to have described them, however generally.
    I should take it as a great favor if you could allow me (in the event of my being unfortunately unable to come myself) to introduce my Sub Editor to your next lecture[4]; for a subsequent comparison of his recollection of it, with your notes, might enlighten us very much.
    Pray let me add, as one who has long respected you, and strongly felt the obligations Society owes to you, that the day on which I took the liberty of writing to you[5] will always be a memorable day in my calendar, if I date from it - as I now hope I shall - the beginning of a personal knowledge of you.
    My Dear Sir | Yours faithfully and obliged | Charles Dickens
Michael Faraday Esquire

1. Faraday's reply to letter 2291 has not survived.
2. See note 2, letter 2291.
3. Faraday's course of six lectures "Upon some points of domestic chemical philosophy" were delivered on 27 April, 4, 11, 18, 25 May and 1 June 1850. Faraday's notes are in RI MS F4 J19.
4. That is on 1 June 1850.
5. Letter 2291.

Letter 2293
Faraday to Charles Rowcroft[1]
3 June 1850
From the original in HLHU

Royal Institution | 3 June 1850

Dear Sir

Let me thank you for your book[2] - still more for your words and most of all for the kindness which has sent both. It is very pleasant to find myself acceptable to those whom one has before hand held in admiration[.]

I am My dear Sir | Your Very Grateful Servant | M. Faraday
Charles Rowcroft Esq | &c &c &c

1. Charles Rowcroft (d.1856, B3). Writer.
2. Rowcroft (1850).

Letter 2294
Faraday to Matthew Cotes Wyatt[1]
3 June 1850
From the original in Royal Institute of British Architects MS Wy Fam /1/5/17

R Institution | 3 June 1850

My dear Sir

I do not think I have thanked you yet for the beautiful engraving you have given me. I do so now heartily. You know the interest I take in the matter[.]

Ever Truly Yours | M. Faraday
M. Wyatt Esq | &c &c &c

1. Matthew Cotes Wyatt (1777–1862, DNB). Sculptor.

Letter 2295
Charles John Huffam Dickens to Faraday
6 June 1850
From Storey et al. (1988), 110

Devonshire Terrace | Sixth June, 1850.

My dear Sir

Very many thanks for your second precious book of notes[1]. I hope, with the assistance of a friend and contributor[2] who has a practical knowledge of chemistry, to convey some very small installment of the pleasure and interest I have in them, to others.
Very faithfully Yours | Charles Dickens
Michael Faraday Esquire.

Address: Michael Faraday | Royal Institution | Albemarle Street

1. See letters 2291 and 2292 and notes therein.
2. According to Storey et al. (1988), 110 this was Percival Leigh (1813–1889, DNB). Writer and former physician.

Letter 2296
Jacob Gisbert Samuel van Breda to Faraday
6 June 1850
From the original in RS MS 241, f.117

Harlem ce 6 Juin 1850

Monsieur!

La Société Royale Hollandaise des Sciences desirant de donner un témoignage de sa haute estime à un savant, qu'elle considère comme le premier des Physiciens de notre époque, Vous a nommé dans sa 98 Séance annuelle, à l'unamimité des voix, un de ses membres associés.

J'ose espérer, Monsieur! que Vous voudriez bien permettre à la Société de se donner l'honneur d'inscrire votre illustre nom parmi ceux, qui en sont deja partie-

J'ai l'honneur d'être avec la plus haute consideration | le Secretaire perpétuel de la Société Hollandaise des Sciences. | J.G.S. van Breda
à Monsieur Faraday | à Londres

TRANSLATION

Haarlem, 6 June 1850

Sir!

The Société Royale Hollandaise des Sciences, desiring to give witness of its high esteem to a savant whom she considers as the first among the physicists of our time has unanimously nominated you at its 98th annual meeting to be one of its associate members.

I dare hope, Sir, that you will be kind enough to allow the Society to honour itself by inscribing your name amongst those who are already members.

I have the honour of being with the highest consideration | the permanent secretary of the Société Hollandaise des Sciences | J.G.S. Van Breda

to Mr Faraday | in London

Letter 2297
Faraday to Jacob Gisbert Samuel van Breda
12 June 1850
From the original in HMW

12 June 1850 | London

Sir

I was honored by the receipt of your letter[1] informing me of the high privilege granted me by the Royal Society of Sciences of Holland of having my name registered among those of the great men whom it has been pleased to place on the list of its foreign associates. I beg most respectfully to offer my thanks for this great distinction and though I can make little promise at my time of life yet I may say it shall be my object to deserve in some small degree the favour conferred upon me.

I have the honor to be | Sir | Your most obliged & Very humble Servant | M. Faraday
A Monsieur | J.G.S. van Breda | Perpetual Secretary | &c &c &c &c

Address: To | J.G.S. van Breda | &c &c &c &c | Perpetual Secretary | Royal Society of Sciences | Harlem | Holland
Postmark: Attleborough[2]

1. Letter 2296.
2. The post town for Old Buckenham.

Letter 2298
Faraday to John Barlow
15 June 1850[1]
From the original in RI MS JB1/1, p.36

Old Buckingham [sic] | Norfolk | Saturday Night

My dear friend

I hope to be at home on Tuesday Evening[2]. As to the Monday meeting of Managers[3] I have not yet been able to obtain estimates from

Newsham[4] for some painting but the whole matter may go over until the Managers meeting in July[5]. Trusting you are quite well

I am | My dear friend | Yours Ever | M Faraday

1. Dated on the basis of the references to the meetings of Managers and on Faraday being in Old Buckenham at this time. For the latter see letter 2297 and records of the Old Buckenham Meeting House, p.40.
2. That is 18 June 1850. See also letter 2299.
3. RI MM, 17 June 1850, **10**: 271 at which Faraday was not present.
4. Richard Newsham. Plumber of 8 Little Stanhope Street, Mayfair. POD.
5. RI MM, 1 July 1850, **10**: 279 when Faraday presented these estimates.

Letter 2299
Faraday to Angela Georgina Burdett Coutts
18 June 1850
From the original in the possession of W.A.F. Burdett-Coutts

Royal Institution | 18 June 1850

My dear Miss Coutts

I have only this moment returned to town[1] or (as I cannot have the honor of being with you on next Monday[2]) I would have returned you the enclosed card before. I hope I am not too late[.]

Though in haste yet not the less your very obliged & | faithful Servant | M. Faraday

1. From Norfolk where the Faradays had visited Old Buckenham from 8 to 18 June 1850. Records of the Old Buckenham Meeting House, p.40; see also letter 2297.
2. That is 24 June 1850.

Letter 2300
Paolo Volpicelli to Faraday
22 June 1850
From the original in IEE MS SC 2
A Monsieur Faraday

Londres le 22 Juin 1850

Les deux anciennes colonnes Trajane et Antonine, qui sont à Rome, ont été en differentes époques frappées par la foudre, car aussi l'une que l'autre est surmonteée d'une statue colossale en bronze. C'est pourquoi Mr Faraday est prié de vouloir bien exprimer son opinion très-appréciable sur les trois articles suivants, par le sousigné.

1° S'il soit nécessaire pour la conservation des deux monuments, de les garnir d'un paratonnerre.
2° Si le fil conducteur puisse être placé en ligne spirale dans l'interieur des deux colonnes.
3° Si des paratonnerres placés *seulement* sur les maisons voisines aux deux colonnes suffisent à produire le même effet.

Votre tres humble | et tres obeiss. Serviteur | Paul Volpicelli

TRANSLATION
To Mr Faraday

London 22 June 1850

The two ancient Trajan and Antonine columns in Rome have at different times been hit by lightning, for both of them are surmounted by a colossal bronze statue. This is why Mr Faraday is kindly asked by the undersigned to express his most valued opinion on the following three points:

1° If it is necessary for the conservation of these two monuments to furnish them with a lightning conductor.
2° If the conductor can be made into a spiral and placed inside the two columns.
3° If lightening conductors placed *solely* on houses surrounding the two columns are enough to produce the same effect.

Your most humble | and most obedient servant | Paul Volpicelli

Letter 2301
Faraday to Charles Richard Weld
24 June 1850
From the original in RS MS MM 13.122

Royal Institution | 24 June 1850

My dear Sir

In reference to your letter I have to inform you & the Council that I have seen Mr. Bate and examined various documents which I called for, in order that I might in some degree audit the account you sent me from him[1]; and have the following remarks to make.

The works referred to were no doubt undertaken at the time; i.e. in the years 1832–1836; and the present Mr. Bate has letters from Mr. Baily[2], who was then Treasurer of the Royal Society, of dates 27 April and 9th May 1836 (which I have seen), requiring that the account should be sent in. I understand the account was not sent in; but of that I have no knowledge.

The first item in the bill is for apparatus ordered on the 13th Jany 1835, as may be seen at page 6 (second paying) of the proceedings of the Excise Committee[3].

The second item is for various fittings of a large balance and other apparatus in a room at Mr. Bate's house, and was money *paid out*; of which I have seen the bill & receipt[.]

The third item is for money *paid out* for spirit and bottles used in the experiments reported in the proceedings of the Committee:- I have seen the bill & receipt. About one half of the spirit remains and all the bottles, and if it be considered as the property of the Royal Society, or rather of the Government, should either be claimed or else disposed of in part liquidation of the debt.

In relation to the fourth item I have examined such documents as were placed before me and find notes of 69 days of experimental working, and 14 days of attendance on the Committee at Somerset house and elsewhere, making 83 days. In a note from the present Mr. Bate to myself he makes the days 94; and there are days mentioned in the records, as those on which instructions were received, making up this difference. The item says "say 6 months from March 1835 to this date" (Jany 27, 1836) - but the 83 or 92 dates I have verified extend from 11 Decr. 1832 to 27 Jany 1836. Mr. Bate assumes that half as much more time would be occupied with calculations, writing out fair &c; in reference to which I ought to remark that the entry for Jany 12, 13, 14, 1836 is for entering experiments and making corrections in the Royal Society's book; which perhaps leads to the inference that time so occupied was always noted. I am unable at this long deferred date, even by the aid of the documents, to say whether the 83 or 94 days might be considered each as a day of paid employment for three persons, or whether there was not in them plenty of time for calculation & notes:- and therefore whether the assumption of half as many more for the latter purpose is right or not.

The fifth item is for four hydrometers ordered by the Committee and the receipt of which is acknowledged by a letter in Mr. Robertons[4] hand writing[.]

The sixth item is for an instrument ordered at the time and recently delivered in (as I am told) to the Royal Society[.]

I am My dear Sir | Very Truly Yours | M. Faraday
C.R. Weld Esq | &c &c &c

---

1. This invoice is in Bate to Secretary of the Royal Society, 13 May 1850, RS MS MM 13.121 and came to £194.15.2. It related to some work that Robert Brettel Bate (1782–1847, DNBmp, scientific instrument maker in the Poultry, London) had carried out for the Royal Society Excise Committee in the mid–1830s. For the complex background to this issue see McConnell (1993), especially pp.54–5 and p.69 where the invoice is printed.
2. Francis Baily (1774–1844, DSB). Astronomer and Treasurer of the Royal Society, 1835–1838.

3. See Minutes of Excise Committee, 13 January 1835, RS MS CMB67, p.30 for this authorisation.
4. John David Roberton (d.1843, age 43, GRO). Assistant Secretary of the Royal Society, 1835–1843. Anon (1940), 344.

Letter 2302
Faraday to Third Earl of Rosse
25 June 1850
From the original in The Birr Scientific and Heritage Foundation MS J/14

Royal Institution | 25 June 1850

My dear Lord

My opinion is not worth your having but bad as it is you have the fullest right to it & I ought not to withhold it. My impression is that either Hansen[1] or Weber (W?) or Wöhler deserve the Copley Medal but as it is for the *Most Worthy* I feel extreme difficulty in forming any idea respecting the preference of one over the others[2]. And my trouble on this point is chiefly if not entirely my infirm memory for I cannot remember that which I have read of what they have written[.] Hence my confidence in myself becomes shaken and I fear to form conclusions least they should be unjust[.]

I told your Lordship that you would have very useless opinion from me & I am sincerely grieved that I cannot make it better or more useful. As a man of true Science & as the Head of the Royal Society Your Lordship deserves the active & working support of every lover of Science and for other relations independant of Science Your Lordship deserve[s] the help of every right minded man[.]

I am My dear Lord | with the Greatest respect Most faithfully | M. Faraday
The Earl Rosse | &c &c &c

1. Peter Andreas Hansen (1795–1874, DSB). Director of the observatory in Seeberg.
2. Hansen was awarded the Copley Medal in 1850. He had been previously nominated by the Committee of Astronomy of the Royal Society. RS CM, 16 November 1849, 2: 123.

Letter 2303
Faraday to Henry Bowie[1]
1 July 1850
From the original in NLS MS 581 499

Royal Institution | 1 July 1850

Sir

I have the honor to acknowledge the receipt of your letter and beg to thank you for the confidence & trust which you are pleased to place in me. I can only say in reply that my health & my duty will not allow me to accept the post you offer me and that I now lecture no where out of the Walls of the Royal Institution except at the Royal Military Academy at Woolwich[.]

I am Sir | Your Obedient humble Servant | M. Faraday
Henry Bowie Esq | &c &c &c

1. Unidentified.

Letter 2304
Charles Barry[1] to Faraday
2 July 1850[2]
From the original in IEE MS SC 2

Royal Institution | Tuesday aft

Dear Faraday
I am very anxious to see you touching the acoustics at the New House of Commons[3][.] I hear you are coming to Town tomorrow for the Levee[4] and so could attend any appointment you would have the goodness to make for tomorrow (as time presses) either to see you here or at my own house and either before or after your Court homages[.]

You see I am bold enough to assume that you will kindly assist me in this matter[.]

Yours very truly | Charles Barry

1. Charles Barry (1795–1860, DNB). Architect of the Houses of Parliament.
2. Dated on the basis of the references cited in notes 3 and 4.
3. On this see Crook and Port (1973), 623–4. Faraday gave evidence to the Select Committee on the New House of Commons on 15 July 1850. *Parliamentary Papers*, 1850 (650) 15, p.iii.
4. Faraday attended the Levée on 3 July 1850. See PRO LC6 / 13.

Letter 2305
Joseph Antione Ferdinand Plateau to Faraday
8 July 1850
From the original in IEE MS SC 2

Gand 8 Juillet 1850

Mon cher Monsieur Faraday.

Recevez d'abord mes bien vifs remerciements pour le morceau de verre pesant dont vous avez eu la bonté de me faire cadeau: je le conserve comme un précieux souvenir de vos bons sentiments pour moi.

Maintenant, je viens importuner, et c'est avec grand regret que je m'y décide; mais voici ce que est arrivé. Au commencement du mois de Janvier, Monsieur Wheatstone m'a écrit que mon Mémoire (*Recherches expérimentales et théoriques sur les figures d'équilibre d'une masse liquide sans pesanteur, deuxième serie*[1]) allait être reproduit dans les *Scientific Memoirs*[2] de Mr. Taylor, et paraitrait probablement le mois suivant, c'est à dire en février; Mr. Wheatstone ajoutait qu'il se chargeait lui même de la traduction, ainsi que de la correction des épreuves. Après avoir attendu plus de quatre mois sans recevoir aucune nouvelle de cette traduction, j'ai écrit à Mr. Wheastone, pour le remercier d'abord de la peine qu'il voulait bien se donner, et pour le prier de me faire savoir à quel point en était l'affaire; mais je n'ai reçu aucune réponse, quoiqu'il y ait près de deux mois de cela. Mr. Wheatstone est il absent, ou bien ma lettre ne lui est elle point parvenue? Dans cette incertitude, j'ai pris le parti de recourir à votre extrême obligeance, et je sais que ce ne sera pas inutilement: car vous, si grand comme physicien, vous êtes encore l'un des hommes les meilleurs. Voici donc ce que j'attends de votre bonté: si Mr. Wheatstone est à Londres, veuillez l'engager à me répondre de suite; s'il est absent, j'espère que vous voudrez bien vous informer si la traduction de mon Mémoire a parue ou va paraître, ou s'il n'en est rien, et me le faire savoir par un mot de réponse. Le Mémoire dont il s'agit m'a couté extrêmement de temps et de travail, et j'y attache une très grande importance: je dois donc faire tout ce qui dépend de moi pour répandre parmi les savants la connaissance des résultats qu'il renferme, et je tiens beaucoup à sa reproduction dans les *Scientific Memoirs*.

J'attends votre lettre avec impatience, et je vous prie de me croire, avec tous les sentiments d'une respectueuse amitié,

Votre entièrement dévoué | Jh Plateau

place du Casino, 22, à Gand.

Address: Monsieur | Faraday, Membre de la Société | Royale, &c A l'Institution Royale, | Albemarle street. | Londres

TRANSLATION

Ghent 8 July 1850

My dear Mr Faraday,

First of all, please accept my very great thanks for the piece of heavy glass of which you had the kindness to make me a gift; I keep it as a precious souvenir of your kind feelings towards me.

Now, I must inconvenience you, and it is with great regret that I have decided to do so; but this is what has happened. At the beginning of January, Mr Wheatstone wrote to me that my Paper (*Recherches expérimentales et théoriques sur les figures d'équilibre d'une masse liquide sans pesanteur, deuxième série*[1]) was going to be reproduced in Mr Taylor's *Scientific Memoirs*[2] and would appear probably the following month, that is to say in February; Mr Wheatstone added that he would himself see to the translation, as well as the correction of proofs. Having waited over four months without receiving any news of this translation, I wrote to Mr Wheatstone, to thank him first of all for his trouble and also to ask him to let me know where the business stood; but I received no reply, although that was over two months ago. Is Mr Wheatstone absent or did my letter never reach him? In this uncertainty, I have taken the course of calling on your extreme kindness, and I know it will not be in vain: for you, so great a physicist, are still one of the best of men. This then is what I ask of your goodness: if Mr Wheatstone is in London, please ask him to write to me by return; if he is away, I hope that you will have the kindness to enquire whether the translation of my paper has appeared or will appear, or if nothing has happened, and to send me word. The paper in question took me a great deal of time and work and I attach enormous importance to it. I must do everything in my power to disseminate amongst my colleagues the knowledge of the results contained in it, and I attach great importance to its publication in the *Scientific Memoirs*.

I await your letter with impatience and I ask you to believe me to be, with all the sentiments of a respectful friendship,

Your entirely devoted | Jh Plateau

place du Casino, 22, in Ghent.

1. Plateau (1849).
2. Plateau (1852).

Letter 2306
Second Marquis of Breadalbane[1] **to Faraday**
10 July 1850
From the printed original in the possession of Rosalind Brennand

The Lord Chamberlain is commanded by the Queen to invite Mr. Michael Faraday to a Ball on Wednesday the 10th of July 1850 at $\frac{1}{2}$ past 9 o'clock[2].
Buckingham Palace.
Full Dress.

1. John Campbell, 2nd Marquis of Breadalbane (1796–1862, DNB). Whig politician and Lord Chamberlain, 1848–1852.
2. This ball would have been cancelled because of the death, on 8 July 1850, of the Queen's uncle Adolphus Frederick, Duke of Cambridge (1774–1850, DNB), field marshal and Viceroy of Hanover, 1816–1837.

Letter 2307
Faraday to Angela Georgina Burdett Coutts
15 July 1850
From the original in the possession of W.A.F. Burdett-Coutts

Royal Institution | 15 July 1850

My dear Miss Coutts

We are living at Norwood & it was only this morning that I found your kind letter and I regret greatly that that circumstance has retarded both my thanks and the return of your privilege in time for it to be of use in any other direction. With warm thanks for your kind remembrances of such an unsocial being as I am

I remain | Your true & Obliged Servant | M. Faraday

Letter 2308
Faraday to John Tyndall
19 July 1850
From a typescript in RI MS T TS, volume 12, p.4125

Upper Norwood, | 19 July, 1850.

Dear Sir,

I am very much obliged to you for the specimens of calcareous spar, and I shall very likely turn to the matter again some day soon[1]. At present I am working upon another part of the great subject[2], and withal am out here for a little change of air and rest[3].

I am, my dear Sir, | Very truly yours, | M. Faraday.
J. Tyndall Esq. | &c. &c. &c.

Address: To J.T. | Spring Bank | Over Darwen | Lancashire

1. See Faraday, *Diary*, 14 August 1850, 5: 11169–70.
2. At this time Faraday was working on the diamagnetism of various substances. Faraday, *Diary*, 16 and 20 July 1850, 5: 10900–46.
3. Tyndall, *Diary*, 7 August 1850, 4: 427–8 recollected his meeting with Faraday of which this letter was clearly an outcome. Faraday and Tyndall differed from each other in their views on diamagnetism but at this meeting Faraday, according to Tyndall, said that it was of "No matter ... you differ not as a partisan but because your convictions compel you". Also quoted in Eve and Creasey (1945), 32.

Letter 2309
Faraday to Benjamin Vincent
25 July 1850
From the original in RI MS F1 C36

Upper Norwood | 25 July 1850

My dear friend

A few words together even on paper (if it cannot be by mouth) are pleasant though I do not find myself good for either just now. Because of much pain in my jaw & the known bad state of my teeth, which I had only hoped to keep through the lectures[1], I went on Monday morning[2] to the dentist[.] He pulled out five teeth & a fang[.] He had much trouble & I much pain in the removal of a deep stump and I think from the feeling then & now he must have broken away part of the jaw bone to get at it, for it is very sore & the head is rather unsteady. On the whole the operation were well & cleverly carried on by the dentist, the fault was in the teeth[.] Just let me say in addition about myself that the cold shiverings which came on on Saturday night[3] are gone & I believe my tendency to chill is very much less[.]

I should like to know how you are & the family but I am at the wrong end of the post for that. Mr Barnard was here yesterday - Mrs Hillhouse[4] left this place yesterday and I hope she reached home in safety[.] Mr Hillhouse[5] was with her so that all that could be done in the way of assistance would be done[.]

On Saturday morning *and Sabbath day* I trust to see you[6]. I am afraid I shall be of no use to others on the Sabbath for my voice is a queer one having lost some of the alphabetical sounds for the present, it is not wonderful that both the dental & labial modifications should be touched: but I hope Mr. H Deacon is in good use.

I am My dear friend | Yours Very affectionately | M. Faraday
Mr. B. Vincent | &c &c &c

Endorsement: 1 Cor 13

1. That is Faraday's course of six lectures "Upon some points of domestic chemical philosophy" which he delivered on 27 April, 4, 11, 18, 25 May and 1 June 1850. Faraday's notes are in RI MS F4 J19.
2. That is 22 July 1850.
3. That is 20 July 1850.

4. Ann Hanbury Hillhouse (d.1860, age 71, GRO). A member of the London Sandemanian Church.
5. John Wilson Hillhouse. Given as an accountant on Ann Hanbury Hillhouse's death certificate.
6. That is 27 and 28 July 1850.

Letter 2310
Faraday to William Whewell
1 August 1850
From the original in TCC MS O.15.49, f.26

Royal Institution | 1 Aug 1850

My dear Sir

I want a distinctive word, it may only be for a time for I am quite in uncertainty as to how finally the subject of Magnetism may settle down[.] But taking the Word *Magnetic* to represent the *general action* of the forces so called, and having already employed the word diamagnetic to represent that part of the general action which is manifest in bismuth phosphorous &c I want a word to represent the other & more know[n] part of the action manifested in Iron, Nickel Cobalt &c. The distinction is very important to me just now & I can hardly write my notes without it[1]. Assuming the Earth as a Planet or a whole to represent one of these actions I have written Terro magnetic or Terra-magnetic & so made the following distinction in my notes

Magnetic $\begin{cases} \text{Terromagnetic} \\ \text{Diamagnetic} \end{cases}$

I feel that *Terro* & *Dia* are not in fair relation. Can you give me a better word, or considering the transitory state of Magnetic language is Terro magnetic or something like it admissible into printed papers for the present?

At present my head is full of visions: whether they will disappear as experiment wakens me up or open out into clear distinct views of the truth of nature is more than I dare say. But my hopes are strong.

Ever My dear Sir | Very Truly Yours | M. Faraday
Revd. Dr. Whewell | &c &c &c

1. See Faraday, *Diary*, 31 July 1850, **5**: 11023–57.

Letter 2311
William Whewell to Faraday
12 August 1850
From the original in TCC MS O.15.49, f.62

Kreuznach, | Rhenish Prussia

My dear Sir

I am always glad to hear of your wanting new words[1], because the want shows that you are pursuing new thoughts, and your new thoughts are worth something: but I always feel also how difficult it is for one who has not pursued the train of thought to suggest the right word. There are so many relations involved in a new discovery and the word ought not glaringly to violate any of them. The purists would certainly object to the opposition or coordination of terro-magnetic and diamagnetic, not only on account of the want of symmetry in the relation of *terro* and *dia*, but also because the one is Latin and the other Greek. But these objections, being merely relative to the form of the words would not be fatal, especially if the new word were considered as temporary only to be superseded by a better when the relation of the phenomena are more clearly seen. But a more serious objection to *terromagnetic* seems to me to be that diamagnetic bodies have also a relation to the earth as well as the other class; namely a tendency to place their length transverse to the lines of terrestrial magnetic force. Hence it would appear that the two classes of magnetic bodies are those which place their length *parallel* or *according to* the terrestrial magnetic lines, and those which place their length *transverse* to such lines. Keeping the preposition *dia* for the latter then the preposition *para* or *ana* might be used for the former; perhaps *para* would be best as the word *parallel*, in which it is involved, would be a mechanical memory for it. Thus we should have this distinction

Magnetic { Paramagnetic: Iron, Nickel, Cobalt &c
          { Diamagnetic: Bismuth, Phosphor &c

If you like *anamagnetic* better than *paramagnetic*, as meaning magnetic *according to* our standard, terrestrial magnetism, I see no objection. I had at one time thought of *ortho magnetic* and *diamagnetic*, directly magnetic and diametrally [sic] magnetic, but here the symmetry is not so complete as with two prepositions.

In considering whether I quite understand the present state of the subject, I have asked myself what would be the effect of a planet made up of bits of bismuth, phosphor &c, of which the general mass had their lengths parallel to a certain axis of the planet. I suppose all *paramagnetic* bodies would arrange themselves transverse to its meridian, and all diamagnetic bodies in its meridian. Am I right?

I rejoice to hear that you have new views of discovery opening to you. I always rejoice to hail the light of such when they dawn upon you.

I have been at the meeting of Swiss naturalists at Aarau, where I met Schönbein who talked much of you, and told me you were going to explain his views of ozone².

I shall be in London in a few days and shall perhaps try to see you when I am there. Letters sent to Cambridge always find me.

Believe me, my dear Sir, yours most truly | W. Whewell

Address: England | Dr Faraday | Royal Institution | Albemarle Street | London
Postmark: 12 August 1850

1. In letter 2310.
2. See letter 2287 and Faraday (1851g), Friday Evening Discourse of 13 June 1851.

Letter 2312
Faraday to Benjamin Vincent
13 August 1850
From the original in RI MS F1 C37

Upper Norwood | 13 Aug 1850

My dear Friend

My wife had your letter yesterday & I am sure will thank you for it as I did for its contents. I called at Wardour St.¹ yesterday but found all but Mrs. Vincents sister² at Gravesend[.] I told her how you were & found she had sent a letter from you on to Mrs Vincent³. There had been a very sad fire at Gravesend⁴ but I trust it was too far from Mrs. Vincent to do more than startle her. Friends appeared pretty well on Sabbath⁵ but a great many were away & we missed both You & Mr. Leighton[.] Mrs. Anderson⁶ also was absent having gone off to Edinburgh where she has safely & well arrived. Her two daughters leave London tomorrow & feel the separation from friends very deeply. Mr Boosey & Mr Leighton both seemed pretty well. On Sabbath Evening I saw Mr Clarke⁷ in bed but looking well in the face & cheery but as he says he is so soon tired he does not know what is the matter with him except it be age.

Dear friend it rejoiced us to hear how you were going on⁸ & I hope you will have a happy time every where. Give our love to Mr Baxter⁹, Mr Maxwell¹⁰, Dr. - but I will not specify - to all the friends who speak of us to you. We were very happy in their company[.]

With love to Mr Leighton & yourself I am my dear friend | Very Truly Yours | M. Faraday

Letter 2313 171

Mr B. Vincent. | &c &c &c
I hope this letter will find you but I do not know how to send it

1. Where James Faraday had his brass foundry.
2. Unidentified.
3. Janet Young Vincent, née Nicoll (1811–1863, GRO). Wife of Benjamin Vincent.
4. On 11 August 1850. See *Ann.Reg.*, 1850, **92**: 97–8.
5. That is 11 August 1850.
6. Elizabeth Anderson. The records of London Meeting House confirm her move to Edinburgh.
7. William Clarke. Cantor (1991a), 300 notes him as a member of the London Sandemanian Church.
8. Vincent was in Scotland. See letter 2315.
9. Possibly the same as Thomas Baxter to whom Faraday wrote on 12 April 1847, letter 1974, volume 3.
10. Unidentified.

Letter 2313
Friedrich Wilhelm Heinrich Alexander von Humboldt to Faraday
13 August 1850
From the original in IEE MS SC 2
Mon cher et illustre Confrere! Les porteurs de ce peu de lignes, les deux freres Docteurs Schlagintweit de Munnich, munis des conniossances les plus variées et les plus valides en Physique, Geologie et Géographie des Plants ont parcouru pendant plusieurs années et avec noble courage la Chaine orientale des Alpes jusqu'aux cimes les plus élevées. Ils vont publier un grand ouvrage[1], semblable à celui que Saussure[2] a donné par des stages de l'ouest[3]. Ce sont des jeunes gens aimables et modestes. J'ose les recommander bien chaudement à Votre bienveillance. Ils sont dignes de Vous approcher, mon illustre ami, je leur envie ce bonheur.

Agreez, je Vous supplie, l'homage respectueux d'un devouement et d'une admiration qui datent de si loin. | Alexandre de Humboldt
Sans Souci ce 13 Aout 1850.

Address: A Monsieur | Mr le Docteur Faraday, | membre de toutes les | Academies des deux | Continens | à | Londres | de la part de | Bn de Humboldt | par Mrs les Docteurs | Schlagintweit

TRANSLATION
My dear and illustrious colleague! The bearers of these few lines, the two brothers Doctors Schlagintweit of Munich, armed with the most varied and valuable knowledge of the Physics, Geology and Geography of Plants,

Plate 4. Friedrich Wilhelm Heinrich Alexander von Humboldt.

have crossed, over several years and with considerable courage, the eastern chain of the Alps, up to the highest summits. They are going to publish a great work[1], similar to that which Saussure[2] wrote in stages on the west[3]. These are very amiable and modest young people. I dare to recommend them warmly to your kindness. They are worthy of approaching you, my illustrious friend, and I envy them their happiness.

Please accept, I beg you, the respectful homage of a devotion and an admiration which go back such a long time | Alexandre de Humboldt Sans Souci this 13 August 1850.

1. Schlagintweit and Schlagintweit (1850).
2. Horace Bénédict de Saussure (1740–1799, DSB). Swiss geologist and naturalist.
3. Saussure (1779–96).

Letter 2314
Harriet Jane Moore to Faraday
14 August 1850[1]
From the original in IEE MS SC 2

August 14th
My dear Mr. Faraday
    Though you and Mrs. Faraday will not try how this pretty spot would suit a philosopher I hope you have found Norwood agreeable, and enjoyed picnics with my friends the Gypsies, whom I never see without thinking it would be pleasant to pass a summer under their tents. We have had a great deal of thunder and lightning this summer; and one storm in

which the hail fell far larger than I had ever seen, the stones were like hazel nuts, and rejoiced the glaziers of the neighbourhood. We had a too short but very agreeable visit from our friends the Lyells[2] lately; he finds books in the running brooks, sermons in stones, and good in every thing; and she is interested in all that interests him. They are now geologising in the Hartz Mountains, or Saxon Switzerland[3]. I often I think over the delightful lectures I heard last spring[4], and only wish I could remember and understand every word of them, for they encrease my stock of happiness, for which I feel grateful to you. I trust that (if not sooner) I shall have the pleasure of seeing you and Mrs. Faraday, to whom I beg my kind regards, next winter in health & happiness.

Believe me dear Mr. Faraday | Your's Most Sincerely | Harriet Moore
The Cedars | Sunning Hill | Berks
My sister[5] I am happy to say has somewhat improved since she left London.

1. Dated on the basis that the Faradays were at Norwood at this time.
2. Charles Lyell and Mary Elizabeth Lyell, née Horner (1808–1873, Burkhardt et al. (1985–94), **4**: 652). Conchologist who married Lyell on 12 July 1832.
3. See Lyell to Moore, 3 August 1850, in Lyell (1881), **2**: 163–4.
4. Possibly Faraday's course of six lectures "Upon some points of domestic chemical philosophy" which he delivered on 27 April, 4, 11, 18, 25 May and 1 June 1850. Faraday's notes are in RI MS F4 J19.
5. Julia Moore (d.1904, age 100, GRO). Sister of Harriet Jane Moore.

Letter 2315
Faraday to John Barlow
16 August 1850
From the original in RI MS F1 N1/21

Upper Norwood | 16 Aug. 1850.
My dear Barlow
If I do not write at once I shall be too late to write at all and so shall send a few scratches. I can make nothing else just now for I am under the dentists hands & what with losing 8 teeth & filings &c &c my head & my hands too are very shaky[.]

All goes on well at Albemarle St where I am 2 or 3 times a week experimenting. Mr. Vincent is taking his holiday in Scotland[.] I have sent him your note[.]

Your description of the route you have taken revives a great many imperfect recollections[.] I could wish we had been with you now & then for the pleasure must have been very great. What does Mrs. Barlow think of it. That glorious scenery & all the fine storms & incidents you have had. I do not mean the Cholera but hail storms & lightning. Perhaps even more striking in the remembrance than at the moment. I shall listen when you come home.

As for me I am weary but hard at work. Magnetism has me altogether & I cannot promise you a thought for machinery until you come home. I think I have hold of some fine things but my head grows weary of the same trains of thought & yet I cannot & dare not throw it off for I would wish to reap my own harvest.

I intend paying Dr Latham a visit as soon as the Dentist has done with me, that he may set me up again.

Mrs. Faraday is out just now but you know what she would say to you both if she could. We live in hopes of having you back again soon well & with a great store of enjoyment in the memory for a continual resource[.]

Kindest remembrances is but poor words to either or both of you[.]
Ever | Most Truly Yours | M. Faraday
Rev. J. Barlow, MA | &c &c &c

Address: Revd. John Barlow MA | &c &c &c | Post Restante | Interlachen | Switzerland

Letter 2316
Faraday to Adolph Ferdinand Svanberg[1]
16 August 1850
From the original in Stockholmes Universitetsbibliotek

Royal Institution | 16 Aug 1850

My dear Sir

I cannot resist my desire to write at once & to thank you for the great pleasure your letter[2] (received yesterday) gave me first as coming from you whom my bad memory well retained in mind[3] and next for the delight which the facts therein described occasioned. They came with the force of truth and are very beautiful & consistent.

How wonderful it is to see the simplicity of nature when we rightly interpret her laws and how different the convictions which they produce on the mind in comparison with the uncertain conclusions which hypothesis or even theory present.

I am not sorry that you find some things unexpected or curious or a little anomalous, for they serve to shew that there are more treasures to be obtained & I see from your letter that you both know how to work for them & will work. The earnest ardent experimentalist is ever rewarded for his labour.

I have got some fancies in my head but they will require a good deal of development & elaboration before I dare venture to trust them forth. Nevertheless I am in hopes they will be fruitful in due time if health be spared me. But the head becomes giddy[.]

Ever my dear Sir | Your Obliged & faithful Servant | M. Faraday
M. Svanberg | &c &c &c

Address: Dr. A.J. Svanberg | &c &c &c | Upsala | Sweden

1. Adolph Ferdinand Svanberg (1806–1857, P2). Professor of Physics at Uppsala University.
2. Not found.
3. Faraday had met Svanberg in 1846. See Berzelius to Faraday, 23 April 1846, letter 1863, volume 3.

Letter 2317
Faraday to William Whewell
22 August 1850
From the original in TCC MS O.15.49, f.27

Upper Norwood | 22 Aug 1850

My dear Sir

I am living and working at Norwood, and so lost the great pleasure of seeing you & what would have been more of hearing you[.] One can consider many things in talking which writing is very unfit for. I received with thankfulness your kind letter[1] and have since, then in my notes & M.S., used the word *Paramagnetic*[2] which will serve my purpose well if after a little further explanation you think it is is [sic] (as I imagine) right. I conclude that a long piece of *soft iron* unable to *retain* magnetism would in a field of equal magnetic force stand in the direction of the lines of force, still the power which can make it do so must be exceedingly small in effect, for I imagine it would require an extremely delicate apparatus to shew the pointing of a bar of such iron under the Earths force which we may take as presenting a field of equal force. The diamagnetic power of bismuth or phosphorous is exceedingly small as compared to the corresponding magnetic power of iron & there is no chance that a bar of either would stand transverse to the Earths lines of force. I have found lately that such a bar does stand transverse in a field of equal force made by two walls of iron 5 inches by 3 inches $\frac{3}{4}$ of an inch apart. But then a piece of Iron[3] destroys such a field as one of equal force for it generates contingent poles in the parts of the iron walls opposed to its ends & the phosphorous as I believe produces a reverse effect equivalent to a destruction of the power there. Hence both the case of the Iron & the phosphorous fall as respects by far the greater part of their effect & perhaps the whole as to position into the law I gave originally - that Magnetic (paramagnetic) bodies pass or tend to pass from weaker to stronger places of magnetic action and diamagnetic bodies from stronger to weaker.

I have been driven to assume for a time especially in relation to the gases a sort of conducting power for magnetism[.] Mere space is Zero. One substance being made to occupy a given portion of space will cause more lines of force to pass through that space than before and another substance will cause less to pass. The former I now call Paramagnetic & the latter are the diamagnetic. The former need not of necessity assume a polarity of particles such as iron has when magnetic and the latter do not assume any such polarity either direct or reverse. I do not say More to you just now because my own thoughts are only in the act of formation but this I may say that the atmosphere has an extraordinary magnetic constitution & I hope & expect to find in it the cause of the *annual & diurnal variations*, but *keep this to yourself* until I have time to see what harvest will spring from my growing ideas.

I am My dear Sir | Most Truly Yours | M. Faraday
Rev. Dr. Whewell | &c &c &c

1. Letter 2311.
2. Faraday first used the term paramagnetic in Faraday (1851c), ERE25, 2790 This paper was dated 2 August 1850. In the manuscript, RS MS PT 40.2, the term "terromagnetic" has been replaced throughout with "paramagnetic".
3. At this point "or a piece of phosphorous" is crossed out.

Letter 2318
Faraday to Harriet Jane Moore[1]
24 August 1850[2]
From Bence Jones (1870a), 2: 256

I have kept your picture to look at for a day or two before I acknowledge your kindness in sending it[3]. It gives the idea of a tempting place; but what can you say to such persons as we are who eschew all the ordinary temptations of society? There is one thing, however, society has which we do not eschew; perhaps it is not very ordinary, though I have found a great deal of it, and that is kindness, and we both join most heartily in thanking you for it, even when we do not accept that which it offers. I must tell you how we are situated. We have taken a little house here on the hill-top, where I have a small room to myself, and have, ever since we came here, been deeply immersed in magnetic cogitations. I write and write and write, until three papers for the Royal Society are nearly completed[4], and I hope that two of them will be good if they justify my hopes, for I have to criticise them again and again before I let them loose. You shall hear of them at some of the Friday evenings[5]; at present I must not say more. After writing, I walk out in the evening, hand-in-hand with my dear wife, to enjoy the sunset; for to me who love scenery, of all that I have seen or can see there is none surpasses that of Heaven. A glorious sunset brings with it a thousand thoughts that delight me.

1. Recipient identified on the basis of the reference in Thompson, S.P. (1898), 207.
2. Dated according to Bence Jones (1870a), **2**: 256 who added that it was written from Upper Norwood.
3. Possibly a reference to the vignette in letter 2314.
4. Faraday (1851c, d, e), ERE25, 26 and 27.
5. Faraday (1851a), Friday Evening Discourse of 24 January 1851 and Faraday (1851f), Friday Evening Discourse of 11 April 1851.

Letter 2319
Brough George Maltby[1] to Faraday
24 August 1850
From the original in GL MS 30108/1/50&51

Trinity House London | 24 Augt 1850.
Sir,
    I am desired to forward to you the accompanying 4 samples of White Lead, marked A, B, D, & E, - and to acquaint you that, the Board are desirous[2] that each Sample should undergo the regular process of Analization, and that you furnish the result thereof, for the information of the Corporation[3].-
    I am, | Sir, | Your most humble Servant | B.G. Maltby pro Secty
M. Faraday Esq.

1. A clerk at Trinity House, 1810–1853. *Imperial Calendar*.
2. See Trinity House Wardens Committee, 20 August 1850, GL MS 30025/20, p.131.
3. Faraday's analyses are in GL MS 30108/1/50&51 together with a note that he replied on 3 September 1850. This was read to Trinity House Court, 3 September 1850, GL MS 30004/24, pp.288–9 and was referred to the Deputy Master (John Henry Pelly), Wardens and Light Committee. A joint meeting on 10 September 1850 (GL MS 30025/20, p.148–9) led to Trinity House placing an order for the white lead.

Letter 2320
William Whewell to Faraday
2 September 1850
From the original in RI MS F3 F187

Lancaster | Sept. 2, 1850

Dear Dr Faraday
　Since I wrote to you I have read your "Twenty Third Series"[1] as well as the letter which you sent me[2]; and the result is that I have some doubts whether the suggestions which I sent you respecting names are sound[3]. I had entertained a view which your paper is employed in refuting, that the magnetism of bismuth &c on the one hand and of iron &c on the other is of a coördinate kind, and that therefore it was desirable to designate the two by two coördinate words, such as paramagnetic and diamagnetic. But as I now understand your view is that bismuth &c have no real polarity but only a seeming polarity arising from each particle having a tendency to go from strong to weak magnetic spaces; while iron &c have a polarity, or at least hard iron is capable of a polarity which may be defined directly and finally as a tendency to arrange its polar axes parallel to the magnetic lines of the earth. This want of exact symmetry in the two kinds of magnetism, and the seeming connection of one with the *earth's* magnetism make me hesitate about the term *paramagnetic*, and reconsider whether your term *terromagnetic*[4], or the corresponding Greek compound, *geomagnetic*, might not be better. But all things considered, I am still disposed to recommend the use of the term *paramagnetic*, advising only that the term should be explained at the outset (if I am right in my view) as implying that iron &c have a (seeming) polarity which places them *parallel* to the magnetic lines of the earth. Perhaps it is still conceivable that there may be planets so constituted that diamagnetic bodies shall arrange themselves parallel to *their* lines of force; but whether or not, the reference to the earth may I think be *understood* in your sense, as safely as *expressed*.
　I am much interested in the hints you give me about the magnetic constitution of the atmosphere and its results. I am exceedingly struck by the ingenuity of the modes of experimenting explained in your 23d Series, but I think you might have given a Diagram to make the construction of the apparatus more evident. For instance, in 2643 what is the position of the brass rod and the core with reference to the wooden lever? I suppose at right angles, and in the plane of the motion; but I think you have not said so.
　I am travelling in the lake country but am always accessible at Cambridge. I was there the other day and found your card[5], as if I had missed seeing you there also[6], which I am sorry for[.]
　Believe me always | Yours most truly | W. Whewell

1. Faraday (1850), ERE23.
2. Letter 2317.
3. In letter 2311.
4. Letter 2310.
5. This suggests that Faraday had visited Trinity College, Cambridge, on his way to (or from) Old Buckenham in June 1850 (see note 1, letter 2299). Whewell was in Germany at the time. See Whewell to Forbes, 25 June 1850, Todhunter (1876), **2**: 360–1.
6. See letter 2317.

Letter 2321
Faraday to George Cramp[1]
3 September 1850
From the original in Historical Society of Pennsylvania, Etting Collection, Scientists, p.26, under Michael Faraday

R Institution | 3 Sep 1850

Sir
    I do not know of any Electro magnetic engine that will do what you require. There is much talk about one in America by Mr Page[2] I think. But I shall not believe in it until I see it[3][.]
    Your Very Obedient Servant | M. Faraday
Geo Cramp Jun Esq | &c &c &c

1. Unidentified.
2. Charles Grafton Page (1812–1868, DAB). Principal Examiner in United States Patent Office, 1841–1852 and proto electrical engineer.
3. On Page's large electric motor see Post (1976), 84–107 and Martin and Wetzler (1887), 19–21.

Letter 2322
Faraday to William Whewell
6 September 1850
From the original in TCC MS O.15.49, f.28

R Institution | 6 Septr 1850

My dear Sir
    I ought to have sent you the enclosed before. Another friend found the difficulty you refer to[1] & this wood cut was the consequence[2].
    I was only waiting to send it with something better. Many thanks for your last letter. I will use Para*magnetic* and restrain its intention as you suggest. The atmospheric magnetism works out beautifully, but my head aches with thinking of it & I am now so giddy I must lay the matter down for a few days[.]

Ever Truly Yours | M. Faraday
Revd. Dr Whewell | &c &c &c

1. In letter 2320.
2. The illustration of the apparatus in Faraday (1850), ERE23, p.188 was omitted from the offprint. See Faraday's copy in RI MS F3 F.

Letter 2323
Lambert-Adolphe-Jacques Quetelet to Faraday
9 September 1850
From the original in IEE MS SC 2

Bruxelles, le 9 Sept. 1850

Monsieur et très illustre confrère,

Je me suis a aperçu avec un regret infini que vous n'aviez pas encore reçu le diplôme d'associé de notre Académie[1]. M. le docteur Pincoffs[2], qui passe par Londres, a bien voulu se charger de réparer cette négligence de la personne qui aurait du vous faire parvenir ce diplôme depuis longtemps. J'y joins deux volumes des dernières publications de notre académie, qui vous sont destinés.

Je suis très-heureux de trouver cette occasion pour vous exprimer mes sentiments de haute estime et ma reconnaissance pour toutes les bontés que vous ne cessez de me témoigner. Vos encouragemens ont toujours été pour moi les plus précieuses recompenses que je pusse ambitionner, et vous me les avez toujours prodigués avec la plus grand bienveillance.

Vous m'avez montré un recueil très-précieux de portraits de personnes aux quelles vous portez de l'affection[3]. Il y aurait, peut-être, de la fatuité à vous demander une place auprès d'elles; si, cependant, vous n'y voyez pas trop d'outrecuidance permettez-moi de vous offrir un exemplaire d'un de mes portraits[4], dessiné par mon beau-frère[5]. Je serais heureux surtout s'il pouvait me valoir en retour le portrait d'un des hommes dont j'estime le plus le talent et le noble caractère.

Agréez je vous prie Mon cher et illustre confrère mes compliments les plus distingués et les plus affectueux. | Quetelet
Monsieur Faraday

TRANSLATION

Brussels, 9 September 1850

Sir and most illustrious colleague,

I realised with infinite regret that you had not yet received the certificate of Associate of our Académie[1]. Dr Pincoffs[2], who is passing

Plate 5. Lambert-Adolphe-Jacques Quetelet by Jean-Baptiste Madou. See letter 2323.

through London, has kindly agreed to put right this negligence on the part of the person who ought to have sent you the certificate long ago. I enclose two volumes of the most recent publications of our Académie, which are destined for you.

I am very happy to find this occasion to express my sentiments of high esteem and my gratitude for the kindness that you never cease to show me. Your encouragement is always one of the most precious rewards that I could wish for, and you have always showered it on me with the greatest kindness.

You showed me a very precious collection of portraits of the people you are fond of[3]. It would, perhaps, be fatuous to ask you for a place beside them; if, however, you do not find me too presumptuous, allow me to offer you a copy of one of my own portraits[4], drawn by my brother-in-law[5]. I would be particularly happy if it gained for me in return the portrait of one of the men whose talent and noble character I admire the most.

Please accept, I beg you, My dear and illustrious colleague, my most distinguished and most affectionate compliments | Quetelet
Mr. Faraday

1. See Quetelet to Faraday, 17 December 1847, letter 2040, volume 3.
2. Peter Pincoffs (d.1872, B2). Dutch born physician who was practising in England at this time.
3. That is RI MS F1 H and I.
4. This portrait, dated 1839, is in RI MS F1 I62. See plate 5.
5. Jean-Baptiste Madou (1796–1877, BNB). Belgian painter.

Letter 2324
Faraday to Smith[1]
10 September 1850
From the original in the Maddison Collection, Templeman Library, University of Kent at Canterbury

Brighton | 10 Sept 1850
Sir
I received your letter only this morning or I would have written sooner. I am not professional and cannot undertake the analyses which you desire[.]
I am Sir | Your Obedient Servant | M. Faraday
Mr. Smith

1. Unidentified.

Letter 2325
Edward Sabine to Faraday
11 September 1850
From the original in IEE MS SC 2

Geneva, Sept. 11h, 1850.

Dear Faraday,

Capt. Younghusband[1] has informed me of the communications which you have lately had with him; I rejoice greatly that any of your investigations should lead you to desire a knowledge of the facts which we have been engaged in obtaining, or give you a prospect of being able to throw light on their causes. I need not add how delighted w⟨e⟩ shall be able to arrange the facts in such ways as shall be either most clear to your apprehension, or illustrative of your opinions. Younghusband tells me you are pleased with the plate in the Hobarton Volume, shewing the diurnal movements of the magnetic direction & force in each month of the year in that quarter of the Globe[2]. I have a precisely similar *drawing* (not yet lithographed) of the diurnal variation in direction & force in the different months at Toronto, as nearly as maybe in the same Latd as Hobarton, but in the other hemisphere. It is in a drawer in my cottage, and I will have it copied for you, if you wish it, immediately I return, which will be early next month. It will be lithographed in the 2d Vol. of the Toronto Obs, of which the printing is now commencing[3]; but you can make any use of it, or indeed of any thing else of mine, that may suit you. I am now engaged on a plate, for which I have the materials with me, which will exhibit the actual curve made by the one extremity of a freely suspended needle in its mean diurnal motion in every month of the year at the 4 observatories of Toronto, Hobarton, St. Helena, & the Cape of Good Hope[4]. I mention this because I think it is exactly the plate that will suit you, and I will endeavour to have it ready for you towards the end of October, if you wish it. It will differ from plates such as the one you have seen for Hobarton, inasmuch as it will shew the actual movement which the end of the needle would itself describe on a [two words illegible] perpendicular to the axis of the needle. The plate you have seen exhibits the movements in *Declination* & *Inclination*. If the 2 kinds of plate belonged to a station on the *magnetic Equator*, they would be identical; but every where else they differ; the movement in declination requiring to be multiplied by the cosine of the Inclination when the representation is to be of the *space* travelled over by the needle.

We are on our way home via Paris, but if you have occasion to write, send your letter to Capt. Younghusband who will forward it.

always truly yours | Edward Sabine

Address: M: Faraday Esq. | Royal Institution | Albemarle St.

1. Charles Wright Younghusband (1821–1899, B3). Officer in Royal Artillery.
2. Sabine (1850–3), **1**: between pp. lxviii and lxix.
3. Sabine (1845–57), **2**: opposite p. xx.
4. *Ibid.*

Letter 2326
Faraday to Edward Sabine
17 September 1850
From the original in BeRO MS D/EBy F48 110

Royal Institution | 17 Septr. 1850

My dear Sabine

I am afraid you will think I am dreaming or else I am sapping the very foundations of terrestrial magnetism as it at present stands but indeed I cannot help it mine are experimental researches and the foundation is a foundation of facts[.] Very likely I may have erred in the conclusions I have raised upon it but the facts cannot alter. I received yours yesterday[1] and thank you for its kindness which I was sure of[.] I intended to have written to you before but have been *so occupied*. The paper which I have now finished and which is in the hands of the copyist that it may go in quickly to the R.S. and take date is above 100 pages[2] & is preceded by one already sent in of more than 40[3] and there is as much more matter waiting till I get my aching head a little rested. I thought it not only a pleasure but a duty to tell you, who have distinguished yourself so much in this branch of Science, first of their facts and except some general remarks which would fall out in communication with Whewell, Captn. Younghusband[4] & some others I have reserved myself for you. I think you had better say nothing about the matter until you come home that the thing may be fairly dated & known at the Royal Society and here first. When the matter is published either by reading or printing it is the property of all but before that I like to have possession of my own game.

Now for a few facts. Oxygen is highly magnetic nitrogen not at all. If one takes a given weight of oxygen and then take *seventeen times* its weight of crystallised proto sulphate of iron which is a good magnetic salt and dissolve that in as much water as will equal the bulk of the oxygen the solution & the gas then have *equal* magnetic power. Oxygen is at ordinary temperatures in the state of iron when by heat it is just upon losing its magnetic force the addition of heat rapidly *diminishes* the magnetic power of oxygen the lowering of its temperature rapidly increases it. Besides that rarefying oxygen alters its magnetic force at half an atmosphere it has, as far as I can tell yet, it has half the power[.] All these qualities it carries into the atmosphere unchanged except that they are diluted to a certain extent by the nitrogen but all that happens to oxygen happens to air[.]

The Sun in his daily course heat[s] & expands the oxygen of the air with the air and is continually changing its magnetic condition hence affections of the lines of force from the earth and as I believe of all the diurnal and annual variations and of many other natural magnetic effects which I have classed together in my paper under the title of Atmospheric magnetism. The results of induction beautifully accord with Observations but I want to know more of the dip than I can get as yet.

Many conclusions flow from these principles which are new & have not occurred to the minds of philosophers. There is one very serious one and when I think of it I am almost sorry for it but I hope it may not be so important as my fears represent it. The *magnetic force* of the Earth is not shewn by the needle: Being in a magnetic medium which can change and does change continually the needle when employed to measure force measure the joint result of the two actions which sometimes *coincide* and sometimes *oppose* each other and of which the needle gives no indication. When by temperature the conducting power of the air or oxygen for the magnetic force is changed the needle shews such change at a higher temperature indicating increased action & at a lower temperature diminished action and yet there has been no change in the amount of power during the whole time either at the Source in the Earth or in the neighbouring space - and many other shapes does this two fold action on the needle take. This for the present you will soon be at home & I shall be anxious to tell you all, even far more than is written[.] I am in hopes that you can get at the paper at the RS as our Secretary[5] & judge it and so help me to a quick publication for I do not think I could wait as I did for the publication of the last paper[5][.]

Many many thanks for your kind offers[.] There will be time enough to consider the matter when you come home. Captn Younghusband has been very kind too and has let me have the MS. data which when you come I shall perhaps ask you about as appendix to my paper[6].

St Helena as you may imagine is a very fine case & I am gradually unravelling it but the statement of principles &c made the last paper so long that I have kept it with the Cape of Good hope the night action & other points for the next paper[7][.]

With kindest remembrances to Mrs Sabine[8] I am my dear friend
    Most Truly Yours | M. Faraday
Coll Sabine RE | &c &c &c

1. Letter 2325.
2. Faraday (1851d), ERE26.
3. Faraday (1851c), ERE25.
4. Charles Wright Younghusband (1821–1899, B3). Officer in Royal Artillery.

5. Faraday (1850), ERE23. See letter 2289.
6. This appendix was published in Faraday (1851d), ERE26, pp.79–84.
7. Faraday (1851e), ERE27.
8. Elizabeth Juliana Sabine, née Leeves (1807–1879, see DNB under Edward Sabine). Scientific translator.

Letter 2327
Faraday to William Whewell
10 October 1850
From the original in TCC MS O.15.49, f.29

Folk[e]stone | 10, Octr. 1850

My dear Sir

I snatch a few moments to send you some account of the points I referred to in my last[1][.] I will not occupy your time by leading you through the successive investigations which are described in two or three papers now sent into the R.S.[2] but go at once to the chief results. Oxygen is a magnetic body in its gaseous state & rather strongly so. The kind of proof is this.

N & S are the poles of an electro magnet & *cc* are piece of soft iron connecting them but turned away in the middle[.] The consequence is that I have a strong field of action all round the attenuated part equal in force & condition[.] If I take two equal diamagnetic bodies

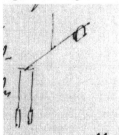

as for instance 2 glass cylinders & suspend these about $1\frac{1}{2}$ or 2 inches apart on a cross bar fixed to the end of a horizontal line supported by a fibre of cocoon silk & then adjust this so that the cylinders hang on opposite sides of the keeper described & then put on the magnetic power immediately the whole arrangement takes up a position in which the glass cylinders are equidistant from the core. If for one of the glass cylinders I substitute a cylinder of bismuth or phosphorous or a tube of water the two stand at different distances & by putting on a force of torsion (by a wire not a cocoon thread) I can measure the relative diamagnetic force of the bodies. I have a *differential balance* with the capability of measuring very accurately & yet of removing or at least balancing & estimating [the] interfering circumstance.

Now I have the power of examining *gases* at *different pressure* & *different temperatures* a power which I have long been searching after. First I prepare flint glass bulbs

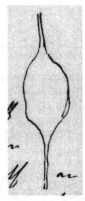

& obtain them so as to be nearly equal in diamagnetic force[.] Then I fill them with gases & seal them up hermetically & I do this when the gas has been by the air pump reduced to half an atmosphere or a third or as nearly as may be to a vacuum, for I reason that if a body (gaseous) were diamagnetic it ought to become less diamagnetic by rarefaction or if magnetic less magnetic by rarefaction[.]

When two bulbs with Oxygen & Nitrogen were put up the oxygen went close up to the angle & the Nitrogen was carried out. With oxygen at 1 atmosphere & $\frac{1}{2}$ atmosphere - the 1 atmosphere went up to the core the other out. A bulb with $\frac{1}{2}$ atmosphere Ox against another of 1 atmosphere went out but against another of $\frac{1}{3}$ of an atmosphere it went inwards. In fine the denser the oxygen the more magnetic it was & when I put a bubble containing $\frac{1}{3}$ of a cubic inch of oxygen against an equal bubble containing only an oxygen vacuum, it went inwards with such power as to require force equal to 0.1 of a grain to carry it back to equidistance. With Nitrogen no such thing happens expansion causes no difference, and three bulbs being Nitrogen 1 atmosphere - Nitrogen Vacuum & Oxygen Vacuum. When compared against each other are alike. Nitrogen is neither Magnetic nor diamagnetic but stands at Zero. Oxygen is magnetic & in proportion to its quantity: Another experiment will give you a notion of the degree[.] I took crystals of proto sulphate of iron which is a pretty magnetic salt &

dissolved it in water & diluted it until it was bulk for bulk equal to oxygen on the differential balance it then contained sulphate of iron equal to $17\frac{1}{2}$ times the weight of the oxygen which was opposed to it[.]

Now for another point in the Magnetism of Oxygen[.] Like iron it loses its magnetism or magnetic condition by heat & is at common temperatures in the condition of iron about to alter. You have a paper of mine about the diamagnetic condition of flame air & gases[3] in which I shew that air heated by a platina helix is rendered more diamagnetic &c &c. Well all the effect there is due to the oxygen by very careful experiments I find that Oxygen rapidly loses its magnetic character by heat whilst Nitrogen Carbonic acid, &c are unchanged.

All these properties of oxygen it keeps in the atmosphere & I formerly shewed that by cooling the air it was made either more magnetic or less diamagnetic as the case might be. Now I know where the Zero is & that it is the oxygen of the air that was altering in its magnetic relation[.] Again I must refer you to the papers which are long but I trust not too long, for the steps of the reasoning by which I found upon this property of oxygen a branch of physics which I have called atmospheric Magnetism. I think I see the cause in the heating power of the Sun in the atmosphere for all the *periodic solar variations* of the needle both annual & diurnal & as far as I have gone the results of observations tally with the theory - besides these variations there are many others due to atmospheric changes & I am in hopes we shall be able to refer so many to this cause as to give us a far clearer view of the power as it exists in the Earth as its origin than we have hitherto had & so help us to discriminate that which belongs essentially to its source[.]

The postman is at the door & I must stop[.]

Ever Your Obliged | M. Faraday

Dr Whewell | &c &c &c

1. Letter 2322.
2. Faraday (1851c, d, e), ERE 25, 26 and 27.
3. Faraday (1847b).

Letter 2328
Julius Plücker to Faraday
10 October 1850
From the original in IEE MS SC 2

Dear Sir!

Doctor Brandis[1] of our University, who particularly applies himself to organic Chemistry is going to visit London, invited by the Prussian

Ambassador, Chevalier Bunsen[2]. He wished for this letter, which would give to him the opportunity to see you.

Since I received your last kind letter[3] I continued my experimental researches about the magnetic axes of cristals. I hope allways to reach finally the true & general law of nature, but my researches, presenting new difficulties are not yet finished. I sent a first paper, six weeks ago, to Poggendorff[4], and will be able in a few days, I hope, to present to you a copy of it. Then I'll give to you a short account of the results it contains.

I had preferred by far to publish all my results at once, but I was obliged to go on by the papers of Knoblauch[5] & Tyndall[6] and by the fact, that two of my papers of a former date and written in french, have by the fault of the Editor, not yet been printed[7].

With my best wishes for your health | Yours | very truly | Plücker Bonn 10th of October | 1850.

Address: Professor Faraday | &c &c | Royal Institution | London

1. Dietrich Brandis (1824–1907, DNB2). Privatdocent at the University of Bonn.
2. Christian Karl Josias Bunsen (1791–1860, NDB). Prussian ambassador in London, 1841–1854.
3. Letter 2250.
4. Plücker and Beer (1850–1).
5. Karl Hermann Knoblauch (1820–1895, ADB). Professor of Physics at Marburg, 1849–1853.
6. Knoblauch and Tyndall (1850).
7. See Plücker (1852a), 1 where he says that this paper, in two parts and in French, was sent to the Haarlem Society of Sciences, but was not published.

Letter 2329
Faraday to John Edward Gray[1]
17 October 1850
From the original in APS MS B G784

R Institution | 17 Octr. 1850

My dear Sir

Could you tell me in three words whether Perry's[2] Conchology 1811[3] is a good & standard work & whether it is cheap at $1.15.0$[.] It is offered to us at that price ought I to authorize that expenditure for it?[4]

Ever Yours | M. Faraday

Gray Esq | &c &c &c

1. John Edward Gray (1800–1875, DNB). Keeper of the Zoological Department of the British Museum, 1840–1874.
2. George Perry. Otherwise unidentified.
3. Perry (1811).
4. There is a copy of this in the Library of the Royal Institution.

Letter 2330
Henry Wentworth Acland to Faraday
17 October 1850
From the original in IEE MS SC 2
My dear Sir,

I scarce know whether, (if you remember my existence here,) that remembrance will justify me in troubling you; the more as I expect you will consider my Question an idle one.

My chief avocation is that of a humble Physician but I am dragged out of the routine of the medical life (if indeed the observation of Human disease allows routine at all) by my Physiological lectures.

And these two twin avocations often lead me of course to reflect on the causes & influences which produce the varied effects seen in our frames. In short one's mind cannot be kept off the consideration of how very little we really know of the relation between the greater Cosmical arrangements, and organic life. I really cannot, (for instance) speak precisely of the amount of humidity which is good for my Consumptive Patient. The Dewpoint, & the Temperature varies perpetually, so does he. He alters in weight & chemical composition before & after every meal, and at every hour - so does the atmosphere in its water, its Electrical state &c.

Now, I have I fear neither knowledge, skill, or leisure to clear up any of what we do not know, or to understand & master half of what is known, and mainly thro' your means; but I have a great fancy for seeing in my minutes of leisure (if I can) something of the workings of the great Atmospheric sea. I live in the middle of the Town - and I should like to know what goes on where I & other people live. I have a little garden - as big as a room with every aspect. I can therefore have and have Barometer, Thermometer, and Hygrometer. But I am utterly in the dark about the Magnetic, & Electrical states. This is what I wanted to ask you very kindly to tell me. Is there means of watching any thing of these, without costly instruments, and a special observatory? and much time & labour?

I must own that I am not 'au courant' with all that is known of the relation of the great Imponderable agents and our Nervous system, & therefore all our faculties of body & mind - but most certainly more has to be understood about this, and one's mind reels in the contemplation of even what we perceive of the at once greatness, & simplicity of the connexion which probably exists between our mind & the material universe thro' the subtle powers which surround the Earth.

I fear you will think me very impertinent, or very stupid - or both. But I must plead my distance from London as excuse for my backwardness.

If you are kind enough to answer my question will you tell me what instruments I could easily manage for observing the Magnetic, & Electric variations - & where I can get them - or whether this is so delicate a matter that it is useless to meddle with it, except by reading results. And is there any way that one can note or measure one's own condition in these respects? I suspect that sleep has something to do with the Electromagnetic condition of the body: only, I beseech you not to think me a Mesmerist in 'disguise', but only,

Your faithful servant | Hy.W. Acland
Oxford, Oct: 17, 1850

Letter 2331
Faraday to Henry Wentworth Acland
18 October 1850
From the original in Bod MS Acland d.63, f.144–5

Royal Institution | 18 Octr. 1850

My dear Sir

I had your letter[1] this morning and am able to say very little that will be satisfactory to you in respect of observations of the Electric & Magnetic condition of the atmosphere. You know the difference between the theorist & the observer & I have been very little of the observer in these matters. This I know that careless or slight observations are of no value & that to make good ones the most perfect apparatus & some practice in observing is needful[.] Mr. Ronalds[2] of the Kew Observatory would be able & I have no doubt happy to tell you all that can be communicated from one to another & could assist you much in that respect[.]

Your letter in some degree jumps with my present pursuits for I have been hard at work on a branch of Science which I have called atmospheric magnetism[.] Only to think of oxygen as a highly magnetic body carrying all its properties in that respect into the atmosphere[.] At common temperatures it is as magnetic as a solution containing 17 times its weight of proto sulphate of iron i.e equal bulks of oxygen & of such solutions have equal magnetic powers. Besides all that there is this beautiful fact[.] As iron at a certain high temperature & Nickel at the heat of boiling oil lose their magnetic forces in part not entirely, so oxygen by heat loses of its magnetic power & natural temperatures occur at that point or within that range which affects the oxygen & hence as I think I shew the greater number of variation of terrestrial magnetism, the annual & diurnal variations &c & a multitude of accidental variations which the photographic registrations shew[.] My papers[3] are already at the RS & in the hands of Mr. Christie & Coll Sabine & you will very shortly hear of them[.]

Ever My dear Sir | Most Truly Yours | M. Faraday
Dr Ackland [sic] | &c &c

1. Letter 2332.
2. Francis Ronalds (1788–1873, DNB). Superintendent of the Meteorological Observatory, Kew, 1843–1852.
3. That is Faraday (1851c, d), ERE25 and 26.

Letter 2332
Henry Wentworth Acland to Faraday
22 October 1850
From the original in IEE MS SC 2

Oxford Oct 22, 1850

My dear Professor Faraday,

I am greatly obliged to you for your kind answer to my vague Question[1]. I feared you would say that nothing is to be done in this matter without great labour & costly instruments. I have learned accidentally what I did not know that there is a large & valuable work of M. Dubois Raymont (?) on the Electric currents of Animals[2]. I must try to become acquainted with this.

Your account of the discovery of the magnetic properties of oxygen is deeply interesting, and I will not say throws light upon the electrical properties of animal life, but adds a marvellous & mysterious character to them. The oxygen which we supposed to play an ordinary chemical part, in the changes of textures, and primarily thro' the blood, is then also the great feeder of the nervous system, mediately thro the blood. At least so it seems to shadow itself forth if it itself bears high magnetic properties with it into the system. However I am out of my depth. But I am very grateful to you for your kind forbearance.

I am My dear Professor Faraday | your faithful & obliged servant | Henry W. Acland

1. Letter 2331.
2. Bois-Reymond (1848–9).

Letter 2333
John Tyndall to Faraday
24 October 1850
From a typescript in RI MS T TS, volume 12, pp.4000–1

Marburg | Hesse Cassel | Oct. 24th. 1850

Dear Sir.

A short time after I had the pleasure of seeing you last June I took the liberty of sending you a few specimens of calcareous spar which from their appearance I judged to be magnetic[1]. I had at the time no means of proving whether they were so or not. Since my arrival in Marburg I have tested specimens of the same spar - the chemical analysis proves the absence of iron and the magnet shews them to be diamagnetic. The wish to furnish you with the means of observing the complementary action of the magnetic and diamagnetic specimens of the crystal being thus far disappointed I have great pleasure in now sending you a sample of the proper kind - a small rhomboid of magnetic spar. You will find that the optic axis of the rhomboid will set from pole to pole, in contra distinction to the diamagnetic crystals of the same form which as you are aware set their axes equatorial.

These effects appear to be capable of the fullest explanation by reference to a principle which you were the first to hint at, that is to say "the action of contiguous particles". Wheat flour for instance is pretty strongly diamagnetic. If we take a little ball of dough made from the flour and squeeze it flat the plate thus formed will when suspended vertically in the magnetic field set its *shortest* dimension equatorial, thus behaving like a magnetic body. Now this is evidently due to the peculiar arrangement of the diamagnetic particles, and if we imagine a similarly formed magnetic mass suspended between the poles it is reasonable to suppose that the arrangement which in the former case caused the *repulsion* of the line of greatest compression would now cause its *attraction*. This conjecture is verified, for if instead of flour we use in the composition of our dough a precipitate of oxide of iron, the squeezed plate formed from the latter will sets its *shortest* dimension *axial*. This action appears to be strictly analogous to that exhibited by the spar. In the diamagnetic specimen the shortest dimension stands equatorial; in the magnetic specimen the same dimension stands axial. The molecules of both crystals are similarly arranged, but in the one case we have to deal with magnetic molecules and the other with diamagnetic.

I remain dear Sir | Most truly and respectfully yours | John Tyndall.
Dr. Faraday. | etc. etc.

Endorsed: Transmitted through Francis Don't crush!

1. See letter 2308 and note 3.

Letter 2334
Faraday to Charles Richard Weld
28 October 1850
From the original in RI MS F1 C38

R Institution | 28 Octr. 1850

My dear Sir

Can you let the bearer have the Parts of the Phil Trans that I signed for some time ago. My last is 1849, part 1.

Can you also let me have my paper series XXVI[1] *the last one* both for the purpose of making an abstract[2] of it & also to make a few alterations in it which Coll. Sabine approves of & recommends.

Ever Truly Yours | M. Faraday
R.C. [sic] Weld Esq | &c &c &c

1. That is the manuscript of Faraday (1851d), ERE26, RS MS PT 40.3, which shows that a number of alterations were made.
2. Published in *Proc.Roy.Soc.*, 1850, 5: 998–1000.

Letter 2335
Sarah Faraday to William Buchanan
31 October 1850[1]
From the original in the possession of Joan Ferguson

3 o clock AM Octr 31st RI | London

My very dear Friend & Brother

I rise from a restless bed to write to you feeling sure of your kind sympathy in my great anxiety & affliction[.]

My beloved husbands mind has been much disturbed in the view the Church takes of not receiving an excommunicant more than once - & he has had some conversation with our Elders which so far does not seem very satisfactory. Now I have heard him say sometimes your clear view of the scriptures struck him. May I ask you to write me by return of post if possible, your view of this subject - from the scriptures alone - he does not know of my writing, & I hope I am not wrong in so doing[.] I do not write to you as an Elder but as to a dear friend & brother with whom we have had many communings & surely if his mind could be set at rest with out sowing division in the church or casting him out it will be a happy thing[.]

Most affectionately your | sister in the truth (I know) | S. Faraday

My husband has been anxious not to involve me in these considerations & I have not seen the Elders[.] I hope if I am wrong in this step my friends will forgive & feel for my great anxiety[.]

1. Dated on the basis that this letter is the first of a sequence of four letters dealing with Faraday's possible second (and thus final) exclusion from the Sandemanian Church. Of these four letters only letter 2340 is dated precisely and the others have been dated accordingly. On this episode see Cantor (1991a), 275–7 and 345.

Letter 2336
Faraday to William Buchanan
3 November 1850[1]
From the original in the possession of Joan Ferguson

R Institution | Sabbath Evening

Most dear & beloved friend

I am in deep distress and I write to you for my heart is full, not with any presumptuous hope, but that my tears may overflow for I know that you love me both in body & spirit and I may perhaps never write to you - more, bear with my anguish & do not refuse to sympathise a little with me by receiving this patiently though I be utterly unworthy[.] I may well fear that a deceived heart hath turned me aside[2] for where my only comfort ought to be there is my sore grief & trouble[.] I have read your words to my dear wife[3] to whom I am a snare, I hope the father of mercies will have her in his keeping, through Christ Jesus our Lord. My great distress is, that the Apostle after say that we have the Spirit of Christ, then tells them, to[4] deliver such an one to satan for the destruction of the flesh that the Spirit may be Saved in the day of the Lord Jesus[5] doing this in the name of our Lord Jesus Christ when gathered together with his spirit by the power of our Lord Jesus Christ and then says, sufficient to such a man is this punishment inflicted of many so that contrawise ye ought rather to forgive him & comfort him lest perhaps such a one should be swallowed up with overmuch sorrow and beseeches them to confirm their love toward him &c, least Satan should get an advantage of us for we are not ignorant of his devices[6]. Then my mind lingers here that the command or rather instruction to separate & to restore may be alike general; that as there is no case of a second restoration so there is not of a second separation, yet the church is surely right in separating whenever the signs which our Lord or the Apostles point out occur and therefore my mind is not fully persuaded from the scriptures that we should limit the continuation of the instruction i.e. the restoration. But well may I fear for myself and hear the precept lean not unto thine own understanding[7]. The Church is the body of Christ the pillar & ground of the truth[8] and the Lord tells the brethren in Matt XVIII to hear the Church[9][.] I have been refusing to hear & have in my heart & to the Elders been lightly esteeming the Church his body speaking of it in all presumption & pride & casting the fear of it away as if God was not in the midst of her & he who walketh amidst the Seven Golden candlesticks[10]. I fear to say what my thought[s] are least I be, as men, in deceit but I hope I could with free love to the brethren subject my thoughts to its voice as the voice of him who speaks by his body. Dear friend forgive me. Your love has drawn me out and I am very thankful for the tender words in your letter which has given me this courage. May the father of mercies put it into my heart next Wednesday[11] to speak according to his mercies which are in Christ Jesus without limitation, he can make reprobates willing in the day of his power[.]

Your afflicted | M. Faraday

1. Dated on the same basis as letter 2335.
2. Isaiah 44: 20.
3. That is Buchanan's reply to letter 2335 which has not been found.
4. The word "separate" is crossed out at this point.
5. 1 Corinthians 5: 5–6.
6. 2 Corinthians 2: 6–11.
7. Proverbs 3: 5.
8. 1 Timothy 3: 15.
9. Matthew 18: 17.
10. Revelation 1: 12.
11. That is 6 November 1850.

Letter 2337
Sarah Faraday to William Buchanan
3 November 1850[1]
From the original in the possession of Joan Ferguson

Thanks many thanks my very dear friend for your letter & its enclosure. My mind is greatly relieved by it & the voice of Church to day & I have hopes that my dear husband will be allowed to see that we have been reasoning beyond what the Scriptures allow, he has been encouraged by your kind expressions to pour out his heart to you but after he had done so he thought he should destroy it, however I venture to enclose it[2] - he has not been heard in the Church yet some & I among them requested patience[.]

We are to meet again on Wednesday evening[3][.]

My fear now is that his mind is quite over taxed & he seems almost as if reason would fail, this subject has pressed with such intense weight upon a brain already worn with much study & he again & again says "I may not be a hypocrite" he seems to me as if he could hardly take in any arguments but I hope he will get sleep & his mind be restored[.]

I am well my dear friend & brother | ever most affectionately yours | S. Faraday
Royal Institution | Novr 3rd

1. Dated on the same basis as letters 2335 and 2336.
2. Letter 2336.
3. That is 6 November 1850.

Letter 2338
Faraday to John Britton[1]
5 November 1850
From the original in National Portrait Gallery, John Britton Collection

R Institution | 5 Novr. 1850

My dear Sir

The point of purchasing books is one the Library Committee decide upon & your application is or will be brought under their consideration they are obliged to be careful of their funds for the deficiencies in the Library are very great. I have *not* presented your works for as I could not decide upon the other part of the letter I thought that would not be right to you but they will go before the Committee as illustrations[.]

Ever Truly Yours | M. Faraday
John Britton Esq | &c &c &c

There is no address to your letter so that this note has been unavoidably delayed. | Yours | MF

1. John Britton (1771–1857, DNB). Antiquary.

Letter 2339
Paul Frederick Henry Baddeley to Faraday
5 November 1850
From the original in IEE MS SC 2

Lahore | 5, Nov 1850

Dear Sir

I have had the honor of receiving your letter of 23 July last[1]; and it gives me great satisfaction to learn, that my observations on "*Dust Storms*," met with your approval.

If permitted to remain at Lahore, I hope to be enabled to continue my observations on the same subject, during next hot season; at which period of the year, Dust Storms are of frequent occurrence. My attention will be chiefly directed to discover if possible, some rule for determining their course, should they be under any fixed law.

Mr. Piddington[2] of Calcutta, who has for many years past, been engaged in collecting and investigating facts which bear on the law of Storms, and has I understand, written valuable memoirs on this important subject - expressed himself much gratified with the published account of D.S's he had just perused in the Philos: Magazine for August - particularly so, with his own ideas as to the probable Electric origin of these Phenomena, seem likely to be confirmed[3].

He says - "In 1848, I published in my Sailors Horn Book[4] the Hypothesis that the Cyclone, is an Electric Meteor composed of one or many close and nearly horizontal, but yet slightly spiral streams of Electric fluid, descending thus from the higher regions, and in its (or their) descent, giving rise to currents in all the air it necessarily passes through, but not carrying that same air along with it"[.]

I have not had the pleasure of seeing this work - but in April & May 1847 while at Lahore, I discovered the Electl nature of Dust Storms or whirlwinds, and reported the same in my Annual return for that year - and again more fully, in 1848/49 - and from the first arrived at almost precisely the same conclusions as Mr. P. respecting the Elecl. Nature of the Cyclone.

Concluding that all like phenomena whether by sea or land, were to be ascribed to one and the same cause.

Should I in the course of inquiries meet with any thing worth of your notice, I will communicate it to you without delay, since you have kindly given me encouragement to do so - and need scarcely add, that any suggestions you may be pleased to offer, to guide me in the proper method of investigating these Phenomena, will be most acceptable.

I have pleasure to enclose 2 rough sketches of approaching Dust Storms - that occurred at Lahore this year[5].

I am | my dear Sir | Yours Faithfully | P. Baddeley | Surgeon Asst
M. Faraday Esq, F.R.S. &c &c | Royal Institution | London.

Address: M. Faraday Esq, F.R.S. | Royal Institution | London.

1. Not found, but in reply to letter 2281.
2. Henry Piddington (1797–1858, DNB). Meteorologist and Curator of the Museum of Economic Geology at Calcutta, 1830–1858.

3. Endorsed by Faraday here "See Phil Mag. 1850 xxxvii, 155". That is Baddeley (1850a).
4. Piddington (1848).
5. Enclosed with this letter are eleven sides of description and sketches of dust storms. See Baddeley (1852).

Letter 2340
Faraday to William Buchanan
6 November 1850
From the original in the possession of Joan Ferguson

Royal Institution | 6, Novr. 1850

My most dear & kind friend & Brother

God has wrought with me when I was against him & broken down my pride & false reasoning and has this evening[1] shewn me what love there is in the Church to an erring brother for Christ sake I believe all my difficulties are taken away. I hope so but fear to say much for though I rejoice it is with great trembling remembering how ready I was to give up & to cast his fear behind me. It is of the Lord's mercies that we are not consumed but how great is the occasion I have to apply that truth to myself. I would venture to thank you my dear Brother for your kindness to me unworthy in this trouble but that I feel unworthy to do so knowing that it was done for his sake whose body the Church I was in the greatest danger of setting at naught but he is able to save unto the uttermost all that come unto God by him and there is hope in your most unworthy but still affectionate

M. Faraday

My dear Wife sends her earnest love to you. She condemns herself very much but it is I who have been a snare to her. Only that almighty power could deliver us. | Yours MF

1. That is at the meeting of the London Sandemanian Church. See letters 2335, 2336 and 2337.

Letter 2341
Faraday to William Robert Grove
16 November 1850
From the original in RI MS G F27

R Institution | 16 Novr. 1850

My dear Grove

Next Tuesday[1] at $\frac{1}{2}$ p 2 oclk I shall be shewing some of the fundamental experiments[2] to Christie & others. I think I promised to let you know[.]
Ever Yours | M. Faraday
My kindest remembrances to Mrs Grove[3].

1. That is 19 November 1850.
2. Presumably these experiments related to Faraday's work on atmospheric magnetism.
3. Emma Maria Grove, née Powles (d.1879, age 68, GRO). Married Grove in 1837 see DNB under his entry.

Letter 2342
Charles Grey to Faraday
17 November 1850
From the original in the possession of Dennis Embleton
Colonel Grey presents his compliments to Mr. Faraday, & is commanded by The Prince Albert to say that if quite convenient to Mr. Faraday to come to Windsor Castle on Wednesday[1] about $\frac{1}{2}$ past 4, it will give His Royal Highness much pleasure to have the opportunity of having some conversation with him[.]
Windsor Castle | Novr. 17, 1850

1. That is 20 November 1850. No evidence has been found to indicate whether or not this meeting took place.

Letter 2343
Faraday to Christian Friedrich Schoenbein
19 November 1850
From the original in UB MS NS 391
Royal Institution | 19, Novr. 1850
My dear Schoenbein
I wish I could talk with you instead of being obliged to use pen & paper[.] I have fifty matters to speak about but either they are too trifling for writing or too important for what can one discuss or say in a letter[.] Where is the question & answer & replication that brings out clear notions in a few minutes whilst letters only make them more obscure one cannot speak freely ones notions & yet guard them merely as notions:- but I am fast losing my time & yours too. I received your complimentary kindness & like it the better because I know it to be as real as complimentary[.]

Thanks to you my dear friend for all your feelings of good will towards me. The bleachings by light & air are very excellent[.] I see a report of part of your paper in the account of the Swiss association but not of the latter part[1]. However a friend has your paper in hand & I hope to have the part about atmospheric electricity soon sent to me. I should be very *glad indeed* to have from any one and above all from you a satisfactory suggestion on that point. I know of none as yet.

By the bye I have been working with the oxygen of the air also. You remember that three years ago I distinguished it as a Magnetic gas in my paper on the diamagnetism of flame and gases[2] founded on Bancalari's[3] experiment[4]. Now I find in it the cause of all the annual & diurnal and many of the irregular variations of the terrestrial magnetism. The observations made at Hobarton Toronto Greenwich St Petersburg Washington St Helena the Cape of Good Hope & Singapore all appear to me to accord with & support my hypothesis. I will not pretend to give you an account of it here for it would require some detail & I really am weary of the subject[.] I have sent in three long papers to the Royal Society[5] & you shall have copies of them in due time & reports probably much sooner in Taylors Magazine[6][.]

I forwarded your packets immediately upon the receipt of them[.]

But now about Ozone. I was in hopes you would let me have a list of points with reference to where I should find the accounts in either English or French Journals and also a list of about 20 experiments fit for an audience of 500 or 600 persons - telling me what sized bottles to make ozone by phosphorus in - the time & necessary caution &c &c &c. My bad memory would make it a terrible & almost impossible task to search from the beginning & read up, whereas you who keep all you read, or discover with the utmost facility could easily jot me down the real points. If you refer to any such notes in your last letter when you ask me whether I have received a memoir on Ozone & *some other things* then I have not received any such notes & I cannot indeed I *cannot* remember about the memoir[7].

I was expecting some such notes & I still think you mean to send me them and though I may perhaps not give Ozone as an Evening *before Easter* still do not delay to let me have them because I am slow - & losing much that I read of, have to imbibe a matter two or three times over & if I do *ozone*[8] I should like to do it well.

My dear wife wishes to be remembered to you & I wish most earnestly to be brought to Madame Schoenbein's mind. Though vaguely I cling to the remembrance of an hour or two out of Bâle at your house[9] & though I cannot recall the circumstances clearly to my mind I still endeavour again & again to realise the idea[.]

Ever My dear Schoenbein | Most Truly Yours | M. Faraday

Address: Dr. Schoenbein | &c &c &c | University | Basle | on the Rhine

1. Schoenbein (1850b).
2. Faraday (1847b).
3. Michele Alberto Bancalari (1805–1864, DBI). Professor of Experimental Physics at the University of Genoa, 1846–1863.
4. Described in Francesco Zantedeschi, "Dei movimenti che presenta la fiamma sottoposta all'influenza elettro-magnetica", *Gaz.Piemontese*, 12 October 1847, no 242, [pp.1–2]. On this work see Boato and Moro (1994).
5. Faraday (1851c, d, e), ERE 25, 26 and 27.
6. Reports of the reading of these papers were published in *Phil.Mag.*, 1851, 1: 69–75.
7. This refers to a letter from Schoenbein to Faraday of 9 July 1850 which Faraday had not yet received and which has not been found. See letter 2353.
8. Faraday (1851g), Friday Evening Discourse of 13 June 1851.
9. In 1841. See Schoenbein to Faraday, 27 September 1841, letter 1364, volume 3.

Letter 2344
Faraday to John Tyndall
19 November 1850
From a typescript in RI MS T TS, volume 12, p.4127

Royal Institution, | 19 Nov. 1850.
Dear Sir,
I do not know whether this letter will find you at Marburg, but though at the risk of missing you I cannot refrain from thanking you for your kindness in sending me the rhomboid of calcareous spar[1]. I am not at present able to pursue that subject, for I am deeply engaged in terrestrial magnetism, but I hope some day to take up the point respecting the magnetic condition of associated particles. In the mean time I rejoice at every addition to the facts and to the reasoning connected with the subject. It is wonderful how much good results from different persons working at the same matter; each one gives views and ideas new to the rest. Where science is a republic, there it gains; and though I am no republican in other matters, I am in that.

With many thanks for your kindness, | I am, Sir, | Your very obliged servant, | M. Faraday
John Tyndall Esq, | &c. &c. &c.

Address: J.T. | Marburg | Hesse Cassel

1. Sent with letter 2333.

Letter 2345
Pietro Odescalchi[1] and Paolo Volpicelli to Faraday
20 November 1850
From the original in RS MS 241, f.118

Accademia Pontificia | DE' NUOVI LINCEI | Roma, 20 Novembre 50.

Num.[blank in MS] di Protocollo | cui si risponde | Oggetto | Nomina di socio corrispondente estero

Chiarissimo signore
    L'Accademia nella tornata del 17 novembre andante nominò Va Sa Chiarissima uno di suoi corrispondenti esteri.
    Perciò col presente riceverà Ella il relativo diploma, unito ad una copia dello statuto che il sommo Pontefice Pio IX[2] assegnò all'accademia quando volle ridurla di governativa istituzione.
    Voglia pertanto Ella gradire questa nomina, dovuta unicamente al suo merito scientifico, e favorire l'accademia nostra coi suoi lumi, e con qualche suo scritto, che riguardato sempre qual dono preziosissimo, sarà tosto pubblicato negli atti della medesima.
    Fra tanto rallegrandomi sinceramente con Lei ho l'onore sommo di assegniarla, e di profferirmi coi sentimenti della maggiore stima
Di Va Sa Chiarissima
il Presidente | D. Pietro Odescalchi
il Segretario | Paolo Volpicelli
Sig. *Faraday* membro della Società Reale di | Londra

TRANSLATION
    Accademia Pontificia | DE' NUOVI LINCEI | Rome, 20 November 50 Reference number | Please reply to | Concerning: | Nomination of foreign correspondent
Dear Sir,
    At its meeting on 17 November last, the Accademia nominated You as one of its foreign correspondents.
    With this letter, therefore, you will receive the relevant certificate, along with a copy of the Statute that the Sovereign Pontiff Pius IX[2] assigned to the Accademia when he wished to change it from a government institution.
    Please accept this nomination, due uniquely to your scientific merit, and favour our Accademia with your learning and with some of your writing, which will always be regarded as a most precious gift, and will be published at the earliest opportunity in the proceedings of the same.

In the meanwhile, in sincerely congratulating you, I have the highest honour of awarding you and of pronouncing myself, with the sentiments of the highest esteem, of your dear self
the President | D. Pietro Odescalchi
the Secretary | Paolo Volpicelli
Sig. *Faraday* membro della Società Reale di | Londra

1. Pietro Odescalchi (1789–1856, AG). President of the Accademia Pontificia de'nuovi Lincei, 1850–1856.
2. Pius IX (Giovanni Maria Mastai-Ferretti, 1792–1878, NBU). Pope, 1846–1878.

Letter 2346
Faraday to Julius Plücker
23 November 1850
From the original in NRCC ISTI

Royal Institution | 23 Novr. 1850

My dear friend

I feel as if I ought to have written to you before but I have been so engaged in researches about Terrestrial Magnetism that I have lost my remembrances of other things & am only slowly coming back to them. I had your kind letter[1] by Dr. Brandis[2] but saw very little of him far less than I desired and I hope you will say so to him. I believe I was in the country the chief part of his time in London but my chief difficulty in the way of intercourse with foreigners & even with society in general is my failing memory. I often intend to do things & then entirely forget their performance until too late.

You still work as I know & cannot work without making discoveries. One of my sorrows is that they are to me concealed as it were in the German language[.] Still by degrees I get hold of the matter. I hope one day to return to the subject of magnetic & diamagnetic bodies and their *inchangeability* or their *convertibility*. I cannot conceive the latter in my mind & think that if it be so there must be some far higher point of philosophy hanging thereby[.]

Ever My dear Sir | Yours Very faithfully | M. Faraday
Dr. Plucker | &c &c &c

Address: Professor Plücker | &c &c &c | University | Bonn

1. Letter 2328.
2. Dietrich Brandis (1824–1907, DNB2). Privatdocent at the University of Bonn.

Letter 2347
Faraday to Lambert-Adolphe-Jacques Quetelet
23 November 1850
From the original in BRAI ARB Archives No 17986 / 989

Royal Institution | 23 November

My dear Sir

I ought long ago to have returned you my heartiest thanks for your very great kindness in sending the Portrait[1] I so much desired for my book. It forms a great addition to the pleasure I take in looking into the volume[2]. The only excuse I have is that I have been deeply occupied and I hope that the subject of my thoughts will be acceptable to you. I am vain enough to think that I have found the true physical cause for the periodical & many of the irregular variations of the magnetic needle and perhaps even in part for the magnetic storms. You remember that three years ago I made known the magnetic characters of oxygen in a letter in the Philosophical magazine devoted to the diamagnetic condition of flame & gases[3] and spoke generally of its effect in the atmosphere. Since then I have continually thought & worked on the subject & of late have devised experimental means of ascertaining the effects of rarefaction and of temperature separately in relation to the different gases. I find that all the effects of these two modes of change are exerted on the oxygen & none on the nitrogen that if oxygen is rarefied by the air pump it loses in magnetic power in proportion, that if it is heated it loses in proportion, but that in regard to the nitrogen neither rarefaction nor change of temperature produces any effect. Then by a chain of reasoning which is given in the three papers that I have sent in to the Royal Society[4] supported by facts drawn from other bodies than oxygen & nitrogen I deduce the effect which the daily changes of temperature ought to produce upon the direction of the lines of force of the earth & as far as I have been able to compare the conclusions with the results obtained at Hobarton, Toronto, Washington, Lake Alhabasen Fort Simpson, Greenwich, St. Petersburgh, Cape of Good Hope, St Helena & Singapore the one accords with the other. You will hear more about them soon.

You desire me to send you a copy of the last portrait that was taken of myself and I shall do so on the first occasion that I can find conveyance perhaps by the Royal Society when the papers are printed. Believe it to represent one who has the highest feelings for your character as a Gentleman a Philosopher & a kind friend[.]

Ever My dear Sir | Most Truly Yours | M. Faraday
à monsieur | Monsieur Quetelet | &c &c &c

Address: A Monsieur | Monsieur Quetelet | &c &c &c &c | Observatory | Bruxelles

Postmark: 1850

1. See letter 2323.
2. That is RI MS F1 I.
3. Faraday (1847b).
4. Faraday (1851c, d, e), ERE25, 26 and 27.

Letter 2348
Christian Friedrich Schoenbein to Faraday
25 November 1850
From the original in UB MS NS 392
My dear Faraday
  Will you be kind enough to forward the parcels inclosed to their places of destination.
  There is no hurry in it, you may deliver them quite leisurely.
  If you should happen to get the parcel with my sulphuret-papers it is very possible that those of lead have turned brown again. I see that by degrees sulphate of lead is acted upon by paper in the dark, so as to become brown i.e. sulphuret of lead.
  I at least cannot account in another way for the fact that sulphuret of lead paper often having been completely bleached by ozonized or insolated oxigen turns gradually brown again in the dark.
  The silhouettes laid by, which except the figures were once quite white will show you that action.
  Yours | very truly | C.F. Schoenbein
Bâle 25, Nov. 1850.

Letter 2349
Faraday to the Editor of the Daily News
29 November 1850
From the original in WIHM MS FALF

R Institution | 29 Novr. 1850
Sir
  I regret I have no time to do more than send you the enclosed rough notes. Use them as you please (*not* mentioning me as the writer) and take out any bad english you may find[1][.]
  I am Sir | Your Obedient Servant | M. Faraday
Editor | &c &c | Daily News

1. These notes were for the article "Professor Faraday on Atmospheric Magnetism", *Daily News*, 30 November 1850, p.6, col. g which reported the readings of Faraday (1851c, d, e), ERE25, 26 and 27 to the Royal Society on 28 November 1850.

Letter 2350
Fabian Carl Ottokar von Feilitzsch to Faraday
3 December 1850
From the original in IEE MS SC 2
Honourable Sir!
If the exhibition of a new theory conditionates likewise a progress in science, because the apparitions alredy known are comprised under one point of sight engaging to new essays to prove or to disprove them - then I dare hope to have made by my efforts one though but little an advance in that branch, that you Sir the great discoverer of diamagnetism have opened. I dared not so assure you of my unbounded esteem, to express you, of what a veneration I am penetrated by following your disinterested indeavours in that science that I love over all, and of which I have made the task of my life,- before I could not put to your feets a little work so unimportant it might be. I wish so vehemently to give you a mark of my deference, that I can not wait longer, to manifest the awe with which I look up to you, might you pardon to such feelings and kindly excuse, that I take now the liberty, to dedicate you this little work.-

I could not be entirely satisfied by that theoretical contemplation of the nature of the diamagnetism, that you and after you M[ess]rs Reich[1], Poggendorff[2], Weber[3] et Plücker[4] have settled[5]. This theory asked the hypothesis: that in every molekule of a magnetical substance by exterial induction the magnetism is in such manner distributed, that to the inducing Southpole is turned a Northpole, and to the inducing Northpole a Southpole; but that in diamagnetical substances the distribution takes in such a manner place that in every molekule to the inducing Southpole a Southpole and to the inducing Northpole a Northpole is turned too. Or what is the same thing, that the currents of the theory of Ampère in the magnetical substances are in a contrary direction moved as in the diamagnetical substances. I tried rather to explicate myself the apparitions by an hypothesis that Mr. van Rees F had explained; consequently I suppose, that in magnetical as in diamagnetical substances the polarity of the molekules have the same direction, but so that all the Northpoles are turned to the Southpole, and all the Southpoles to the Northpole of the inducing Magnet, only whith that difference, that in a bar of *magnetical* substance the *intensity* of the distribution of the molekules *increases* from the ends to the midst, while it *decreases* in a bar of *diamagnetical* substance from the ends to the midst. The currents of Ampère in magnetical substances would be consequently more feebly directed in every particles that is situated next one of the centres of excitation, as in one more distant, but in diamagnetical substances they would be more strongly directed.

These suppositions are permitted, if we attribute to the two groups of substances a diverse resistance against the magnetical excitation (a different Coërcitiv-power). The particles of a magnetical body have a very little Coërcitiv-power, thus the distribution of magnetism must take place in such a manner, that the magnetism dispensed by the primitive excitation in every particle acts distributing on his part on the others, and particulary on the neighbouring particles. Because the molekules are situated very near one to the other, it is to be thought that this part of magnetism is stronger, than that of the primitive excitation. But in the diamagnetical bodies the Coërcitiv-power is so important, that this portion of magnetism, that takes place by the excitation of the molekules one to the other, is more feeble, than that which is produced by the primitive excitation. A bar of magnetical or diamagnetical substance can we excite in two manners, either from the ends to the midst, or from the midst to the ends

A. The excitation from the ends to the midst is done usually thereby, that a bar is suspended between two magnetpoles.

1., To its Coërcitiv-power so important that the effect of the molekules to each other can be neglected, then every particle that is nearer to the magnetpole will be more strongly excited than that next neighbouring and more distant particle.

Do we such observe two neighbouring particles near the exterial Southpole then will the more near exert a Southpole with the intensity s, the more distant will turn to a Northpole with the intensity n', but in such a manner that n' < s. But outwardly works this two excited magnetisms with the difference of their power s-n', but this is in our case *southpolar*, consequently of the same kind, as the exciting Southpole. The contrary will take place near the Northpole, so that *the disengaged magnetism, extended over the bar, grows southpolar on that half which is turned to the Southpole, but northpolar on the other half, that is turned to the Northpole*. A substance, where this takes place is *diamagnetical*, it puts itself equatorial.

2., Is the bar of a magnetical substance therefore so qualified, that the separating action of the molekules on each other must be taken in consideration than can it grow so strong, that the molekules in the midst of the substance are more strongly magnetical, than towards the ends. Do we observe once more two such particles near the exterial Southpole, of the intensity s_1 so will the next avert a Southpole of the intensity s_1 from this exterial Southpole, but the more distant turn towards it a Northpole of the intensity n_1'; but in such a manner, that $n_1' > s_1$ outwardly works both with the intensity $n_1'-s_1$, but this is northpolar, therefore of a contrary nature, as the excitating Southpole. The contrary shall take place near the Northpole, so that *the disengaged magnetism, extended over the bar, grows northpolar on the half that is turned towards the Southpole, but southpolar on that half, that is turned towards the Northpole.* A substance, where this takes place is *magnetical*, it puts itself *axial*.

3., Besides of this observed disengaged magnetism must be yet considered that magnetism, that grows disengaged on the final surface of the bar and that can not be compensated by the neighbouring particles. But this is always of a contrary nature than the excitating neighbouring pole. For magnetical bodies does it support the effect of the disengaged magnetism extended over the bar; in diamagnetical bodies acts it in the contrary, and it is to be thought, that even it is preponderating. Perhaps might this be cause of the feeble magnetism, that you were finding in the Platin, Paladium and Osmium[6].-

B., An excitation from the midst to the ends takes place, if we do lay a bar in an electrical spiral. But in this case all the substances must gain an equal polarity as the iron[.]

To prove that, I puted a thick bar of Bismouth in a very strongly acting spiral, which were excitated by 4 cells of Mr. Grove every one of twelve □ inch of platin plate. I set this spiral on a side near a little declinations-needle suspended on a silk-thread and I compensated its effect by a magnet of steel, that I dislocated as long on the other side of the needle, as it was returned to its first place. Did I withdrew the bismuth bar out of the spiral, than the needle declined in favour of the compensating magnet, but if I puted it again in the spiral, then the needle declined in favour of the spiral. Unlucky the poor fortune of the physical establishment of our university did not allow, to prove also other substances, than the bismouth, but I schall supply this defect as soon as it is do be done; but you will allow me the consequence:

> that the diamagnetismus and the magnetismus are only modifications of the same power, that are produced partly by the different Coërcitiv power of the substances, partly by the different manner of excitation.

Transporting the former in the theory of Ampère I startled, because it has teached hitherto only: Currents, that are parallel and directed in the same way, attracts themselves, but if they are parallel but not directed in the same manner they are repulsive; therefore that a current, moving in the sense of a hand of a watch, in a spiral produces a Southpole on the entrance point in the spiral, but a Northpole on the egression point. But hitherto one has only constructed spirals, where the currents in every winding schows an equal intensity.

But I tried, to construct spirals in such a manner as these, which I have adjointed[7]. One of them is in such a way constructed that on two copper wires are solderd to each of them 15 thin threads spined over with silk. With all this 15 threads is the first winding layed backwards over the copperwire; the second winding is only winded with 14 threads, during that the fifteenth moves itself along the axis etc. Consequently has every of the 15 windings a thread less, and the ends of all the other threads have the direction of the axis. Are soldered in the midst the lasted ends of the 2. 15 threads and are the two thick threads without touching themselves, in each manner bowed, that they can be suspended in the little cups of the apparatus of Ampère, than a current, passing by the spiral, will divide itself in such a manner, that it is the most strong on the exterial ends of the spiral, but decreases more and more to the midst. If the winding of the spiral took place in the direction of the hand of a watch, then must the end of it, where the current enters, grow a Southpole: *but a Northpole*, kept parallel of the spiral, will it repulse. Only the final winding will be attracted and is representing this disengaged magnetism of the final surface. The second spiral is winded like this, only with that difference, that the strongest windings are laying in the midst and the feelblest near the ends. This spiral will be attracted of the Northpole of a magnet over the half in which the current moves at first, but the other half will be repulsed by it. Of the third spiral at last have all the windings the same strength, over all the extension, it is indifferent against a magnetpole, that is not to[o] near, and only their last ends are attracted or repulsed.

Therefore it is permitted to enlarge the theory of Ampère in this manner

If an electrical current passes through a spiral in the direction of a hand of a watch, and

a., if the current is more feeble in every winding that is nearer to the midst of the spiral, then that half is *attracted* by a Southpole, in which enters the current excepted the first winding,

b., but if the current is more strong in every winding that is nearer to the midst of the spiral, then that half is repulsed by a Southpole, in which enters the current including the first winding[.]

The contrary will be adopted for that half in which deserts the current and likewise for the Northpole of the magnet opposed.

In consequence of this extension of the theory of Ampère, it is easy to transfer them in the opinions above produced. In the molecules of magnetical and diamagnetical bodies are to find electrical currents: By the magnetism they will be in such a way directed, that they put themselves parallel of the exterial acting currents. In the diamagnetical bodies is opposed a very great resistance to the direction of these currents of molekules, therefore will their intensity decrease from the centre of the excitation; these bodies will comply with the opinion that is given in a., they will be repulsed. But in magnetical bodies acts the currents of the molekules, that are diverted by exterial influence, on their part also directing on the neighbouring currents of molekule, and in such a manner, that these currents are the most energical directed in the midst of the bar, but are more feebly directed near the ends. These bodies comply with the opinion given in b., they will by attracted and puts themselves axial.

But I fear to tire you, would I transfer the opinions that I have explained on the different apparitions, that followed to the your discovery of the diamagnetism. The apparitions of the minglings of magnetical and diamagnetical substances; the predominating attraction or repulsion of the axis of the crystals; the apparitions of the magne-crystallic axis; the currents of induction that gives a bar of Bismuth exhibited by Mr Weber:- all this apparitions follows by themselves by plain adaptions.-

I ask your pardon, Sir, that I have dared to write you in your own language, that I know so very imperfectly: but I hope I have not to much deformed the sense of what I would express.

I have the honour to be Sir, your most humble servant | Dr. von Feilitzsch | Professor of the university of | Greifswald
Greifswald in Prussia, | 3 Dec. 1850
F cfs. Memoirs of the Netherlandish Institution Vol 12[8].

1. Reich (1849).
2. Poggendorff (1848).
3. Weber (1848, 1849).
4. Plücker (1849e).
5. Endorsed here by Faraday: "I have not adopted the view referred to[.] See Phil Trans. 1850 p.171 M.F." This note was printed in Feilitzsch (1851), 47 and refers to Faraday (1850), ERE23.
6. Faraday (1846c), ERE21, 2379-82, 2385.
7. See Faraday, *Diary*, 15 January 1851, 5: 11274 for Faraday's experiments with these helices.
8. Rees (1846).

Letter 2351
John Phillips to Faraday
5 December 1850
From the original in IEE MS SC 2

St Mary's Lodge | York | Dec. 5 1850

My dear Faraday

Many thanks for your kind explanations of the meaning which ought to be attached to your lines of magnetic force, which I quite comprehend and (I think) see the bearings of.

Believing from the interest you appear to take in the question of Aurora - that you may be induced to try to bring that *recusant* within the pale of your magnetism, I will venture to send you my little Hypothesis derived from my many nights of gazing, & not a few of needle scrutinizing.

In my little page[1] only the peculiar Aurora of 1847 is described, but to comprehend my view fully it is necessary to take two cases.

1. The *Auroral Arch, really weak*, viz a narrow white 'ring of light' stretched across the dipping needle like the ordinary conjunctive wire. This moves from NNW to SSE, at a pretty regular rate. I esteem it to be a simple Electrical Current.

2. *The Beam*, which is commonly seen at a great distance & seen edgeways, appears like a brush or pencil of light, *parallel* to the Dipping Needle, of the plane or nearly so. But I apprehend this appearance is not to be trusted altogether. I suppose that the *beam* when it is viewed from beneath has a different aspect, & such a structure as to justify me in regarding it as an Electrical Magnet. When the sky is so favorably covered by Aurora as in 1847 & 1834, the great area of light is found to be subject to intermittence, as

This is as the swift flashing, or curtain waving.
or

This is the wonderful 'Pulsation' (so called by Airy[2] & self independently) and I conceive that the Electric discharges thus subject to fits of alternate brightness & darkness, in a line

parallel to dipping needle may have the effect of interrupted spirals - or continuous spirals, & affect the Needle - though when looked at edgeways they seem (at 70 miles distance) to be like travelling pencils of light.

They certainly lie parallel to the dipping needle, & some (or probably all) affect the Compass; this could hardly be if they were simple brushes of light - for in that case they have no business to look to the same Star as the Dipping Needle & the Compass ought not to *see* them. The simple *Arch* or ring of Auroral Light is rather uncommon.

See what a penalty you pay for being famous!

I have just refixed my old Suspension Magnet, in hopes to get some further evidence of the influence of Aurora on it this winter.

My Sister[3] joins in kind Comps to Mrs Faraday with your faithful friend | John Phillips

1. Phillips (1847). An offprint is in IEE MS SC 2.
2. According to John Phillips, "Aurora Borealis", *Yorks.Gaz.*, 19 October 1833, [p.2, col. f], this was in a letter from Airy to Phillips.
3. Ann Phillips (1803–1862, Private communication from Jack Morrell). She kept house for John Phillips from 1829.

Letter 2352
Faraday to John Phillips
7 December 1850
From the original in University Museum Oxford MS John Phillips Papers

53 King's Road | Brighton | 7 Decr. 1850.

My dear Phillips

It rejoices me to think that you are still looking for Aurora and are thinking about that of which so much is said with assurance & yet so little assuredly know[n][1]. It and the physical cause of the electric condition of the atmosphere are two things which I am persuaded are naturally linked together and yet we may say we know little or nothing of either of them. Now I have fancies but they are as yet not (*to me*) better they are even worse than other peoples still they return but as you know in London there is no chance of doing any thing with aurora and I am unfortunate in not being able to see even a common chemical experiment with other peoples eyes.

As regards the relation of Electricity & Magnetism in the Aurora there are some observations which if they could be made surely would be of the utmost importance to a true theory. The Aurora affects the Magnet - but how does it affect it? When the needle end goes west is it not rather suddenly? and is that always as the beams mount? Is the *first* effect on the needle *always* in the *same direction* the return effect being merely cessation & is that first effect always in the same direction? that point is I think very important. Or if it be always in the same direction in a station of low latitude is it sometimes the other way in a station of high latitude? or is it at such a station sometimes one way & sometimes the other.

Then in the Southern hemisphere what is the direction of motion & how does it correspond to the movements of the north? Does the S end there agree with the N end in the north or the contrary?

Do your two kinds of Aurora the *Arch* & *beam* both affect the needle & both in the same direction?

Oh that we could know a little more about the Aurora & the magnetic needle. Really it is the free needle that we ought to interrogate for the inclination might tell us even more than the declination what is going on if we could obtain it. But it must be a very difficult matter to obtain such indications freely & above all to separate them from other sources of local action some of which though not auroral may be dependant on aurora action[.]

See what outpouring I have given you. If we were together we would have a long talk about the physics of the Aurora borealis[.]

Kindest remembrance to Miss Phillips[2] from Yours Ever (& wife) | M. Faraday

1. See letter 2351.
2. Ann Phillips (1803–1862, Private communication from Jack Morrell). She kept house for John Phillips from 1829.

Letter 2353
Faraday to Christian Friedrich Schoenbein
9 December 1850
From the original in UB MS NS 393

Brighton | 9 December 1850

I have just read your letter dated July 9, 1850 exactly *Six months* after it was written[1]. I received the parcel containing it just as I was leaving London and I do not doubt it was in consequence of your moving upon the receipt of my last to you a few weeks ago[2]. Thanks thanks my dear friend, for all your kindness. I have the Ozonometer and the summary &

all the illustrative packages safe and though I have read only the letter as yet and that I may acknowledge your kindness write before I have gone through the others yet I see there is a great store of matter & pleasure for me. As to your theory of atmospheric electricity I am very glad to see you put it forward[.] Of course such a proposition has to dwell in ones mind that the idea may be compared with other ideas and the judgment become gradually matured: for it is not like the idea of a new compound which the balance & qualitative experiments may rapidly establish still as I study & think over your account of Ozone & insulated oxygen so I shall gradually be able to comprehend & imbibe the idea. Even as it is I think it is as good as any and much better than the far greater number of hypotheses which have been sent forth as to the physical cause of atmospheric electricity - and some very good men have in turns had a trial at the matter. In fact the point is a very high & a very glorious one:- we ought to understand it & I shall rejoice if it is you that have hold of the end of the subject[.] You will soon pull it clearly into sight.

The German account you sent me of insolated oxygen & your theory of atmospheric electricity[3] is in the hands of a young friend who is translating it - whilst it is going on & also in reading your letter a question arises in my mind about the *insolated oxygen* which perhaps I shall find answered when I come to read the paper. It is whether the oxygen having been insolated is then for a time a different body out of the presence of light as well as in it. I think an American[4] (I forget who) says that Chlorine after being exposed to the Sun is of a brighter colour & acts far more readily than such as has been kept in the dark for a time[5]. Suppose a little box blackened inside with two little glass windows that a ray of sun light could be passed through it & the box filled with oxygen & a proper test paper put up in the dark part of the box would it show change or must the test paper be in the ray to be acted upon. Of course Ozone would act upon it in the dark place is insolated oxygen like ozone in that respect? - I do not doubt that I shall find the answer amongst the data that I am in possession of and so do not trouble yourself for a reply just now[.] As told you in my last I must talk about Atmospheric *Magnetism* in my Friday evenings before Easter[6] & I am glad that Ozone will fall in the Summer months[7] because I should like to produce some of the effects here. I think I told you in my last how that oxygen in the atmosphere which I pointed out three years ago in my paper on flame & gases[8] as so very magnetic compared to other gases is now to me the source of all the periodical variations of terrestrial magnetism and so I rejoice to think & talk at the same time of your results which deal also with that same atmospheric oxygen. What a wonderful body it is[.]

Ever My dear Schoenbein | Yours faithfully | M. Faraday

Address: Dr Schoenbein | &c &c &c &c | University | Bâle | on the Rhine

1. Not found.
2. That is letter 2343.
3. Schoenbein (1850c).
4. John William Draper.
5. Draper (1844).
6. Faraday (1851a), Friday Evening Discourse of 24 January 1851 and Faraday (1851f), Friday Evening Discourse of 11 April 1851.
7. Faraday (1851g), Friday Evening Discourse of 13 June 1851.
8. Faraday (1847b).

Letter 2354
Faraday to Edward Sabine
11 December 1850
From a microfilm in BL RP 243

Brighton[1] | 53 King's Road | 11 Decr. 1850

My dear Sabine

I received your packet with many thanks to you for it, this morning all seems quite safe though the parcel came open in the Mail. The little magnet appears to be in perfect order[.] I perceive it is for me and I must thank you as well as M. Haecker[2] for it[.] Will you do me the favour when you next write to say how much I feel indebted to M. Haecker for it. As to the amount of strength in large magnets I still have the impression that the person at Harlaem [sic] produces them of twice the strength given by the formula of M Haecker but I cannot tell until I reach home[3][.]

In reference to Mr. Thomson's letter I gave Newman the instructions for preparing the Glass globe as we agreed & I hope he will soon have it ready for the oscillations at Woolwich as you proposed. My own power of working will now be limited as the Season comes on but still I am pushing forward the construction of a differential torsion balance & making drawings for it here & I hope by that to settle many points that are of importance to me[.]

As regards the apparent intensity of the needle that I think ought to be affected by the matter gas or solution surrounding the needle but as respects the variations I think it is very rarely the direct action which affects them but an indirect one like very many of the actions which we have in Static electricity where power exerted in one place is made to affect indirectly the results on another. All forces of tension shew this kind of action which is *remarkably in contrast* in this respect with the force of *Gravity*[4].

I presume that Mr. Thompson [sic] may refer to any thing in the papers. It is probably the magnecrystallic results which I shewed him that he wants[5].

Ever my dear Sir | Yours faithfully | M. Faraday
Coll. Sabine RA | &c &c &c

1. "Royal Inst" crossed out above "Brighton".
2. Paul Wolfgang Haecker. Jungnickel and McCormmach (1986), 123 identify him as a scientific instrument maker of Nuremberg.
3. See letter 2278.
4. See Faraday (1851b), ERE24.
5. Presumably a reference to Faraday (1851d), ERE26, 2836–46. See Thomson, W. (1851).

Letter 2355
Charles John Huffam Dickens to Faraday
11 December 1850
From the original in IEE MS SC 2

Devonshire Terrace | Eleventh December 1850.
My dear Sir
    Will you do me the favor to accept the accompanying book[1] - a poor mark of my respect for your public character and services, and my remembrance of your private kindness in so generously lending me your valuable notes[2].
    Concerning which, let me say that I have them in safe keeping, and will shortly return them. The gentleman[3] who has them to refer to, still tells me when I ask if he has done with them, "that they are not so easily exhausted, and that they suggest something else."
    My Dear Sir | Yours faithfully and obliged | Charles Dickens
Professor Faraday.

1. Possibly Dickens (1850).
2. See letters 2291, 2292 and 2295.
3. According to Storey et al. (1988), 110 this was Percival Leigh (1813–1889, DNB). Writer and former physician.

Letter 2356
Faraday to Christian Friedrich Schoenbein
13 December 1850
From the original in UB MS NS 394

Brighton | 13 Decr. 1850

My dear Schoenbein

It will be very strange if I do not make your subject interesting[1]. I have gone twice through the MS. & the illustrations both are beautiful[2]. As soon as I reach home I shall begin to prepare for ozone making & repeating your experiments. This morning I hung out at my window one of the Ozonometer slips that was about 2 hours ago. Now when I moisten it, a tint of blue comes out between Nos 4 and 5 of the Scale. Though I face the sea & have the wind on shore still I am not aware that the spray can do this or any thing that comes from the Sea water but before I send off this letter I shall go down and try the sea myself[.]

Well! I have been to the Sea side & the sea water does nothing of the kind - nor the spray - but as I walk on the shore holding a piece of the test paper in my hand for a quarter of an hour, at the end of that time it by moistening shews a pale blue effect.

That which is up at my window has been out in the air 4 hours & it when wetted comes out a strong blue tint about as No. 6 of the scale. The day is dry but with no sun the lower region pretty clear but clouds above[.]

After reading your notes & examining the illustrations I could not resist writing to you though as you see I have nothing to say[.]

Ever truly yours | M. Faraday

Address: Dr. Schoenbein | &c &c &c | University | Bâle | on the Rhine

1. Faraday (1851g), Friday Evening Discourse of 13 June 1851.
2. Referred to in letter 2353.

Letter 2357
John Phillips to Faraday
14 December 1850
From the original in IEE MS SC 2

14 Dec. 1850 | York

My dear Faraday

It is indeed too true that the *exact* relation of the effect of an Auroral Arch or beam to its position, is not to be stated, except with extreme reserve[1]. The Phenomena are normally of too exciting a kind, & come on us when we are too unprepared, to allow of all being done which should be done. In 1833 it was almost a *novelty* to assure the Meteorologist that the needle was powerfully affected at all. Few persons have a needle fit for the purpose of observing *horizontal* deviation, still fewer a *Dipping* needle worthy of the name: & of these almost 0 has the needle or needles mounted & conveniently placed for use.

In 1839 I changed house, & now it is 1850. In these 11 years I have never had time & power to mount either of my needles. *Thank God!* both are now mounted, in my study & *so good*, that I intend to devote some time to them, & to make such regular Observations as to give me well their normal state (or rather the *Mean* state) & be prepared against all Auroras. If we have any good Arches or Beams I hope to see them with my needles[.]

In turning over old papers I find a duplicate of some notes of the great aurora of 12 Oct 1833[2]. Pray accept it - as being pretty full of Magnetical notes. The *displacements* of the needle *air* [sic] always rather sudden, by little starts.

Ever Yours truly | John Phillips

1. See letter 2352.
2. John Phillips, "Aurora Borealis", *Yorks.Gaz.*, 19 October 1833, [p.2, col. f]. This offprint is in IEE MS SC 2.

Letter 2358
Faraday to Charles Dickson Archibald[1]
17 December 1850
From the original in WIHM MS FALF

Brighton | 17 Decr. 1850

My dear Sir

I expect to be in Town at the end of this week & shall be very happy to see you & your friend next week at your convenience.

I think I told you that your friend *could not* be the discoverer that hydrogen was water & Negative Electricity & Oxygen water & P. Electricity for the idea is old & belongs to De la Rive and moreover it has been very fully canvassed by philosophers & rejected[.]

If your friend did not know this I am afraid that there is a great deal more in the common stock of knowledge that he is not acquainted with for it was published in all the French & English Scientific Journals of the time[.]

Ever Truly Yours | M. Faraday

C.D. Archibald Esq | &c &c &c

Address: C.D. Archibald Esq | &c &c &c | 15, Portland Place | London

1. Charles Dickson Archibald (1802-1868, B1). Writer on colonial topics.

Letter 2359
Faraday to John Phillips
21 December 1850
From the original in Bod MS dep BAAS 59, f.47-8

R Institution | 21 Decr. 1850

My dear Phillips

If I were with you I would ask a word or two about Arches[1] but my thoughts are not definite enough to make me put them down in ink in association with No. 2. The luminous arch & the dark circular segment though not (apparently) concentric have as far as I have seen the same relation to a vertical plane passing North [of] their centres & the observer but is this center of both a fixture for the same evening & the same as regards astronomical North on different nights. No. 3 would perhaps answer the question but if the plane seems to shift at all though slowly it might be observed if *looked for*[.]

If I had opportunities of observing aurora I should make this experiment. A horizontal needle hung say by cocoon silk. A bar of thoroughly soft iron about 2 feet long in the position of the dip fixed with one end so near one end of the magnetic needle as to affect it a certain amount deflecting it for instance 45° or more. It seems to me likely that any horizontal current of Electricity or equivalent derangement of general magnetism might be better shewn by its effect on this bar & through it on the needle than on the needle alone[.] I think also that a similar bar of iron fixed horizontally & perpendicular to the line of the dip & plane of Mag meridian acting on a *horizontal* needle hung by silk - might tell about streams of Electricity vertical or approaching to verticality - and that both as to the upping or downing of the discharges.

Ever Yours | M. Faraday

John Phillips Esq | &c &c &c

I have made no marks on the paper but intend to keep that MF

1. See letter 2357. This letter refers to Phillips (1847) (sent with letter 2351) where the paragraphs are numbered.

Letter 2360
Faraday to Richard Taylor
24 December 1850
From Feilitzsch (1851), 46

Royal Institution, | Dec. 24, 1850.

My dear Sir,
I have just received the inclosed letter[1]; and though I have not had time to consider the view experimentally, I think it such an important contribution to the philosophy of magnetic and diamagnetic bodies, and am, as always, so anxious to establish the date of a new theory or fact, that I send it to you at once for publication if you think fit. I have left it almost in the author's language, that I might not misstate his view.
Ever, my dear Sir, | Very truly yours, | M. Faraday
Richard Taylor, Esq., | &c. &c.

1. Letter 2350, published as Feilitzsch (1851).

Letter 2361
Faraday to John Frederick William Herschel
27 December 1850
From the original in MU MS 173

R Institution | 27 Decr. 1850

My dear Herschell
Your note gave me great pleasure & when you think of joining the R Institution you encourage us greatly in our exertions. Since I called Barlow tells me in a note he has written to you suggesting some change. Let us know what you would but like & it shall be done[1]. As to coming in I hope you feel that you are always free here and as to any body else I am always at your service[.]
Ever Most Truly Yours | M. Faraday

1. Herschel did not become a member of the Royal Institution but RI MS GM, 3 February 1851, 5: 547 noted the nomination by Faraday, among others, of Margaret Brodie Herschel, née Stewart (d.1884, age 73, GRO. Married Herschel in 1829, see DNB under John Frederick William Herschel) to be a member. Her election is recorded in RI MS GM, 3 March 1851, 5: 550.

Letter 2362
John Stevens Henslow to Faraday
27 December 1850
From the original in IEE MS SC 2

Hitcham | Hadleigh | Suffolk | 27 Dec 1850

My dear Sir,

Tho' it is a shame to apply to one so occupied as yourself, I know there is nothing like appealing to the highest authority. We have determined on forming a collection of objects for the Ipswich Museum which may serve as *types* of the principal groups under the 3 Kingdoms of nature - & to keep such collections well labelled & illustrated by models & drawings apart from the ever shifting series of specimens arranged in the cases of a continually increasing collection. At the head of the Mineral Kingdom I wish to illustrate the Elementary substances by exhibiting as many of them as possible contenting ourselves with the names of those which cannot be exhibited. My idea is to give the S.G. & Chemical Equivalent where known, & to place the specimens under glasses. Circular discs of the metals, one half polished & the other not, with an exhibition of their *fracture* seem to me the best way of showing these. Though I cannot ask you to assist further than by offering any suggestion that may occur, perhaps you can inform me where I am likely to pick up examples of the more uncommon substances properly put up (as Iodine, Bromine &c) or of such metals as may require rather severer treatment to reduce them into the required shape than I am able to bestow on them. As I am asking on account of the public, I feel less scrupulous than if it were on my own account alone, though I shall not be less obliged.

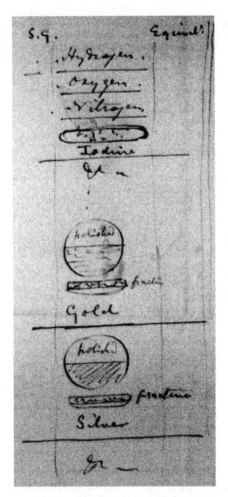

Some such arrangement as this is what I am thinking of - a small watch glass over the metals. Where can I find the best arrangement? Will that in Gregory[1] do?[2]

Believe me | very truly yours | J.S. Henslow

Address: Prof. Faraday | Royal Institution | Albemarle St | London

1. William Gregory (1803–1858, DNB). Professor of Chemistry at Edinburgh University, 1844–1858.
2. Gregory (1845), **1** where he divided the metals into various classes and orders depending on their properties.

Letter 2363
Antoine-César Becquerel to Faraday
27 December 1850
From the original in IEE MS SC 2
Faraday number 1

Monsieur et très illustre confrère,

oserai-je vous prier de rendre une service à mon fils Edmond, qui se présente comme candidat à la place laissée vacante dans le Sein de l'Académie des Sciences par le décès de Mr Gay-Lussac[1]; en lui étant utile vous m'obligerez personnellement et je vous en aurai beaucoup de reconnaissance. voici ce dont il s'agit: mon fils a lû un premier mémoire à l'Académie des Sciences le 21 mai 1849 (voir les comptes rendus mai de la même année[2] et annales de physique et de chimie nouvelle série T.28 p 283[3]) dans le quel il a étudié l'action des milieux environnants gazeux sur les substances soumises à l'action du magnétisme.

Il a découvert entre outres propriétés que l'oxygène est magnétique, ou susceptible d'aimantation par l'influence à un dégré - assez énergique. Les autres gaz ne lui ont pas manifesté cette action et se sont comportés comme diamagnétisme.

il a démontré cette puissance magnétique 1° en mesurant les effets produits sur de petits barreaux placés successivement dans le vide et dans l'oxygène; 2° en condensant ce gaz dans des barreaux de charbon, puis il a évalué le rapport de l'action à celle qui est exercée sur le fer doux.

Dans un 2ème mémoire lû à l'académie le 12 août 1850 (voir comptes rendus de l'académie des sciences[4]) il a étudié de nouveau les differentes circonstances de ces phénomènes et les effets produits par la condensation des gaz. il est parvenu à démontrer en outre l'action magnétique exercée sur l'oxygeène, en l'enfermant dans des petits tubes de verre fort mince, qui se trouvent ainsi constituer sous l'influence des aimants de petits barreaux aimantés.

Les détails de ces expériences sont décrits dans le mémoire qui se trouve entre les mains des membres de l'académie chargés de faire un rapport.

Je désirerai que vous voulassiez bien avoir l'obligeance de me dire si - vous vous êtes occupé de cette question c'est-à-dire du magnétisme de l'oxygène, à quelle épôque et dans quels recueils scientifiques se trouvent consignés vos recherches? auriez-vous fait par hazard une lecture Bakerienne sur le même sujet avant la publication de mon fils dont je vous ai cité les epôques? Les renseignements que je prends la liberté de vous demander sont d'une grande importance pour lui et serviront à établir l'historique des recherches entreprises pour mettre en évidence le magnétisme d'un gaz, qui joue un grand rôle dans la nature.

Comptez moi toujours, monsieur, en rang de vos admirateurs et des personnes qui vous sont le plus dévouées | Becquerel

Paris. Jardin des plantes | ce 27 décembre 1850

Address: Monsieur | Monsieur Faraday, professeur | à l'institution royale, | Londres | (angleterre)[5]

TRANSLATION
Sir and most illustrious colleague,
Dare I ask you to render a service to my son, Edmond, who is presenting himself as a candidate for the place left vacant at the heart of the Académie des Sciences by the death of Mr Gay-Lussac[1]; in helping him, you would oblige me personally and I would be very grateful to you. This is what it concerns: my son read his first paper to the Académie des Sciences on 21 May 1849 (see the *Comptes rendus* of May of the same year[2] and the *Annales de physique et de chimie* new series volume 28, p.283[3]) in which he studied the effects of gaseous atmospheres on materials subjected to the action of magnetism.

He discovered, amongst other properties, that oxygen is magnetic or susceptible to magnetism to a high degree. Other gases did not show him this action and behaved diamagnetically.

He demonstrated this magnetic power 1st by measuring the effects produced on small bars placed successively in a vacuum and in oxygen; 2nd by condensing this gas in bars of charcoal, then he compared the relationship between that action and that which exerted itself on soft iron.

In the second paper read to the Académie on 12 August 1850 (see *Comptes rendus de l'académie des sciences*[4]) he studied once again the different circumstances of these phenomena and the effects produced by the condensation of gases. He was able to show moreover the magnetic action exerted on oxygen by enclosing it in little tubes made of very thin glass, which found themselves constituting under the influence of magnets little magnetised bars.

The details of these experiments are described in a paper which is currently in the hands of the members of the Académie, entrusted with making a report.

I would very humbly ask you to have the kindness to tell me if - you worked on this question, that is to say on the magnetism of oxygen, on which date and in which scientific collection were your findings published? Would you by chance have given a Bakerian lecture on this subject before my son's publications, of which I cited the dates above? The information that I take the liberty of asking you is of great importance to him and will serve to establish the history of research undertaken to show the magnetism of a gas which plays such a great role in nature.

Please count me always, Sir, among the rank of those of your admirers who are most devoted to you | Becquerel
Paris, Jardin des plantes | this 27 December 1850

1. Joseph Louis Gay-Lussac (1778–1850, DSB). French chemist and co-editor of the *Ann.Chim.* who died on 9 May 1850. Becquerel was not elected to the Adadémie until 1863. Crosland (1992), 182.
2. Becquerel (1849a).
3. Becquerel (1850a).
4. Becquerel (1850b).
5. The envelope is endorsed by Faraday "Nos. 1, and 2.".

Letter 2364
Faraday to Antoine-César Becquerel
30 December 1850
From the original copy in IEE MS SC 2
Faraday number 2

Royal Institution | 30 Decr. 1850

My dear Sir

It is with great pleasure that I receive a letter from you[1], for much as I have thought of your name and the high scientific labours connected with it I do not remember that I have seen your handwriting before. I shall treasure the letter in a certain volume of Portraits & letters that I keep devoted to the personal remembrance of the eminent men who adorn science whom I have more or less the honour & delight of being acquainted with.

In reference to the Queries in your letter I suppose the following will be sufficient answer[.] I developed and *published* the nature & principles of the action of magnetic & diamagnetic media upon substances *in* them more or less magnetic or diamagnetic than themselves, in the year 1845, or just *five* years ago. The paper was read at the Royal Society[2] 8 Jany 1846, and is contained in the Philosophical Transactions for 1846 p.50, &c[3][.] If you refer to the numbered Paragraphs *2357*, *2363*, *2367*, *2400* &c, *2406*, *2414*, *2423*, *2438* you will see at once how far I had gone at that date. The papers were republished in Pogendorf[f]s Annalen[4] & I believe in the Geneva[5], the Italian & German journals - in one form or another[.]

In reference to the Magnetism of Oxygen *three* years ago i.e in *1847*, I showed its high magnetic character in relation to nitrogen & all other gases & that Air owed its place amongst them to the oxygen it contained. I even endeavoured to analyse the air, separating its oxygen & nitrogen by magnetic force for I thought such a result possible. All this you will find in a paper published in the Philosophical Magazine for 1847, vol xxxi, page 401, &c[6]. This paper was also published at full length in Poggendorf[f]s Annalen 1848, vol lxxiii, page 256, &c[7]. I shall send you a copy of it immediately by M. Bailliere[8] who has undertaken to forward it to you. I have marked it in ink to direct your attention[9]. In it also you will find the effect of *heat on oxygen, air &c*. The experiments were all devised & made upon the principles before developed concerning the mutual relation of substances & the media surrounding them[.]

This year I have been busy extending the above researches & have sent in several papers to the Royal Society[10] & have also given a Bakerian lecture in which they were briefly summed up[11]. I fortunately have a copy in slips of the Royal Society's Abstract[12] of these papers & therefore will send it with the paper from the Philosophical Magazine. I suppose it will appear in the out coming number of the Philosophical Magazine[13][.] The Papers themselves are now in the hands of the Printer of the Transactions[.]

I was not aware until lately of that Paper of M. Edmond Becquerel to which you first refer. My health & occupation often prevent me from reading up to the present state of science[.] Immediately that I knew of it I added a note (by permission) to my last paper Series xxvi, in which I referred to it & quoted at length what it said in reference to atmospheric magnetism calling attention also to my own results as to oxygen three years ago & those respecting media five years ago[14]. I have no copy of this note or I would send it to you. It was manifest to me that M. Edmond Becquerel had never heard of my results and though that makes no excuse to myself I hope it will be to him a palliation that I had not before heard of his. The second one I had not heard of until I received your letter the day before yesterday. I was exceedingly struck with the beauty of M. E. Becquerels experiments and though the differential balance I have described in my last papers will I expect give me far more delicate indications when the perfect one which is in hand is completed still I cannot express too freely my praise of the apparatus & results which the first paper describes & which is probably surpassed by those in the second.

I know the severe choice of Your Academy of Sciences and I also know that France has ever been productive of Men who deserve to stand as candidates whenever a vacancy occurs in any branch of knowledge & though as you perceive I do not know all that M. E. Becquerel has done, I know enough to convince me that he deserves the honour of standing in that body & to create in me strong hopes that he will obtain his place there.

Ever My dear M. Becquerel, Your faithful admirer, | M. Faraday

Address: A Monsieur | Monsieur Becquerel | Membre de l Academie des Sciences | &c &c &c &c | Jardin des Plantes | à Paris

1. Letter 2363.
2. "18 Decr. 1845 &" is crossed out here.
3. Faraday (1846c), ERE21.
4. Faraday (1847a).
5. Faraday (1846d).
6. Faraday (1847b).
7. Faraday (1848b).
8. Hippolyte Baillière (d.1867, age 58, B1). French bookseller in London.
9. In IEE MS SC 2 there is a copy of Faraday (1847b) which he numbered as "2" and which is also marked up by him.
10. Faraday (1851c, d, e), ERE25, 26 and 27.
11. Although Faraday (1851b), ERE24 was designated as the Bakerian Lecture, *Proc.Roy.Soc.*, 1850, **5**: 994 stated that "Dr. Faraday the delivered the Bakerian Lecture, which in substance was a résumé of the following papers", that is Faraday (1851b, c, d, e), ERE24, 25, 26 and 27.
12. *Proc.Roy.Soc.*, 1850, **5**: 995–1001.
13. *Phil.Mag.*, 1851, **1**: 68–75.
14. Faraday (1851d), ERE26, pp.42–3 cited Becquerel (1850a). This note was dated 28 November 1850.

Letter 2365
Edward Adolphus Seymour[1] to Faraday
31 December 1850
From the original in IEE MS SC 2

Office of Woods | 31st Dec / 50

My dear Sir

I shall be glad to see you because Mr Barlow seemed to consider it desirable for the interests of the Royal Institution that I should have a few minutes conversation with you. I shall be here tomorrow from 12 until 2 o'clock & probably disengaged.

Yours very faithfully | Seymour

1. Edward Adolphus Seymour (1804–1885, DNB). Commissioner of Woods, 1849–1851.

Letter 2366
Faraday to John Stevens Henslow
3 January 1851
From the original in IEE MS SC 3

R Institution | 3 Jany 1851

My dear Sir

I found that if I did not prepare the tubes at once[1] I might not be able to do so for months so you will shortly receive a box with the four named[.] I have tried to put up the phosphorous dry it was very good but I fear the surface will alter it does so either in or out of water. The chlorine colour is scarcely sensible unless you look along the tube. The bromine atmosphere is far more sensible or rather visible. The sealing of the Chlorine & bromine tubes is rather a ticklish matter for both act on the glass & the latter is very volatile. Both are safe now & I trust will reach you so. Remove the cork protections carefully. Beware of the bromine if the tube should break at the end for the vapour is very deleterious[.]

  Ever Most Truly Yours | M. Faraday
Professor Henslow | &c &c &c

1. See letter 2362.

Letter 2367
John Stevens Henslow to Faraday
5 January 1851
From the original in IEE MS SC 2

            Hitcham Hadleigh | Suffolk | 5 Jany 1851
My dear Sir,
  I shall be most careful in unpacking the box[1]. I know of old what a disagreeable inmate is Bromine. I had some in a glass stoppered bottle which gradually evaporated & attacked several minerals in the same drawer. I do not mention it that you should think of repeating the preparations of this & Chlorine - but the thought strikes me a *globe*, rather than a tube, would be the best way of securing mass enough to see the colour of the gases.

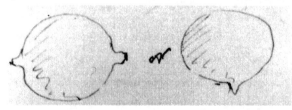

I have had some correspondence with the assayer, & have asked his advice about the mode of securing the metal disks.

My idea is to procure a number of square blocks of wood, all of the same size, & coat them with some (?) varnish, which may serve as a cement to the disks, the glass covers, & keep out the air. This would admit of ready re-arrangement in an additional covering of a glass case large enough to contain the whole series. Perhaps one of the simplest arrangements would be to take either the S.G's, or the Atomic Weights, leaving such as are not known at the bottom of the series. According to intrinsic values I have suggested disks of 4 sizes suitable to 1.oz - $\frac{1}{2}$oz - $\frac{1}{4}$oz - $\frac{1}{8}$oz. I never saw Silicium, & do not know whether it can be thus exhibited. As for carbon I suppose a very *small* diamond would be the best specimen - with a sample of charcoal? or plumbago? - or these latter are hardly pure enough. I see the Assayer has some of the Earths also - & it will be as well to have a selection of such as are common, as an introduction to the geological series.

Believe me | Very sincerely Yours | J.S. Henslow

Address: M. Faraday Esq | Royal Institution | Albemarle | London

1. See letter 2366.

Letter 2368
William Vernon Harcourt[1] to Faraday
7 January 1851[2]
From the original in IEE MS SC 2

Residence | York | Jan. 7.

Dear Faraday

I cannot refrain from sending you my congratulations at this commencement of a new year on the great discovery you have made in placing so universal & important an element as oxygen side by side with iron in the *"Paramagnetics"*.

May I enquire whether you obtain a neutral point at which there is neither Dia- nor Para-magnetism by mixing oxygen & nitrogen in a certain proportion, & whereabouts that proportion is? also whether any of the chemical compounds of these gases exhibit such *neutrality*, or whether the most oxygenous are themselves in any degree *para* magnetic at common temperatures? I am curious also to know whether Chlorine Bromine or Iodine, have, or approach to, this property and whether their admixtures or combinations with oxygen counteract its *para*magnetism in the same degree with other substances? I would not ask these questions if I supposed it would give you any trouble to answer: but they are points which I dare say you have ascertained.

Believe me | Yours sincerely | Wm. Vernon Harcourt

1. William Vernon Harcourt (1789–1871, DNB). One of the founders of the British Association. Rector of Bolton Percy, 1837–1861.
2. Dated on the basis that Faraday first used the term paramagnetic in Faraday (1851c), ERE25, 2790. The reading of this paper to the Royal Society was reported in *Phil.Mag.*, 1851, 1: 69–71 where the term was used on p.71.

Letter 2369
Faraday to William Vernon Harcourt[1]
9 January 1851[2]
From the original in private possession

R Institution | 9 Jany. 1850 [sic]

My dear Sir

Thank you for your kindness[.] I am in hopes you will not think it less fitly bestowed at the end of the year than the beginning for I have full confidence at present in the deductions which are drawn from the new facts regarding oxygen & gases. Space being my zero. My present means of measuring gave me Nitrogen at zero so that a mixture of oxygen & nitrogen never can arrive at zero. Then as to *compounds*, of Oxygen & Nitrogen, Nitrous oxide appears to be at zero - Nitrous acid is Magnetic - Nitric oxide perhaps so. A better apparatus is in the course of construction which will settle all these points minutely. Chlorine & Bromine appear to be close to zero[.]

Oxygen in combining generally loses all its power. Carbonic acid is close to zero - Per oxide of Iron is scarcely magnetic. Per oxide of Lead & many other things is diamagenetic[.]

In haste | but Ever Truly Yours | M. Faraday
Revd. W.V. Harcourt | &c &c &c

1. William Vernon Harcourt (1789–1871, DNB). One of the founders of the British Association. Rector of Bolton Percy, 1837–1861.
2. Dated on the basis that this is the reply to letter 2368.

Letter 2370
Faraday to Charles Manby
9 January 1851
From a photocopy in RI MS

Royal Institution | 9 Jany 1851

My dear Manby
 If I can help you I will i.e. I will set down any book that I think you ought to have[.] But my memory is so bad that I could not tell you any thing at once - & the same circumstances prevents me reading any books but those which belong especially to my present and particular pursuits. But what I can do I will.
 Ever Truly Yours | M. Faraday

Letter 2371
Faraday to Mary Fox
14 January 1851
From the original in RI MS F1 D1

Royal Institution | 14 Jany 1851

My dear Mrs. Fox
 I understand you wish for a ticket for my next Evening here[1]. I hasten to send one & only wish I could offer you such assurance of accommodation as would be certain to make you comfortable. I cannot give you power to go into the inner circle with Mr Fox for that belongs to Members or else to their Lady friends which *Lady Members* invite and as my wife is not a member she cannot take you there. She will be most happy to take charge of you - but I say all this because fearing a crowd I fear to make you uncomfortable at [sic] unawears - and would rather therefore mention the matter before hand.
 Ever My dear Mrs Fox | Yours faithfully | M. Faraday

1. Faraday (1851a), Friday Evening Discourse of 24 January 1851.

Letter 2372
Antoine-César Becquerel to Faraday
14 January 1851
From the original in IEE MS SC 2
Faraday number 5

Monsieur et illustre confrère,

L'aimable reproche que vous m'adressez de ne point posseder de lettre de moi[1], me fait sentir mes torts envers vous de ne pas vous avoir encore donné signe de vie ne fut-ce que pour vous féliciter des brillantes découvertes dont vous avez enrichi la physique. L'autuer de la découverte de l'induction, d'un ordre aussi élevé que celle d'Oersted, qui n'a su tirer aucun parti de la sienne ne pouvait m'être indifferent; j'applaudissais en secret à les succès à l'avenir. je serai moins discrèt et vous apprendrez de moi directement la haute estime que je professe pour vos travaux et votre personne.

J'ai lû très attentivement les diverses publications que vous m'avez indiquées relatives au diamagnétisme de l'oxygène, mais je ne sais si j'ai bien compris, il me semble qu'il y a une difference entre vos découvertes et les travaux de mon fils; permettez moi de vous soumettre mes réflexions à cet egard. Si je me trompe vous aurez la bonté de rectifier ma manière de voir.

Le mémoire que vous avez publié (trans. phil. 1846, p.50 et suivantes[2]) a été inséré par extrait dans la bibliothèque de Genève t.2 1846[3]. il est relatif à l'action du magnétisme sur differents corps solides et sur diverses dissolutions. il n'est pas question dans ce travail de gaz et de vapeur. vous avez été conduit par ces premières observations aux conséquences suivantes: (bibliothèque de Genève t.2 p.157, 1846) "on voit que, dans toutes ces expériences, les gaz et les vapeurs occupent une position intermédiaire entre les substances magnétiques et diamagnétiques. quelles que soyent leurs propriétés chimiques ou autres, leur difference de densité ou leur dégré de raréfaction, tous les fluides élastiques paraissent être parfaitement identiques dans leurs rapports magnétiques et être à cet égard semblables au vide"[4].

Autre travail. Vous me mandez dans votre lettre que vous avez montré la haute puissance magnétique de l'oxygène, relativement à l'azote et aux autres gaz et que ces faits se trouvent consignés dans le philosophical magazine 1847 vol. 31 p.406[5].

mon fils n'a point eu connaissance de ce dernier travail avant la rédaction de ses deux mémoires[6], mais je ne vois rien qui établisse nettement la puissance magnétique exercée sur chaque gaz isolé, en particulier sur l'oxygène.

dans ce travail remarquable vous êtes parti de l'action exercée sur la flamme et tous les gaz échauffés ou refroidis, pour examiner les effets produits sur des courants de gaz, qui circulent dans d'autres gaz. Vous avez établi qu'il existait des effets produits sur des courants gazeux circulant dans d'autres gaz, mais sans décider comme on va le voir, si cet effet est dû à la difference du diamagnétisme du gaz, ou s'il y en a - quelque uns de magnétiques. en outre il me semble que vous n'avez pris aucune mesure.

page 410[7], après avoir parlé de l'action exercée sur l'oxygène dans l'air, on lit: so oxygen appears to be magnetic in common air. Whether it be really so, or only less diamagnetic than air &c &c en français: ainsi l'oxygène semble être magnétique dans l'air ordinaire. mais nous ferons place à mince d'examiner ci-après, si cela a lieu en effet, où si l'oxygène est moins magnétique que l'air (l'air étant un mélange d'oxygène et dazote[)].

page [4]12 art 6 &c[8]:
of all the vapours and gases yet tried, oxygen seems to be that wich [sic] has the least &&c i [sic] believe it to be diamagnetic (oxygen).

c'est à dire: de tous les gaz et vapeurs examinés jusqu'à présant l'oxygène semble être celui qui a la moindre force diamagnétique. il est encore douteux de la place que l'on doit assigner à l'oxygène, car il peut être quelque fois aussi peu diamagnétique que le vide, ou même passer au côté magnétique et l'experience ne résoud pas encore ce problème. *moi, je le crois, diamagnétique* (l'oxygène).

plus loin p.413[9], en haut, je trouve: Still i think the results are in favour of the idea that oxygen is diamagnetic. ou bien: je pense que les résultats sont en faveur de la supposition *que l'oxygène est diamagnétic.* [sic]

dans le cours du mémoire l'oxygèné [sic] s'est présenté comme magnétique, Lorsqu'un courant de ce gaz passait dans l'acide carbonique ou dans l'hydrogène. mais comme vous n'avez observé que des differences, on ne peut, je crois, en rien conclure relativement aux effets exercés sur chaque gaz; un effet vous ajoutez, page 420[10], à la fin du mémoire: now, until it is distinguished &c c'est à dire, maintenant jusqu'a cela soit démontré, on ne peut pas dire lesquels des corps gazeux doivent être rangés parmi les corps diamagnétiques et les quels doivent être considérés comme magnétiques et en outre s'il en existe qui n'éprouvent aucune action.

ces expériences très importants, quant aux actions relatives des differents gaz, à diverses températures, d'après votre opinion même, ne décident rien quant aux effets produits individuellement, sur les substances gazeux, tandis que la méthode d'experimentation de mon fils lui a permis de mesurer l'action directe exercée par le magnétisme sur les gaz considérés isolement et particulièrement sur l'oxygène qu'il a démontré être fortement magnétique et dont il a évalué numériquement la puissance par rapport au fer doux; en outre les autres gaz se sont présentés á lui comme diamagnétiques.

telles sont les observations que je soumets à votre excellent jugement, dont je vous prie de me faire connaitre le résultat.

Veuillez bien, monsieur et illustre confrère, agréer l'assurance de mes sentiments les plus dévoués et les plus affectueux | Becquerel
jardin des plantes | ce 14 janvier

Endorsed by Faraday: *Nos 5 and 6*
Address: Monsieur | Monsieur Faraday, professeur | à L'institution royale | Londres | (angleterre)
Postmark: 1851

TRANSLATION
Sir and illustrious colleague,
The gentle reproach that you have addressed to me of not having had a letter from me[1], makes me feel the guilt of giving no sign of life and not congratulating you on your brilliant discoveries with which you have enriched physics. The author of the discovery of induction of an order as high as that of Oersted, who was unable to draw any conclusions from his, could not be indifferent to me, and I applauded in secret his success and the future. I shall be less discreet and you shall learn directly the high esteem that I profess for your work and your person.

I read with great attention the various publications that you indicated in connection with the diamagnetism of oxygen, but I do not know if I have understood correctly, as there seems to me to be a difference between your discoveries and the work of my son; allow me to submit to you my thoughts on the subject. If I am mistaken, you will have the kindness to rectify my way of seeing things.

The paper you published (*Phil Trans* 1846, p.50 and following[2]) was inserted in sections in Vol 2, 1846 of the *Bibliothèque* of Geneva[3]. It concerns the action of magnetism on different solids and their various dissolutions. This paper does not deal with gases or vapours. You were led by these first observations to the following conclusions: (*Bibliothèque* of Geneva Vol 2, p.157, 1846) "one sees that in all these experiments, gases and vapours occupy an intermediary position between magnetic substances and diamagnetic substances. Whatever their chemical or other properties, their difference in density or their degree of rarefaction, all elastic fluids appear to be perfectly identical in their magnetic ratio and seem in this regard similar to a vacuum"[4].

Other work. You say in your letter that you have shown the relatively high magnetic power of oxygen in comparison to nitrogen and other gases and these facts are consigned to the *Philosophical magazine* 1847 vol 31, p.406[5].

My son did not know of this latter work before writing his two papers[6], but I see nothing that establishes clearly the magnetic power exerted on each isolated gas, in particular on oxygen.

In this remarkable work you begin by showing the action exerted on a flame and on all heated or cooled gases, in order to examine the effects produced on currents of gases circulating in other gases. You have established that there existed effects produced on currents of electricity circulating in other gases, but without deciding as one shall see, if this effect is due to the difference of the diamagnetism of the gas or if there are some which are magnetic. Moreover, it seems to me that you have taken no measurements.

Page 410[7], having spoken of the action exerted on oxygen in air, one reads: "so oxygen appears to be magnetic in common air. Whether it be really so, or only less diamagnetic than air" - &c &c in French "so oxygen appears to be magnetic in common air. But we shall examine later if that is really the case or if the oxygen is less magnetic than air (Air being a mixture of oxygen and nitrogen"[)].

Page 412 para 6 &c[8]:
"of all the vapours and gases yet tried, oxygen seems to be that which has the least &c &c. I believe it to be diamagnetic (Oxygen)." That is to say of all the gases and vapours examined up till now, oxygen seems to be the one that has the least diamagnetic force. The place one should give to oxygen is still doubtful, as it can sometimes be as little diamagnetic as a vacuum or even pass on to the side of the magnetic and experimentation has not yet resolved this problem. *I myself believe it to be diamagnetic* (oxygen).

Further on, p.413[9], at the top, I find: *"Still I think the results are in favour of the idea that oxygen is diamagnetic."* Or, I think the results are in favour of the idea *that oxygen is diamagnetic*.

In the course of the paper, oxygen was presented as magnetic, when a current of this gas was passed through carbon dioxide or hydrogen. But as you have observed only differences, one cannot, I believe, conclude anything relative to the effects exerted on each gas; in fact you add at the end of the paper, p.420[10]: "Now, until it is distinguished" &c, that is to say now until it is distinguished one cannot say which of the gases must be listed as diamagnetic and which as magnetic and moreover if there are any that are not subject to any action.

These very important experiments, as to the relative action of different gases at different temperatures, according to your own opinion, decide nothing as to the effects produced individually on the gaseous substances, whilst my son's experimental method has allowed him to measure the direct action exerted by magnetism on isolated gases, particularly on oxygen which he found to be strongly magnetic and of which he has numerically calculated the power in comparison to soft iron; moreover the other gases showed themselves to be diamagnetic.

These are the observations that I submit to your excellent judgement, of which I ask you to let me know the result.

Please accept, Sir and illustrious colleague, the assurance of my most devoted and most affectionate sentiments | Becquerel
Jardin des plantes | this 14 January.

1. In letter 2364.
2. Faraday (1846c), ERE21.
3. Faraday (1846d).
4. Ibid., 157.
5. Faraday (1847b).
6. Becquerel (1849a, 1850b).
7. Faraday (1847b), 410. This passage is marked in the copy in IEE MS SC 2.
8. Ibid., 412. This passage is marked in the copy in IEE MS SC 2.
9. Ibid., 413.
10. Ibid., 420. This passage is marked in the copy in IEE MS SC 2.

Letter 2373
Faraday to Antoine-César Becquerel
17 January 1851
From the original copy in IEE MS SC 2
Faraday number 6

Royal Institution | 17 Jany 1851

My dear Mr. Becquerel

I received your letter of the 14th instant[1] yesterday & hasten to reply to it as you desire: first however thanking you for your kind[2] expressions which will be a strong stimulus to me coming as they do from a Master in Science. I would not have you for a moment think that I put my paper of three years ago[3] & that of M. E Becquerels of last year on the same footing, except in this that we each discovered for ourselves at those periods the high magnetic relation of oxygen to the other gases[.] M. E. Becquerel has made excellent measurements which I had not & his paper[4] is in my opinion a most important contribution to science[.]

I am not quite sure whether you are aware that in my paper of 1847 the comparison of one gas with another is always at the *same* temperature i.e at common temperatures and it was a very striking fact to me to find that oxygen was magnetic in relation to hydrogen to such an extent as to be equal in attractive force to its force of gravity, for the oxygen was suspended in the hydrogen by magnetic force alone. Phil. Mag. xxxi, pp. 415, 416[5]. I do not think that much turns upon the circumstance of calling oxygen magnetic or diamagnetic in 1847 when the object was to shew how far oxygen was apart from the other gases in the magnetic direction these terms being employed in relation to other bodies and with an acknowledgement that the place of zero was not determined[.] If I understand rightly M Edmond Becquerel still calls bismuth & phosphorus magnetic whilst I call them diamagnetic. He considers space as magnetic I consider it as zero. If a body should be found as eminently diamagnetic in my view as iron is magnetic; still I conclude M. Edmond Becquerel would consider it magnetic. He has not yet adopted the view of any zero or natural standard point[.] But this does not prevent us from fully understanding each other and the facts upon which the distinction of oxygen from nitrogen and other gases are founded remain the same and are just as well made known by the one form of expression as the other. It was therefore to me a great delight when I first saw his paper in last November[6] to have my old results confirmed and so beautifully enlarged in the case of Oxygen & Nitrogen by the researches of M. E Becquerel and beyond all to see the beautiful system of measurements applied to them which is described in his published paper. Pray present my kindest remembrances & wishes to him and believe me to be with the highest respect My dear M. Becquerel

Your faithful Obliged Servant | M. Faraday

1. Letter 2372.
2. "& encouraging" crossed out here.
3. Faraday (1847b).
4. Becquerel (1850a).
5. Faraday (1847b), 415–6.
6. Becquerel (1850a).

Letter 2374
R. Smith[1] to Faraday
25 January 1851
From the original in IEE MS SC 2

Blackford Perthshire | 25 January 1851

Honoured | Sir

I hope that you will excuse the liberty I have taken in troubling you with this communication, I have enclosed a sketch and short description of a telegraph which I have in progress, & would consider it a great favour, acquisition and honour were you to have the kindness to give me your opinion in regard to this kind of telegraph, I sent a description of it to Mr Clark[2] engineer to the London telegraph company, he returned the paper with the following remark "I have no doubt that your ingenious telegraph will work well on short lines, but for long distances it is my opinion that

it will not do so well". I do not see why it would not work on long lines as well as short ones. I have the honour to be
    revered | Sir | Your most | obedient servant | R. Smith
Prof. Faraday | Royal Institution | London

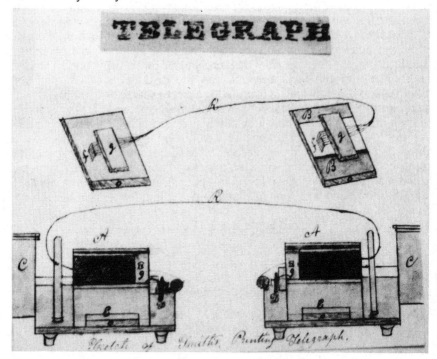

Description of Smiths New Electro Printing Telegraph,
This kind of Telegraph is so constructed that the same instrument both communicates and indicates. One of the instruments A.A. is placed at each station on the line, a communication is set up with types as f. and placed upon the board Fig. 1. and pute into the gro[o]ve E. of the instrument at one station. And a sheet of papir [sic] which is keept constantly wet with a solution of ferro-cyanide of potassium, to which has been added a few drops of Nitric Acid is laid upon the lead block B.B. Fig 2, and pute [sic] into E of one of the instrument at a distant station, The cathode electrode of the Voltaic battery communicates with the leaden black, and the anode with the types. The slides g.g. carries the opposite ends of a number of wires insulated and placed together in the form of a rope, and which passes along the line C.C. is boxes containing clock work, or a system of pulleys D.D. an electro-magnet connected with the battery. As soon as the electric circuit is Formed, the guard is attracted by the magnets the motion

of which lifts a small lever which holds the slides containing the wires and as soon as the electro-magnet liberates the slides, the clock-work or the arrangement of pulleys moves the slides in the gro[o]ves of the instrument at the same time. When the wires are in contact with the surface of the types the electricity passes along to the other instruments containing the moistened paper. Consequently the ferro-cyanide of potassium is decomposed, the acid attacks the iron, and cyanurate of iron is formed, which produces blue lines upon the paper, but when the points of the wires are not in contact with the types the electric current is interrupted and the points ceases to mark so that a positive copy or impression of the types is produced upon the paper. Instead of the types a message may be written upon paper with a kind of metallic ink, and which may be transmitted in the same way. As soon as the wires passes over the types and paper the motion is reversed, the Slides are taken back to there [sic] former position, the boards E.E. are then drawn out till a new surface of the types and paper is presented to the wires, and which is regulated by means of a scale, the circuit is again closed, and the machine operates as before, and so on, till the whole of the message is printed. Any number of wires may be used from one to fifty, and all intermediate. Stations may be supplied with branch wires in the usual way. In order to return an answer to a communication a board with a leaden block and paper is placed into the instrument where the types were, and of course types or hand writing into the other, and the electric current reversed.

The specimen sent[3] was produced by a small coursely constructed model, in the presence of Lord Berriedale[4] and Captain Duncan[5][.]
R. Smith

1. Unidentified.
2. Edwin Clark (1814–1894, DNB1 under Josiah Latimer Clark). Civil engineer and, from 1850, Engineer to the Electric Telegraph Company.
3. This is a slip of paper in IEE MS SC 2 with the word "Telegraph" printed on it (shown with the figure).
4. James Sinclair, styled Lord Berriedale (1821–1881, DNB). Amateur man of science and heir to the Earldom of Caithness.
5. Unidentified.

Letter 2375
Arthur-Auguste De La Rive to Faraday
27 January 1851
From the original in IEE MS SC 2
Mon cher Monsieur,

J'ai essayé depuis quelques jours de me remettre au travail & j'ai commencé par achever de corriger les épreuves du premier volume de mon ouvrage¹ dont l'impression est complètement achevé. Mais pendant l'intervalle qui s'est écoulé depuis le moment òu j'ai suspendu tout travail jusqu'à ce jour, il s'est fait beaucoup de choses nouvelles que j'ai été encore à temps d'introduire dans ce volume. Parmi ces choses la plus importante est, je crois, les nouvelles recherches que vous venez de faire & dont le *Phil Mag* de Janvier donne un extrait malheureusement bien abrégé². Il est en particulier un point sur lequel je desirerais vivement avoir plus de détails, c'est le *magnétisme* de l'oxygène, la manière dont vous le prouvez & la conséquence que vous en tirez - Vous serait-il possible de me donner une notice abrégé sur ce point si intéressant afin que je ne commette point d'erreur dans ce que j'en dirai dans mon ouvrage & que je puisse en parler avec un peu de détails? J'ai fait suspendre le tirage de la feuille où je traite ce sujet afin de pouvoir y faire l'addition par laquelle je réclame de votre bonne amitié quelques renseignements³. -

Excusez, Monsieur & très cher ami, mon indiscretion & noyez aux sentiments affectueux de votre devoué & malheureux ami | Aug de la Rive

Geneve, le 27 janv 1851.

Address: Profr Faraday F.R.S. &c &c | Royal Institution | Albemarle Stt | Londres

TRANSLATION
My dear Sir,

I have been trying for several days to get back to work and I began by finishing the correction of the proofs of the first volume of my work¹ of which the printing has been completed. But in the time that has elapsed between when I ceased working and today, many new things have been discovered which I was just in time to include in this volume. Amongst these things the most important is, I believe, the new research that you have just completed and of which the January edition of the *Phil.Mag.* contained an all too brief extract². There is a point on which I would particularly like to have more details, that is the *magnetism* of oxygen, the method by which you can prove it and the corollaries that you draw from this. Would it be possible for you to send me an abridged note on this point of such great interest, so that I may commit no errors in what I shall say in my work and so that I can speak of this with a few details? I have held the printing of the page where I deal with the subject so that I can make the necessary additions and I am relying on your friendship for a little information³.

Please excuse, Sir and very dear friend, my indiscretion and drown in the affectionate sentiments of your devoted and unhappy friend | Aug de le Rive

Geneva, this 27 January 1851

1. De La Rive (1853–8), 1.
2. *Phil.Mag.*, 1851, 1: 68–75 which reported the reading of Faraday (1851b, c, d, e), ERE24, 25, 26 and 27.
3. De La Rive (1853–8), 1: 468–72.

Letter 2376
Mervyn Herbert Nevil Story-Maskelyne[1] to Faraday
1 February 1851
From the original in IEE MS SC 2

The Ashmolean Museum, Oxford. | February 1, 1851.
My dear Mr. Faraday
 I have been considering about a subject for my Friday Evening[2], and I have come to the conclusion that one which I thought of some months ago will make a very good one, and though it has been alluded to by Grove in an Evening he gave[3], I am not aware that it has been at all enlarged on in any of our Friday Evenings. It is the discovery by Pasteur[4] of the two acids in Paratartaric acid [(]see Annales de Chimie 1850 p.56[5][)], the one turning the Polarised ray to the right, the other Polarising it circularly to the left. I think with Darkers[6] Lime Light Apparatus I could make the Phenomena sensibly visible to the audience.
 But my difficulty will be in getting the salts of the acid to exhibit. The Paratartaric acid has only once been produced & though Brodie has a little, it is but a little.
 Might I ask if you are acquainted with Pasteur, and if so would you do me the great kindness of writing to him to ask if he could give me a little of the acid, or specimens of his salts? If this is imposing a task in any way unpleasant to you, I beg you will not think of doing so, but I apply to you as thinking that you are the most probable person to be acquainted with him whom I know.
 The title of my Paper I suppose should be something of this sort "On the connection of chemical Forces with the Polarization of Light" - some facts might be added about Turpentine &c which would give the Lecture a general character such as would be represented by the above title, or perhaps I had better adhere to the single fact of Pasteur & call the Evening "On the mode in which certain Organic acids affect Polarized Light".
 Would you do me the kindness of selecting or altering in any way one of these titles. I feel that I am intruding on your kindness rather, in asking you for so much advice & assistance in a matter which ought to be entirely my own, but I know your interest in our Friday Evenings and I

need not to say how I share it with you. I must make this my excuse for asking you for your aid in procuring specimens or information from M. Pasteur, and can only promise in return that I will do all in my power to make the Evening as little unworthy as I can of the Royal Institution.

I dare say you will do me the kindness of communicating the title of my paper, to Mr. Barlow.

I remain, Ever, | My dear Mr. Faraday, | Yours most faithfully | Nevil Story Maskelyne
To Michael Faraday Esq | &c &c &c

1. Mervyn Herbert Nevil Story-Maskelyne (1823–1911, DNB2). Mineralogist at Oxford.
2. Story-Maskelyne (1851), Friday Evening Discourse of 28 March 1851.
3. See *Athenaeum*, 26 January 1850, p.106 for an account of Grove's Friday Evening Discourse of 18 January 1850 "On some recent Researches of Foreign Philosophers".
4. Louis Pasteur (1822–1895, DSB). Professor of Chemistry at Strasbourg University, 1849–1854.
5. Pasteur (1850).
6. William Hill Darker. Scientific instrument maker of 9 Paradise Street, Lambeth. Clifton (1995), 76.

Letter 2377
Faraday to Thomas Bell
3 February 1851
From the original in RS MS RR 2.214

Royal Institution | 3 Feby 1851

My dear Sir

As far as I have have any power of judgment I find reason to commend Coll Sabines paper[1] as a most important contribution to Magnetic science. It resolves a multitude of apparently wild extravagances of action into an important degree of order and develops a new law of periodicity amongst terrestrial effects which can be explained only after it is (as in this paper) made known. I cannot doubt that Terrestrial magnetism is rapidly advancing when such a paper as Sabines comes forth and not only explains but provokes others also to give explanations[.]

Ever Truly Yours | M. Faraday
Thos. Bell Esq | &c &c &c

1. Sabine (1851a).

Letter 2378
Faraday to Arthur-Auguste De La Rive
4 February 1851
From the original in BPUG MS 2316, f.67–8

Royal Institution | 4 February 1851

My dear De la Rive,

My wife & I were exceedingly sorry to hear of your sad loss[1]: it brought vividly to our remembrance the time when we were at your house; and you, & others with you, made us so welcome[2]. What can we say to these changes but that they shew by comparison the Vanity of all things under the Sun[3]. I am very glad that you have spirits to return to work again; for that is a healthy & proper employment of the mind under such circumstances[.]

With respect to my views & experiments[4] I do not think that anything shorter than the papers (& they will run to above 100 pages in the Transactions[5]) will give you possession of the subject; because a great deal depends upon the comparison of observations in different parts of the world, with the facts obtained by experiment and with the deductions drawn from them: but I will try to give you an idea of the root of the matter. You are aware that I use the phrase *line of magnetic force* to represent the presence of magnetic force & the direction (of polarity) in which it is exerted; and, by the idea which it conveys one obtains very well, and I believe without error, a notion of the distribution of the forces about a bar magnet or between near flat poles presenting a field of equal force; or in any other case. Now if circumstances be arranged so as to present a field of equal force, which is easily done as I have shewn by the Electro magnet, then if a sphere of iron or nickel be placed in the field it immediately disturbs the direction of the lines of force for they are concentrated within the sphere. They are, however, not merely concentrated but *contorted*; for the sum of forces in any one section across the field is always equal to the sum of forces in any other section; and therefore their condensation in the iron or nickel cannot occur without this contortion. Moreover the contortion is easily shewn by using a small needle, ($\frac{1}{10}$ of an inch long) to examine the field: for, as before the introduction of the sphere of Iron or nickel, it would always take up a position parallel to itself: afterwards it varies in position in different places near the sphere. — That being understood let us then suppose the sphere to be raised in temperature; at a certain temperature it begins to lose its power of affecting the lines of magnetic force, and ends by retaining scarcely any; so that as regards the little needle mentioned above, it now stands every where parallel to itself within the field of force. This change occurs with iron at a very high temperature & is passed through within the compass, apparently, of a small number of degrees: With Nickel it occurs at much lower temperatures being effected by the heat of boiling oil.

Now take another step. Oxygen as I shewed above three years ago in the Philosophical Magazine for 1847, vol. 31, p.p. 410, 415, 416[6], is magnetic in relation to Nitrogen & other gases. E. Becquerel, without knowing of my results, has confirmed & extended them in his paper of last year[7], and given certain excellent measures. In my paper of 1847 I shewed, also, that oxygen (like iron & nickel) lost its magnetic power & its ability of being attracted by the magnet when heated p.417; and I further shewed that the temperatures at which this took place was within the range of common temperature; for the oxygen in the air i.e the air altogether increased in magnetic power when cooled to 0°F. page 406.

Now I must refer you to the papers themselves for the (to me) strange results of the incompressibility (magnetically speaking) of oxygen and the inexpansibility of nitrogen & other gases:- for the description of a differential balance by which I can compare gas with gas: or the same gas at different degrees of rarefaction:- for the determination of the true zero or point between Magnetic & diamagnetic bodies:- and for certain views of magnetic conduction & polarity. You will there find described certain very delicate experiments upon diamagnetic and very weak magnetic bodies concerning their action on each other in a magnetic field of equal force: the magnetic bodies repel each other and the diamagnetic bodies repel each other; but a magnetic & a diamagnetic body *attract* each other: and these results combined with the qualities of oxygen as just described convince me that it is able to deflect the lines of magnetic force passing through it just as iron or nickel is; but to an infinitely smaller amount; and that its power of deflecting the lines varies with its temperature & degree of rarefaction.

Then comes in the consideration of the atmosphere and the manner in which it rises & falls in temperature by the presence & absence of the Sun. The place of the great warm region nearly in his neighbourhood;- of the two colder regions which grow up & diminish in the northern & southern hemispheres as the sun travels between the tropics;- the effect of the extra warmth of the Northern hemisphere over the Southern;- the effect of accumulation from the action of preceding months;- the effect of dip & mean declination at each particular station;- the effects that follow from the noncoincidence of magnetic and astronomical conditions of polarity, meridians and so forth;- the results of the distribution of land & water for any given place;- all these and many other things I must refer you for, to the papers. I could not do them justice in any account that a letter could contain, and should run the risk of leading you into error regarding them. But I may say that, deducing from the experiments & the theory what are the deviations of the magnetic needle at any given station, which may be expected as the mean result of the heating & cooling of the atmosphere for a given Season & hour, I find such a general accordance with the results of observations, especially in the direction & generally in the amount for different seasons of the *declination* variation, as to give me the strongest hopes that I have assigned the true physical cause of those variations and shewn the modus operandi of their production.

And now, my dear De la Rive I must leave you and run to other matters. As soon as I can send you a copy of the papers I will do so and can only say I hope that they will meet with your approbation. With the kindest remembrances to your Son[8] Believe me to be My dear friend
Ever Truly Yours, | M. Faraday

1. That is the death of De La Rive's wife, Jeanne-Mathilde De La Rive on 18 August 1850. Choisy (1947), 50.
2. In 1835. See Faraday to Magrath, 19 July 1835, letter 807, volume 2.
3. See Ecclesiastes 1:2-3.
4. Which De La Rive had asked for in letter 2375.
5. Faraday (1851c, d, e), ERE25, 26 and 27.
6. Faraday (1847b).
7. Becquerel (1850a). See also letters 2363, 2364, 2372 and 2373.
8. William De La Rive.

Letter 2379
John Tyndall to Faraday
4 February 1851
From a typescript in RI MS T TS, volume 12, pp.4002-5

Marburg | Feb. 4th. 1851

Dear Sir,

Your last kind letter informed me that you were occupied with terrestrial magnetism[1]. I had however read with deep interest previously that you had arrived at the probable origin of that hitherto enigmatical phenomenon - the variation of the needle[2].

During the last three or four months I have worked at the hem of the same great garment. My belief in the living interest you feel in the progress of science encourages me to lay a brief abstract of my investigation before you. It relates however, not to terrestrial magnetism but to electromagnetism.

The subject of the investigation embraces the following four propositions:-

1. To determine the general relation of the strength of an electromagnet and the mutual attraction of the magnet and a mass of soft iron when both are in contact.

2. A constant force, being opposed to the pull of the magnet being applied to the mass of soft iron, to determine the conditions of equilibrium between this force and magnetism, when the distance between the magnet and the force varies.

3. To determine the general relation between force and distance, that is to say, the law according to which the magnetic force decreases when the distance between the magnet and mass of soft iron is increased.

4. To determine the general relation between the strength of a magnet and the mutual attraction of the magnet and a mass of soft iron, when both are separated by a fixed distance.

The first proposition relates to the so called "lifting power" of the magnet which, as you are well aware, has been the subject of manifold investigation. The results however heretofore obtained are incapable of being reduced to anything like law.

To avoid the causes of divergence complained of by previous experimenters a peculiar method of experiment has been adopted, and instead of irregular masses of iron I have made use of spheres. The coincidence of the results is truly surprising.

The reply to the first proposition is, *that the force with which the magnet and the sphere cling together is directly proportioned to the strength of the magnet.*

The "strength of the magnet" is measured by the intensity of the current which circulates in the surrounding helix and the current was measured by means of a galvanometer of tangent[.] In the investigation the tangent of the angle of deflection is taken as the measure of *the strength of the magnet*.

The reply to the 2nd. proposition is, that when the distance between the magnet and the sphere varies, and a constant force opposed to the magnet is applied to the latter, to hold this force in equilibrium *the strength of the magnet must vary as the square root of the distance*.

I ought to mention that the "distances" are very small - the unit of distance is $\frac{1}{1000}$ of an inch being the thickness of a leaf of foreign post paper; by placing a number of such leaves between the sphere and magnet the distance could be varied at pleasure.

The reply to the 3rd. proposition is, *that the force varies inversely as the distance*.

You may perhaps find some little difficulty in separating the 2nd. proposition from the 3rd. This will vanish when you consider, that, in the former case a constant force (a weight) operated against the magnet, and the question was one between *magnetism and distance*:- in the latter case, the magnetism is preserved constant, and the question is one between *weight and distance*.

The 4th. proposition embraces the rather celebrated law of Lenz[3] and Jacobi[4] - who solved it by direct experiment[5]. It can however be deduced *á priori* from the 2nd. and 3rd. proposition just noticed - I have submitted the deduction to experimental test and found the coincidence remarkably close.

The answer to the 4th. proposition is, that *the attracting force is directly proportional to the square of the strength of the magnet.*

This latter law holds good when a distance of little more than $\frac{1}{1000}$ of an inch separates sphere and magnet. Is it not most singular that this small distance should so entirely change the nature of the law? In *contact*, as before remarked, the attracting force is proportional *to the strength of the magnet simply.*

A most remarkable analogy exists between some of the results established and the formulae which Poisson has developed for electrified balls[6]. I am not at all surprised that Prof. Barlow[7] arrived at the notion, that magnetism is a surface phenomenon[8]. As far as I am able to judge at present the whole might be explained on this supposition.

A memoir containing an account of the investigation accompanies this letter to the office of the Philosophical Magazine. The memoir will, I trust, appear on the 1st. March[9].

When Science is a Republic as you say it gains, and yet I dare affirm that no living man knows better than yourself how little benefit is to be derived in this way in comparison with that which results from the solitary communion of the individual with nature. There are people in the world who are very fond of what they call "composite ideas". They imagine, if six men come together and talk on a matter, that more will be elicited than if one man held his tongue and simply *thought* over it. I must say that I have little faith in the proceeding, and if I needed an authority to confirm me in my scept[ic]ism I should without hesitation turn to Professor Faraday.

I remain dear Sir | Most faithfully yours, | John Tyndall.
Professor Faraday | etc. etc.

1. Letter 2344 which Tyndall received on 23 November 1850. Tyndall, *Diary*, 23 November 1850, **5**: 12.
2. No evidence has been found which suggests that Faraday's work was publicly known before its reading to the Royal Society on 28 November 1850. Faraday (1851c, d, e), ERE25, 26 and 27. Possibly Tyndall had read the summary in *Phil.Mag.*, 1851, **1**: 68–75 and forgotten when he had received Faraday's letter.
3. Heinrich Friedrich Emil Lenz (1804–1865, DSB). Russian physicist.
4. Moritz Hermann von Jacobi (1801–1874, DSB). Member of the Imperial Academy of Sciences at St Petersburg.
5. Jacobi and Lenz (1839).
6. Poisson (1811).
7. Peter Barlow (1776–1862, DSB). Professor of Mathematics at the Royal Military Academy, Woolwich.
8. Barlow (1823), 181.
9. Tyndall (1851a).

Letter 2380
Faraday to John Frederick William Herschel
8 February 1851
From the original in RS MS HS 7.191

Royal Institution | 8 Feby 1851

My dear Sir John

I find I made an error last night which is almost inexcusable for me & would be if not considered as a result of my most treacherous memory[.] But as I suppose Mr Barlow had said it is when Lady members *come* that they can take their two friends in with them to the reserved seats[.] Our president the Duke [of Northumberland] was asking for you last night I trust he met with you after Owens discourse[1] was over[.]

Ever My dear Sir John | Your faithful Servant | M. Faraday
Sir John Herschell Bart | &c &c &c

1. Owen (1851), Friday Evening Discourse of 7 February 1851.

Letter 2381
Faraday to Richard Owen
8 February 1851
From the original in RI MS F1 D2

8 Feby 1851

Oh Owen! I got into a sad hobble last night[1]. All the time I was talking with you they were searching for me (and you & Mrs. Owen[2]) to take you up stairs[.] My wife was half angry with me but she is a good creature as yours is also & I hope you will all forgive me[.]

Ever Yours | M. Faraday

1. Owen (1851), Friday Evening Discourse of 7 February 1851.
2. Caroline Owen, née Clift (d.1873, age 70, GRO). Married Owen 20 July 1835. See DNB under his entry.

Letter 2382
Richard Adie[1] to Faraday
19 February 1851
From the original in RI MS F2 J120

Liverpool 19 February | 1851

Sir

I take the liberty to intrude upon you with a small specimen of a preparation from the acetate of zinc in the hope that it and a brief notice of some of its properties may not be uninteresting to you. The acetate of zinc is ignited on a plate of silver copper or any other convenient metal[.] The residue contains a few darker coloured particles which for magnetic properties are very little inferior to similar portions of iron, with a Common Sheffield magnet they can be drawn along the paper and examined in the microscope, the magnet to be used underneath the paper only[.] Besides zinc I have obtained similar compounds varying in degree from all the heavy metals met with in the arts, that is all I can get to try yeild [sic] particles that will move to a magnet on paper several of them are the partially reduced oxides merely but the more difficult ones tin, silver, mercury, platina, require the Carbon from the decomposition of a vegetable acid to assist them, for bismuth and zinc also a vegetable acid is used but the particles of these two when obtained are strongly attracted when compared with the others.

Perhaps I should mention that the account of these I have submitted to Professor Jameson[2] in the hope that he may find them worthy of a place in his journal[3], indeed it was in making some expts expressly for a paper for him that the fact of so many metals yeilding magnetic products came out. The experiments in question were to show that among bodies of like constitution the darker coloured ones are the most attracted by a magnet, latterly the coralla of flowers promise to give evidence of this fact, at least where they do not contain too much water or odoriferous matter to interfere, for example the pale yellow everlasting flower and the pile most coralla with varnished like surfaces are dia-magnetic, while common gorse and dandelion two similar yellow colours without the glossy surface are attracted freely, save the everlasting flower the others contain their natural quantity of moisture[.]

Yours very respectfully | Richard Adie

Endorsed by Faraday: The particles contained Iron. On digesting the grey powder in pure dilute MA the zinc deposited & left a few black particles & a little black powder - this well washed & digested in strong ether acid - dissolved in part forming a yellow solution as if iron & this being diluted & tested by ferro-pruss potash gave prussian blue.
22 Feby 1851 MF
Wrote Mr Adie word on the Monday 24th

1. Richard Adie (1810–1880, P3). Chemist and meteorologist in Liverpool.
2. Robert Jameson (1774–1854, DNB). Editor of the *Edinb.New Phil.J.*
3. Adie (1851).

Letter 2383
Pieter Elias[1] to Faraday
24 February 1851
From the original in IEE MS SC 2

Haarlem | ce 24 Février 1851.
Monsieur!

Monsieur Logeman, mécanicien d'ici s'est chargé de vous transmettre ces lignes.

Puisque vous avez bien voulu faire quelque attention aux aimants travaillés et aimantés à ma manière par MM Logeman et van Wetteren[2], j'espère que vous me permettrez de vous adresser quelques renseignements au sujet des instruments magnétiques exposés par ces Messieurs.

Il y a d'abord trois aimants en forme de fer à cheval, le premier pesant environ 90 livres (av.d.p.), le second 6 livres et le troisième une livre. Le premier est capable de supporter un poids de 500 livres à 550 livres, le second 100 livres et le troisième 29 livres. Quant au grand, il est assez difficile de le charger comme il faut, de manière que la pièce de fer doux ne glisse pas, mais soit arrachée tout droit.

Ensuite M. Logeman expose une machine magnéto-électrique d'une construction assez simple et d'une force assez grande pour être employée à quelque fin utile, au télégraphe p.e., à la dorure ou à autre chose.

Enfin ces Messieurs exposent une boussole de marine ordinaire. Quant à celle ci, veuillez bien me permettre d'entrer en quelque détails. Les marins, comme on sait, jugent de la bonté d'une boussole par la vitesse de ses oscillations, et ils ont raison: à diamètre égal de la rose des vents, la boussole, qui dans un tems donné fera un plus grand nombre d'oscillations qu'une autre, sera retenue à sa place avec plus de force, sera donc plus indépendante des mouvements du navire, aura en un mot plus de stabilité (steadiness). La rose des vents étant donnée, la vitesse des oscillations doit dépendre de la force et des dimensions de l'aiguille magnétique. Une aiguille trop petite ne sera pas capable de vaincre assez le moment d'inertie de la rose; elle lui imprimera un mouvement lent. Une aiguille trop lourde aura par elle même un grand moment d'inertie, qui accroit en plus grande portion que son magnétisme. Donc elle fera de lentes oscillations, d'autant plus quand elle est grevée de la rose des vents. Il faut donc qu'il y ait entre deux une grandeur de l'aiguille qui pour une rose des vents donnée soit la plus convenable. Laquelle?

Afin de pouvoir répondre à cette question il faudrait 1° connaitre la meilleure forme de l'aiguille (les meilleures proportions de ses dimentions linéaires), et 2° la loi selon la quelle le *moment magnétique* accroit avec l'accroissement égal de toutes les dimentions linéaires des aiguilles.

Or, j'espère pouvoir prouver, quoique je n'aye encore rien publié la dessus, que les moments magnétiques des barreaux, *dont toutes les dimentions linéaires sont entre elles dans la même proportion*, sont en raison des cubes des dimentions linéaires (ce qui diffère entièrement de la loi des attractions). La meilleure forme de l'aiguille étant supposée comme, et la mécanique donnant les moyens de trouver le moment d'inertie tant de l'aiguille elle même que de la rose des vents, on peut par le calcul des maxima et minima trouver les dimentions d'une aiguille qui communiquera à une rose des vents donnée les oscillations les plus rapides.

L'aiguille de la boussole exposée par MM. L. et v.W. à [sic] été faite d'après mes calculs. La rose des vents est de masse ordinaire ($\frac{1}{32}$ livre av.d.p. environ) et son diamètre est d'un peu plus que $6\frac{1}{3}$ pouce Anglais. Son tems d'oscillation est ici de 4,8 secondes.

Je me flatte qu'elle pourrait encore être utile aux marins dans des régions où des boussoles médiocres refusent leurs services.

Pardonnez moi, Monsieur, de vous avoir occupé si longtems. C'est que notre plus grand désir serait que vous daigniez jeter un regard sur ces instruments et qu'ils puissent être trouvés pas entièrement indignes de votre attention[3].

Agréez l'expression de mes sentiments les plus respectueux, | P. Elias
Monsieur M. Faraday à Londres.

TRANSLATION

Haarlem | this 24 February 1851

Sir!

Mr Logeman, an engineer from here, has agreed to convey these few lines to you.

Since you have expressed some interest in magnets manufactured and magnetised using my methods by Messrs Logeman and van Wetteren[2], I hope that you will allow me to address some points on the subject of the magnetic instruments demonstrated by these gentlemen.

There are first of all three horse shoe magnets, the first weighing around 90 pounds (avoirdupois), the second 6 pounds and the third one pound. The first is capable of withstanding a weight between 500 and 550 pounds, the second 100 pounds and the third 29 pounds. As to the large one, it is quite hard to position it properly, in such a way that the piece of soft iron does not slide, but is lifted straight out.

Then M. Logeman will show a magneto-electric machine of a fairly simple construction and of a force large enough to be employed to some useful end, the telegraph, perhaps, gilding or some other thing.

Finally, these gentlemen will demonstrate an ordinary marine compass. As to this, please allow me to go into a little detail. Sailors, as one knows, judge the quality of a compass by the speed of its oscillations and they are right: given a compass card of equal diameter, the compass that in a given time will make a larger number of oscillations than another, is the one that will be kept in its place with more force, will therefore be more independent of the movements of the ship, and will, in one word, have more steadiness. Given a particular compass card, the speed of the oscillations must depend on the force and the dimensions of the magnetic needle. A needle that is too small will not be able adequately to overcome the moment of inertia of the compass; it will give it a slow movement. A needle that is too heavy will of itself have a great moment of inertia, that will grow disproportionately to its magnetism. Thus it will make slow oscillations, all the more so when it is hampered by the compass card. There must therefore be an ideal size of needle for a given compass card. But which?

In order to be able to reply to this question one must 1st know the best form of needle (the best proportions and its linear dimensions) and 2nd the law by which the *magnetic moment* increases in line with the linear dimensions of needles.

Now I hope to be able to prove, although I have not yet published anything on this, that the magnetic moments of bars, *of which all the linear dimensions are in proportion to each other*, are in fact cubes of the linear dimensions (which differs entirely from the law of attraction). Assuming one knows the best form of a needle and the mechanics giving the means of finding the moment of inertia both of the needle itself and of the compass card, one can by calculating the maxima and minima find the dimensions of a needle which will give to a compass card the fastest oscillations.

The needle of the compass demonstrated by Messrs. L. and v.W. was made following my calculations. The compass card is of ordinary mass ($\frac{1}{32}$ pound avoirdupois approximately) and its diameter is a little over $6\frac{1}{3}$ English inches. Here it oscillates in 4.8 seconds.

I flatter myself that it could still be useful to sailors in the regions where mediocre compasses refuse to work.

Forgive me, Sir, for having occupied you for so long. This is because our greatest desire would be that you deigned to cast an eye on these instruments and that they would be found not entirely unworthy of your attention[3].

Please accept the expression of my most respectful sentiments | P Elias

Mr M. Faraday in London

1. Pieter Elias (1804–1878, NNBW). Scientific instrument maker in Haarlem.
2. Unidentified.
3. See Anon (1851), **3**: 1147.

Letter 2384
Henry Thomas De La Beche to Faraday
26 February 1851[1]
From the original in the possession of Elizabeth M. Milton

Wednesday Morng. | Jermyn St. | 26 Febr.

My dear Faraday

Are you inclined for the levée this fine morning - if you are, and can get ready in time, I will, if you would approve, call for you at the R. Institution at 1 o'clock[.] What say you?

Snow Harris, Playfair and self are going in a Clarence, and there is, therefore, a seat vacant, if you will take it.

Ever sincerely | H.T. De la Beche

1. Dated on the basis that De La Beche, Harris and Playfair attended the levée (PRO LC6 / 13). Faraday did not attend as he was at a meeting of the Senate of the University of London (ULL MS ST 2/2/3, p.1).

Letter 2385
Jacob Herbert to Faraday
27 February 1851
From the original in GL MS 30108/2/88

Trinity House | 27, Febry 1851.

My dear Sir,

I have been instructed to request that you will favor me, for the Board's information, with your Opinion whether fresh Water, deposited in Leaden Cisterns is likely to obtain any quality rendering it prejudicial if used for culinary purposes, or for drinking.-

I remain, | My dear Sir, | Your very faithful Servant | J. Herbert

M. Faraday Esq.

Letter 2386
Faraday to Jacob Herbert
28 February 1851
From the original copy in GL MS 30108/2/88

Royal Institution | 28 Feb 1851

My dear Sir

In reply to your letter of yesterday[1] containing the enquiry whether fresh water deposited in leaden cisterns is likely to obtain any quality rendering it prejudicial if used for culinary purposes or for drinking; I may state that there are cases where such injury has occurred & may occur. The liability is proportionate to the purity of the water in respects of *Saline* matter. Thus pure distilled water will act freely on lead - pure rain water will act sometimes - some spring waters are so pure that they act on the lead of pipes & cisterns. In thus acting they tend to coat the lead with an insoluble compound that serves more or less as a protection, and the protection is more perfect as the water is less pure in salts; especially of a certain kind. The consequence is that *new* lead cisterns & pipes injure such waters as act upon them *more at first* than afterwards - and again that a water which being not quite pure can be injured by a new cistern or pipe, ceases to be injured after some time[.]

The cases of injury in practice are not numerous, though they occur. The multitude of instances in which the water is *not* injured is manifest in London where leaden pipes & cisterns innumerable are properly & safely employed for the conveyance and retention of fresh water. Doubtful cases have to be examined each by itself[2][.]

I am My dear Sir | Your faithful Servant | M. Faraday

1. Letter 2385.
2. This letter was read to Trinity House Court, 4 March 1851, GL MS 30004/24, pp.331-2.

Letter 2387
Faraday to an unidentified correspondent
1 March 1851
From the original in HCL, Quaker Collection

Royal Institution | 1 Mar 1851

Sir

I hasten to inform you that I do not take pupils nor am I in any way professional. I have not published any work on Electro-magnetism other than my papers in the Philosophical Transactions[.] Such of these as were collected some years ago into two volumes 8vo under the original title of Experimental Researches were published by Taylor of Red Lion Court Fleet Street[1][.] I do not think they or the Transactions are out of print but do not know.

I am Sir | Your Obedient Servant | M. Faraday
Geo [name illegible] Esq | &c &c &c

1. Faraday (1839b, 1844b).

Letter 2388
Faraday to Christian Friedrich Schoenbein
5 March 1851
From the original in UB MS NS 395

Royal Institution | 5 March 1851.

My dear friend

I had your hearty Christmas letter in due time[1] - and was waiting for the papers referred to in it when lo! they arrived about 4 days ago and your friend Professor Bolley[2] called and left them & his card but not his address. I was ill & I believe in bed and could not see him. I have not been out of the house for a week or more because of inflamed throat & influenza - being unable to speak & obliged to give up lecturing but I am now improving & trust I shall see the Professor soon[.] The papers & the specimens of oil of turpentine are all quite safe & most valued treasures[.] I have read the papers through and I think you must now begin to rejoice in ozone for though it has cost you a great deal of trouble & work still it has surely made wonderful way & what is more is progressing & will progress. Though you may sometimes get tired of it still I think you never take it up afresh without being rewarded. I have been consulting with a medical friend about the medical paper & he (Dr. Bence Jones) recommends that it be sent to the Medico chirurgical Society - where it will be introduced *at once* into the minds of the Medical Profession and appear in the transactions[3][;] tomorrow we shall meet again when he will have read the paper & we shall decide. The chemical paper I have sent off at once to the Chemical Society[4] - it will appear there in time for me to have access to & use of it on my or rather your evening which I expect will be the 13th June[5] or the middle of our great exhibition. When I drew out a sort of preliminary sketch of the subject I was astonished at the quantity of matter - real matter - and its various ramifications - and it seems still to grow upon one. What you will make it before I begin to talk, I do not know[.]

I do not as yet see any relation between the magnetic condition of oxygen & the ozone condition but who can say what may turn up? I think you make an inquiry or two as to the amount of magnetic force which oxygen carries into its compounds. This is indeed a wonderful part of the story for magnetic as *gaseous oxygen* is the substance seems to lose all such force in compounds. Thus water which is $\frac{8}{9}$th oxygen contains no sensible trace of it: and peroxide of iron which itself consists of two most magnetic constituents - is scarcely sensibly magnetic so little have either of these bodies carried their forces into the resulting compound. Sometimes I think we may understand a little better such changes by thinking that magnetism is a physical rather than a chemical force but after all such a difference is a mere play upon words & shews ignorance rather than understanding. But you know there are really a great many things we are as yet ignorant of - and amongst the rest the infinitesimal proportion of our knowledge to that which really is *to be known*. I have a copy of my last papers[6] ready for you & if Professor Bolley can take charge of it shall give it into his hands.

I read your theory of the pile in the Geneva journal with great pleasure and go with you I think to the full extent[7]. My mind was quite prepared for the view years ago. I do not suppose you ever see the back numbers of an old work which still drags its slow length along or else you would see that at Paragraph 949, 950[8], and again 1164[9] and 1345, 1347[10], and elsewhere that I was ready to agree with you 10 or 15 years back.

I have no doubt I answer your letters very badly but my dear friend do *you remember* that *I forget*, and that I can no more help it than a sieve can help the water running out of it. Still you know me to be your old & obliged & affectionate friend and all I can say is the longer I know you the more I desire to cling to you[.]

Ever My dear Schoenbein | Yours affectionately | M. Faraday

Address: Professor Schoenbein | &c &c &c | University | Bâsle | on the Rhine

1. Not found.
2. Alexander Pompius Bolley (1812–1870, P1, 3). Professor of Chemistry in Aarau, 1838–1854.
3. Schoenbein (1851b).
4. Schoenbein (1851a). This was read to the Chemical Society on 7 April 1851. *J.Chem.Soc.*, 1851, 4: 190.
5. Faraday (1851g), Friday Evening Discourse of 13 June 1851.
6. Faraday (1851b, c, d, e), ERE24, 25, 26 and 27.
7. Schoenbein (1850a).
8. Faraday (1834), ERE8, 949–50.
9. Faraday (1838a), ERE11, 1164.
10. Faraday (1838b), ERE12, 1345, 1347.

Letter 2389
Faraday to Edward William Brayley
5 March 1851
From the original in RI MS F1 D3

R Institution | 5 Mar 1851

My dear Sir

Herewith the ticket. Herewith also a copy of the proceedings[1] which I find on my table. Mr Barlow has managed the matter well I think. I will see what I can do with the committee as to sending a copy, it should come. I do not think there is any idea at present of re-collecting together the Old data.

I hope soon to send you my last four papers[2]. The subject is a very fine one & *I* am quite satisfied with its state. There is far more to do.

Ever Truly Yours | M. Faraday
E.W. Brayley Esq | &c &c &c

1. That is the first issue of *Proc.Roy.Inst.*
2. Faraday (1851b, c, d, e), ERE24, 25, 26 and 27.

Letter 2390
Thomas Graham to Faraday
5 March 1851[1]
From the original in the possession of Elizabeth M. Milton

4 Gordon Sqr | Wednesday March 5.

Dear Mr. Faraday
Our friend Hofmann has consented at the eleventh hour to be proposed at the Royal Society & will I have no doubt be highly gratified by your name to his certificate[2].

Very truly Yours | Tho. Graham
M. Faraday Esq | &c &c

1. Dated on the basis that Hofmann's certificate for the Royal Society was first read on 6 March 1851.
2. Faraday did sign Hofmann's certificate. RS MS Cert 9.275. Hofmann was elected on 5 June 1851.

Letter 2391
William Scoresby[1] **to Faraday**
6 March 1851
From the original in IEE MS SC 2

Torquay, 6th March 1851.

Dear Dr. Faraday,
Whilst *you* are working so successfully on great laws & phenomena, I am, as time admits, doing a little, or rather trying to do something, though in the matter to which I refer in this note, the results are merely negative.

I have been trying the effect of the magnetic condition, in iron, as also of a galvanic current, with a view to the inquiry, whether any alterations in *dimensions* are thereby occasioned? My experiments, within the limits of observation in my apparatus, indicate no sensible effect, except what belongs to temperature. The value of my scale, however, observed by reflection in a mirror, is not well determined on account of the different points of contact & leverage - the lengths not being easily determinable. I wish therefore to verify my present impression, as to the value by the interposition of a disc or slip of metal of *known* thickness. I thought that you, perhaps, might be able to advise me how I could procure such a disc or plate? The Mint has a fine rolling apparatus, which I believe has values attached? But you, possibly, might have some little bit of sheet metal of known dimensions. What I require is a disc or "blank" or slip of small size - that of a shilling[2] is large enough, and of a thickness of $\frac{1}{50}$th or any where from $\frac{1}{50}$th to $\frac{1}{100}$th of an inch. This I could interpose between two of my bars which would verify or correct my present estimations.

The apparatus which I constructed myself here, works so beautifully, that *a* degree of temperature produces near $\frac{1}{4}$ inch movement in the reflected scale - consequently the *quarter of a degree of temperature*, or the alteration in length due to that small quantity, is easily determinable.

Thinking the experiments might not be altogether uninteresting, though negative as to results, I am preparing a paper for the Royal Soc. thereon[3].

I am, my Dear Dr. Faraday, | Yours very faithfully, | Willm Scoresby

1. William Scoresby (1789–1857, DNB). Retired clergyman who lived in Torquay. Worked on magnetism.
2. That is 2.4cms.
3. No evidence has been found which suggests that Scoresby submitted such a paper.

Letter 2392
Faraday to William Scoresby[1]
c7 March 1851[2]
From the original in Whitby Literary and Philosophical Society MS

Royal Institution

My dear Scoresby

I have applied to Mr. Brande as clerk of the Irons at the Mint who tells me that they have not measured plates of metal - for though they gauge them they do it to weight & not to measure.

I have applied to Brockedon[3] who says he will if possible supply me with a piece of measured plate that I may send it to you by post - but I do not know when[.]

I have applied to Cowper & De la Rue both of whom refer me to Mr Whitworth[4] Engineer of Manchester who has perfect gauges &c &c &c[.]

They say he would supply such a thing instantly but I do not know him - perhaps you will write to him yourself[.]

I could find no change in the absolute bulk of Iron bismuth or other bodies under intense magnetic power[.] See Par 2752[5] of recent Experimental researches a copy of which I am just setting aside for you. It is quite a volume[6] or I would send it by post[.]

I suppose you remember Joules results in the Phil Mag for 1848 or 1849[7]. I think he makes out iron to become larger & narrower but cannot remember[.]

Ever Very Truly Yours | M. Faraday
Revd Dr. Scoresby | &c &c &c

1. William Scoresby (1789-1857, DNB). Retired clergyman who lived in Torquay. Worked on magnetism.
2. Dated on the basis that this is the reply to letter 2391.
3. William Brockedon (1785-1854, DNB). Painter, author and inventor.
4. Joseph Whitworth (1803-1887, DNB). Mechanical engineer in Manchester.
5. Faraday (1851c), ERE25, 2752.
6. Faraday (1851b, c, d, e), ERE24, 25, 26 and 27.
7. Joule (1847).

Letter 2393
William Spence[1] to Faraday
10 March 1851
From the original in RI MS Conybeare Album, f.37

18 Lower Seymour | St. | Portman Square | March 10, 1851

My Dear Sir,

As an addition to your series of Ransome portraits[2], I beg your acceptance of a lithograph profile of my dear old friend Mr. Kirby[3], from a pencil sketch of my eldest son[4] when on a visit at Sir W. Middleton[5] near Barham two years ago, & which having been taken when Mr. Kirby was not aware of what was going on, is an excellent likeness[6].

I am | My dear Sir | Yours very truly | W. Spence
M. Faraday Esq

1. William Spence (1783-1860, DNB). Entomologist.
2. That is for the portraits contained in RI MS F1 K.
3. William Kirby (1759-1850, DNB). Entomologist and incumbent of Barham, 1782-1850.
4. William Blundell Spence (1815-1900, TBKL). Artist and engraver.
5. William Fowle Middleton (1786-1860, AC). Country gentleman.
6. This portrait is in RI MS F1 I149.

Letter 2394
John Frederick William Herschel to Faraday
12 March 1851
From a copy in RS MS HS 23.102

March 12, 1851.
My dear Sir,
Many thanks for your *most liberal* provision of the healing liquid - the Caledonian water - which I shall make last me my life, and which for the donor's sake shall never be applied to meaner uses. I think I got a quieter night last night by it's application.

May I ask you for *two* admission tickets to the lecture of Friday next[1] for 2 friends (*not* for reserved seats but for the general circle.)

Your speculations on the diurnal & annual variations[2] as referable to the magnetization of oxygen took me quite by surprise. I had no Idea you had gone so far into the matter. I find them difficult to follow from the habitual use of so totally different a mode of "envisaging" the magnetic forces.

Believe me &c | (sd) J.F.W.H

1. Gull (1851), Friday Evening Discourse of 14 March 1851.
2. In Faraday (1851c, d, e), ERE25, 26 and 27.

Letter 2395
Faraday to John Frederick William Herschel
12 March 1851
From the original in RS MS HS 7.192

My dear Sir John
Herewith the tickets[1][.] I hope the Whisky will deserve a good character[.]

Ever Yours | M. Faraday
12 Mar 1851

1. Asked for in letter 2394.

Letter 2396
Lambert-Adolphe-Jacques Quetelet to Faraday
20 March 1851
From the original in IEE MS SC 2

Bruxelles | le 20 mars 1851
Mon cher ami,
J'ai reçu par Mm. Schlagintweit[1] le beau présent que vous m'avez fait: rien ne pouvait m'être plus agréable que de recevoir Votre portrait avec une inscription de Votre main qui témoigne de l'Amitié dont Vous

Voulez bien m'honorer. Je veux désormais avoir toujours ce portrait sous mes yeux; il se trouve dans le lieu le plus apparent de mon cabinet de travail, dont il fait le principal ornement.

Je Vous félicite sur les résultats remarquables que Vous avez obtenus au sujet des propriétés magnétiques du gaz oxygène. Notre Academie a qui j'ai cru pouvoir communiquer ce que Vous m'en avez dit dans votre dernière lettre[2], en a pris connaissance avec un Vif intéret[3]. j'aurais du vous exprimer mes remerciments dupuis longtemps; mais je voulais, en vous écrivant, Vous transmettre la suite de mes travaux sur l'électricité de l'air, travaux auxquels Vous avez bien Voulu prêter quelqu'attention[4]. Je trouve enfin, aujourd'hui, un moment pour satisfaire à ce désir.

Mon principal but en vous fesant cette communication est de vous prier d'examiner si vos découvertes récentes sur l'oxygène peuvent aussi rendre compte des Variations périodiques de l'électricité. Je ne doute aucunement que cet important sujet n'ait déjà fixé Votre attention. Mon[5] premier travail sur l'électricité contenait les résultats de quatre années et demie d'observations[6]; je puis y joindre aujourd'hui, une des deux années 1849 et 1850, et du commencement de 1851: peut être ne les verrez vous pas sans intêret. Ils confirment en général les principaux faits déduits de mes observations antérieures; et cependant l'année 1849 a présenté une anomalie très sensible pendant les sept premiers mois: l'électricité a été notablement inférieure à ce qu'elle est dans son état ordinaire: vous pourrez en juger par le *tableau* cijoint qui, outre les Valeurs moyennes et extremes de chaque mois *observés directement*, contient aussi les moyennes mensuelles, en ramenant à une même échelle les déterminations de chaque jour.

on y retrouve aussi la même loi de continuité; les mêmes différences entre l'hiver et l'été, pour autant, bien entendu, que le permettent les années considérées individuellement.

Rien n'est plus propre à faire apprécier la lacune qui existe encore dans nos connaissances relativement à l'électricité de l'air, que le doute qui entoure les anomalies que j'ai observées pendant la première partie de l'année 1849. Il m'a été impossible de trouver des observations qui pussent servir de contrôle aux miennes. Aucun observatoire, du moins à ma connaissance, ne publie jusqu'à ce jour des observations quotidiennes régulières sur l'électricité atmosphérique: c'est là une lacune trés facheuse, comme on ne tardera pas à le reconnaître.

Quand une année est remarquable par une température anormale, par des pluies excessives ou des sècheresses, tous les météorologistes s'accordent pour constater ces irrégularités dans le cercle où elles se sont manifestées: malheureusement il n'en est pas de même ici. J'ai consulté des recueils d'observations des plus estimés et je n'y ai trouvé aucun renseignement qui peut me satisfaire.

Il en résulte qu'on peut me demander si l'affaiblissement dans l'état électrique de l'air était un fait bien réel, ou s'il n'était qu'apparent; s'il n'était point du, par exemple, à un dérangement de mon instrument? et en supposant cet affaiblissement d'électricité bien constaté, était il purement local? toutes ces questions peuvent prendre d'autant plus d'importance que l'anomalie signalée coincidait à peu près avec le retour du fléau qui a si cruellement éprouvé nos populations[7]. ces difficultés m'ont porté à rechercher s'il n'y aurait pas moyen de rendre *un électromètre comparable à lui même à différentes époques.*

J'avais montré déjà, dans mon premier travail[8], qu'il est facile de comparer entre eux deux électromètres de Peltier[9], et de construire des tables d'équivalents pour leurs indications mais, quand une comparaison a été faite, et que l'instrument comparé a été transporté dans un autre pays; ou même, sans qu'il y ait eu transport, après qu'il s'est écoulé un certain temps, il importe de s'assurer que les indications n'ont pas Varié, qu'elles ont bien conservé leurs Valeurs absolues.

Cette Vérification peut se faire d'une manière très facile. il suffit en effet de s'assurer que l'aiguille a conservé toute sa mobilité et que sa force directrice est restée la même. or, cette force directrice, ici, est donnée par la petite aiguille aimantée attachée à l'aiguille indicatrice de l'électromètre. il suffira alors de soumettre la petite aiguille aimantée aux procédés ordinaires qui servent à constater son énergie magnétique; c'est à dire de la faire osciller librement dans un plan horizontal et de constater si son état magnétique est resté le même. On doit tenir compte, bien entendu, des corrections ordinaires employées en pareil cas pour la température, la torsion des fils, la variation de l'intensité horizontale du magnétisme tettestre, &c.

On conçoit que, par des procédés analogues, on peut faire dépendre aussi la détermination de la force électrique absolue de la terre, de celle de son magnétisme absolu: problème important, mais dont je n'ai point à m'occuper pour le moment. Il me suffit d'avoir établi qu'on peut par un procédé fort simple reconnaître qu'un électromètre de Peltier est resté comparable à lui même. Bien que la chose soit très simple en soi, je m'étonne que cette précaution n'ait pas été indiquée encore. Si l'idée m'était Venue de Vérifer ainsi l'aiguille directrice de mon électromètre, en 1849, et aux époques qui ont precédé et suivi, je n'aurais pas à rechercher aujourd'hui, si mon instrument a pu subir un dérangement temporaire, ni à m'occuper d'observations étrangères qui puissent contrôler les miennes.

Si vous pensez que ces remarques si simples puissent être de quelqu'utilité, ou qu'elles puissent porter d'autres physiciens à entreprendre des séries d'observations électriques si nécessaires à la Science, je vous prie de faire de cette lettre tel usage que vous jugerez convenable. Je me soumets entièrement à Vous, comme au juge le plus compétent en ces matières.

Je vous prie de vouloir bien présenter mes hommages à Madame Faraday, et de recevoir, mon cher et illustre confrère, mes compliments les plus respecteux,
Votre ami | Quetelet

Je ne vous parle pas de la mort d'Oersted[11]: cette nouvelle m'est d'autant plus penible que ce grand physicien m'honorait de son amitié. Oersted et Schumacher[12] à des termes si rapprochés!

Degrés d'électricité aux différents mois (observatoire royal de Bruxelles)

Moyenne des degrés observés à l'électromètre

	1844	45	46	47	48	49	50	51	1844–50
Janvier	..	50	50	63	50	39	49	52	40
Février	..	55	45	45	44	36	38	52	44
Mars	..	44	26	47	36	27	36	31(1)	35
Avril	..	27	23	30	27	20	19		24
Mai	..	26	19	21	18	16	22		20
Juin	..	18	18	18	18	13	14		17
Juillet	..	21	14	18	22	14	12		17
Aout	28	27	22	6	24	21	19		21
Septembre	29	29	23	17	24	24	28		25
Octobre	31	42	26	30	32	31	36		33
Novembre	33	44	41	35	36	45	34		38
Décembre	46	53	57	48	45	38	50		48
Année		36	30	31	31	27	30		31

(1) jusqu au 20 mars

Moyenne des nombres proportionnels

	1844	45	46	47	48	49	50	51	1844–50
		471	562	957	487	219	507	462	534
		548	256	413	295	163	180	532	309
		262	95	282	164	90	194	125	173
		93	94	221	155	132	70		128
		163	49	67	59	32	220		98
		51	39	47	48	27	24		39
		58	33	43	61	25	21		40
	90	89	57	11	64	92	55		65
	91	95	62	39	63	69	69		74
	110	299	98	107	120	122	172		147
	127	334	274	160	152	364	155		224
	340	742	799	356	281	304	451		468
	..	267	202	225	162	137	179		192
Année		49	44	46	39	36	41		43

Maxima

Janvier		65	71	77	76	65	74	75	71
Février		70	60	73	62	62	55	74	64
Mars		64	56	62	47	55	66		58
Avril		48	40	48	51	35	65		48
Mai		41	33	41	40	25	75		42
Juin		48	30	24	36	39	22		35
Juillet		43	32	31	44	24	25		33
Aout	36	45	37	23	38	34	44		37
Septembre		42	29	30	22	44	50		40
Octobre	48	67	55	48	54	57	67		56
Novembre	51	60	65	53	57	77	64		61
Décembre	67	73	74	66	65	74	71		70
[Année]		55	49	49	50	49	56		51

Minima

	32	8	38	19	0	0	12	16
	28	0	23	11	17	0	31	13
	25	0	21	19	0	13		13
	10	0	0	8	0	3		4
	0	0	0	0	0	0		0
	0	3	0	0	0	0		0
	3	0	4	0	0	0		1
4	2	9	0	12	0	0		4
	15	8	0	0	13	11		8
6	0	0	12	22	16	0		8
13	24	18	4	9	11	11		14
21	30	24	27	7	5	20		19
	15	6	11	9	5	5		8

TRANSLATION

Brussels | 20 March 1851

My dear friend,

I received from Messrs Schlagintweit[1] the beautiful present that you sent: nothing could be more agreeable than to receive a portrait with an inscription in your own hand, testifying to the friendship with which you wish to honour me. I wish from now on to have this portrait always before my eyes; I have put it in the most obvious place in my study, where it is the principal ornament.

I congratulate you on the remarkable results that you have obtained on the magnetic properties of oxygen. Our Académie to which I believed myself able to communicate what you told me in your last letter[2] learned of it with great interest[3]. I should have thanked you a long time ago; but I wanted, in writing to you, to send you the results of my work on atmospheric electricity, work in which you have expressed some interest[4]. I have finally found time, today, to satisfy this desire.

My foremost reason for sending you this information is to ask you to examine if your recent discoveries on oxygen can also account for the periodic variations in electricity. I have no doubt that this important subject has already awakened your interest. My[5] first memoir on the electricity of the air contained the results of four and a half year's observations[6]. I can now add to them those of 1849, 1850, and the beginning of 1851; you will see them, probably, not without interest. They confirm generally the principal facts deduced from former observations;

nevertheless the year 1849 has presented a very sensible anomaly during the first seven months: the electricity is then less in a remarkable degree than the ordinary proportion, as you will perceive by the accompanying tables, which, besides the mean and extreme values of each month *observed directly*, contain also the monthly means obtained in reducing the determinations of each day to the same scale.

As far as is permitted by the consideration of the years individually, we meet again with the same law of continuity and the same differences between summer and winter as before.

Nothing is more fitted to make one appreciate the deficiency which still exists in our relative knowledge of the electricity of the air, than the doubt which surrounds the anomalies I have observed during the first part of the year 1849. I have been unable to find any observations which may serve to control my own. No observatory, at least as far as I know, has published up to this time regular daily observations of the electricity of the air; and this constitutes a deficiency most annoying, as we shall not be long in perceiving.

When a year is remarkable by an abnormal temperature, by extreme rainy or dry periods, all meteorologists agree in determining its irregularities in the district where they are manifested; unfortunately it is not the same here. I have consulted the most esteemed collections of observations, and I have found nothing there which could satisfy my want.

Consequently it may be demanded of me whether the deficiency in the electricity of the air is real or only apparent; whether the result is not due, for example, to a derangement of my instrument? or, supposing this weakening of the electricity well ascertained, was it local or not? All these questions derive still higher importance from the fact that the anomaly remarked coincided very nearly with the return of that scourge which caused our population to suffer so cruelly[7]. These difficulties have led me to search whether there were not means of rendering *an electrometer comparable to itself at different times*.

I have already shown in my first memoir[8] that it is easy to compare two Peltier's electrometers[9] with each other, and to construct tables of equivalents for their indications; but when a comparison has been made and the compared instrument has been carried into another country, or even without, after it has been employed for a certain time, it is important to be assured that its indications have not varied, but have preserved their absolute values.

This verification may be very easily made; it is enough to ascertain that the needle has preserved all its mobility, and that its directive power has remained unchanged. Now the directive force is given by the little magnetic needle attached to the indicating needle of the electrometer; it is sufficient therefore to submit this little needle to the ordinary processes employed to determine magnetic energy, that is to say, to make it oscillate freely in a horizontal plane, and to ascertain that its magnetism has remained unaltered. It should be well understood that account is kept of the ordinary corrections employed in such cases for temperature, torsion of thread, variations of the horizontal intensity of the earth, &c.

One can understand how by analogous processes we may also make the determinations of the absolute electric force of the earth depend on that of its absolute magnetism; an important problem with which I cannot occupy myself at this moment. It is sufficient for me to establish, that by a very simple process we may ascertain whether a Peltier's electrometer has remained comparable with itself. The thing is so simple that I am astonished the precaution has not as yet been indicated. If the idea had come into my mind of thus verifying the needle of my electrometer in 1849 and at preceding and following epochs, I should not now have had to search out whether my instrument could undergo a temporary derangement, nor to look after foreign observations to control my own.

If you think that these simple remarks may be of any service or induce other philosophers to undertake those series of electric observations so important to science, I beg you to make such use of this letter as may seem to you expedient[10].

I submit entirely to you as the best judge in these matters.

I ask you kindly to present my homage to Mrs Faraday and to receive, my dear and illustrious colleague, my most respectful compliments

Your friend | Quetelet.

I will not talk of the death of Oersted[11]: this news is all the more painful since this great physicist honoured me with his friendship. Oersted and Schumacher[12] so close together!

1. See letter 2313.
2. Letter 2347.
3. On 30 November 1850. See *Bull.Acad.Sci.Bruxelles*, 1850, **17** (2): 371-1.
4. See letter 2263.
5. From here the translation is taken from Quetelet (1851a).
6. Quetelet (1849).
7. That is cholera.
8. Quetelet (1849).
9. Peltier, J.C.A. (1842).
10. End of translation from Quetelet (1851a).
11. Oersted died on 9 March 1851.
12. Heinrich Christian Schumacher (1780-1850, DSB). German astronomer who worked largely in Denmark. Died 28 December 1850.

Letter 2397
Faraday to Richard Taylor
22 March 1851
From Quetelet (1851a), 329

My dear Sir,

I think M. Quetelet's observations so important[1] that I hope you will publish them in the next Number of your Journal[2], and I doubt not they will have their due effect in inducing many to join the very small band of those who at present observe and study the phaenomena of atmospheric electricity.

Very truly yours, | M. Faraday.
22nd March 1851.

1. In letter 2396.
2. That is *Phil.Mag.* which Taylor did as Quetelet (1851a).

Letter 2398
Faraday to Jacob Gisbert Samuel van Breda
22 March 1851
From the original in HMW

Royal Institution | 22, March 1851

Sir

I have been greatly honored by the receipt of Your letter[1]; and though proud of such a mark of confidence fear that I shall hardly prove deserving of it. If I understand rightly you desire me to say what great experiment or research there is, which, being important to science, is in the matter of expence beyond the reach of an individual, and deserves the assistance of the Royal Society of Sciences. For my own part I am so happily situated in this Institution, that I am sure if I could say that I needed even a very large sum up to £1000 or even £2000 to decide a great enquiry; I could raise it amongst our members, for my personal use, in a week. I therefore have every want satisfied; and am not able to mention a subject from amongst those which form the object of my own especial studies.

But there is a research which I will venture to suggest to your consideration for its importance which is to a large extent proved;- because of the character of the philosopher who has already entered into it;- and because of the circumstances which, as I have reason to suppose, have in part caused its cessation:- I refer to that on *Respiration* by Regnault - which he has carried so far already; and the first fruits of which he has published[2][.]

I cannot doubt that your meeting of 1852 will be important for Science, not merely as respects your own country but as regard[s] the world. Such determinations for the advancement of knowledge do good every where not only by the direct fruits but by the example. I doubt not the meeting will be a happy one for the Society as a body & for its members individually: and in that happiness though at a distance I shall feel that I have the honor to share[.]

I am Sir | With every respect | Your Most Obliged & Humble Servant | M. Faraday
To | J.G.S. Van Breda | &c &c &c &c

Address: A Monsieur | Monsieur J.G.S. Van Breda | Secretaire Perpetuelle | &c &c &c &c | Societe Royale Hollandaise des Sciences | Haarlem

1. Not found.
2. Regnault and Reiset (1849).

Letter 2399
Paul Frederick Henry Baddeley to Faraday
22 March 1851
From the original in IEE MS SC 2
My dear Sir M. Faraday,
The information contained in the accompanying paper[1], will not be communicated to anyone for publication, till after you have received it[.]
I am, with respect | Yours faithfully | P. Baddeley
Lahore | March 22nd 1851
I send this thro' my Brother in law - Mr. J. Fleming[2] of Liverpool.

Address: Sir M. Faraday F.R.S. | Royal Institution | London

1. Enclosed with this paper are thirteen sides of descriptions and drawings of dust storms. See Baddeley (1852).
2. Liverpool POD gives John Fleming in charge of a boarding school in Bootle.

Letter 2400
Gideon Algernon Mantell[1] to Faraday
27 March 1851[2]
From the original in RI MS Conybeare Album, f.45

My dear Dr Faraday,

I thank you very, very much, for your kindness is presenting me with a separate copy of another series of your splendid researches[3]; believe me I appreciate your remembrance of me most highly.

Prof. Silliman[4] & his son Dr. B.S.[5] (of Connecticut) come to town tomorrow on their way to the Continent, and they are most anxious to see you if but for a few minutes. They will therefore attend the meeting of the R.I.[6] tomorrow evening, if I can get tickets for them[7]. I have written to Mr Barlow on their behalf; but as the wife[8] & sister[9] of Dr B.S. & two friends[10] accompany them, I must beg, borrow, or steal, tickets for the whole if possible.

Prof. Silliman comes to me from Oxford tomorrow & stops at my house: on Monday he & his son go to Paris.

Ever dear Dr Faraday | with the greatest regard | Gideon Algernon Mantell
19 Chester Square | Thursday

1. Gideon Algernon Mantell (1790–1852, DSB). Geologist.
2. Dated on the basis of the reference to Silliman mentioned in Curwen (1940), 266.
3. Faraday (1851b, c, d, e), ERE24, 25, 26 and 27.
4. Benjamin Silliman (1779–1864, DSB). Professor of Chemistry and Pharmacy at Yale Medical School from 1813. Founder and editor of *Am.J.Sci.*
5. Benjamin Silliman (1816–1885, DSB). Professor of Medical Chemistry and Toxicology at the University of Louisville, 1849–1854 and an editor of *Am.J.Sci.*
6. Story-Maskelyne (1851), Friday Evening Discourse of 28 March 1851.
7. Faraday took the tickets to Mantell himself on 28 March 1851. See Curwen (1940), 266.
8. Susan Silliman, née Forbes. She had married Benjamin Silliman Jr in 1840. Fulton and Thomson (1968), 222.
9. Actually the sister of Susan Silliman. Fulton and Thomson (1968), 222.
10. According to Fulton and Thomson (1968), 222 these were Walter S. Church, a grandson of Benjamin Silliman Sr, and George Jarvis Brush (1831–1912, DAB), an assistant of Benjamin Silliman Jr at the University of Louisville.

Letter 2401
Faraday to George Merryweather
28 March 1851
From p.1 of a printed version tipped in Merryweather (1851)

Royal Institution, March 28, 1851.

I beg to thank you most heartily for the copy of your work on the "Tempest Prognosticator,"[1] which I have read with great interest.

I perceive you have read an account of an Evening that I gave on the Researches of Quetelet with Peltier's instrument[2]. I have no doubt you have quoted the "London Medical Gazette"[3] accurately; nevertheless I thought you would rather know of certain errors it contains before a new edition of your work comes forth. In the first place, it was not *Pelletier*[4], but *Peltier*, who invented the Electro-meter. It is curious that both men lived at the same time in Paris, and both were scientific.

At page 50, line 3, the name should be Quetelet, not Peltier.

At page 51, bottom line, the name should be *Peltier*, not Quetelet. Quetelet carefully refrained from putting forth any theory.

At page 52, line 9, for "they," it should be "the former."

M. Faraday

1. Merryweather (1851).
2. Peltier, J.C.A. (1842). See *Lond.Med.Gaz.*, 1850, **10**: 255–6 for an account of Friday Evening Discourse of 1 February 1850 "On the Electricity of the Air".
3. This account was quoted in full in Merryweather (1851), 49–52.
4. Pierre-Joseph Pelletier (1788–1842, DSB). French pharmaceutical chemist.

Letter 2402
Faraday to Benjamin Silliman[1]
29 March 1851
From the original in Boston Public Library MS Ch.I.6.32

R Institution | 29 Mar 1851

My dear Sir

I have this morning found a friend who will take charge of *all* my packets to Paris. I am happy to relieve you from the trouble but most thankful for the readiness with which you offered to do me the favour[2][.]

Ever Yours Most Truly
Dr. Silliman | &c &c &c

1. Benjamin Silliman (1779–1864, DSB). Professor of Chemistry and Pharmacy at Yale Medical School from 1813. Founder and editor of *Am.J.Sci.*
2. Presumably made at the Royal Institution the previous evening. See letter 2400.

Letter 2403
George Merryweather to Faraday
31 March 1851
From p.2 of a printed version tipped in Merryweather (1851)

Whitby, March 31, 1851.

I do indeed feel myself exceedingly obliged to you for pointing out to me a grave error that I have copied from the "London Medical Gazette."[1] Although it was with perfect innocence on my part, yet, at the same time, as I appear as the ostensible party, in charging M. Quetelet with an hypothesis that I myself should not wish to be charged with, I shall certainly take immediate measures to correct this most unfortunate mistake.

Your kind letter has just arrived in time to save me from a world of trouble. As it is my intention to present a copy to each of the learned societies on the Continent, for which object they are now binding, I should esteem it a great favour if you would allow me to publish your letter, or to quote those parts of it relating to the errors, which I would prefix to each Essay, and thereby avoid a multiplicity of explanation. I could also pursue those copies already delivered to the learned institutions of this country, and neutralize at once the errors.

George Merryweather

1. See letter 2401 and notes 2 and 3.

Letter 2404
Faraday to George Merryweather
1 April 1851
From p.2 of a printed version tipped in Merryweather (1851)

Royal Institution, April 1, 1851.

My dear Sir, - I can have no objection to your making any use of the corrections in my letter[1], &c. &c.

Very truly yours, | M. Faraday

1. That is letter 2401. See letter 2403 for the request.

Letter 2405
Lyon Playfair to Faraday
7 April 1851
From the original in IEE MS SC 2

[Royal Coat of Arms] | Great Exhibition of the Works of Industry of all Nations, 1851. | President, | His Royal Highness Prince Albert, K.G. | &c. &c. &c. | Office for the Executive Committee, | Exhibition Building, | Kensington Road, London. | 7th April / 51
Dear Sir
    Many thanks for your consent. I am sure it will delight the Prince much. You may have your choice of Classes I, III or IV. II I can scarcely afford two scientific chemists on as I must have Pharmaciens & Manufacturers. Perhaps you would like I, as you would then only require to give aid as a Chemist on Metallurgical subjects & require less constant attention[1].
    Truly Yours | Lyon Playfair
M. Faraday Esq | Royal Institution, Albemarle St

1. Faraday was appointed a member of the Jury of the Great Exhibition for Class I, "Mining, Quarrying, Metallurgical Operations, and Mineral Products". Anon (1852a), 1: xxvi.

Letter 2406
Faraday to John Frederick William Herschel
8 April 1851
From the original in RS MS HS 7.193

Royal Institution | 8 April 1851
My dear Sir John
    I have no doubt you are aware of the lines on glass produced by M. Nobert. I have received from him some rulings & a paper which he wishes me to present to the Royal Society & I propose doing so by next Thursday[1] but I thought you would like in the mean time to see an attempt to measure the length & velocity of the undulations of a ray of light by the direct application of a scale to them[.]
    Ever Yours Truly | M. Faraday
Sir John F.W. Herschell Bart | &c &c &c

1. That is 10 April 1851. Faraday did this. See F.A. Nobert, "Description and purpose of the glass plate which bears the inscription 'Interferenze-spectrum. Longitudo et celeritas undularum lucis relativa cum in aëre tum in vitro'", *Proc.Roy.Soc.*, 1851, 6: 43–5. The manuscript is in RS MS AP 33.21.

Letter 2407
Faraday to Thomas John Fuller Deacon
10 April 1851
From the original in RI MS F1 H70

Royal Institution | 10 April 1851

My dear Friend

I must thank you before I get into the full business of the day: when that begins I am pulled from Pillar to post at this part of the Season. That is doubly the case just now because tomorrow evening I must talk about atmospheric magnetism[1]. Caroline knows a little of that matter[.]

Your translation was just what I wanted[.] I have altered a few technicalities & sent it in to the Royal Society[2][.] The subject is an exceedingly curious one but rather out of my way but it was sent to me by Nobert from Pomerania with a highly characteristic English German letter[3] which needed no translation[.]

I hope you [are] now in comfortable lodgings & feeling occupied & cheerily so[.] Sometimes we have too much to do & sometimes too little - and we should go strangely wrong if there were not one over us who is always caring for us even for the most unworthy & the most rebellious, who is the preserver of all men. Give my Love to your wife & to the little impertant [sic] Miss Constance - also to Mr Paradise[.]

Ever Your affectionately | M. Faraday
Mr. Deacon

1. Faraday (1851f), Friday Evening Discourse of 11 April 1851.
2. That is Nobert's paper cited in note 1, letter 2406.
3. Not found.

Letter 2408
Faraday to John Frederick William Herschel
15 April 1851
From the original in RS MS HS 7.194

Royal Institution | 15 April 1851

My dear Sir John

The accompanying book was addressed to me on the outside but to you on the inside. I consider it as yours.

Ever Truly Yours | M. Faraday

Letter 2409
Matthew Fontaine Maury[1] and George Manning[2] to Faraday
16 and 23 April 1851
From the original in IEE MS SC 2

National Observatory, | Washington D.C | April 16th 1851

My dear Sir,

I have recd your letter of the 24th Ulto with the Copy of the "Experimental researches in Electricity 24th 25th 26th & 27th Series"[3] which you had the kindness to send, and for which I thank you most heartily[.]

I need not say how eagerly these several papers were devoured; I have studied them with profit and pleasure; and admired, step by step as I went along, the true spirit of philosophical research that pervades them.

To me, some of your results look very much like a cleu [sic] which you have placed in the hands of physicists to guide them through dark and doubtful places of research.

It may lead through many tortuous windings before it brings us to the end, but we look to you still further; You must not tire.

Will you not therefore embrace in your researches the electric or magnetic properties of Sea water at various temperatures? for it appears to me that our philosophy is as much at fault in accounting for the velocity of the Gulf Stream, as it is for the cause of magnetism.

The Mississippi river, where its fall has been computed at $2\frac{1}{2}$ inches to the mile, is said to have an hourly velocity of $1\frac{1}{2}$ miles. *But*, the Gulf Stream, running on a *water level*, maintains a velocity of 4, and reaches occasionally a velocity of 5 miles, an hour.

I am prepared to show that the Trade Winds have very little to do with giving direction and force the currents of the sea, and there is room for the conjecture that the submarine currents are as regular if not as active as those at the surface.

What is the cause of this great velocity of the Gulf stream on a *water level*? Not gravitation certainly[.]

The difference of Specific gravity between the hot waters of the Gulf stream and the cold waters of the poles would cause *motion*; but unless some other agent were concerned, not with such velocity as that of the Gulf Stream; nor can it be comprehended how, without the help of some other agent, the waters of the Gulf Stream should collect themselves together, and flow to the North in a body, as they do.

That the waters of the Gulf stream do not readily mingle with those of the ocean about them, we know; and if you will do me the favor to look at pp 15–20 "Investigations of the Winds and Currents of the Sea"[4] several copies of which I have caused to be placed at your disposal, you will find some striking evidence of this fact.

Why is this? the very waters through which the Gulf Stream is flowing, are themselves bound down to the Trade Wind regions, to supply the air with vapor, to have their temperature raised, to enter the Gulf, to perform their circuit, and to issue thence as Gulf Stream water, invested with antagonistic principles - so to speak. Whence is this antagonism derived? meaning by Antagonism, in this place, simply the indisposition of the two waters - those flowing from the Gulf and those through which they flow, to commingle.

The reluctance of the waters of any two streams when they meet, to mingle is often manifested, and perhaps the display of this propensity by those of the Gulf Stream may not be considered remarkable, except as to distance and extent.

If you will look at the shape of the Gulf Stream you will be struck with its cuniform proportions; From the Straits of Bemini to the Grand Banks it is like a wedge, with its apex cleaving the Straits[.]

What has this form and pressure of the cold and therefore heavier waters which make the bed and banks of the Gulf Stream, to do with its velocity? Striking analogies might be pointed out in this connexion.

You will observe too, that according to the Storm tracks which meteorologists assign to the West India Hurricanes, those storms manifest by their course, a tendency towards the waters of the Gulf Stream, conforming with it in their general direction[.]

Can therefore the waters of this Stream, and the air above them, be more or less paramagnetic or diamagnetic, than the sea waters generally?

Why may not the oxygen of the water be paramagnetic, as well as the oxygen of the air? or are the gasses and salts of sea water capable of any magnetic influences? If the oxygen of the water be magnetic will not that of its vapor be magnetic also? If an affirmative reply be given to these interrogations, would it not be suggestive of the agency through which, or by which, the Gulf Stream rules the course of the Storm, and preserves its own peculiarities?

You will, I hope, understand me in making the above remarks, and in propounding the above interrogatories.

I make the remarks with the hope of interesting you, and of engaging your thoughts upon these captivating subjects; and I ask the questions, in the true spirit of philosophical inquiry, so well described in your 2702 et seq[5], as moving you to your beautiful train of researches.

Will you not therefore - or is it your intention to do so, or would it be useless - to extend your "Researches in Electricity" to the sea, its gasses, & its salts, and its vapors, and determine as to the degree and nature of "Aqueous Magnetism" if there be such a thing.

Reading your remarks as to the great "Magnetic lense"[6] which follows the sun through the atmosphere and studying the diagrams of magnetic declination at Toronto & St. Petersburg I was led to ask myself the question, What have the Gulf Stream and the mantle of warmth which it spreads over the extra tropical North and places between St. Petersburg and Toronto to do with the time and period of the great "sun swing"[7] of the needle at the two places. One place has a great extent of land surface to the East - the other to the West and the two are separated by the warm waters of the Gulf Stream.

I hope you will pardon me for trying so earnestly to tempt you out to sea. I am there, and often find myself bewildered, and shall be most happy to come within your hail, and get fresh points of departure from you now and then[.]

Respectfully &c | M.F. Maury
To | M. Faraday Esq | Royal Institution | London
The charts &c referred to by Lieut Maury, will be forwarded to you in a package to John W. Parker[8], Bookseller, if possible by next London Packet sailing of 24 inst -or 1st Prox.

Very truly &c &c | Geo Manning
New York April 23 / 51

1. Matthew Fontaine Maury (1806–1873, DSB). American naval officer and oceanographer.
2. According to Williams (1963), 260 George Manning was a distribution agent in New York.
3. Faraday (1851b, c, d, e), ERE24, 25, 26 and 27.
4. Maury (1851).
5. Faraday (1851b), ERE24, 2702 *et seq*.
6. Faraday (1851d), ERE26, 2892 and 2920.
7. Faraday (1851d), ERE26, 2909 and Faraday (1851e), ERE27, 3031.
8. John William Parker (1792–1870, DNB). Printer and publisher in London.

Letter 2410
Faraday to John Barlow
18 April 1851
From the original in RI MS JB1/2, p.47

Hastings | 18 April 1851

My dear Barlow
Look at the accompanying sketch of regulations[.]
Supposing them fit & proper or altered until they are so what do you say to having them printed in a proper form as a page of post paper and kept lying upon the tables in the library on Friday nights[.] A thousand or two being printed so that they might be given to the members reading there & that those which are removed or dirty may be continually replaced[.]
They ought of course to be sanctioned by the Managers & signed by you as Secretary: *but*

Would it not be important to have them on the tables before the first Friday Evening i.e the 2nd of May[1] - and as no Managers meeting will occur before that would you mind our taking the responsibility on ourselves for that time. I have no objection to taking as much as you please[2][.]

I do not wait until we meet at home but send this that there may be plenty of time for consideration. If we had thought of it sooner some such regulations might have been sent to each member with the report[3] &c for the 1st of May[4][.]

Ever Truly Yours | M. Faraday

1. Airy, G.B. (1851), Friday Evening Discourse of 2 May 1851.
2. The regulations for the Friday Evening Discourses were approved by the Managers at RI MM, 2 June 1851, **10**: 333 and their earlier implementation was also approved.
3. That is *The Annual Report of the Visitors of the Royal Institution of Great Britain for the Year 1850*.
4. That is the Annual Meeting of the Royal Institution. RI MS GM, 1 May 1851, **5**: 564–9.

Letter 2411
Faraday to John Tyndall
19 April 1851
From the original copy in Tyndall, Diary, 28 April 1851, 5: 52-3

Hastings 19th April 1851.

Dear Sir,

Whilst here resting for awhile I take the opportunity of thanking you for your letter of the 4th of February[1] and also for the copy of the paper in the phil. Magazine which I have received[2]. I had read the paper before and was very glad to have the development of your researches more at large than in your letter. Such papers as yours make me feel more than ever the loss of memory I have sustained, for there is no reading them or at least retaining the argument under such a deficiency. Mathematical formulae more than anything requires quickness and surety [in receiving and retaining the true value of the symbols used,][3] and whilst one has to look back at every moment to the beginning of a paper, to see what H or α or β mean there is no making way. Still though I cannot hold the whole train of reasoning in my mind at once I am fully able to appreciate the value of the results you arrive at, and it appears to me that they are exceedingly well established and of very great consequence. These elementary laws of action are of so much consequence in the development of the nature of a force which, like magnetism is as yet new to us.

My views with regard to the cause of the annual, diurnal, and [some][4] other variations are not yet published though printed[5]. The next part of the philosophical transactions will contain them. I am very sorry I am not able to send you a copy from those allowed to me, but I have had so many applications from those who had some degree of right that they are all gone. I only hope that when you see the Transactions you may find reason to think favourably of my hypotheses. Time does not lessen my confidence in the view I have taken but I trust when relieved from my present duties and somewhat stronger in health to add experimental results regarding oxygen so that the mathematicians may be able to take it up.

As you say in the close of your letter I have far more confidence in the one man who works mentally and bodily at a matter than in the six who merely talk about it - and I therefore hope and am fully persuaded that you are working. Nature is our kindest friend and best critic (exciter?)[6] in experimental science if we only allow her intimations to fall unbiassed on our minds. Nothing is so good as an experiment which whilst it sets an error right gives us a reward for our humility in being refreshed by an absolute advancement in knowledge[.]

I am my dear Sir | your very obliged and faithful Servant | M. Faraday
Dr. J. Tyndall | &c &c

1. Letter 2379.
2. Tyndall (1851a).
3. There is an ellipsis in the MS here. The passage in square brackets is taken from the typescript in RI MS T TS volume 12, p.4128.
4. As in typescript. See note 3.
5. Faraday (1851c, d, e), ERE25, 26 and 27.
6. "critic" is given in the typescript.

Letter 2412
Faraday to Lambert-Adolphe-Jacques Quetelet
19 April 1851
From the original in BRAI ARB Archives No 17986 / 989

Royal Institution | 19 April 1851

My dear Quetelet

Directly that I received your letter[1] I translated it and sent it to the Philosophical Magazine where I have no doubt you have seen its insertion[2] and I trust it will induce some to join with you in the observation of atmospheric electricity. Your observations regarding the first part of the year 1849 are most interesting and I have full confidence in them so that though it may in some degree (& a large one) be unfortunate that there are not other observations made elsewhere to compare with yours still it is of the utmost consequence that yours have been made & I hope they will awaken the sleepy observers.

Your flatter me by the manner in which you receive my portrait:- I do not think much of my own face but I have very great pleasure in looking upon yours[3] & it brings by association all your kind feelings towards me back to my mind and very pleasant they are[.]

I sent you a long paper or rather several papers[4] a little while ago - containing some on atmospheric magnetism[.] I have no fear as to the experimental part & I entertain hopes that the hypothetical part may find favour before your philosophic mine. My hopes in it are not as yet any less than at any former time - but I shall leave the paper to tell its own story[.]

I am not quite right in health & though I have from habit dated from the Royal Institution am really at Hastings on the Seashore waiting on rest & fresh air[.]

Ever My dear M. Quetelet | Your faithful Servant | M. Faraday

Address: A Monsieur | Monsieur Quetelet | &c &c &c | Observatory | Bruxelles

1. Letter 2396.
2. Quetelet (1851a).
3. See letter 2323.
4. Faraday (1851b, c, d, e), ERE24, 25, 26 and 27.

Letter 2413
Faraday to Christian Friedrich Schoenbein
19 April 1851
From the original in UB MS NS 396

Hastings | 19 April 1851.

My dear Schoenbein

Here we are at the seaside and my mind so vacant (not willingly) that I cannot get an idea into it. You will wonder therefore why I write to you since I have nothing to say but the fact is I feel as if I owed you a letter and yet cannot remember clearly how that is. Still I would rather appear stupid to you than oblivious of your kindness and yet very forgetful I am. In 6 or 7 weeks I shall be talking of ozone[1]. I hope I shall not discredit you or fail in using well all the matter you have given me abundant & beautiful as it is. But I feel that my memory does not hold things together in hand as it used to do. Formerly I did not care about the multiplicity of items, they all took their place & I picked out what I wanted at pleasure[.] Now I am conscious of but few at once & it often happens that a feeble point which has present possession of the mind obscures from recollection a stronger & better one which is ready & waiting:- but we must just do the best we can - and you may be sure I will do as well for you as I could for myself.

I set about explaining the other evening my views of atmospheric magnetism[2] & found when I had done that I had left out the two or three chief points. I only hope that the printed papers[3] contain them & that they will be found good by the men who are able to judge. The copy for you is either with you or on the way for the Gentleman whom you introduced to me[4] whose name I forget (from Aarau?) kindly took charge of it.

And now my dear Schoenbein with kindest remembrances to Madame Schoenbein (and my wife joins all she can to you & yours)

I am as Ever | Most truly yours | M. Faraday

Address: Professor Schoenbein | &c &c &c | University | Basle | on the Rhine

1. Faraday (1851g), Friday Evening Discourse of 13 June 1851.
2. Faraday (1851f), Friday Evening Discourse of 11 April 1851.
3. Faraday (1851c, d, e), ERE25, 26 and 27.
4. Alexander Pompius Bolley (1812–1870, P1, 3). Professor of Chemistry in Aarau, 1838–1854. See letter 2388.

Letter 2414
Faraday to John Barlow
21 April 1851
From the original in RI MS F1 N1/22

R Institution | 21 April 1851

My dear Barlow

I am glad you think the regulations are not improper & may be useful - alter them at your pleasure[1][.] As to the title of Secretarys list, it can be called the *President's*[2] or the *Evening* or *the* list - only care must be taken from the first to prevent folks thinking that *any body* may write a name down on it. Even managers *individually* ought to have no power of the kind but just one or two persons as the president or Secretary. The numbers of the list combined with that of the managers ought perhaps to be limited so that on occasions even managers might have to give place & I am sure in case of need would do so. When we come together we will see how many the reserved bench seats of the front row can contain and then that number with something additional for chairs upon very especial occasions ought to make up the whole number.

Perhaps a good additional regulation would be to place no chairs except for distinguished *visitors* or if that be too strong we can talk of such a rule.
Ever Yours | M. Faraday

1. See letter 2410.
2. It was called the President's list. RI MM, 2 June 1851, **10**: 333.

Letter 2415
William Snow Harris to Faraday
25 April 1851
From the original in IEE MS SC 2

Plymouth, 25 April 1851

My dear Faraday
I am in possession of your late researches in Electy printed in the First Part of the Philosophical Transactions for this present year[1], with which I am greatly charmed and instructed, and for which I feel deeply indebted to you[.] When I look at what you have done in Science, and your still brilliant progress, I am inclined to say of you, as the Marquis de L'Hospital[2] said of Newton "Does Faraday eat drink and sleep like other men"?[3] Your new discovery of Atmospheric Magnetism has opened a vast field of further research and impressed upon every new fact in this department of Science a peculiar Interest.
Having lately been myself engaged in the further prosecution of inquiries concerning the nature, laws, and mode of action of this species of force, and having arrived at several conclusions, apparently novel and important I have assembled them under the form of a Communication to the Royal Society[4]. My first impulse however was to send the Paper to you, with a view of your looking it over, and presenting it, should you have thought it worthy of that honor - that would have been the course which my great admiration of your Labours and I will add my sincere personal regard toward you, would have led me, to adopt - but I considered, that I ought not to trouble you on such an occasion seeing how much you are hourly occupied in matters of such vast importance to the interests of Science. Thinking however, that the experimental part of my paper, may possibly be interesting to you, seeing that it is involved in those very researches in which you are engaged, and that it embraces new phenomena on Magnetism requisite to be considered, & accounted for, in any further view, we may be led to take of this mysterious power, I have determined on forwarding for your information a rough copy of my paper, with the few following notices, so that you may have no trouble about it,

and refer to the facts in it in such way as may suit your convenience. I trust you will not think me intrusive in having ventured on this step - for after all I am not sure whether what I have done may be esteemed of sufficient importance to merit your consideration or not[.]

The First pages consist of Introductory remarks, and a reference to Instruments principally those you saw at the Geological Museum - but with improvements. This is followed by a particular view of the nature and mode of Operation of Magnetic Force, which is considered to arise out of what I have called waves of Induction or Magnetic reverberations set up between the surfaces of the opposed bodies, this may after all come to your theory of Polarized particles - however if you think it worthwhile you may see what is said sec 4 and 5.

The laws of these inductive forces, as given in sec 11, may perhaps be found very well worth consideration - more especially when taken in connection with secns 23, 24 &c in which various laws of reciprocal force are fully elucidated - and all the several results of experiments by Hauksbee[5], Brook Taylor[6], Muschenbroek[7] and others quite reconciled with the primary or more elementary laws of Magnetic force. I think this is so far interesting inasmuch, as it helps to throw light on the nature of Magnetism - more especially in any reference of this peculiar Physical Force, to a general or universal principle as you seem disposed to do[.] So far as I see - we have in the phenomenon of a Magnet attracting Iron or that of two Magnetic poles attracting each other, one end only of a chain in our grasp, the other end of which is out of sight. Taking the reciprocal force near the magnet, it is certainly in no inverse ratio greater than that of the simple distance - as we recede from this into Space - the force becomes as the $\frac{3}{2}$ power of the distance inversely - then as the Squares, next as the $\frac{5}{2}$ power then as the cubes of the distances inversely as stated by Newton[8] - and to what other inverse powers of the distances, the force may extend as the action fades away in distance, it is almost impossible to say, without instruments of extreme sensibility such as we can not at present boast of, it is not possible to further investigate this point. You see what I have said about this at Sec. 28. Observe I do not say that you obtain the above laws in every instance successively with the same Magnet so much depends on the stability of the Inductive force. But I have no doubt whatever, not the slightest, of the facts, or that the experiments, can in any way, either Mathematical or Physical, be called in question.

In Sec 14 and following I think you will find a new class of Phenomena, unless I am unacquainted with all you & others have effected - e.g. you will find a curious and interesting example of *a diminished Magnetic Intensity by an increase of surface*, being in magnetism, precisely the same experiment and class of fact, which Franklins[9] Expet. of the Can and Chain is in Electricity[10], or of any analogous experiment in which by extending the Surface the Electrometer falls see Fig 7 Sec (15). You will further see (16) that a hollow tempered steel Cylinder becomes equally if not more powerfully magnetic than a similar solid-tempered steel Cylinder of the same diameter - shewing that it is the *Surface* and not the *Mass* which is concerned. That a soft Iron Cylinder made to fit, and passed into a tempered hollow magnetic Cylinder operates as a sort of discharging rod, and permanently discharges as it were the opposite polarities, leaving a residuum as in the Leyden Jar &c &c. There are a class of facts here which perhaps may be interesting to you[.] We have not I think as yet done enough in the way of investigating *internal* magnetism, as in your Electrical cage, you suspended and lived in[11].

Sec. 34 - which treats of magnetic *quantity*, and *its law* of *measurement* is necessarily in association with some of your views sec. 2870[12]. I have not the slightest doubt but the law and condition of charge, is identical with Electricity the force of *attraction is as the square of the quantity*. Sec 35 contains my apology for terms, and refers to an Experiment Fig 8. with which I have been much satisfied & pleased from the great, I may say the *extreme precision* of its results. I do not think that such a combined action for the measurement of Volta magnetic force has been as yet effected. It served me in magnetism as the Unit Jar did in Electricity - and we have now I think a means at command for determining how much more magnetism we have in the pole of any one magnet, or in any other point, than in the pole &c of any other magnet. The Expt in detail is given (36) the Law for quantity is in (34) afterward proved to be true.

In sec. (42) is the problem of the magnetic development in different points of a Magnetic Bar, which has not always been accurately and definitely stated[.] Finally I have thought it worth while to refer in the end of this Paper to the peculiar parallism in the phenomena of the *Electrical Jar* and the *Magnetic Bar*, and to what Newton and others of the more remote periods of the R.S have advanced on this subject, and which instead of being open to the severe animadversions and unmerited criticisms of many eminent mathematicians[13], are I firmly believe perfectly true in all their details and results, and quite consistent with demonstrable laws of magnetic force - you will see my views on this subject (45)[.]

I do not see how it is possible to consider Magnetism as a central force or *emanation from a magnet* spreading out into space and getting weaker in Proportion to the amount of space spread over:- all our experience is to my humble apprehension against such a deduction - the action must be after all referred to an action between terminating planes; much after the fashion you have described in 1299, 1302, 1301 1163[14] &c - and in various other parts of your researches. Two terminating planes or surfaces with *something going on* or *established* between them is the immediate feature to me of both Electrical and magnetic force thus

In the attraction of a magnetic Pole n and a similar mass of Iron m, end on, as it were, as in the above Diagram I do not at all believe, that the forces uniting in the other parts & about the center of the Bar m enter into the surface action at the pole -

and for the reasons I have given (7) (18) (44) &c. It is not therefore as stated by Robison[15] a sort of balance of Electrical attractions & repulsions we have to consider, as referred to all points of the mass of the opposed bodies in such a case[16]; but an *exclusive* action between the surfaces. The *something which goes on between* the surfaces you call polarization of particles. So far as I see the general principle is true, but I think we shall have to refer this polarization to some medium different from that of air for I have no doubt whatever but that the force between a plane magnetic pole and an Iron Plate which goes on in vacuo in no sense differs in its nature from the force between a plane polar electrified surface and a neutral conducting plate. I am quite sure that both these actions are identical all that the air does in Electy of high Intensity may be to arrest the passage of the Electy, thus in magnetism is not wanted. Robison seems to think (Mech. Phil. vol iv p273) that the law of attraction as observed between magnets or between magnets & Iron, must be different from the *real* law of magnetic action - because he says the magnetism is always increasing or decreasing with the distance[17] - but the ground I take in my paper is, that it is *really* this *increasing* or *decreasing* magnetism as depending on induction which *constitutes magnetic action* - it is in fact the combined effect of induction between the opposite poles at different distances which gives the law of the force; this *is* the *real* magnetic action, there is *no other*. It appears to me that there has been a great bias with many profound mathematicians in favor of a certain law of magnetic force, which they think must be the same as the law of Gravity, and if we do not find it so by Expt. the Experiments are false, we *ought to find it so*. The more modern & French Theorys of Electricity & Magnetism depend much on the truth of this position all their fine mathematical superstructures are built on it - and I really do think in many instances, that if it were a point at issue between Poissons Mathematics[18] and the course of Nature, very many would give it in favor of Poisson[.] It is quite curious to see how severely every Experimental result is handled which does not coincide with the law of the inverse duplicate ratio of the distance. Robison cuts down at one sweep all the valuable expts of Hauksbee[19], Brook Taylor[20] Whiston[21], Muschenbroek[22] & others[23], says in as many words they are worthless - that Electrical attractions & repulsions are not the most proper phenomena for declaring the precise law of variation - yet was it from *those same attractions and repulsions* that both *Coulomb*[24] and *Lambert*[25] deduced their law of magnetic force - Lambert especially resorted to the method of Hauksbee & Brook Taylor, certainly in a more refined way - now *their experimental inquiries* are quoted with confidence; and in no way objected to, in fact they were considered to have arrived at the *true law of the Force*: then again Newton having observed that the Magnetic attraction decreases in a certain case in the Inverse triplicate ratio of the distance - we are told in as many words by Biot, that Newton was ignorant of the whole matter[26], and had not accurate Ideas of magnetic Phenomena, and so in a variety of other instances. In fact without in any way entering as you have done into a severe and close investigation of nature by Experiment many profound Mathematical men have been content to adopt a certain set of principles, derived it seems to me, not always from very unexceptionable experiments - and they seem determined to bend every thing to those principles. For my self I have no belief in Electrical & Magnetic fluids in their hypothetical density & distribution, in virtue of the assumed law of their constitution &c. There is something to come yet far beyond all this - however we are still in the dark about it. I have read over your attempt to establish a relation between Gravity & Electricity[27]. I have not myself much hope of success by any direct experiment. The results of the two forces being so very different - I can conceive no identity in these forces except through a sort of Aethereal medium in which all matter may be conceived to float and different relations or affections of which to the particles of common matter is in one case Gravity in another Electricity: but it is in vain for me to speculate on such things at this moment. You are the only Philosopher on Earth likely to throw light on the question[.]

I had always great misgivings relative to the changes which might ensue in vibrating a magnetic needle in air, and I pointed out some of these changes in my papers in the Edinb. Phil. Transactions for 1834 vol xiii, "On the Investigations of Magnetic Intensity &c-"[28] and I felt quite assured at that time, that Christies Expts. relative to the Influence of Light Phil. Trans. 1825[29] were disturbed by taking the oscillations in air - all of which I have enlarged on at the close of that Paper sec 33. And I think you may find some Experiments there immediately coinciding with and bearing on what you say 2871[30] and the direct connection of the two is really very striking see also sec 34 and 35 of my paper above quoted. Your observations on Atmospheric Magnetism confirm my view of the propriety of observing Intensity oscillations in vacuo - which I have always done as in the long series of Expts in the Phil. Trans for 1831 p69[31].

I take it for granted, that by this time I have tired you out, but I find the subject so very interesting that I could continue to write about it to almost any extent, especially to you. I will however now bid you adieu & will subscribe myself your very faithful & affect[ionate] friend
    W. Snow Harris

1. Faraday (1851b, c, d, e), ERE24, 25, 26 and 27.
2. Guillaume-François-Antoine de L' Hospital (1661–1704, DSB). French mathematician.
3. Quoted in, for example, Hutton (1795), **2**: 150. On this quotation see Westfall (1980), 473.
4. William Snow Harris, "On Induced and other Magnetic Forces", *Proc.Roy.Soc.*, 1851, **6**: 87–92. The manuscript is in RS MS AP 33.15.
5. Francis Hauksbee (c1666–1713, DSB). Experimental natural philosopher.
6. Brook Taylor (1685–1731, DSB). Mathematician.
7. Petrus van Musschenbroek (1692–1761, DSB). Dutch natural philosopher.
8. Newton (1726), 403. [Book 3, prop. 6, corr. 5].
9. Benjamin Franklin (1706–1790, DSB). American natural philosopher.
10. Franklin (1751–4), 121–2.
11. See Faraday (1838a), ERE11, 1174.
12. Faraday (1851d), ERE26, 2870.
13. Possibly a reference to Whewell (1835) which criticised Harris. See Harris to Faraday, 28 April 1839, letter 1166, volume 2.
14. Faraday (1838a), ERE11, 1163, 1299, 1301, 1302.
15. John Robison (1739–1805, DSB). Professor of Natural Philosophy in the University of Edinburgh, 1773–1805.
16. Robison (1822), **4**: 272.
17. *Ibid.*, 273.
18. Poisson (1811).
19. Hauksbee (1712).
20. Taylor, B. (1721).
21. William Whiston (1667–1752, DNB). Theologian and natural philosopher. On this work see Whiston (1719).
22. Musschenbroek (1725).
23. Robison (1822), **4**: 217.
24. Coulomb (1789).
25. Johann Heinrich Lambert (1728–1777, DSB). German natural philosopher. On this work see Lambert, J.H. (1766a, b).
26. Biot (1830), 270.
27. Faraday (1851b), ERE24.
28. Harris (1834b).
29. Christie (1826).
30. Faraday (1851d), ERE26, 2871.
31. Harris (1831).

Letter 2416
Faraday to Joseph Henry
28 April and 2 May 1851
From the original in SI A, JHC, Box 10

Royal Institution | April 28, 1851

My dear Henry

The instant that I received your letter I applied to an Architect whom we employ Mr. Vulliamy to make me the necessary drawings of our Lecture room. I have only just received them. I do not know what has occasioned the delay but I could not help it. Why did you not come over yourself & see the room & hear in it; & if needful speak in it (we should have been very glad to listen to you) so that you might have been well able to judge of your own knowledge how far it was worthy to suggest any thing or serve for imitation in your own great room? We should have been very glad to see you for we have not forgotten the pleasure we received at Your last visit[1].

Besides the Exhibition is coming on & though I am very little moved or excited by such things yet it would have been pleasant to see you here for any reason and as it is we do expect a number of very good things from your Country which we reckon also as half ours[.]

I wonder whether I shall ever see America - I think not - the progress of years tells & their effect on me is to blot out many a fancy which in former days I thought might perhaps work up into realities - and so we fade away. Well I have had & have a very happy life at home nothing should make me regret that I cannot leave it & indeed when the time for decision comes - home always has the advantage[.] Mrs. Faraday wishes to be kindly remembered to you[.] We look at your face painted in light by Mayall[2] & I dare say it is like He & nature together have made you look very comfortable & I suspect that we have both altered much since last we saw each other[.] My wife mourns with half mimic half serious countenance over my changes & chiefly that a curly head of hair has become a mere unruly grisly mop. I think that is on the whole the worst part of the change that 60 years nearly have made[.]

And now having scratched this letter I merely wait for the Architects charge which is — and shall then send it & the drawing off.

Ever My dear Henry | Most truly Yours | M. Faraday
Prof. Joseph Henry | &c &c &c
2 May. I have just obtnd the bill 5.16.6 which I have enclosed | Yours Ever MF

Address: Professor Joseph Henry Esq | &c &c &c | Smithsonian Institution | Washington | United States

1. In 1837. See Reingold and Rothenberg et al. (1972–99), **3** *passim*.
2. John Jabez Edwin Mayall (1810–1901, Reingold and Rothenberg et al. (1972–99), **6**: 418). American photographer who spent most of his life in England.

Letter 2417
Faraday to Novello[1]
1 May 1851
From the original in Knox College Galesburg MS

Mr Faraday would be obliged to Mr. Novello if he would do him the favour to address & forward the accompanying letter[.]
Royal Institution | 1 May 1851

1. Unidentified.

Letter 2418
Faraday to Henry Thomas De La Beche
6 May 1851
From the original in British Geological Survey MS GSM1/507

Royal Institution | 6 May 1851

My dear De la Beche
    I think I am right in trusting to you for the carriage for tomorrow night. I do not know how late it comes for me - will 11 o'clk do? I will come away whenever you like. I suppose we shall meet[1].
    I do not know whether heavy glass falls within the scope of your collection[2]. I send the best piece I now have to spare[.] If you do not want it let me have it back again. It is a *silico-borate of lead*.
    Ever Yours | M. Faraday

1. This was the State Ball held at Buckingham Palace on 7 May 1851. For an account see *Times*, 8 May 1851, p.5, col. a-c. Faraday's attendance is noted in col. b.
2. At the Geological Museum.

Letter 2419
William Spence[1] to Faraday
7 May 1851
From the original in RI MS F1 K39

18 Lower Seymour St. | Portman Square | May 7, 1851

My dear Sir

Dr & Mrs Percy of Birmingham take tea with us in a free way (no large party) on Saturday evening next[2] at half past eight, & if you & Mrs Faraday chance to have no better enjoyment, we shall be very glad if you will favour us with your company to meet them.

I am | my dear Sir | Yours very truly | W. Spence
M. Faraday Esq

1. William Spence (1783–1860, DNB). Entomologist.
2. That is 10 May 1851.

Letter 2420
Faraday to John Ridout[1]
8 May 1851
From the original in WIHM MS FALF

R Institution | 8 May 1851

My dear Sir

Did you ask me for an admission to the Lecture on Saturday[2] or has my bad memory led me to some mistake?

Ever Truly Yours | M. Faraday
J. Ridout Esq | &c &c &c

Address: John Ridout Esq | &c &c &c | 10 Montague Street | Russell Square

1. John Ridout (1784–1855, B3). Surgeon and member of the Senate of the University of London, 1836–1855.
2. That is 10 May 1851 when Faraday gave his second lecture in a course of six lectures "On some points of Electrical Philosophy". His notes are in RI MS F4 J11.

Letter 2421
Alfred Swaine Taylor to Faraday
13 May 1851
From the original in RI MS Conybeare Album, f.17

3 Cambridge Place | Regts Park May 13 | 1851

Dear Faraday

It has occurred to me that the enclosed recent case of death from lightning might be of interest to you. I therefore send a copy of it for your acceptance.

It was formerly thought that all nervous and muscular power was so completely destroyed by the passage of an electric current through the body, that in death from lightning, the limbs *never became* rigid. John Hunter[1] held this opinion. The fallacy, however, arose from the rigidity having supervened and passed away before the body was seen. The enclosed case shows that the opinion is erroneous[.] This might have been inferred from the fact that the sutere destruction of nervous and muscular power by disease during life (as in paralysis) does not present that extraordinary condition of the muscles after death which is known as rigidity or rigor mortis - the last indication of actual power in the body.

In the enclosed case, the electric fluid appears to have burst the blood vessels and led to considerable effusion of blood. The red stripes with an arboriferous disposition is an appearance which has given rise to many superstitious notions[.] They have been supposed to represent the impression of a tree. The soldiers of Titus[2] engaged in the siege of Jerusalem, who were struck during a storm presented the mark of *a Cross* on the back - so it is said.

The man who lost his life in this case appears to have first received the shock, - and to have been more elevated than the lad: but both were on the low level of the river.

I am | Your's very truly | Alfred S. Taylor
Prof Faraday

1. John Hunter (1728–1793, DNB). Anatomist and surgeon.
2. Titus (39–81, NBU). Roman military commander in Palestine. Emperor, 79–81.

Letter 2422
Faraday to Alfred Swaine Taylor
14 May 1851
From the original in DUL, Great Britain, papers (Misc) Vol 1 (1795–1869), p.64

R Institution | 14 May 1851

My dear Taylor
    Let me thank you at once for your kindness[1]. I have read the letter. The slip I am going to take with me to read in our minutes.
    Ever Truly Yours | M. Faraday

1. In letter 2421.

Letter 2423
Faraday to Thomas Bell
15 May 1851
From the original in RS MS RR 2.215

Royal Institution | 15 May 1851

My dear Sir
    Herewith I return Coll Sabines paper[1] with congratulations that we have such an one to work in such a manner for the cause of science in relation to terrestrial magnetism. There are very few who would undertake the labour of such comparative & analytical labours and none who could do them better[.] As far as I may be permitted to judge I consider his investigations as of the utmost value to science.
    Ever My dear Sir | Very Truly Yours | M. Faraday
Thomas Bell Esq | &c &c &c

1. Sabine (1851b).

Letter 2424
George Ransome to Faraday
17 May 1851
From the original in RS MS 241, f.118

Ipswich Museum | [Coat of Arms] | Ipswich 1851 5mo 17
To Prof Faraday F.R.S. | &c &c &c
    It gives me much pleasure to inform thee, thou art unanimously elected an Honorary Member of the Ipswich Museum in testimony of the

high appreciation the Members of this Institution entertain of the valuable services thou hast rendered to the cause of Science, and I am further directed to forward the enclosed card of Honorary Membership for thy acceptance,
Believe me with high esteem | Thine faithfully | Geo Ransome | Hy Secy

Letter 2425
Faraday to Mrs Smyth[1]
22 May 1851
From the original in WIHM MS FALF

Royal Institution | 22 May 1851

My dear Madam
I am made proud by your request and hasten to enclose an order[2]. Will you do me the favour to introduce the names[.]
Ever Sincerely Yours | M. Faraday
Mrs. Smyth

1. Unidentified.
2. Probably to attend Hosking (1851), Friday Evening Discourse of 23 May 1851.

Letter 2426
Faraday to George William Blunt[1]
22 May 1851
From the original in Massachusetts Historical Society MS Washburn 273

Royal Institution | 22 May 1851.

Sir
I beg to acknowledge the receipt of your letter accompanied by Professor Bache's[2] report[3] & also by the three bottles of water and return you very sincere thanks for your kindness. I have not heard any thing from Lieut Maury[4] in relation to the waters but I will take care of them. My health & occupation is such that I am unable to enter into any investigation of them for my own researches are suspended from day to day & month to month for want of physical power - but I dare say all will come right in time & the waters will find their true destination.
I am Sir | Your Very Obliged Servant | M. Faraday
Geo W. Blunt Esq | &c &c &c

1. George William Blunt (1802–1878, DAB). First Assistant to the United States Coast Survey, 1833–1878.
2. Alexander Dallas Bache (1806–1867, DSB). Superintendent of the United States Coast Survey, 1843–1867.
3. "The report of the Superintendent of the Coast Survey, showing the progress of that work during the year ended November, 1850", Executive Document number 12, volume 1, of the 2nd session of the 31st Congress (1850–1). Laid before the House on 19 December 1850 and ordered to be printed.
4. Matthew Fontaine Maury (1806–1873, DSB). American naval officer and oceanographer.

Letter 2427
John Tyndall to Faraday
26 May 1851
From a typescript in RI MS T TS, volume 12, pp.4006–8

95 Dorothreen-Strasse | Berlin | 26th. May 1851

Dear Sir,

Shall I thank you for your last encouraging letter?[1] By thus doing I should imply that a kind of equilibrium might be established between my thanks and your kindness - this cannot be done and I therefore refrain from making the attempt, appealing rather to my future actions to testify the effect which your inspiring words have had upon me.

I write now just to mention that I had the honour of an interview to-day with Humboldt. I introduced your recent investigations in terrestrial magnetism. "I have read them[2] with astonishment" was his remark "I do not imagine that very little variation can be thus accounted for but in the main he is correct"[.] His last words to me were - "Tell Mr. Faraday that I am quite convinced of the validity of his hypothesis."[3] Dove[4] has expressed the same opinion to me.

Could you not pay Berlin a visit? there is no place on earth where you would be more enthusiastically welcomed. Your presence would call forth an exhibition of Hero-worship with which even Thomas Carlyle[5] himself would be satisfied[6].

I have been working for the last five weeks at diamagnetism. Prof. Magnus has been kind enough to place the necessary space and apparatus at my disposal - indeed I cannot speak too highly of the kindness of the men of science of Berlin generally.

My results I hope will interest you but I forbear mentioning them as they are not yet complete. It has been again my misfortune to arrive at conclusions very divergent from those of Prof. Plücker. A paper[7] on the subject shall be ready for the British Association at its next meeting[8].

Believe me dear Sir, | Most truly and respectfully Yours | John Tyndall

Prof. Faraday | etc. etc. etc.

Since writing the above I have spent a few hours with Du Bois-Raymond and have succeeded completely in developing a current by muscular contracted [sic] - I obtained a deflection of about 30° - right or left to the arm contracted. The experiment requires a delicate apparatus and some care. Du Bois' multiplying Galvanometer contains 24.000 windings[.]

1. Letter 2411.
2. Faraday (1851c, d, e), ERE25, 26 and 27.
3. See Tyndall, *Diary*, 26 May 1851, 6: 59–60 for an account of his meeting with Humboldt.
4. Heinrich Wilhelm Dove (1803–1879, DSB). Professor of physics at Berlin.
5. Thomas Carlyle (1795–1881, DNB). Writer and historian.
6. A reference to Carlyle (1841).
7. Tyndall (1851b).
8. In Ipswich.

Letter 2428
Arthur-Auguste De La Rive to Faraday
27 May 1851
From the original in IEE MS SC 2

Vichy (Dept. de l'Allier) | le 27 mai 1851.

Mon cher Monsieur,

Je ne veux pas tarder plus long-temps à venir vous remercier de votre bonne & aimable lettre que vous m'avez écrite il y a quelques mois[1]. Je suis, comme vous le verrez par la date de cette lettre à Vichy où m'ont envoyé les Médecins à la suite d'une indisposition assez grave que j'ai eue ce printemps. Je me trouve bien de cette cure & je crois qu'elle me guérira. Mais qu'est que la guérison du corps quand le coeur est malade? Sous ce dernier rapport je ne vais guère mieux & il n'y a guère de chance que mon état s'améliore; je ne le voudrais même pas; car, chose singulière, dans la triste situation où je suis, on déteste sa souffrance & pourtant on ne voudrait pas ne pas l'avoir. Il n'y a, je le sais, qu'un seul remède à ce genre de douleurs morales; je le cherche autant que je puis, mais je suis homme & très homme, parconséquent faible & très faible & parconséquent j'ai des moments de désolation où il me semble que je n'ai plus la foi & que Dieu m'abandonne. Je suis sur que vous me comprenez & que vous me plaignez; je voudrais bien être soutenu d'en Haut comme vous l'êtes.

Je vois beaucoup Arago qui est aussi ici pour sa santé; il est arrivé bien malade, mais il semble qu'il est déjà un peu mieux. Vous savez combien sa conversation est intéressante; c'est incroyable combien il sait de choses & comme il les sait bien; aussi sa conversation est pour moi d'un immense interet. Nous parlons souvent de vous, car il est aussi l'un de vos grands admirateurs. Nous attendons avec une grande impatience votre mémoire sur les variations diurnes du magnétisme terrestre[2]. Ce sujet nous intéresse d'autant plus que nous nous en sommes tous les deux beaucoup occupés. Je vous avertis que, jusqu'à ce que je connaisse votre théorie, je tiens assez à celle que j'ai donnée de ces phénomènes & que je n'entrevois pas encore trop bien comment vous pouvez rendre compte des anomalies du phénomène (*Cap de Bonne Espérance*) & de la différence qui existe entre l'hémisphère Sud & l'hémisphère Nord quant autour dans lequel a lieu la variation diurne. Mais tout cela sera probablement éclairci quand on aura lu votre mémoire.

Je suis bien sur que vous avez quelque chose de nouveau sur le métier; car vous êtes infatigable. Vous savez l'intérêt que je mets à vos recherches, aussi j'ose vous prier de m'écrire quelques lignes quand vous avez quelque chose de nouveau. L'intérêt de la Science est le seul parmi les intérêts humains avec celui de ma famille que j'essaie de conserver encore; & rien ne peut m'y aider davantage que quelques communications de votre part.

J'ai eu bien de la peine à me remettre à mon ouvrage; c'est pour moi un travail plein de souvenirs doux & amers en même temps. Enfin j'ai pris courage & j'avance; mais, j'ai de temps à autre des moments de découragement qui m'ôtent toute possibilité de m'en occuper; alors je suis obligé momentanément de faire autre chose. Cette inconstance jointe à la maladie qui j'ai eue ce printemps vous expliquera tous les retards qu'éprouve la publication de cet ouvrage[3] qui aurait dû paraître, il y a plus d'un an.

J'espère que votre santé est passablement bonne dans ce moment & qu'elle ne souffre pas de vos travaux multipliés. J'espère également que Madame Faraday est bien; veuillez avoir la bonté de me rappeler à son bon souvenir. Pardonnez moi cette mauvaise lettre; j'ai obéi en vous écrivant au désir que j'avais de me rapprocher de vous par la pensée pendant quelques moments. Croyez moi, mon cher Monsieur, votre bein dévoué & affectionné | A. de la Rive

Si vous avez le temps de m'écrire quelques lignes, je dois vous prévenir que je serai à *Vichy* jusqu'au 8 juin seulement & que delà je retournerai à *Genève* d'où je ne bougerai plus.

Address: Monsieur Faraday | Royal Institution | Albemarle Stt | Londres

TRANSLATION

Vichy (Dept. del'Allier) | 27 May 1851

My dear Sir,

I do not want to delay any longer in coming to thank you for the good and kind letter that you wrote to me a few months ago[1]. I am, as you shall see by the date on this letter, in Vichy, where the doctors have sent me following a rather serious illness that I had this spring. I feel better for the treatment and I believe it will cure me. But what is the health of the body when the heart is sick? Concerning the latter I am barely better and there is little chance of any improvement in my condition; neither would I wish it; for, strange as it may seem in the sad situation in which I find myself, one detests suffering and yet one would not be without it. There is, I know, but one remedy to this kind of grief; I seek it as much as I can, but I am human and very human at that, thus weak and very weak and therefore I have moments of desolation where it seems to me that I no longer have faith and that God has abandoned me. I am sure that you understand me and that you pity me; I would like to be sustained from Above as you are.

I see a lot of Arago who is also here because of his health; he arrived extremely ill, but he seems a little better already. You know how interesting his conversation is; it is incredible how much he knows and how well he knows it; also his conversation is of immense interest. We often speak of you, for he is also one of your great admirers. We await with great impatience your paper on the daily variations in terrestrial magnetism[2]. This subject is of interest to us all the more because both of us have done a lot of work on it. I am warning you that, until I know your theory, I am keeping to the one I have given to these phenomena and I cannot very well see as yet how you can account for the anomalies of the phenomenon (*Cape of Good Hope*) and of the difference which exists in the Southern hemisphere and in the Northern hemisphere as to around which the daily variation takes place. But all that will be clear when we have read your paper.

I am sure that you have done some new work as you are indefatigable. You know the interest that I take in your research, also I ask you to write to me briefly when you have something new. Interest in Science is the only thing amongst human interests, alongside my family, that I try to keep up; and nothing can help me better than some communication from you.

I have had a lot of trouble in getting back down to my work; it is for me a work full of sweet and bitter memories at the same time. Finally I have taken heart and I advance; but I have from time to time such moments of discouragement that take from me any possibility of occupying myself, then I am obliged momentarily to do something else. This inconstancy added to the illness that I had this spring will explain all the delays I have experienced in the publication of this work[3], which ought to have appeared more than a year ago.

I hope that your health is tolerably well at the moment and that it is not suffering from your multiple works. I hope also that Mrs Faraday is well; please have the kindness to remember me to her good memory. Please excuse this bad letter; in writing to you, I have acquiesced to the desire of bringing myself closer to you through my thoughts for a few moments. Believe me, my dear Sir, your most devoted and affectionate | A. de la Rive

If you have the time to write a few lines to me, I must warn you that I shall be in *Vichy* only until 8 June and thereafter I shall return to *Geneva* from where I shall not be moving.

1. Letter 2378.
2. Faraday (1851c, d, e), ERE25, 26 and 27.
3. De La Rive (1853–8), **1**.

Letter 2429
Faraday to Charles Babbage
3 June 1851
From the original in BL add MS 37194, f.541

Royal Institution | 3 June 1851

My dear Babbage
 I have received & heartily thank you for your book[1]. I have begun to read it. I dare say if I had seen it before it was in print I should have been amongst those who would have tried to persuade you from publishing it[.] The fact is I grieve that your powerful mind ever had cause to turn itself in such a direction and away as it were from its high vocation & fitting occupation[.] Still I know that we cannot avoid the checks & jars of a naughty world and that at times we are driven from our most direct courses by very unworthy objects under our feet[.]
 Ever My dear Babbage | Very Truly Yours | M. Faraday

1. Babbage (1851).

Letter 2430
Joseph Henry to Faraday
4 June 1851
From the original in IEE MS SC 2

Smithsonian Institution | June 4th 1851

My Dear Dr. F.

I owe you many thanks for your prompt attention to my enquiries relative to the lecture room of the Royal Institution and also for your very interesting letter of the 28th of april[1]. It was received in Washington while I was absent on a visit to Cincinnati to attend an extra meeting of the American Association for the advance of science. This was my first visit to the great basin of the Mississippi and I have returned with ideas much expanded of the fertility, resources, and extent of this country. The site of Cincinnati was about fifty years ago a wilderness and it is now occupied by a city which contains 120,000 inhabitants. The houses are built of brick and free stone and the whole country around has the appearance of a long settle place[.] We, (Mrs. H[2], Bache[3] and myself) returned by the way of Niagara Falls, a distance, from Cincinnati, of upwards of 1100 miles, which can now be travelled in three days by rail road and steam boat. Steam as a locomotive power is of great importance in every part of the world, but is no where of as much value as in this country. A friend of mine, one of the Judges of the supreme court of the U.S. has just returned to Washington after deciding 200 law cases and travelling 9000 miles since the 1st of March. I have also been much impressed with the effects produced by the telegraph; lines of which are now forming a net work over the whole inhabited part of the United States. In my late tour I could every where immediately communicate with Washington. On one occasion I held a conversation with a gentleman at a distance of 600 miles; the answers were immediately returned. In dry winter weather I am informed that communications can be sent immediately from Philadelphia to Louisville through 1200 miles of wire[.] This is effected by calling into action a series of batteries distributed along the line.

It would give me much pleasure to see Mrs. F. and yourself in this country and though in comparison with england we could show you nothing of much interest in the way of art, I think you would be gratified with the objects of nature. In our boasted improvements you might possibly be disappointed. An englishman and an american look on these things from different points of view & arrive at very different conclusions. The european finds cities here which in comparison with those he is familiar with in the old world often appear in no respect remarkable while the american after a few years absence returns to a place which he left a wilderness and finds it the site of a large and prosperous city. The one compares this country with the conditions of Europe as they now are the other this country with itself at different times and is astonished at the change.

I am pleased to learn that I still hold a place in the memory of Mrs. F. and yourself[.] The presentation of the daguerrotypes was a proposition of Mayall[4] himself which I did not know he had carried into execution. I had some thoughts of going to London this summer but felt myself too poor to bear the expense of the voyage besides this I would not care to be there during the excitement of the great exhibition and indeed I would by far prefer seeing one of your recent experiments on Diamagnetism than all the contents of the crystal palace.

I send you a copy of an engraving of the Smithsonian building and also of a portrait of Smithson[5] with whom I believe you were acquainted. I regret to say that the building, though picturesque, is not well adapted to the uses of the Institution. The architecture of the 12th century is not well adapted to the wants of the 19th. I consider the crystal palace the true architectural exponent of the feelings and wants of the present day.

Please give the accompanying draft to Mr Vulliamy and ask him to send me a receipt for the amount which may serve as my voucher. Please also to request the secretary of the Royal Institution[6] to address any communications intended for me to Washington instead of Princeton my former residence.

With my best wishes for your | continued health and prosperity | I remain very Truly | your friend & servant | Joseph Henry
M. Faraday LLD

1. Letter 2416.
2. Harriet Henry, née Alexander (1808–1882, Reingold and Rothenberg et al. (1972–99), 1: 211). Married Henry in 1830.
3. Alexander Dallas Bache (1806–1867, DSB). Head of the United States Coast Survey, 1843–1867.
4. John Jabez Edwin Mayall (1810–1901, Reingold and Rothenberg et al. (1972–99), 6: 418). American photographer who spent most of his life in England.
5. James Louis Macie Smithson (1765–1829, DSB). English chemist.
6. John Barlow.

Letter 2431
Paul Frederick Henry Baddeley to Faraday
4 June 1851
From the original in IEE MS SC 2

Lahore June 4th 1851

My dear Sir Michael

I think it practical to lay a line of Electric Telegraphs in the air by means of Baloons.

The line should be floated to at least one and a half miles above the surface of the earth, so as to be beyond the influence of the dense atmosphere and the region of the active portion of storms. At that height, the air is so rarefied that the force or power of the wind upon floating bodies, must necessarily be very considerably less than at the earth's surface, and for the same reason friction would be much less - and progressive motion would not be so much impeded.

The plan seems to preserve many and great advantages over a line laid in the sea, or even on the earth especially for long distances.
1 In the first place, it would be much cheaper.
2. Capable of being carried to an unlimited extent.
3 More easily constructed, and arranged than the sea line.
4 More out of the reach of harm - than either - not so liable to derangement or injury from various causes.
5 And moreover, it might be made a means of safe and expeditious transit.

To explain in general terms the plan I propose I must remark that a strong two inch rope might be easily boyed up to the higher regions of the atmosphere by means of small Baloons placed at intervals of five miles and numerous smaller ones of about 3 feet or more in diameter, at every 100yds along the line - composed of silk varnished with gutta percha or some such impermeable substance to prevent the escape of the Hydrogen gas. A line of this kind would be so boyant as to be capable of sustaining considerable weights, and it might therefore easily carry a line of wire. It might be secured by ropes at every 10 miles and communication established at such points by means of small baloon cars conducted up the communication lines. Frequent fixed points at the earth's surface would not however be absolutely necessary - but on the contrary, it is probable that the line might be carried more than a hundred miles straight and without any communication with the earth.

Transit might be effected by means of baloon cars attached to the projected line that being nearly of the same sp. gravity as the line itself - might be sustained & guided in their course, without much stress upon the floating line. Mechanical & philosophical science would be brought to bear upon the undertaking, so as to secure success.

Believe me | Yours Faithfully | P. Baddeley | Surgeon Asst

I have many interesting points to tell you about storms & the nature of Dust storms for I cannot find time to arrange my notes - or even to peruse them. In the dust storms there is a mass of Elecl. spirals of limited weight I think limited at least as far as the Dust is concerned, to $\frac{1}{2}$ or $\frac{3}{4}$ of a mile above the earths surface. This mass may be a broad strip extending for many miles and going at the rate of from 10 to 50 or 60 miles an hour and directed down to the surface from the higher regions.

This mass is composed of innumerable spirals and the gusts and in all cases of storms are occasionally the passage a greater number of these Elecl spirals & in greater force at this particular period & they come on sometimes in regular intervals - thus - In spirals shape so that the part in contact with the earth is behind & the whole slopes from below upwards and forwards ///// thus

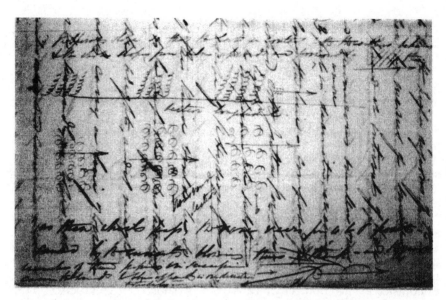

As these spirals pass, the vane veers from 4 to 6 points caused by the currents blowing thus I think - and the great number of these passing in succession cause the winds to blow apparently in one direction & steadily

Letter 2432
Faraday to Arthur-Auguste De La Rive
5 June 1851
From the original in BPUG MS 2316, f.71–2

Royal Institution | 5 June 1851

My dear De la Rive
Your last letter[1] has affected me deeply and has renewed my sorrow on your account. I knew of your sad loss[2] and had heard also of your personal illness & its very serious character: but I knew also that you had within that might sustain you under such deep trials. Do not be discouraged - remember - wait patiently. Surely the human being must suffer when the dearest ties are rent but in the midst of the deepest affliction there is yet present consolation for the humble minded which (through the power that is over us) may grow up and give peace & quietness & rest. Your letter draws me out to say so much for I feel as if I could speak to you on account of something more than mere philosophy or reason. They give but a very uncertain consolation in such troubles as

yours and indeed nothing is more unsatisfactory to me than to see a mere rational philosopher's mind fighting against the afflictions that belong to our present state & nature - as on the contrary nothing is more striking than to see such afflictions met by the weakest with resignation & hope. Forgive me if my words seem to you weak and unfit for the occasion. I speak to you as I have felt and as I still hope to hope to feel to the end and your affectionate letter has drawn me forth[.]

I have nothing to say to you about Philosophy. For the duties of the season & the exhibition Jury work[3] occupy all my poor ability which through a failing memory becomes less & less[.] As to my last papers[4] I hope you will find them at home & there you will see how I consider the Cape of Good Hope & other cases.

I intended to have written to you at Vichy but I think now I had better send to Geneva. Indeed I ought to apologize for this letter both for its bad writing & its unconnected matter but if I wait to write a better it may turn out worse for I have let my thoughts run on though unconnectedly and as to my hand writing I find the muscles will not obey my will as they used to do and that I cannot write as well as I would[.] The machine is becoming rusty. Let it, there is a time for all things[5] - and there is a time without end[6] coming[.]

Ever My dear De la Rive | Your affectionate friend | M. Faraday

Address: A la Professor | Auguste de la Rive | &c &c &c | Geneva

1. Letter 2428.
2. The death of Jeanne-Mathilde De La Rive. See letter 2378.
3. That is of the Jury of the Great Exhibition. See letter 2405.
4. Faraday (1851c, d, e), ERE25, 26 and 27.
5. William Shakespeare, The Comedy of Errors, ii, 2, 65.
6. See Ephesians 3:21.

**Letter 2433
George Ransome to Faraday
6 June 1851
From the original in RI MS F1 K35**

Ipswich 1851 6mo 6

My dear Friend

I feel truly grateful for thy kindness - but it would be quite out of the question my presuming to include my own portrait in the series I am bringing out[1][.] However it strikes me thy wish can be complied with in another way and which will not interfere with thy kind concession to my wishes. Last year some kind friends requested me to submit to the same ordeal I had induced so many to undergo and the Committee presented me with a few surplus copies one of which is quite at thy service and shall be forwarded today² - I can assure thee I feel most grateful for thy kindness not only on my own account but also for the sake of others to whom I know thy Portrait will be valuable.

Believe me my dear friend | Ever truly thine | Geo Ransome

P.S. I will write to Maguire³ that he may arrange to have all ready for thee⁴[.]

1. That is of portraits of members of the Ipswich Museum of whom Faraday was one. See Letter 2424.
2. This lithograph is in RI MS F1 K35.
3. Thomas Herbert Maguire (1821–1895, Wood (1995), 337). Portraitist.
4. That is for his portrait of Faraday. See plate 6.

Letter 2434
Faraday to Augusta Ada Lovelace
10 June 1851
From the original in Bod MS dep Lovelace-Byron 171, f.50–1

Royal Institution | 10 June 1851

My dear Lady Lovelace

You see what you do - ever as you like with me. You say write & I write - and I wish I had strength & head rest enough for a great deal more for it would give me very great pleasure to move more earnestly for those young creatures whom I rejoiced to know as your children[.] Their intelligence was astonishing, their manners kind & themselves in every way most interesting[.]

But for myself whilst the objects I should like to pursue remained before me - the weight of years & of a failing memory retard me more & more in their pursuit so that from day to day I have to circumscribe the space I destine for my own activity. Else I should most assuredly call on you. I met the Earl¹ one evening I think at Lord Rosse's where I went as in duty bound & he told me I think that you were not there & went very little out because of ill health & I was very sorry to hear it. I hope yours will be but a short restraint, mine is a growing defect and can only be expected to end with life[.]

Ever My dear Lady Lovelace | Your Grateful Servant | M. Faraday

1. William King, Earl Lovelace (1805–1893, B6). Lord Lieutenant of Surrey, 1840–1893.

Plate 6. Faraday by Thomas Herbert Maguire. See letter 2433.

Letter 2435
George Wilson to Faraday
12 June 1851
From the original in IEE MS SC 2

24 Brown Square Edinburgh | June 12th 1851

Dr Faraday
Dear Sir

Observing from the Athenaeum[1] that you are about to Lecture on Ozone[2], I take the liberty of addressing a few lines to you on the subject, though afraid that they may not reach you in time.

Schönbein has suggested that Ozone may be the cause of Influenza[3], & Mr Robert Hunt has drawn attention to the probable increased presence of the substance in question in the atmosphere at the period when the last Epidemic of Cholera was passing away. What I wish to suggest in the way of a query is that if Ozone can produce influenza it should be possible to induce that artificially by causing Ozonised Air to be breathed, as Berzelius produced in himself severe bronchitis, by inspiring air containing Seleniuretted Hydrogen[4], which Dr Prout[5] thought *might be* the cause of Influenza[6]. Experiments, however, on the Subject seem scarcely needed, since for more than a Century Electricians have undesignedly developed around their persons large quantities of Ozone, and have been exposed for hours together to Air, much more highly charged with Ozone, than upon any hypothesis the atmosphere can ever be. I should like to put this question to you, in case you think it worth while answering publicly. Has it ever happened in your own very extensive experience of the action of the largest Friction Hydro-Electric - & Galvanic Machines, & batteries to observe in yourself or your assistants *or audience*, any development of a disease resembling influenza? Surely within a Century we should have heard of a *Morbus Electricus*, if Electricians had been sufferers from the action of Ozone?

Conclusions have been drawn as to the relative quantity of Ozone in the air, from the development of a blue colour of varying intensity in Starch-paste mixed with Iodide of Potassium, but according to M. Chatin's[7] communication made to the Academy of Science at Paris (Séance du 5 Mai 1851 L'Institut pour 7 Mai 1851, p 148[8]), the atmosphere always contains free Iodine, so that experiments thought to prove the presence of Ozone in the Air are exposed to a serious fallacy. In the case also of Iodide of Potassium *used alone*, I venture to suggest that the same Agency, Electricity of High Tension which developes Ozone in the Air, also developes at least when undergoing discharge, Nitric Acid, which by liberating Iodine from Iodide of Potassium may give a deceptive appearance as to the presence of free Ozone. I do not wish to call in question the presence of Ozone in the atmosphere, but merely to refer to the doubtful character of some of the supposed proofs of its presence, and to suggest the improbability on the ground of the impunity with which Electricians breath Ozone, (if I am right in believing in this impunity) of the Ozone-theory of Influenza.

I trust you will excuse this intrusion. Your Courtesy on former occasions induces me to send this hasty scrawl to catch the post.

I Remain | Yours very truly | George Wilson M.D

1. *Athenaeum*, 7 June 1851, p.608 announced Faraday's lecture.
2. Faraday (1851g), Friday Evening Discourse of 13 June 1851.
3. Schoenbein (1851b).
4. Berzelius (1819), 101.
5. William Prout (1785–1850, DSB). Physician and chemist.
6. Prout (1834), 350–2.
7. Gaspard-Adolphe Chatin (1813–1901, DBF). Professor of Botany at the Ecole de Pharmacie, 1848–1873.
8. Chatin (1851).

Letter 2436
Anne Isabella Noel King[1] to Faraday[2]
15 June 1851[3]
From the original in RI MS F1 N3/26

6 Gt Cumberland Place | June 15th

Dear Sir

I hoped to find an opportunity last Friday evening[4], to thank you for your kind letter, but I was obliged to give up all attempts to reach you. I have got Harris's book on electricity[5], and shall read it with great interest; it is to you that I owe my desire to become more nearly acquainted with this interesting branch of science.

Believe me yours truly | & much obliged | Annabella King

1. Anne Isabella Noel King (1837–1917, DNB4 under Wilfrid Scawen Blunt). Only daughter of Ada Lovelace.
2. Recipient established on the basis of provenance and content. See letter 2434.
3. Dated on the basis that this letter follows on from letter 2434.
4. Faraday (1851g), Friday Evening Discourse of 13 June 1851.
5. Probably Harris (1851).

Letter 2437
William Henry Fox Talbot to Faraday
15 June 1851
From the original in RI MS Conybeare Album, f.37
M. Faraday Esq

Lacock Abbey, Chippenham | June 15/51

Dear Sir

The important experiment tried yesterday at the Royal Instn. succeed[e]d perfectly. A printed paper was fastened to a disk, which was then made to revolve as rapidly as possible. The battery was discharged, and on opening ye Camera it was found to have received an impression. The image of the printed letters was just as sharp as if the disk had been motionless. I am not aware of this experiment having ever been made before. I should be happy to repeat it in the presence of some of our Scientific friends, but I wish first to obtain effects on a greater scale of development and therefore I should be very glad if you would devise means of augmenting the brilliancy of the flash. Two methods occur to me which I submit for your judgment. (1) Professor Hare[1] of America says that if a flat coil of copper ribbons is placed in the circuit the spark from a Leyden jar is greatly increased in brilliancy. Supposing this to be equally true of a battery, would it not be desirable to adopt it?

(2) If the battery discharge were taken thro' a tube lined spirally with pieces of tinfoil (one of which is at the R. Instn.) would not the total effect of these numerous sparks light up the room more than the single discharge does?

If a truly instantaneous photographic representation of an object has never been obtained before (as I imagine that it has not) I am glad it should have been first accomplished at the Royal Instn[2].

Believe me | Dear Sir | Ever truly yours | H.F. Talbot

1. Robert Hare (1781–1858, DSB). Professor of Chemistry in the University of Pennsylvania Medical School, 1818–1847.
2. This experiment was reported briefly in *Athenaeum*, 28 June 1851, p.688. See also William Henry Fox Talbot, "Note on instantaneous Photographic Images", *Proc.Roy.Soc.*, 1851, 6: 82 and Talbot (1852).

Letter 2438
Faraday to William Henry Fox Talbot
16 June 1851
From the original in Lacock Abbey MS 51-22

R Institution | 16 June 1851

My dear Sir

I do not know of Hares[1] Experiment[2][.] We have no flat coil of copper: but perhaps an Electromagnetic coil would do. When you pursue your subject you can try such as we have and also the spiral tube. I should not expect much from either process: the latter you cannot put into the focus of a lens or mirror as you can the former i.e the single spark[.]

I do not know what mirror you used but I should have thought that a deep one would answer better than a shallow one. Such as they use at the lighthouses collects $\frac{2}{3}$ or even $\frac{3}{4}$ of the rays which issue from the light in the focus.

There is no occasion for any one to think much for you are well able to think your own thoughts to an end. For my part the moment I can get clear of the run of ordinary business all my thought will given to the Magnetism of Oxygen[.]

Ever Truly Yours | M. Faraday
H.F. Talbot Esq | &c &c &c

1. Robert Hare (1781–1858, DSB). Professor of Chemistry in the University of Pennsylvania Medical School, 1818–1847.
2. See letter 2437.

Letter 2439
Julius Plücker to Faraday
18 June 1851
From the original in IEE MS SC 2
Dear Sir!

A longe time ago I got your last kind letter[1]. Soon after receiving it I sent you "sous bande" the first two papers on the magnetic axes of cristals[2] (a third and last one is not yet finished). I paid them at the post office, but four months after I got them back, with the words "Rowland refused". Newspapers are sent to England in the same way for a trifle of money. When there is a mistake, I would be angry if by my fault. You will get the papers now by a gentlemen returning to London. With pleasure I had brought them myself, in this time of the Exhibitions fever and the meetings of the British Association[3]; but I cannot leave Bonn befor[e] August.

I have joined to the papers a new metal, named Donarium (Do) discovered by professor Bergeman[4] of Bonn[5], thinking it may interest yourself and perhaps also the chemical section of the Association[6]. M. H Rose confirmed prof. Bergeman's results. Fearing the reduced metal might be altered by crossing the Channel, he (B) prefers to present the hydrat of the oxyd ($\overline{Do}$ + Aq.) You may reduce it easily, better by potassium or natrium than by hydrogenium. I give also a specimen of the mineral (Orangit), which is very rare, containing the new metal.-

Last sumer unhappily I did not succede in repeating E Becquerels experiments with oxigenium absorbed by charcoal[7]. At the same time I convinced myself that the results of my former experiments which induced me to admit the dilation of air in the neighbourhood of a magnetic pole, were illusorious. It may be proved mathematically, that there can be in such cases no sensible effect at all. In that desolate state of my head, regarding the magnetisme of the gases, I received with great satisfaction the first notice of your recent experiments with oxigenium[8]. I instantly undertook a long series of experiments, making use of a very fine balance of glass, indicating with full certainty $\frac{1}{10000}$ of a gramme. In that way I may determine the magnetic power nearly with the same accuracy, then the weight of a body. I concluded first from my experiments that the specific magnetism of a body is a quantity as constant as the specific heat &c &c. The specific magnetism of oxigenium is exactly proportional to its density; it is not changed in a sensible way, when this gas is mixed with hydrogenium nitrogenium, carbonic oxyd, chlore &c &c. The specific magnetism of pure iron being 1000000, that of oxigenium is very near 3500. Hydrogenium, Chlore, Cyanic acid, carbonic oxyd, carbonic acid, $\mathcal{K}$ (protoxide d'azote) &c &c were not affected, the first one showing only a trace of diamagnetic action, but certainly not amounting to $\frac{1}{200}$ of that on oxigenium.

Two months ago I sent two elaborate papers to Poggendorff but till now I got no copies of them[9]. Two weeks ago I sent a new paper[10], schowing by experiment the coercitiv force of the gases, similar to that of steel.

By far the most curious and allways unexpected results presented Nitrogenium in its different combinations with oxigenium. I undertook about it a laborious series of experiments not yet quite finished. But I hope to overpower these days the last difficulty by condensing $\mathcal{K}$ (acide hyponitrique) in its purity. I expect the condensed gase will be strongly magnetic, as it is in its aëriform state. Then I shall be able to answer in a most accurate way one important question, by comparing the specific magnetism of the same substance in the two different states.

Within the narrow limits of a letter it seems to me rather impossible to give a true account of my researches. I'll spare some of the results already obtained to a next comunication.

I hope your health be quite good now; with great satisfaction at least I may conclude it from what you worked in science. Tis only the impulse given by you, from which originates my recent researches.

With all my heart, Sir I am | Yours very truly | Plücker
Bonn, 18th of June | 1851.

Address: Professor Faraday | &c &c | London | Royal Institution

1. Letter 2346.
2. Plücker and Beer (1850-1).
3. At Ipswich.
4. Carl Wilhelm Bergemann (1804-1884, P1, 3, 4). Professor of Chemistry at the University of Bonn.
5. Bergemann (1851).
6. See letter 2449 and note 7.
7. Becquerel (1850a).
8. Faraday (1851c, d, e), ERE25, 26 and 27.
9. Plücker (1851a, b).
10. Plücker (1851c).

Letter 2440
Augusta Ada Lovelace to Faraday
20 June 1851[1]
From the original in RI MS Conybeare Album, f.44

Gt C. Place | Friday Evg | 20th June

Dear Mr Faraday

Many thanks for y[ou]r kind & prompt fulfilment of my request.

Annabella[2] has just carried off the Book[3] with her to the country, where *I* shall join her tomorrow. She said she should write to you herself from there.

I shall be back here before *Wed[nes]d[a]y*[4], - & shall have indeed great delight in expecting you on Weddy evg.

Very truly & very gratefully yours | A.A. Lovelace

1. Dated on the basis that this letter follows on from letters 2434 and 2436.
2. Anne Isabella Noel King (1837-1917, DNB4 under Wilfrid Scawen Blunt). Only daughter of Ada Lovelace.
3. See letter 2436.
4. That is 25 June 1851.

Letter 2441
Christian Friedrich Schoenbein to Faraday
25 June 1851[1]
From the original in UB MS NS 397

My dear Faraday

I think an excellent likeness of our illustrious countryman Euler will prove acceptable to you it was made at the expense of Basle and I am charged by the Council of our Museum to send you a copy of it as an humble hommage they desire to render you. There are some other copies joined and intended for the Royal Society &c. and I beg you to be kind enough to forward them quite leisurely to their respective places of destination. From Mr. Burckhardt[2] I learned that you are doing well, he was highly pleased with the Lion of the Royal Institution.

I am continually riding my hobby horse and now and then pick up something new. I am very sorry I did not sooner ascertain some facts; they would have made a good figure in your experiment on Ozone. You shall before long have details about [them].

By this time I think your lecture on that subject will be over[3], let me know something of the matter. In the beginning of August I intend to go to Glarus, where the meeting of our association will take place. Have you no mind to come over and ramble about a little with me?

Pray present my best compliments to Mrs. Faraday and believe me
Quite in a hurry | Your's | very truly | C.F. Schoenbein
Basle Aug. [sic] 25, 1851.
P.S. Mr. Sarasin[4] a young friend of mine has the kindness to take charge of the parcel; should he happen to deliver it in person pray receive kindly[.] | S.

1. Dated on the basis that Faraday's reply, letter 2453, is dated Tynemouth, 1 August 1851 (where he had been since 12 July 1851 (Faraday, *Diary*, 5: 11322)) and in which he referred to taking this letter with him. Schoenbein may have had August in mind as he had just referred to the meeting of the Swiss Association in that month.
2. Unidentified.
3. Faraday (1851g), Friday Evening Discourse of 13 June 1851.
4. Unidentified.

Letter 2442
Faraday to George Ransome
28 June 1851
From the original in CITA

R Institution | 28 June 1851

My dear friend

I had your kind note and also the one from the Lodging Committee[1] which I gave to M. Dumas & as he was about to write I asked him to acknowledge it on my part[.]

I expect we shall leave London on Wednesday[2] by the 11 o clk train. De la Beche, Dumas, Barlow & several of us.

Ever Truly Yours | M. Faraday

Mr Geo Ransome

1. Of the British Association at Ipswich.
2. That is 2 July 1851.

Letter 2443
César Mansuète Despretz[1] to Faraday
29 June 1851
From the original in IEE MS SC 2
Mon cher Confrere
    Mr. Chuard[2] qui a plusieurs appareils à l'exposition de Londr[e]s, et notamment une nouvelle lampe de surete pour les min[e]s, et un Gazoscope, me demande une lettre pour vous. C'est un simple et bon garcon connu par plusieurs inventions. Permett[e]z moi de vous le recommander. J'ai lu avec bien del'interet vos nouvelles experiences sur les gaz[3].
    Recev[e]z l'assurance des sentimens distingués avec lesquels je suis | votre devoué confrer[e] | C. Despretz | m. del'Institut
Le 29 Juin 1851

Address: A Monsieur | Mr, Faraday de la | the royale de Londr[e]s | &c &c à Londres

TRANSLATION
My dear colleague,
    Mr Chuard[2] who has several instruments at the London exhibition, notably a new mining safety lamp and a Gasoscope has asked me for a letter of introduction. He is a simple and good lad known for several inventions. Please allow me to recommend him to you. I read with a lot of interest your new experiments on gases[3].
    Please accept the assurance of my distinguished sentiments with which I am | your devoted colleague | C. Despretz | member of the Institut.
29 June 1851

1. César Mansuète Despretz (1792–1863, DBF). French chemist.
2. Chuard was a French scientific instrument maker of 6 Rue Carnot, Paris. Anon (1851), **3**: 1177.
3. Faraday (1851c, d, e), ERE25, 26 and 27.

Letter 2444
Charles-Nicolas-Alexandre Haldat Du Lys[1] to Faraday
July 1851
From the original in IEE MS SC 2

Nancy, | le July 1851. | Société des Science, Lettres et Arts | de Nancy Le Secrétaire perpétuel | De la Société des Sciences, Lettres et Arts de Nancy, | Membre correspondant de l'Institut & de la Société National de Médecine.
Monsieur Faraday
Monsieur et illustre promoteur de la Science magnetique &c.

Je profite avec bien de l'empressemant du voyage de L'un de mes compatriotes qui va visiter lexposition Europeene pour me rapeller a votre bienveillant Souvenir et vous remercier de votre bonne reception lan dernier. j'étais alors plein de vigueur quoique fort agé mais les fatigues de mon voyage qui ont éccedé mes forces m'ont donné une gastralgie des plus peinibles et des plus menacantes. toutes fois je ne me repentirai jamais d'avoir visité un eminant physicien, d'avoir vu lune des plus belles ville [sic] du monde et d'avoir eu des relations avec des personnes dun peuple qui brille aux premiers rangs - par son industrie, *son bon sens son attachemant a Ses loix et Son respect pour lautorité de Son gouvernemant*. je Suis parti de ce pays avec le desir de le voire de nouveau et plus en detail mais ma Santé Sy oppose encore plus que mes 82 ans et c'est un grand Sujet de regret.

j ai lu dans les ouvrages periodiques quelques notices Sur vos ingenieux travaux relatifs au magnetisme des gaz mais ils me Sont insuffisants je serais bien reconnaissant Si vous pouviez me donner le memoire qui les contient[2] - ou quelquextrait, quelquanalyse bien faite. le jeune becquerel fils de mon ami a ausi travaillé Sur ce Sujet mais que dire du vide ou il fait jouer à lether le meme role quaux gaz. lEther est une hipothese commode mais....

j ai lhonneur de vous faire hommage de quelques petits travaux extraits des memoires de l'academie de Nancy ou j'habite. veuillez aggréer avec indulgence et les communiquer a Mrs vos confreres aux quels je les adresse *parmi ces memoires il en est un Sur le quel je desire extremement fixer de nouveau votre attention*[3]. cest celui ou jai etabli l'universalité du magnetisme[4] depuis rigoureusement prouvé par vos decouvertes mon but principal est de vous demander *franchemant* votre opinion Sur la pretention que j'eléve d'avoir le premier depuis votre illustre compatriote Gilbert[5] qui en avait deja donné quelqu'idee; d'avoir, dije, contre lopinion de Coulomb alors g[e]nerallemant adoptée proclamé au nom de tous les physiciens qui m'ont fourni les preuves l'universalite de la force, ou puissance &c magnetique. *Votre opinion Sera pour moi d'un grand poid veuillez ne pas me refuser cette grace*. Si vous avez la bonté de menvoyer quelques uns de vos memoire en original ou en extrait vous pouvez me les adresser par l'intermédiaire de la maison de librairie J.B. Ba[i]lliere⁶ regent Street et a paris rue de lEcole de médecine[.]

aggreéz [sic] Monsieur les humbles temoignages de la plus haute consideration de Votre Serviteur | C.N. Haldat

TRANSLATION

Nancy, | le July 1851. | Société des Science, Lettres et Arts | de Nancy Le Secrétaire perpétuel | De la Société des Science, Lettres et Arts de Nancy, | Membre correspondant de l'Institut & de la Société National de Médecine.
Mr Faraday
Sir and illustrious promoter of magnetic science &c,

I am hurrying to take advantage of a trip by one of my compatriots, who is going to visit the European exhibition, in order to bring myself to your kind memory and to thank you for your good reception last year. I was then full of vigour, although I was rather aged, but the tiring effects of the journey which exceeded my strength, have given me the most acute and threatening stomach pains. Nevertheless I shall never regret visiting an eminent physicist, seeing one of the most beautiful cities in the world and having met people from a nation of the first order that stands out because of its industry *its common sense, its attachment to its* laws and its respect for the authority of its government. I left your country with the desire to see it again and in more detail, but my health is against it even more than my 82 years and this is of great regret to me.

I read in some periodicals notices of your ingenious work regarding the magnetism of gases but they are insufficient. I would be extremely grateful if you could give me the paper that contains it[2] - or some extract or good analysis. Young Becquerel, the son of my friend, has also worked on this subject but what can you say of a vacuum where he allows aether to play the same role as gases? Aether is a convenient hypothesis but ....

I have the honour of paying you homage through a few small works extracted from papers of the académie of Nancy where I reside. Please accept them kindly and communicate them to your colleagues to whom I address them. *Amongst these papers is one to which I would like to ask you to turn your attention once more*[3]. It is the one where I established the universality of magnetism[4] since then rigorously proved by your discoveries. My principal reason is to ask you *frankly* your opinion on the pretension that I raise that I was the first, after your compatriot Gilbert[5], who had already given some notion of this, to have, as I say, against the generally accepted theory of Coulomb, proclaimed in the name of all the physicists that furnished me with proofs of the universality of magnetic force, or power &c. *Your opinion will be for me of great weight. Please do not refuse me this grace.* If you have the kindness to send me some of your papers in the original or in extracts, you can address them to me through the bookshop of J.B. Baillière[6] in Regent Street or in Paris at the rue de l'ecole de médecine.

Please accept, Sir, the humble testimony of the highest consideration of your servant | C.N. Haldat

1. Charles-Nicolas-Alexandre Haldat Du Lys (1769–1852, DBF). Physician in Nancy.
2. Faraday (1851c, d, e), ERE25, 26 and 27.
3. See Haldat Du Lys to Faraday, 26 May 1846, letter 1880, volume 3.
4. Haldat Du Lys (1845).
5. Presumably William Gilbert (1544–1603, DSB). Physician and natural philosopher. See Gilbert (1600).
6. Jean-Baptiste-Marie Baillière (1797–1885, Vapereau (1880), 104, (1893), 75). Bookseller in London and Paris.

Letter 2445
Nathaniel Wallich[1] to Faraday
8 July 1851
From the original in RI MS F1 K40

5 Lower Gower St | 8 July 1851

A thousand thanks, My Dear Dr Faraday, for your most kind and generous note, the contents of which shall immed[iatel]y be communicated to Mr Hjorth[2], his address shall in one line be forwarded.

Believe me with | profound respect, | My dear Dr Faraday | Your most obliged | N. Wallich
M. Faraday | &c &c &c

1. Nathaniel Wallich (1786–1854, DNB). Superintendent of the Calcutta Botanic Gardens, 1815–1850.
2. Søren Hjorth (1801–1870, DBL). Danish inventor. There is a description of his electric motor in Anon (1851), 1359–60.

Letter 2446
Faraday to Henry Ellis
16 July 1851
From the original in BM CA

Tynemouth | North Shields | 16 July 1851

My dear Sir Henry

I received your note here and am sorry you could not have an answer by the day you wished. I should have been happy to have assisted in a

good cause but I shall not be in town perhaps for some months & therefore am unable to meet the wishes of the Trustees[1][.]
Ever My dear Sir Henry | Your faithful Servant | M. Faraday
Sir Henry Ellis Bart | &c &c &c

1. The Trustees of the British Museum had requested, on 12 July 1851, Faraday's advice on cleaning metal objects discovered by the archaeologist Austen Henry Layard (1817–1894, DNB1) in the Middle East.

Letter 2447
Tom Taylor[1] to Faraday
17 July 1851[2]
From the original in IEE MS SC 2

3 Figtree Court | Temple | July 17

Dear Professor Faraday

Am I the most impudent man in the world, in presuming so far on your kindness, (shown at the time I gave once a Friday evg lecture at the Royal Institution on a subject very little in the ordinary track of the lectures there[3]) as to present to you Il Dottere Polli[4], Professor of Chemistry at the Scola Mechanica (or School of Industrial Arts) at Milan & compiler of the "Annali di Chimica Applicata alla medicina" who is here observing all that London & the Great Exhibition have to show bearing on his speciality, or chemistry applied to the arts & industrial Processes.

He is most anxious for the honour of an introduction to you & for an opportunity if possible of hearing one of your lectures. Forgive me, if I am taking an unwarrantable liberty - but I have so vivid a recollection of your great kindness, that I am emboldened to do what I am doing. I have written to Mr Barlow to ask for a ticket, in the event there being an opportunity of hearing a lecture.

But I thought I might make the request to you also, in the event of Mr Barlow not being able.

Most truly yours | Tom Taylor
Professor Faraday | &c &c &c | Royal Institution

1. Tom Taylor (1817–1880, DNB). Assistant Secretary to the Board of Health, 1850–1854.
2. Dated on the basis of the reference to the Great Exhibition.
3. See *Athenaeum*, 15 May 1847, p.525 for an account of Taylor's Friday Evening Discourse of 7 May 1847 "On the Saxon Epic Beowulf".
4. Giovanni Polli (1812–1880, EI). Professor of chemistry at the Scuola technica, Milan, 1849–1851.

Letter 2448
Faraday to Joseph Henry
23 July 1851
From the original in SI A, JHC, Box 10

Royal Institution London | 23 July 1851.

My dear Dr. Henry
    I received your last letter[1] and herewith enclose Mr Vulliamy's receipt, I thought I had sent it in the former letter[2] but found out my mistake[.]
    Your account of the country you have been through excites me far more than Palaces or Exhibitions. The beauties of nature are what I most enjoy. Scenery and above [all] the effects of light & shadow Morning & Evening & Midday or a storm or a cloudy sky[.] My predilection is for out of door beauties & just now I and my wife have run away from London to the seaside[3] to get quiet & rest. My head even now aches & I feel very weary[.]
    When I left London I had not received the Engraving of the Smithsonian Building but I dare say I shall find it upon my return[.] I thank you very heartily for it. I did not know Mr Smithson[4] though I think I used to hear his name I was then of no consequence[.] My wife send[s] her kindest remembrances with mine.
    Ever My dear friend | Yours Very Truly | M. Faraday

Endorsed: Recd 8 Aug

1. Letter 2430.
2. Letter 2416.
3. That is Tynemouth. See letter 2446.
4. James Louis Macie Smithson (1765–1829, DSB). English chemist.

Letter 2449
Faraday to Julius Plücker
23 July 1851
From the original in NRCC ISTI

23 July 1851

My dear Plucker
    I received your kind letter of last month[1]: but I know nothing about the papers which you refer to as having been sent to me "sous bande". I do not think they were refused at the Royal Institution for although miscellaneous papers do not come as newspapers do through the post on moderate terms yet I never refuse them. I am always glad to see your

papers for though I cannot make out the German language they are ever very pleasant remembrances of you and of the happy hours we have had together. I am in hopes that before this you will have received a copy of my papers from the Philosophical Transactions[2]. It was a great pleasure to me so to confirm & enlarge the results of the magnetic character of oxygen which I had obtained in 1847[3]. Your results with this body agree entirely with my own[.] I shall be curious to know what are your results with the hyponitric acid to which you refer, that is, when you have obtained them. My memory is so bad that I cannot be quite sure but I think I tried the condensed nitrous acid & did not find it sensibly magnetic and that I stated the result in the letter on the Magnetic characters of flame & gases[4] - but as I am not at home but write from Tynemouth where my wife & I have gone out of the turmoil of London for rest & fresh air I cannot refer to either papers or notes[.]

The papers on Donarium[5] and also the specimens of the oxide & mineral came quite safe and I took them to the meeting of the Association at Ipswich where they were received with great interest by the Chemical section & philosophers in general[6]. I was about to ask you to thank Professor Bergeman[7] for me but I think I will write a few words on the next page for though my head aches and is giddy I am anxious to express to the Professor my great sense of his kindness[.]

Ever My dear Plucker | Yours Most Truly | M. Faraday

Address: Dr Plücker | &c &c &c | University | Bonn | Germany

1. Letter 2439.
2. Faraday (1851b, c, d, e), ERE24, 25, 26 and 27.
3. Faraday (1847b).
4. *Ibid.*, 411.
5. Bergemann (1851).
6. See *Athenaeum*, 12 July 1851, p.750 for an account of this which occurred on 4 July 1851.
7. Carl Wilhelm Bergemann (1804–1884, P1, 3, 4). Professor of Chemistry at the University of Bonn.

Letter 2450[1]
Faraday to Benjamin Vincent
27 July 1851
From the original in RI MS F1 D4

Tynemouth | 27 July 1851

My dear friend

I purposed writing to you about this time but did not expect to do so under such circumstances. It is the Sabbath day & yet I am confined to this place and unable to meet with the very few brethren who are at Newcastle. My system had sunk too low and last week it settled into an attack of sore throat so that I speak with labour can hardly swallow even the Saliva because of the pain & have at last called in the Dr. We are glad of it for he has put us on the right track (and we were not quite right in that which we had been pursuing) and I hope in a day or two to get out of the house again[.] Mrs. Geo Buchanan[2] came to us here & last Sabbath day[3] we had the very unexpected pleasure of her husbands company in his way home[.] She has been very poorly and so little benefitted by Tynemouth that on Thursday[4] she went away to Newcastle or rather to Rye Hill. We had a letter from her this morning in which she says she is better. She tells us of the death of Mrs. Macnaughton[5] the wife of a brother at Edinburgh - and also the death of Mr Pratts[6] grandson Thomson[7] very suddenly.

Mr. Paradise seems fairly well[.] We had an outing one day (last Tuesday[8]) he, D. Reid & I, and enjoyed it very much together though I rather think I was not in a state to be braced up but rather knocked down by it. Mr. & Mrs. Deacon have arrived but I have not yet been able to see them perhaps in two or three day[s] I may get into Newcastle.

We do not think of coming home until about Wednesday week[9] - if all goes on well[.] Now on looking at my list of remembrance dates yesterday I found that would be after your holiday had began. I do not expect that will be any inconvenience to you since the arrangements are all made not to interfere one with another and Anderson is fully instructed & trusted by me in his duty, but if there is any thing I have not perceived or that you wish attended to write me a line[.] In any case I should be very glad to see a word from you. I feel very weary as to letter writing myself but I am glad when others write to me[.] My wife sends her love to you and yours with mine[.]

Ever My dear friend | Yours affectionately | M. Faraday
Mr B. Vincent

1. This letter is black edged.
2. Charlotte Buchanan, née Barnard (1805–1866, GRO). Sister of Sarah Faraday and wife of George Buchanan.
3. That is 20 July 1851.
4. That is 24 July 1851.
5. Unidentified.
6. Unidentified.
7. Unidentified.
8. That is 22 July 1851.
9. That is 6 August 1851.

Letter 2451
John Tyndall to Faraday
30 July 1851
From a typescript in RI MS T TS, volume 12, pp.4009–12

Queenwood College, | Stockbridge, | Hants. | 30th. July 1851

Dear Sir,

It would have afforded me great gratification to have made these magnetic experiments before you; but as the destinies seem disposed to deny me this pleasure, I take the liberty of forwarding you a few substances which you perhaps will be good enough to suspend in the magnetic field when your leisure permits.

No. 1. contains two cylinders of bismuth; between the *flat poles* one sets axial, the other equatorial. Between the pointed poles when *near*, the former sets equatorial also, but on withdrawing the poles to a distance from each other it turns into the axial position. In this case the principal cleavage is transverse to the axis of the cylinder, as you can prove by trial with a knife; and I believe you will conclude that the cause of its setting equatorial between the near poles is due to the fact that the tendency of the mass to pass from stronger to weaker places of magnetic force overcomes the directive power of the crystal. If the mass were a sphere the powerful cleavage would always set equatorial, between both points and flat faces. This turning towards the axial line, as you are aware, is attributed by Plucker to a triumph of magnetism over diamagnetism; but I think you will see that it is on account of the more equable condition of the magnetic field permitting the directive action of the crystal to assert itself.

No. 2. contains two bismuth plates; the magne-crystallic axis is perpendicular to the flat surface of each, but it will set equatorial. This has been brought about by compression between two plates of copper in a hard vice. Before compression the magne-crystallic axis set in both cases strongly axial.

No. 3. contains a plate of compressed wax; it is diamagnetic, and the line of compression sets equatorial, thus causing the plate to set its length from pole to pole.

No. 4. contains a crystal of carbonate of iron, and a model of the same, formed from the crystal when pounded into dust and sifted through linen. I think you will see that the deportments of both are so far similar as to be suggestive of a common origin. The dough from which the model is made was compressed in the direction of the optic axis.

No. 5. contains a prism of carbonate of iron artificially made. suspended from the centre of one of its sides it will set axial - change the point of suspension 90° in the equatorial plane; it will set strongly equatorial. Between the near *points*, when they are not too strongly excited the prism suspended as in the latter case will set axial, between the distant points it will turn into the equatorial position. This is the phenomenon which Plucker attributed to the triumph of the optic axis force at a distance in the case of tourmaline, idocrase, and other magnetic crystals. In explanation of this I will merely say, that when the prism set equatorial between the excited poles, the mass of dough from which it was taken was compressed in the horizontal line perpendicular to the axis of the prism - the line of compression sets axial and hence the prism itself equatorial. With regard to the turning round I believe you will see the analogy between this case and that of the bismuth cylinder, the one being *recession* and the other *approach*.

No. 6. contains a bit of shale; suspended from one of its edges it sets axial, from the edge adjacent, it sets equatorial. This is similar to Plücker's first experiment with the plate of tourmaline. I had a much finer piece of shale to show this experiment but it has unfortunately got broken.

It is with considerable reluctance that I introduce any personal affair of mine to your notice, and yet I am induced to do so at present. I returned from Germany, as you are aware, a few weeks ago and entered a situation here. I am teacher in this college and when our work commences I expect to be occupied about 7 hours a day. My salary is £150 a year, rooms in the college and board; and I find the Principal[1] and his family willing to do all they can to render my life agreeable. The institution is a private one, and with regard to its durability I am unable to utter a word. The boys whom I have to teach are young, and I shall be engaged with them at the lower mathematics, such as Euclid, Algebra perhaps to quadratic equations, and a little trigonometry, and elementary physics. I shall have to lecture twice a week, but to suit my audience must confine myself to elementary matters. There is a danger in retrogression in such a position, and there is also the everlasting consciousness of want of permanancy. I think this is a sufficient account of my present position.

In the last two numbers of the Athenaeum I find an advertisement stating that in the university of Toronto there are 6 professorships now vacant at the annual salary of £350 Halifax currency[2]; one of these is the professorship of Natural Philosophy[.] It occurred to me as I read the advertisement that I ought to make application for the post, and after turning this matter round in my mind I concluded that the wisest plan would be to solicit your advice, and this conclusion was confirmed by my friend Dr. Francis of the Philosophical Magazine. Nobody can be more sensible than I am how slender are the grounds upon which I can justify this liberty, and had I been left to my own unaided cogitations I should hardly have attempted it. But to return; if both choices were before me I should prefer remaining in this country to going to Canada - not that I lay stress upon locality any further than that I like to be near men of science. But I dont know my chances in England and hence I think it unwise to allow an opportunity like the present to escape me. To settle myself down at Queenwood, even granting it permanent, would be to sacrifice an object for which I have battled harder than anybody knows, and that is to approve myself a worker in science. Seven hours plus meal times and other contingencies, plus the time to depolarize the intellect after having been engaged with other matters is a heavy subtraction from the day. I ask your counsel in this state of things - something doubting I must confess, for I know I have no right to expect it - I have already written to Magnus and Poggendorff for testimonials so that if you advise the step I shall be ready to take it promptly.

    I remain dear Sir | Most faithfully & respectfully yours | John Tyndall

Prof. Faraday F.R.S. | etc. etc.

1. George Edmonson (1798–1863, DNB). Headmaster of Queenwood College. On the College see Thompson, D. (1955).
2. *Athenaeum*, 19 July 1851, p.761 and 26 July 1851, p.793.

Letter 2452
Faraday to John Tyndall
1 August 1851
From a typescript in RI MS T TS, volume 12, p.4129

    Tynemouth, | 1 August, 1851.

My dear Sir,

    Your letter[1] finds me here ill of a quinsey but now recovering, and though I cannot write much, I determined to answer you at once. In the first place many thanks for the specimens which I shall find presently at home. I was very sorry not to see you make your experiments but hope to realise the profound results which interest me extremely. I want to have a very clear view of them.

    But now for the Toronto matter. In such a case private relationships have much to do in deciding the matter, but if you are comparatively free from such considerations and have simply to balance your present power of doing good with that you might have at Toronto, then I think I should (in your place) choose the latter. I do not know much of the university but I trust it is a place where a man of science and a true philosopher is required, and where in return such a man would be nourished and cherished in proportion to his desire to advance natural knowledge. I cannot doubt indeed that the University would desire the advancement of its pupils and also of knowledge itself. So I think that you would be exceedingly fit for the position, and I hope the position fit for you. If I had any power of choosing or recommending, I would aid your introduction into the place, both because I know what you have already done for science, and I heard how you could state your facts and treat your audience.

Now I do not, for I cannot, proffer you a certificate, because I have in every case refused for many years past to give on the application of candidates. Neither indeed have you asked me for one. Nevertheless I wish to say that, when I am asked about a candidate by those who have the choice or appointment, I never refuse to answer; and indeed, if my opinion could be useful and there was a need for it, you might use this letter as a private letter, shewing it or any part of it to any whom it might concern[2].

And now you must excuse me from writing more, for my muscles are stiff and weak, and my head giddy.

Ever, my dear Dr. Tyndall, | Yours most truly, | M. Faraday

1. Letter 2451.
2. In *Testimonials of John Tyndall, Ph.D., Candidate for the Professorship of Natural Philosophy in the University of Toronto*, (copy in RGO6 / 373, f.414–9), Tyndall stated (p.9), "I am permitted to state that Dr. Faraday and the Astronomer Royal are prepared to respond to any personal reference made to them respecting my qualifications for the Professorship in question".

Letter 2453
Faraday to Christian Friedrich Schoenbein
1 August 1851
From the original in UB MS NS 398

Tynemouth | 1 August 1851

My dear Schoenbein

On running away from the bustle and weariness of London I brought your letter[1] here intending to answer it long before now and lo! I have been attacked by inflammation of the throat have had a quinsy and been held in much pain and debility until now. I will not longer delay believing that a few words are better than none. I have not yet received the portrait of Euler but doubt not it is at home will you do me the favour to return my most sincere thanks to the Council of the Museum for the great [honour] they have done me in favouring me with a copy which I shall ever look upon with great pleasure. The others I will deliver according to their addresses.

The Ozone Evening went off wonderfully well[2][.] Our room overflowed and many went away unable to hear (my account at least) of this most interesting body. Through your kindness the matter was most abundant & instructive & the experiments very successful. The subject has been sent into the world so much piecemeal that many were astonished to see how great it became when it was presented as one whole, and yet my whole must have been a most imperfect sketch for I found myself obliged to abridge my thoughts in every direction. Many accounts were printed by different parties & some very inaccurately since they had to catch up what they could. A notice of four pages appeared in the proceedings of the Royal Institution and though I think that has appeared in the Athenaeum or the Philosophical Magazine[3] yet I shall send you copies of it when I can. The subject excited great interest and from what the folks said I had no reason to be ashamed either for the subject or myself[.]

and now my dear Schoenbein I am very weary[.] Perhaps to day you are at Glarus[4]. I was two days at Ipswich at our meeting[5] - no more for want of strength[.] Queens balls[6] - Paris fetes - &c &c &c I am obliged (& very willing) to leave all to others[.]

With kindest remembrances to Made Schoenbein & Yourself in which my wife has full part

I am ever yours | M. Faraday

Address: Dr. Schoenbein | &c &c &c | University | Basle | Switzerland

1. Letter 2441.
2. Faraday (1851g), Friday Evening Discourse of 13 June 1851.
3. No report appeared in either *Athenaeum* or *Phil.Mag.*
4. For the meeting of the Swiss Association.
5. Of the British Association.
6. See letter 2418.

Letter 2454
John Tyndall to Faraday
c3 August 1851[1]
From a typescript in RI MS T TS, volume 12, pp.4013

Aug. 1851.

Dear Sir,

I have to return you my sincere thanks for your kind and valuable letter. I was already aware of your dislike to giving written testimonials and therefore took care not to ask for anything of the kind. Indeed the manner of recommendation which you allude to would be more agreeable to me than that by certificate, and if the parties have an agent in this country (of which I am not yet aware) I doubt not it would be the most practical and effectual.

With regard to my paper at Ipswich[2] there is one matter which I should like to call to your mind. During the reading of it I was haunted by the consciousness that the Section was already tired, and this caused me,

I doubt not, to appear somewhat hurried. Had it occurred earlier in the day, as I hoped it would I think I should have satisfied you better[3].

I believe you will find the experiments to be as I have described them. If you once turn the light of your intellect upon this matter I think you will not long halt between two opinions - that you will once more take the thought to your bosom which suggested that significant note to a passage in the Bakerian Lecture for 1849 "Perhaps these points may find their explanation hereafter on the action of contiguous particles."[4]

With best thanks, dear Sir | Most faithfully yours | John Tyndall
Prof. Faraday | etc. etc.

1. Dated on the basis that this is the reply to letter 2452.
2. Tyndall (1851b), given at the meeting of the British Association.
3. See *Athenaeum*, 12 July 1851, p.748 for Faraday's comments on Tyndall's paper.
4. Faraday (1849b), ERE22, note to paragraph 2586.

Letter 2455
Lambert-Adolphe-Jacques Quetelet to Faraday
3 August 1851
From the original in IEE MS SC 2
Bruxelles, le 3 Aout 1851. | Académie royale | des | Sciences, des Lettres
et des Beaux Arts | de Belgique.
Mon cher et illustre confrère,

Mon ami M. Stas[1] qui se rend à Londres a desiré obtenir quelques mots d'introduction auprès de vous. Je me suis empressé de diférer à sa demande, non parceque je pense qu'il ait besoin d'être recommandé mais pour avoir moi même le plaisir de vous présenter mes amitiés et de vous remercier pour le bienveillant accueil dont vous m'avez honoré pendant mon sejour à Londres[2].

Je ne vous parlerai pas des travaux de M. Stas, vous les connaissez probablement aussi bien que moi, et vous avez plus de connaissances pour les apprécier. M. Stas vous dira que nous avons repété ensemble et avec M. Melsens, Vos belles expériences sur le diamagnetisme: elles sont font bien réussi et j'ai lieu d'être très satisfait de l'appareil que m'a fourni M. Ru[h]mkorff.

Je vous envois la continuation de mon ouvrage sur le climat de la Belgique[3]. Je n'ose me flatter que vous ayez le temps d'y jeter les yeux, cependant je crois que la partie des *ondes atmosphériques*[4] n'est pas tout à fait indigne de votre attention.

Veuillez présenter mes compliments les plus respectueux à Madame Faraday, et agréez pour vous même les assurances réiterées de mes sentiments d'amitié et de devouement.

Tout à vous | Quetelet
Monsieur M. Faraday, &c

TRANSLATION
Brussels 3 August 1851. | Académie royale | des | Sciences, des Lettres et
des Beaux Arts | de Belgique
My dear and illustrious colleague,
My friend M. Stas[1], who is travelling to London, has come to ask me for a few words of introduction to you. I hastened to comply with his wishes, not because he needs to be recommended, but for the pleasure of presenting my own compliments to you and to thank you for the kind reception with which you honoured me during my stay in London[2].
I shall not speak to you of the works of M. Stas, you know them probably as well as I do myself and you are better able to appreciate them. M. Stas will tell you that, together with M. Melsens, we repeated your beautiful experiments on diamagnetism. They were very successful and I was most satisfied with the instrument provided by M. Ruhmkorff.
I am sending you the continuation of my work on the Belgian climate[3]. I do not flatter myself that you have the time to glance over it, however I believe that the section on *atmospheric waves*[4] is not entirely unworthy of your attention.
Please be kind enough to present my compliments to Mrs Faraday and please accept for yourself the reiterated assurances of my sentiments of friendship and devotion.
Yours | Quetelet
Mr M. Faraday, &c

1. Jean-Servais Stas (1813–1891, DSB). Professor of Chemistry at the Military School in Brussels, 1840–1868.
2. Probably a reference to Quetelet's visit to London when he was an Associate of the Jury of the Great Exhibition dealing with Class X, Philosophical Instruments. Anon (1852a), 1: xxviii.
3. Quetelet (1851b).
4. *Ibid.*, 73–104.

Letter 2456
Harriet Jane Moore to Faraday
4 August 1851[1]
From the original in IEE MS SC 2

The Spring Aug 4th

Many thanks my dear Mr. Faraday for your kind note. I am much vexed that you should have suffered so distressing a malady, and that Mrs. Faraday should have felt the anxiety your illness must have caused her: I trust that you will continue to improve, and will be careful not to catch cold, relapses are generally so much more formidable than the first attack. I am so pleased that I could send any thing you liked; as soon as I hear of your return Mrs. Faraday shall have a nosegay with long stalks, at least as long as I can contrive to procure them, but our garden is not too well stocked with flowers. The humble river Brent is now full of the most elegant plants; weeds, fashionable gardeners would call them, but really very beautiful, rather too much so for a lady who came down to spend the evening with us last week, in attempting to reach some white water lilies fell into the water, & was most completely drenched. I hope you have had some fine sunsets in the north; I always wish you were here when there is one more beautiful than usual. The weather is now favourable for the harvest, & wheat & oats are being cut in this neighbourhood. I do not think you can say as much for Northumberland. My people join in kindest remembrances to yourself & Mrs. Faraday,

& believe me Most truly Yours | Harriet Moore

---

1. Dated on the basis of the reference to Faraday's illness and to his being in Northumberland at this time. See letters 2450, 2452 and 2453.

Letter 2457
Faraday to Royal Society
19 August 1851
From the original in RS MS RR 2.89
I consider Harris' paper[1] as an important experimental paper and one which should be printed in the Transactions not only because of the character of the author & the right he has to speak as a philosophical authority - but for its own manifest merit. It contains the proofs of what I believe to be facts in magnetism and which as they may not be set aside must be admitted & finally explained in any view of the nature of magnetic force which can hope to keep its place in the future progression of science[.] | M.F.
Royal Institution | 19, Aug 1851.

1. William Snow Harris, "On Induced and other Magnetic Forces", *Proc.Roy.Soc.*, 1851, **6**: 87–92. The manuscript is in RS MS AP 33.15. This paper had been referred on 22 May 1851 (RS MS CMB90c) and postponed on 19 June 1851. Despite this report of Faraday's, Harris's paper was again referred on 30 October 1851 since an unfavourable report by Humphrey Lloyd dated 1 October 1851 (RS MS RR 2.90) had also been received. See letter 2489.

Letter 2458
Faraday to Heinrich Rose
23 August 1851
From the original in SI D MS 554A

R Institution | 23 Aug 1851

My dear Rose
    I quite grieve that I did not see you to day. I cannot call on Monday[1] before 1 or 2 o'clk because I must be here to meet Magnus - as I think you perhaps may have heard. You will be so full of occupation that I have no right to think it probable you may drop in on Monday. If you do not I shall come to you[.] This is the more necessary on my part because on Tuesday & Wednesday in part I must be out of town[.]
    Yours Affectionately | M. Faraday

1. That is 25 August 1851.

Letter 2459
Faraday to Andrew Reid[1]
26 August 1851
From the original in the possession of Rosalind Brennand

R Institution | 26 Aug 1851

My dear Andrew

I received the balance & the note in due course:- and though I write in haste when I reached home let me again thank You for your kindness in taking care of us. The trip was a source of very great pleasure & as to health I believe did more good than all Tynemouth put together[.]

Your Very Affectionate Unkle | M. Faraday

1. Andrew Reid (1823–1896, Reid, C.L. (1914)). A nephew of Sarah Faraday and a printer in Newcastle.

Letter 2460
George Biddell Airy to Faraday
2 September 1851
From the original press copy in RGO6 / 373, f.410

Royal Observatory Greenwich | 1851 Sept. 2

My dear Sir

Dr. Tyndall writes to me[1] in reference to his candidature for a Toronto professorship. I had just read his last paper in the Phil. Mag.[2] & thought it good. But he refers me distinctly to you - and thus I am as it were compelled to trouble you with the question whether *you* do not think him a good man?

Yours very truly, | G.B. Airy
Michael Faraday Esq | &c &c &c

1. Tyndall to Airy, 1 September 1851, RGO6 / 373, f.402. Airy replied that he would be willing to give Tyndall a reference rather than a testimonial. Airy to Tyndall, 10 September 1851, RGO6 / 373, f.412. See note 2, letter 2452.
2. Tyndall (1851c).

Letter 2461
D.F. Van der Pant[1] to Faraday
2 September 1851
From the original in RS MS 241, f.129

Rotterdam September 2d 1851

The Directors of the Batavian Society of experimental Philosophy at Rotterdam in their meeting of August 25 a.c.[2] have decreed to request you to honour the Society with your acceding to it as a corresponding member.

The Directors trusting they will receive a favourable answer to this proposal beg you to acquaint them with your different titles.

In the name of the said Directors | D.F. van der Pant | 1st Secretary Michael Faraday, Prof. of Chemistry &c. | London.

1. Unidentified.
2. Anno corrente. Of the current year.

Letter 2462
Faraday to Caroline Deacon
4 September 1851
From the original in the possession of Elizabeth M. Milton

Thursday 4 Septr | 1851 | Park end Cottage | Lee Road | Blackheath
My dear Caroline

I am about to write a few words I hardly know why except that your last letter moves me in its mention of the troubles of one I love namely yourself; and truly you have much to make you sorrowful seeing that though in possession of the surpassing gift of hope in Christ you with us all are still in the flesh and our Lord in his mercy testified that even when the spirit was willing the flesh was weak[1]. When I think of you & Thomas I try to cheer myself by what I doubt not also sustains you namely that all is under the ruling guidance of a merciful God who knows best what is needed by those whom he has chosen[.] All the scriptures of the New Testament & therefore all the body of the word in its true meaning & intent shew that for the people that are of God in Christ this world is not their rest implying I think that it must be different to them to what it will be to those who are of it. In the parable Abraham says that when the rich man had his good things in the world Lazarus was receiving evil things[2] - *but now* he is comforted & thou art tormented[.]

And then again there is that part of the Scripture (Hebrews) where the Apostle speaks of those who are without chastisement as being out of the Fathers love & its exercise;- as bastards & not sons[3]. And I do not understand this chastisement as implying merely some evident form of rebuke like that which I trust has been in mercy manifested towards Elizabeth[4]; but also such dealing of God with us (very often in private) as shewing us how soon we are beset by repining & impatience may lead us to see how weak our faith & trust in him is, and how little we understand of the true value of the prize of the high calling of God in Jesus Christ. And indeed such chastisements may & do come as much to those who seem at ease in respect of such matters as you are greatly tried, with, as to those who are surrounded by them: for we make our own trouble by our folly, and when temporal things are easy to us then the world is as a snare, and we are perhaps in more danger of letting go the faith and saying where is the presence of his coming for since the fathers all things continue as they were[.] What a happy thing it is my dear Caroline that if we are moved by him to think of these things; then we are encouraged & *urged* to go to a throne of Grace for the remedy and ask for Mercy & Grace to help us in time of need.

I write to you as if I were forgetting Thomas but it is not so. Whatever the nature of our trial we *all* have access by the same door to him who is a very present help in every time of trouble. Only I think you can better choose the time of giving my earnest love to him than I can by the Postman & so I write to you.

Give the little one[5] a kiss for me[.] I think when she was looking at the setting sun she probably saw a phenomena which you may perhaps never have distinguished - a certain reflection of the light from the eye lashes which gives the appearance of a ray elongating & darting forth. Ever my dear Caroline | Your affectionate Unkle M. Faraday[.]

I am very glad to hear what you say of Elizabeth[.] We shall see if the effect continues. It certainly has not continued since when the expression were so sober that it could not stand for a moment against them[.]

1. Matthew 26: 41.
2. Luke 16: 19–31.
3. Hebrews 12: 8.
4. Elizabeth Reid.
5. Constance Deacon.

Letter 2463
Faraday to George Biddell Airy
5 September 1851
From the original in RGO6 / 373, f.411

Royal Institution | 5 Septr. 1851

My dear Sir

I think so well of Dr Tyndalls papers that I should be very glad to hear he had the Toronto Professorship because I think it would give him the power of working & that he would work to the honor of the University[1][.]

He spoke so well & clearly at Ipswich[2] in explaining his results that I think he would make a very good Lecturer[.] With kindest remembrances to Mrs. & Miss Airy

I am as Ever | Yours Most Truly | M. Faraday
The | Astronomer Royal | &c &c &c

---

1. See letter 2460.
2. Tyndall (1851b), given at the meeting of the British Association. For Faraday's comments on Tyndall's paper at the time see *Athenaeum*, 12 July 1851, p.748.

Letter 2464
John Barlow to Sarah Faraday
15 September 1851[1]
From the original in RI MS F1 I38

Rectory | Buxted | Uckfield Sep 15

My dear Mrs Faraday,

This note is indited on the pure principles of Hibernian reciprocity viz to "get from the lady all I can and give her nothing in return"[.]

I want you to tell me how you and Faraday are. It is quite three weeks since I heard about you.

My life gives no materials. My wife and I and our friend Miss Grant are living in this quiet parsonage, and I go every day, on some errand or other, to Uckfield where I was curate nearly 30 years ago.

The condition of the rural labourers in this part of England is infinitely better than I have ever known it. Wages 10s a week, and every body employed. Then the harvest has been got in so cheaply & so well that the Farmer cannot lose by the low prices. I hear that the millers prefer the corn of this year for immediate grinding to the old wheat[.]

Spend five minutes and one penny in giving me a bulletin & believe me
Always yours | John Barlow

1. Dated on the basis that Barlow was Rector at Uckfield in 1822, that no evidence has been found that he went abroad in the summer of 1851 (unlike other summers in this period) and because of the reference to Faraday's illness.

Letter 2465
George Biddell Airy to Faraday
23 September 1851
From the original press copy in RGO6 / 373, f.199

Royal Observatory Greenwich | 1851 September 23

My dear Sir

Allow me to introduce to you Professor Listing[1] of Göttingen (Professor of Experimental Philosophy) who brings letters from Gauss, and whose acquaintance I think you will find very agreeable.

I am, my dear Sir, | Yours very truly | G.B. Airy
Michael Faraday Esq | &c &c &c

1. Johann Benedict Listing (1808–1882, P1, 3). Professor of Physics at University of Göttingen.

Letter 2466
Faraday to Edward Hawkins[1]
11 October 1851
From the original in BM DWAA MS Correspondence 1826–1860, volume 5, 1722

Park End Cottage | Lee Road | 11, Octr. 1851

My dear Hawkins

Those who make casts do it so often from Sulphate of lime in the form of plaster that they ought to be far better practical judges than I am but I should incline to avoid the use of plaster against the slabs. There would however be no difficulty I suppose in taking wax impressions from the slab, and then making the plaster casts in these. In a small way the process is common enough[2].

Ever Truly Yours | M. Faraday
Edw. Hawkins Esq | &c &c

1. Edward Hawkins (1780–1867, DNB). Keeper of Antiquities at the British Museum, 1826–1860.
2. Hawkins needed this information before proceeding to make casts from Assyrian and Lycian sculptures in the British Museum for the French government. See minutes of the British

Museum Trustees, 11 October 1851 and of its Standing Committee, 13 December 1851 which authorised the making of the casts. The request to Faraday is not noted in either minute.

Letter 2467
Faraday to John Barlow
14 October 1851[1]
From the original in RI MS F1 D5

Park end Cottage | Tuesday 14 Oct

My dear Barlow

I was talking with Scott Russel[l][2] at the Exhibition about the American Yacht her sailing[3] &c and said I thought he should give us an Evening on the subject of sails & their principles in that instance. He assented far more freely than I expected. Do you approve or not if so good if not let me know & I will back out properly. I think it would be a good subject & that he would do it well[4].

Ever Yours | M. Faraday

1. Dated on the basis that Faraday was staying in this cottage at the time. See letter 2462.
2. John Scott Russell (1808–1882, DNB). Naval architect.
3. This was the schooner 'America' which won the prize of the Royal Yacht Squadron at the Cowes Regatta in August 1851. See Folkard (1901), 373–4.
4. Russell (1852), Friday Evening Discourse of 6 February 1852. At this Discourse a model of the 'America' was exhibited, p.119.

Letter 2468
Faraday to John Tyndall
21 October 1851
From a typescript in RI MS T TS, volume 12, pp.4130

Lee Road, | 21 Oct. 1851.

My dear Sir,

Many thanks for your note and the paper[1] and the testimonials[2]. I hope you will obtain your desire at Toronto. I have read the paper briefly, but must do so again: at present I am writing, or rather copying, a paper on *lines of force*, which touches the point of *Polarity*, so that it is only hereafter I shall be able to collate all together. I propose giving the paper in to the Royal Society today[3].

Ever truly yours, | M. Faraday.
Dr. J. Tyndall, | &c. &c. &c.

1. Probably Tyndall (1851c).
2. That is *Testimonials of John Tyndall, Ph.D., Candidate for the Professorship of Natural Philosophy in the University of Toronto*, (copy in RGO6 / 373, f.414–9).
3. Faraday (1852b), ERE28. Faraday sent the paper in the following day.

Letter 2469
Joseph Toynbee[1] to Faraday
24 October 1851
From the original in IEE MS SC 2

16 Savile Row, Octr 24, 51

Dear Sir,
I beg to thank you very much for your autograph which I shall value greatly as long as I live.
I am dear Sir, | Yours very faithfully | Joseph Toynbee
M. Faraday Esq

1. Joseph Toynbee (1815–1866, DNB). Ear surgeon.

Letter 2470
Johan Rudolf Thorbecke[1] to Faraday
27 October 1851
From the printed original in RS MS 241, f.127
No. 100 | V Division

La Haye, le 27 Octobre 1851.

Monsieur,
J'ai l'honneur de vous faire parvenir ci-joint un exemplaire de l'arrêté de Sa Majesté le Roi[2] du 26 de ce mois, par lequel l'Institut Royal des Sciences, des Lettres et des Beaux-Arts est supprimé et remplacé par une Académie Royale des Sciences, dont il a plu au Roi de vous nommer membre[3].

Vous trouverez également sous ce pli un exemplaire du règlement pour la nouvelle Académie.

Je saisis cette occasion, Monsieur, pour vous offrir l'assurance de ma considération très-distinguée.
Le Ministre de l'Intérieur, | Thorbecke
A Monsieur M. Faraday | à Londres

Endorsed by Faraday: See page 89

TRANSLATION
No. 100 | V Division
                                            The Hague, 27 October 1851
Sir,
I have the honour of enclosing a copy of the decree of His Majesty the King[2] of the 26th of this month, by which the Institut Royal des Sciences, des Lettres et des Beaux-Arts is superseded and replaced by an Académie Royale des Sciences, of which it has pleased His Majesty to nominate you a member[3].
You will also find enclosed a copy of the constitution of the new Académie.
I take this opportunity, Sir, to offer you the assurance of my most distinguished consideration.
The Minister of the Interior, | Thorbecke
To Mr M. Faraday | in London

1. Johan Rudolf Thorbecke (1798–1872, NNBW). Netherlands Minister of the Interior, 1849–1853.
2. William III (1817–1890, NNBW). King of the Netherlands, 1849–1890.
3. See letter 2180.

Letter 2471
Faraday to Henry Bence Jones
27 October 1851
From the original in RI MS F1 D6
                                            Royal Institution | 27 Octr. 1851.
My dear Dr Jones
My wife & I have been looking forward to our return home[1] & to a hoped for conversation or consultation with you in regard to her health at any convenient opportunity. If I knew how you were circumstanced I would make one[.] May I ask you when you next come to the Institution to tell the Porters to inform me & I will come to you[.]
Ever My dear Dr | Most truly Yours | M. Faraday

1. From their stay in Blackheath. See letters 2462, 2466, 2467 and 2468.

Letter 2472
Prince Albert to Faraday
31 October 1851
From the original in RS MS 241, f.120

Windsor Castle | October 31st 1851

Sir

I have the honor, as President of the Royal Commission for the Exhibition of 1851 to transmit to you a Medal that has been struck by order of the Commissioners, in commemoration of the valuable services which you have rendered to the Exhibition, in common with so many eminent men of all Countries, in your capacity of Juror[1].

In requesting your acceptance of this slight token on our part of the sense entertained by us of the benefit which has resulted to the interests of the Exhibition from your having undertaken that laborious office, and from the zeal and ability displayed by you in connection with it, it affords me much pleasure to avail myself of this opportunity of conveying to you the expression of my cordial thanks for the assistance which you have given us in carrying this great undertaking to its successful issue.

I have the honor to be | Sir | Very gratefully yours | Albert | President of the Royal Commission
Professor Faraday F.R.S.

1. See letter 2405.

Letter 2473
Nathaniel Bagshaw Ward[1] to Faraday
3 November 1851
From the original in IEE MS SC 2

Clapham Rise | 3 Nov 1851.

Dear Sir,

I feel most reluctant to trouble you about any personal matters, but I am in a strait, and know not how to act. Paxton[2] - not contented with his reputation as an Architect, a Botanist & Horticulturist - has come forward as a Medical Champion for the relief of consumptive cases by advocating the erection of a Sanatorium in connexion with the Consumption

Plate 7. Prince Albert.

Hospital³. Now this idea most certainly did not originate with him as I advocated it most strongly in a little work published in 1842⁴.

It is just possible that Paxton may never have read my book - although the magnific conservatory of his noble patron at Chatsworth⁵ was filled with its choicest vegetable treasures by means of the closed cases sent expressly to the East Indies for that purpose under the care of a gardener chosen by Paxton. Relying on the boasted impartiality of the Times I wrote a letter to the Editor, simply stating the above facts - but this letter has not been noticed.

Now may I take the liberty of asking you whether (in the conversation I had with you - resp[ectin]g the Lecture you did me the honor to deliver at the R.I.⁶ & which was prior to the publishing of my book) I stated my conviction that the same principle which had proved so beneficial with regard to the vegetable kingdom was likewise applicable to the animal.

I feel that I have no right to trouble you with such a question - but a Portfolio in which the minutes of your Lecture were preserved - has been lost or mislaid in my removal from Wellclose Sqr to this place. I can hardly however expect that you can have recollected what I - who am personally interested in the matter have nearly forgotten. What I now wish to do is to publish a plain statement in one of the medical journals - leaving the public to draw their own conclusions. I need not of course add that I shall not print any statement from you without your consent.

Hoping you will pardon my thus intruding upon you.

Believe me to be Dear Sir | Very truly yours | N.B. Ward

1. Nathaniel Bagshaw Ward (1791–1868, DNB). Botanist.
2. Joseph Paxton (1801–1865, DNB). Superintendent of the Gardens at Chatsworth from 1826 and architect of the Crystal Palace.
3. On Paxton's ideas for a sanatorium in Victoria Park see "Foreign Airs and Native Places", *Household Words*, 1851, 3: 446–50. It was not built.
4. Ward, N.B. (1842), 71–2.
5. William George Spencer Cavendish, 6th Duke of Devonshire (1790–1858, DNB). Collector of coins, books and paintings whose seat was Chatsworth.
6. See *Lit.Gaz.*, 14 April 1838, p.233 for an account of Faraday's Friday Evening Discourse of 6 April 1838 "On Mr. Ward's Method of preserving Plants in limited Atmospheres".

Letter 2474
Faraday to Jean-Baptiste-André Dumas
6 November 1851
From the original in AS MS

Royal Institution | 6 November 1851

My dear & kind friend

I hope you will not be startled at my presumption but I had formed the rather ambitious thought of endeavouring (if you should sanction it) to convey to our Members at one of the Friday Evening meetings an idea of the remarkable and important views which you developed to us in some degree at Ipswich[1] and first I have to ask you whether such a proceeding would be agreeable to you or whether for any reason or feeling you would rather I should not do it.

But in the next place if you see no reason against it but on the contrary are willing to let me touch so fine a subject before our members then I am obliged to confess to you that I feel greatly startled in finding how much of that which you communicated to us & to myself personally my decaying memory has allowed to escape and though I have the sheet of paper on which you wrote me down a few pencil figures & a few rough lines and also the journal accounts of your discourse at Ipswich yet they do not sufficiently clear up my recollections to enable me to do that which I want to do well.

And so my boldness extends to this. If in the first place it is agreeable to you that I should do it then have you any papers MS or other that you could lend to me giving me an account of the results whether deduced from change of volume or of soluble or progression of character or equivalent numbers &c &c and the probabilities arrived at as the conclusion. As our audience though they contain I am happy to say many high philosophers consist chiefly of persons who though gentlemen of high & liberal education are still not exclusively scientific (500 or 600 persons being present perhaps) so I generally introduce an experiment or two to make them quickly comprehend any *point* which is under consideration:- As for instance in speaking of the progression of Chlorine Bromine & iodine in reference to your views I should shew them these bodies but such helps I could arrange and do not wish to trouble you about experiments unless indeed you have some which have occurred to yourself as fit illustrations. As to diagrams or curves I am at a loss for the few lines I have do not now recall my memory except very vaguely to that which they represent[.]

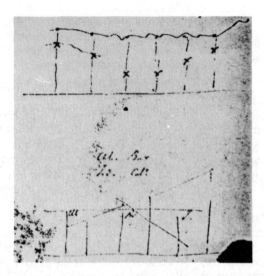

Now I think I have said enough to frighten you[.] If I could obtain possession of the matter and give the subject well I should be greatly honoured in the doing of it but I should not like to put you to too much trouble. Any papers you may trust me with I will most carefully return and use them with every reservation that you may desire. If there is any thing yet published about them and you favour my proposition send me a reference to it. The evening would be in the middle or end of next January, but I should like to have possession of the matter (if I gave it) by or before the beginning of the next year that I may study it well. Our kindest remembrances to Madam Dumas also to M. Dumas your Son & to Madam Edwards[2].

Ever My dear friend, Yours faithfully | M. Faraday
Monsieur Dumas | &c &c &c

Address: A Monsieur | Monsieur Dumas | &c &c &c | Rue de Vaugirade | Faubourg St. Germain | à Paris

1. At the meeting of the British Association. See *Athenaeum*, 12 July 1851, p.750 for an account of Dumas's paper "Observation on Atomic Volumes and Atomic weights, with considerations of the probability that certain bodies now considered elementary may be decomposed". Faraday did not lecture on this topic. See letter 2493.
2. Laure Edwards, née Trézel. Married Henry Milne-Edwards in 1823. See his entry in DBF.

Letter 2475
Faraday to John Frederick William Herschel
14 November 1851
From the original in RS MS HS 7.195

R Institution | 14 Novr. 1851

My dear Herschell
    I did with the papers as you wished.
    Being busy with Magnetism & lines of force I send for your present amusement a couple of printings from nature[1][.] The disposition of the magnets is marked at the back[.]
    Ever Yours | M. Faraday

1. See Faraday, *Diary*, 11, 12, 13 November 1851, 6: 11666–11703 for Faraday's production of iron filing diagrams.

Letter 2476
John Tyndall to Faraday
28 November 1851
From a typescript in RI MS T TS, volume 12, p.4014

Queenwood College | Stockbridge, | Hants. | 28th. Nov. 1851.
Dear Sir,
    From a letter received yesterday from Col. Sabine I conclude that you are already aware of his kind intentions towards me[1][.] In that letter he tells me that your signature will be one of those from personal knowledge required by the Royal Society previous to the admission of a new member. For this allow me to return you my sincerest thanks; I do most fervently hope that the great kindness of which I have of late been the recipient from you and others will bear its proper fruit in keeping me faithful to my work. Few have had greater encouragement than I have had, and the greater will be my condemnation if it be not turned to suitable account. I owe you much - more perhaps than you are aware of. However sweet the instances of personal good will with which you have favoured me may be,- and sweet they are beyond a doubt - you have been my unconscious benefactor in other ways to an extent which, as far as the cultivation of my proper manhood is concerned outweighs those instances of private friendship. To me you have been a preacher of toil, courage, and humility - a preacher in the highest sense of the term, demonstrating by the silent power of example the possibilities which are open to the honest worker. This has been your relation to me and may it long continue so.
    I remain dear Sir, | Your faithful and obliged servant | John Tyndall
Professor Faraday | etc. etc. etc.

1. See Tyndall, *Diary*, 27 November 1851, **5**: 79 which noted that Sabine had obtained Faraday's signature on Tyndall's nomination for Fellowship of the Royal Society from personal knowledge. RS MS Cert 9.300.

Letter 2477
Henry William Pickersgill[1] to Faraday
30 November 1851
From the original in IEE MS SC 2

14 Stratford Place | 30 Novr. 1851

My dear Faraday

I should long 'ere this have replied to your very kind note had I not hesitated whether I ought to accept your generosity towards my Daughter for I assure you I consider myself a great intruder and have strong feelings of compunction upon entering that Building and I should long ago have subscribed had not my industry been heavily taxed to assist those whose claims are all powerful and in the present case of my enquiry I was in hopes there was a means of admission within my grasp.

I cannot expect you will point out a mode of easing my conscience of the debt of gratitude but I must endeavour to find out myself the means.

With the highest considerations, esteem, and gratitude

Believe me | Ever your most obliged | friend & servant | H.W. Pickersgill
To Ml. Faraday Esq DCL | &c &c &c

1. Henry William Pickersgill (1782–1875, *DNB*). Portrait painter.

Letter 2478
Faraday to Joseph Ellison Portlock[1]
1 December 1851
From Bence Jones (1870a), 2: 289–92

December 1, 1851.

My dear Portlock, - ... As one of the Senate of the University of London, and appointed with others especially to consider the best method of examination[2], I have had to think very deeply on the subject, and have had my attention drawn to the practical working of different methods at our English and other Universities; and know there are great difficulties in them all. Our conclusion is that examination by papers is the best, accompanied by *viva voce* when the written answers require it. Such examinations require that the students should be collected together, each with his paper, pens, and ink; that each should have the paper of questions (before unknown) delivered to him; that they should be allowed three, or any sufficient number of hours to answer them, and that they should be carefully watched by the examiner or some other officer, so as to prevent their having any communication with each other, or going out of the room for that time. After which, their written answers have to be taken and examined carefully by the examiner and decided upon according to their respective merits. We think that no numerical value can be attached to the questions, because everything depends on how they are answered; and that is the reason why I am not able to send you such a list at the present time.

My verbal examinations at the Academy go for very little, and were instituted by me mainly to keep the students' attention to the lecture for the time, under the pressure of a thought that inquiry would come at the end. My instructions always have been to look to the note-books for the result[3]; and so the verbal examinations are only used at last as confirmations or corrections of the conclusions drawn from the notes.

I should like to have had a serious talk with you on this matter, but my time is so engaged that I cannot come to you at Woolwich for the next two or three weeks, so I will just jot down a remark or two. In the first place, the cadets have only the lectures, and no practical instruction in chemistry, and yet chemistry is eminently a practical science. Lectures alone cannot be expected to give more than a general idea of this most extensive branch of science, and it would be too much to expect that young men who at the utmost hear only fifty lectures on chemistry, should be able to answer with much effect in writing, to questions set down on paper, when we know by experience that daily work for eight hours in *practical laboratories* for *three months* does not go very far to confer such ability.

Again: the audience in the lecture-room at the Academy always, with me, consists of four classes, i.e. persons who have entered at such different periods as to be in four different stages of progress. It would, I think, be unfair to examine all these as if upon the same level; they constitute four different classes, and we found it in our inquiries most essential to avoid mixing up a junior and a senior class one with the other. Even though it were supposed that you admitted only those who were going out to examination, and such others from the rest as chose to volunteer, yet as respects them it has to be considered that I may not go on from the beginning to the end of their fifty lectures increasing the importance and weight of the matter brought before them, for I have to divide the fifty into two courses, each to be begun and finished in the year, and I ever have to keep my language and statements so simple as to be fit for mere beginners and not for advanced pupils.

I have often considered whether some better method of giving instruction in chemistry to the cadets could not be devised, but have understood that it was subordinate to other more important studies, and that the time required by a practical school, which is considerable, could not be spared. Perhaps, however, you may have some view in this direction, and I hasten to state to you what I could more earnestly and better state by word of mouth, that you must not think me the least in the way. I should be very happy, by consultation, in the first instance, to help you in such a matter, though I could not undertake any part in it. I am getting older, and find the Woolwich duty, taking in as it does large parts of two days, as much as I can manage with satisfaction to myself; so that I could not even add on to it such an examination by written papers as I have talked about: but I should rejoice to know that the whole matter was in more practical and better hands[4].

Ever, my dear Portlock, yours very truly, | M. Faraday.
I refused to be an examiner in our University. | M.F.

1. Joseph Ellison Portlock (1794–1864, DNB). Inspector of Studies at the Royal Military Academy, Woolwich, 1851–1856.
2. This was a long running issue in the University of London (see Faraday to Rothman, 9 June 1843, letter 1501 and note 2, volume 3) which had re-emerged in the late 1840s. Brande raised the issue by letter at a meeting of Senate (at which Faraday was not present) on 19 July 1848 (Minutes of Senate, ULL MS ST2/2/2, pp.61-2), but discussion was deferred. At its meeting on 20 June 1849 (at which Faraday was not present) the Senate referred the matter to a committee of the whole Senate (ULL MS ST2/2/2, p.94). This committee met on 27 June 1849 (at which Faraday was not present) and it adopted Brande's wording for examinations (ULL MS ST3/2/4, pp.30-1). The report of this committee was accepted by the Senate at its meeting on 1 August 1849 at which Faraday was, again, not present (ULL MS ST2/2/2, pp.105-6).
3. See Faraday to Drummond, 11 January 1836, letter 873, volume 2.
4. See Abel to Gladstone, nd, Gladstone (1874), 30 where Abel wrote "But for some not ill-meant, though scarcely judicious, proposal to dictate modifications in his course of instruction, Faraday would probably have continued for some years longer to lecture at Woolwich". See letter 2495.

Letter 2479
Faraday to Edward Magrath
6 December 1851
From the original in RI MS F1 D7

Royal Institution | 6 Decr. 1851.
My dear Magrath

The progress of Old time bringing with him in my case as in all others the usual effects tends with me to the diminution of income and therefore necessarily the diminution of pleasures depending upon it. One of the first of these which I am constrained to give up is my Membership

at the Athenaeum. Will you therefore have the goodness to communicate to the Committee my resignation[.]
Ever My dear Magrath | Very Truly Yours | M. Faraday

Letter 2480
Faraday to Edward Magrath
8 December 1851
From the original in RI MS F1 D8

8 Decr. 1851

My dear Magrath
I send you another note[.] I do not wish the Club to think that I withdraw from any pique or coolness and therefore prefer assigning a reason which I have made as general as possible. You need not be sorry for the reason of my withdrawal there are plenty of means left for all that is needful.
Ever Truly Yours | M. Faraday

Letter 2481
Warren De La Rue to Faraday
16 December 1851
From the original in IEE MS SC 2
[De La Rue's embossed heading] | 7 St Mary's Road | Canonbury |
December 16th 1851

My dear Sir
In reference to our conversation, I must inform you that immediately after leaving you I took home the iron core which has been prepared now about a twelvemonth or perhaps more and placed it on the stage of my microscope, and was glad to find that it would answer well. Being desirous, however, that no slight defect should cause the waste of your time, I determined on making the core part of the instrument, so as to ensure steadiness and to avoid its sliding about in screwing up the ends. A little consideration enabled me to decide on the plan, which I drew out before I went to bed and the next morning I placed the drawing in the hands of Mr Ross[1] who pledged himself not to divulge what was in preparation or to allow the core to be seen by his workmen or any other person. Tonight at $9\frac{1}{2}$ o'clock I received back the instrument.
As I conceive it to be very important to get the points as nearly in the same plane as possible with the object to be operated upon so as to enable

us to bring them very close and to still have the lines of force in the most favourable direction, I intend taking my instrument to Bunhill row so as to turn up the moveable end high enough in order that the thinnest film of mica may be used instead of glass on some occasion; and also to see if our workmen can make make [sic] *microscopic points* as at present they look like clumsy fractures of stout wire. Thirdly to see that there is conducting communication from end to end, that is that the wire has not been injured.

I am having constructed 6 holders of glasses so that one experiment may be prepared whilst another is observed; at present I have only one but I have desired that its thickness might be guaged in order that the top surface of the plate in all should be the same height from the stage[.]

As there is a nichols prism P which fits under the stage experiments on polarized light or the change in the properties of crystals with respect to it may be observed with facility.

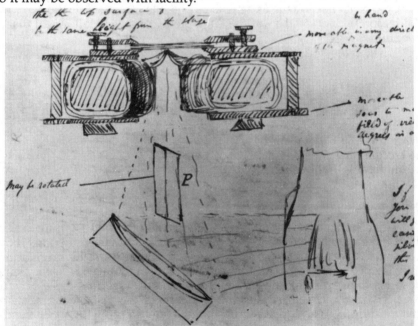

If there is any other preparation which you would like to have made I will get it done. A contrivance might easily be made by means of a slip of silver to heat the glass plates if you think this provision desirable.

I remain | Yours Very truly | Warren De la Rue
Michael Faraday Esq | &c &c &c

1. Andrew Ross (d.1859, age 61, GRO). Scientific instrument maker of 2 Featherstone Buildings, Clerkenwell. Clifton (1995), 238.

Letter 2482
Faraday to Christian Friedrich Schoenbein
16 December 1851
From the original in UB MS NS 399

Royal Institution | 16 Decr 1851

My dear Schoenbein

If I do not write at once (and even though I may seem to have but little to write about yet if I delay) all that I have to say passes from my remembrance and I involuntarily become remiss in my duty. Dr Bence Jones has just called on me to say that the Society having printed the paper you sent to me in their Transactions[1] have sent 25 copies of it to h⟨is⟩ house for you. It occupies 16 pages. Can you help me in telling me how I shall send these to you? I will do whatever you may instruct me in. I have besides a formal letter of thanks to you from the British Museum for the Portrait of Euler which I will send at the same time.

I keep working away at Magnetism[2] whether well or not I will not say. It is at all events to my own satisfaction. Experiments are beautiful things & I quite revel in the making of them. Besides they give one such confidence and as I suspect that a good many think me somewhat heretical in magnetics or perhaps rather fantastical I am very glad to have them to fall back upon[.]

Remember me very kindly to Madam Schoenbein & believe me to be

Ever most truly yours M. Faraday

Address: Dr. Schoenbein | &c &c &c | University | Basle | Switzerland

1. Schoenbein (1851b).
2. See Faraday, *Diary*, 11 November to 20 December 1851, 6: 11666–11928 which was primarily concerned with mapping magnetic fields.

Letter 2483
John Percy to Faraday
16 December 1851
From the original in IEE MS SC 2

Museum Dec 16 / 51.

My dear Dr Faraday,
I regret that I was absent when you called yesterday. I herewith send a piece of nickel such as I understand you to require[1]. A small cube may easily be made out of it by filing. I have prepared it today by fusing Evans & Askin's[2] nickel with $\frac{5}{100}$ of oxide under plate-glass. If you require a larger piece I shall have great pleasure in making one 1lb weight or more. I have a flat ingot cast weighing 14ozs which is quite at your service; but it has not been fused with oxide. I shall always be happy to prepare anything in the metallurgical line you may need.

Ever yours sincerely, | & with great respect, | John Percy
Dr Faraday

1. See Faraday, *Diary*, 20 December 1851, **6**: 11928 for Faraday's experiment mapping the magnetic field of this nickel.
2. Evans presumably took over running the metal supply firm established in Birmingham by Charles Askin (d.1847, *J.Chem.Soc.*, 1847, **1**: 149–50). See also Faraday to Lloyd, 14 October 1848, letter 2114, volume 3.

Letter 2484
Paul Frederick Henry Baddeley to Faraday
23 December 1851
From the original in IEE MS SC 2

My dear Sir
I have the pleasure to acknowledge receipt of your letter of 4th of November last[1].

I shall take your kind [advice] about fixing my observations on the Dust Storms of India and now that I have had more time to consider the subject, and to add to and confirm my former results, those that may be now published, will appear in a more useful and kind form than this could now done before[2].

I have just invented an instrument for enabling ships to navigate with greater safety out of danger when involved in these rotatory Storms - and it will moreover tend to elevate the subject, by explaining the real power of the Cyclone, and cause of the veering of the winds, and confused sea, and in fact to answer for all or most of the phenomena connected with it. I have sent a model, with diagrams and description, the first one I have made to Lord Palmerston[3] by this mail or the following.

As to the navigation of the aerial regions, I would not venture to propose such an apparently extravagant idea, save in a private letter to yourself[4].

I am however surprised that the elecl state of the higher regions of the atmosphere has not been sought or *probed*, by means of a small baloon carrying a small wire covered with gutta percha and on a strong ship cord.

I must now disabuse you of the idea of my intention to flatter. I had good reason for supposing I addressed you correctly, for your name appears twice in Mr. Piddington's[5] book - the "Sailor's Horn Book" 1851 at page 259 and again in the index, page 335[6] - as I designated you - so that I am clear for this - Believe me

With great respect | Yours Faithfully | P. Baddeley
Lahore | Dec. 23rd 1851.

Address: M. Faraday Esq | &c &c | Royal Institution | London

1. Not found, but the reply to letter 2431.
2. See Baddeley (1852).
3. Henry John Temple, 3rd Viscount Palmerston (1784–1865, DNB). Foreign Secretary, 1846–1851.
4. See letter 2431.
5. Henry Piddington (1797–1858, DNB). Meteorologist and Curator of the Museum of Economic Geology at Calcutta, 1830–1858.
6. Piddington (1851), 259, 335 which gives Faraday's title as "Sir".

Letter 2485
Faraday to George Biddell Airy
24 December 1851
From the original in RGO6 / 404, f.35

Royal Institution | 24 Decr. 1851

My dear Airy

I send you by way of remembrance at this Season with our best wishes to You & Mrs. Airy a few forms of iron filings over magnets[1]. You will see the places of the magnets &c on the under side[.] When you told me about the total Eclipse you did not repress my hope that we should hear you tell the results in our Lecture room this season on a Friday Evening[2]. How I should rejoice to hear them & a few of your thoughts about the physical constitution of the Sun or of the bodies about him. May I nurse this hope up and give it more & more strength? Tell me Yes[.]

Ever Yours Truly | M. Faraday

1. See Faraday, *Diary*, 11, 12, 13 November 1851, **6**: 11666–11703 for Faraday's production of iron filing diagrams.
2. Airy, G.B. (1851), Friday Evening Discourse of 2 May 1851. Airy did not give a Discourse during 1852.

Letter 2486
Albert Richard Smith[1] to Faraday
24 December 1851
From the original in IEE MS SC 2
Mr. Albert Smith presents his compliments to Mr Faraday, and with every apology for troubling him, requests the favour of an answer, in one word, to the question over leaf. The only excuse he can plead for taking this liberty is that there is the question of a 'dinner' depending on it, and the other party will not be satisfied with any other decision than that of Mr Faraday.
12 Percy St | Bedford Square. | Dec 24, 1851

At a certain height above the level of the sea - say 10,000 feet - would a barometer stand at the same degree in a balloon as it would on a mountain at the same elevation?

1. Albert Richard Smith (1816–1860, *DNB*). Journalist and lecturer.

Letter 2487
George Biddell Airy to Faraday
29 December 1851
From the original copy in RGO6 / 404, f.36
Playford near Ipswich | 1851 Dec 29
My dear Sir
Thank you for your magnetic remembrance, which has followed me hither[1]. I wish I could instantly convey your kind wishes to Mrs. Airy: she is I trust safe either on the sea or in Madeira, for which place she has embarked with our eldest daughter, whose health has given us much uneasiness[2].
About the Eclipse[3]. First, pray let me yet believe for some time that I am under no engagement of any kind. I have been sitting, this morning, three hours by the fire without doing or thinking of anything, and begin to understand in what happiness consists. In the next place, the Eclipse

will be a very meagre subject: ten minutes would suffice for it: but I will in some more active state turn it over and report to you.

I am, my dear Sir, | Yours very truly | G.B. Airy

Michael Faraday Esq | &c &c &c

1. Letter 2485.
2. Elizabeth Airy (1833–1852, Airy, W. (1896), 98, 212). She died on 24 June 1852.
3. See note 2, letter 2485.

Letter 2488
Faraday to Miss Murray[1]
8 January 1852
From the original in WIHM MS FALF

Mr Faraday presents his compliments to Miss Murray and will have the honor of paying his respects tomorrow at $\frac{1}{2}$ past 10 o'clk when he hopes to place half an hour if needful at Miss Murrays service[.]

Royal Institution | 8 Jany 1852.

1. Unidentified.

Letter 2489
Faraday to Thomas Bell
13 January 1852
From the original in RS MS RR 2.91

Royal Institution | Jany. 13, 1852

My dear Bell

I do not see that I need alter my report[1]. Lloyd I perceive gives up practically[2], i.e as respects real magnets, both the idea of resultant poles and also the law of the inverse square of the distance, and if all were like him, then indeed Harris' paper would be superfluous, except as a proof that experiment & theory agreed. But there are many who are of opinion that the law of the inverse square of the distance is the true law of magnetic action[.] Tyndall & others for instance who have been working and publishing very lately[3]. Now Harris probably feels that that is no law which does not apply to the near intervals, which are perhaps the most important intervals in determining the true nature of magnetic action and that one might as well say that the revolution of a parabola produces a cone neglecting the consideration of all the parts about the focus which are

Plate 8. Faraday in his laboratory by Harriet Jane Moore, 1852.

the best fitted to make manifest the truth. As long, therefore, as the law of the inverse square of the distance is assented in an unqualified manner he may feel bound to offer experimental proof to the contrary. In these remarks I am not putting myself either in the place of Harris or Lloyd and I think if I were Harris I should like to know of Lloyd's letter and reconsider what modification such admissions as it contains might require but whether that is proper or not I cannot say. I perceive that the hope I have expressed in the first of the two papers of mine that you now have[4], that the apparently contrary results of Harris and others will be reconciled & coalesce[5] is near upon being fulfilled[.]

Ever My dear Bell | Truly Yours | M. Faraday

1. That is letter 2415 which refereed William Snow Harris, "On Induced and other Magnetic Forces", RS MS AP 33.15.
2. Lloyd's unfavourable report of 1 October 1851 (RS MS RR 2.90) on Harris's paper caused consideration of the paper to be postponed on 18 December 1851 (RS MS CMB 90c) and again on 15 January 1852. Following this letter of Faraday's, another report was obtained from William Thomson dated 6 February 1852 (RS MS RR 2.92) which was also unfavourable and on 19 February 1852 the paper was archived (RS MS CMB 90c).
3. Tyndall (1851a), 284.
4. Faraday (1852b, c), ERE28 and 29.
5. Faraday (1852b), ERE28, 3075.

Letter 2490
Joseph Antione Ferdinand Plateau to Faraday
16 January 1852
From the original in IEE MS SC 2

Gand, 16 Janvier 1852

Mon Cher Monsieur Faraday.

J'ai l'honneur de vous adresser, par la poste, un exemplaire de la première livraison d'un petit traité de physique[1] à l'usage des gens du monde, auquel je travaille depuis une couple d'années. Si vous daignez jeter les yeux sur l'avant-propos, vous comprendrez que ce petit ouvrage qui, au premier abord, peut paraître aisé à composer, a dû cependant m'offrir des difficultés considérables; on m'en a chargé à peu près malgré moi, et c'est lui qui, depuis deux ans, m'a empêché de publier la suite de mes recherches sur les figures d'équilibre liquides. Je n'ai pu mettre de suscription sur la couverture de l'exemplaire que vous recevrez, parce que la poste ne le permet pas.

Permettez-moi maintenant de vous demander un petit service. Je ne le fais qu'avec grand regrêt, parceque je vous ai déjà importuné pour la même chose[2]; mais vous me pardonnerez, j'espère, en réfléchissant que mon infirmité rend pour moi les voyages impossibles, et que, ne pouvant ainsi aller moi-même à Londres m'occuper de mes affaires, je suis contraint de recourir à l'obligeance des personnes qui veulent bien m'honorer de leur amitié; vous, qui êtes si bienveillant pour moi, je suis certain que vous ne rejetterez pas ma requête. Voici ce dont il s'agit: il y a plus de six mois que la traduction anglaise de mon mémoire sur les figures d'équilibre liquides[3], traduction destinée à paraître dans les Scientific Memoirs de Mr. Taylor, est imprimée, je n'ai aucune nouvelle de sa publication, et je désirerais vivement savoir si je dois renoncer à l'espoir de cette publication; n'est-ce point trop abuser de vos bontés pour moi que de vous prier de vouloir bien prendre des informations à cet égard, et me les transmettre?

J'ai encore une petite demande à vous faire. Mr. Wheatstone a eu l'obligeance d'envoyer, il y a environ sept mois, au Journal des Sciences d'Edimbourg, une analyse de mon mémoire rédigée par moi. Si vous recevez ce journal, puis-je espérer que vous voudrez bien vous assurer si l'analyse en question y a été insérée?

Je le répète, Monsieur, c'est avec grand regrêt que je viens vous prier de perdre ainsi pour moi une partie d'un temps aussi précieux que le Vôtre, et je vous supplie de nouveau de me pardonner mon importunité.

J'ai reçu les exemplaires de vos dernières séries que vous m'avez fait l'honneur de m'envoyer[4]. J'ai admiré vos belles expériences sur de légères boules de verre pleines d'oxigène ou d'un autre gaz à différentes densités et soumises à l'action d'un électro-aimant; j'ai admiré aussi, en particulier, les moyens ingénieux que vous avez employés, indépendamment de celui dont il a déjà été question entre nous[5], pour montrer que les gaz n'éprouvent ni condensation ni dilatation dans le voisinage des pôles[6], et je suis encore à me demander comment cela est possible.

Agréez, Monsieur, l'assurance de tous mes sentiments de respectueuse affection. | Jh Plateau

P.S. J'ai envoyé également à Mr. Wheatstone un exemplaire de mon petit ouvrage; veuillez avoir la bonté de lui dire, à la première occasion, pourquoi je n'ai pas mis de suscription sur la couverture.

Address: Monsieur | Faraday, membre de la Société Royale, &c | à l'Institution Royale, Albemarle Street | Londres

TRANSLATION

Ghent | 16 January 1852

My Dear Mr Faraday,

I have the honour of sending, by post, a copy of the first edition of a little treatise on physics[1] for the use of the people of the world, on which I have been working for a couple of years. If you deigned to cast your eyes on the foreword, you would understand that this little work, which at first might seem easy to write, must have given me considerable difficulty; I was charged with it a little against my will, and it has, for two years, stood in the way of the publication of my further research on liquid equilibrium figures. I have not been able to put any inscription on the cover of the copy you will receive, because the post office does not allow it.

Allow me, now, to ask you a small favour. I do it with enormous regret, since I have already troubled you once about this[2], but I hope you will forgive me, when you realise that my infirmity makes travel impossible for me, and being thus unable to go to London myself to see to my affairs, I am obliged to rely on the generosity of people who wish to honour me with their friendship; and someone who is as kind towards me as you are, will, I am certain, not reject my request. This is what it concerns: it is now more than six months that the English translation of my paper on liquid equilibrium figures[3], the translation destined to appear in Mr Taylors *Scientific Memoirs*, was printed, and I do not have any information on its publication, and I would really like to know if I should renounce any hope of ever seeing it printed; is it too much of an abuse of your kindness towards me to ask you to make some enquiries about this and let me know the outcome?

I have another small request to make. Mr Wheatstone very kindly sent, about seven months ago, to the *Edinburgh Journal of Science*, my own analysis of the paper. If you receive this journal, could you please check if the analysis in question was indeed inserted?

I repeat, Sir, that it is with great regret that I come to ask you to waste on me time as precious as yours, and I beg you again to forgive my importunity.

I received the copies of your latest series which you honoured me by sending[4]. I admired your beautiful experiments on the light balls of glass filled with oxygen or other gases of different densities and subjected to the action of an electro-magnet; I admired in particular the ingenious methods that you have used, independently of that which we have already once discussed[5], to show that gases to not condense or dilate near the poles[6], and I am still asking myself how that is possible.

Please accept, Sir, the assurance of all my sentiments of respectful affection | Jh Plateau.

P.S. I have sent also to Mr Wheatstone a copy of my little work; please kindly tell him, at the first opportunity, why I have not written anything on the cover.

1. Plateau and Quetelet (1851–5), part 1.
2. See letter 2305.
3. Plateau (1852).
4. Faraday (1851b, c, d, e), ERE24, 25, 26 and 27.
5. Letters 2164 and 2241.
6. Faraday (1851c), ERE25, 2718–56.

Letter 2491
Faraday to Edmund Belfour[1]
22 January 1852
From the original in BL add MS 42240, f.23

Royal Institution | 22 Jany 1852

Sir

I have the honor to acknowledge the very kind invitation for Saturday the 14th of next Month and beg to return my grateful thanks for it. I shall hope to avail myself for the Oration[2] at three o clk but am obliged to decline the Evening Hospitality.

I have the Honor to be | Sir | Your Very Humble Servant | M. Faraday

Edw Belfour Esq | &c &c &c

1. Edmund Belfour (d.1865, age 75, B1). Secretary of the Royal College of Surgeons, 1814–1865.
2. That is the Hunterian Oration of the Royal College of Surgeons.

Letter 2492
Roderick Impey Murchison[1] to Faraday
29 January 1852
From the original in IEE MS SC 2

16 Belgrave Square | Jany 29 1852

My dear Faraday,

Before I make my appeal to the board of Managers of the Royal Institution as I am authorized to do by the Royal Geographical Society over which I preside, I wish to consult you as 'amicus curiae'[2] & grand Master of all that pertains to your admirable establishment.

The history of our case is this. We are a thinking, active body now, to whose geographical memoirs & discussions great interest is attracted & we literally have not sitting room.

We are memorializing the Government to do something in their way for us, as we do much for them; but in the mean time it would be an enormous boon & advantage, if we could have the use of your theatre on a Monday Evening (every other Monday) when nothing is done with it.

Any additional expence or even a moderate consideration (if requisite) we would pay, & it would be quite understood that every Member of the R. Institution could have free access to our *performances*.

In short, it would be one attraction more added to your good bill of fare, & thus instead of having me once on mountainous matters you could be bored with me very often.

Your President³ is one of *my* Council & if you & Barlow give me any hopes of success I will write to His Grace to have my letter laid before the Managers next Monday⁴.
Yours most sincerely | Roderick Murchison
M. Faraday Esq

1. Roderick Impey Murchison (1792–1871, DSB). Retired Army officer and geologist.
2. A friend of the court.
3. The Duke of Northumberland.
4. See RI MM, 2 February 1852, **10**: 364–5 where this proposal was agreed.

Letter 2493
Faraday to Jean-Baptiste-André Dumas
3 February 1852
From the original in AS MS

Royal Institution | 3 Feby 1852

My dear Friend
I would not on any account that you should have the least anxiety added to your present heavy charge on my account[1], I would far rather if that were possible help to remove some from you, but each of us have our burden in life and though mine be a light one it has not rendered me unable to sympathize with my friends. As to the special subject I would on no account desire that it should be injured by hasty production and I hope you will in due time find leisure enough to develop it in all its beauty[.] Our kindest remembrances to Madame Dumas and to yourself my kind friend.
Ever Most Truly Yours | M. Faraday
A Monsieur | Monsieur Dumas | &c &c &c

Address: A Monsieur | Monsieur Dumas | &c &c &c | Rue de Vaugirade | coin de la Rue du Pt de Fer[ou] | Faubourg St. Germain | Paris

1. A reference to the proposal made by Faraday in letter 2474.

Letter 2494
William Whewell to Faraday
7 February 1852
From the original in RI MS F3 G73[1]

Trin. Lodge, Cambridge | Feb. 7, 1852

My dear Sir

I am much obliged to you for the specimens of magnetic curves[2] which I received two days ago. I have been thinking what word will best answer your purpose, but it is difficult to decide such a question without knowing the kind of connection in which it is to be used. *Spheroid* would describe the surface which you wish to express but is not mechanical enough. You might perhaps get on in English by calling it the *spindle shaped* surface or *fusiform* surface, but a new word would be better. I should recommend you to call it the *sphondyloid* surface, and then, the *sphondyloid* simply, making it a substantive. *Sphondylos* in Greek is a pulley or socket which turns on an axis, a spindle, a vertebra, and the like, and is already familiar in anatomy and botany. Used as a substantive *sphondyloid* will group well enough with *solenoid* which has been adopted by English writers.

It is rash to suggest anything to you in the way of manipulation; but would not some magnetic curves come out more neatly if instead of filings you were to use fine wire cut into minute lengths.

I am glad you are going on with your magnetic speculations and am always

Yours Truly | W. Whewell

Dr Faraday

Endorsed by Faraday: *Sphondyloid*[3].

1. This letter is mounted in Faraday's offprint of Faraday (1852d), [ERE29a] opposite paragraph 3271.
2. That is Faraday's iron filing diagrams. The accompanying letter has not been found.
3. Faraday used this term in Faraday (1852d), [ERE29a], 3271.

Letter 2495
Faraday to Joseph Ellison Portlock[1]
9 February 1852
From the original in PRO WO44 / 523

Royal Institution | 9th February 1852

My dear Portlock

As I have already intimated to you in conversation it is my wish to retire from the duty of delivering the chemical lectures in the Royal Military Academy at Woolwich provided that can be without inconvenience to the Establishment. For Twenty two years I have been honored with a hearing there but my memory is failing me & I feel it to be right that I should restrict the field of my exertions. I have long felt that if it were possible the Gentlemen Cadets should have practical instruction combined with the lectures and as far as I can judge of your propositions for a certain amount of change[2] (though of course I cannot tell how far they consist with the other arrangements of the Academy) they have my full approbation[.] Hoping that you will succeed in arranging all to your own satisfaction & for the good of the Academy and with kindest wishes & remembrances to my friends there

I remain | My dear Portlock | Ever Very Truly Yours | M. Faraday
Coll Portlock RE | Inspector | &c &c &c | Royal Military Academy | Woolwich Common

1. Joseph Ellison Portlock (1794–1864, DNB). Inspector of Studies at the Royal Military Academy, Woolwich, 1851–1856.
2. See note 4, letter 2478.

Letter 2496
Faraday to William Whewell
9 February 1852
From the original in TCC MS O.15.49, f.30

Royal Institution | 9, Feby 1852
My dear Sir

I hasten to thank you for your kind suggestions[1]. The term Sphondyloid will I suppose answer every purpose I may want it for[.] I just enclose a figure[2] to make sure that I have not deceived[.] Supposing c.c a magnet as marked on the back of the paper, then various lines of force are shewn at the axis AB, and the lines DD.E.F. Now I have occasion to consider the solid which would be generated by the revolution of the area between F & C round the axis AB, or that produced by the like revolution of the area enclosed by E and C or by D and C, or even that enclosed between the lines of force D and E and such like. But in saying this I do not wish to give you the trouble of answering this letter. If you should however desire to do so tell me at the same time what the fair English meaning of the word *Solenoid* is meant to be[.]

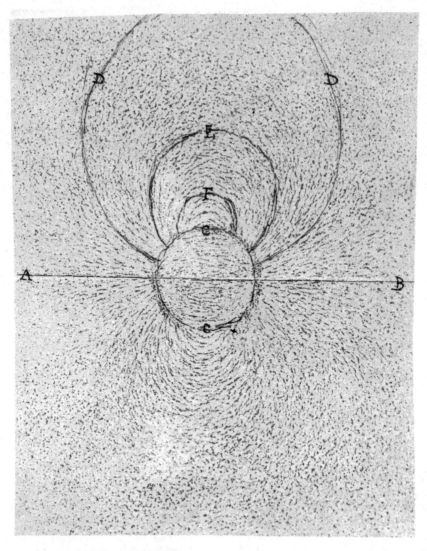

  Ever Your Very Obliged | M. Faraday
Revd. Dr. Whewell | &c &c &c

1. In letter 2494.
2. This figure is mounted at the beginning of TCC MS O.15.49.

Letter 2497
William Whewell to Faraday
10 February 1852
From the original in RI MS F3 G73[1]

Trin. Lodge, Cambridge | Feb. 10, 1852

My dear Sir

I think *sphondyloid* will answer your purpose. I have just stumbled on a passage in Jeremy Taylor[2] which will show you that the term sphondyl is not new in English. It is at the beginning of the Dedication to the Cases of Conscience. "The circles of Divine Providence turn themselves upon the affairs of the world so that every *sphondyl* of the wheels may mark out those virtues which we are then to exercise."[3] *Spondyl* is the Ionic, *Sphondyl* the Attic form. Perhaps you may think it wise to introduce the word in some such way as this. ["]The sphondyloid body contained between the two surfaces of revolution E and F, which for the sake of brevity I shall call simply *the sphondyloid*"[4].

Solen in Greek means a pipe or gutter. I think Ampere and his followers use it for a figure generated by a ring formed curve sliding along any other line and so with the termination *oid* it may mean any pipelike surface[5][.]

Believe me yours very truly | W. Whewell
Dr Faraday

1. This letter is mounted in Faraday's offprint of Faraday (1852d), [ERE29a] opposite paragraph 3271.
2. Jeremy Taylor (1613–1667, DNB). Anglican theologian.
3. Taylor, J. (1660), dedication gives "spondel" in the passage quoted.
4. A modified version of this sentence was included in the note to Faraday (1852d), [ERE29a], 3271.
5. See Ampère (1823), 279.

Letter 2498
Faraday to William Whewell
11 February 1852
From the original in TCC MS O.15.49, f.31

R Institution | 11, Feby 1852

My dear Sir

Many thanks for your last letter[1] which is curiously to the point in respect of Jeremy Taylor[2][.]

I enclose a notice of certain views[3].

Ever Truly Yours | M. Faraday
Revd. Dr Whewell | &c &c &c

1. Letter 2497.
2. Jeremy Taylor (1613–1667, DNB). Anglican theologian. See letter 2497.
3. Possibly Faraday (1852a), Friday Evening Discourse of 23 January 1852.

Letter 2499
Faraday to Edward William Brayley
2 March 1852
From the original in RI MS F1 D9

R Institution | 2 March 1852

My dear Sir

I send you the ticket. With regard to the researches I have not the slightest objection to their being in the hands of other person[s] because there are corrections in them. All these things I leave to take a natural course.

They seem however to be upon your hands. Now I am often asked for them and have no doubt I could meet with a purchaser for them some time or other if you have not parted with them before[.] You can let me know whether I shall look out for one when we next meet[.]

Ever Yours Truly | M. Faraday
E.W. Brayley | &c &c &c

Letter 2500
George Butler[1] to Faraday
3 March 1852
From the original in IEE MS SC 2
Confidential and immediate

Office of Ordnance | 3rd March 1852

Sir

I am directed by the Board of Ordnance to acquaint you that certain French shells containing combustible matter of a peculiar nature have recently been sent to England from Gibraltar, having been picked up after the French attack on Salee[2], unexploded.

And the Board being desirous of ascertaining the nature of this composition, I am to request that you will favor them with your advice as to the best way of proceeding in the matter.

Should you consider it necessary to employ some practical chemist, the Board will thank you to name a person to whom immediate application may be made, as the Committee of Artillery Officers at

Woolwich (to which Committee such matters are always referred) is already assembled at Woolwich.

Mr. Tozer[3], of the Royal Laboratory Woolwich, has been directed by the Committee to wait on you for the purpose of affording any information in respect to the beforementioned composition which it may be in his power to give.

The Board, I am to add, fully appreciates your liberality and kindness on all occasions on which it has been necessary to apply to you.

I am Sir | Your most obedient | humble servant | G. Butler
Professor Faraday

Endorsed by Faraday:

The Exploding power of the fuze - in small quantity - detonated by blow or friction - contains *Chlorate of potassa* & *Sulphuret of antimony*.

The star or loose vessel is about this size

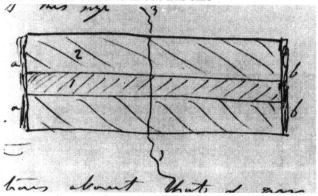

1 is a portion of black mixture seen here in section. It is apparently a coarse meal powder or powder composition consisting simply of nitre Sulphur & charcoal in proportions about that of gun powder. As the ends of the case are open at aa bb and are covered over with meal powder no doubt this is to be inflamed on the bursting of the shell which inflame the composition 1 which inflame composition 2 being aided in that by the quick match 3 which goes right through all[.]

3. is a yellowish brown mass - breaking under the pestle granular in appearance as if heterogeneous - smelling a little of camphor, softening & melting by heat[.] It fumes rises in part containing resin - chars - bursts into a flame - gives a vivid combustion at last as if nitre present - & finally acts on the platina by antimony in it. It takes fire by a flame but not very readily when hit goes on burning & burns well having *access of air*. It contains a resin as common *resin* - sulphur - nitre - no chlorate potassa - metallic antimony - and a wax. The proportions present are nearly

Nitre	—	47.5
Sulphur	—	20.5
Resin	—	25.0
Wax	—	2.5
Antimony	—	4.5
		100.0

} but is irregular

Replied 5[4] March 1852[5]

1. George Butler. Secretary to Board of Ordnance. *Imperial Calendar*, 1852, p.228.
2. Salé in Morocco was bombarded by the French in November 1851. On this see Brown (1976), 177–80.
3. J.S. Tozer. See Tozer to Faraday, 30 December 1846, letter 1943, volume 3. Otherwise unidentified.
4. Faraday clearly wrote a "5" over a "4" here.
5. Letter 2501.

Letter 2501
Faraday to George Butler[1]
5 March 1852[2]
From the original in PRO WO44 / 644

Royal Institution | 4 March 1852

Sir

I have the honor to acknowledge the receipt of your letter[3]; and also, by Mr. Tozer[4], of the substances referred to in it; I am happy that I could find time to make an examination; and beg you to offer the results to the Board of Ordnance. There were specimens of *three* substances[.] One of these was called *friction composition*: it proved to be a mixture of pulverized chlorate of potassia and sulphuret of antimony. The quantity was too small for me to determine the proportions;- but there can be no difficulty in preparing such a mixture, for the materials, & their effects when mixed, are well known here; and there was nothing particular in the proportion[.]

The cylindrical missile contained two mixtures, besides the ordinary powder about it; the central or axial portion, which was black, & the surrounding or chief part in bulk, which was brownish yellow. The central portion consisted of nitre, charcoal, and sulphur in proportions to form a gunpowder. I do not think it has ever been granulated; for the mixture appears to be but coarsely made. It may however have been meal powder or else a green mixture of the materials of powder. It has all the qualities of powder; & not having time, I did not ascertain its proportions.

The chief mass of the missile; i.e the yellow substance has only such hardness as to be cut or impressed by the finger nail. It breaks down under pressure, yet clings a little together. It softens & melts by heat and has probably been softened when put into its place, being adjusted there by pressure. It is granular in its structure; the sulphur & other materials used having evidently not been in a state of fine division or very perfect mixture; and if it has been softened by heat when put into its place, it has not been melted. It does not burn very vividly, that probably not being desired; but its habits in this respect will be better understood by trial & experiment. It contains the following substances:- Nitre - Sulphur - Resin, perhaps common resin - Wax - Metallic Antimony - No chlorate of potassia. The following are the results as to proportions which one portion gave me; but from the appearance I should conclude that the mixture is not accurately made, & that the proportions are likely to differ in different parts.

Nitre	—	47.5
Sulphur	—	20.5
Resin	—	25.0
Wax	—	2.5
Antimony	—	4.5
		100.0

I have reserved as much of the composition as I could and return it herewith.

I am Sir | Your Very Obedient humble Servant | M. Faraday
Geo. Butler Esq | Secretary | &c &c &c &c

Endorsed: 5 March 1852 | Ordered to be transmitted, with the accompanying Packet, to the Director of the Royal Laboratory[5]. Acknowledge the receipt to Dr. Faraday, and express to him the thanks of the Board for his communication and for the trouble he has taken in this matter. | TH[6] Wrote Dir Rl Laboratory and Professor Faraday | 5th[7]

1. George Butler. Secretary to Board of Ordnance. *Imperial Calendar*, 1852, p.228.
2. Dated on the basis of the evidence referred to in note 4, letter 2500.
3. Letter 2500.
4. J.S. Tozer. See Tozer to Faraday, 30 December 1846, letter 1943, volume 3. Otherwise unidentified.
5. Richard Hardinge (1790–1864, B1). A Royal Artillery officer.
6. Thomas Hastings (1790–1870, DNB). Storekeeper to the Ordnance Office, 1845–1855.
7. This line is in Butler's hand.

Letter 2502
Griffith George Lewis[1] to Faraday
9 March 1852
From the original in RS MS 241, f.127

Royal Military Academy | Woolwich; March 9th 1852

Dear Sir,

Having submitted to His Lordship the Master General[2] Your communication of the 9h Ultimo[3], expressing a wish to retire from the Office of Chemical Lecturer to the Royal Military Academy as soon as Convenience would permit, - I have the honor to state that a person Competent to fill the situation of Chemical Lecturer to this Institution having been found in the person of Mr. Abel - the Master General has been pleased to accept your resignation and to appoint Mr. Abel Your Successor.

I beg to subjoin an Extract from the Official Letter of the Master General's Secretary expressive of His Lordships high sense of the advantage the Ordnance Service & Gentleman Cadets of the Royal Military Academy have gained from your instruction for so many Years.

I have the honor to be | Dear Sir | Your most obedient Servant | G.G. Lewis | Major General | Lt. Governor

Extract

Office of Ordnance | 8th March 1852

Sir,

"The Master General cannot allow Mr. Faraday to withdraw from the Ordnance Service, without expressing his high sense of the advantage which it has derived from the Professor's instruction. The Lectures of so distinguished a Chemist cannot fail to have encouraged amongst the Cadets a taste for a science intimately Connected with professional objects, and to have laid the foundation of practical attainments which must be eminently useful in their future career."

I have the honor to be | Sir | Your obedient Servant | Signed / Edward Elliot[4] | Pro: Secty
M. General Lewis C.B. | &c &c &c

1. Griffith George Lewis (1784–1859, DNB). Governor of the Royal Military Academy, Woolwich, 1851–1856.
2. Henry Hardinge (1785–1856, DNB). Master-General of the Ordnance, March to September 1852.
3. Letter 2495. The delay in reply was due to the change in government.
4. Edward Elliott. Chief Clerk to the Master General of the Ordnance Office. *Imperial Calendar*, 1852, p.228.

Letter 2503
Faraday to Thomas John Fuller Deacon
16 March 1852
From the original in the possession of Elizabeth M. Milton

Royal Institution | 16 Mar 1852

My dear friend

Perhaps you may have heard from Margery[1] a rumour of an apparent situation here perhaps not. I have seen Sir John Boileau[2] today & as far as I can remember the points are as follows. At the Royal Society of Literature there is an Honorary Secretary and a Clerk, who acts as Foreign Secretary but they want an Assistant Secretary of gentlemanly manners and good education for a comparatively small time & for a small Salary. The pay is £80 annually. The duty would be to attend fortnightly meetings of the members - also meetings of the council or committee, to keep minutes of proceedings to take charge of correspondence answer letters, examine memoirs sent in, select such as are fit for reading at the meetings, to read them at the meetings to the Members, make notes of proceedings, superintend the printing of papers &c &c. I asked Sir John Boileau about Greek & Latin he seemed to think the party should know enough to read papers in them if needful, or at least to be able to manage them in some way[.]

The connexion might be very good for a Young man of Education starting in life, but I have my fears about its offering much that would suit you. I cannot find that Sir John Boileau has the place to offer he says that Mr Hallam[3] would chiefly dispose of it. I think he felt there could be no deciding about it without seeing the party, but he has promised to keep it open as far as he can whilst I wrote to you. If circumstances are such that you would wish to use or examine the opportunity, you must let me know so consult together & let me hear quickly what your thoughts are. The point I feel most anxious about is the responsibility of deciding upon the many circumstances that would be left to such a person to settle in relation to the papers, the selections, the correspondence &c &c which the Council I suppose would delegate[.] It is true that much ought not to be expected for such a Salary as £80 but now a days high qualifications & continual responsibilities are paid for by poor salaries[.]

I dare hardly say how my own feelings go. I think on the whole that a sturdy ordinary light business occupation would perhaps be better and yet I cannot but feel that for a young, vigorous determined man the present might by the connexion it would give afford an excellent opportunity of entering on the world[4].

Here is a business like letter and I must not wait to alter it[.] So I will merely say My love to you & Caroline & Constance from

Yours affectionately | M. Faraday
Mr Thos. Deacon

Sir John Boileau[5] thinks it might not occupy more than a couple of days a week[.] I cannot tell the employment out of time often mounts up. MF

1. Margery Ann Reid.
2. John Peter Boileau (1794–1869, DNB). Archaeologist.
3. Henry Hallam (1777–1859, DNB). Historian.
4. Deacon was not appointed to the Royal Society of Literature of which Boileau and Hallam were both Vice-Presidents. According to their *Annual Reports* there was no change in staff in this period.
5. Boileau attended the Managers' Meeting that day. RI MM, 16 March 1852, 10: 376.

Letter 2504
George Richmond[1] to Faraday
21 March 1852
From the original in IEE MS SC 2

10 York Street | Portman Sq | Mar 21, 1852.
My dear Sir

I am very sorry to hear from Dr. Bence Jones that you are not well enough to keep your appointment with me for tomorrow. Perhaps it will be better to wait for your last sitting[2] till these cold winds leave us and if they should take to doing so before Wednesday the 31st inst and you should be equal to a sitting then, would you kindly come to me on that morning at the old hour half past nine only please not to trouble yourself the least about it on my account if when the day comes, you should not feel quite up to sitting, but I hope you will be quite well again long before then

and have the honor to remain | My dear Sir | Your very truly obliged Servant | Geo Richmond

1. George Richmond (1809–1896, DNB). Portrait painter.
2. For Richmond's portrait of Faraday, commissioned by Bence Jones, see plate 9.

Plate 9. Faraday by George Richmond, 1852. See letters 2504 and 2523.

Letter 2505
Faraday to Charles Mansfield Clarke
22 March 1852
From the original in Museo di Storia della Fotografia Fratelli Alinari

R Institution | 22 Mar 1852

Many thanks my dear Sir Charles for your note. I found after I wrote that a form of rate had been entered & reported. We wanted it for the printed notices for the Members[.]
     Ever Your Obliged | M. Faraday
Sir Charles Clarke Bart | &c &c &c

Letter 2506
Faraday to Gideon Algernon Mantell[1]
22 March 1852
From the original in National Library of New Zealand MS Papers 83 (Mantell family) folder 38

R Institution | 22 Mar 1852

My dear Mantell
     I am ever your debtor. Many thanks on the present occasion for the Notornis[2]. When shall you have your own bird home I mean he who sends you all these most interesting things?[3] He must be a source of great pleasure & pride to you[.]
     Ever Yours | M. Faraday

1. Gideon Algernon Mantell (1790–1852, DSB). Geologist.
2. A bird from New Zealand. See Mantell (1850) on this. A drawing had been displayed at Mantell (1852), Friday Evening Discourse of 5 March 1852, p.146.
3. That is Walter Baldock Durrant Mantell (1820–1895, DNZB). Civil servant in New Zealand.

Letter 2507
Jacob Herbert to Faraday
25 March 1852
From the original in GL MS 30108/1/50&51

Trinity House, London, | 25 March 1852.
My dear Sir,
     The Elder Brethren being desirous that the White Lead supplied for the Corporation's Service, for the present year, - of which the quantities sent herewith are Samples (mark'd No 1 & 2 respectively) should be

subjected to test similarly to that forwarded to you for that purpose in the month of May 1850[1].- They request you will favor them by testing the same, and reporting the result, as upon that occasion[2].
 I remain | My dear Sir, | Your very faithful Servant | J. Herbert
M. Faraday Esq. D.C.L. | &c &c &c.

1. Letter 2288.
2. Faraday's analyses are in GL MS 30108/1/50&51 where he noted that he replied on 30 March 1852.

Letter 2508
Faraday to Duchess of Northumberland[1]
26 March 1852
From the original in the possession of Wayne Lee Radziminski
                                                      Royal Institution | 26 Mar 1852
Madam
 I had hoped to have the honor personally of offering my sincere thanks tomorrow Evening for Your Grace's kindness in remembering one who is so little known to your Grace but regret to say that circumstances have arisen which preclude even a hope of, at any hour being able to make my appearance for the performance of that most agreeable duty[2][.]
 I have the honor to | Madam | Your Graces most | humble faithful Servant | M. Faraday
Her Grace | The Duchess of Northumberland

1. Eleanor, Duchess of Northumberland, née Lupus (1820–1911, CP). Married the Duke of Northumberland, 25 August 1842.
2. This appears to be Faraday declining an invitation to attend a soirée at Northumberland House. He had attended a soirée there on 13 March 1852. See Curwen (1940), 284.

Letter 2509
Faraday to Charles De La Pryme[1]
3 April 1852
From the original in RI MS F1 N1/23
                                                      Royal Institution | 3 April 1852
Sir
 Your note & its contents are very interesting. It is to be hoped that the lad will be able to pass safely through the period that must intervene between the present & that time when he shall have found his true

position according to the powers of his mind amongst his fellow men[.] This will depend very greatly upon the friends around him. If he could have for half an hour the experience which is sure to be his ten years hence I do not doubt he would be able to steer his immediate course aright. I trust he will even now know how to value courteous commendation and disregard flattery and the sentiment of the two first specimens and more especially of the mind makes me hope this.

I should be happy to take a copy of the volume but I cannot let my name appear. For years past I have declined. It must therefore be simply M.F. with no address[2][.] I should have sent you a post office order for the 5/- but am refused at the office for want of your Christian name[.]

I am Sir | Your Very Obedient Servant | M. Faraday
C. de la Pryme

1. Charles De La Pryme (1815–1899, B4). Lawyer and author.
2. Realf (1852) which has a preface by De La Pryme. There is no reference to Faraday, by initials or otherwise, in the volume.

Letter 2510
Faraday to Charles Lyell
3 April 1852
From the original in SI D MS 554A

3 April 1852

Dear Lyell
Would you send me back the enclosed Envellope only as a sign that the Post has been faithful in its performance. With Sincere thanks for the Evening[1] & all else
Ever Yours | M. Faraday

1. Lyell (1852), Friday Evening Discourse of 2 April 1852.

Letter 2511
Thomas Andrews to Faraday
8 April 1852
From the original in the possession of Elizabeth M. Milton

Queen's College Belfast | April 8, 1852.
My dear Sir

Our arrangements for the meeting of the British Association which you are aware is to be held in this place in the first week in September are going on very favorably; and we have good reason to expect a successful meeting. We find it necessary however to apply to you to assist us in this good work by favoring us with your presence on the occasion, and I am only expressing the universal sentiment here when I add that an unfavorable reply from you will throw a gloom over our proceedings, while even a conditional promise that you will attend will have the best possible effect.

I may farther undertake to assure you that you will not be called upon to take a greater share in the public proceedings of the meeting than you may feel inclined to do; as I am fully aware that relaxation & not excitement is what you have a right to expect during your short intervals of labour.

But I have other inducements to offer; the month of September is here the finest season in the year and a tour round the Antrim Coast as far as the Giants Causeway will I am sure gratify your love of coast scenery & afford even an opportunity of examining without labour the extensive & varied basaltic formations which are equally interesting to the geologist & to the admirer of picturesque nature.

Graham has promised to attend and Liebig (who was here last autumn) has given me a conditional promise. They will be both inmates of my house, and as I am perhaps a little inclined to monoplolize the élite of the meeting, I wish to add you to the number of my guests. Mrs. Andrews desires me to express how much she will be gratified if Mrs. Faraday can accompany you.

Believe me to be | My dear Dr. Faraday | Yours most truly & obliged | Th Andrews

M. Faraday Esq | &c &c &c

Letter 2512
Faraday to Thomas Bell
10 April 1852
From the original in RS MS RR 2.83

Royal Institution | 10 April 1852

My dear Bell

Having carefully read Mr. Grove's paper[1] I arrive at the conclusion that it is correct in the experiments and sound in the deductions and touching as it does upon the unity of electrical phenomena, whether in the static or dynamic form, is of great value in the development of that part of

science. My voice therefore is for the paper. Personally I hail it as a great aid in the expansion of that subject which I love[.]
    Ever Very Truly Yours | M. Faraday

1. Grove (1852a).

Letter 2513
Jacob Herbert to Faraday
10 April 1852
From the original in GL MS 30108/1/50&51
                          Trinity House, London, | 10 April 1852.
My dear Sir,
    It being deem'd necessary that a further Test of the White Lead, samples of which were analysed by you, & reported upon in your Letter dated 31 Ult:-, should take place[1];- Herewith I beg to forward to you 6 additional Samples thereof, letter'd A. to F. and to request that you will favor the Elder Brethren by ascertaining the contents of each separately, and communicating the results at your early convenience[2].
    I remain | My dear Sir, | Your very faithful Servant | J. Herbert
M. Faraday Esq. DCL | &c &c &c

1. See letter 2507.
2. Faraday's analyses are in GL MS 30108/1/50&51.

Letter 2514
Faraday to Thomas Andrews
13 April 1852
From the original in SM MS 350/1, f.76
                          Royal Institution | 13, April 1852
My dear Andrews,
    Your kind letters[1] delight & melt me and raise many pleasant thoughts which may perhaps lead to realization. At this moment my desire is strongly to be with you[2] and if I can I will. As to my wife she is most grateful for the kindness which Mrs. Andrews and you so freely direct towards us & I shall leave the idea to work both in her mind and mine for a little while. I have been thinking of you again & again lately for having had some copies of a late research to send to my friends I find the last three or four Researches for *you* still in my desk directed but not sent[3][.]

Either I have never had the opportunity or if I have had it my treacherous memory has at the time deceived me. *Now,* I do not know whether it is worth while looking out for an immediate means, because in five or six weeks I hope for one or two more papers[4] to add to the packet. What is the most expedient means of sending such things to you? Has your College any bookseller in London or other channel?

I am suffering just now under a heavy cold which has been on me for weeks & which the East wind keeps on me so that I feel somewhat confused and had better say no more than that with most grateful respects to Mrs Andrews I am My dear Friend

Very Truly Yours | M. Faraday

Address: Dr. Andrews | &c &c &c | Queens College | Belfast | Ireland

1. Including letter 2511.
2. For the Belfast meeting of the British Association.
3. Faraday (1851b, c, d, e), ERE24, 25, 26 and 27.
4. Faraday (1852b, c), ERE28 and 29.

Letter 2515
Faraday to Thomas Andrews
20 April 1852
From the original in SM MS 350/1, f.35

Royal Institution | 20 April 1852

My dear Andrews

Receive my sincere thanks for your kind letter[1] and for the paper[2] & for all your good will & favours to me: tell Mrs. Andrews I hope to thank her personally & trust when the time comes she will put up with me & not be disappointed.

As to your proposed experiments, it will be of extreme interest to me to know how electrified bodies behave in a perfect vacuum. Perhaps you know that Masson[3] not long since concluded that Electric currents or discharges *cannot exist* in an absolute vacuum[.] As an introduction to his memoires of which there are four or five I may refer you to the Annales de Chimie 1851, Vol xxxi, p125. At pp.149, 150 you will see some of his conclusions[4][.]

I have sent my papers to Williams & Norgate[5].

Ever Truly Yours | M. Faraday

1. Not found.
2. Probably Andrews (1852).

380   Letter 2515

3. Antione-Philibert Masson (1806–1860, DSB). Professor of Physics at the Lycée Louis-le-Grand.
4. This would seem to refer to a separately paginated offprint of Masson (1845, 1850, 1851). For his conclusions see Masson (1851), 326. On Masson's work see James (1983b), 146–8.
5. Williams and Norgate. Booksellers of 14 Henrietta Street. POD.

Letter 2516
Charles Mansfield Clarke to Faraday
23 April 1852
From the original in IEE MS SC 2
Faraday numbers 4 and 5
My dear Dr Faraday
    I hardly know how to answer your note: as there was much talking and little regularity at the meeting alluded to - no written motion was made but I believe that the substance of it corresponded with the enclosed[1].
    Yours very sincerely | Charles M. Clarke
15 Berkeley Square | April 23, 1852

That the Gentleman and Ladies who have attended the Lectures given by Mr Brande in the Theatre of the Royal Institution desire to tender him their best acknowledgements for the kind and admirable manner which he has displayed in illustrating the subjects which have been brought before them in the various courses of lectures on Chemistry delivered at the Institution[.]

1. This refers to Brande's announcement of his resignation as Professor of Chemistry at the Royal Institution following new regulations issued by the Royal Mint which prohibited its employees from holding other positions. He announced his resignation following a lecture at the Royal Institution on 3 April 1852. See *Proc.Roy.Inst.*, 1852, 1: 168–71 and letters 2524, 2525, 2527 and 2529.

Letter 2517
Faraday to George Grove[1]
27 April 1852
From the original in SI D MS 554A

                                                            R Institution | 27 April 1852
My dear Sir
    I can only say in your own words that I appreciate very highly the privilege of subscribing the address[2].

Ever Truly Yours | M. Faraday
G. Grove Esq | &c &c &c

Endorsement: His name may be appended to the Dedic[ati]on. April 27, 1852

1. George Grove (1820–1900, DNB2). Secretary of Society of Arts, 1849–1852.
2. This was the dedication to Anon (1852a). The address, to the Queen, is on pp.iii-iv with the names of the subscribers, including Faraday, on p.iv.

Letter 2518
Jacob Herbert to Faraday
29 April 1852
From the original in GL MS 30108/1/50&51

Trinity House | 29 April 1852.
My dear Sir,
I am directed to forward to you Three farther parcels of White Lead, letter'd respectively G, H, & I, and to request you will favor the Elder Brethren, by making an analysis thereof and communicating the results,- as you have done with reference to the Six Samples, reported upon in your Letter to me of the 20th Instant[1].
I remain | My Dear Sir | Your very faithful Servant | J. Herbert
M. Faraday Esq. D.C.L

1. See letter 2513. Faraday's analyses are in GL MS 30108/1/50&51. On the basis of Faraday's report the Trinity House By Board, 11 May 1852, GL MS 30010/37, p.422 placed an order.

Letter 2519
Faraday to Frederick Augustus Abel
May 1852[1]
From Gladstone (1874), 30–1

I hope you feel yourself happy and comfortable in your arrangements at the Academy[2], and have cause to be pleased with the change. I was ever very kindly received there, and that portion of regret which one must ever feel in concluding a long engagement would be in some degree lessened with me by hearing that you had reason to be satisfied with your duties and their acceptance. - Ever very truly yours, M. Faraday

1. Date given in Gladstone (1874), 30.
2. That is the Royal Military Academy, Woolwich.

Letter 2520
Faraday to William Robert Grove
4 May 1852
From the original in RI MS G F28

R Institution | May 4, 1852

My dear Grove

M Dubois Reymond will be making his experiments *here* on Thursday[1] next beginning about 11 o'clk & going on for 2 or 3 hours. I thought you would like to know I have seen one or two - beautiful[2][.]

Ever Yours | M. Faraday

1. That is 6 May 1852.
2. These experiments were noted briefly in *Lit.Gaz.*, 22 May 1852, p.435. These were on animal electricity as described in Bois-Reymond (1848–9).

Letter 2521
Faraday to William Whewell
4 May 1852
From the original in TCC MS O.15.49, f.32

R Institution | 4 May 1852

My dear Sir

Dr. Dubois Reymond of Berlin will be engaged here next Thursday[1] the 6th from 11 o'clk onwards in shewing his fine experiments on the Muscular & Nervous Electricities at *this house*[.] Thinking you would like to see them if in town or at least to know of them I send this hasty note[2][.]

Ever Truly Yours | M. Faraday
Revd. Dr. Whewell | &c &c &c

1. That is 6 May 1852.
2. See note 2, letter 2520.

Letter 2522
Faraday to Charles Wheatstone
4 May 1852
From Gladstone (1874), 46

Royal Institution, 4th May, 1852.

My dear Wheatstone,

Dr. Dubois-Raymond will be making his experiments *here* next Thursday, the 6th, from and after 11 o'clock. I wish to let you know, that you may if you like join the select few[1].

Ever truly yours, | M. Faraday

1. See note 2, letter 2520.

Letter 2523
Faraday to George Richmond[1]
4 May 1852
From the original in the possession of Jorge C.G. Calado

Royal Institution | 4 May 1852

My dear Sir

Dr. Bence Jones has left it with me to ask your convenient time[2]. Would either next Friday or Monday or Tuesday[3] suit you at the former time or any other morning hour?

Ever Truly Yours | M. Faraday
G. Richmond Esq | &c &c &c

1. George Richmond (1809–1896, DNB). Portrait painter.
2. That is 7, 10 or 11 May 1852.
3. Presumably for a sitting for Richmond's portrait of Faraday commissioned by Bence Jones. Plate 9.

Letter 2524
William Thomas Brande to Faraday
5 May 1852
From the original in IEE MS SC 2
Faraday number 1

Royal Mint | 5 May 1852

My dear Faraday

I have been much annoyed at two independent communications which have reached me since we met on Monday last[1] from two members of the Institution, stating the extreme regret they felt "in consequence of Mr *Barlow* having prevented the members expressing their feelings of thankfulness to me for my long services" and have heard for the first time that such a desire on their part was expressed after I had left the room on the occasion of my last lecture "but that Mr Barlow was instrumental in crushing it"[.]

I am convinced that there must be some mistake somewhere but the statement has been so unequivocally made to me that I think it my duty to ascertain from you what was really said upon that occasion[2] - and further, whether any thing transpired upon the subject at the members meeting either on Saturday or Monday last, as I am told is the case.

I am sure you will at once tell me what passed on these several occasions as I am sadly perplexed by the reports which have reached me. I have not said a word to any other than yourself upon the subject and of course the matter must be quite confidential between us.

Yours my, dear | Faraday always & sincerely | Wm Thos Brande.

1. That is when the both attended the meeting of Managers. RI MM, 3 May 1852, **10**: 392.
2. See letter 2516.

Letter 2525
Faraday to William Thomas Brande
6 May 1852
From the original copy in IEE MS SC 3
Faraday number 2

Royal Institution | 6 May 1852

My dear Brande

Your informants have not done Barlow justice perhaps they mistook him[1]. I will endeavour to give you such an account as you ask me for though doubting my memory[.] I was in the Gallery at your last lecture[.] Your audience were taken much by surprise by your farewell[2] and when you left the room a member I think Sir H. [sic] Hall[3], called on Mr. Barlow to take the chair that the audience might give an expression of their feelings. Other persons spoke and I think that several members thought they ought to have been informed by the Managers of the coming resignation and my impression was that they considered the Managers ought to have done & said somewhat & were hurt by the neglect[4]. This as *you* know was impossible because of the recent date of your announcement to the board for the Monday following your lecture, was the first monthly meeting after it. Mr Barlow ventured to mention the recent circumstances and the managers intention to report on the following

Monday your resignation & their proposition to express their feelings by taking precisely the same steps as in the case of the resignation of Davy. Then expressions very kind to you were uttered as was most natural after such a long term of what I may truly call affectionate relationship, accompanied by some vague propositions of a fellowship a bust or some other mark to be awarded as by a vote of those present. Mr. Barlow endeavoured to explain that that mixed meeting could not act or vote on a lecture day as a body of members the act charter & bye laws being against it but that they could give the expression of their conjoined opinion in any form they thought proper whereupon a vote of thanks moved & seconded by Sir Charles Clarke & Mr. John Pepys[5] was carried & communicated immediately by the former to you. The vote is also recorded in the printed notices & I believe elsewhere as in a report to the Managers but I am not sure about that. I cannot remember. On Monday 5th April the Managers made their report to the monthly meeting and I had the honor of proposing you as the Honorary Professor of Chemistry[6]. Several then spoke in the very highest terms of your long connexion with the R Institution and were glad to hear what the managers recommended. Several proposed some token of their feelings in which they could be joined personally, and Mr. John Pepys' generous mind was very forward in this; but a real obstruction was thrown in the way by one member proposing so many things that nothing was distinct, a chair, a scholarship, a bust presented to yourself, a portrait, a medal, were amongst them, and some members including myself had to remark upon the fitness of things. I recommended a committee & it was understood as I believe that any thing of the kind ought to be done not as an act of the meeting of members acting for the *whole body* of members, but by a committee & subscription as in other like cases and I have been waiting to hear of the formation of such a Committee by those who seemed earnest for it. May 1st was the Annual meeting: then also many kind expressions were uttered during the hour of waiting for the Election of Officers, your name being in the Managers list but, as you know, nothing but the Election & the Visitors Report *could* then be taken as the business of the day. May 3 was the next monthly meeting and then the Election as Honorary Professor occurred[7]. You ask me whether any thing transpired[.] I cannot call to mind that any proposition (beyond what the Managers had recommended) or any hint was made. I was still expecting the formation of a Committee but those who said most on the first occasion were not present.

 I have thus endeavoured to answer your enquiries but feel I have not remembered the order of things clearly: Sir Charles Clark[e] was present on all the occasions[8] and he is one who could tell you what occurred & whom I think you would feel you could trust. On the whole I do not see how Mr Barlow when called upon could act otherwise, and I know the impression on the minds of several who were present is *not* that which you have received. I shall say nothing to him or any body else about your letter but consider it at present quite confidential as you desire and I trust that you will soon hear enough from other parties to remove altogether & entirely the impressions you have received. It would be indeed a sad pity if after Forty years of kind & active association between the Royal Institution & yourself the least uncomfortable feeling should remain as its result, and I cannot help saying that if I knew your informants I should feel very much inclined to speak to them as a justice due to Mr Barlow & yourself conjointly.

Ever My dear Brande | Yours faithfully | M. Faraday

1. See letter 2524.
2. That is Brande's announcement of his resignation as Professor of Chemistry at the Royal Institution following new regulations issued by the Royal Mint which prohibited its employees from holding other positions. He announced his resignation following a lecture at the Royal Institution on 3 April 1852. See *Proc.Roy.Inst.*, 1852, **1**: 168–71 and letters 2516, 2524, 2527 and 2529.
3. John Hall (1787–1860, *Ill.Lond.News*, 1860, **36**: 381). Member of the Royal Institution.
4. Brande had previously informed the Managers of his resignation. RI MM, 16 March 1852, **10**: 379.
5. John Pepys (d.1866, age 90, GRO). Member of the Royal Institution, 1800–1866.
6. RI MS GM, 5 April 1852, **6**: 26–8.
7. RI MS GM, 3 May 1852, **6**: 39.
8. See letter 2516.

Letter 2526
Christian Friedrich Schoenbein to Faraday
7 May 1852
From the original in UB MS NS 400
My dear Faraday

What may be the cause of the very long silence kept by your friend on the Rhine? This question has perhaps more than once been asked in the Royal Institution these last six months. First of all, let me assure you that that somewhat strange taciturnity has nothing to do with any thing being in the remotest degree akin to forgetfulness. Why, I don't know, but the fact is, that Mr. Schoenbein has of late conceived an almost invincible dislike to pen and ink so that nothing but the most cogent reasons can force him to make use of them. He therefore has become a most lazy correspondent to all his friends. Whether that antipathy be a symptom of advanced age or only one of those unaccountable fits and whims, which even the strongest minds are now and then liable to, I cannot say, but this I know, that he trusts your inexhaustible kindness will grant full pardon and indulgence to this piece of human frailty of his. Though strongly disinclined to handle his pen, he has not yet lost his relish for scientific pursuits and, as far as I know, was rather active last winter. It cannot be unknown to you that our mutual friend entertains very curious and even highly strange notions regarding oxigen, which he considers as the first-rate Deity not only of the chemical but of the whole terrestrial world[.] He is indeed a most enthusiastic devotee to that Deity, talking and thinking of nothing but of her, praising and exalting her glory wherever he can. He pretends that our philosophers, much as they think to know of oxigen, are as yet blinded and ignorant of the omnipotence of that mighty ruler of the elementary world. Upon many agents considered as equal to oxigen he looks down as upon upstarts and usurpers, assuming powers and privileges to which they have no right and declares that an infinite number of glorious deeds ascribed to the agency of inferior deities, are in fact the work of what he calls the "Jove of the philosophical Olympos."

As a matter of course, our friend entertains feelings of peculiar love and esteem towards those whom he considers as high-priests to his Jupiter and who tend to increase the authority and glory of the King of elements. He asserts that you are the leader of those chosen adepts; that you more than any other have unravelled the mysteries of the wonderful workings of oxigen in nature and that you are the man who first has brought to light that the influence of our friends' favorite deity reaches far beyond the limits of the chemical world. He goes even so far as to maintain that upon your discovery of the magnetical powers of oxigen a new philosophical era will be founded.

Having said so much about our queer and enthusiastic friend you will not be surprized when I tell you that he is continually worshipping his goddess in a little smoky room, which he calls "Jove's Temple" and if I be not misinformed, there upon a sort of "tripod" he asks all sorts of questions with the view of getting as deep as possible into the mysteries of his deity. The other day he hinted at very strange answers having received from his oracle. Oxigen, he says, is the lord and master even of the most subtle and all pervading beings in existence, destroying and creating light, making and unmaking colors at pleasure &c. Indeed, he showed me some very strange tangible substances exhibiting in a most extraordinary manner the nature of a Chameleon, for within a few minutes I saw the very same thing assuming white, green, yellow, orange, light-red, dark-red and even black colors. Heaven knows how such a wonderful change was brought about; our friend says that his oxigen and nothing but his oxigen had been the Charmer; but being afraid that he is a little cracked, I am rather sceptical about his assertions. He also talks now and then of oxigen being closely allied to the great powers of Electricity and Magnetism and gives to understand that their apparent might and force are only borrowed from his sovereign's.

I wonder whether he will divulge his queer Ideas to the world[.] I should like to see them kept back from the philosophers of our days for these people are too sober and rational as to relish the extravagant notions of our hot-headed friend.

Mrs Schoenbein and the Children are well and have not forgotten their English friend to whom they beg to be kindly remembered. Mrs Faraday, I hope, recollects still the writer of these lines and will be indulgent enough as to accept friendly his compliments.

Pray let me soon hear of your doings and believe me

Your's | most truly | C.F.S.

Bâle Mai 7, 1852.

Address: Doctor Michael Faraday | &c &c &c &c | London | Royal Institution

Letter 2527
William Thomas Brande to Faraday
8 May 1852
From the original in IEE MS SC 2
Faraday number 3

Tunbridge Wells | 8 May 1852

My dear Faraday

I am sincerely obliged by your kind and candid reply to my selfish queries[1], which however, have only part to enable me thoroughly to divest my mind of the impression it had received from the reports of two old friends of mine, and of the Institution.

Of the bust, picture, medal, &c I had heard nothing, and am very far from aspiring to, or even desiring any thing of that kind. But some public expression of the kindly feeling of a *Body* with which I had been intimately, and I may truly say, affectionately associated, for what I may call the whole of my life, would have been more than acceptable; and this is what I was told was prevented. I have however written, as you suggested, to Sir Charles Clarke, and if he tells me that such an expression as I have adverted to was not so opposed as has been represented, I shall be at ease. I have received an *official* intimation of my Election as Honorary Professor, to which I have returned an official answer - and very glad I was to find from your letter, that you were my proposer. All this is all I would have wished.

Ever my dear Faraday | Sincerely yours | Wm Thos Brande

1. Letter 2525.

Letter 2528
Jacob Herbert to Faraday
10 May 1852
From the original in GL MS 30108/1/52

Trinity House, | 10 May 1852.
My dear Sir,
Herewith I send you a sample of Oil taken from the Iron Tank, from which the Light Houses &c belonging to the Corporation are supplied, in which Tank it has stood for some months,- and having recently been burnt at this House, has not been found to burn so well as heretofore; it appearing also, in some degree, clouded, And I am thereupon instructed to request you will favor the Elder Brethren, by stating your opinion, whether vegetable Oil, of this description, deposited in an Iron Tank, would probably possess such chemical affinity for the metal as to take up a sufficient portion of it, so as to injure it's qualities of combustion[1].
I remain | My dear Sir, | Your very faithful Servant | J. Herbert
M. Faraday Esq. D.C.L. | &c &c &c

1. Faraday's notes on this are in GL MS 30108/1/52. He concluded that the iron tank had exerted no effect.

Letter 2529
William Thomas Brande to Faraday
11 May 1852[1]
From the original in IEE MS SC 2
Faraday number 6

Royal Mint | Tuesday Morning
My dear Faraday
I have this morning Sir Charles Clarkes reply to my queries - and I am happy to say that his version of the Transaction entirely agrees with yours[2] - so that my informant must certainly have misunderstood or misinterpreted Mr Barlows interference.
Yours my dear Faraday | very sincerely | W T Brande

1. Dated on the basis that this letter follows letter 2527 in sequence.
2. Letter 2525.

Letter 2530
Faraday to John Potter[1]
15 May 1852
From the original in Manchester Public Library MS Potter 26

Royal Institution | 15 May 1852

Sir
 I regret that I am unable to accept the kind invitation to be present at the opening of the Manchester Free Library on the 9th of next month[.]
 I am Sir | Your Very Obliged & faithful servant | M. Faraday
John Potter Esq | &c &c &c

1. John Potter (1815–1858, B2). Manchester Alderman, 1845–1858.

Letter 2531
Faraday to Constance Deacon
19 May 1852
From the original in Vassar College

Royal Institution | 19 May 1852

My dear Constance
 First a kiss-s-s-s-ss. Next thank you for Your good letter - very well written and very pleasant - and now thanks for the letter you are going to write to me in which you must tell me how Papa[1] & Mamma[2] do - and what you are about. I went this morning to see a fish like a great eel take his breakfast. This morning he had three frogs for breakfast - yesterday he eat [sic] 9 fish in the course of the day each as large as a sprat and the day before 14. When the fish are put into the water he electrifies and kills them & then swallows them up - and if a man happens to have his hands in the water at the same time the fish that is the eel, electrifies the man too. The eel is now above 12 years old and is heavier I think than you are[.]
 Yesterday I saw the Royal children the Prince of Wales[3] & the duke of York[4]. Such nice children they would make famous playmates for you but I do not know whether Princes do play much[.] I do not think they can be so happy in their play as you are[5][.]
 As to the magnet when you & I meet we will have a long talk about it and make some *experiments*.
 and so with my love to Papa & Mamma and curious Constance with a kiss for each I am
 Your loving old Uncle | M. Faraday

1. Thomas John Fuller Deacon.
2. Caroline Deacon.

3. Albert Edward, Prince of Wales (1841–1910, DNB2). Eldest son of Queen Victoria and Prince Albert. Prince of Wales, 1841–1901.
4. Prince Alfred (1844–1900, DNB1). Second son of Queen Victoria and Prince Albert. Faraday seems to have been confused by this title which Prince Alfred did not bear.
5. *Times*, 19 May 1852, p.5, col. c noted that the Royal princes visited the Polytechnic Institution the previous day.

Letter 2532
Alexander William Williamson to Faraday
22 May 1852
From the original in RI MS

University College, London | 22nd May 1852

Dear Sir,

Having been informed, that in consequence of the resignation of Prof. Brande at the Royal Institution, the Managers contemplate making arrangements for the provisional discharge of the duties hitherto appertaining to that gentleman, I am desirous of stating that I would feel much honoured by obtaining leave to deliver a course of chemical lectures in the Institution.

I am sufficiently aware of the difficulty of presenting the higher scientific truths in an easy and striking form, to feel that much indeed is needed of a lecturer at the Royal Institution, but the difficulty of the task would rather stimulate than discourage me, and no exertion should be wanting on my part to render the course as effective as possible.

By submitting my qualifications to the consideration of the Managers[1], you will greatly oblige

Yours sincerely and respectfully | Alexd. W. Williamson
To Professor Faraday

1. The Managers accepted Williamson's proposal and invited him to deliver nine lectures in the term before Easter 1853. RI MM, 7 June 1852, **10**: 401.

Letter 2533
Jacob Herbert to Faraday
29 May 1852
From the original in GL MS 30108/2/88

Trinity House, London, | 29th May 1852.

Sir,

I am instructed to forward to you the accompanying Jar of Water, which has been received from the Longships Light House, together with a

Copy of a Letter, dated 18th Inst, from the Agent for that Establishment in relation thereto[1]; And I am to signify the request of the Elder Brethren, - that you will favor them by examining the Water refer'd to, and furnishing them with your opinion as to any impurities which, on analysis, it may be found to contain with respect to it's wholesomeness for drinking purposes[2].

I am also directed to transmit to you herewith, a Sample of White Lead which it is desirable should be tested as on previous occasions[3]; and to request that you will accordingly favor the Elder Brethren by doing so, and reporting the result, for their information[4].

I am | Sir, | Your most obedient Servant | J. Herbert
M. Faraday Esq. D.C.L. | &c &c &c

1. This letter is in GL MS 30108/2/88.
2. Faraday's analyses are in GL MS 30108/2/88.
3. See letters 2288, 2507, 2513 and 2518.
4. Faraday's report on both these matters, dated 2 June 1852, was read to Trinity House By Board, 8 June 1852, GL MS 30010/37, p.431. It was referred to the Light Committee.

Letter 2534
Faraday to Christian Friedrich Schoenbein
2 June 1852
From the original in UB MS NS 401

Royal Institution | 2 June 1852

My dear friend

Though very stupid & weary yet I write, chiefly for the purpose of thanking you for your last very kind letter[1] - it was quite a refresher and did me good. I wish more had such power, then I should think I might be of some little use amongst my friends by cheering them up.

Your paper in the Chirurgical Transactions[2]. I think I asked you what I should do with some copies that were printed off[3][.] However I forget whether you told me any thing about them - and I find by enquiring that Dr. Bence Jones has sent them to you by a friend that hoped to see Basle perhaps you have them already[.]

Presently you will have three papers of mine all at once. Two from the Phil Trans[4] & one from the Phil Mag[5]. They all relate to one subject i.e the lines of magnetic force.

Every now & then I stir my audience by talking about your Ozone[6] - and then there are many enquiries[.] I wish we had a good general English account of it both as to its preparation actions and history[.] An acquaintance of mine the Revd Mr. Sidney[7] is busy putting slips from your ozonmeter which I have supplied him with through the cleft stems of vegetable & says he procures many effects just like those of ozone. In such cases however there is a great deal to eliminate as due to other actions of the ozonometrical strip & the juices before he will have his subject clear[.] Still experimentation is always useful[.]

What are your mysterious results - or what the results of your mysterious friend. Have you made gold - or iron rather for it is a more useful metal[.] Or have you condensed oxygen. I wish you could tell me what liquid or solid oxygen is like. I have often tried to coerce it & long to know[.]

With kindest remembrances to Mrs. Schoenbein

I am My dear Schoenbein | Your lazy friend | M. Faraday

Address: Professor Schoenbein | &c &c &c | University | Bâle | on the Rhine

1. Letter 2526.
2. Schoenbein (1851b).
3. In letter 2482.
4. Faraday (1852b, c), ERE28 and 29.
5. Faraday (1852d), [ERE29a].
6. Faraday's first lecture, on 24 April 1852, in his course of six lectures on the non-metallic elements was on oxygen and hydrogen. See RI MS GB 2: 72 and also Faraday (1853b), 104–11.
7. Edwin Sidney (d.1872, age 74, B6). Rector of Little Cornard, 1847–1872 and lecturer at the Royal Institution and elsewhere.

Letter 2535
George Biddell Airy to Faraday
7 June 1852
From the original in IEE MS SC 2

Royal Observatory | Greenwich | 1852 June 7

My dear Sir

At the late meeting of the Board of Visitors, there was a wish expressed by almost every member of the Board that I should confer with you respecting the reduction of our Magnetic Observations. The object - at least the ultimate object - of the reductions being understood to be, the reference of the phaenomena to their physical causes.

I do therefore by this writing take the first step in the conference, and I earnestly hope that you will respond to it.

Assuming that you will do so, I now proceed to consult you on the best method of proceeding. And I will beg leave to commence with the following statement.

We have, as you probably know, very full and accurate records (such as I firmly believe to be unequalled in the world) of the constant changes

of magnetic elements, formed by our photographic apparatus. Without examination of these, I do not think that any body can form an accurate notion of the phaenomena that stand to be explained. There are, as is commonly believed, the regular phaenomena of
> Annual variation
> Diurnal variation

but there are also the anomalous phaenomena of
> Fret
> Irregularity
> Storms

which are so much greater in magnitude than the others, that I have no confidence (and in this I was supported by several members of the Board) in deductions applying to the so called regular phaenomena until the irregular ones are cleared out of the way.

Now would you like to look over a volume or two (say a years collection) of our Photographs, and form a general notion of the things? On receiving your assent, I will send you the bound collections. And then, in due time, would you make an appointment for me to wait on you and talk over the matter? - Or do you see any better course?

I am, my dear Sir, | Yours very truly | G.B. Airy
Michael Faraday Esq | &c &c &c

Letter 2536
William Whewell to Faraday
8 June 1852
From the original in TCC MS O.15.49, f.63

[Athenaeum letterhead] | June 8, 1852

My dear Dr Faraday,

You find such admirable fields of research for yourself, and work them so well that it is not a light matter to offer to you any suggestion in that way: but what you have already done with regard to the local peculiarities of the declination needle in your most important paper on that subject[1], points out you as the person most fit to undertake the remainder of the problem. We want some explanation or at least some beginning of explanation of the *magnetic storms* which agitate the needle from time to time, and of which the mean results affect the daily and monthly changes of declination. The Astronomer Royal has a large mass of accumulated photographic observations of the daily changes of the needle, and among them, of course are included the records of various kinds of storms; from a mere magnetic *fact* which takes place in some degree to a state of frantic oscillating leaps which occur at other times. He will, I know, be very glad if you will look at them and suggest any probable or even possible connexion of these with other physical phenomena[2].

I saw your "sphondyloid" paper in the Phil. Mag[3]. and I saw that I had led you into saying a sphondyloid *body*[4] instead of *solid*; which would have been more proper as it is a geometrical solid, not a mechanical body which you speak of.

I leave England today: but I shall be glad to hear from you at Kreuznach, Rhine-Prussia; and am always
Yours very truly | W. Whewell

1. Faraday (1851d), ERE26, 2929–46.
2. Letter 2535.
3. Faraday (1852d), [ERE29a].
4. *Ibid.*, paragraph 3271. See letters 2494, 2496, 2497 and 2498.

Letter 2537
Faraday to William Whewell
10 June 1852
From the original in TCC MS O.15.49, f.33

Royal Institution | 10 June 1852

My dear Whewell

I received your pleasant letter[1] which I take as a great compliment; or rather (for I do not care for compliments) as a great encouragement. Now I do not know that I have the ability to enter upon the irregular variations of the Earths magnetism, but I have the wish; and shall some day be encouraged to do so, but not immediately, for the following reasons. The *seat* of the terrestrial magnetic force is (for us at present) in the earth:- and *within* the earth there may be causes of variation, not in the distribution of the power only, but also in its amount; the latter varying not simply for the whole mass but probably in different parts at different times. Such variations (*within the earth*) are I suspect the chief, perhaps the only, cause of magnetic storms, frets &c &c; and as yet we have scarcely any other hold than that of the imagination upon them. On the other hand considering the seat of the force to be *in the earth*, the force itself extends externally around the planet & its *distribution* is, as I believe, affected by the medium which is there present, and which, under the influence of the sun, is the chief cause of the daily and of some other regular variations. Now I think I have a hold, by experiment, on this cause of variation; & I further think that if I succeed in making that clear, I shall do far more good to the whole cause & aid more rapidly in the elucidation of the other set of variations, than by going to them at once. For this reason I mean to devote all my thoughts & means to a determination of the rate of variations of the paramagnetic force of oxygen from summer heat or higher, down to 0° or if possible 20°, 30° or 40° below 0°. If by such experiments my expectations are confirmed, and we should be able to know surely what is the amount of atmospheric variation; then, subtracting this from the whole amount of effect, how much better we should be able to comprehend the proportion & also the nature of the effect due to internal causes of variation? - and we might then hope that not one or two minds only, but many would start on the search after the cause of the irregular phenomena with far greater advantage, because from a far more assured position than at present. Whatever hopes we have of solving the riddle of terrestrial magnetism must depend upon our successive elimination of the causes of variation; and any one which presents itself in a tangible shape should be pursued as far as possible and estimated in value so that it may be taken from the rest, that they may be studied in a simple form. If we could begin with a magnetic storm the process would be the same as regards the residual phenomena:- but we have as yet little or no hold ⟨on⟩ the storms, & I think we have a grasp on the annual & daily variation.

When we really come to magnetic storms I think we shall want needles far lighter than such as are now used for the observations. Needles for such a use ought to be without weight, so as to have neither inertia or momentum[.] How can a bar of some pounds weight tell us the frequency or the extent of rapid vibrations, or give us anything more than a slurring mean? Yet it may be that these variations in their full development are essential to the explication of the mystery. But we shall see by degrees what is wanted & I dare say at last obtain it[2][.]

Ever My dear Sir | Yours Most Truly | M. Faraday

Address: Revd. Dr. Whewell | &c &c &c | Kreuznach | Rhine | Prussia

1. Letter 2536.
2. This letter was discussed in Whewell to Airy, 15 June 1852, RGO6 / 694, f.60.

Letter 2538
Faraday to George Biddell Airy
10 June 1852
From the original in RGO6 / 694, f.58–9

Royal Institution | 10 June 1852

My dear Sir

Your letter[1] stirs up a great many thoughts in my mind which I have been obliged to leave at rest for a time, simply because experimental work is slow work, & strength & health are limited. My next occupation must be the determination, if possible, of the amount of change in the paramagnetic force of oxygen (or air) by depression of temperature to 0°, and if possible to –40°. or –50°; and at different degrees of pressure or condensation. These data are essentially necessary for the development of the effects which occur under the head of atmospheric magnetism; and which as I believe include a large part if not the whole of the daily variation & much of the annual variation. In the mean time it is a great thing that other persons should be moved to collate and analyse the results of observation; & separate as much as possible the compound result which appears in the continuous stream of a continued set of perfect observations, into such subordinate parts as the periodical variations and the irregular or anomalous phenomena:- for until the great result is in some measure analysed & referred to the various parts that make it, there can be but little hopes of clearing up the scheme of these most complicated phenomena. Therefore I think that such analyses as Sabines[2] of the larger variations will be very useful in due time.- M. Lamont has already been moved to search, & has, he says, found a recurring cycles of ten years[3][.]

To me, the phenomena of fret, storms &c have far greater attraction than the periodical variations already referred to: but we have now some hold of the latter and little or none of the former. They, i.e the former appear to me as probably due chiefly to actions within the magnet i.e the earth; and may have little or no relation to the action of the surrounding medium. If it be so, to determine the condition of a magnet (as the earth) producing variations of such large extent, with such rapidity, and so general in their influence, is a problem of such difficulty that we can only hope for its solution by degrees and knowing so little of the interior of the earth as we do, at first seems almost *hopeless*. Nevertheless to know, as I now do, that a perfectly invariable magnet, as regards the amount of force[4], may have the external *disposition* of its power varied to a very large extent, either suddenly or slowly, without the least change in the *sum* of power externally, is one point of rest and one ground of hope for the mind. Another is the power which, I hope, the mathematician has of pointing out what sources of change or irregularity are external to the magnet (or earth) and what internal. I think I have heard that Gauss has done something of this kind[5]. If we could separate the variations due to external causes from those caused by internal action, it would certainly clear up the subject in some degree. Hence it is that I must, (being already in the path,) work out the effect of temperature on air as far as I can, before I attempt to meddle with the causes of action within the earth; and, indeed, if I succeed in this first object there are then others waiting for me, which seem to present better handles for experiment than the internal state of the earth[.]

But I hope in a week or two to have another paper for you[6], and then if you like I will come some day to the Observatory. It would be a great treat to me, and a source of great instruction, to have a talk with you over some of the records of irregularity, storms &c and over the instruments[.]

I am My dear Sir | Most Truly Yours | M. Faraday
G.B. Airy Esq | &c &c &c

1. Letter 2535.
2. Sabine (1851a, b).
3. Lamont (1852).
4. "power" is crossed out here and "force" written above.
5. Gauss (1841), 229.
6. Faraday (1852e), Friday Evening Discourse of 11 June 1852.

Letter 2539
Jacob Herbert to Faraday
10 June 1852
From the original in GL MS 30108/1/53

Trinity House, London, | 10 June 1852.
My dear Sir,

The accompanying Bottle should contain Olive Oil, but a statement having been made, that it is mixed with another kind of vegetable Oil (Rape), I am desired to send it to you, and to request to be favor'd with your opinion, as to it's being genuine Olive Oil.

I know not if you have the means of determining this point, both Oils being vegetable; but am sure, you will kindly give the Board all the information which may be in your power[1].

I remain | My dear Sir, | Your very faithful Servant | J. Herbert
M. Faraday Esq, D.C.L. | &c &c &c

1. Faraday's analysis is in GL MS 30108/1/53. He concluded that there was not much olive oil in the sample and replied to this effect on 15 June 1852. This letter was read to Trinity House By Board, 22 July 1852, GL MS 30010/37, pp.440–1. They instructed that the offending oil be removed.

Letter 2540
Faraday to Third Earl of Rosse
12 June 1852
From the original in RI MS F1 A22
Thanks my dear Lord Rosse for a sight of the note & for freedom[.]
    Ever Yours | M. Faraday
12 June 1852

Letter 2541
Henry Bence Jones to Faraday
15 June 1852
From the original in RI MS Conybeare Album, f.7
Dear Mr Faraday
    His Grace the Duke of Northumberland was so good as to say that he would attend to the note which I wrote to you regarding the Electric Eels. If I could inform the Governor of Demerara[1] what his Graces directions have been it would perhaps render him better able to assist me in this matter. If you can obtain this information for me I shall be still more your much obliged
    H. Bence Jones MD, FRS.
30 Grosvenor Street | June 15, 1852.

1. Henry Barkly (1815–1898, DNB1). Governor of Demerara, 1848–1853.

Letter 2542
Faraday to John Barlow
16 June 1852
From the original in RI MS F1 D10
                                                  R Institution | Wednesday 16 June | 1852
My dear Barlow
    I cannot be at the Managers meeting to day but think there is nothing of consequence which requires me. I hope by degrees to obtain some good notions from Vulliamy & others about W. Closets - Shelves & Cases but they are not ready yet[1]. Such repairs as must be done are of the ordinary kind and the managers will I have no doubt trust them to my discretion & care[2].
    Ever Yours | M. Faraday

1. These plans were not laid before the Managers until RI MM, 7 February 1853, **10**: 421 and not approved until RI MM, 6 June 1853, **11**: 20.
2. This was agreed. RI MM, 16 June 1852, **10**: 403.

Letter 2543
Jacob Herbert to Faraday
16 June 1852
From the original in GL MS 30108/1/53

Trinity House | 16, June 1852

My dear Sir,
Many thanks for your Letter of the 15 - reporting on the (so call'd) Olive Oil[1].
May I ask you to give me a Line stating the price of each of the pure samples which you obtain'd, - per imperial Gallon - that is of Olive & of Rape.
I remain, | My dear Sir, | Faithfully Yours | J. Herbert
M. Faraday Esq

1. See letter 2539.

Letter 2544
Faraday to Edward Elliott[1]
17 June 1852
From the original in CITA

Mr Faraday presents his compliments to Mr. Elliott & is greatly obliged by the copy of the Warner[2] papers[.]
Royal Institution | 17 June 1852

1. Edward Elliott. Chief Clerk to the Master General of the Ordnance Office. *Imperial Calendar*, 1852, p.228.
2. Samuel Alfred Warner (c1794–1853, private communication from Anita McConnell). A charlatan who persuaded successive governments to examine various weapons that he had claimed to have invented.

Letter 2545
Faraday to Francis Beaufort[1]
18 June 1852
From the original in Hydrographic Office MS

Royal Institution | 18 June 1852

My dear Sir Francis

I mentioned to you the kindness of the Duke of Northumberland in the matter of the Gymnoti but Dr. Bence Jones hardly knows how to profit by it (as you will see by the enclosed note[2]) because of uncertainty. Could you help us in this matter by ascertaining at some convenient opportunity what his Grace is able to do for us?

Ever Your Obliged | M. Faraday

1. Francis Beaufort (1774–1857, DNB). Hydrographer to the Navy, 1829–1855.
2. See letter 2541.

Letter 2546
William Thomson to Faraday
19 June 1852[1]
From the original in IEE MS SC 2

32 Duke Street, St. James' | Saturday June 19.

My dear Sir

After our conversation today I have been thinking again on the subject of a bar of diamagnetic non-crystalline substance, in a field of magnetic force which is naturally uniform, and I believe I can now show that your views lead to the conclusion I had arrived at otherwise, that such a bar, capable of turning round an axis, would be set stably with its length along the lines of force. As I may not have another opportunity of seeing you again before you leave town[2], I hope you will excuse my continuing our conversation by writing a few lines on the subject.

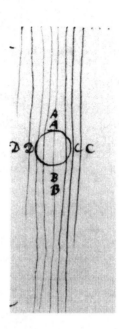

Let the diagram represent a field of force naturally uniform, but influenced by the presence of a ball of diamagnetic substance. It is clear that in the localities A and B the lines of force will be less densely arranged, and in the localities D and C they will be more densely arranged than in the undisturbed field. Hence a second ball placed at A or at B would meet & disturb fewer lines than if the first ball were removed; but a second ball placed at D or C would meet and disturb more lines of force than if the first ball were removed. It follows that two equal balls of diamagnetic substance would produce more disturbance on the lines of force of the field if the line joining their centres is perpendicular to the lines of force than if it is parallel to them. But the disturbance produced by a diamagnetic substance is an effect of worse "conducting power", and the less of such an effect the better.

Hence two balls of diamagnetic substance, fixed to one another by an unmagnetic framework, would, if placed obliquely and allowed to turn freely round an axis, set with the line joining their centres, *along* the lines of force.

The same argument, for a contrary reason, shows that two balls of soft iron similarly arranged, would set with the line joining their centres, also *along* the lines of force. For in this position *more* disturbance is produced on the lines of force of the field than in any other. But now, the *more* disturbance (being of better "conduction") the better. Hence the conclusion. Of course similar conclusions follow for bars, or elongated masses, of the substances.

I can never however make my assertion regarding the tendency of a diamagnetic bar in a uniform field without repeating that I believe no experiments can make it sensible. I doubt even whether the corresponding tendency in the case of a neutral bar in as strong a solution of sulphate of iron as could be got, could be rendered sensible by actual experiments, as excessively slight deviations from uniformity in the field would entirely mask the results of this tendency even if by themselves they might produce appreciable effects.

I remain, My dear Sir, | Your's very truly | William Thomson
M. Faraday Esq

---

1. Dated on the basis that Thomson's meeting with Faraday referred to in this letter is the same one as that discussed in Thomson to James Thomson, 21 July 1852, ULC MS add 7342, T441, which was written from the same address. Also Tyndall, *Diary*, 19 June 1852, **5**: 119–20

noted that Tyndall interrupted this meeting. The content of this letter is discussed in Gooding (1982) and Smith and Wise (1989), 274–5.
2. Faraday was about to leave for a holiday in Norfolk and Suffolk. See letters 2548, 2549 and 2551.

Letter 2547
Faraday to Henry Bence Jones
21 June 1852
From the original in RI MS F1 D11

Royal Institution | 21 June 1852.

My dear Sir

My earnest thanks for your kindness in sending me the copy of the work[1] also for the further proof of kindness contained in it and for the many corresponding manifestations of feeling on your part upon different occasions of all of which I am very sensible[.]

Next I enclose a note I have had from Admiral Beaufort[2] in relation to your last enquiry[3]. I hope it will be useful & the instructions he speaks of efficient[.]

Ever My dear friend | Yours faithfully | M. Faraday
Dr. Bence Jones | &c &c &c

1. Probably Bence Jones (1852) dedicated to Faraday.
2. Francis Beaufort (1774–1857, DNB). Hydrographer to the Navy, 1829–1855.
3. See letter 2545.

Letter 2548
Faraday to Benjamin Vincent
28 June 1852[1]
From the original in RI MS F1 D12

Old Buckenham | Monday Morning | June 28th

My dear friend

Your note was indeed very acceptable & I hope that the cheering account you gave of our friend is now backed up by continual improvement on his part so that our next letters may leave us to rejoice without anxiety as regards this attack[.] When I tell the friends that you say Mr Whitelaw is better they naturally say & how is himself, but there is no word of that in your note so we trust that no news is good news[.] Our friends here are much as usual[.] All were at meeting yesterday except Mrs. Thos. Loveday[2] Junr. who draws near to her confinement & Mrs. Bigsby[3] whom as yet we have not been able to see or even hear of, but to day or tomorrow will I trust do something in that way[.]

The weather here is a great mixture of fine sunshine & rain. Yesterday was a beautiful day until tea time when a Thunder storm came on with floods of rain & every part of the surface of the earth streamed with water[.] It tied up many friends but about 10 o'clk it cleared up. The Loveday family have a clamp of bricks in hand & the rain of Saturday Friday & yesterday was very inopportune for them. I shall see presently what harm it has done.

With respect to our movements I cannot tell you certainly what we shall do. We intend to go to Lowestoff some day about the middle of the week and it is just possible we may be home on Saturday next[4] so that I stand some chance of being at the Monday meetings of Managers & Members[5][.] But if I find Lowestoff takes off the languor (which this place does not remove) then we may be tempted to stop a few days longer there. I have written Anderson a note telling him to stop letters until he hears further. Will you have the kindness to give it to him[.]

Friends here are earnestly enquiring after you & the London friends. They are looking forward with expectation to your fathers[6] visit & we encourage their hopes. I hope we do not do wrong. They send their love to you & the brethren generally. I put in that of my wife (who is just gone shopping) and self to you & yours.

I am My dear friend | Yours affectionately | M. Faraday
Mr. B. Vincent.

1. Dated on the basis of the reference to the visit to Lowestoft. See letter 2551.
2. Elizabeth Loveday, née Crick (GRO certificate of son's birth).
3. Unidentified.
4. That is 3 July 1852.
5. Faraday was not present. RI MM, 5 July 1852, **10**: 404.
6. Thomas Vincent (1783–1854, GRO). Accountant.

Letter 2549
Sarah Faraday to Benjamin Vincent
9 July 1852[1]
From the original in RI MS F1 G29

Friday July 9th | High Street | Lowestoft
My very dear Friend

Tho' you are so kind as to say we need not answer your letter I think I may write a few lines to thank you in both our names, for it was most acceptable & perhaps you will be so kind as to say to our Mary Ann[2] that we do not intend being at home before Wednesday or Thursday next[3][.] I promised to let her know if we did not mean to be at home this week.

We are tempted by the beautiful weather & fine sea breezes to stay a little longer, as we find we can spend the Sabbath with our Brethren at Old Buckenham taking a return ticket tomorrow but though I tell you this, we do not wish the time of our return to be known to people in general - for sometimes it has been quite over powering to Mr Faraday to have so many callers immediately on his arriving[.]

Mr Faraday joins me in kind love to Mrs. Vincent[4] & yourself & I remain my dear Friend

Yours very affectionately | S. Faraday

Affectionate remembrance to enquiring friends, particularly to Mr Whit[e]law who we rejoice to hear is so much recovered[.]

1. Dated on the basis that the Faradays were visiting Lowestoft at this time and on the reference to Whitelaw's illness. See letters 2551 and 2548 respectively.
2. Mary Ann Champ (age 22 in 1851 Census returns PRO HO107 / 1476, f.64). Maid at the Royal Institution.
3. That is 14 or 15 July 1852.
4. Janet Young Vincent, née Nicoll (1811–1863, GRO).

Letter 2550
John Tyndall to Faraday
11 July 1852
From a typescript in RI MS T TS, volume 12, pp.4015–6

Queenwood College | near Stockbridge Hants. | 11th, July 1852.
Dear Sir,

I return, with many thanks, the memoir which you were kind enough to lend me[1]. I believe I must plead guilty to the charge of keeping it a day beyond the time permitted me, and the reason of my doing so is, that during my stay in London (which I quitted yesterday) I was so tossed about, in mind and body, that the tranquility necessary to the profitable reading of such a paper was denied me. If I dared I would keep it a day or two longer, for a man requires to brood over such a subject, to get as near as possible to the writers point of view, and to familiarize himself with the peculiar form under which the matters treated of present themselves to the writers mind. The constancy of the magnetic force regardless of distance, as indicated by the moving wire, is an exceedingly striking and significant result:--I am still a little in the dark as regards your remark in 3090, that "the system of power about the magnet must not be considered as revolving with the magnet."[2] Now in your bar magnets the force emanating from one of the corners is, I imagine, more intense than that emanating from any other point; in the rotation of the magnet this corner moves, and it appears to me that as the line of force emanating from the corner will always preserve the same position with regard to the corner, when the latter revolves the line of force peculiar to it must revolve also. In the case supposed by yourself in 3110^3, where the wire and the line of force are conceived to coincide, if the wire and the magnet revolves together I should imagine that the wire and the line of force would continue to coincide, that one would be perfectly motionless with regard to the other, and that hence no current could be formed.

Most heartily do I subscribe to the sentiment expressed in 3159^4. The object of the analyst in many cases seems to be to draw boundaries and limitations which the advancing experimentalist ever tends to break through. Far be it from me to decry the labours of the analyst - they are the necessary complement to experimental knowledge; but our knowledge is a thing *flowing* and not a thing *fixed*, and as long as this is the case the widest generalization which man has yet made is in danger of being swallowed up by one still more comprehensive. The analyst is the exponent of the centripetal tendency of the human mind, while the experimentalist is *centrifugal*, and seeks continually for still wider expansions.

I am making preparations at present to enter a struggle for a post in one of the Queen's Colleges in Ireland. The chair of Natural Philosophy in Galway College has become vacant and I intend to become a candidate for the post. Some of my friends have already commenced operations on my behalf - Col Sabine has written to a very influential quarter and other agencies are also at work. The election will probably not be made till towards October; in the mean time I will invoke all the aid I can, and if I fail, the consciousness of having done all it behoved me to do in the matter will be sufficient to render me content.

believe me dear Sir | most faithfully Yours | John Tyndall
Professor Faraday | &c &c &c

1. Faraday (1852b), ERE28 which he lent Tyndall on 19 June 1852 when Tyndall called on him. Tyndall, *Diary*, 19 June 1852, **5**: 119–20.
2. Faraday (1852b), ERE28, 3090.
3. *Ibid.*, 3110.
4. *Ibid.*, 3159.

Letter 2551
Faraday to Caroline Deacon
12 July 1852
From the original in the possession of Elizabeth M. Milton

Lowestoft | 12 July 1852

My dear Caroline

I write with a mind thoroughly empty of news but still containing a little affection for my friends and some for you. Writing seems but a very poor way of communication for thoughts which are scarcely representations of facts but concern the feelings & though I should dearly like to have a long & close chat with you about many things I cannot replace that by a letter & I really find the difficulty greater in trying to do so because I find that I am now rather apt to forget what the former part of a sentence is and so often make it awkward in its construction & what is worse confused in its meaning & mere nonsense[.] But though we cannot see as yet how things will turn, I hope that in one way or another we shall have the opportunity of personal chat & then we will talk. In the mean time let me say cheer up. It is a very common place phrase but still is not quite so common place with us as it is in the world:- or ought not to be. We were at Old Buckenham yesterday and our thoughts were led to the *prize* of the high calling of God in Christ Jesus and the infinitely great compensation it gives for the *tribulations* of this life and though we must feel the tribulations whilst in the flesh, yet still the prize set before us enables one of Christs people to say to another Cheer up with far greater effect & power than could accompany such words and without reference to the prize[.]

I hear of your worldly concerns by the letters of one & another & think I understand them as far as they are settled. I hope matters go on somewhat as you would wish. Remember me affectionately to Thomas and the little one[1] and think me.

Your Very Affectionate Unkle | M. Faraday

1. Constance Deacon.

Letter 2552
Faraday to George Grove[1]
27 July 1852
From the original in Royal Society of Chemistry MS

Royal Institution | 27 July 1852

Sir

I am very thankful for the kind invitation[2] you have sent me and hasten to say that I have the full intention of availing myself of the kindness of the Directors & Messrs Fox & Henderson[3][.]
I am Sir | Your Obedient Servant | M. Faraday
Geo Grove Esq | &c &c &c

1. George Grove (1820–1900, DNB2). Secretary of Society of Arts, 1849–1852.
2. This would appear to be for the erection of the first column of the new Crystal Palace at Sydenham on 5 August 1852. See *Athenaeum*, 7 August 1852, p.848.
3. John Henderson (1811–1858, *Ill.Lond.News*, 1858, **32**: 38). Engineer on the Crystal Palace.

Letter 2553
Robert John Kane[1] to Faraday
31 July 1852
From the original in RI MS Conybeare Album, f.30

51 Stephens Green | Dublin | 31 July 1852
My dear Dr Faraday
Your kind note and the parcel of memoirs[2] have been sent to me here from Cork and I beg to return to you my most sincere acknowledgements for your kind remembrance.
I should have called to pay my respects to you lately when I was in London but that I was aware of how much your time is occupied and I did not like to disturb you. I hope that I may be able to attend the Belfast meeting[3], but there are some official engagements which I cannot neglect but which may possibly interfere.
Believe me to be | My Dear Dr Faraday | Most Faithfully Yours | Robert Kane

1. Robert John Kane (1809–1890, DSB). President of Queen's College Cork, 1845–1873.
2. That is Faraday (1852b, c, d), ERE28 and 29.
3. Of the British Association.

Letter 2554
Johann Rudolf Wolf to Faraday
2 August 1852
From the original in IEE MS SC 2

Bern | 2/8 1852
Hochgeehrter Herr.

Ich habe soeben eine Entdeckung gemacht und veröffentlicht (Mittheilungen der Naturforschenden Gesellschaft in Bern Nro. 245–247[1]), die für Sie so viel Interesse haben dürfte, um mich zu entschuldigen, weñ ich mir erlaube Sie in einigen Worten damit bekañt zu machen. Ein vergleichendes Studium der von Herrn Hofrath Schwabe[2] beobachteten jährlichen Anzahl von Fleckengruppen der Soñe, und der von Herrn Professor Lamont[3] mitgetheilten Jarhresmittel für die Declinationsvariationen der Magnetnadel (oder auch der mittlern täglichen Bewegung der Horizontalintensität) hat mich nämlich zu folgendem Gesetze geführt: *Die Anzahl der Fleckengruppen und die mittlere Variation sind nicht hur derselben Periode von* circa $10\frac{1}{3}$ *Jahren unterworfen, sondern diese Perioden correspondieren bei beiden bis ins kleinste Detail; in demselben Jahre, wo die Anzahl der Fleckengruppen ihr Maximum oder Minimum erreicht, hat dasselbe auch für die mittlere Variation statt.* Es geht wohl mit Evidenz daraus hervor, daß die Soñenflecken und die Declinationsvariationen sich auf dieselbe Endursache zurückführen lassen müssen, und es eröffnet sich ein weites Feld der Speculation. Möchte es Ihnen gefallen an der Hand dieses Gesetzes neue Felder der Wissenschaft zu erleuchten, so würde sich seiner kleinen Mitwirkung herzlich freuen

Dero hochachtungsvollst Ergebensten | Rudolf Wolf | Director der Sternwarte

Address: Monsieur le Professeur Faraday | Royal Institution | Londres.

TRANSLATION from the original in IEE MS SC 2
Highly honoured Sir

I have just made a discovery & explained it (communications of the natural philosophical society in Bern Nos. 245–247[1]) which will have so much interest for you as to excuse me if I permit myself to make it known to you in a few words.

A comparative study of the yearly number of groups of spots in the sun observed by Herr Hofrath Shwabe[2] ([word illegible] of [word illegible]) and of the communicated yearly mean of the variations of declination of the magnet needle by Professor Lamont[3] (or also of the mean daily movement of the horizontal intensity) has led me to the following view. *The number of the groups of spots & the mean variation are not only subject to the same period of about* $10\frac{1}{3}$ *years but these periods will correspond in both to the smallest detail in the same year when the number of the groups of spots reaches its maximum or minimum Has the same also for the mean variation*[1]. It agrees with the evidence therefore that the sun spots & the variations of declination must allow themselves to be referred to the same - cause & a wide field for speculation opens itself. If you were pleased, on the suggestion of this view, to enlighten new fields of science Of his small share in the work would heartily rejoice.

Devoted | Rudolf Wolf | Director of the Observatory
Bern | 2/8 1852

1. Wolf (1852a).
2. Schwabe (1844).
3. Lamont (1852).

Letter 2555
Faraday to William Henry Sykes
6 August 1852
From the original in RI MS F1 D13

R Institution | 6 Aug 1852

My dear Sir

Yesterday a friend[1] & a friend of Liebigs told me that Liebig having the desire of sending his son[2] (who has now completed his medical education) to India has been advised to apply to you as able to assist him being both a man of Science & having the power to direct an appointment his way[.] I was asked to speak to you but said I could scarcely make so free: that if such a thing were fit in itself, I was sure you would do what you could for Liebigs sake that if not duty would come first. As far as a love for science might lead you I thought that Liebigs name ought to have more weight with you than mine.

At first I thought I would say nothing to you about it, but on consideration concluded that I would let you know. To oblige Liebig would be to oblige myself - but other considerations must come before. Do not trouble yourself to answer this note[.]

Ever Yours | M. Faraday
Coll. Sykes | &c &c &c

1. Possibly Alexander William Williamson. See letter 2557.
2. Georg Liebig (1827–1903, Brock (1997), 316). He was appointed to be an Army doctor in India.

Letter 2556
Faraday to Thomas Andrews
14 August 1852
From the original in SM MS 350/1. f.7

Royal Institution | 14 August 1852

My dear Andrews

I look rather anxiously at my remembrancer as the time passes on & the 1st Septr. draws nigh, and the sanguine hopes I entertained in the distance fade much as the period approaches when I thought of enjoying myself with you & Mrs Andrews. There is no chance of my dear wife being with you for her inability to walk much increases and without the power of moving about rather freely there could be no ability to use the privileges of the meeting[1]. A walk even of a few hundred yards makes her require a little rest. Our medical friends think it is rheumatism, but Sir B Brodie & some others think it rather a deficiency of energy in the nerves of the limbs[.] She is very thankful for your kind invitation but dares not accept it[.] For myself my obstruction is just the old one and as I had to run away from Oxford[2] after three days so I fear it would be at Belfast[.] I find in myself an illustration of one of the chapters of Dr. Hollands late volume on physiological subjects namely that on the *time* essentially required in mental operations[3][.] That time is now with me considerable in proportion to what it was naturally & the consequence is that I can only hold my way in a quiet progression of things[.] When I am involved in rapid changes of thoughts or persons then I have to use extra exertion mentally & then confusion & giddiness comes on. All this I forget when I have been in the country for a few weeks doing nothing, & then I think myself as able as ever to race with others. But having come home and gone to work upon oxygen & a magnetic torsion balance for a little while[4] I find the old warning coming on & I have to suspend my occupation. Formerly I thought that to enter into such a thing as the Association meeting would be rest in comparison but found that it was equivalent to work & that under pressure, and so may not look to my visit to Belfast in that point of view.

However I mean to do the best I can between this and the end of the Month, and have not yet given up hopes of seeing you[.] Do not suppose however that the association would be so chief a reason for my coming as the earnest desire to thank you for your kindness by enjoying it; and when I think of giving up Belfast my regret is to lose the pleasure of seeing you & Mrs. Andrews at home[.] Remember that if I come any closet or corner will do for me to sleep in[.] I will write again in the course of 12 or 14 days[.]

Ever My dear Andrews | Affectionately Yours | M. Faraday

1. That is the meeting of the British Association in Belfast.
2. In 1847.
3. Holland (1852), chapter 4.
4. See Faraday, *Diary*, 28 July to 13 August 1852, 6: 12034–12223.

Letter 2557
Faraday to Justus Liebig
14 August 1852
From the original in Liebig Museum, Giessen
Royal Institution | 14 August 1852

My dear Liebig
    I profit by the occasion of a visit of a friend of mine to Munich to send you a note by way of hearty remembrances. it seems strange to associate you with any place but Giessen: still I hope you will find a happy home in the great city and that you will now & then think of me there. I have been within its walls but now I forget it altogether & cannot recall it to my mind[1]. Mr. Blaikley[2] my friend whom I beg to introduce to you hopes to see its pictorial treasures well. Any aid which in that respect you can give him will be esteemed a great favour both by him & me. If you have heard from Dr Williamson lately he may have said something to you about Coll. Sykes. I wrote to Coll. Sykes a week or two ago[3] & his answer was very kind in respect of both the names of Liebig & Faraday.
Ever My dear Friend | Yours Truly | M. Faraday
My wife's kind remembrances. She often speaks of York[4].

Address: Baron Liebig | &c &c &c | Munich

1. In 1814. See Faraday to Margaret Faraday, 10 November 1814 and Faraday to Abbott, 26 and 30 November 1814, letters 38 and 40, volume 1.
2. Alexander Blaikley (1816–1903, Cantor (1991a), 299). Sandemanian and artist.
3. Letter 2555.
4. A reference to the York meeting of the British Association. See Liebig to Faraday, 19 December 1844, letter 1660, volume 3.

Letter 2558
Faraday to George Venables Vernon[1]
22 August 1852
From the original in JRULM Unitarian A2 Woodhouse Collection
Royal Institution | 22 Aug 1852
Sir

As regards Electrical observations I suppose Peltiers instrument as used by Quetelet[2] would suit your purpose best. They are made by Mr Becker[3] whose card I enclose and he could inform you of their price & where you could find instructions how to use it.

I do not know of any portable magnetometer that would be of any use in determining the magnetical elements. At the same time I should say I have not given my attention to the practice of observation or the use of the instrument & am not aware of all the varieties[.]

I am Sir | Your Very Obedient Servant | M. Faraday
G.V. Vernon Esq

1. George Venables Vernon (1831–1878, P3). Meteorologist in Manchester.
2. Peltier, J.C.A. (1842). On this see *Athenaeum*, 9 February 1850, pp.161-2 for an account of Faraday's Friday Evening Discourse of 1 February 1850 "On the Electricity of the Air".
3. Carl Ludwig Christian Becker (1821–1875, *Month.Not.Roy.Ast.Soc.*, 1876, **36**: 136–7). German born scientific instrument maker who moved to London in 1849.

Letter 2559
John Somerset Pakington to Faraday
22 August 1852
From the original in IEE MS SC 2

Westwood Park | Driotwich | Aug 22 / 52

Dear Sir,

In the storm last Tuesday[1] an oak tree in this park of large size was destroyed by lightning in a very extraordinary manner and it now presents the most wonderful illustration of the power of lightning that I ever saw.

If you think it worth your while to come to me here to look at it, I shall be very happy to receive you.

I shall be at home tomorrow, and Tuesday, and Saturday this week[2].

Your best train is via Birmingham & thence by Worcester line to the Driotwich Station - train at 10a.m. brings you here at 3 - train at $5\frac{1}{4}$ p.m. brings you home at 9.

I beg to remain | faithfully yours | John S. Pakington
M. Faraday Esq.

1. That is 17 August 1852.
2. That is 23, 24 and 28 August 1852.

Letter 2560
Faraday to Johann Rudolf Wolf
27 August 1852
From Wolf (1852c), 16

I am very greatly obliged and delighted by your kindness in speaking to me[1] of your most remarkable enquiry, regarding the relation existing between the condition of the Sun and the condition of the Earths Magnetism. The discovery of periods and the observation of their accordance in different parts of the great system, of which we make a portion, seem to be one of the most promising methods of touching the great subject of terrestrial magnetism. The power is wonderful;- and the whole problem set before philosophers, very complicated; - whilst at the same time our opportuniti[e]s of access to the power are very few and imperfect:- for, what is going on in the bowels of the earth who can tell. But we have first to discriminate amongst the great things which can or do, affect the power; and knowing, first of all that much concerning its seet [sic] must be within the earth, it is then a great advance to know that the Sun has also much to do with it;- and again (as I beli[e]ve fully) that the atmosphere by its oxygen has a great deal to do with it also. These division though at first imperfect, are the beginning of the analysis of the power; by which kind of analysis we can alone hope ultimat[e]ly to understand its nature and natural arrangement and variations.

1. See letter 2554.

Letter 2561
Jacob Herbert to Faraday
27 August 1852
From the original in GL MS 30108/1/54

Trinity House | 27, Aug: 52

My dear Sir,

It is the intention of the Board in the ensuing Spring to improve the Lights on the Casket Rocks near Alderney, by enlarging the Lanterns, giving them an encreased elevation of 20 Feet, and very probably replacing the present Reflecting Apparatus by Lenses, - and the needful preparations will progress here during the Winter that the Work may be commenced at the proper time next Year.

The Lights burn at equal Heights above the Level of the Sea in 3 distinct Lanterns and are revolving, - the arrangement is the same in each, viz 8 Lamps & parabolic Reflectors placed on a Circle, - the period of revolution bring 2 minutes, - & the present range of the Lights 15 miles - the additional Height will add something more than a mile to that range.

It is important that the present character or appearance of the Lights should remain unchanged, - and the Deputy Master[1] has instructed me to say that he will be obliged if you will consider how this may be best effected by dioptric apparatus of the Second Order & favor him at any time convenient to yourself by calling here & talking the matter over with him.

Believe me, | My dear Sir, | Very faithfully Yours | J. Herbert
M: Faraday Esq.
I think it will be desirable that you should let me know a day or two before hand, when you propose to come. | JH

1. John Shepherd.

Letter 2562
Christian Friedrich Schoenbein to Faraday
29 August 1852
From the original in UB MS NS 402
My dear Faraday!

To give you a sign of life I write these lines quite in a hurry. They will be delivered to you by the kindness of our mutual friend Dr. Whewell. Your last letter[1] shall be answered at a more convenient time and so as it merits, for your friend is in this present moment not in his writing-mood. He has continued to ride his hobby-horse and found out different little things. If you have got a friend knowing german, he will perhaps give you the substance of the papers, I have published in Erdmann's[2] Journal for practical Chemistry[3].

Tuas literas expectabo, quum ut, quid agas, tum, ubi sis sciam, et cura, ut omnia sciam, sed maxime ut valeas. Tuae uxori carissimae salutem[4]

C.F.S.
Bâle Aug. 29, 1852.

Address: Doctor M. Faraday | &c &c &c | Royal Institution | Albemarle street | London

1. Letter 2534.
2. Otto Linné Erdmann (1804–1869, DSB). Professor of Chemistry at Leipzig and editor of *J.Prak.Chem.*
3. Schoenbein (1852b, c, d, e, f, g, h).

4. "I shall await your letter, in order that I may know what you are doing and where you are, take care that I know everything, but most of all that you are well. I hope your wife is well".

Letter 2563
Faraday to Thomas Andrews
30 August 1852
From the original in SM MS 350/1, f.9

Royal Institution | 30 August 1852

My dear Andrews

When I received your last letter[1] & the instructions from the Secretaries as to the Routes[2], I made up my mind to leave London on the Sabbath Eveg[3] &, passing by Dublin, to be with you on the Monday evening, for the rest of the time. But since that I have been so weak & depressed as to rejoice that I was not with you in that state, as I should have been a mere burden; & shackle on your kind exertions to others:- indeed home & even the bed room were the only fit places. I do not know whether it was the season or not - I cannot tell - it was one of the old time depressions[.] I am better now but not in a fit state for the excitement of the meeting. I may go on improving & if I feel able enough mean to fulfil the intentions:- still I hardly expect it;- I could not come as I am; and I do not know which way matters will go this week. I shall be sorry, & much disappointed, to be away; but I ought not to complain, and when I do so, & murmuring thoughts arise, it seems to me that I am very ungrateful to a kind providence.

Wishing you a happy and prosperous meeting in every case and with most earnest thanks to Mrs. Andrews & yourself for your great kindness I am

My dear Andrews | Yours gratefully | M. Faraday

1. Not found.
2. That is to the Belfast meeting of the British Association.
3. That is 5 September 1852.

Letter 2564
Faraday to Henry Stevens[1]
1 September 1852
From the original in UCLA UL Henry Stevens collection #801, Box 39, folder 6

Hampstead | 1 Septr. 1852

Sir
  I beg to acknowledge with many thanks your kindness in forwarding me Vols II, III, & IV of the Smithsonian Contributions. I have just sent some papers[2] to the United States or I would have intruded on your kindness for conveyance[.]
  I have the honor | to be Your Obliged Servant | M. Faraday
Henry Stevens Esq | &c &c &c

Address: Henry Stevens Esq | &c &c &c | Morleys Hotel | Trafalgar Square | London

1. Henry Stevens (1819–1886, B3, DAB). American born bookdealer who worked in London from 1845.
2. Probably Faraday (1852b, c, d), ERE28, 29 and [29a].

Letter 2565
Jacob Herbert to Faraday
1 September 1852
From the original in GL MS 30108/1/54

Trinity House | 1, Sept: 52

My dear Sir,
  I thank you for your note of yesterday[1] - and now send you an extract from the French official publication of Light Houses in France[2], which as it describes the practical Operation of a dioptric Light of the Second Order revolving, may be of some use in your Consideration of the subject of the Caskets.
  Upon the whole perhaps it may be better that you should make the proposed enquiry of the Construction of the dioptric Lights in Paris, - and, if you see no objection, perhaps you will do so.
  The Deputy Master[3], is disposed to think that, if there be any doubt of preserving the present appearance of the Lights, by substituting a dioptric arrangement instead of the present circle of reflectors, - that it may be as well to consider whether it would be advisable to encrease the power of the Light, by having two circles of reflectors in each Lantern, - they

being placed vertical, so that the axis of two shall be precisely on the same Line.

I remain, | My dear Sir, | Very faithfully yours | J. Herbert
M: Faraday Esq

1. That is the reply (not found) to letter 2561.
2. This paper is in GL MS 30108/1/54.
3. John Shepherd.

Letter 2566
Jacob Herbert to Faraday
1 September 1852
From the original in GL MS 30108/1/54 $\frac{1}{2}$

Trinity House | 1, Sept. 52.

My dear Sir,

The enclosed papers[1] will I think explain the Nature of an alteration which our Committee have adopted in the burning of the Concentric Wick Lamps, and which is consider'd an improvement as respects the power of the Light from the Lamp, and in other respects - but I think it has not yet been brought under your notice.

It has been suggested that notice should be taken of it in my Letter to Col: Aspinwall[2] - but as it reduces the number of Wicks by the discontinuance of the Central or $\frac{7}{8}$ths Inch Burner, - I have been unwilling to do so without first submitting the matter to you; - and I have therefore been instructed to make this Communication to you, and to request to be favor'd with any observations which may occur to you on this point.

Mr. Wilkins will attend you, if you desire him, - and give you any further information which you may require[3].

I remain, | My dear Sir, | Always faithfully yours | J. Herbert
M: Faraday Esq

1. These papers are in GL MS 30108/1/54 $\frac{1}{2}$.
2. Thomas Aspinwall (1786–1876, Smith, C.C. (1891)). United States Consul in London, 1816–1854.
3. Faraday's reply of 3 September 1852 was read to Trinity House Court, 7 September 1852, GL MS 30004/25, p.139 which reported Faraday's approval of the plan.

Letter 2567
John Somerset Pakington to Faraday
2 September 1852
From the original in SM MS 350/1, f.95

Westwood Park | Septr. 2 / 52

My dear Sir,

Lord Lyttelton[1] & other friends who are now staying here, agree with me in thinking that for the sake of Science it is very desirable that either you or some other Scientific man of eminence should see the action of the lightning on the tree about which I wrote to you[2].

Though perhaps you might not think it so extraordinary as we do - I should mention that what we consider is wonderful is the *explosive action from below*, which can, apparently, alone account for the manner in which the tree was *blown out* of the ground roots & all, leaving a socket or cup where it stood - one portion of the stem *with the earth* attached being moved 3 or 4 *yards* from the socket - any one leaving London by an early train (via Birmingham) might return by the evening Express. I mention this is case you should like to send anybody else.

Yours very faithfully | John S. Pakington
M. Faraday Esq.

1. George William, 4th Baron Lyttelton (1817–1876, DNB). Principal of Queen's College, Birmingham.
2. Letter 2559.

Letter 2568
Faraday to Thomas Andrews
6 September 1852
From the original in SM MS 350/1, f.78

Hampstead | 6 Septr. 1852

My dear Andrews

I am sorry & glad. Sorry not to be with you[1] but glad I am here at home - for my head aches enough to make me know I am right in remaining quiet[.]

Give my kindest thanks to Graham for his persuasive note[.] It wanted nothing to be added to your own words if I could have come[.]

I enclose a letter from Sir John Pakington[2][.] I cannot go to him as I cannot go to you but I send it hoping that some one will call on their way back from Belfast[.] I have written to him to say so & he would be very glad to see any one[.]

Ever My dear friend | Yours | M. Faraday

Address: Dr. Andrews | &c &c &c | Queen's College | Belfast | Ireland

1. At the Belfast meeting of the British Association.
2. Letter 2567.

Letter 2569
Sarah Faraday to Thomas Andrews
6 September 1852
From the original in SM MS 350/1, f.9
My dear Dr. Andrews

I have delayed answering your kind & cordial invitation[1] till now, hoping that my dear Husband would be able to accept it though I could not; but he has been so depressed & unable to bear a part, even in casual conversation that I felt at last I could not urge him to undertake such a journey & excitement without me, when he himself felt so unfit for it, and we have reluctantly given it up[.]

Now though I am disappointed in not being able to be with you at the British Association I still hope we may see you & Mrs. Andrews in your hospitable home some future time & this notwithstanding my philosophical husband tells me that I must not expect to be strong again, that old age is coming upon us &c &c &c.

I can bear travelling though unable to walk, and at another time I might not be such a burden to friends as I should on such an occasion as the present[.]

With our best thanks to Mrs Andrews for all her intended kindness[.]

I am my dear Sir | Yours very sincerely | S. Faraday
Hampstead | Sept 6th 1852
Dr. Andrews | &c &c &c

1. To attend the Belfast meeting of the British Association. See letter 2511.

Letter 2570
Jacob Herbert to Faraday
8 September 1852
From the original in GL MS 30108/1/54

Trinity House | 8 Septem. 1852.

My dear Sir,

In reply to the latter paragraph of your note, dated 3d Inst[1]: - in relation to the Burner of the Dioptric Lights, - I beg to state, that the Addresses of the manufacturers of Dioptric Lenses in Paris are Mons. H. Lepaute[2]
    No 247, Rue St. Honoré &
    The successor of
M. Létourneau[3]
    No. 37, Allée des Veuves
        Champs Elysées
    I am | My dear Sir | Your very faithful Servant | J. Herbert
M. Faraday Esq | &c &c &c

1. Not found but presumably the reply to letter 2565. See also note 3, letter 2566.
2. Augustin-Michel Henry Lepaute (b.1800, Glaeser (1878), 446–7). Clockmaker and manufacturer of lighthouse equipment in Paris.
3. Létourneau. According to Paris POD a manufacturer of lighthouse equipment at avenue Montaigne 37.

Letter 2571
John Barlow to Faraday
13 September 1852[1]
From the original in IEE MS SC 2

Paris | 7 Rue 29 Juillet | Septr. 13h

My dear Faraday

Since I wrote to you we have gone over much space, & have had many things to think about, recalling associations with yourself... From Berne we went to Interlaken, where I found your letter at the Poste Restante. We remained at Interlaken 3 weeks, and had $2\frac{3}{4}$ days of available weather. These small morsels of time we devoted to the Wengen Alp Reichenbach & Rosenlaui. From Interlaken we went to Vevay [sic] (& thence to Geneva) (by the Simmenthal). The weather was just beginning to clear up: luckily one is not so dependent on the clearness of the atmosphere for valley as for mountain scenery.

At Geneva we were very kindly received by De la Rive Marcet &c and enjoyed ourselves for 11 days getting many glimpses of Mont Blanc, and witnessing De la Rive's phenomenon of the second pink which comes over the mountain, on a fine evening[2]. After eleven agreeable days we quitted Geneva, & travelled post hither, coming over the Jura to Dijon... We have taken lodgings for a month and I hope that we may meet at the beginning of October...

While at Berne I was invited to attend the meeting of a philosophical Society. I was not able to do so. I regretted this afterwards as Wolf? read two papers on interesting subjects.
1. Proving from a series of observations a connection between magnetic variations and the spots on the sun's disc[3].
2 Describing a method for making platinum malleable at a very little cost.
If there really is any thing in these communications, they are very important. You will of course hear more about them if you have not done so already.

The political state of Switzerland is any thing but satisfactory. The 'Black' & 'White' parties (i.e Conservatives & Radicals) abuse each other vehemently, but we cannot easily see what is the point of real principle at issue. The Black Conservative denounces the White Radical as a source of all sort of moral corruption - imputes to him the increase of drunkenness, & the growing depravity des moeurs. The White rejoins by impeaching the Black as a conspirator against the rights of the poor. One effect of the triumphs of the White party at Geneva is sufficiently deplorable. They have, as you well know, ostracized the De la Rives Marcets &c. I see that all the old families regard Geneva as no longer the place of their interests. And yet notwithstanding that the intellectual Glory of the University has departed, the material prosperity of that town never could have been greater than it is now. One cannot help thinking what a bribe it offers every day to France or to its starving neig[h]bour, Savoy.... Marcet told me that, in the month of August, 6 inches of rain ($\frac{1}{5}$th of the average of the year) had fallen at Geneva[4].

Delarive is going to write to you on one point in your recent researches on which he is not sure that he has got your exact thoughts. I told him that I had heard from you, and that you were working satisfactorily - by the bye, in reference to another sentence in your letter, Bence Jones & Dubois Reymond are to meet at the Hotel des Bergues at Geneva on the 15th.

I met Dr. Webster[5] in the street on Friday. He is full of the sea-serpent, which he declares that he & half a dozen other persons saw, going at the rate of 10 miles in 2 minutes, between Dieppe & Newhaven. He is ready to make any amount of declarations that it was no porpoise, grampus or anything of the kind. He has written to the Times, & will write to the Royal Society - so let Owen beware[6].

I found luckily two very good examples of your *"filings"*[7] in my blotting book. I gave one to De la Rive, & the other to Dumas, with whom I dined yesterday.. He supports his station very elegantly. His son[8] is Directeur de la Monnaie at Rouen.

And now good bye. If you can find 5 minutes to tell me of your welfare & that of Mrs. Faraday I shall be thankful - so will my wife. Miss Grant offers her best remembrances[.]

Ever yours | John Barlow

1. Dated on the basis of the reference to the weather and to Wolf's paper.
2. See De La Rive (1839).
3. Wolf (1852a).
4. See *Bibl.Univ.Arch.*, 1852, **21**: 82 for details of the August rainfall in Switzerland.
5. John Webster (1795–1876, Munk (1878), **3**: 233). Scottish physician in London specialising in medical institutions and sometime a Manager of the Royal Institution.
6. For Owen's view of sea monsters see Rupke (1994), 324–32.
7. That is Faraday's iron filing diagrams. See letters 2475 and 2494.
8. Ernest-Charles-Jean-Baptiste Dumas.

Letter 2572
Henry Clutterbuck[1] to Faraday
15 September 1852
From the original in RI MS F1 H34
My dear Sir

Allow me to begin my note, with expressing the pleasure I feel in hoping that it leaves you in good Health & strength.

I reply to your query respecting Mr Monpriut[2] I can only say I remember him as an old pupil, but I have not had the pleasure of knowing more of him. Very shortly after your note reached me, I received the enclosed from that gentleman, and which I send you for your perusal.

Believe me, my dear | Sir, one of your very | sincere well-wishers & admirers | H. Clutterbuck
New Bridge Street | Sept 15, 1852.

Address: Mr Faraday.

1. Henry Clutterbuck (1767–1856, DNB). Physician and medical writer.
2. Unidentified.

Letter 2573
Faraday to John Scoffern[1]
22 September 1852
From the original in SI D MS 554A

Royal Institution | 22 Septr 1852

My dear Sir

I beg to thank you most sincerely for the copy of your work on Gold[2] and for the kind feeling which has prompted you to put my name in it.

I am Very Truly Yours | M. Faraday

Dr Scoffern

Let me have my lecture notes as soon as you have done with them[3] | MF

1. John Scoffern (1814–1882, B3). Physician, chemist and scientific writer.
2. Presumably a reference to an autograph dedication in an advance copy of Scoffern and Higgins (1853) which does not mention Faraday by name in the text. However, the calculations of the amount of oxygen in the world on p.82 are taken directly, and without explicit acknowledgement, from Faraday (1853b), 113–4.
3. That is Faraday's notes (RI MS F4 J13) of his lectures on the non-metallic elements which he had delivered after Easter and from which Scoffern edited Faraday (1853b).

Letter 2574
Faraday to Jacob Herbert
27 September 1852
From Symons (1882), 187–89

Royal Institution, | 27th September, 1852.

My dear Sir,

I fortunately reached the Nash Low Lighthouse[1] last Thursday[2], before any repairs were made of the injury caused by the discharge of lightning there, and found everything as it had been left: the repairs were to be commenced on the morrow.

The night of Monday, 30th August, was exceedingly stormy, with thunder and lightning; the discharge upon the lighthouse was at six o'clock in the morning of the 31st, just after the keeper had gone to bed. At the same time, or at least in the same storm, the flagstaff between the upper and lower lights was struck, and some corn stacks were struck and fired in the neighbourhood. It is manifest that the discharge upon the tower was exceedingly powerful, but the lightning conductor has done duty well - has, I have no doubt, saved the building; and the injury is comparatively slight, and is referable almost entirely to circumstances which are guarded against in the report made by myself and Mr. Walker 22nd September, 1843[3].

The conductor is made fast to the metal of the lantern, descends on the inside of the tower to the level of the ground, and passes through the wall and under the flag pavement which surrounds the tower. It is undisturbed everywhere, but there are signs of oxidation on the metal and the wall at a place where two lengths of copper are rivetted together, which show how great an amount of electricity it has carried.

A water-butt stands in the gallery outside the lantern. A small copper pipe, 1 inch in diameter, brings the water from the roof of the lantern into this butt; it does not reach it, but terminates 10 or 12 inches above it. A similar copper pipe conducts the surplus water from the butt to the ground, but it is not connected metallically with the other pipe, or with the metal of the conductor, or the lantern. Hence a part of the lightning which has fallen upon the lantern has passed as a flash, or, as we express it, by disruptive discharge from the outside of the lantern to this tub of water, throwing off a portion of the cement at the place, and has used this pipe as a lightning conductor in the rest of its course to the ground. The pipe has holes made in it in three places, but these are at the three joints, where, it being in different lengths, it is put together with tow and white lead, and where of course the metallic contact is again absent; and thus the injury there (which is very small) is accounted for. The pipe ends below at the level of the ground in a small drain, and at this end a disruptive discharge has (naturally) occurred, which has blown up a little of the cement that covered the place. Some earth is thrown up at the outer edge of the pavement round the tower over the small drain, which tends to show how intense the discharge must have been over the whole of the place.

Instead of the lantern there are traces of the lightning, occurring at places where pieces of metal came near together but did not touch, thus at the platform where a covering copper plate came near to the top of the stair railing, but the effects are very slight. All the lamps, ventilating tubes, &c., remained perfectly undisturbed, and there was no trace of injury or effect where the conductor and the lantern were united.

Inside of the tower and the rooms through which the conductor passes there were and are no signs of anything (except at the rivetting above mentioned) until we reach the kitchen or living-room which is on a level with the ground, and here the chair was broken and the carpets and oil-cloth fired and torn. To understand this, it must be known that the separation between this room and the oil-cellar beneath is made by masonry consisting of large stones, the vertical joints of which are leaded throughout, so that the lead appears as a network upon the surface, both of the kitchen floor above, and the roof of the oil cellar beneath, varying in thickness in different places up to $\frac{1}{3}$ or more of an inch, as in a piece that was thrown out. The nearest part of this lead to the conductor is about 9 inches or a little more distant, and it was here that the skirting was thrown off, and the chair broken; here also that the fender was upset and the little cupboard against the skirting emptied of its articles. If this lead had been connected metallically with the conductor, these effects would not have happened.

The electricity which in its tendency to pass to the earth took this course, naturally appeared in the oil-cellar beneath, and though the greater portion of it was dissipated through the building itself, yet a part appeared in its effects to have been directed by the oil cans, for though they were not at all injured or disturbed, the wash or colour in the wall above four or five of them was disturbed, showing that slight disruptive connections or sparks had occurred there.

At the time of the shock, rain was descending in floods, and the side of the tower and the pavement was covered with a coat of water. This being a good conductor of electricity has shown its effects in connection with the intense force of the discharge. A part of the electricity leaving the conductor at the edge of the pavement and the tower, broke up the cement there, in its way to the water on the surface, which for the time acted to it as the sheet of copper - which I conclude is at the end of the conductor - does, *i.e.*, as a final discharge to the earth. Also on different parts of the external surface of the tower near the ground, portions of cement, the size of half a hand, have been thrown off by the disruptive discharges from the body of the tower to this coat of water: all testifying to the intensity of the shock.

I should state that the keeper says he was thrown out of bed by the shock. However, no trace of lightning appears in the bedroom, still there are evidences that powerful discharges passing at a distance, and on the other side of the thick walls may affect bodies and living systems, especially by spasmodic action, and something of the kind may have occurred here. It may be as well for me to state that the upper floors are *leaded* together like that of the kitchen. The reason why they did not produce like effect is evident in that they from their position could not serve as conductors to the earth as the lower course could.

The keeper said he had told the coppersmith to make the necessary repairs in the pipe, and I instructed him to connect the waste pipe and the upper pipe by a flat strap of copper plate. I would recommend that the lead of the lower floor be connected metallically with the conductor to a plate of copper in the earth. I could not see the end of the present conductor, not being able by any tools at the lighthouse to raise the stonework, but I left instructions with the keeper to have it done, and report to me the state of matters.

I am, &c., | (signed) M. Faraday
The Secretary, | Trinity House.

1. In Glamorganshire.

2. That is 23 September 1852.
3. Faraday to Herbert, 25 September 1843, letter 1526, volume 3.

Letter 2575
Faraday to George Biddell Airy
4 October 1852
From the original in RGO6 / 694, f.63

Hampstead | 4 Octr. 1852

My dear Airy

I am greatly obliged by the volumes I have received[1]. I long to look at them but whether I shall ever do so with understanding I cannot tell. I have much work to do in making my torsion balance right, and the power itself even in the most stable cases is such a restless fidgetly thing that it is difficulty [sic] to catch wholly & rightly though it ever presses definite in the end if well pursued. I turn many sometimes in my present course but I have learned to know that patient perseverance in experimental researches ever leads to some good or another and therefore I resist the desire I have to turn away to some other point before the one I have taken be brought to something like a resting place indicative[.]

With kindest respects to Mrs & Miss Airy I am | Most faithfully Yours | M. Faraday

1. Presumably the volumes referred to by Airy in letter 2535.

Letter 2576
Justus Liebig to Faraday
6 October 1852
From the original in RI MS F1 I48

Giessen 6 Oct. 1852

My dear Faraday

Perhaps you know already, that my son[1] has been fortunate enough to obtain the appointment as Assistant Surgeon in the Indian Army, which I had endeavoured to gain for him. Colonel Sykes wrote me that he would appoint him if he passes the necessary examinations.

To you chiefly my dear Faraday, I am no doubt indebted for the fulfilment of this favourite wish of mine and I find it difficult to say how sensible I am of the proof you have thus given me of your friendship by using your interest in behalf of my son on this occasion[2]. Never will I forget this service.

Hoping that you will continue in your kindness towards my son I remain with all my heart
Yours very sincerely | Dr Justus Liebig
M. Faraday Esq. | London.

1. Georg Liebig (1827–1903, Brock (1997), 316). Army doctor in India.
2. See letter 2555.

Letter 2577
Faraday to Arthur-Auguste De La Rive
16 October 1852
From the original in BPUG MS 2316, f.73–4

Royal Institution | 16 Octr. 1852

My dear De la Rive

From day to day and week to week I put off writing to you, just because I do not feel spirit enough; not that I am dull or low in mind, but I am as it were becoming torpid:- a very natural consequence of that kind of mental fogginess which is the inevitable consequence of a gradually failing memory. I often wonder to think of the different courses (naturally) of different individuals, and how they are brought on their way to the end of this life. Some with minds that grow brighter & brighter but their physical powers fail; as in our friend Arago, of whom I have heard very lately by a nephew[1] who saw him on the same day *in bed* & at *the Academy*: such is his indomitable spirit. Others fail in mind first, whilst the body remains strong. Others again fail in both together; and others fail partially in some faculty or portion of the mental powers, of the importance of which they were hardly conscious until it failed them. One may, in one's course through life, distinguish numerous cases of these and other natures; and it is very interesting to observe the influence of the respective circumstances upon the characters of the parties and in what way these circumstances bear upon their happiness. It may seem very trite to say that *content* appears to me to be the great compensation for these various cases of natural change; and yet it is forced upon me, as a piece of knowledge that I have ever to call afresh to mind, both by my own spontaneous & unconsidered desires and by what I see in others. No remaining gifts though of the highest kind; no grateful remembrance of those which we have had, suffice to make us willingly content under the sense of the removal of the least of those which we have been conscious of. I wonder why I write all this to you: Believe me it is only because some expressions of yours at different times make me esteem you as a thoughtful man & a true friend:- I often have to call such things to remembrance in the course of my own self examination and I think they make me happier. Do not for a moment suppose that I am unhappy. I am occasionally dull in spirits but not unhappy, there is a hope which is an abundantly sufficient remedy for that, and as that hope does not depend on ourselves, I am bold enough to rejoice in that I may have it.

I do not talk to you about philosophy for I forget it all too fast to make it easy to talk about. When I have a thought worth sending you it is in the shape of a paper before it is worth speaking of; and after that it is astonishing how fast I forget it again; so that I ⟨have⟩ to read up again & again my own recent communications and may well fear that as regards others I do not do them justice. However I try to avoid such subjects as other philosophers are working at; and for that reason have nothing important in hand just now. I have been working hard but nothing of value has come of it[.]

Let me rejoice with you in the Marriage of Your daughter[2]. I trust it *will be* as I have no doubt it *has been* a source of great happiness to you. Your Son[3] too whenever I see him makes me think of the joy he will be to you. May you long be blessed in your children and in all the things which make a man truly happy; even in this life. Ever My dear friend, Yours Affectionately, M. Faraday

Address: A Monsieur | Auguste De la Rive | &c &c &c &c | Presenge | Geneva | Switzerland.

1. Frank Barnard. See letter 2589.
2. Jeanne-Adèle Tronchin, née De La Rive (1829–1895, Choisy (1947), 51). Married 1852.
3. William De La Rive.

Letter 2578
Christian Friedrich Schoenbein to Faraday
17 October 1852
From the original in UB MS NS 403
My dear Faraday

I trust you received in due time the letter I sent you through Dr. Whewell some months ago[1]. Now I avail myself of a friend going to London to forward to you a paper of mine[2], which I hope will not remain a sealed book to you. If you should feel curious to decipher that whimsical letter I once wrote you about oxigen[3], get the memoir translated by some friend of your's and you will perhaps be interested in the matter as it regards some of your most important discoveries.

Entertaining the notion that in many if not in all cases the color exhibited by oxycompounds is due to the oxigen contained in them, or to express myself more distinctly, to a peculiar chemical condition of that body, I have continued my researches on the subject and obtained a number of results which I do not hesitate to call highly curious and striking. Far be it from me to think, on that account, my hypothesis correct and proved; but the fact is that I owe the discovery of a number of remarkable phenomena solely and exclusively to the conjecture mentioned. I am nearly sure that you will be pleased to repeat the experiments, for either by mere physical means or by chemical ones you may make and unmake or change the color of certain substances without altering the chemical constitution of those matters. To my opinion, that wonder is performed by changing the chemical condition of the oxigen of the oxycompound.

I cannot help thinking that the colors of substances, which up to this present moment have been very slightly treated (in a chemical point of view) will one day become highly important to chemical science and be rendered the means to discover the most delicate and interesting changes taking place in the chemical condition of bodies. In more than one respect the color of bodies may be considered the most obvious "signatura rerum"[4], as the revealer of the most wonderful actions going on in the innermost recesses of substances, as the indicator of the most elementary functions of what we call ponderable matter. But alas! Whilst we are pleased with and wonder at that rich field of chromatic phenomena, which continually strikes our eye, we know as yet little or nothing of the connexion which certainly exists between the chemical nature of bodies and the influence it exerts upon light. We must try to dissipate that thick darkness which still hangs about and obscures the most luminous phenomena. Clearing up but the smallest part of that vastly important subject would be of more scientific value, I think, than discovering thousand and thousand new organic compounds, things which I cannot help considering in the same light as I do the infinite number of figures which may [be] produced by the caleidoscope.

What would the world say of a man, who should take the trouble to shake for whole years that plaything and de[s]cribe minutely all the shapes (pretty as they might be) he had obtained from his operation!

You know, I am no great admirer of the present state of Chemistry, and of the Ideas leading the researches made upon that field. Atoms, weight, ratio of quantities, endless production, and formula of compounds, i.e. the "caput mortuum"[5] of nature are the principal if not the only subjects with which the majority of our Chymists know to deal. Force power, action, life in fact, are as it were phantoms to them, disliked if not hated. The world being a system of Ideas, its very essence power and intellect, how can we expect great things from men who so much mistake the nature of nature.

In perusing what is written above I find it is not worth of being sent over the water, but having no more time to write another letter, you must take it as it is and excuse my random talking. Mrs. Schoenbein and the Children are well and beg to be kindly remembered to you. My best compliments to Mrs. Faraday and to you the assurance that I shall for ever remain

Yours | most truly | C.F. Schoenbein
Bâle Oct. 17, 1852.

1. Letter 2562.
2. Schoenbein (1852a).
3. Letter 2526.
4. A reference to the doctrine of signatures.
5. "A dead head". That is the residuum left by a process of chemical analysis.

Letter 2579
William Withey Gull[1] to Faraday[2]
22 October 1852
From the original in IEE MS SC 2

8 Finsbury Sq | Oct 22 1852

My dear Sir,

I am almost ashamed of the time I have kept the abstract of Du Bois Reymond's labours[3]. The book was laid on one side and overlooked. I am much obliged to you for its perusal.

Yours very faithfully | William W. Gull

1. William Withey Gull (1816–1890, DNB). Professor of Physiology and Comparative Anatomy at Guy's Hospital, 1846–1856.
2. Recipient established on the basis of provenance of this letter.
3. Bence Jones (1852).

Letter 2580
Faraday to Jacob Herbert
26 October 1852
From the original copy in GL MS 30108/1/54[1]

Royal Institution | 26 Octr 1852.

My dear Sir

In reference to the matter of the Caskets lights[2] I wrote some time since to M. Lepaute[3] for any experimental results he could supply me with as to the relative power of the beams of light sent forth by lenses of the 2nd order. He was so kind as to answer my letter, and stated that 2nd order lights, revolving in one minute, had 8 lenses, and those revolving in 30 seconds had 12 lenses; both kinds having the same radius of 70 centimeters:- but he did not give me their area or their power; both of which I had asked for. I wrote to him again for the latter data on the 27th Septr. but have received no answer. Probably he is not in possession of experimental comparisons of the beams from the two different sized lenses. I enclose his letter that you may see exactly what he says.

I therefore venture some considerations which are necessarily imperfect, since I have not had the opportunity of examining a 2nd order light with either 8 or 12 lenses: you will easily see how far they offer safe suggestions to the mind. The quantity of light evolved by combustion will be, with equal care, nearly as the oil consumed; as former experiments have shewn. I have tried to ascertain the quantity of oil burnt in a second order lamp, but Mr Wilkins has not been able to supply me with the result[.] I therefore take the proportion of cotton in the burners as my guide; and I find that the cotton in a 2nd order lamp is six times that in a reflector lamp, whilst the cotton in a 1st order lamp is ten times that of the reflector lamp. I think I have never seen the light of the first order lamp exceed that of fourteen argand lamps: and though the 2nd order lamp may not give so intense a light in proportion, as the 1st order, yet, taking that proportion, then its light will be eight times (or a little more) that of the Argand.

Of the light thus generated, probably $\frac{3}{4}$ falls on the surrounding lenses; but as I have not the area or measurement of the lenses I cannot be sure; the rest go off above & below:- indeed less than this must be intercepted by the lenses, because much that would proceed to the lower part of the lenses, is intercepted by the burner. About the same proportion of light is also received by the reflectors. The quantity of light extinguished respectively by the lenses and the reflectors, I do not know: it was in reference to that point that I wished to know the experimental results as to the power of the beams, but judging (from the comparison at Purfleet of a good reflector & a first order lens) that the loss was about equal: then it would seem probable that eight lamps and their reflectors, would give beams of *equal power* to a 2nd class light with eight lenses; or might probably surpass it. I have no doubt they would much surpass the same 2nd class light with twelve lenses. It is needless for me to say that sixteen lamps & reflectors would (in my opinion) far surpass any arrangement of the 2nd class lights[4].

I am My dear Sir | Most truly Yours
Jacob Herbert Esq | &c &c &c

1. This unsigned copy is not in Faraday's hand.
2. See letters 2561, 2565 and 2570.
3. Augustin-Michel Henry Lepaute (b.1800, Glaeser (1878), 446–7). Clockmaker and manufacturer of lighthouse equipment in Paris.
4. This letter was read to Trinity House Court, 2 November 1852, GL MS 30004/25, p.155. It was referred to the Deputy Master (John Shepherd), Wardens and Light Committee. It was read to the Trinity House Wardens Committee, 8 November 1852, GL MS 30021/21, pp.275–6. It was agreed that the Caskets light would be of twelve argand lamps and reflectors.

Letter 2581
Faraday to Anne Brodie[1]
28 October 1852
From the original in the possession of K.W. Vincentz
Dear Lady Brodie

My wife called in Savile Row to day, but she has not courage to write to you; I therefore do so knowing that you will excuse me for the objects sake. A case has come before us much recommended by circumstances; which is entirely fit for the Widows relief Society[2]; and believing that you belong to it, and if you have no other claim before you, would be glad to help it, I take the liberty of enclosing a note of the circumstances. Knowing your kindness I will waste no words in apologies but use them rather in thanks. My wife joins me in kindest remembrances to you in many acknowlegeds [sic] to Sir Benjamin for his consistent endeavours to do us both good[.]

Ever Dear Lady Brodie | Your faithful Servant | M. Faraday
28 Octr. 1852

Endorsement: To be kept as an autograph of Faraday

1. Anne Brodie, née Sellon (d.1861, age 64, *Gent.Mag.*, 1861, **11**: 218). Philanthropist and wife of Benjamin Collins Brodie whom she married in 1816 (see under his DNB entry).
2. That is the Society for the Relief of Widows and Orphans of Medical Men. Benjamin Collins Brodie was a Vice-President. See Taylor, J.L. (1988), 21.

Letter 2582
Faraday to Cecilia Anne Barlow
3 November 1852
From the original in RI MS F1 D15

R Institution | 3 Novr. 1852

My dear Mrs. Barlow

A Proxy which I had mentally assigned to you & Lady Shelley[1] (you are both I think in the same case) has been lost but the enclosed paper has been sent to me as a substitute and will serve instead[2][.]
Ever Yours | M. Faraday

1. Jane Shelley, olim St John, née Gibson (d.1899, age 79, GRO, B6 under Percy Florence Shelley). Writer and editor of the works of Percy Bysshe Shelley (1792–1822, DNB), poet.
2. Neither Barlow nor Shelley are listed as subscribers to the London Orphan Asylum on which see note 1, letter 2173.

Letter 2583
Henry Hart Milman[1] to Faraday
3 November 1852
From the original in RI MS Conybeare Album, f.26

Deanery, St Pauls, Nov 3.
My dear Professor Faraday,
The Dean and Chapter of St Pauls propose to offer a few seats at the Funeral of the Duke of Wellington[2] to gentlemen most eminent in letters and science. Your name of course, occurs among the very first. As these are personal tickets, you will oblige me by informing me whether you wish to be present.
Ever with great respect | faithfully &c | H.H. Milman
Professor Faraday

1. Henry Hart Milman (1791–1868, DNB). Dean of St Paul's, 1849–1868.
2. Arthur Wellesley, 1st Duke of Wellington (1769–1852, DNB). Field Marshal and politician. Prime Minister, 1828–1830. Master of Trinity House, 1837–1852, Chaplin [1850], 30. He died on 14 September 1852 and his funeral took place in St Paul's Cathedral on 18 November 1852.

Letter 2584
Faraday to Henry Hart Milman[1]
3 November 1852
From the original in WIHM MS FALF

Royal Institution | 3 Novr. 1852
My dear Dr Milman
I am surprized by and *very grateful* for your kindness most unexpected on my part but most acceptable as a mark of the good will of one whom I so highly respect. I shall be *unable* to avail myself of it and think it a duty to give you the earliest intimation that I can that I may

restore into your hands a prize of such value which you were willing to bestow on me[2].
Ever Most truly Yours | M. Faraday
The | Revd. Dr. Milman | &c &c &c

1. Henry Hart Milman (1791–1868, DNB). Dean of St Paul's, 1849–1868.
2. See letter 2583.

Letter 2585
William Crane Wilkins to Faraday
3 November 1852
From the original in GL MS 30108/1/54

[Wilkins's letterhead] 24 Long Acre | 3rd Novr 1852
Sir
Pray pardon my not having sent you the undermentioned particulars before[.]
I am | Sir respectfully | Your Obedient Servant | W.C. Wilkins
Consumption of Lamps[1]
[7.625 pints] Lamps with 4 Wicks 3qts $1\frac{1}{2}$ Pint & $\frac{1}{2}$ Gill
[7.875] ditto with breaker without a centre wick - 3qts $1\frac{1}{2}$ Pint & $1\frac{1}{2}$ Gill
[4.0] Lamps with 3 Wicks [2 quarts]
[0.4375] Argand Lamps - $1\frac{3}{4}$ Gills
The above consumption is for 8 hours -
M Faraday Esq | &c &c &c

1. Text in square brackets is in Faraday's hand.

Letter 2586
Johann Rudolf Wolf to Faraday
4 November 1852
From the original in IEE MS SC 2
Monsieur.
Je ne saurais Vous remercier d'une manière plus convenable de Votre lettre bienveillante du 27 Août passé[1], qu'en Vous indiquant les progrès que je viens de faire dans l'étude des relations entre les tâches solaires et la terre. Je terminerai sous peu un mémoire[2], qui les développera en détail, et je prendrai la liberté de Vous en présenter un exemplaire; mais en

attendant je Vous en doñe le résumé suivant. Mon mémoire se divisera en six parties:
Dans le premier chapitre je démontrerai, appuyé sur 16 époques différentes, établies pour le Minimum et le Maximum des taches solaires, que la dureé moyenne d'une période des taches solaires doit être fixée à
$$11{,}111 \pm 0{,}038 \text{ années}$$
de sorte que 9 périodes équivalent justement à un siècle.
Dans le second chapitre j'établirai que dans chaque siècle les añées
  0,00 11,11 22,22 33,33 44,44 55,56 66,67 77,78 88,89
correspondent à des Minimums des taches solaires. L'intervalle entre le Minimum et le Maximum suivant est variable; la moyenne en est de 5 années.
Le troisième chapitre contiendra l'énumération de toutes les observations des taches solaires depuis Fabricius[3] et Scheiner[4] jusqu'à Schwabe, continuellement mise en parallèle avec ma période. L'accord est surprenant.
Le quatrième chapitre établira des analogies remarquables entre les taches solaires et les étoiles variables, par les quelles on peut présumer une liaison intime entre ces phénomènes singuliers.
Dans le cinquième chapitre je démontrerai que ma période de 11,111 añées coincide encore plus exactement avec les variations en déclinaison magnétique que la période de $10\frac{1}{3}$ añées établie par Mr Lamont[5]. Les variations magnétiques suivent même les taches solaires non seulement dans leurs changemens réguliers, mais aussi dans toutes les petites irrégularités,- et je pense que cette dernière remarque suffira pour avoir prouvé définitivement cette relation importante.
Le sixième chapitre traitera d'une comparaison entre la période solaire et les indications météorologiques contenues dans une chronique Zuricoise sur les añées 1000–1800. Il en résulte (conformément aux idées de William Herschel[6]) que les années où les taches sont plus nombreuses sont aussi en général plus sèches et plus fertiles que les autres[7],- ces dernières au contraire plus humides et plus orageuses. Les aurores boréales et les tremblements de terre indiqués dans cette chronique, s'accumulent d'une manière frappante sur les années de taches.

Si ces petites découvertes Vous semblent avoir assez d'importance pour en faire le sujet d'une communication à la Société royale ou à la Société astronomique et surtout à Mr Herschel, je Vous prie d'en disposer, Vous m'obligeriez infiniment par une telle comunication.

Agréez, Monsieur, l'assurance de la plus haute considération de Votre très dévoué serviteur | Rodolphe Wolf | Directeur de l'Observatoire de Berne
Berne 4/11 1852.

Endorsement: 812 recd Nov 25, 1852 TB[8]. Letter from Mr Wolf Director of the Berne Observatory to Mr Faraday Dated 4th Novr 1852.
Address: Monsieur le Professeur Faraday | Institution royale | Londres

TRANSLATION
Sir.

I would not know how to thank you in a more fitting way for your kind letter of 27 August last[1], than to share with you the progress that I have just made in the study of the relationship between sunspots and the earth. I shall shortly finish a paper[2] which will develop this in detail and I shall take the liberty of presenting you with a copy; but in the meantime, I give you the following summary. My paper will be divided into six parts:

In the first section I shall show, based on sixteen different periods established for the Minimums and Maximums of sun spots that the average duration of a period of sun spots must be fixed at

$$11.111 \pm 0.038 \text{ years}$$

so that 9 periods are exactly equal to a century.

In the second section I shall establish that in each century the years

 0.00 11.11 22.22 33.33 44.44 55.56 66.67 77.78 88.89

correspond to the Minimums of sun spots. The interval between the Minimum and Maximum that follows is variable; the average is 5 years.

The third section will contain the enumeration of all the observations of sun spots from Fabricius[3] and Scheiner[4] to Schwabe, continually placed in parallel with my period. There is surprising agreement.

The fourth section will establish remarkable analogies between sun spots and variable stars, by which one can assume there is an intimate link between these singular phenomena.

In the fifth section I shall show that my period of 11.111 years coincides even more exactly with the variations in magnetic declination than the period of $10\frac{1}{3}$ years established by Mr Lamont[5]. The magnetic variations even follow the sun spots not only in their regular changes, but also in all their small irregularities, - and I think that this last remark will have been enough to prove for definite this important relationship.

The sixth section will consider a comparison between the solar period and the meteorological indications contained in a Zurich chronicle of the years 1000–1800. It transpires (in accordance with the ideas of William Herschel[6]) that the years when the spots are more numerous are also, in general, drier and more fertile than others[7],- the latter being wetter and more stormy. The aurora borealis and earthquakes indicated in this chronicle, occur in a striking way in the years of the spots.

If these little discoveries seem to you to have enough importance to make them the subject of a communication to the Royal Society or to the

Astronomical Society and above all to Mr Herschel, I beg you to do so,- You would oblige me infinitely by such a communication.

Please accept, Sir, the assurance of the most high consideration of your very devoted servant | Rodolphe Wolf | Director of the Observatoire de Berne.
Berne 4/11 1852.

1. Letter 2560.
2. Wolf (1852c).
3. Johannes Fabricius (1587–1615, NDB). German astronomer.
4. Christoph Scheiner (1573–1650, DSB). Jesuit astronomer.
5. Lamont (1852).
6. Frederick William Herschel (1738–1822, DSB). Astronomer and natural philosopher.
7. Herschel, F.W. (1801), 316–7.
8. Thomas Bell. This endorsement implies that this letter was sent to the Royal Society (and somehow found its way back to Faraday). No firm evidence has been found that it was ever read to the Royal Society. See, however, letter 2592.

Letter 2587
Faraday to John Frederick William Herschel
8 November 1852
From the original in RS MS HS 7.197

Royal Institution | 8 Novr. 1852

My dear Herschell

I send you a letter I have received from M. Wolf[1] as he especially desires. After you have seen it I will send it if you approve according to his wish to the Royal Society or to the Astronomical[2] if you think it better[.]

Ever faithfully Yours | M. Faraday

Endorsed by Herschel: Period of solar spots 9 Periods in 100 y or 11 y 11 Min to Min
The commencement 1800, &c. of each century is a *minimum*

1. Letter 2586.
2. See note 8, letter 2586 for evidence that letter 2586 was sent to the Royal Society. There is no indication one way or the other that it was sent to the Royal Astronomical Society. The topic was raised there but based on Wolf (1852d).

Letter 2588
Frederick Madan to Faraday
8 November 1852
From the original in GL MS 30108/1/54

Trinity House | Novr: 8th:/52

My dear Sir

We have a considerable difference of opinion here relative to the expressions in your letter of the 26th: ulto:[1] stating the conclusion you come to as to the power of a second order annular Lense. You say "it would seem probable that 8 lamps & their reflectors would give beams of equal power to a second class light with 8 lenses": or might probably surpass it. Now the Deputy Master[2] & some others understand you to mean from this, that one lamp & its reflector would be equal to one 2d order annular Lense, but this is so contrary to all I have hitherto heard, that I think they must misunderstand you. I have always heard that even Cookson's[3] 1st order Annular Lenses with the 4 wick lamps were about equal to *ten lamps* & Reflectors, & that the improved French Lense, which is 8 or 10 inches deeper, would be fully 10, or even 20 more; & from that I had concluded that a 2d order Lense would be equal to 6 or 8 Lamps & Reflectors. Pray let me know how this is. The area or measurement of the 2d order annular Lense will probably be 20 inches wide by 2F 6I deep, that is, of 8 to the circle, the same as the fixed Refractors. Those of 12 to the circle will of course be less wide.

Your's sincerely | Fredk Madan

The Light Apparatus at the Caskets at present consists of 8 Reflectors in 8 faces, that is, only one Reflector on a face.
M Faraday Esq | &c &c

1. Letter 2580.
2. John Shepherd.
3. William Isaac Cookson (1812–1888, Morris et al. (1988), 11). Newcastle glass manufacturer.

Letter 2589
Faraday to Frank Barnard
9 November 1852
From Gladstone (1874), 51–2

Royal Institution, 9th Nov., 1852.

My dear Nephew,

Though I am not a letter-writer and shall not profess to send you any news, yet I intend to waste your time with one sheet of paper: first to thank you for your letter to me, and then to thank you for what I hear of your letters to others. You were very kind to take the trouble of executing my commissions, when I know your heart was bent upon the entrance to your studies. Your account of M. Arago was most interesting to me[1], though I should have been glad if in the matter of health you could have made it better. He has a wonderful mind and spirit. And so you are hard at work, and somewhat embarrassed by your position: but no man can do just as he likes, and in many things he has to give way, and may do so honourably, provided he preserve his self-respect. Never, my dear Frank, lose that, whatever may be the alternative. Let no one tempt you to it; for nothing can be expedient that is not right; and though some of your companions may tease you at first, they will respect you for your consistency in the end; and if they pretend not to do so, it is of no consequence. However, I trust the hardest part of your probation is over, for the earliest is usually the hardest; and that you know how to take all things quietly. Happily for you, there is nothing in your pursuit which need embarrass you in Paris. I think you never cared for home politics, so that those of another country are not likely to occupy your attention, and a stranger can be but a very poor judge of a new people and their requisites[2].

I think all your family are pretty well, but I know you will hear all the news from your appointed correspondent Jane, and, as I said, I am unable to chronicle anything. Still, I am always very glad to hear how you are going on, and have a sight of all that I may see of the correspondence.

Ever, my dear Frank, | Your affectionate Uncle, | M. Faraday

1. See letter 2577.
2. A reference to the events leading to the establishment of the Second Empire on 1 December 1852. See *Ann.Reg.*, 1852, **94**: 261–6.

Letter 2590
Faraday to Frederick Madan
9 November 1852
From the original copy in GL MS 30108/1/54

R Institution | 9 Novr. 1852

My dear Sir

My letter[1] with its reservations does mean what the Deputy Master[2] understands "- that one lamp & its reflector would probably be equal to one Second Order lens." Your note[3] seems to me to imply that you think one first order lens sends forwards as much light as ten lamps & reflectors. In reference to this point you no doubt remember the experiments made at Purfleet and Blackwall I think in September 1840[4], when both the French

& English first class refractors were compared with a single lamp in its reflector, and the latter was found to surpass *both*. Now the refractors though they had concentrated the rays from above & below had not done so for the horizontal line and therefore to get an idea of what might be the increase of effect in that respect we may assume that as each of the eight refractors receives 45° horizontally of the light issuing from the lamp; & that as this light in the *lens* form is reduced from a divergence of 45° to that of 15°, so such a lens would be equal to three such lamps & reflectors. But the lamp then used and was one of *four* wicks burning above $7\frac{1}{2}$ pints of oil (7.625) in 8 hours[.] Whereas the 2nd order or three wicked lamp as Mr. Wilkins now tells me burns only 4 pints per 8 hours[5]. Therefore one of the eight lenses of the 2nd order might be expected to equal about $1\frac{1}{2}$ such lamps & reflectors - and one of twelve lenses only 1 such lamp i.e a lamp & reflector not so good as compared with the refractors at Purfleet[.]

I think the lenses generally, especially the excellent ones of France, compress the chief part of the beam into a less divergence than 15° at certain distance: & then of course the effect on the axis of the beam is brighter but that means the edge of the 15° dimmer than the mean: But I have not seen a second class lens and therefore my reply[6] to Mr Herberts letter[7] is worded with much diffidence & reservation. Still you will perceive that the above mode of viewing the question founded on experiment & observation does not lead to a result much different from the former, and in the absence of a more direct comparison, I dare not venture to go farther[.]

Yours Very Truly | M Faraday
Captn Madan | &c &c &c

1. Letter 2580.
2. John Shepherd.
3. Letter 2588.
4. See Faraday to Trinity House, 28 August 1840, 16 September 1840, 16 October 1840, 30 October 1840, GL MS 30108A/1, pp.109–29, 130–43, 144–56, 162–5. (These letters will be published in the addenda to the final volume).
5. See letter 2585.
6. Letter 2580.
7. Letter 2561.

Letter 2591
John Frederick William Herschel to Faraday
10 November 1852
From the original in IEE MS SC 2

Harley Street | Nov. 10 1852

My dear Faraday

Wolf's letter[1] excites great expectations[2]. A law of perioding in the recurrence of the Solar Spots seems to be established by Schwabe's[3] (?) observations referred to by Sabine in his recent paper on Magnetic disturbances[4], and the period (of somewhere about 10 or 11 years if I remember right) agrees with Wolff.- Sabine has in that paper (whether originally or not I know not) distinctly connected the two Phaenomena - extraordinary Magnetic disturbances and great Solar Spots. - their identity of period.- If all this be not premature we stand on the verge of a vast cosmical discovery such as nothing hitherto imagined can compare with. Confer what I have said about the exciting cause of the Solar light - referring it to Cosmical electric currents traversing space and finding in the upper regions of the Suns atmosphere matter in a fit state of tenuity to be *auroralized* by them (Astron. Note on Aur. 400)[5][.]

(Query the red Clouds seen in Solar Eclipses - are they not reposing auroral masses)[.]

As Sabines paper was read to the R.S. and the subject is one of quite as much physical as purely astronomical Interest I should think the RS would be the fit point of delivery of Mr Wolfe's ideas - only in what form I know not - perhaps in some conversational form - or in that of a statement from the chair (but ? as this would be a precedent[6]).

However I would not preclude the Astronomical Society from learning and discussing it and I don't see why it might not pass on from one to the other in the way of Scientific News or Gossip[.]

Yours very sincerely | J.F.W. Herschel
Dr Faraday
PS. What treatise on Chemistry (not organic) should I put into my Son's[7] library as a text book in furnishing him with matter for meditation in India?

1. Letter 2586.
2. See letter 2587.
3. Schwabe (1844).
4. Sabine (1852), 121.
5. Herschel, J.F.W. (1849), paragraph 400 (p.238).
6. See note 2, letter 2587.
7. William James Herschel (1833–1917, WWW2). Civil servant in India, 1853–1878.

Letter 2592
Edward Sabine to Faraday
11 November 1852
From the original in IEE MS SC 2

Woolwich, Nov. 11, '52

Dear Faraday

I return Mr Wolf's *german* letter[1] with thanks. The french letter[2] shall be given to Mr Christie for communication[3]. Humboldt wrote me some time since an account of Mr. Wolfe's discovery of the connexion between the Magnetic variations & the solar spots, & remarked that I had preceded him in publication by between 4 & 5 months[4]: and I have reason to believe that he will notice my priority in his forthcoming vol. of Cosmos[5], which I am glad of, as the thing itself has excited but little interest in this country, & foreign countries are not always ready to do justice to a man of a country which is comparatively regardless, of a matter in which they take a far greater interest. I believe that I have a claim to be considered as the first announcer of the probable existence of a *secular magnetic period in the Sun.*

Mr. Wolf has been forestalled by Lamont in the period of the variations of the magnetic declination[6]. By myself in the period of the variations of the magnetic dip & total force, & in the period of the supposed irregular disturbances or storms - and in the coincidence of these periods with that of the solar spots. There remains to him of original suggestion therefore the examination of the earlier solar spot observations from the time of Fabricius[7], & their connection with the recent far more ample research of Schwabe a supposed correction of the *period* from 10.33 to 11.11 years - the suggestion of a connexion between the *solar spots & variable stars*, which would be indeed surprising, as one cannot well see how the solar spots are to affect beyond the solar system, & it would seem to make the solar magnetic period only one phenomenon of a general cosmical magnetic period. The proof of the years of maximum solar spots being years of dryness & fertility would seem to be very difficult to establish if the proof is to extend, as it ought to do, *over the surface of the Globe generally.* I should doubt greatly whether this will prove more than a mere speculation.

The connexion between the appearance of Aurora & the magnetic storms has long since been established - the greater frequency of the storms carries with it therefore the greater frequency of Aurora.

Robinson[8] at Belfast[9] suggested that the variable light of the stars might be analogous to the solar spots, as indicative of a magnetic period, but not the *same* period for *sun* & *stars*[.]

Sincerely yours | Edward Sabine

1. Letter 2554.
2. Letter 2586.
3. That is to the Royal Society. See note 8, letter 2586.
4. See Sabine (1852).
5. Which he did in Humboldt (1846–58), 4: 81. The reason why Sabine was aware of this was because Elizabeth Juliana Sabine, née Leeves (1807–1879, see DNB under Edward Sabine) was the translator.
6. Lamont (1852).
7. Johannes Fabricius (1587–1615, NDB). German astronomer.
8. Thomas Romney Robinson (1792–1882, DNB). Director of the Armagh Observatory, 1823–1882.
9. That is at the Belfast meeting of the British Association.

Letter 2593
Frederick Madan to Faraday
11 November 1852
From the original in GL MS 30108/1/54

Trinity House | Novr: 11th: /52

My dear Sir

I am much obliged to you for your explanation[1].

I was absent in the yacht during nearly the whole of the month of Septr 1840, & was not present at the trials you mention[2], neither do I remember hearing the result of them. I had always understood, I think it must have been from my late Friend Drew[3], that our Dioptric apparatus at the Start Point was considered equal to those at Beachy Head & St Agnes, the one being 1st order revolving, with Cookson's[4] Lenses & Mirrors, & the other having ten Reflectors on a face. This must be my excuse for the mistake I have made, but still I am inclined to think that my impression is more in accordance with Alan Stevenson's[5] experiments, & as it is an important question which are most eligible for a revolving Light, Reflectors or Lenses[6], I shall observe to the Deputy Master[7], when I show him your letter, that it would be very desirable for you to be furnished with a 2d order Lense, & whatever you may require to come to a correct conclusion.

In your letter you estimate the divergence of the Lense to be 15°, the same as the Reflector;- I thought it had been 5° or 6° degrees.

I think also that Wilkin's information as to the consumption of oil of the 3 & 4 wick lamps is not correct, & that our Mean here would make the one about $\frac{2}{3}$ds of the other.

I should like to have the whole subject thoroughly investigated.

Your's very sincerely | Fredk Madan

M Faraday Esq: | &c &c

1. Letter 2590.
2. See note 4, letter 2590.
3. Richard Drew (1787–1843, *Gent.Mag.*, 1843, **20**: 329). An Elder Brother of Trinity House, 1826–1843, Chaplin [1950], 84.
4. William Isaac Cookson (1812–1888, Morris et al. (1988), 11). Newcastle glass manufacturer.
5. Alan Stevenson (1807–1865, DNB). Scottish lighthouse engineer.
6. Stevenson (1850), 2: 1–148 described work which demonstrated the superiority of lenses over refractors.
7. John Shepherd.

Letter 2594
Faraday to Frederick Madan
12 November 1852
From the original copy in GL MS 30108/1/54

R Inst | 12 Novr. 1852

My dear Sir

You know far more about lenses practically than I do & I do not know that you are under any mistake[1]. *Practically* I know nothing of them and cannot have the opportunity, and it strikes me as very serious that there have not been practical comparisons of the lenses (as well as of the refractors) with an Argand & reflector at Purfleet & Blackwall[.] It would be very easy so to shade the parts of a lens as to make a comparison with a reflector that should be very useful. The axial part of the beams could be compared and the divergence might be laid hold of by some contrivance so as to make the comparison more complete[.]

I referred to a divergence of 15° because I understand it is *essential* that the character of the Caskets lights should not be altered[.] If you were to contract a divergence of 15° into one of 5° or 6° and so make the duration of the light only about one third what it was before I concluded that could not be allowed[.]

Ever Truly Yours | M. Faraday
Captn Madan | &c &c

1. Letter 2593.

Letter 2595
Faraday to John Frederick William Herschel
12 November 1852[1]
From the original copy in RS MS HS 7.196

Royal Institution | 12 Mar. [sic] 1852

My dear Herschel

I have not read up the different elementary Treatises on Chemistry and should rather ask advice than give it but I will mention three, Fowne[s]'s[2], Graham's[3], and Brande's[4]. The first is the smallest but excellent. The third is the largest.

Ever truly yours | M. Faraday

I have sent Wolf's letter[5] to Sabine for I thought he was far the best judge as to fitness[6]. | MF

Endorsement in Herschel's hand: The original given by J.F.W.H. to Is. H[7].

1. Dated on the basis that this is the reply to letter 2591.
2. George Fownes (1815–1849, DSB). Professor of Chemistry to the Pharmaceutical Society, 1842–1846. The text was Fownes (1850).
3. Graham (1842).
4. Brande (1848).
5. Letter 2586.
6. See letter 2592.
7. Isabella Herschel (1831–1893, Buttmann (1970), 72, Evans et al. (1969), 6). Second daughter of John Herschel.

Letter 2596
Faraday to Edward Sabine
13 November 1852
From the original in BeRO MS D/EBy F48 108/1

R Institution | 13 Novr. 1852

My dear Sabine

You say there is not so much interest felt here as there ought to be in the probable joint magnetic relations of the Sun & the Earth[1]. I think that ought not to be so and you know how much I have striven & Barlow also to make our Friday Evenings useful in that respect as concerns a highly intelligent & somewhat influential audience. The subject is a fit one for such an audience if treated in the manner which would make the ideas comprehensible to those who being intelligent still are hearing for the first time. What do you say? Would you like them i.e. the ideas put forth here on such an occasion & will you give us an account of them some Friday Evening?[2]

Or if you would rather not give an evening & *still would like them noticed* it seems to me that I shall be giving an evening on some points relating to the magnetic force as the amount of action at different distances and it is very probable I could devote a quarter of an hour or 20 minutes to your subject and should wish to do so if you wished it and would undertake to instruct me a little in the general vein of the matter[3].

However I speak very much at random until I know your inclinations. The thing I should like most would be to hear you here[.]
Ever Truly Yours | M. Faraday

1. Letter 2592.
2. Sabine never delivered a Friday Evening Discourse.
3. Faraday (1853a), Friday Evening Discourse of 21 January 1853, pp.237-8 is devoted to this topic.

Letter 2597
Faraday to Isambard Kingdom Brunel
13 November 1852
From the original in BrUL MS

Royal Institution | 13 Novr. 1852

My dear Brunel
Dampness is probably as bad for you as oxygen so that I suppose moist matters are to be avoided[.] But as against both dampness and oxygen why not use Potassium or Sodium. Select either, and then *just before you change the cup*, remove the napth[a]line from the metal by dry blotting paper or a piece of linen, cut the metal into three or four pieces by a sharp clean knife put them instantly into the cup and the cup into the cavity & close all in[.]
Ever Truly Yours | M. Faraday

Letter 2597
George Herbert to Faraday
22 November 1852
From the original in GL MS 30108/1/54

My dear Mr. Faraday
I can hardly summon up sufficient courage to intrude upon your already overoccupied time but your great kindness upon all occasions to me makes me hope you will look leniently upon the Intruder. I have made the enclosed estimate (based as you will see upon your observations) of the comparative amount of light thrown upon the horizon from the different apparatus in use by this Board[1][.] The conclusion arrived at in regard to the 1st order revolving (French) light being so nearly that arrived at by Mr. Drummond[2] leads me [to] think that the estimate may not be altogether incorrect, and if you will do me the favor of returning me the

enclosed with just a word to say whether I have understood the matter aright or not, you will greatly oblige my dear Mr. Faraday
   Your very faithful | George Herbert
Trinity House | 22d Nov 1852

1. This is in GL MS 30108/1/54.
2. Thomas Drummond (1797–1840, DNB). Officer in Royal Engineers. Worked on Ordnance Survey of Ireland. This work is in Drummond (1830), 390.

Letter 2599
Jacob Herbert to Faraday
24 November 1852
From the original in GL MS 30108/2/64.2
                    Trinity House London | 24 Novr. 1852.
My dear Sir,
   Enclosed I forward to you Copy of a Letter from Dr. Joseph J. W. Watson, dated 20th Instant:- relative to Electricity as a source of Illumination[1].
   The Board have no hesitation in complying with Dr. Watson's request, and are solicitous that you should be present whenever the Light and it's Apparatus are exhibited at this House,- I am directed therefore to enquire what days and hours will best suit your convenience to attend, either in the next or following weeks;- I am also to suggest, that should you not already have some acquaintance with Dr. W.'s invention, - you might probably have an opportunity of learning something on the subject prior to such exhibition.
   I understand that the Light will be shewn from the top of a House at Charing Cross on Thursday (tomorrow) from 8 to 11 P.M.
   I remain, | My dear Sir, | Your very faithful Servant | J. Herbert
Mr. Faraday Esq

1. Watson to Herbert, 20 November 1852, GL MS 30108/2/64.1. This was read to Trinity House By Board, 23 November 1852, GL MS 30010/37, p.571 who ordered that Watson's light be tested.

Letter 2600
Faraday to George Herbert
25 November 1852
From the original copy in GL MS 30108/1/54

Brighton 25, Novr. 1852

My dear Sir

Your figures[1] are I think all right but as they are founded on my own, which as I have said in the letter containing them[2] are quite insufficient for any serious conclusions, so of course they must partake of their imperfection; and I have no confidence in the results wherever they may be opposed to or want the confirmation of observation. There are too many data necessarily left out. For instance we do not know correctly the comparative loss of light by reflection from the silver and transmission & reflexion by the glass. The loss by silver has been thought to be the greatest but recently it has been found to be only one tenth of the whole quantity[3] a result which Lord Rosse appears to have confirmed[4]:- Another point is that the oil that is burnt must be burnt in or near the focus to be useful; if burnt out of it it only gives as the result a larger divergence but not a brighter ray:- The divergence in any arrangement is chiefly dependant on the extent of the place of combustion[.] Another point is that the burner obstructs the progress of a very large portion of the light that is produced in or near the focus. In the four wicked lamp probably one half of the flame is hidden from the bottom of the refractor or the lens by the great breadth of the burner. It was this circumstance which to a large extent rendered the Gurney oxy lamp so deficient in effect: the light was intense but it was set down low upon the burner & much of it thus stopped in its course. Again in assuring a divergence of 6° or 15° the area cannot for a moment be supposed to be uniformly illuminated & the light in the axis of the beam does not represent the intensity of the light generally. Neither does it represent a *given ratio* to the area. I expect that the light in the axis of a good French lens is far more intense in relation to the light generally over the 6° of divergence than the light in the axis of a reflector is to the general light over its 15° of divergence[.] The extent of the divergence of 6° & 15° is only a general assumption approaching to the truth; and the progressive disposition or distribution of the light over the areas so represented is I believe even less clearly known to us than the extent of divergence. The brightness of the revolving lens light coexists with briefness the dullness of a single reflector with comparative duration and I suppose that this difference of duration which may be said to be as 6 to 15 must be of much importance in the character of a given light[.]

So after all, & as always, I think that *observation* is the only useful way of arriving at a good practical conclusion[.] It was in 1842 [sic] I believe that the experiments at Purfleet & Blackwall were made[5]. A first order French refractor - a first order English refractor and a reflector were compared and the results were such as to make me *very cautious* in drawing conclusions except from practice: for the English refractor, though not at all comparable to the French refractor in workmanship & the form of the ribs, was but little inferior to it in power & that chiefly because of the colour of the glass. Further the Parabolic reflector clearly surpassed *both*. I think your father[6] was there with the Deputy Master[7] & the Elder Brethren but at all events he has some account of the results.

I am away from home and my bad memory causes that I cannot give Wilkins results as to oil burnt[8]. That same bad memory makes me desire to preserve papers on which I have ventured an opinion even though the opinion be private. So though I send you your paper[9] with my marginal notes, yet I will thank you to let me have both back again (you can copy my notes if you like) that I may store them away in their place.

Ever My dear Sir | Very Truly Yours | M. Faraday
Geo Herbert Esq | &c &c

1. Sent with letter 2598 and which are in GL MS 30108/1/54.
2. A copy of these figures, in Faraday's hand, is in GL MS 30108/1/54 and give the consumption of oil at the Start Point, St Catherine's and South Foreland lighthouses. They seem to have been sent with a letter that has not been found.
3. Jamin (1848).
4. See *Rep.Brit.Ass.*, 1851, pp.12–14.
5. See note 4, letter 2590.
6. Jacob Herbert.
7. John Henry Pelly.
8. Letter 2585.
9. This is in GL MS 30108/1/54.

Letter 2601
George Gabriel Stokes to Faraday
26 November 1852
From the original in IEE MS SC 2

Pembroke College Cambridge | Nov 26th 1852
My dear Sir,

I received your letter yesterday, and am much obliged to you for having given directions to your assistant to have both the voltaic and the electric apparatus ready, as I should like to try both, though I am not very sanguine as to the possibility of making the fundamental experiment visible to so large an audience[1].

As regards my lecture, it would be decidedly an advantage if the audience were previously acquainted with the singular phenomenon which Sir John Herschel described in a dilute solution of sulphate of quinine[2]; and if there should be time, without interfering with more important matters, to mention it, at the previous meeting, I should be much obliged by your doing so.

The nature of the phenomenon is a follows. If a dilute solution of sulphate of quinine be poured into a tumbler and placed in a window (it is well to place it on a blank cloth or velvet) a blue gleam of light is seen near the surface of the glass (the common surface of the glass & fluid) *on the side next the incident light and on that side only*. If the eye be placed so as to look down in a direction parallel to the surface of the glass, i.e. to the sill of the vessel, the blue gleam is foreshortened into a very narrow blue arc. A test tube containing a portion of the same solution exhibits the same phenomenon; but if the test tube be placed in the vessel with the solution of quinine the blue colour seen in test tube disappears.

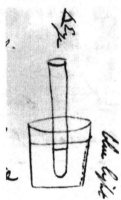

If the fluid in the tumbler be water instead of a solution of quinine, the fluid in the test-tube exhibits the blue light in perfection[3].

It appears then that the light which has once passed through a thin stratum of a solution of sulphate of quinine, although to all appearance unaltered, has yet undergone some mysterious analysis whereby it has been deprived of those portions, whatever they be, which produce the blue light in a solution of quinine.

A solution of 1 part of the common disulphate of quinine in 200 parts of water acidulated with sulphuric (or citric, tartaric, phosphoric, nitric, acetic) acid gives a medium very well suited to the experiment. However it will succeed within wide limits as to the strength of the solution.

I am dear Sir | Yours very truly | G.G. Stokes
M. Faraday Esq | &c &c &c

1. Stokes (1853), Friday Evening Discourse of 18 February 1853. See James (1985), 151–2 for a discussion of this lecture.
2. Herschel (1845a, b).
3. This was Stokes's discovery of fluorescence. See Stokes (1852). For a discussion of this work see James (1983a).

Letter 2602
Faraday to John Barlow
1 December 1852[1]
From the original in RI MS F1 G11

Brighton | Wednesday, 1 Decr

Many thanks My dear Barlow for your letter. I am encouraged by it to think of returning here until tomorrow week[2]. But I shall be up on Saturday[3]. I am very lazy or inert & have not been able yet to write my lecture notes[4]

Ever Truly Yours | M. Faraday

1. Dated on the basis that Faraday was in Brighton at this time. See letters 2600 and 2607.
2. That is 9 December 1852.
3. That is 4 December 1852.
4. These were Faraday's Christmas lectures on Chemistry. His notes are in RI MS F4 J14.

Letter 2603
Faraday to Jabez Hogg[1]
5 December 1852
From the original in HLHU b MS Eng 1009 (82)

Royal Institution | 5 Decr. 1852

Sir
I beg to return you my sincere thanks for your kindness in sending me Part LVI of the Instructor[2][.]
I am Sir | Your Very Obedient Servant | M. Faraday
Ja Hogg Esq | &c &c &c

1. Jabez Hogg (1817–1899, DNB1). Journalist and ophthalmologist.
2. That is Hogg's *Weekly Instructor*, 21 March 1846, **3**: 57–8.

Letter 2604
Faraday to Christian Friedrich Schoenbein
9 December 1852
From the original in UB MS NS 404

Brighton | 9 Decr. 1852

My dear friend
If I do not write to you now I do not know when I shall - and if I write to you now I do not know what I shall say - for I am here sleeping eating & lying fallow that I may have sufficient energy to give half a dozen

juvenile Christmas lectures[1]. The fact is I have been working very hard - for a long time to no satisfactory end - all the answers I have obtained from nature have been in the negative and though they shew the truth of nature as much as affirmative answers yet they are not so encouraging and so for the present I am quite worn out. I wish I possessed some of your points of character. I will not say which for I do not know where the list might end and you might think me simply absurd and besides that ungrateful to Providence[.]

I had your letter[2] by Dr. Whewell and I have received also your last of the 17th October[3] and the paper[4] and I hope when I return home to get the latter done into English. It is a very great shame to us that such papers do not appear at once in English but somehow we cannot manage it. Taylor appears to be much embarrassed in respect of the Scientific memoirs[.] I hope now that they have changed their shape & are to appear in two series physical & chemical that they will be more servicable to such as I am[.]

Your letter quite excites me and I trust you will establish undeniably your point. It would be a great thing to trace the state of combined oxygen by the colour of its compound not only because it would shew that the oxygen had a special state which could in the compound produce a special result - but also because it would as you say make the optical effect come within the category of scientific appliances and serve the purpose of a philosophic induction & means of research whereas it is now simply a thing to be looked at. Believing that there is nothing superfluous or deficient or accidental or indifferent in nature I agree with you in believing that colour is essentially connected with the physical condition and nature of the body possessing it and you will be doing a very great service to philosophy if you give us a hint however small it may seem at first in the development or as I may even say in the perception of this connexion.

As I read your letter I wondered whether there was any connexion between your phenomena and those recently investigated by Stokes[5]. I do not mean any immediate likeness but distant connexion. He has been rendering the invisible chemically acting rays visible - that is to say he has been converting them into visible rays. You by giving a given condition to a substance make it when in compounds send one ray to the eye - and then by giving it another condition cause it to send other rays to the eye the body being chemically the same. Both these are phenomena of radiation & both are connected with chemical agencies or forces. If they could be connected what a heap of harvests would spring up between the two. I do not know enough yet of Stokes phenomena to form any thing but a crude idea and I know nothing of yours yet so that you will think me very absurd to write such stuff but then it is only to a friend[.]

You are very amusing with your criticisms on Organic chemistry. I hope that in due time the chemists will justify their proceedings by some

large generalisations deduced from the infinity of results which they have collected[.] For me I am left hopelessly behind & I will acknowledge to you that through my bad memory organic chemistry is to me a sealed book. Some of those here Hoffman for instance consider all this however as scaffolding which will disappear when the structure is built. I hope the structure will be worthy of the labour. I should expect a better & a quicker result from the study of the *powers* of matter but then I have a predilection that way & am probably prejudiced in judgment. My wifes kindest remembrances to you & yours. My earnest wishes for the happiness of you all[.]
    Ever | my dear Schoenbein Your Affectionate friend | M. Faraday

Address: Dr. Schoenbein | &c &c &c | University | Basle | on the Rhine | Switzerland

1. These were Faraday's Christmas lectures on Chemistry. His notes are in RI MS F4 J14.
2. Letter 2562.
3. Letter 2578.
4. Schoenbein (1852a).
5. That is fluorescence. Stokes (1852).

Letter 2605
Faraday to Edward Solly[1]
11 December 1852
From the original in SI D MS 554A

                                      Royal Institution | 11 Decr. 1852
My dear Solly
    I really am not well enough to undertake duty at the Adelphi and you must excuse me. I have been trying to settle my head at Brighton this last three weeks that I might be able to get through my small proportion of work here and yet I am by no means sure that I shall be able to do so. Whilst I give up all friendly society the bodies must excuse me.
    Ever Truly Yours | M. Faraday

1. Edward Solly (1819–1886, DNB). Secretary of Society of Arts, June 1852 to May 1853.

Letter 2606
Jacob Herbert to Faraday
16 December 1852
From the original in GL MS 30108/2/64.3

Trinity House | 16 Decemr. 52

My dear Sir, -
I learn with regret that your absence from London has not been attended by all the benefit we might have desired - but I hope you will Continue to improve in health, and that when I next have the pleasure of seeing you, it will be with your usual energy & spirits.

I have written this day to Dr. Watson requesting him to put himself in Communication with you, in order that such arrangements may be made with the object of testing his Electric Light, as respects it's application for the Illumination of Light Houses,- as you may deem most advisable[1].

I remain, | My dear Sir, | Very faithfully Yours | J. Herbert
M: Faraday Esq.

1. It was agreed at Trinity House Court, 7 December 1852, GL MS 30004/25, p.174 that Faraday would be asked to test Watson's light.

Letter 2607
Christian Friedrich Schoenbein to Faraday
18 December 1852
From the original in UB MS NS 405

My dear Faraday
I had already given up the hope of my paper[1] having reached you, when I was most agreeably undeceived by your kind letter from Brighton[2]. I am really curious to know what you will think about my notions on the relations of the different conditions of oxigen to the voltaic, magnetic and optical properties of that body. The conviction of their being correct has by no means been shaken by my recent experimental results of which you shall hear before long. But however they may turn out I trust they will at any rate draw the attention of philosophers to a most important set of phenomena.

I am not acquainted with the experiments of Stokes[3] but from what you say about them I am inclined to believe that they are closely connected with my subject. I am just now working upon the optical action of nitrous gas (NO^2) upon the solutions of the protosalts of iron, which, as you are well aware, is so very striking. As I entertain the notion that the deep coloring of those solutions produced by NO^2 is due to a change of the condition of the oxigen being contained in the base of the ironsalt i.e. to the transformation of the inactive state of that oxigen into the active one, I suspect that the paramagnetic force of the black liquid is smaller than the sum of the paramagnetic forces of its constituent parts. You know that by uniting two equiv of inactive i.e. paramagnetic oxigen to one equiv of paramagnetic deutoxide of Nitrogen a diamagnetic compound is produced and you are likewise aware, that the two eq. of oxygen united to NO^2 exist in hyponitric acid in the ozonic or excited condition. Again by associating 2 equiv. of the highly paramagnetic protoxide of iron to one equiv. of paramagnetic oxigen a compound is obtained being, according to your own experiments, magnetically indifferent. I have shown in my paper that $Fe^2O^3 = 2FeO + \overset{\circ}{O}$ that is to say that the third equiv. of the peroxide of Iron exists in the exalted condition. From these facts I infer that in the first case the diamagnetism of 2 equiv. of ozonic Oxygen is stronger than the paramagnetism of the two equiv. of inactive oxigen contained in NO^2; and that in the latter case the diamagnetism of one equiv. of ozonic Oxigen neutralizes the paramagnetism of 2 equiv. of protoxide of Iron. Now I conjecture that by uniting the two paramagnetic compounds: a proto iron salt to NO^2 either a diamagnetic or a less paramagnetic fluid will be obtained. I should consider it as a great favor, if you would settle that point by experiment.

I trust the bracing air of Brighton will refresh your body and mind so much as to enable you not only to resume your Lectures, but what is more important your scientific labors. We cannot spare you, our present age being so woefully deficient of original thinkers and experimental Philosophers. There are indeed but a very few to whom I might say: You are the salt of the Earth, but if the salt have lost his savour, wherewith shall it be salted?[4] Permit me to tell you that I count you amongst those few.

Mrs. Schoenbein and the Children are well. My eldest daughter is now rather a big child i.e. a grown up Lady. They charge me with their best compliments to you and Mrs. Faraday, to whom you will reme⟨m⟩ber me in particular and in the most friendly mann⟨er.⟩ Excuse my badly written letter, which I was obliged to scribble down in a great hurry and believe me, my dear Faraday

Your's | most truly | C.F. Schoenbein
Bâle Dec. 18th 1852.

Address: Doctor Michael Faraday | &c &c &c | Royal Institution London
Postmark: Melksham[5], 27 December 1852

1. Schoenbein (1852a).
2. Letter 2604.
3. On fluorescence. See Stokes (1852).

4. Matthew 5: 13.
5. In Wiltshire.

Letter 2608
Faraday to Joseph John William Watson
20 December 1852
From the original in SI D MS 554A

Royal Institution | 20 Decr. 1852

Sir

On Saturday[1] I was at the Trinity House & consulted with the Secretary[2] who agreed with my general statement[3] of what I had said to you & Mr. Presler[4]. It will therefore be expedient first to have a day of 8 hours at the Trinity house placing the Electric light in the Dioptric light room in the corner by the steps so as to have it as far from the French light as possible but with the sight clear from one to the other between the Iron pillars[.]

If the days trial is encouraging then a daily trial for a fortnight excepting Sunday of at least 8 hours each day will be required - also at the Trinity house. Upon the result of these trials further steps can be taken[.]

Now the single days trial might be this week excepting Saturday[5] - but the successive trials could not well come on until the week beginning 10th January because of the possible absence of some who would have to regard the light from day to day. I leave it with you therefore to decide whether the single trial shall take place this week or at that time - & would be glad if you would let me know at once on what you decide and when the day is to be[.] Above all I most earnestly beg of you to produce no imperfectly prepared apparatus or any thing that requires excuse. If the apparatus is not yet complete keep it back until it is. The Trinity house has not time to Witness incomplete trials and it would only do harm to call their attention to any thing avowedly imperfect[.]

I think I have said nothing here to which you did not freely assent[.] If you make a trial this week let me know the day as soon as possible[6][.]

I am Sir | Your Obedient Servant | M. Faraday

Dr. Watson | &c &c &c

1. That is 18 December 1852.
2. Jacob Herbert.
3. Faraday put this statement into a memorandum on 20 December 1852 which was approved by Trinity House By Board, 21 December 1852, GL MS 30010/37, p.600.
4. Unidentified.
5. That is 25 December 1852.
6. Watson's agreement to the proposals in this letter were noted in Trinity House Court, 4 January 1853, GL MS 30004/25, p.182.

Letter 2609
Joseph John William Watson to Faraday
20 December 1852
From the original in GL MS 30108/2/64.4

11, Adam St | Adelphi | Decr. 20: 1852

Sir,
As you have kindly permitted me to select the time for the trials to be made of my lamp[1] if you have no objection I should prefer after the 10th of January as I fully anticipate the first experiment to be successful and then the after-trials, for 14 days may succeed as rapidly as you may wish.

I will endeavour to profit by your kind advice and I trust that you will not have to complain of my having brought before you an imperfect or impracticable invention[.]

I am | Sir | Your obliged and obedient Servant | Joseph J.W. Watson
M. Faraday Esq. D.C.L | &c &c &c
Since Writing the above I have received your notes and I shall make it a special object of care to observe rigidly the directions therein contained | J.J.W.W.

1. In letter 2608.

Letter 2610
Arthur-Auguste De La Rive to Faraday
24 December 1852
From the original in IEE MS SC 2

Genève | le 24 Xbre 1852

Monsieur & très cher ami,
Je n'ai pas répondu plus tôt à votre bonne & amicale lettre[1] parceque j'aurais voulu avoir quelque chose d'intéressant à vous dire. Je suis peiné de ce que votre tête est fatiguée; cela vous est déjà arrivé quelquefois à la suite de vos travaux si nombreux & si persévérants; mais vous vous rappelez qu'il suffit d'un peu de repos pour vous remettre en très bon état. Vous avez ce qui contribue le plus à la sérénité de l'âme & au calme de l'esprit, une foi pleine & entière aux promesses de notre Divin Maitre & une conscience pure & tranquille qui remplit votre coeur des espérances magnifiques que nous donne le Evangile. Vous avez en outre l'advantage d'avoir toujours mené une vie douce & bien réglée exempte d'ambition & parconséquent de toutes les agitations & de tous les mécomptes qu'elle entraine après elle. La gloire est venue vous chercher malgré vous; vous avez su, sans la mépriser, la réduire à sa juste valeur. Vous avez su vous

concilier partout à la fois la haute estime & l'affection de ceux qui vous connaissent. Enfin vous n'avez été frappé jusqu'ici, grâce à la bonté de Dieu, d'aucun de ces malheurs domestiques qui brisent une vie. C'est donc sans crainte comme sans amertume que vous devez sentir approcher la vieillesse, en ayant le sentiment bien doux que les merveilles que vous avez su lire dans de livre de la nature doivent contribuer pour leur bonne part à en faire encore plus admirer & adorer le Suprême Auteur.

Voilà, très cher ami, l'impression que votre belle vie m'a toujours fait éprouver. Et quand je la compare à nos vies agitées & si mal remplies, à tout cet ensemble de mécomptes & de douleurs dont la mienne en particulier a été obscurcie, je vous estime bien heureux surtout parceque vous êtes digne de votre bonheur. Tout cela m'amène à penser au malheur de ceux qui n'ont pas cette foi religieuse que vous avez à un si haut degré, & en particulier à la pauvre Lady Lovelace[2]; avez-vous su quelque chose sur ses derniers moments & sur ses dispositions morales & religieuses à cette heure suprême?

Mon fils[3] qui vous remettra cette lettre est dans ce moment à Londres avec sa jeune femme[4]; ils ont été voir leur excellente grand-mère Madame Marcet qui semble être mieux maintenant qu'elle ne l'a été depuis longtemps.

Mon premier volume sur l'Electricité doit avoir paru dans ce moment[5]; le second avance & ne tardera pas à suivre son ainé[6]. Si vous le lisez (je vous en enverrai un exemplaire aussitôt qu'il aura paru), faites moi la grande amitié de me présenter toutes vos observations, afin que j'en fasse encore profit pour l'édition française[7].- Je suis bien sur que malgré ce que vous me dites, vous saurez encore trouver quelque belle mine à exploiter dans ce riche domaine que vous cultivez avec tant d'ardeur & de succès. N'oubliez pas de m'en faire part, car vous savez tout l'intéret que je mets à ce qui vient de vous parceque c'est de vous avant tout & ensuite parceque c'est toujours original & remarquable.

Merci de toutes vos précédentes communications. Votre affectionné & bien dévoué | A. De la Rive

TRANSLATION

Geneva | 24 December 1852

Sir and very dear friend,

I did not reply earlier to your kind and friendly letter[1] because I should have preferred to have something interesting to tell you. I am saddened that your head is tired; it has troubled you before, as a result of your numerous and persevering works; but you have that which most contributes to the serenity of the soul and peace of mind, a full and complete faith in the promises of our Divine Master and a pure and tranquil conscience which fills your heart with the magnificent hopes that the Gospel gives us. You have moreover the advantage of having always led a gentle and well ordered life, free from ambition and consequently of all the restlessness and all the disappointments that it brings in its train. Glory has come to find you despite yourself; you have known, without despising it, to reduce it to its true value. You have been able to reconcile, everywhere and at all times, the high esteem and the affection of all those who know you. Finally, thus far you have not been hit, thanks to the grace of God, by any of those domestic calamities that shatter a life. It is thus without fear and without bitterness that you must sense old age approaching, having that agreeable feeling that the marvels that you have been able to read in the book of nature must contribute in their own way to make the Supreme Author even more admired and adored.

There, my dear friend, is the impression that your beautiful life has always made on me. And when I compare it to our agitated and unfulfilled lives, and all the disappointments and pains that have overshadowed mine in particular, I regard you as happy above all because you are worthy of your happiness. All this brings me to think of the unhappiness of those that do not have that religious faith that you hold so deeply, and in particular of poor Lady Lovelace[2]; did you know something of her last moments and of her moral and religious disposition at that supreme hour?

My son[3], who will give you this letter, is at this moment in London with his young wife[4]; they went to see their excellent grandmother, Madam Marcet, who seems better now than she has been for a long time.

My first volume on Electricity should appear any day now[5]; the second is advancing and will not be long behind the first[6]. If you read it (I shall send you a copy as soon as it has appeared), please do me the great friendship of sending me all your observations, so that I can still profit from them for the French edition[7]. I am quite sure that despite everything you tell me, you will be able to find some good mine to exploit in the rich domain that you cultivate with such ardour and success. Do not forget to include me, for you know all the interest that I take in everything you send, first of all because it comes from you and then because it is always original and remarkable.

Thank you for all your preceding communications. Your affectionate and very devoted | A. De la Rive.

1. Letter 2577.
2. Lovelace died on 27 November 1852.
3. William De La Rive.
4. Cécile-Marie De La Rive, née De La Rive (1831–1893, Choisy (1947), 51). Married William De La Rive on 15 March 1852.
5. De La Rive (1853–8), **1**.
6. De La Rive (1853–8), **2**.
7. De La Rive (1854–8).

Letter 2611
Faraday to George Cargill Leighton[1]
29 December 1852
From the original in RI MS F1 D15a
Portraits
To be half bound - free in the back. The Portraits to be in the present alphabetical order. They are numbered 1, 2, 3, 4 &c in the lower right hand corner. Not to be cut down any more than may be quite needful[.]
M. Faraday
Decr. 29, | 1852

Endorsed: Ipswich Museum[2]
Address: Mr Leighton

1. George Cargill Leighton (1826–1895, Cantor (1991a), 301). Printer.
2. That is RI MS F1 K.

Letter 2612
George Gabriel Stokes to Faraday
3 January 1853
From the original in IEE MS SC 2
Pembroke College, Cambridge | Jan 3d 1853
My dear Sir,
I have no particular engagements here before the last week of the month. I can go to London any time next week or the week after. The day is a matter of perfect indifference to me, and I beg you will fix according to your own convenience. There are one or two things I should like to have ready, and on that account perhaps it would be well to fix on a late day. However there is no occasion to fix on any day at present; if you send me notice a day or two before the time you fix that will be sufficient[1].
The things to which I alluded, which I should like to have ready before I go to London, are
(1) The glass vessel which has been already ordered at Newman's.
(2) A pair of deal stands of a kind useful in optical experiments. For want of such I felt like a fish out of water the night I was at the Royal Institution[2]. Perhaps you will have the goodness to give the orders for them. It is common carpenter's work. I send the description on a separate sheet.
I understand the Institution possesses the apparatus for a revolving mirror. If so, I hope the Managers will kindly allow me the use of it for an

experiment of a good deal of scientific interest in connexion with the subject of the phenomena exhibited by solutions of quinine &c. I am writing to Mr. Barlow, as I suppose he is the proper person to apply to.

I am dear Sir | Yours very truly | G.G. Stokes
(Turn over)
P.S. I put up several letters per post in a hurry, and just after they were gone I found on my table this letter which ought to have gone to you. I must either have sent you a wrong letter or an empty envelope[3].

Two sliding tables of deal to be made according to this pattern.

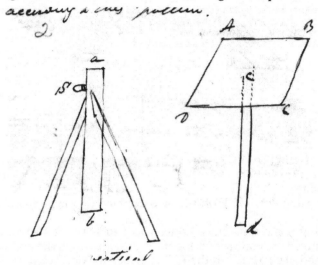

ab is a vertical hollow square tube of board resting firmly on three legs nailed to it (only two are represented in the figure). The base b not to reach quite to the floor on which the whole stands. The top a may stand 2ft. 9in. from the floor. S is a coarse wooden screw for clamping, ABCD is a square table 1ft. 4in. square, fixed to a thick square rod cd, which is attached at the centre, and stands perpendicular to its plane. The rod is of the same size as the interior of the tube ab, so as to slide within it, and to be clamped by means of the screw S to any height that may be desired. The length cd is not quite 2ft. 9in, so that when S is not screwed home d does not quite touch the ground.

The bases of the legs ought to form pretty nearly an equilateral triangle with the base b of the tube nearly over its centre.

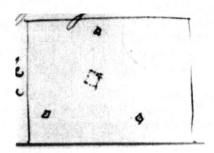

1. That is to prepare for Stokes (1853), Friday Evening Discourse of 18 February 1853. See James (1985), 151–2 for a discussion of this lecture.
2. That is the visit referred to in letter 2601.
3. From "(Turn over)" to this point has been crossed through.

Letter 2613
Edward Sabine to Faraday
4 January 1853
From the original in IEE MS SC 2

11 Old Burleigh S. | January 4, 1853.
Dear Faraday
    I have just read Mr. Wolf's little pamphlet[1] on the period of the sun's spots.
    Having some time since noticed to you that I felt some degree of disappointment in contrasting the interest with which the announcement which I made last March to the R.S. of the connexion between the period of the sun's spots & that of the magnetic variations[2] had been received on the Continents of Europe & America as compared with this country[3] - a remark which perhaps you thought might stand in need of some justification, - Mr. Wolf's pamphlet furnishes me I think with a very fine illustration of it, which I notice at once to yourself because you are concerned, & to mark more particularly that what I regret is rather the *general indifference of Englishmen to the scientific discoverys of their own countrymen*, than *individual* indifference; for from no Englishman should I look for more consideration & regard than from you. But now to my point. In page 16 Mr Wolf quotes in two notes, long extracts from letters from Humboldt & yourself, Humboldt's dated Sept. 10, yours Augt. 27[4]; both acknowledging letters from him announcing the coincidence of the 2 periods, the magnetic variations & the solar spots. Humboldt does me the justice to inform Mr. Wolf of my previous announcement of the same first & refers to date & place of publication. You writing from England take no

notice whatsoever of the previous publication of the very same coincidence which had been communicated to you some months before by a friend & countryman of your own!-

The fact is a remarkable one; and will no doubt be adverted to by Mr Wolf in the reply which he will probably make to the remarks which I must make on the injustice which overlooking Mr. de Humboldts letter he has done to me in placing the date of my publication in *September*, instead of March or May (at which latter date the printed copies were in general circulation), and thus countermanding his own claim to a nearly simultaneous publication in the Comptes Rendus for September 13[5].

Sincerely yours | Edward Sabine

1. Wolf (1852c).
2. Sabine (1852).
3. See letter 2592.
4. Letter 2560.
5. Wolf (1852a).

Letter 2614
Joseph John William Watson to Faraday
6 January 1853
From the original in GL MS 30108/2/64.5

11 Adam Street | Adelphi | January 6th. 1853

Sir,

In reply to your letter just received, I beg to say that I shall be prepared to exhibit my light at the Trinity House on any day agre[e]able to yourself after Wednesday next the 12th instant. I am now making my arrangements for that purpose and shall send some of the requisite materials to the Trinity House in the course of a day or two.

I shall be quite at your service on any day after Wednesday.

I have the honor to be | Sir, Your obedient servant | Joseph J.W. Watson

M. Faraday Esqre | DCL | &c &c

1. Faraday reported these arrangements in two letters of 6 and 10 January 1853 (not found) which were read to Trinity House By Board, 11 January 1853, GL MS 30010/38, pp.6–7 who ordered the necessary arrangements to be made.

Letter 2615
Faraday to Edward Sabine
7 January 1853
From the original in BeRO MS D/EBy F48 109/2

Royal Institution | 7 Jany 1853

My dear Sabine

I have read your remarks[1] with great interest & am much obliged for them. I was puzzled a little at first, for from habit when thinking of a free needle in each hemisphere, I always compare their upper ends, and forget continually that you were comparing the north ends i.e. the upper end of one & the lower end of the other. I seem to have had the variation, which you so properly distinguish as an annual variation, in my mind before but then I had had it as *part* of a larger effect i.e. the increase of the *whole extent of the daily variations* at Hobarton[2] or Toronto[3] as the Sun approaches either one or the other; or as the solstice comes on. The eastern effect at Toronto in its summer morning appears to me to be half of a large result of which the other half is the western effect in the evening - and both these *diminish* together as the Sun goes South & as the correspondent effect at Hobarton *increases*.

Whatever the physical nature of the suns action may be, as he comes on from the East, at the time of the equinox, we may suppose him to act equally on the needles at Hobarton & Toronto. But as he goes north towards the June Solstice the daily variation increases in extent at Toronto & diminishes at Hobarton. This natural action will produce precisely the variation which you describe i.e. the northern end of the needle at Toronto will at 7AM go most east in June or thereabouts because of an *increase* in the extent of daily variation and the North end of the needle at Hobarton will at the same time go *least west* or as you put it (if I am not mistaken) *most east* for the evening hour because the extent of variation is the *least* at that time. When the Sun goes South the other part of the change occurs in its due order[.]

Perhaps I have not caught the whole of your meaning and I think that is more likely because I have only given above the views which I had in 1850[4] and the endeavour I then made to shew that the annual increase & diminution of the extent of the daily variation coincided with the action of the cause which I then assigned as a probable one. That the sun is the cause I do not doubt; that he acts in the way I supposed may be true only for *a part* and even a *very small part* of the whole effect[.] But the other part of the action if large must be in the same direction as my hypothetical action; for otherwise the effect you refer to would not and could not as I believe occur. Your deductions therefore confirm my expectations and I am also very glad to find that your plate II[5] agrees with my plate of curves in the 1850 paper[6]. The *amount* of daily variation & its annual change are noted in paragraphs 2948[7], 3009, 3027[8] & elsewhere. I do not intend to refer to this subject in my evening but mean to keep myself in the latter part of it entirely to your magneto-solar coincidences[9]. I find that my last letter to Wolf in which I referred to your publication was written 13 Novr[10]. If he had it in time he ought to have referred to it in his pamphlet[11][.]

Ever My dear Sabine | Very Truly Yours | M. Faraday

1. Letter 2613.
2. Sabine (1850–3).
3. Sabine (1845–57), **1, 2**.
4. Faraday (1851d, e), ERE26 and 27.
5. Sabine (1845–57), **2**, opposite p.xx.
6. Faraday (1851e), ERE27, opposite p.96.
7. Faraday (1851d), ERE26, 2948.
8. Faraday (1851e), ERE27, 3009, 3027.
9. Faraday (1853a), Friday Evening Discourse of 21 January 1853, pp.237–8.
10. Not found.
11. Wolf (1852c).

Letter 2616
George Gabriel Stokes to Faraday
7 January 1853
From the original in IEE MS SC 2

Pembroke College, Cambridge | Jan 7th 1853

My dear Sir,

I found your letter on my return to Cambridge this evening. As you seem to be disengaged in the early part of next week, I propose to go to Town on Monday[1], so as to be ready to try experiments on Monday evening. Please to leave word with the Porter, or with your assistant, what hour you would like to begin. It is a matter of perfect indifference to me.

By way of fixing on something, I would propose on Monday evening to try the effects of different flames, especially on sulphur burning in oxygen, the effect of which, from your account[2], must be very powerful. I should be glad also to try the experiment you mentioned respecting the light given by the explosion of oxygen and hydrogen in a glass vessel, and some others which that suggested to me.

The glass vessel which I ordered is destined to hold a very weak solution of chromate or bichromate of potash. I am not sure which will answer best, but I am inclined to think the chromate. It is no consequence whether the salt be or be not chemically pure.

I should be glad to have some Canton's[3] phosphorous[4]. I have got some here, but as I made it myself merely in my fire I am afraid it may not be good.

If there be an Argand lamp at the R.I. I should be glad to repeat the experiment Draper's which I have referred to at p.547 of my paper l.10 from the bottom[5]. According to Becquerel, the rays of low refrangibility cause Canton's phosphorus *if previously excited by rays of high refrangibility* to give out more quickly than it otherwise would the light which it is capable of giving out[6]. It strikes me as possible that the phosphorous with which Draper worked when he obtained this result may have been previously excited.

I dare say the experiments I have mentioned, and some things I should like to show you, (such as the absorption-bands of permanganate of potash mentioned in Note D of my paper[7]) will afford work enough for one evening without the electrifying machine.

Art. 224 of my paper[8] will explain what I wanted with a revolving mirror. The experiment could be performed in any place where there are these two things, a revolving mirror apparatus, and an electrifying machine[.] If Prof. Wheatstone will kindly lend his apparatus, or else undertake the experiment himself, it might be performed at his house or at King's College as might be convenient.

Yours very truly | G.G. Stokes

1. That is 10 January 1853 when he wanted to prepare for Stokes (1853), Friday Evening Discourse of 18 February 1853. See James (1985), 151–2 for a discussion of this lecture.
2. Faraday had found this when working on fluorescence. See Faraday, *Diary,* 14 December 1852, 6: 13011–23, especially 13015.
3. John Canton (1718–1772, DSB). Natural philosopher.
4. Canton (1768).
5. Stokes (1852), 547 referred to Draper (1845), 436.
6. Becquerel (1847).
7. Stokes (1852), 558–9.
8. Stokes (1852), 548.

Letter 2617
George Gabriel Stokes to Faraday
8 January 1853
From the original in IEE MS SC 2

Pembroke College Cambridge | Jan 8th 1853

My dear Sir,

A passage in my last letter[1] must have puzzled you if you have referred to art. 221 of my paper[2]. Writing from memory I spoke of Dr. Draper's experiment as having been prepared with an *Argand lamp*, whereas I see it is *incandescent lime*[3]. Draper speaks of the effect of an Argand lamp in the same paper. I am nearly sure I had it originally in my M.S. an Argand lamp, but corrected it[4] on referring to the original paper in the Phil.Mag. I merely mention this to excuse (as far as it is excusable) my mistake, which I fear may have caused you needless trouble. As to the experiment itself, I should like very well to perform it along with you if you feel an interest in this particular subject; but if getting up the lime light should involve any trouble, or if you wish to reserve your time for more important matters, there will be no occasion to try the experiment at the Royal Institution, for a friend of mine in Cambridge has a lime light apparatus, so that I could try the experiment here.

I am dear Sir | Yours very truly | G.G. Stokes

1. Letter 2616.
2. Stokes (1852), 547–8.
3. Draper (1847).
4. The manuscript of Stokes's paper, RS MS PT 45.2, shows this correction at art. 221.

Letter 2618
Edward Sabine to Faraday
8 January 1853
From the original in IEE MS SC 2

11, Old Burleigh S. | January 8, '53

Dear Faraday

I think that the facts point to a more distinct recognition of distinct causes than appears in your note[1]: that we must separate the effects which are occasioned by the earths revolution in its orbit, from those which are occasioned by its revolution on its axis: the one producing a variation in the magnetic direction which is properly called "annual" because its period is a *year*, and the other "diurnal" because its period is a *day*. Now with respect to the first, or annual variation, & confining ourselves to *our* element, viz the Declination: and (the better to fix our ideas) to *one* hour, viz 8 am, (as that is one of the hours followed out in the dates which I sent to you). Now, as far as we have yet experience, if the declination be observed at the same hour of 8 am in *any part of the world*, i.e. in *any* meridian, or in *any part of any meridian*, whether north or south of the Equator, or on the Equator, the declination so observed will shew an annual variation, the *direction and amount of which will be every where the same*: i.e. at the northern solstice the needle will be at its eastern extreme, at the southern solstice at its western extreme, and it will pass through its mean position at the two equinoxes: in the one case when proceeding from east to west, & in the other when returning from west to east. The direction of the motion is every where the same, i.e. of the north end of the needle from east to west from June 21 to December 21, and from west to East from December 21 to June 21, without reference to whether the one end of the needle be directed upwards (by reason of the Dip) or the other end of the needle be directed upwards, or whether the needle be horizontal:- and the amount is also every where the same, or very nearly so, viz five minutes of Arc.

Now, having found & measured this annual variation, we can eliminate it; and when it is eliminated, the needle will have its mean place for 8 am all the year round, *so far as the annual variation is concerned*. Now let the same thing be done for each of the 24 hours separately, (assuming that you have hourly observations to deal with), and let it be done for three places, one in Northern Latitude, a second in Southern Latitude, and a third on the Equator; (the annual variation so eliminated will have been the same amount, & have had the same direction at each of the three locations as already stated). Now then the comparison of the direction of the needle at the different *hours* of the day will give us the *diurnal* variation; and, as the result of the diurnal variation, we find that at 8 am the north end of the needle is considerably to the *East* of its mean position in the 24 hours at the station in *North* Latitude, - considerably to the *West* of its mean position at the station in South Latitude; whilst at the Equator the diurnal variation is null; the needle at 8 am pointing the same as it does on the mean of the 24 hours. The *character* of the diurnal variation at 8 am then is that the north end of the needle is affected in opposite senses in opposite hemispheres, passing thro' a Zero on the Equator - and that its amount varies from 0 at the Equator to several minutes in the high Latitudes being continually variable according to some law into which we need not now enter - whereas the Character of the *annual* variation is that the same end of the needle is affected in the same sense, and to the same amount, equally whether the locality be in the North or South hemisphere or on the Equator.

There is another feature in the Annual Variation which is a very remarkable one, which is shown I think well by the Plate I sent you. I have said that the annual variation passes thro' its mean position at the

Equinoxes; it not only does so on the very day of the Equinox, but in one week after the Equinox is passed, it has reached *in amount* the full extent (or very nearly so indeed) of the half amplitude of the whole movement; in the same way that a week before the Equinox the position of the needle depending on the annual variation is almost if not quite as distant from the mean position in the other direction. This phenomena is most markedly the same at St. Helena & the Cape of Good Hope - whilst at St. Helena the dip is 21° & at the Cape 54'; the Lat of the one 17°, & ⟨o⟩f the other 34°.5.

You can lay bye this note till you have more leisure to look into the subject, and then I shall be very glad to receive your remarks[.]

Very truly yours | Edward Sabine

1. Letter 2615.

Letter 2619
Faraday to Joseph John William Watson
10 January 1853
From the original in the possession of Roy G. Neville

Royal Institution | 10 Jany 1853

Sir

I am much obliged by your note. I will be at the Trinity House at or before 11 o clk on Friday[1][.]

Yours Very Truly | M. Faraday

Dr. J.J. Watson | &c &c &c

1. That is 14 January 1853.

Letter 2620
Faraday to Edward Sabine
12 January 1853
From the original in BeRO MS D/EBy F48 109/1

Royal Institution | 12 Jany 1853

My dear Sabine

I am sorry for your sake that I gave you the trouble of writing again[1]; but not for my own, for now I understand clearly the part of the effect to which you wished to draw attention, and am much struck with it, especially with the equality of amount in the different parts of the earth, and still more especially, with the fact as you state it that in a single week after the sun has passed the equator a full half of the *whole amount* of this annual effect is produced. What with a bad memory & the many calls upon my attention in other directions I find it difficult & headachey work at times, to realize the positions due to this or that particular kind of variation. I wish you could devise such a mode of representing them, either by coloured areas or otherwise, that one could look at any one simply, and then superpose one on another so as to obtain the *sum* of effect for any given time. My plate of curves[2] will not help me do this, because the means of the months are made to coincide; and therefore do not help me to see how the variation you now distinguish, combines with that I referred to of the amount of variation at a north or south place at the time of the one or other solstice. If I could succeed in catching experimentally the amount of change in the paramagnetic character of oxygen by a given difference of temperature, as that of 60° and 0°, then I should be encouraged to try & compare the phenomena & the facts again: but I have not succeeded in that yet, and yet worked so hard for that purpose in the autumn, that at last I was obliged to relinquish it from the effect of mere weariness & fatigue. I dare say I shall be tempted to try again after the season is over, if all remains pretty well.

I do not know that I have any right to pretend to speak with any authority, but it seems to me that after your observation of the Solar spots and variation periods;- the determination - separately, & distinction one from another, of such particular phenomena as the variation you now refer to, that presented by the larger disturbances, & the several others which I may express by &c &c, are of the utmost consequence to the development of the physical condition of the magnetism of the earth. And although one may be tempted by the coincidence of sun spots & variations, or by the magnetic condition of oxygen, to try ones powers now & then; yet in fact it is working without the necessary data that are within our reach, if one sets to work before all these distinct variations, or features of variation, are developed, which such comparisons & collections as you have made & are making, can supply.

I am | My dear Sabine | Ever Yours | M. Faraday

Mr. Stokes left me to day after a most beautiful series of demonstrations on Monday & yesterday[3] regarding his discoveries. The finer experiments cannot be made in public but some beautiful ones are possible | MF

1. Letter 2618.
2. Faraday (1851e), ERE27, opposite p.96.
3. That is 10 and 11 January 1853.

Letter 2621
Lyon Playfair to Faraday
18 January 1853
From the original in IEE MS SC 2

[Athenaeum Letterhead] | Jermyn St. | 18th Jany 53

My dear Sir

I am sure that you will have no objection to sign Franklands certificate[1]. I believe him to be one of the most rising Chemists of the day & I am sure that he would be very proud of having your Signature attached to his certificate.

Yours Truly | Lyon Playfair

1. Faraday did sign Frankland's certificate for Fellowship of the Royal Society, to which he was elected on 2 June 1853, "From Personal Knowledge", RS MS Cert 9.319.

Letter 2622
Faraday to Lyon Playfair
19 January 1853
From the original in the possession of Mrs Raven Frankland

Royal Institution | 19 Jany. 1853

My dear Playfair

I am very much obliged to you for the opportunity of signing Franklands certificate[1].

Ever Truly Yours | M. Faraday

1. See letter 2621.

Letter 2623[1]
Sarah Faraday to an unidentified correspondent
19 January 1853
From the original in RI MS

RI Jany 19th 53

My dear Sir

I am afraid you will think me very negligent of your request for some records of my husbands *birth parentage & education* but I was not sure whether what *I* could say would be satisfactory & really he has not had a moment to spare this last fortnight. I have now got him to find me a paper on which he had put down some dates a few years ago - and if you can call

& tell me what you want we will look it over together or I will send you date of birth, 22 Sepr 1791 Father, James Faraday[2] a smith, from Yorkshire, &c &c or any thing you wish to know for as he says though not anxious to place his private life before the public, he is quite willing to answer any questions[.]

I am My dear Sir | Yours very sincerely | S. Faraday

1. This letter is black-edged.
2. James Faraday (1761–1810, information from the family). Blacksmith who had moved from near Kirkby Stephen to London before Faraday's birth.

Letter 2624
Samuel Warren to Faraday
19 January 1853
From the original in IEE MS SC 2

Inner Temple | 19th January 1853

Dear Dr. Faraday,

Will you send me an order for admission to the R.In. on Friday next, to hear your Lecture on *"Magnetic Force"*?[1]

I have just sent to press a work on *"The Intellectual & Moral Development of this Age"*[2] (of which I shall send you a copy); & in it I speak not a little about your discoveries[3] - & represent you, as the head of the Chemists of the Age, or on the eve of making some prodigious discovery: as I truly believe you are.

I am sorry that I missed you so soon after we had begun to talk together, at the Duke of Northumberland's, some few months ago[4]. I looked for you, within 5 minutes afterwards, everywhere & you were gone!

Believe me | My dear Professor Faraday | Very faithfully Ever | Your's | Samuel Warren

P.S. I presume the Lecture begins at 8pm?
Professor Faraday DCL

1. Faraday (1853a), Friday Evening Discourse of 21 January 1853.
2. Warren (1853).
3. *Ibid.*, 69–70.
4. See note 2, letter 2508.

Letter 2625
Faraday to Samuel Warren
20 January 1853
From the original in MU MS 173

R Institution | 20 Jany 1853

My dear Sir
 The discourse[1] will begin at 9 o'clk but the Library is open at 8 o'clk & the Lecture room at $\frac{1}{2}$ p 8 o'clk. I am proud of the thought of your being there[2] but have not much to say and having an oppressive cold am in a poor state to say it. I am pretty well worn out and you must not either in your thoughts or your writings expect much more from a pitcher that has been often to the well & has received some fractures[.]
 Ever Very Truly Yours | M. Faraday
Saml. Warren Esq | &c &c &c

1. Faraday (1853a), Friday Evening Discourse of 21 January 1853.
2. See letter 2624.

Letter 2626
Faraday to Thomas Hodgkin[1]
22 January 1853
From the original in Durham County Record Office MS D/HO/C38/10

Royal Institution | 22 Jany 1853

My dear Sir
 There must be some mistake in your impressions. I gave a discourse here last night[2] but have no other evening lecture to give. Neither have I any day lectures on hand perhaps I may be so engaged after Easter[3] but that is not arranged and want of health may prevent me[.] However I am much better than I have been & hope to go through the season. If the American Minister should be in London & willing to honor me with his presence I should be very glad to send you the pass & very proud of his presence.
 Ever Truly Yours | M. Faraday
Dr. T. Hodgkin | &c &c &c
 I send you a Friday Evening ticket *undated* & *unnamed* that if you should like to bring Mrs. Hodgkin[4] on any Evening you may be able to do so | MF

1. Thomas Hodgkin (1798–1866, DNB). Physician.
2. Faraday (1853a), Friday Evening Discourse of 21 January 1853.
3. Faraday delivered a set of six lectures on static electricity after Easter. His notes are in RI MS F4 J15.

4. Sarah Frances Hodgkin, olim Scaife née Callow (d.1875, age 71, DQB). Married Hodgkin in 1850. See his DNB entry.

Letter 2627
Faraday to William De La Rive
24 January 1853
From the original in BPUG MS 2316, f.77

Royal Institution | 24 Jany 1853

My dear Sir

I have received from Longman's[1] the copy of the First Vol of the Work[2] of my friend your father of which you spoke to me. I thank you first for it as it comes through your intervention and I shall ever value it highly first & foremost as an association with & remembrance from him whom I esteem so highly and next as a great contribution to science[.] Though I have only looked into it as yet, I am able to see it has a claim to such character. When I have had time to read it I shall then write to your father himself and acknowledge both the book & his last kind letter[3][.]

Ever Very Truly Yours | M. Faraday

M. de la Rive Esq | &c &c &c

1. That is Longman, Brown, Green and Longmans, publishers in Paternoster Row. Wallis (1974), 40.
2. De La Rive (1853–8), 1.
3. Letter 2610.

Letter 2628
Faraday to Jacob Herbert
24 January 1853
From Symons (1882), 189–90

Royal Institution, | 24th January, 1853.

My Dear Sir,

In reference to the remarkable stroke of lightning which occurred at the Eddystone Lighthouse, at midday on 11th January of this year, and made itself manifest by a partial flash discharge in the living rooms, I have to call your attention to the drawing herewith returned, and to the circumstances which appear (from it) to have accompanied and conduced to the discharge[1].

In the body of the stone work above the store-room exist eight rings of metal; each going round the building, and each being four inches square of solid iron and lead. Also, latterly the bed-room and sitting-room have been lined with a framework of iron bars, situated vertically, and pinned by long bolts into the stonework.

The part of the tower above the floor of the living-room is, therefore, filled with a metallic system, which, with the metal lantern, gives a very marked character to the upper half of the structure.

The recent metallic arrangements (but not the rings) are connected with the lightning rod; and the copper part of this rod, beginning at the floor of the living-room, then proceeds downwards by the course which can be followed in the drawing, and terminates on the outside of the rock between high and low water marks.

Considering all these circumstances, I was led to conclude that the conductor was in a very imperfect condition at the time of low water; and I had little doubt that I should find that the discharge had taken place when it was in this state, and very probably with a spring tide.

The day of the stroke was the 11th January - a new moon occurred on the 9th, so that it was at a time of spring tide.

The occurrence took place at midday; and, according to the tide tables, that was close upon the time of low water at Devonport. The end of the conductor would then be 6 feet from the water, if the latter were quiescent, and I cannot doubt that this circumstance gave rise to that diverted discharge which became so manifest to the keepers. Mr. Burges[2], with whom I have conversed about the matter, thinks it probable that, through the violence of the waves, the conductor does not now descend so much as is represented in the drawing.

I think it essential that the lower end of the conductor be made more perfect in its action; and I should prefer this being done on the *outside* of the tower and rock, if the rod can be rendered permanent in such a situation.

If it be impossible to prolong and fix the lower end of the conductor where it now is, so that it shall have large contact with the sea at low water, then I would suggest, whether or no, on the more sloping part of the rock, about midway between high and low water, three or four holes could not be sunk to the depth of 3 feet, and about 3 or 4 feet apart, and that copper rods being placed in these, they should be connected together, and the lightning rod continued to them.

If this *cannot* be done, then it might be right to consider the propriety of the making a hole through the centre of the building and rock, about 2 or more inches in diameter, and 30 feet deep, and continuing the conductor to the bottom.

A conversation with Mr. Burges regarding the present state of the Bishop's Rock Lighthouse, now in the course of construction, induces me also to suggest the propriety of making provision for the lightning conductor as the work proceeds.

It would be easy now to fix terminal rods of copper, and to combine them upwards with the work. Considering the isolated and peculiarly exposed condition of a lighthouse on this site, I would propose that there be *two* conducting rods from the lantern, down the outside on opposite sides of the tower, each terminating below in two or three prolongations, entering as proposed into the rock, or into fissures below low water mark, so as to be well and permanently fixed[3].

I am, &c. | (signed) M. Faraday
The Secretary, | Trinity House.

1. The report of this lightning strike was noted in Trinity House By Board, 18 January 1853, GL MS 30010/38, pp.16–17 who ordered that a copy be sent Faraday. A copy of the drawing of the Eddystone lighthouse is given in Symons (1882), 191.
2. Alfred Burges (d.1886, age 84, GRO). Engineer and a partner of James Walker from 1829. Smith, D. (1998), 24.
3. This letter was read to Trinity House By Board, 25 January 1853, GL MS 30010/38, p.31. They ordered that information on the lightning conductor be provided to them. This was given at Trinity House By Board, 15 February 1853, GL MS 30010/38, p.47 and sent to Faraday.

Letter 2629
Charles Wheatstone to Faraday
26 January 1853
From the original in IEE MS SC 2

Lower Mall | Hammersmith | Jany 26th 1853

My dear Faraday

If you will send to King's College Dr Miller[1] will let you have the original revolving mirror with the whirling table to which it is fitted; I have spoken to him about it. I have at home a small revolving mirror ($\frac{1}{4}$ of an inch square) with a watch movement making 200 revolutions per second with a tolerably accurate means of measuring the angular elongation of a spark, which I have used to measure the duration of sparks in electro-magnetic coils; Mr Stokes can also have this if it will be of any service to him[2].

Yours very truly | C. Wheatstone

1. William Allen Miller (1817–1870, DSB). Professor of Chemistry at King's College, London, 1845–1870.
2. See letter 2616.

Letter 2630
John Tyndall to Faraday
26 January 1853
From the original in APS MS

Queenwood College | near Stockbridge Hants | 26th Jan. 1853.
Dear Prof. Faraday,
Will you allow me to introduce to your notice an experiment I made in Berlin, which I thought very remarkable at the time, but which I have had no opportunity to follow up since. It appears to me to lead to the same conclusion, if it be not the same in principle as those you have recently made. - It proves that the law of decrease for iron is different from that of a salt of the metal. Do you not consider it probable that a difference of this kind between water and bismuth is the cause of the greater repulsion of the latter at an increased distance. As the bismuth is carried to a greater distance it may in fact be regarded as immersed in a new fluid altogether, and if the difference of repulsion between this fluid and the bismuth be greater than the corresponding difference *near the poles* the facts observed would of course follow.

I confess I never entertained the thought of applying the experiments described on the accompanying leaf as you have done, and though startled by the profound ingenuity of your argument on Friday night[1], I did not imagine that the conclusion could be at all approached by the route chosen, the old magnetic law has still I confess a hold upon my convictions.

Believe me dear Sir | Most faithfully Yours | John Tyndall

1. Faraday (1853a), Friday Evening Discourse of 21 January 1853. See Tyndall, *Diary*, 21 January 1853, 5: 184–5 for Tyndall's description of the lecture.

Letter 2631
Faraday to William Whewell
29 January 1853
From the original in TCC MS O.15.49, f.34

Royal Institution | 29 Jany 1853
My dear Sir
You frighten me; for if I have not conveyed to you a clear idea of my meaning by my papers, how can I expect I have succeeded in relation to others. And then how *obscure & confused* my three last papers must have seemed[1]. As you have thought it worth while to look at them once I hope you will again:- for a great deal must have appeared to you to be utter nonsense, which I trust will now have a new meaning & some value in

your eyes. My ideas of lines of force as closed curves passing through the magnet is, that they pass *from end to end*, or in a direction coinciding therewith: *every* line of force, wherever it may

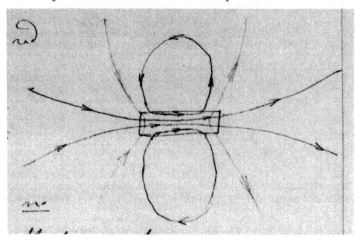

issue from the regular magnet, is *within* it at the magnetic equator; and passes through its *section* at that place. The black lines represent my curves *not* the red lines: the latter could not form continuous lines of force because their polarities would be opposed at the poles of the magnet. I recognize *no poles or centres of action* in that respect. It is curious that I do not find in all my figures one in which the lines of force are marked within the magnet. Of course all those given by filings could give only the direction of the external parts of the lines. But in (3116[2]) I speak of the full action of the moving internal radius wire alone (b.d. fig 6. par 3095. 3098[3]) and in 3117[4], say that the curved portions of the lines are continuations of the outer portions *absolutely unchanged in their nature*: which would not be true of the red lines on the previous page, where the polarity is suddenly inverted. In (3120[5]) I compare the magnet to an Electromagnetic helix in regard to the direction of the lines; a comparison further illustrated by figg. 5 and 20 of the plate. In 3276[6] of the paper on the physical lines of force I compare it to a Voltaic battery immersed in a decomposable or conducting fluid; & to a Gymnotus surrounded by water; and I am in hopes the Paragraphs 3277. 3278. 3283. 3287. 3288. 3295[7] will have a new interest to you, now that you will have caught my idea. 3271. 3265[8]. &c. 3231[9]. also: in fact I think you will be induced to read the papers again.

Dr. Rogets[10] determination of the forms of the lines of magnetic force, is given in the Journal of the Royal Institution 1831, vol I, page 311[11]: and *his* further account of them in the Treatise on Magnetism published in the Library of Useful Knowledge Natural Philosophy vol. II, page 19 of the particular treatise[12]. Wheatstone promised to shew me a folio volume containing demonstrations & forms by a German author but has never kept his word[.]

In a few days I hope to send you Newtons testimony in favour of physical lines of force as regards Gravity[13]. If that idea is necessary to gravitation, how much more so it must be in relation to the dual powers of Magnetism and Electricity which act in curve lines, and which in the case of a magnet act at all times and when only a single system is concerned[.]

Ever My dear Sir | Truly Yours | M. Faraday
Revd. Dr. Whewell | &c &c &c

Address: Revd. Dr. Whewell | &c &c &c | Trinity College | Cambridge

1. Faraday (1852b, c, d), ERE28, 29 and [29a].
2. Faraday (1852b), ERE28, 3116.
3. Faraday (1852b), ERE28, 3095, 3098.
4. Faraday (1852b), ERE28, 3117.
5. Faraday (1852b), ERE28, 3120.
6. Faraday (1852d), [ERE29a], 3276.
7. Faraday (1852d), [ERE29a], 3287, 3288, 3295.
8. Faraday (1852d), [ERE29a], 3271, 3265.
9. Faraday (1852c), ERE29, 3231.
10. Peter Mark Roget (1779–1869, DNB). Physician and Secretary of the Royal Society, 1827–1848.
11. Roget (1831).
12. Roget (1832), 19.
13. Possibly a reference to Newton to Bentley, 25 February 1692/3, Turnbull (1961), letter 406 which is quoted to this effect in Faraday (1855b), [ERE29b], 3305. Faraday could have read this letter of Newton's in a number of places including the entry on Newton in *Biographia Britannica*, 6 volumes, London, 1747–1766, 5: 3244.

Letter 2632
Faraday to John Tyndall
29 January 1853
From the original in APS MS

Royal Institution | 29 Jany 1853

My dear Sir

I have only at this moment been able to read your letter[1], having laid all I could aside that I might write some simple plain notes of last F. evening[2]: and now that I take it up, I cannot venture to judge of your experiments or draw conclusions from them not having seen the experiments or the apparatus employed. I never can *decide* upon a result until I have seen the experiment it depends upon made and then never until I have considered repeated & varied the experiment[.] Pluckers conclusion is one that I do not know of my own knowledge to be correct[.] I mean the varying rate of change of the Paramagnetic & diamagnetic form of force though I have seen many experiments. I know that some of the forces of his experiments are truly stated and think he may be right but would not vouch for it or give it as my own decided opinion. There is no end to the cases in which I think it safest not to come to a conclusion least it should be erroneous[.]

Ever Truly Yours | M. Faraday
Dr. J. Tyndall | &c &c &c

1. Letter 2630.
2. Faraday (1853a), Friday Evening Discourse of 21 January 1853.

Letter 2633
William Whewell to Faraday
30 January 1853
From the original in TCC MS O.15.49, f.64

Trin. Lodge, Cambridge | Jan. 30, 1853
My dear Dr Faraday,

As I expected, a single line[1] from you has turned my darkness into day. I ought never to have been *so* dark; and I believe I had formerly understood you better; but I have not attended to such subjects for a long time; and turning to them again for a few hours, I found myself in a puzzle of which you had the result, and which is now quite dissipated. I think, at the same time, that it may be useful to others of your readers that you should draw some figures in which you delineate the currents inside the magnets; for readers are much led by the eye. To explain my getting into such perplexity, I may say that I have been generally in the habit of thinking of your curves as identical with the old magnetic curves, which only give the active part of the curves. Also, when I gave you the word *sphondyl*[2] I thought rather of a solid generated by the revolution of this figure

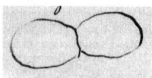

than this.

though the word may describe either. The sphondyl is in fact an annulus described by the revolution of a closed curve round an axis.

You will find a calculation of the properties of the magnetic curve, besides the places you have mentioned, in Leslie's[3] Geometrical Analysis (1821)[4] but undoubtedly I have seen it in much older books.

With regard to Newton's views of lines of force, I have given some account of them, as well as of other ways of expressing gravity, in the Philosophy of the Inductive Sciences. B. III. Ch. IX, Sect. 7[5][.]

I know that you try carefully to keep yourself free from all assumptions as to the mode and machinery of the actions which you investigate: but I think the incautious followers of the same line of speculation may run into something very like Descartes'[6] system of *vortices*[7]. I do not at all mean that because it is a system of vortices it must be wrong; but if it be that it must be an assumption which requires a good deal of proof.

Believe me dear Dr Faraday | Yours very truly | W. Whewell

1. Letter 2631.
2. Letter 2494.
3. John Leslie (1766–1832, DSB). Professor of Natural Philosophy at Edinburgh University, 1819–1832.
4. Leslie, J. (1821).
5. Whewell (1840), 1: 250–1.
6. René du Perron Descartes (1596–1650, DSB). French philosopher.
7. Also discussed in Whewell (1840), 1: 250, 256.

Letter 2634
Julius Plücker to Faraday
9 February 1853
From the original in IEE MS SC 2
Dear Sir!

After a longer silence, allow me at first to thank you for your last paper[1], I received two months ago. From your admirable activity in science I may conclude, that your health be improved, and I am happy to do so.

I take the liberty to send you my last mémoir[2] on the way of library, two former ones[3] will have reached you, six months ago, on the same way.

The indications you gave of the contents of your lecture at the Royal Institution on magnetic lines of force[4], engaged myself to send a paper of mine, unfinished as it was, to Poggendorff. When I was last time at London, I intended to explain to you what it contains, but I desisted to do so in that troublesome period, not thinking then I might meet you in these researches.

I part from the law of the action of an element of the current on a magnetic pole, as given by Biot, immediately after Oerstedt's discovery[5]. From that law I deduce the law of the induction of a current by a revolving pole, admitting the Newtonian principle, of the equality of action and reaction, as existing between both classes of phenomena. A pole turning round a conducting wire is acting on this wire as an "electromotrice force", I deduce the laws of its action, both experimentally and mathematically. In the case of a revolving magnet bar (if you take the case of the Earth) I conclude that there is a tension of negative Electricity in the equatorial zone and a tension of positive one in the arctic regions. &c &c &c.

Mr. Geissler[6], our clever artist, constructed a year ago, for the use of wine-makers', an apparatus, indicating the quantity of alcool, by the tension of the vapours of the mixture mixed with a certain quantity of air, all at the temperature of 100C. This apparatus improved on my advice by excluding all air, is that which Dr. Waller[7] bought from Mr. Geissler and for which he took a "brevet" for England[8]. Dr Hofmann, as it is written to Mr. G., intends to present it at the R. Institution. This apparatus startled me by the *regularity* of its indications, non [sic] only on mixtures containing alcool but also in many other cases, namely in the case of solutions. I left to Mr Geiss[l]er the care to work out (exploiter) his technical apparatus, reserving to myself, after that was done, to take use of the principle, well known before, but not at all sufficiently appreciated, for scientific researches on the affinity of vapours, on Dalton's[9] law, modified for the case of mixed vapours &c &c. After Dr Waller's departure I took up the question and I intend now to send to Poggendorff the first part of the results I obtained, assisted by Mr Geissler[10]. For this purpose new apparatus were constructed, the "Vaporimeter" not being able for scientific researches.

I beg you Sir to present my respects to Mad. Faraday.

Yours very truly | Plücker

Bonn, 9th February | 1853.

Address: Professor Faraday | &c &c | Royal Institution | London.

1. Probably Faraday (1852d), [ERE29a].
2. Possibly Plücker (1852b).
3. Possibly Plücker (1852a) and Plücker and Geissler (1852).

4. Faraday (1852a), Friday Evening Discourse of 23 January 1852. See Plücker (1852b) which cites this lecture on p.352. Page 107 of the lecture is quoted *in extensio* by Plücker on p.371.
5. That is Biot's comments on Oersted's discovery of electro-magnetism. Oersted (1820) and Biot (1821). For a discussion of Biot's views see Gooding (1990), 36–43. See Plücker (1852b), 376–8 for his discussion of Biot's work.
6. Johann Heinrich Wilhelm Geissler (1815–1879, DSB). Technician at the University of Bonn from at least 1852.
7. Augustus Volney Waller (1816–1870, DNB). Physiologist.
8. Patent 1852/288.
9. John Dalton (1766–1844, DSB). Chemical philosopher who lived in Manchester. Developed a version of the atomic theory of matter.
10. This was published as Plücker (1854b) which described the apparatus on pp.199–200.

Letter 2635
Samuel Warren to Faraday
11 February 1853
From the original in RI MS Conybeare Album, f.37

Inner Temple | 11th Febry 1853.

My dear *Philosopher,*

Pray drop me a line when you shall have read my little book & tell me what you think of it - particularly the note at p69[1].

Your's with hearty regard & | respect, ever, | in the bonds of true philosophy | & friendship, | Samuel Warren
Mr T. [sic] Faraday Esq | D.C.L., FRS

1. Warren (1853), 69–70 referred in glowing terms to Faraday's recent work on magnetism.

Letter 2636
Faraday to William Francis
12 February 1853
From the original in St Bride Printing Library, Taylor and Francis Collection: author's letters

R Institution | 12 Feby 1853

My dear Sir

I send you a copy of a report on a Friday evening subject[1]. As the matter has not been communicated to the Royal Society or elsewhere I have drawn it up more carefully that usual & at some length. I thought perhaps you might like it for the Phil.Mag. but use your own pleasure entirely in respect of it. If you insert it you may either have it as a report or as communicated from myself - or give it any shape you please[2][.]

I ought to say that though drawn up only for the use of the Members of the R Institution these reports get out & appear occasionally in the Athenaeum or elsewhere[3]. I do not know that it will appear any where prior to the 1st of March or even after, but I thought it right to mention the possibility.

Tyndall gave us an excellent discourse last night delivered in an admirable manner[4][.]

Ever Truly Yours | M. Faraday
Dr. Francis | &c &c &c

1. Faraday (1853a), Friday Evening Discourse of 21 January 1853.
2. This was put in as a report, *Phil.Mag.*, 1853, **5**: 218–27.
3. See *Athenaeum*, 19 February 1853, pp.230–1 for an account of Faraday's Friday Evening Discourse of 21 January 1853 "Observations of the Magnetic Force".
4. Tyndall (1853), Friday Evening Discourse of 11 February 1853. For accounts of this Discourse see Tyndall, *Diary*, 11 February 1853, **5**: 191–3, Tyndall (1868), 243–4, Eve and Creasey (1945), 39–41.

Letter 2637
George Gabriel Stokes to Faraday
14 February 1853
From the original in IEE MS SC 2

Pembroke College, Cambridge | Feb 14th 1853

My dear Sir,

You will receive I hope some time tomorrow a tin case containing some diagrams. I enclose you the key. Will you have the goodness to order them to be laid flat? I do not think that I shall use more than the 4 outside ones; I merely send the rest for fear I should wish for them and regret having left them behind. They are what I used at Belfast[1], and they have got a good deal dirtied and rubbed. Will you have the goodness to look at the four outside ones, to see if you think they are decent to produce before a London audience? If not, perhaps there would be time to have them copied or retouched.

I intend to go to the Royal Institution on Thursday evening[2]. I should be glad to have some oxygen ready, as I wish to try some more experiments with the sulphur light. I think it would be possible, and interesting, to show the audience some chemical reactions observed by means of the effects produced by the media on the invisible rays. For this purpose I should wish to have a little of a solution of quinine, (not a quinine salt) in alcohol.

I have found an easy way of purifying the horse-chestnut solution. I mean to bring some of the purified solution with me, but in case of any accident please tell Anderson to make a decoction of a good part or the whole of the bark which is left, to the fluid decanted or filtered, to add a little carbonate of ammonia, and then leave it in a shallow open vessel, so as to be exposed to the air. It requires a day or two's exposure. It will be time enough when I come to go on with the process.

I should be very sorry that you should forego any engagement on my account. Anderson can attend me. Indeed you have always got *one* engagement, namely to go on with your most important investigations.

I should like a little hydrogen as well as oxygen, to try the effect of the lime light once more with absorbing media.

I am dear Sir | Yours very truly | G.G. Stokes

1. A reference to the evening lecture Stokes gave on fluorescence to the Belfast meeting of the British Association on 3 September 1852. *Rep.Brit.Ass.*, 1852, p.xl.
2. That is 17 February 1853 for Stokes (1853), Friday Evening Discourse of 18 February 1853. See James (1985), 151–2 for a discussion of this lecture.

Letter 2638
John Tyndall to Faraday
15 February 1853
From a typescript in RI MS T TS, volume 12, p.4022

Queenwood, 15th, Feb, 1853.

Dear Prof Faraday

I never calculated on my expenses being paid, but am perfectly content to abide by your usual practice on such occasions - My travelling expenses are the only ones worth naming and they amounted to about thirty shillings - This sum I shall be equally contented to receive or not to receive.

When a man feels deeply on any point I believe as a general rule he will not talk much about it. I once heard the silence of Goethe[1] for a year during which he was in love with a girl in Frankfort accounted for in this way, and I have always thought the hypothesis a probable one. Friday night[2] is to me what his sweetheart was to Goethe - I shall never forget it but cannot well write about it. The memory of it I trust will serve to keep me more loyally to my task, to assure me when I doubt, to strengthen me when I falter and to establish my faith in a sentiment which I have met somewhere in your own writings, that the patient and conscientious investigator of natural truth is ever sure of his reward[.]

Believe me dear Professor | Most sincerely Yours | John Tyndall

1. Johann Wolfgang von Goethe (1749–1832, DSB). German writer and philosopher.
2. Tyndall (1853), Friday Evening Discourse of 11 February 1853.

Letter 2639
Faraday to John Tyndall
16 February 1853
From a typescript in RI MS T TS, volume 12, p.4132

Royal Institution, | 16 Feb. 1853.

My dear Tyndall,

    I do not know what post office near you would be convenient for a money order, so I send a cheque for the 1.10.0, with many thanks on the part of the Institution, and I may truly say of many members who were at your discourse[1]. If the cheque is inconvenient to you, return it to me by post, and tell me what I should do.

    Ever truly yours, | M. Faraday.

1. Tyndall (1853), Friday Evening Discourse of 11 February 1853. See letter 2638.

Letter 2640
Carlo Matteucci to Faraday
19 February 1853
From the original in IEE MS SC 2

Pise 19. Fev.1853

Mon cher Faraday,

    J'ai mis à la poste pour vous une lettre imprimée[1] qui est addressée au Dr Jones a propos son livre sur m D B Reymond[2] - je suis tres-content que vous ayez offert a mon adversaire (il faut bien que je l'appelle malgré moi ainsi) les moyens de vous montrer ses expériences -. J'ai été faché que vous ayez accepté le dedicace d'un livre qui est plein de calomnies et de mensonges contre mon caractere et aux experiences d'electricité amicale. Vous savez que l'un et les autres sont *vraies*. Je me vengerai en vous priant (j'ai presque dit) en exigeant de vous, que vous lisiez cette lettre. Donnez moi ces 15 minutes - vous avez fait un si beau usage de votre temps que ce n'est pas le perdu que en employer si peu pour connaitre la verité sur un homme qui a pour vous vener[a]tion et grande amitié. Je fais partir demain par Mr Arago l'extrait de mes expériences sur le magnétisme de rotation et sur le diamagnétisme[3] - La plus grande curiosité c'est votre opinion -

    Agreez les hommages | d un respect et | attachement | tout de | C Matteucci

TRANSLATION

Pisa 19 February 1853

My Dear Faraday,

I have put in the post for you a printed letter[1] which is addressed to Dr Jones concerning his book on Mr D.B. Reymond[2] - I am very pleased that you have offered my rival (as I am obliged to call him, despite my self) the opportunity to show you his experiments. I was angry that you accepted the dedication of a book that is full of calumnies and lies about my character and my experiments on animal electricity. You know that both are *true*. I shall avenge myself by asking, (I almost said) by demanding, that you read this letter. Give me those 15 minutes - you have made such good use of your time that it is not lost by using so little of it to know the truth on a man who has respect and great friendship for you. I am sending tomorrow, through Mr Arago, the extract of my experiments on the magnetism of rotation and on diamagnetism[3]. I am most curious of your opinion.

Please accept the homage | of the complete respect and | attachment | of | C. Matteucci.

1. Matteucci (1853a).
2. Bence Jones (1852) which was dedicated to Faraday.
3. On this see Matteucci (1853b).

Letter 2641
Faraday to William Gilpin[1]
20 February 1853
From the original in Birmingham Public Library MS 135

Royal Institution | 20 Feby 1853

Sir

I regret that I cannot comply with your request but I am constrained by circumstances to give one common answer in the negative to all the applications made to me to act as Steward at Public dinners or meetings[.]

I am Sir | Your Very Obedient Servant | M. Faraday
William Gilpin Esq | &c &c &c

1. Unidentified.

Letter 2642
Faraday to George Biddell Airy
23 February 1853
From the original in RGO6 / 405, f.145

Royal Institution | 23 Feby. 1853.

My dear Airy

May I ask you to send the enclosed to your Carpenter perhaps he will return me the bill with *"Paid"* by post[1][.]

The kindest remembrances of my wife & self to Mrs. Airy[.]

Ever Truly Yours | M. Faraday

---

1. This refers to an invoice for some work carried out for Airy, G.B. (1853), Friday Evening Discourse of 4 February 1853. See Airy to Barlow, 21 February 1853, RGO6 / 405, f.144.

Letter 2643
Faraday to Jacob Herbert[1]
24 February 1853
From the original copy in GL MS 30108/1/56

Royal Institution | 24 February 1853

Dear Sir

On Tuesday last[1] an examination was made at the Trinity House into the relative merits of the French lens, belonging to the frame in the experimental room, and a corresponding lens recently manufactured by the Messrs Chance of Birmingham; which for the purpose of comparison was attached to the same frame. According to my judgment the *forms*, and *adjustment* of the different portions of glass of Mr Chance's lens are as perfect as in the French lens, and the light from the lamp therefore as well and properly dispersed on the distant screen. The *colour* of all parts of Mr Chances lens appeared to me by artificial light the same and very good; and equal to the great central part of the French lens. The upper and lower ribs of the French lens were slightly green, and therefore slightly inferior in that respect. On the whole I consider the lenses as being very equal in power and quality[3].

I am | My dear Sir | Your very faithful Servant | M. Faraday
Jacob Herbert Esqr | &c &c &c

---

1. Apart from the signature, this copy is not in Faraday's hand.
2. That is 22 February 1853.
3. This letter was read to Trinity House Court, 1 March 1853, GL MS 30004/25, p.217. The Court expressed its satisfaction with the contents.

Letter 2644
George Gabriel Stokes to Faraday
25 February 1853
From the original in IEE MS SC 2

Pembroke College Cambridge | Feb 25th 1853

Dear Prof. Faraday,

I deferred answering your letter in hopes of being able to say that I *had* sent the M.S. of the abstract of my lecture[1]. However I do not wish to delay longer, though I can only say that it is in progress.

With respect to your other question, I have not been at any expense about instruments, as what I used I had previously, with the exception of a single piece of coloured glass. As to my travelling expenses, which do not come to much, it is not my wish to be repaid, I should prefer leaving it as it stands.

Believe me | Yours very truly | G.G. Stokes

1. Stokes (1853), Friday Evening Discourse of 18 February 1853. See James (1985), 151–2 for a discussion of this lecture.

Letter 2645
John Forbes Royle[1] to Faraday
1 March 1853
From the original in RI MS Conybeare Album, f.38

Acton 1st March 1853

My Dear Faraday

I was much obliged by your kind note, of which the tenure [sic] is exactly what I expected & is exactly the cause I myself perceive in all elections at Kings College. But having at the same time that I wrote to you addressed others of the several whom I personally knew I was surprised to find that your rule was not observed in all cases. So much so as to make me doubt whether I should become a Candidate[2]. But as I want to present you with or persuade you to look at the chemical part of my Manual of Materia Medica[3] as connected with the subject I will do myself the pleasure of calling one of these days & telling you more particularly to what I allude.

Believe me | Yours Very Truly | J.F. Royle

1. John Forbes Royle (1799–1858, DNB). Surgeon, naturalist and Professor of Materia Medica at King's College London, 1836–1856.

2. For the Professorship of Physiology at King's College London. On this see Hearnshaw (1929), 232. The physician and microscopist Lionel Smith Beale (1828–1906, DNB2) was appointed to the chair.
3. Royle (1847).

Letter 2646
William Stevenson to Faraday
1 March 1853
From Proc.Roy.Soc., 1853, 6: 291

Dunse (N. Britain), March 1, 1853.

Dear Sir, - In the report in the Athenaeum of your lecture at the Royal Institution on the 21st of January[1], I observe that you refer to the highly interesting observations of Schwabe[2], Sabine[3], Wolf[4], Gautier[5], &c., from which it would appear that a connection exists between the solar spots and the variations of the terrestrial magnetic forces. Since a connection has been demonstrated to exist between the latter and auroral phenomena, I was induced to look over my notes relating to the aurorae observed at this place, with a view to ascertain whether these also exhibited maxima and minima, and if so, whether the periods of such agreed with those of the solar spots and of the magnetic variations. The subjoined table shows the distribution of the aurorae seen here in the years 1838 to 1847 inclusive:-

	Jan.	Feb.	Mar.	Apr.	May.	June.	July.	Aug.	Sept.	Oct.	Nov.	Dec.	Sum.
1838	5	3	4	3	...	...	...	2	4	1	2	3	27
1839	9	1	2	4	1	...	...	...	11	7	2	1	38
1840	5	5	2	4	...	...	...	3	7	6	6	5	43
1841	6	3	4	4	2	...	...	3	3	3	7	7	42
1842	2	2	...	...	...	...	1	...	3	...	1	...	9
1843	2	1	1	1	...	...	...	2	...	3	...	...	10
1844	1	...	2	...	...	...	...	...	1	3	4	2	13
1845	1	2	...	1	...	...	...	1	2	1	1	1	10
1846	...	1	...	...	...	...	1	2	7	4	1	...	16
1847	2	2	3	1	...	...	...	1	5	6	6	4	30
	33	20	18	18	3	...	2	14	43	34	30	23	238

These figures speak for themselves. I may remark that the returns for 1842 are incomplete, as I was absent from home during March and April of that year. In 1848 I was also absent for some months, but from the number of aurorae which I have noted during that year, I am satisfied that a maximum then occurred, both as regards the number and intensity of auroral displays. The present winter has been very barren in auroral phenomena.

Of crimson aurorae I find I have noted two in 1837, one in 1839, one in 1846, three in 1847, and no less than six in 1848.

A discussion of the aurorae seen in North America and the North of Europe during a series of years would be interesting with reference to the points in question.

Apologizing for troubling you, I am, dear Sir, | With the greatest respect, yours faithfully, | Wm. Stevenson

1. See *Athenaeum*, 19 February 1853, pp.230–1 for an account of Faraday's Friday Evening Discourse of 21 January 1853 "Observations on the Magnetic Force". Faraday (1853a).
2. Schwabe (1844).
3. Sabine (1852).
4. Wolf (1852a).
5. Jean Alfred Gautier (1793–1881, P1, 3). Professor of Astronomy in Geneva and Director of the Observatory. See Gautier (1852), 189.

Letter 2647
Faraday to Carlo Matteucci
3 March 1853
From the original copy in IEE MS SC 2

Royal Institution | 3 Mar 1853

My dear Matteucci

I was quite startled the other day by the receipt of your letter (I mean the MS. one to myself[1]) for my imperfect memory made me quite unaware that there was any thing in Dr. B Jones *translation* of Müller's[2] account of Du bois Reymonds experiments[3], which, could make it any source of annoyance or irritation beyond the original. I knew from matters reported in the Comptes Rendus & otherwise, that you & Du Bois Raymond were in some degree antagonistically placed[4]; a thing very much to be regretted, but which often happens amongst the highest men in every department of Science, and more often when there are two or three only that really pursue the subject, than when there are many. Still I may truly say that when Du bois Reymond was here[5] he never spoke of you in hard terms or objectionably to me; probably he avoided the subject, but he did not embitter it. Dr. Bence Jones translation was not completed I think in print until after he was gone; but of that I am not quite sure. Being entirely unacquainted with German I do not know what either Du bois Reymond or Dr. Muller may have said controversially, but I concluded you had borne with the work of the latter with that patience which most men of eminence have to practice. For who has not to put up in his day with insinuations & misrepresentations in the accounts of his proceedings given by others, bearing for the time the present injustice which is often unintentional and often originates in the haste of temper; & committing his fame & character to the judgment of the men of his own & future time[6][.]

I see that that moves you which would move me most namely the imputation of a want of good faith; and I cordially sympathise with any one who is so charged unjustly. Such cases have seemed to me almost the only ones for which it is worth while entering into controversy. I have felt myself not unfrequently misunderstood, often misrepresented, sometimes passed by; as in the cases of Specific inductive capacity[7], magneto electric currents[8], definite electrolytic action[9], &c &c but it is only in the cases where moral turpitude has been implied that I have felt called upon to enter on the subject in reply[10]. I can feel with you in the regret which you express (pp. 14, 15[11]) at having to write such a letter and employ time in such a manner and looking again at the abstract can see how p23[12] & some other parts have made you think it necessary to do so, but the letter being written it will at all events have the good effect of collating dates both before and after the year 1842. Ultimately this collation of dates is every thing for in all matter of scientific controversy the dates form the data upon which that final umpire is appealed to (i.e the scientific world) will judge[.]

I am sorry the dedication annoys you[.] I suppose the Italian & the English feeling must differ in that respect. I do not like dedications but I look upon them as Honorary Memberships and not to be refused without something like an insult to the other parties concerned. In the chief number of cases in which I have been concerned, I have not been *asked* before hand & in all cases would rather not. We are bound by our duty to the Members and to Science to let Du bois Reymond (or any other person) make his experiments here[13] and to the accident of his making them here is due the dedication itself as the book says.

These polemics of the Scientific world are very unfortunate things they form the great stain to which the beautiful edifice of scientific truth is subject: *Are they inevitable?* They surely cannot belong to science itself but to something in our fallen natures. How earnestly I wish in all such cases that the two champions were friends[.] Yet I suppose I may not hope that you & Du bois Reymond may some day become so. Well let me be your friend at all events & with the kindest remembrances to Madame Matteucci[14] & yourself believe me to be, my dear Matteucci,
    ever Very Truly Yours | M. Faraday

1. Letter 2640.
2. Johannes Peter Müller (1801–1858, DSB). Professor of Anatomy and Physiology at Berlin University, 1833–1858.
3. Bence Jones (1852) which was dedicated to Faraday.

4. Matteucci (1850b), Bois-Reymond (1850a, b), Matteucci (1850c), Bois-Reymond (1850c).
5. In 1852. See letters 2520, 2521 and 2522.
6. The following passage is crossed through here: "to whom all the necessary *dates* are made known in a manner that cannot be altered".
7. See Gooding (1978).
8. See Steinle (1996).
9. See Faraday to Solly, 23 November 1836, letter 952, volume 2.
10. See Faraday to Wollaston, 30 October 1821, letter 154, volume 1, relating to priority over electro-magnetic rotations.
11. Matteucci (1852a), 14–15.
12. Bence Jones (1852), 23.
13. See letters 2520, 2521 and 2522.
14. Otherwise unidentified.

Letter 2648
William Stevenson to Faraday
5 March 1853
From the original in RI MS Conybeare Album, f.38
M. Faraday Esq | &c &c &c

Dunse, (N.B.) 5 March 1853

Dear Sir,

I have no objection to your bringing the substance of my communication under the notice of the Royal Institution[1]. On the contrary I feel very much gratified that you consider it of so much interest.

I have for many years paid a good deal of attention to the phenomena of the Aurora & have often sent notices of the more remarkable displays to various public journals. Some of these have probably come under your notice. Mr. Glaisher[2] & I have had a good deal of correspondence regarding the connection of the Auroral displays seen here, with the disturbances of the magnetic instruments at Greenwich.

In 1840, I communicated to Sir D. Brewster some observations on the connection of *cirri* with Auroral phenomena. These were laid by him before the Phil. Society of St. Andrews[3] & I have often since then sent similar communications to that body. As many of my observations bear upon the subject of atmospheric magnetism, and, on that account, would probably be of some interest to you, I shall have much pleasure in looking over my notes when I have a little leisure & extracting such obs[ervatio]ns as are likely to be of interest.

I may state that the communication referred to above as made in 1840, related chiefly to the tendency of cirri to effect a linear arrangement in the direction of the Magnetic meridian. For two or three years previous to 1840 I had been particularly struck by this tendency, which was all the more remarkable since the direction of the motion of these clouds was generally at *right angles* to that of the Mag. meridian. Exceptions, due principally to cyclonic movements of the atmosphere, were of pretty frequent occurrence, but from what I observed I was satisfied that the normal or undisturbed position of the long parallel lines of cirrus & cirrostratus coincided with that of the Mag. meridian. For some years I supposed that I had been the first to detect this magnetic tendency of these clouds. I was however surprised & gratified (tho' it took from me all title in the eyes of the world to claim originality of discovery,) to find in the "Cosmos", that its illustrious Author[4] had remarked in South America, (under circumstances certainly much more favorable than are presented in our climate) the same tendency of cirri to an arrangement in "meridional bands"[5].

I have at this moment an impression that for two or three years somewhere between 1840 & 1848, fine, regular displays of cirri were much rarer than for a year or two prior to 1840; but I must look into this matter.

I Remain | Dear Sir | With the Greatest Respect | Ever Faithfully Yours | Wm. Stevenson

1. Faraday read letter 2646 to the Royal Institution on 4 March 1853. *Proc.Roy.Inst.*, 1853, 1: 275.
2. James Glaisher (1809–1903, DSB). Superintendent of the Magnetic and Meteorological Department at the Royal Greenwich Observatory, 1838–1874.
3. For an account of this see "Literary and Philosophical Society of St Andrews", *Fife Herald*, 5 November 1840, p.2, col. f.
4. Friedrich Wilhelm Heinrich Alexander von Humboldt.
5. Humboldt (1846–58), 1: 182–3.

Letter 2649
Faraday to William Stevenson
8 March 1853
From Duns (1883), 291

Royal Institution, | 8 Mar. 1853.

My Dear Sir,

I am very much obliged by your kind note[1] and shall be most grateful for any data you can favour me with in relation to atmospheric magnetism. The observations about the Cirri are exceedingly interesting, and if at your leisure you devote me a few minutes I shall be very thankful. Captn. [sic] Sabine was very much struck by your note to me (the first one[2]) and asked me to allow it to be read at the Royal Society next Thursday Evening[3], and I have sent it to him for that purpose.

Your Very Obliged Servt., | M. Faraday

1. Letter 2648.
2. Letter 2646.
3. That is 10 March 1853 when it was read.

Letter 2650
Faraday to Albert Way[1]
10 March 1853
From the original in FACLM H MS c1

Royal Institution | 10 March 1853

Dear Sir

I have seen & talked with Mr Medlock[2]; I have received from him an account of the composition of *the glass* and specimens of the glass itself. I have also made certain brief experiments on the glass with strong chemical agents which though they cannot pretend to produce in a short period the pure effect of time & exposure to air, seemed the most fitted to help in guiding the mind. The result is that with respect to your question "whether I should apprehend from the mode of fabrication or materials used, that this impressed glass may be more liable to risks of deterioration than ordinary glass"? I may freely answer that I do not see any reason to expect inferiority on that point[.] The consideration of the composition of the glass and its behaviour with powerful chemical agents suggest no suspicion of that kind[.]

I am | My dear Sir | Very Truly Yours | M. Faraday
Albert Way Esq | &c &c &c

1. Albert Way (1805–1874, DNB). Antiquary.
2. Henry Medlock (1825–1875, B6). Industrial chemist who also worked on the chemistry of medieval artefacts.

Letter 2651
Faraday to Edward William Brayley
10 March 1853
From the original in the possession of Elizabeth Baird

Royal Institution | 10 Mar 1853

My dear Sir

I owe you many thanks for your kindness in sending me the extract from O'Shaughnessy's[1] report. By degrees the evidence of facts accumulates and I hope we shall before very long be able to form a much better notion of the Electro magnetic or Magneto electric state of our earth than we have at present.

Your kindness in sending me the extract reminds me of a little chat we had once about my title "On the *Magnetization of light*". When you told me (what I knew) of the objections of others I held my own & said I was looking a little beyond their views - and I added to the title *a note*[2][.] If you happen to look at Walker's[3] translation of De la Rive['s] recent work on Electricity p522 at the bottom - p.523 and page 524 bottom[4] where De la Rive is speaking his own mind on the facts, you will see that I have reason to be satisfied.

My lectures after Easter will be very common place & old in matter for I have had no health or strength to construct or devise new matter but such as they may be I send you an order[5][.]

Ever Truly Yours | M. Faraday
E.W. Brayley Esq | &c &c

Endorsed by Brayley: Answered June 2nd. See p3.
"De la R's expression, p524[6], at the last par. "the Phen." to "action...." "*on the manner*" to "*the ether*" is exactly equivalent to those in yr. note."
"It appears to me, however, that the *resultant* of all yr researches in E. is the establishment or at the least the impossible indication of the existence of the ether"

1. William Brooke O'Shaughnessy (1809–1889, DNB). Physician and Professor of Chemistry at Calcutta.
2. Faraday (1846a), ERE19, p.1.
3. Charles Vincent Walker (1812–1882, DNB). Electrician.
4. De La Rive (1853–8), 1: 522–4.
5. Faraday delivered six lectures on static electricity after Easter. His notes are in RI MS F4 J15.
6. De La Rive (1853–8), 1: 524.

Letter 2652
Faraday to Dr M
12 March 1853
From Bence Jones (1870a), 2: 322

Royal Institution: March 12, 1853.

Dear Sir, - My words are *simple* and *correct*. I know that there are plenty of portraits: I do not know that there is a single likeness. I have compared the portraits with my face in the glass, and I cannot see a likeness in any one of them. Therefore, if I wished, I could not send you one. But as I never help to publish either portrait or likeness, I cannot in

any manner accede to your request. I think we may now consider this matter as finished.
Very truly yours, | M. Faraday

Letter 2653
Mary Somerville[1] to Faraday
12 March 1853
From the original in RI MS F1 I56b

Genoa 12th March 1853

Dear Dr Faraday
Your papers on electricity which you so kindly send to me from time to time give me infinite pleasure because they are most valuable in themselves, and because they show me that your health permits you to continue your invaluable experiments[.] Besides it is most gratifying to be remembered by you and to find that time and absence makes no change to your friendship.
I sincerely trust that Mrs Faraday is well, we all pass our best wishes to her, to you and to your niece who I hope is as successful as she used to be in raising exotic plants.
We go to Florence in April and shall be glad if we can do anything for you there[.]
Yours ever sincerely | Mary Somerville

1. Mary Somerville (1780–1872, DSB). Scientific writer.

Letter 2654
Faraday to Charles Manby
14 March 1853
From the original in WIHM MS FALF

Royal Institution | 14 Mar 1853

My dear Sir
I think Regnaults researches are in the Memoires of the Academie des Sciences[1]:- perhaps you will find what you want in Dixons[2] Treatise on heat 8vo (Dublin)[3][.]
Ever Truly Yours | M. Faraday
Chas Manby Esq | &c &c &c

1. Regnault (1847).

2. Robert Vickers Dixon (c1812–1885, B1, AD) Professor of Natural Philosophy at Trinity College, Dublin from 1848.
3. Dixon (1849).

Letter 2655
Jacob Herbert to Faraday
16 March 1853
From the original in GL MS 30108/2/89

Trinity House, London, | 16th. March, 1853.

Sir,

I am directed to send you the accompanying Sample of White Lead, belonging to one of the Tenders received for the supply of that article for the Corporation's Service, - and have to request, that you will analyze the same, and report the result, for the Board's information[1].

I am, | Sir, | Your most humble Servant | J. Herbert
M. Faraday Esq.

1. Faraday's analyses are in GL MS 30108/2/89 with a note that he reported on 26 March 1853. This was read to Trinity House Wardens Committee, 29 March 1853, GL MS 30025/22, p.71 who based their order on it.

Letter 2656
Jacob Herbert to Faraday
19 March 1853
From the original in GL MS 30108/2/60c

Trinity House,- | 19 March 1853.

My dear Sir,

I am instructed to transmit to you the enclosed Copy of certain Observations made before the Select Committee of the House of Commons on Light Houses in 1845[1], in respect to a plan for reflecting Sound; to acquaint you, that the Corporation propose erecting a Fog Bell at the South Stack Light House Establishment,- and to request that you will favor the Elder Brethren by stating whether, in your opinion, the apparatus described in the paper referred to, would be applicable to Sound,- and could be adapted with advantage to the purposes of the Bell in question.

I remain | My dear Sir, | Your very faithful Servant | J. Herbert
M. Faraday Esq. D.C.L. | &c &c &c

1. This is in GL MS 30108/2/60b and is from *Parliamentary Papers*, 1845 (607) 9, p.274 which was the evidence of the lighthouse engineer Alexander Gordon (1802–1868, B1).

Letter 2657
Faraday to Charles Manby
22 March 1853
From the original in RI MS F1 D16

22 Mar 1853 | 118 Kings road | Brighton

My dear Manby

I have not worked specially on the subject of the enclosed letter and therefore must decline to answer the enquiry[.] Indeed my memory is becoming so treacherous that I should not know without much search where to find the safest answer derived from the trials of others. I am sorry I cannot send you this note back in time as you wish[1][.]

Ever Truly Yours | M. Faraday

1. This presumably refers to some unidentified piece of business connected with the Institution of Civil Engineers.

Letter 2658
Faraday to Albert Way[1]
22 March 1853
From the original in FACLM H MS c1

Royal Institution | 22 Mar 1853

Dear Sir

I am very happy to hear that the window is to be as you desired and thank you for your kind letter. I will keep you continually in my debt for a little good will but nothing more[2][.]

Your Very faithful Servant | M. Faraday
Albert Way Esq | &c &c &c

1. Albert Way (1805–1874, DNB). Antiquary.
2. See letter 2650.

Letter 2659
William Stevenson to Faraday
23 March 1853
From the original in IEE MS SC 2
M. Faraday Esq | &c &c &c

Dunse, 23d March 1853

My dear Sir,
I have now the pleasure of enclosing the Abstract of my Observations which I promised to send you[1]. You are heartily welcomed to make *any use* of the paper that you may think fit. If it shall prove of any importance to Science I shall consider myself amply repaid for the labor.

The Extracts from my journals have extended I am afraid, to a tedious length, altho' I have abridged them as much as I could by leaving out the less important details. I have also omitted a great number of instances wherein the phenomena exhibited were very similar to those shown in the cases I have quoted. Of course I have the means of furnishing, if necessary, a great quantity of addt details & instances, but those now given may serve as a sort of general example. With regard to special points of enquiry which may suggest themselves to you, I shall be happy, on hearing from you, to furnish all the assistance in my power.

I intended to have added some general remarks to the paper now sent, but on consideration, thought it better to defer this lest I might do injury by premature attempts at theorizing.

I was much gratified to learn that Colonel Sabine felt so much interested in my note[2]. There are perhaps some things in the abstract sent which will also be interesting to him.

I Remain | My Dear Sir, | With the greatest respect | Yours Faithfully | Wm Stevenson

P.S. I should mention that *at night* there is often much difficulty in distinguishing between cirri & cirrostrati, so that it is possible that in some of my observations I may have mistaken the one form of cloud for the other. | W.S.

1. In letter 2648.
2. See letter 2649.

Letter 2660
Faraday to Jacob Herbert
24 March 1853
From the original copy in GL MS 30108/2/60c

Royal Institution | 24 March 1853.

My dear Sir

In reference to the proposition made on the evidence which you sent to me given by Mr. Gordon on the subject of the fog signal[1] I think I understand that the proposed arrangement has not been really tried but is suggested on the score of the analogy between light and sound. In my opinion it would not be prudent to substitute an untried method in a case of actual service for such as have been tried, the best of which would of course be selected.

If I were asked what results I should expect from experimental trials; with a view of testing the proposition and producing a good & useful apparatus in conformity thereto, I should say that the results are doubtful and that it would require many trials & reconstructions of the apparatus before either an affirmative or a negative result would be obtained. I do not as yet know of any thing approaching to a steam whistle either in power or arrangement of parts which has been obtained by a bellows yet something like such a source would be required not merely in respect of strength of sound but also of form and consequent distribution of the sonorous undulations. A steam whistle sends sounds of nearly equal intensity in all directions around it; any arrangement having the form of an ordinary whistle does not do so; and would not be so fit for a central source as the former.

Then as respects the reflexion of sound the origin of which is in a focus:- though the physical analogy of light and sound is very great, yet there is an infinite difference in degree between them, and we have no material which is to sound what polished silver is to light. In the absence of previous experience & trials it is even possible and most probable that the first acoustic reflectors found for such a purpose might deaden & destroy sound and that to a large extent. On the other hand if by investigation a constant & useful mode of construction could be discovered and the best material be ascertained, still the curvature and the size of the reflectors which would be required are altogether unknown, & would need further experiments for their determination[.]

So that at present if any one were to propose to work out such a system as that described by Mr. Gordon I should not object on the score that the proposition was unphilosophical or against natural principles but I should say experimental investigation would in the first place be required & I should not be surprized if he failed to eliminate any practical or applicable result[2][.]

I am My dear Sir | Very Sincerely Yours | M Faraday

1. See letter 2656.
2. This letter was read to Trinity House By Board, 29 March 1853, GL MS 30010/38, p.93 who referred it to the Lights Committee.

Letter 2661
Jacob Herbert to Faraday
28 March 1853
From the original in GL MS 30108/2/89

Trinity House London | 28 March 1853.

Sir,

I duly received your Letter of the 26th Instant reporting on the Sample of Mr. Grace's[1] White Lead[2],- and am directed to send you a Sample of another Party who tender'd for the same Contract, which you will please analize and report upon[3].

I am, | Sir, | Your most humble Servant | J. Herbert

M. Faraday Esq. D.C.L. | &c &c &c

1. There are several white lead manufacturers by the name of Grace listed in POD.
2. See letter 2655.
3. Faraday's notes of this analysis are in GL MS 30108/2/89. His report, dated 4 April 1853, was read to Trinity House Wardens Committee, 5 April 1853, GL MS 30025/22, p.75 who confirmed their order.

Letter 2662
Faraday to Charles Fellows[1]
4 April 1853
From the original in WIHM MS 5634
    With M. Faradays Compliments to Sir Charles Fellows.
Royal Institution | 4 April 1853

Address: Sir Charles Fellows Bart | &c &c &c | 4 Montague Place

1. Charles Fellows (1799–1860, DNB). Traveller and archaeologist.

Letter 2663
Faraday to William Frederick Pollock
4 April 1853
From the original in the possession of George W. Platzman
                              Royal Institution | 4 April 1853
My dear Pollock
    I enclose an Anderson[1] card and venture through you to recommend it to Mrs. Pollocks favour. As to the other & more masculine matter I assure you I will not forget your letter when the time comes[.]
    Ever Very Truly Yours | M. Faraday

1. Unidentified.

Letter 2664
Faraday to Thomas Newborn Robert Morson[1]
7 April 1853
From the original in SI D MS 554A
                              Royal Institution | 7 April 1853
My dear Morson
    I need for my lecture on Saturday[2] some films or sheets of Collodion[.] I once had some about the size of my hand & cannot tell where I should go for them[.] Have you such a thing or can you tell me where I can get them? They were quite transparent & very electric and it is to illustrate that point that I want them i.e. one or two such sheets[.]
    Ever Truly Yours | M. Faraday

1. Thomas Newborn Robert Morson (1799–1874, Morson (1997)). Pharmaceutical chemist.
2. That is 9 April 1853 when Faraday delivered his first lecture of his course of six lectures on static electricity. Faraday's reference to the use of collodion in this lecture is in his notes, RI MS F4 J15, p.1.

Letter 2665
Edward Augustus Inglefield[1] to Faraday
7 April 1853
From the original in IEE MS SC 2

H.M.S. Phoenix | Woolwich | 7th/4/53

My dear Sir

You may possibly be aware that another Expedition is about to proceed to the Arctic Seas, and I have been appointed to the command[2]. It is not impossible that I may have to spend a winter in those high latitudes[3] & should that be the case, I shall have much leisure to carry on philosophical experiments. If I can make any observations or experiments for you, I shall be too glad of the occasion and if there are any suggestions you would kindly give me as to magnetical or electrical research I shall be ever

Yours very gratefully | E.A. Inglefield

P.S Perhaps it would save your time if I might be allowed an interview with you some morning early. Mr. Gassiot informs me about 9 is generally the most agreeable hour to you[.]

9 Portsea Place | Connaught Sq

1. Edward Augustus Inglefield (1820–1894, DNB1). Commander in the Royal Navy.
2. This was yet another expedition to search for the Arctic explorer John Franklin (1786–1847, DNB).
3. The expedition did not winter in the Arctic.

Letter 2666
Faraday to Charles Tomlinson[1]
8 April 1853
From the original in University of Texas at Austin Charles Darwin letters

Royal Institution | 8 April 1853

Dear Sir

I am very happy to inclose the pass for the Porter and only hope Mr Cunningham[2] may be interested in the simple matters which I shall have to talk about tomorrow[3][.]

Ever Truly Yours | M. Faraday
C. Tomlinson Esq | &c &c &c

1. Charles Tomlinson (1808–1897, DNB). Scientific writer.
2. Unidentified.
3. That is when Faraday delivered his first lecture of his course of six lectures on static electricity. His notes are in RI MS F4 J15.

Letter 2667
William Stevenson to Faraday
8 April 1853
From the original in IEE MS SC 2
M. Faraday Esq | &c &c &c

Dunse, 8th April 1853
Dear Sir,
I am favoured with your letters of the 30th ult & 6th inst.
It is very gratifying to me to learn that such eminent judges as Colonel Sabine & yourself have considered my observations to be of such Scientific importance as to render their publication, in some form, desirable.
I have not the slightest objection to the publication of my observations, wholly or partially, in any journal or in any form, so that they may be made accessible to those who take an interest in the Subject. I have as yet published nothing on Meteorological Subjects except in the shapes of newspaper paragraphs, slips for circulation among my friends, & the like. Perhaps as you suggest the Philosophical Magazine would be a suitable vehicle for the publication of the observations contained in the abstract which I sent you. They would however of course require to be put into a proper form, for I need scarcely say that in making up the abstract I did not think it necessary to attend to anything like literary arrangement. I am quite willing to entrust the Editors of the Phil.Mag. to put the paper into proper shape for their journal. In this case I should like to see what alterations they make upon the M.S. before it is in the hands of the Printer, or, at least, to have an opportunity of looking over the proofs, that any clerical or other errors may be rectified. Or, I could alter the form myself in accordance with any suggestions which you might make, in conjunction (perhaps) with the Editor of the Phil.Mag. I should however rather prefer the former method, for, among other reasons, I have not hitherto been a reader of that journal, and am therefore not a proper judge of style &c suitable to its readers. All I wish is that whatever is of value in my

observations should be made available to the Scientific public. The paper could be much shortened by striking out many of the less important notes[1].

I may perhaps at some future time, publish a small book upon the Subject, but as I am not yet prepared for this, I think it much better not to detain from the public what may be of use in the meantime.

The results of your investigations respecting the effect of low temperatures & rarefaction upon the magnetism of oxygen, will be looked for with great interest[2].

If I recollect right I stated somewhere in the "Abstract" sent you, that I had been led to the opinion that the directions of the progressive motions of cyclones compared, at least approximately, with those of the upper atmospheric currents as indicated by the motions of cirri. Within the last two months I have observed some very striking instances confirmatory of this opinion, and I should like much that the attention of meteorologists was directed to the Subject, as any addition to our knowledge of the laws & indications of the approach of these destructive visitants, must be of great value to humanity, as well as of much Scientific importance. The following is a Short abstract of some of my notes regarding the cases referred to:

On the 18th Febry the cirri moved from N to S, their bands ranging NNE to SSW, wind light from about WNW & barometer declining. I noted in my journal "this seems to indicate the passage of a cyclone from N to S. Perhaps a storm from NNE may be expected with a rise of the barometer". Next day we had Severe Storm of snow & hail from NNE with a rapid rise of the barometer.

On the 21st Feby. at 3 P.M. cirri extended from NE to SW, *moving from about NW*. In the evening the barom. began to decline & the wind changed from N to W. I noted in my journal "a cyclone appears to be coming over from NW, the centre passing to Eastward". This cyclone proceeded according to the "law of Storms" during the 22nd & 23rd. On the 24th before its course was fully completed, it was interfered with by a Second cyclone, which in its turn, was (on the 25th) interfered with by a third,- the progressive motion of all the 3 Storms being from NW. On the 25th about noon, cerri & cirro-cumuli were seen moving from NW, wind then N, and the barom. at its culminating point. About 3 P.M. the wind backed to W, and during the evening the barom. fell rapidly. At 12 PM. the wind was SW, gusty, with Snow. I noted in my journal "a third cyclone has come on before the Second was completed & moving in the Same direction (from about NW). From its commencing at SW, we may expect a heavy Storm from NE as this cyclone appears to be central here". Next day we had a very Severe Storm from NNE with a rising barom. & on the 27th the storm was finished by the wind Sinking at NE, the barom. having risen to fully its average height.

On the 30th March the wind was Easterly, weather fine & barometer which had been high, gradually declining. On the 31st at 8 A.M. I observed numerous cirri moving very rapidly from SE, wind then E & barom. still slowly declining. I inferred a cyclone moving from SE, the centre passing to Westward of this place. This was fully confirmed, for at 1 AM. of the 1st inst a high wind Sprung up from E, the barom. falling very much. During the day the wind veered round to SW, the barom having fallen until the wind passed the SSW point, when it began to rise. On the 2nd at 2 PM. I observed irregular cirri at a lower elevation than usual & passing into cirro-cumuli, moving rapidly from SW, wind then SW. 9 PM. barom still rising, wind high from WSW. The cyclonic movement on this occasion would thus seem to have affected to a certain extent the higher regions of the atmosphere. It would be important to ascertain the exact nature of the disturbances of the meteorological elements in the upper regions during, as well as before & after the passage of cyclones. The Subject tho' difficult, would, I have no doubt yield good results to careful & long continued observation.

Now that we really know Something about Storms, and now that we have the electric telegraph ready to convey intelligence almost instantaneously to and from remote parts of the islands of Britain & Ireland, it is, I think, high time that our Government should establish meteorological observatories at several important & distant stations now connected by telegraph, so that the indications of approaching storms, as well as observations made during their progress should be *"flashed"* over the country, especially to the principal Sea ports, without loss of time. If this were done, with the addition of a system of Storm Signals at prominent points round our coasts, I need scarcely say, that in all probability, a vast number of lives would be saved yearly, as well as an amount of property, compared with which the cost of the proposed establishments would be a mere trifle.

This is Surely a Subject well worthy of being seriously considered by the Government of a country so eminently maritime as our own, and from the highly liberal & Scientific character of Several members of the present Government, I have little doubt that an application to them, urging the great importance of the proposal both as regards the interests of humanity & of Science, would not be made in vain.

I Remain | My Dear Sir | With the greatest Respect | Your's Ever Faithfully | Wm Stevenson

1. Stevenson, W. (1853).
2. Faraday had recently been working on this topic. See Faraday, *Diary*, 3–16 November 1852, 6: 12908–13009 and 15 February 1853, 6: 13029–13038.

Letter 2668
Faraday to Peter Theophilus Riess
11 April 1853
From the original in SPK DD

Royal Institution | London | 11 April 1853

My dear Sir

I cannot deny to myself the pleasure of expressing to you my most grateful thanks for your kind present of the two volumes on Electricity[1], which I received safely a week ago. The pleasure is mingled with a deep regret that I cannot profitably use them, because of my ignorance of your powerful, well used, & highly productive, language; for I know, through translations, of some of your papers[2], the high qualities of your powerful mind, and I feel that your book is a rich mine of wealth, freely opened to others but shut to me[.] Still the kindness which has caused you to send it to me is a proof that I have your good opinion; & that I value most highly. I can appreciate the value of the commendation of men themselves worthy of all praise: and were it not that I am now growing old & memory is failing I would promise that your labours should have their influence a second time by stimulating me to further work. As it is accept my most grateful thanks & the expression of my highest esteem & respect and believe me to be

Most Truly Yours faithfully | M. Faraday
Professor P. Th. Riess | &c &c &c

1. Riess (1853).
2. For example Riess (1846, 1852).

Letter 2669
Faraday to Robert Warington[1]
15 April 1853
From the original in Northeastern Science Foundation, Brooklyn College of the City University of New York

Many thanks my dear Warington for your kindness in the matter of the handkerchief & all else[.] Thank them for me who were so good as to put up the packet.

Yours Truly | M. Faraday
15 April 1853

1. Robert Warington (1807-1867, DNB). Chemist to the Society of Apothecaries, 1842-1867, and Secretary of the Chemical Society, 1841-1851.

Letter 2670
Faraday to William Mure
19 April 1853
From the original in NLS MS 4953, f.49

Mr Faraday presents his Compliments to Coll Muir [sic] and would be much obliged to him if he could arrange Mr. Faraday's examination either for Thursday of this week or for Monday or Tuesday of next week[1] as he is anxious to engage in some important investigations with Mr. Brodie on Friday next[2][.]
Royal Institution | Tuesday 19 April 1853

1. That is 21, 25 and 26 April 1853. Faraday did not give his evidence to the Select Committee on the National Gallery until 10 June 1853. *Parliamentary Papers*, 1852–3 (867) 35, pp.373–83. See letter 2684.
2. That is 22 April 1853.

Letter 2671
Faraday to James Heywood Markland[1]
22 April 1853
From the original in SAU MS QC F2 L2, f.185

Royal Institution | 22 April 1853
Sir
I shall be very happy to talk over the matter of the Bath Waters on Monday next[2] at $\frac{1}{2}$p 9 o clk or 10 o'clk if that would suit your convenience[.]
I am | Your Very faithful Servant | M. Faraday
J.H. Markland Esq | &c &c &c

1. James Heywood Markland (1788–1864, DNB). Bath antiquarian.
2. That is 25 April 1853.

Letter 2672
Faraday to John Tyndall
4 May 1853
From a typescript in RI MS T TS, volume 12, p.4133

R. Institution, | 4 May, 1853.
My dear Tyndall,
Anderson has asked me about the enclosed tables. My pencil notes will point out the points questioned. Will you settle them and let us have

the tables as soon as may be, that he may be employing his spare portions of time on them[1].

Ever yours, | M. Faraday.

1. This note is mentioned in Tyndall, *Diary*, 7 May 1853, 5: 204, but gives no indication as to what the subject of the note was.

Letter 2673
Faraday to Jacob Herbert
5 May 1853
From the original copy in GL MS 30108/2/60c

5 May 1853 | Royal Institution

My dear Sir

I have seen and heard Mr Wells Fog signal[1] and cannot help thinking that its deserves a fair trial:- as to distant effect over the water. The sound is very distinct & I think in character more available than that of a bell or a gong. It will require greater power than either of these but still the power of a man appears to be sufficient - even in the present comparitively [sic] rough state of the apparatus[2][.]

I am My dear Sir | Yours Very faithfully | M Faraday
Jacob Herbert Esqr | &c &c &c

1. A printed advert of G. Wells's (of 15 Upper East Smithfield) apparatus is in GL MS 30108/2/60a.
2. This letter was read to Trinity House By Board, 10 May 1853, GL MS 30010/38, p.130. It was referred to the Light Committee.

Letter 2674
George Wingrove Cooke[1] to Faraday
12 May 1853
From the original in IEE MS SC 2

Sir,

Will you pardon the intrusion of a stranger who really has no excuse for his inroad upon your time except the claim which literature sometimes is allowed to have on science[.]

I am writing an article for the New Quarterly Review[2] upon the history of the Rapping Spirit humbug and I am disturbed by facts or pretended facts like the inclosed. The little scientific knowledge which I do possess tends to the conclusion that these "facts" are utterly inconsistent with all the laws that make up Electricity & Magnetism[.]

Might I trespass upon you as far as to ask whether there is any thing at all *possible* in these statements. Of course I shall make no use of any answer which you may favor me except to take from it my own tone in dealing with this department of the subject - unless indeed you should expressly authorise me to make this use of it.

Perhaps you would kindly return me the extracts enclosed[.]

I remain Sir | Your very faithful servant | Geo Wingrove Cooke
2 Brick Court, Temple | 12 May 1853

1. George Wingrove Cooke (1814–1865, DNB). Lawyer and writer.
2. [Cooke] (1853).

Letter 2675
John Allen[1] to Faraday
16 May 1853
From the original in IEE MS SC 2

Prees Vicarage, Shrewsbury, 16 May 1853

Sir,

Will you pardon me for asking your attention to the phenomenon of *table moving*, which, unless I am greatly deceived, will be ranked among the most astonishing discoveries of this age. No one could as I think have been less disposed to believe the accounts given in the newspapers than I was, but having seen a private letter from Mrs. H.G. Bunsen[2] describing what took place at the Prussian minister's[3] house on Monday evening last[4] I was induced to try the experiment on Saturday[5] with some friends whose good faith I entirely confide in.

1. 7 or 8 persons stood round a small circular table, resting on wooden feet on a carpeted floor, they having previously laid aside metal ornaments (as chains, rings, watches) & taking care that their dresses should not touch, the left hand thumb being under the right hand thumb, and the right hand little finger of each resting under the left hand little finger of the neighbour on the right side, so as to form a continuous chain, the tips of the fingers resting lightly on the table, and the parties exercising to the best of their power a wish that the table should move round in one direction; after about 20' the table began to move round, the velocity of the motion appearing to increase equally so that the experimenters were obliged to run round rapidly with it.

At first I was simply a spectator but being assured that the experimenters were not intentionally deceiving me, I joined (2) a company of nine persons who repeated the experiment round a larger table on iron castors on a carpeted floor; after about 10' it began to move. Fearing lest there might be some deception, the parties (possibly unconsciously) aiding the motion mechanically, I without apprising the rest broke the continuity of the chain more than once while the table was in motion, and each time the table stopped.

I should have mentioned that, in the first experiment, on the experimenters expressing a volition that the table should stop and move round in the opposite direction, the table stopped, and then moved almost immediately in conformity with the wish so expressed.

It may be said that there was some clever person in the party who managed to push the table round, or that unconsciously the persons pushed it in the direction in which they wished it move: but the sensation, the equability of the motion, its cessation when the continuity of the contact was interrupted, and the length of time that elapsed before the motion began, are conclusive arguments to myself who am well acquainted with the parties experimenting, that there is something in the matter which hitherto has not been explained.

I should feel greatly indebted to you sir if you would deem the matter worthy of your attention.

There was a tingling felt at the tips of the fingers and at the elbows previous to and during the motion of the table.

Sir I am with true respect | Your faithful Servant | John Allen | Archdeacon of Salop, dioc. Lichf.

Professor M. Faraday | L.L.D.

1. John Allen (1810–1886, B4). Vicar of Prees, 1846–1883 and Archdeacon of Salop, 1847–1886.
2. Mary Louisa Bunsen, née Harford-Battersby (d.1906, age 84, GRO). Married Henry George Bunsen (d.1885, age 66, GRO, CCD, Vicar of Lilleshall, 1847–1869) in 1847, Bunsen (1868), 2: 78.
3. Christian Karl Josias Bunsen (1791–1860, NDB). Prussian ambassador in London, 1841–1854 and father of Henry George Bunsen.
4. That is 9 May 1853.
5. That is 14 May 1853.

Letter 2676
Faraday to Thomas Challis[1]
17 May 1853
From the original in WIHM MS FALF

Royal Institution | 17 May 1853

My Lord

I exceedingly regret that a rigorous prior engagement will prevent me from profiting by Your Lordships kindness on the evening of the 8th of

June[2] but am unable to remove it[3]. With many thanks for the honor & kindness I have the honor to be
   Your Lordships | Most Obliged & Grateful Servant | M. Faraday
The Right Honorable | The Lord Mayor | &c &c &c

1. Thomas Challis (1794–1874, B1). Lord Mayor of London, 1852–1853.
2. This was a reception following a meeting to discuss the best means of presenting knowledge of science and art in various industrial centres. See *Athenaeum*, 21 May 1853, p.618 and 11 June 1853, p.705.
3. This was possibly the Wednesday meeting of the London Sandemanian Church. See Cantor (1991a), 65.

Letter 2677
William Edward Hickson to Faraday
17 May 1853
From the original in IEE MS SC 2

Fairseat, Wrotham | Kent | May 17/53

My dear Sir,
   Are we not on the eve of some new discovery in Dynamics?
   I allude to the curious phenomena of 'table moving' which I am anxious to see our scientific men take out of the hands of Electro biologist charlatans & spirit manifesters.
   The facts to be investigated appear to me so important, and incredulity respecting them from this singularity is so natural, that I am induced to send you an account of what has passed under my observation within the last two days.
   1st Experiment - Three persons sitting round a small round drawing room work table with their hands upon it & fingers touching, for an hour & a quarter. | *No result*.
   2nd Experiment of *four* persons round a kitchen table for an hour & a half. *No result*.
   In both the above cases of failure the experimenters were on the shady side of forty.
   3rd Experiment. Seven young ladies of from 12 to 16 yrs of age standing in a schoolroom, about a round kitchen table with their hands upon it & fingers joined. In half an hour the table began turning round with considerable rapidity - the feet (without castors) grating harshly on the floor. I then joined the circle, when it paus'd a little, as if to re-establish an equilibrium,- resum'd, tilted itself up with considerable force & fell over on one side.
   These phenomena were repeated several times during the same evening,- changing the circle; and I noticed at each change a pause, and a

pause of longer or shorter duration according to the apparent amount of vitality possessed by the different persons who join'd. One young lady of 16 of a florid complexion & full habit of body seemed to have more power than the rest; so much so that the table almost immediately began spinning when she joined the circle, but with or without her the ultimate result was always the same.

4th Experiment. I placed two needles at right angles on the table while in motion, but could perceive no deflection. They seem'd however to *stick* to the table more than usual, but roll'd off when it tipt up.

5th Experiment. A hat was plac'd on the table, and the fingers join'd (of the last circle) on the hat. Hat & table in a few minutes began moving round together. Afterwards the hat by itself.

6th Experiment. Four of us from the last circle, (myself one) took the hat to another table with a screen top, still keeping up the chain. The hat again began to move, and with it the top of the table, which unscrew'd itself & came off.

Returning home I call'd at my friend Fowler's[1] the architect, whose family had been busy all the evening trying similar experiments. They had fail'd to move a table but were succeeding with a hat when I entered the room. The circle was formed of three young ladies of 20, & Charles Fowler[2] (25) whose fingers were joined & lightly resting on the brim of the hat,- the hat moving round as if turning on a pivot, first from left to right & then, after about $\frac{1}{2}$ a dozen turns, *going back again from right to left*.

This latter circumstance stopt my theorizing, for in Miss Grants'[3] schoolroom the motions, whether of hat or table, had always been from left to right, as if following some tellurian circuit.

But query, may not all *force*, that of magnetism included, be deriv'd from the original motive power of centrifugal & centripetal action?

Yours very truly | W.E. Hickson
M. Faraday Esq

1. Charles Fowler (1792–1867, B1). Architect.
2. Otherwise unidentified.
3. According to Kent POD a Miss Grant ran a school at Stanstead which is three miles from Wrotham.

Letter 2678
William Edward Hickson to Faraday
19 May 1853
From the original in IEE MS SC 2

Fairseat, Wrotham | Kent | May 19/53

My dear Sir,

Your note[1] is just the kind of answer I should have written myself a week ago when I shd certainly not have stirr'd an inch out of my way to verify the phenomena describ'd.

The facts of the Biologists as far as they are facts are explicable by metaphysical laws, & the rest is fraud; but here we have a class of phenomena belonging to the mechanics of chemistry; too easily tested for the scientific world to ignore, and apparently affording a clue to laws worth finding out.

Let four persons in the prime of life & robust health stand round a hat, with their fingers lightly resting on the brim & so that the little fingers of each join, and, if the door be clos'd & the room warm, *in 20 minutes the hat will turn*, first from left to right, then from right to left.

The hat will even continue to turn for a time when the chain is broken & left with one person having his fingers on the brim, but will presently stop.

This experiment I have seen repeatedly, since I wrote, & without a single failure.

The table experiment requires more patience & more power; but there is no mistake about it. You feel at once that it moves as independently of any *pressure* from your fingers as the fly wheel of a steam engine, & sometimes with a rapidity that makes the head giddy in keeping up with it.

Now, if Newton was wise in asking himself why does the apple fall[2], may we not, with due modesty, ask his successors, why does the hat or table turn?

Will you, while this fine weather lasts, take a day or twos holiday with us, & talk the matter over while inhaling the fresh air of our chalk hills?

We have half a dozen beds at your service, and a beautiful country to shew you, worth visiting. A country too that affords to a botanist & geologist numerous points of interest.

My pony chaise shall meet you at the Gravesend station any day you may appoint.

Say you will come & you will confer a great favour on myself and Mrs Hickson[3][.]

Yours truly | W.E. Hickson

M. Faraday Esq

1. Faraday's reply to letter 2677. Not found.
2. See McKie and De Beer (1951–2).
3. Jane Hickson, née Brown. Married Hickson in 1830, see under his DNB entry.

Letter 2679
William Edward Hickson to Faraday
22 May 1853
From the original in IEE MS SC 2

Fairseat | Wrotham | Kent | May 22/53

My dear Sir

Some time at least in the course of the summer I hope you will be more at leisure. Pray consider the invitation a standing one, and that when ever you can command a day or two's holiday you may be sure here of a hearty welcome.

If you should see Mr Carpenter[1] will you make my compts to him & say that I quite agree with his biological exposition[2]. Two years ago in a paper I wrote in the Westmr[3] I went over the same ground.

Our present facts however have nothing to do with the laws of suggestion. The rotation when it commences more generally *interrupts* than allows the order of conversation.

If Mr Carpenter intend following up his experiments, will you ask him if he will run down here as yr *locum tenens*.

With the assistance of Miss Grants[4] establishment in wh there are 60 inmates we can produce all the phenomena any evening with as much certainty as making the wheels of a clock turn by winding it up.

Yours truly | W.E. Hickson

1. William Benjamin Carpenter (1813–1885, DSB). Professor of Forensic Medicine at University College London, 1845–1856.
2. See Carpenter (1852), Friday Evening Discourse of 12 March 1852.
3. [Hickson] (1851).
4. According to Kent POD a Miss Grant ran a school at Stanstead which is three miles from Wrotham.

Letter 2680
Faraday to Frederick Gye
24 May 1853
From the original in SI D MS 554A

Royal Institution | 24 May 1853

My dear Sir

You are most kind - on returning home from the Geographical dinner[1] (the first dinner of the kind I had been to for years) I found your enclosure[.] I would much rather have been with you[2] than where I was but such was my fate. I did not leave home until 20' to 7 o'clk but your note had not then arrived[.] This must be my excuse for not returning it last night[.]

If you have any further intention towards us, do not let it be this week or before Thursday week i.e the 2nd of June: it so happens that I am *never* at liberty on Saturdays and the other opera evenings before then are subject to appointments[.]

With kindest remembrances to Mrs. Gye[3].

I am yours faithfully | M. Faraday

F. Gye Esq | &c &c &c

1. The Anniversary Meeting of the Royal Geographical Society had been held at the Royal Institution prior to the dinner. See *J.Roy.Geogr.Soc.*, 1853, **23**: liii-lv.
2. Faraday would have heard Meyerbeer's "Roberto di Diavola". See *Times*, 23 May 1853, p.4, col. c.
3. Mrs Frederick Gye, née Hughes. Otherwise unidentified. See DNB under Frederick Gye.

Letter 2681
Jacob Herbert to Faraday
26 May 1853
From the original in GL MS 30108/1/58

Trinity House London | 26 May 1853.

My dear Sir,

I am directed to transmit for your consideration and any observations which you may be pleased to make thereon,- the accompanying extract from a Letter from the Principal Keeper of the Light House at St. Catherine's[1], in relation to the defective state of the ventilation of the Lantern thereof.

I beg to add that should you deem it advisable that you should visit that Establishment, to make personal enquiry into the matter, - the Board will quite approve your so doing.

I am, | My dear Sir, | Your very faithful Servant | J. Herbert

M. Faraday Esqr. | &c &c &c

1. This extract is in GL MS 30108/1/58.

Letter 2682
Faraday to an unidentified correspondent
31 May 1853
From the original in SM MS 1324

Royal Institution | 31 May 1853.

Dear Sir

I cannot find a copy of the notes of which I spoke to you[.] I will therefore say in brief that when I examined the books at the Athenaeum it was my opinion that they suffered from the Sulphurous & Sulphuric acid produced by the burning gas and that as it appeared to be impossible so to purify coal gas as to remove every source of these substances from it, so it would be wise to ventilate the gas lamps. This was done according to my suggestion & I understand has perfectly removed the evil[1][.]

I am | My dear Sir | Very Truly Yours | M. Faraday

I may add that I have found no reason to change my opinion of the cause of the evil and I doubt not that the remedy would be as efficient in Lord Spencer's[2] Library as at the Athenaeum[.] | MF

1. See Cowell (1975), 24–5.
2. Frederick, 4th Earl Spencer (1798–1857, CP). Retired Admiral.

Letter 2683
Faraday to Elizabeth Reid
7 June 1853
From the original in RI MS Conybeare Album, f.1

Royal Institution | 7 June 1853

My dear Eliza

There remain three lectures here one on Thursday at 3 o'clk by Dr. Frankland on *Chemistry of food*[1] one by myself on Friday evening at 9 o'clk on *Oxygen*[2] - & one on Saturday at 3 o clk by Dr. Tyndall on *Water*[3]. I send your admissions for all & shall be very happy to see Mr. Davis[4] here[.]

Your Affectionate Unkle | M. Faraday

1. This was Frankland's tenth and final lecture in his course on "Technological Chemistry" delivered on 9 June 1853. See RI MS GB **2**: 77.
2. Faraday (1853c), Friday Evening Discourse of 10 June 1853.
3. This was Tyndall's fourth and final lecture in his course on "Air and Water" delivered on 11 June 1853. See RI MS GB **2**: 78.
4. Richard Hayton Davis (d.1911, age 78, GRO). Chemist who married into the Barnard family and lived mostly in Harrogate. Davis's admission card for Faraday's Friday Evening Discourse is in RI MS Conybeare Album, f.1.

Letter 2684
William Mure to Faraday
7 June 1853
From the original in RI MS F2 J185

Un. Services Club, | June 7/53

Dear Mr Faraday,

Mr Charteris[1], who is more in the way of such curiosities than myself, has promised to procure & send to you the requisite fragment of old oil painting, in the course of tomorrow or next day early.

If your receive it, and are able to operate on it[2], I hope you will favour the Committee with your attendance on Friday next[3] between 12 and 3; bringing with you your various experimental specimens. If you have not yet had the requisite opportunity of performing the spirit of wine experiment, your evidence had better be postponed to a future meeting.

I go to Oxford tomorrow - but hope to be back at latest to breakfast on Friday morning.

Your very truly | Will Mure
M. Faraday Esq | &c &c

1. Francis Richard Charteris (1818–1914, CP under Wemyss). Conservative MP for Haddingtonshire, 1847–1883 and a member of the Select Committee on the National Gallery.
2. Faraday was able to operate (that is clean) on this picture on 9 June 1853. His notes on this work are in RI MS F2 J182–5. He reported this work to the Select Committee on the National Gallery on 10 June 1853. *Parliamentary Papers*, 1852–3 (867) 35, p.376.
3. That is 10 June 1853.

Letter 2685
Faraday to William Mure
7 June 1853
From the original in NLS MS 4953, f.100

Royal Institution | 7 June 1853

Dear Sir

On looking out some old pictures painted for me (very common place) by Penry Williams[1] when he was young i.e about 33 years ago[2], & working with them & alcohol, I found them exceedingly affected; & I suspect they are, as to vehicle either the mixture of oil & varnish or all varnish. On the other hand working with a white lead surface containing no varnish but only oil, which I applied 3 years ago, I found it to resist altogether the alcohol:- I think therefore I have the two extreme cases which illustrate the effect of alcohol on pictures[3] & in that case I am ready for the Committee on Friday[4]. If you think fit I will come; but as Friday

time is valuable⁵, would be glad to know the time accurately & not to be away from home longer than is needful[.] Whatever you desire I will do[.]
I am | Dear Sir | Most truly Yours | M. Faraday
Coll Mure M.P. | &c &c &c

1. Penry Williams (1798–1885, DNB). Painter.
2. See Faraday to Guest, 20 July 1819, letter 103, volume 1.
3. Faraday's notes on this work are in RI MS F2 J182–5. He reported this work to the Select Committee on the National Gallery on 10 June 1853. *Parliamentary Papers*, 1852-3 (867) 35, p.376.
4. That is 10 June 1853. See letter 2684.
5. Due to Faraday delivering the Friday Evening Discourse, Faraday (1853c), that day.

Letter 2686
F.W.M.¹ to Faraday
8 June 1853
From the original in IEE MS SC 2

June 8h 1853.
Sir
I ought to apologise for the liberty I am taking in writing to you, particularly as altho' much interested in Scientific discoveries, I am very ignorant upon these subjects, & I fear it will be thought great presumption in me to offer any suggestions. I am told that when lecturing upon subject of "Table moving" you stated that there was not enough electricity in the human body to act upon the table.

Is it not possible (I ask with the *greatest deference*) that the electricity may be, as it were, renewed or accumulated in our bodes as fast as we part with it? if so it *may* take 20 minutes or $\frac{1}{2}$ an hour for a sufficient quantity to have passed through our hands to charge the table. *Supposing* this to be true, I venture to hope that an idea which occurred to me on first seeing this extraordinary phenomenon may also have some truth in it. My idea is this. 1st It is the *room* which moves & *not the table*. The room goes round with the Earth from West to East, & thus accounting for the apparent movement of the table from East to West. 2dly the electricity in our bodies acting on the table, *lifts it slightly from the ground*, (the 100th part of an inch would do), and 3dy this being the case, as we *use no force*, but passively submit to follow what I must call the power of Electricity, we are obliged to rise from our chairs as soon as the room *goes on without the table* (which has been released from the power of attraction by the power of electricity), & we continue to follow it or rather to keep pace with the Earths movement.

It is difficult to realise the abstraction (or insulation) of the table; but the air *quite straight* comes down a little further on one side or the other, & not on the exact spot it started from *owing to the Earth's having moved on*. Of course the attraction of gravity is only *slightly* overcome, & the *least* movement of our hands may give a check to the table, & even send it round a little the other way - this is no doubt done when people "will" it to go this way or the other. The pressure of the hands may be almost involuntary & yet produce this effect. The "Tipping over" may be caused in the same way or more probably by the weakening of the power of electricity in that direction (the table always appears to move towards the west till influenced). I am so ignorant about electricity that I feel great diffidence in making these observations, but my anxiety that the reason of this extraordinary movement should be discovered, has led me to wish that one capable of judging, should tell me whether there is any truth in them, if there is, perhaps you will kindly write at your leisure a few lines directed to F.W.M. | Post Office | Marlow | Bucks. as I dont like to sign my name. Once more apologising for taking up your valuable time by so long a letter,

I am Sir | Your Obedient &c | F.W.M.

I ought to have said the electricity might be renewed (*possibly*) because I imagine tho' perhaps very ignorantly, that it is supplied from the atmosphere as we breath in.

1. Unidentified.

Letter 2687
Jacob Herbert to Faraday
9 June 1853
From the original in GL MS 30108/2/64.6

Trinity House, London | 9th June 1853.

Sir,

Adverting to our previous correspondence, and communications on the subject of the Electric Light, as proposed by Dr. Watson, for the purposes of Light House Illumination[1],- I have to acquaint you, that at the request of the Electric Power and Colour Company, to whom it appears Dr. Watson has transferred his patent[2], a conference took place at this House on Tuesday 31st. ultimo, between a Deputation of the provisional Directors of that company, and this Board[3], when it was arranged, that, when prepared, Dr. Watson shall place himself on behalf of the said Company, in communication with yourself, with the view of the practical qualities of the Electric Light being tested, in respect of the several points specified in your letter of the 20th December last[4].

I remain, | Sir, | Your most humble Servant | J. Herbert[5]
M. Faraday Esqre.

1. See letters 2599, 2606, 2608, 2609, 2614 and 2619.
2. Patent numbers 1852/211, 1852/595 and 1853/570.
3. This meeting is noted in Trinity House By Board, 31 May 1853, GL MS 30010/38, pp.148–9.
4. For this letter see letter 2608 and note 3.
5. Faraday's reply, dated 13 June 1853 (not found), is noted in Trinity House By Board, 14 June 1853, GL MS 30010/38, p.159 which stated that he was ready to meet Watson.

Letter 2688
F.W.M.[1] to Faraday
c11 June 1853
From the original in IEE MS SC 2

Miss FWM was not aware that she was committing a breach of etiquette in writing anonymously to Mr. Faraday[2], & she is sorry to have done so. What she intended to say was, not of course that the earth revolved round a table leg, but that the table partially released from the power of attraction *might possibly* show the movement of the earth on its axis much in the same way as a ball thrown up in the air; or as the pendulum experiment tried two or three years ago did[3]; especially as the table until influenced or hurried on by those around it invariably seems to go towards the west.

Miss FWM is however she confesses much too ignorant on these subjects to have interfered in them, & she regrets having done so. Mr. Faraday will she trusts excuse her still remaining unknown as it is disagreeable for a lady to make her name public under these circumstances[.]

1. Unidentified.
2. Letter 2686.
3. A reference to Foucault (1851).

Letter 2689
Faraday to Mary Buckland
16 June 1853
From a typescript in RI MS

Royal Institution | 16 June 1853.

My dear Mrs. Buckland

I have not yet seen Lord John Thynne[1] but we have written once or twice & I expect we shall meet tomorrow. I shall be very happy to be of the least service and should be in any case but it is very pleasant to be asked by You[.]

My wife is not strong and I have undertaken to acknowledge your note for her. We both think of the sight of Oxford in vacation time with very great pleasure but I do not know how our ability may turn out. Your kind invitation is a most pleasant thought to us and is & will be a great source of enjoyment whether we profit by it in the body or not:- for I hold kindness intended as kindness already conferred and we both value your offer as such. Perhaps one of us may have to go abroad, but I do not as yet know how that may be[.] We are

Ever - My dear Mrs. Buckland | Most Truly Yours | M. Faraday

1. John Thynne (1798–1881, B3). Sub-Dean of Westminster Abbey, 1835–1881.

Letter 2690
Faraday to James Walker
25 June 1853
From the original in Yale University Library, Joseph Bradley Murray Collection, Box 1/6.1

R Institution | 25 June 1853

My dear Walker

Thanks for the sight of the Chance mirror: it is a very good one. My nephews account is generally correct though of course imperfect. It would take a good many pages to describe the whole case. Ask him to shew you the piece of chilled cast iron & then the whole thing will be clear to you[.]

Ever Yours | M. Faraday

Letter 2691
Faraday to the Editor of the Times
28 June 1853
From Times, 30 June 1853, p.8, col. d

Sir, - I have recently been engaged in the investigation of table-turning[1]. I should be sorry that you should suppose I thought this necessary on my own account, for my conclusion respecting its nature was soon arrived at, and is not changed; but I have been so often misquoted, and applications to me for an opinion are so numerous, that I hoped, if I enabled myself by experiment to give a strong one, you would consent to convey it to all persons interested in the matter. The effect produced by table-turners has been referred to electricity, to magnetism, to attraction, to some unknown or hitherto unrecognized physical power able to affect inanimate bodies - to the revolution of the earth, and even to diabolical or supernatural agency. The natural philosopher can investigate all these supposed causes but the last; that must, to him, be too much connected with credulity or superstition to require any attention on his part. The investigation would be too long in description to obtain a place in your columns. I therefore purpose asking admission for that into the *Athenaeum* of next Saturday[2], and propose here to give the general result. Believing that the first cause assigned - namely, a *quasi* involuntary muscular action (for the effect is with many subject to the wish or will) - was the true cause; the first point was to prevent the mind of the turner having an undue influence over the effects produced in relation to the nature of the substances employed. A bundle of plates, consisting of sand-paper, millboard, glue, glass, plastic clay, tinfoil, cardboard, gutta-percha, vulcanized caoutchouc, wood and resinous cement, was therefore made up and tied together, and being placed on a table, under the hand of a turner, did not prevent the transmission of the power; the table turned or moved exactly as if the bundle had been away, to the full satisfaction of all present. The experiment was repeated, with various substances and persons, and at various times, with constant success; and henceforth no objection could be taken to the use of these substances in the construction of apparatus. The next point was to determine the place and source of motion - *i.e.* whether the table moved the hand, or the hand moved the table; and for this purpose indicators were constructed. One of these consisted of a light lever, having its fulcrum on the table, its short arm attached to a pin fixed on a cardboard, which could slip on the surface of the table, and its long arm projecting as an index of motion. It is evident that if the experimenter willed the table to move towards the left, and it did so move *before* the hands, placed at the time on the cardboard, then the index would move to the left also, the fulcrum going with the table. If the hands involuntarily moved towards the left *without* the table, the index would go to the right; and, if neither table nor hands moved, the index would itself remain immoveable. The result was, that when the parties saw the index it remained very steady; when it was hidden from them, or they looked away from it, it wavered about, though they believed that they always pressed directly downwards; and, when the table did not move, there was still a resultant of hand force in the direction in which it was wished the table should move, which, however, was exercised quite unwittingly by the party operating. This resultant it is which, in the course of the waiting time, while the fingers and hands become stiff, numb, and insensible by continued pressure, grows up to an amount sufficient to move the table or the substances pressed upon. But the most valuable effect of this test-apparatus (which was afterwards made more perfect and independent of the table) is the corrective power it possesses over the mind of the table-turner. As soon as the index is placed before the most earnest, and they perceive - as in my presence they have always done - that it tells truly whether they are pressing downwards only or obliquely, then all effects of table-turning cease, even though the parties persevere, earnestly desiring motion, till they become weary and worn out. No prompting or checking of the hands is needed - *the power is gone;* and this only because the parties are made conscious of what they are really doing mechanically, and so are unable unwittingly to deceive themselves. I know that some may say that it is the cardboard next the fingers which moves first, and that *it* both drags the table and also the table-turner with it. All I have to reply is, that the cardboard may in practice be reduced to a thin sheet of paper weighing only a few grains, or to a piece of goldbeater's skin, or even the end of the lever, and (in principle) to the very cuticle of the fingers itself. Then the results that follow are too absurd to be admitted: the table becomes an incumbrance, and a person holding out the fingers in the air, either naked or tipped with goldbeaters' skin or cardboard, ought to be drawn about the room, &c.; but I refrain from considering imaginary yet consequent results which have nothing philosophical or real in them. I have been happy thus far in meeting with the most honourable and candid though most sanguine persons, and I believe the mental check which I propose will be available in the hands of all who desire truly to investigate the philosophy of the subject, and, being content to resign expectation, wish only to be led by the facts and the truth of nature. As I am unable, even at present, to answer all the letters that come to me regarding this matter, perhaps you will allow me to prevent any increase by saying that my apparatus may be seen at the shop of the philosophical instrument maker - Newman, 122, Regent-street.

Permit me to say, before concluding, that I have been greatly startled by the revelation which this purely physical subject has made of the condition of the public mind. No doubt there are many persons who have formed a right judgment or used a cautious reserve, for I know several such, and public communications have shown it to be so; but their number is almost as nothing to the great body who have believed and borne testimony, as I think, in the cause of error. I do not here refer to the distinction of those who agree with me and those who differ. By the great body, I mean such as reject all consideration of the equality of cause and effect, who refer the results to electricity and magnetism - yet know nothing of the laws of these forces; or to attraction - yet show no phenomena of pure attractive power; or to the rotation of the earth, as if the earth revolved round the leg of a table[3], or to some unrecognized physical force, without inquiring whether the known forces are not sufficient; or who even refer them to diabolical or supernatural agency, rather than suspend their judgment, or acknowledge to themselves that they are not learned enough in these matters to decide on the nature of the action. I think the system of education that could leave the mental condition of the public body in the state in which this subject has found it must have been greatly deficient in some very important principle.

I am, Sir, your very obedient Servant, | M. Faraday
Royal Institution, June 28.

1. See Faraday's notes of seances at John Barlow's house on 20 and 27 June 1853, IEE MS SC 2.
2. "Professor Faraday on Table-Moving", *Athenaeum*, 2 July 1853, pp.801–3.
3. A reference to letter 2688.

Letter 2692
Faraday to Jacob Herbert
29 June 1853
From the original copy in GL MS 30108/1/58

Royal Institution | 29 June 1853

My dear Sir

I was in the Isle of Wight on the 21, 22, & 23 instant examining the condition of the lighthouses at St Catherines and the Needles in respect of condensation & ventilation[1]; and now beg to offer you the result of my enquiries. In reference to the St. Catherines light, I will remind you of a report which I made dated 10, Feby 1841[2], in which the damp state of the tower and of the air in it was adverted to; the description there given still applies; & I have no doubt in certain states of the weather with great force. The SW side of the tower is stained with dampness within from the top to the bottom as if the prevalent wet weather of that quarter could penetrate the stone work. The keeper says that there is frequently a haze in the tower when there is none outside. He reports also that he has occasionally taken up three or even four buckets full of water which have condensed within the tower & run down the walls. These circumstances show that in certain conditions of weather & temperature the tower & its walls must be quite damp, & the air within it saturated with moisture. Yet this is the only air which has access to the lanthorn to ventilate it and when the glass or roof of the lanthorn are by any external change of temperature rendered colder than the tower at such times as these condensation must ensue there. I found at my visit on the morning of the 22nd with moderately fine weather that the dryness within the the [sic] tower was indicated by a hygrometer temperature difference of only 7°, whereas outside the difference was 11°.5. The keeper reports that occasionally when symptoms of condensation have come on in the lanthorn he has hung up a piece of dry flannel or a sheet of blotting paper in the current of air entering from the tower & they have become quite damp in a short time. The condensation is still very much better than it was before the lamp ventilation flue was erected[.] It begins first on the cold metal of the roof & from that runs down: sometimes it occurs also on the glass.

All these results prove what I have urged on former occasions that a lanthorn should have its own ventilators or air passages[.] If this lanthorn were supplied with eight good large ventilators one in each face or a larger number of smaller one[s] equally disposed on different sides, the tower might be continually shut off by the doors (which are up) and I am prepared to expect that the evil now existing would be in a great degree or entirely removed. The ventilators would have to be fixed in the iron sides of the lanthorn; there would indeed be room in the glass without interfering with the light, for the glass extends $8\frac{1}{2}$ inches below the lowest reflecting zone, but their position would be better lower down & in the iron than in the glass, & the appearance also better[.]

The cowl is so formed & placed as to produce I suspect an effect here like that discovered & remedied at the Needles light house[.] It is part of a ball 2 feet in diameter revolving on a truncated cone about 2 feet high which stands on the summit of a conical & ribbed roof. The wind which drives against this roof must on the windward side be driven up & enter into the cowl between it & the cone and in such cases will retard the exit of lanthorn air there and may even make a return into the lanthorn in which case condensation is almost sure to occur. The keeper reports that he has occasionally traced the air of the lanthorn backwards into the upper part of the tower which I can only account for by some such return effect as that I am describing. I verified the facts at the Needle light house (See report 30 June 1849[3]) and the evil was perfectly remedied by the elevation of the cowl & the application of a deflector plate or ledge round the opening. I could not get to the outside of the cowl of the St. Catherines light to make a similar examination there. I should be very glad to see the cowl a couple of feet higher from the roof[.]

The very form of the cowl & cone on which it stands tends to make the wind drive in at the angle or opening *a*

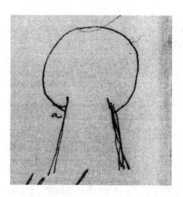

at which they meet, and it is not easy to apply a deflector plate here: for if it be applied on the cone the globular form of the cowl would tend to direct air in between; & if applied on the cowl it would only make matters worse. I should be very glad if a wind guard could be tried once practically instead of a cowl. I believe it would much surpass the cowl in action and help powerfully in windy weather to make a draught through the lanthorn. Such is the case every where with the many smaller wind guards put up in connection with the stove pipes.

The lanthorn stove pipe here is 50 feet long of which 25 feet are horizontal. Yet having no fire in it, there was an excellent draught in every part & at the ash pit of the stove because of the wind guard with which it is furnished. I was able to discover the very part of the pipe within which the keeper had stuffed a duster & then forgotten it; by the action of this cold draught[.]

The last 20 feet of this pipe is vertical and because of the warming power of the previous horizontal part on the air of the lanthorn, it is comparitively cool when the fire is alight, & the water produced by the combustion of the coals condenses in it. The joints are made by putting the upper pieces over the lower & this causes the condensed water to be conducted outwards at every point. The whole of the vertical part should be inverted that the water may run inwards; and the bottom stopper there should be a vessel to catch this water, that it may be removed daily or as it is produced.

The keeper here has ventured as I understand without authority to alter the ventilation pipe associated with the burner; he has turned one of the funnels upside down and then closed the aperture. By this he has increased the length of the lamp chimney to 8 feet 6 inches whereas the Standard or Fresnel[4] height (which I have never interfered with) is only 5 feet. In trying to find any sound or practical reason for this departure from the use of every other like light house I could not find that the keeper had any distinct reason but he considered the result good: but the following occurred to myself. Whenever the condensing condition comes on the keeper shuts off the air of the tower as much as he possibly can: he tells me the light then turns red & foxy and smokes, and I believe he gets the lanthorn into such a state that there is an important increase of carbonic acid & diminution of oxygen. Under such circumstances, to lengthen the chimney would be to increase the draught & to brighten the flame; and I suspect this is what has led to his change. Of course at other times he is using this increased draught & I conclude the consequence is that he is burning more oil than other three wicked lamps. The fourth or middle wick is taken away; but there could be no occasion to remove it from a lamp having a chimney 8 feet 6 inches high. I have no objection in the present case to this increase in regard to ventilation but I would much rather that the funnel or expansion were removed & the whole length of pipe made straight[.]

At St. Catherines all was right in the lanthorn & the ventilation perfect. At the last painting even the slates of the roofs were covered with white lead[.] I think this is not expedient. Above five sixths of the paint has been washed off by the rain and carried into the tanks to the injury of the water and the roofs look very bad[5][.]

I am My dear Sir | Very Truly Yours | M Faraday
Jacob Herbert Esqr | &c &c &c

1. See letter 2681. Faraday's notes of his visit are in GL MS 30108/1/58.
2. Faraday to Trinity House, 10 February 1841, letter 1337, volume 3.
3. Letter 2204.
4. Léonor Fresnel (1790–1869, Tarbe de St.-Hardouin (1884), 186–7). Secretary and Director of the French lighthouse service.
5. This letter was read to Trinity House Court, 5 July 1853, GL MS 30004/25, pp. 255–6. They instructed that Faraday's suggestions be executed. The matter of the paint was referred to the Light Committee.

Letter 2693
John Frederick William Herschel to Faraday
1 July 1853
From the original in RI MS Conybeare Album, f.11

July 1 1853.
My dear Faraday
M. Regnault of Paris is here and wishes much to meet you. Can you dine here on Monday[1] at $7\frac{1}{2}$ or if you do not like to dine come in the evening[.]

Yours very truly | J.F.W. Herschel
PS. All here delighted with your letter in the Times[2] - specially the concluding part, but is this deeper than the system of education.

1. That is 4 July 1853.
2. Letter 2691.

Letter 2694
Third Earl of Rosse to Faraday
1 July 1853
From the original in RI MS Conybeare Album, f.11

13 Connaught Place | July 1st 1853

Dear Faraday

Can you tell me where the best published account is to be found of the process for precipitating silver by grape sugar, essential oils &c[1].

Every process which affords the slightest prospect of facilitating the construction of true brilliant surfaces of silver, deserves to be examined[.]

Truly Yours | Rosse

1. This method had been invented by Liebig but was not published until Liebig (1856). On this see Vaupel (1991).

Letter 2695
Alexander Bath Power[1] to Faraday
1 July 1853
From the original in IEE MS SC 2

Norwich | July 1st 1853.

Dear Sir,

I rejoiced extremely to observe your letter in 'The Times' of yesterday on the 'Table Moving' subject[2] and more particularly as I had intended to express a decided opinion, in an address of which I send a copy. I take the liberty of presenting this to you as I feel it will be satisfactory to you to know, that there are those scattered through the country who are labouring to direct the public mind to worthier objects. I have again and again referred to the indication to be gathered from the history of 'table moving' of the deficiency which exists in our general routine of education; and I have advocated on every opportunity the introduction of more physical science into the schools of the middle and upper classes. It will find its place as a matter of necessity in elementary education, but it will require to be pressed upon the notice of those who conduct higher schools. The new Government Department of Science & Art will be something, but not all that is needed.

I remain | Yours very faithfully | A. Bath Power

1. Alexander Bath Power (d.1872, age 61, GRO, AC). Principal of Norwich Diocesan Normal School, 1840–1857.
2. Letter 2691.

Letter 2696
Robert Stephenson[1] to Faraday
4 July 1853
From the original in RI MS Conybeare Album, f.38

4 July 53

My dear Sir

I have to apologize to you for not sending you an introduction to the Engineer of the Holyhead Railway before this time, the fact is, it escaped my recollection[2].

Yours faithfully | Robt Stephenson
Professor Faraday

Address: Professor Faraday FRS | Royal Institution | 21 Arlbemarle Street

1. Robert Stephenson (1803–1859, DNB). Railway engineer.
2. Faraday visited Wales during July 1853. See letter 2703.

Letter 2697
Thomas Andrews to Faraday
4 July 1853
From a copy in Queen's University Belfast MS 2/13

Queen's College Belfast | 4h July 1853

My dear Dr Faraday

I feel greatly obliged by your kind note but regret you should have considered it necessary to explain when the explanation should have rather come from myself. In calling the first evening I enquired particularly both for Mrs Faraday and yourself, and when I learned to my regret next day that you had left London I took it for granted that Mrs. Faraday had accompanied you. My visit to London was a very short one, and I hope on the next occasion to be more fortunate.

Since my return I have read your very interesting letter in the Times on the table turning exhibitions[1]. I am afraid even the example you have given of the mode of investigating such a question will scarcely check the desire for marvels of this kind which under different forms has of late been manifested by a large portion of the public mind. The extension of education has apparently had very little influence in preventing these delusions: but has it not limited their duration. I mean the period during which they have been received as quasi-established truths? The belief in witch-craft maintained its ground for centuries; while clairvoyance has already run its brief career.

Excuse these speculations which have sometimes occupied my thoughts and were recalled by the concluding observation in your letter.

Please give my kind regards to Mrs. Faraday & express the regret I have experienced in not having seen her & believe me to be yours very truly & obliged | (signed) Th. Andrews.

1. Letter 2691.

Letter 2698
John Tyndall to Faraday
9 July 1853[1]
From a typescript in RI MS T TS, volume 12, p.4018

London, Saturday

Dear Prof Faraday

I regretted on reaching London that I had started two days too late to find you here. I should gladly have had a little conversation with you regarding lectures and concerning a few instruments which I think we shall find useful. On the latter point however I will write to you again from Berlin.

It struck me while looking over the apparatus that a course on heat might be rendered interesting. If you have no objection therefore I would propose a course of 12 on this subject[2]. Should it interfere at all with your arrangements I shall be most happy to choose something else.

Your table-turning article I read with great relish[3] - every body has read it. It has worked wonderful change in the minds of many. I believe you will find the ladies most impenitent upon the subject.

Should you desire to say any thing to me soon a letter addressed to the Post office Berlin will find me. As soon as I arrive however I will write to you and send you my address.

I hope you will find your vacation pleasant and profitable and that you will return to town with renewed vigour. My head is still a little wary

which you will probably have already inferred, but I am getting stronger daily and shall soon be in good working order.
My kind respects to Mrs Faraday
Most sincerely Yours | John Tyndall

1. Dated on the basis that this was the first Saturday after Tyndall had reached London. See Tyndall, *Diary*, 10 July 1853, **5**: 217-20. On p.218 he noted his visit to the Royal Institution on 8 July 1853.
2. Tyndall delivered a course of seven lectures on "Some Phenomena of Heat" after Easter 1854. RI MS GB 2: 83.
3. Letter 2691.

Letter 2699
Christian Friedrich Schoenbein to Faraday
11 July 1853
From the original in UB MS NS 406
My dear Faraday,
Many months ago I sent you a letter[1] and some papers of mine without having received from you any answer since. Being afraid of my parcel having been miscarried I forward to you another by the kindness of Mr. Drew of Southampton and hope you will get it in time.
The single paper treats of a subject of a general nature, and if you should feel curious to get acquainted with certain views of your friend's Schoenbein, you will perhaps find some body translating it for you.
The question of the nature of Ozone seems to have been settled in the laboratory of Mr. Bunsen[2] at Heidelberg and it appears that both views hitherto entertained about that subtle agent are correct[3]; there is one sort of Ozone containing nothing but Oxigen and another that contains some hydrogen. Common oxigen being absolutely anhydrous is transformed into the first one by electrical discharges, as de la Rive[4] and Berzelius[5] maintained some years ago. The odoriferous principle disengaged at the positive Electrode on electrolysing water is a compound consisting of two Eq. of pure Ozone or allotropic oxigen and one Eq. of water = HO^3.
How such a wonderful change of properties can be effected in oxigen without adding to or taking away any ponderable substance from that body is indeed very difficult to say; I, at least, know nothing about it but suspect that something very fundamental is at the bottom of that fact. It is a riddle to be solved by you only.
Just preparing for a journey to Vienna and Munich I am in a great hurry and you will therefore excuse the emptiness of this letter. I promise you to write a better one after my return, which will not be prolonged

beyond four weeks. I intend to go down the Danube, the scenery of which is as yet entirely new to me.

Pray transmit leisurely the volume laid by to Mr. Grove, who I think now and then sees you in the Royal Institution.

In asking you the favor to present my best compliments to Mrs. Faraday I am

My dear Faraday | Your's | most truly | C.F. Schoenbein
Bâle July 11, 1853.

1. Letter 2607.
2. Robert Wilhelm Eberhard Bunsen (1811–1899, DSB). Professor of Chemistry at the University of Heidelberg, 1852–1889.
3. A reference to Baumert (1853) who did this work in Heidelberg. See p.55.
4. See Marignac (1845).
5. See Berzelius, *Jahres-Bericht*, 1847, **26**: 58–64.

Letter 2700
Giovanni Antonio Amedeo Plana[1] to Faraday
11 July 1853
From the original in RS MS 241, f.130

Turin, le 11 Juillet 1853
LE PRESIDENT DE L'ACADEMIE ROYALE DES SCIENCES
A Monsieur le Prof. Michel Faraday, | Membre de la Société Royale de Londres, &c, &c.

Monsieur,

L'Académie Royale des Sciences, dans sa séance du 26 du dernier mois, Vous a nommé Académicien Etranger pour la classe Physico-Mathématique, et Monsieur le Ministre de l'Intérieur vient de transmettre à l'Académie l'ampliation de l'Ordonnance Royale, qui approuve cette nomination.

Je saisirai la première occasion favorable pour Vous envoyer le Diplôme Académique; en attendant cette lettre d'office que je m'empresse de Vous adresser, pourra Vous en tenir lieu.

Je Vous prie, Monsieur et très-honoré Collègue, de vouloir en agréer mes sincères félicitations, ainsi que l'assurance de la haute estime, et de la considération très-distinguée, avec laquelle j'ai l'honneur d'être,

Monsieur, | Votre Dévoué Collègue | Jean Plana

Address: A Monsieur | Monsieur le Chev. Michel Faraday, | Membre de l'Académie Royale des Sciences | de Turin, et de la Société Royale de | Londres | (21, Albemarle-Street)

TRANSLATION

Turin, 11 July 1853

THE PRESIDENT OF THE ROYAL ACADEMY OF SCIENCES
To Mr Prof Michael Faraday | Member of the Royal Society of London &c &c
Sir,

The Royal Academy of Sciences, at its meeting on the 26th of last month, has nominated you a Foreign Member of the Academy in the Physics & Mathematics section, and the Minister of the Interior has just transmitted to the Academy a copy of the Royal Decree that approves this nomination.

I shall seize the first favourable opportunity to send you the Academic Diploma; in the meantime, may this letter of office which I hasten to send to you, serve in its place.

I ask you, Sir and most honoured colleague, to accept my sincere congratulations, and the assurance of my high esteem, of the very distinguished consideration, with which I have the honour of being,
Sir, | Your Devoted Colleague | Jean Plana.

1. Giovanni Antonio Amedeo Plana (1781–1864, DSB). Italian astronomer.

Letter 2701
Robert Espie[1] to Faraday
19 July 1853
From the original in IEE MS SC 2

Sydenham Kent | July 19th 1853
"The errors of great men are the Triumph of dunces, and dunces alone" so says Theodore Parker[2]. I am precluded from uttering even a slight sneer at you for your slip about Table Turning[3], but I confess I am not so lenient with those fools who jumped headlong into your mistake. Coleridge[4] said a rogue was a fool with a circumbendibus[5], a philosopher may be said to be a fool by simply turning him out of his wonted groove - a magnet has an atmosphere of its own playing round it spirally - is this so? is man a magnet? if so what lets but that atmosphere join'd with others in a circle might turn a table or anything in the vortex if you had ever read any of Swedenborgs[6] works you wd not have miss'd your footing as you manifestly have done[.] I would rather you had been right, simple and unhark'd as I am[.] I have the means in my House of putting you to the blush, but philosophy is too proud to look in any direction, but its own[.]

I am Sir | Your very humble servant | Robert Espie Surgeon R.N.

1. Robert Espie (d.1870, age 79, GRO). Retired naval surgeon.
2. Theodore Parker (1810–1860, DAB). American preacher and social reformer. The source for this quotation has not been located.
3. See letter 2691.
4. Samuel Taylor Coleridge (1772–1834, DNB). Poet and philosopher.
5. Coleridge (1835), 18 (4 January 1823).
6. Emanuel Swedenborg (1688–1772, DSB). Swedish theologian.

Letter 2702
Faraday to George Towler
23 July 1853
From the original in SI D MS 554A

R Institution | 23 July 1853

Sir

I cannot give the opinion you ask for. I can only say that I should not publish such a paper in *my own name*. My reasons are that being an experimentalist I do not think theory alone of much force in Magnetism; since many theories as to the natural mode of action, are apparently equally probable. When we have devised new experimental proofs, so as to separate these theories from each other, then we shall be better able to judge which is the nearest to the truth of nature. Hence for my own part I refrain from forming any conclusion as to the real nature of magnetic action & am content for the present in trying to find out the *laws* which govern this extraordinary exertion of force[.]

I am Sir | Your Obedient Servant | M. Faraday
G. Towler Esq

Letter 2703
Faraday to Caroline Deacon
23 July 1853
From the original in the possession of Elizabeth M. Milton

Royal Institution | 23 July 1853

My dear Caroline

Yours to me arrived yesterday and was very pleasant in the midst of the serious circumstances which had come over us and of which you have no doubt heard by this time. I am always cheered by your words & it is well for us to have a remembrance of our hope rebounding from one to another in these latter days[1] when the world is running mad after the strangest imaginations that can enter the human mind. I have been shocked at the flood of impious & irrational matter which has rolled before me in one form or another since I wrote my times letter² and am more than ever glad that as a natural philosopher I have borne my testimony to the cause of common sense & sobriety[.] I have received letters from the most learned & from the highest thanking me for what I did. I cannot help thinking that these delusions of mind & the credulity which makes many think that supernatural works are wrought where all is either fancy or knavery are related to that which is foretold of the latter days & the prevalence of unclean spirits³ - which unclean spirits are waking in the hearts of men & not as they credulously suppose in natural things[.] There is a good hope however which has no relation to these things except by its perfect separation from them in all points & which will not fail those who are kept in it[.]

Poor Mary⁴. But why poor? she is gone in her hope to the rest she was looking for & we may rejoice in her example: as a case of the power of God who keeps those who look to him in simplicity through the faith that is in Christ. But her poor husband⁵ & her many children are deeply to be felt for & you also & her father⁶ we join in deep sympathy with you all. It would be a sad shock to him coming so suddenly upon the cheerful events he had been concerned in at Newcastle. And the Young ones too it must make them grieve. Your Unkle Edward⁷ was here just now & observing how soon all their gaity of appearance & (for the event itself, the fit) bravery would have to disappear in a sober form. I saw Mrs. J. Boyd⁸ this evening & the two elder boys - all are as you might expect. Margery [Ann Reid] was there yesterday & will be there today. Elizabeth⁹ dined with us & was then going to the house. Your Aunt¹⁰ was there yesterday, today she is gone to see her father¹¹ & there I shall be this Evening. Whilst in Wales we were talking about you & settling to have the pleasure of helping you to settle a little in the house & it ended in our putting apart £5 for the purpose. You will receive half of it *with our love* in the present sheet & the other part I will send in a few days[.]

Do you see how crabbed my handwriting has become? the muscles do not obey as they used to do but trip up or fall short of their intended excursions and so parts of letters are wanting or whole letters left out. You must guess it & I know you have a good will for the purpose. We had not heard of Mr. Paradise's serious illness - for serious it must have been to keep him at home. *Give our love to him earnestly.*

And now dear Caroline with kindest remembrances to your husband I must conclude. *Not forgetting the Maiden of the house*¹².

Ever Your Affectionate Unkle | M. Faraday

1. A reference to 1 Timothy 4: 1.
2. Letter 2691.
3. Revelation 16: 13.

4. Mary Boyd, née Ker-Reid (1813–1853, Reid, C.L. (1914)). A niece of Sarah Faraday and a sister of Caroline Deacon. She died on 22 July 1853.
5. Alexander Boyd. Blacksmith. Cantor (1991a), 300.
6. William Ker Reid (1787–1868, Reid, C.L. (1914)). Silversmith and a brother in law of Sarah Faraday.
7. Edward Barnard (1796–1867, GRO). A brother of Sarah Faraday. Silversmith. See Grimwade (1982), 431.
8. Unidentified.
9. Elizabeth Reid.
10. Sarah Faraday.
11. Edward Barnard.
12. Constance Deacon.

Letter 2704
John Tyndall to Faraday
24 July 1853
From a typescript in RI MS T TS, volume 12, pp.4019–21

Berlin | Universitals [sic] Strasse No 4 | July 24th, 1853
Dear Professor Faraday

In commencing to write this letter I am violating to some extent an article of my own creed, that no man is at liberty to write or speak without some necessity. I do not clearly see the necessity at present, I have no distinct purpose in view, but still I write. Perhaps such writing is of a meteorological character depending on the state of the weather, and indeed there is a calmness and a beauty in this morning which might suggest an occupation of the kind. I have been here for the last 9 or 10 days and employ myself in extracting as much information as possible from my friends. A doubt of the value of such information is however often present with me: its value is chiefly conversational, it fills a gap in a lecture sometimes, but beyond this I believe it is very unprofitable. It tends to weaken the fixity of the mental axis and to keep it vibrating amid the multitude of facts. I sometimes think that education may be too liberal, that our books and our acquaintances may weaken our alliance with nature and thus do more harm than good. People often speculate on the possible achievements of an original thinker who has had to struggle with difficulty, had such difficulty not existed, forgetting that the private effort enforced by his circumstances may be the very thing to which his development is due. The tendency of our so called advantages is sometimes simply to dilute a man. Society is the enemy of work, and here I see a danger which lowers upon the foreground of my own future. I must circumscribe myself in London: must set my face against visiting if I would get any work done. I know that some who are now very friendly with me will dislike peculiarity, but I must trust to time for a true verdict. Here however my course is comparatively clear. You have already hewn your way through this jungle and I have nothing to do but follow your steps.

I have been repeating Du Bois experiments thinking possibly that they might be made use of on some Friday evening. Another subject out of which a Friday evening might be manufactured[1] is the old one of Trevelyan's[2] experiment[3]. Since Forbes's papers[4] I believe nothing has appeared on the subject in England and these papers cast a doubt upon your explanation of the matter[5]. Forbes seems to consider it essential that different metals must be used, I have had an instrument made since I came here and obtained it with a distinct tone yesterday, which continued a quarter of an hour, from copper *on copper*. The artifice by which it is obtained is suggested by your explanation. I will try other metals this week and am not without hope of obtaining the action with quartz: at all events I am getting a crystal cut with the view of trying the experiments.

Another point to which I have turned my attention is the generation of cold at a bismuth and antimony joint. Up to the present time the experiment has only succeeded when either bismuth or antimony is in the circuit. It is a most suggestive experiment. I have ordered an apparatus with which I hope to generalize the fact and which may lead to something else.

I have ordered a few instruments which will be required next year in London: but have restricted the order to a sum which will be no means render me bankrupt provided you do not consider the instruments necessary for the institution. There is a most convenient form of the galvanometer used here: it consists of a magnetized polished steel disk with coils. I think it would be very useful both for private experiment and for the lectures - the deflection of the disk might be rendered very evident by the motion of a ray of light reflected from it. It would cost about three pounds. If you had time I should feel thankful if you would let me know whether I might devote 20 or 25 pounds to the purchase of apparatus.

Believe me dear Sir | Most faithfully Yours | John Tyndall

1. Tyndall (1854a), Friday Evening Discourse of 27 January 1854.
2. Arthur Trevelyan (1802–1878, B6). Scientific writer.
3. Trevelyan (1831, 1835).
4. Forbes, J.D. (1833, 1834).
5. Faraday (1831), Friday Evening Discourse of 1 April 1831.

Letter 2705
Faraday to Christian Friedrich Schoenbein
25 July 1853
From the original in UB MS NS 407

Royal Institution | 25 July 1853

My dear Schoenbein

I believe it is a good while since I had your last letter[1] i.e the one previous to that I received by the hands of Mr Drew[2] - but consider my age & weariness & the rapid manner in which I am becoming more & more inert - and forgive me. Even when I set about writing I am restrained by the consciousness that I have nothing worth communication. To be sure many letters are written having the same character; but then there is something in the manner which makes up the value: and which when I receive a letter from a kind friend such as you often raises it in my estimation far above what a mere reader would estimate it at. So you are going down the Danube one point on [sic] which I once saw[3] and are about enjoying a holiday in the presence of pure nature. May it be a happy & a health giving one and may you return to your home loving it the better for the absence & finding there all the happiness which a man sound both in mind & body has a right to expect on this earth.

I have not been at work except in turning the tables upon table turners - nor should I have done that but that so many enquiries poured in upon me that I thought it better to stop the inpouring flood by letting all know at once what my views & thoughts were. What a weak credulous incredulous, unbelieving superstitious, bold, frightened, what a ridiculous world ours is, as far as concerns the mind of man. How full of inconsistencies contradictions & absurdities it is. I declare that taking the average of many minds that have recently come before me (and apart from that spirit which God has placed in each) and accepting for a moment that average as a standard, I should far prefer the obedience affections & instinct of a dog before it. Do not whisper this however to others. There is one above who worketh in all things and who governs even in the midst of that misrule to which the tendencies & powers of man are so easily perverted.

The Ozone question appears indeed to have been considerably illuminated by the researches in Bunsens[4] laboratory[5]. But why do you think it wonderful that Oxygen should assume an allotropic condition? We are only beginning to enter upon the understanding of the philosophy of molecules & I think by what you say in former letters that you are feeling it to be so. Oxygen is of all bodies to me the most wonderful as it is to you. And truly the views & expectations of the philosopher in relation to it would be as wild as those of any table turner &c &c &c were it not that the philosopher has respect to the *laws* under which the wonderful things that he acknowledges come to pass and to the never failing recurrence of the *effect* when the *cause* of it is present. At the close of our Friday Evenings, I gave a little account to our members of Fremy[6] & Becquerels expts.[7] in producing Ozone by Electricity[8] - and I confess myself glad that whilst at Heidelberg they have shewn an HO^3 they have also proved the existence of a true Ô.

My dear Schoenbein, I really do not know what I have been writing above & I doubt whether I shall reread this scrawl least I should be tempted to destroy it altogether. So it shall go as a letter carrying with it our kindest remembrances to Madam Schoenbein and the sincerest Affection and Esteem of
Yours Ever Truly | M. Faraday

Address: Dr. Schoenbein | &c &c &c | University | Bale | on the Rhine

1. Letter 2607.
2. Letter 2699.
3. An occasion when it would appear that Faraday might have seen the Danube would have been on his return from Italy in 1815. See Faraday to Margaret Faraday, 16 April 1815, letter 50, volume 1.
4. Robert Wilhelm Eberhard Bunsen (1811–1899, DSB). Professor of Chemistry at the University of Heidelberg, 1852–1889.
5. In Heidelberg. See Baumert (1853).
6. Edmond Frémy (1814–1894, DSB). Professor of Chemistry at Muséum d'Histoire Naturelle, 1850–1879.
7. Becquerel and Frémy (1852).
8. Faraday (1853c), Friday Evening Discourse of 10 June 1853.

Letter 2706
Faraday to John Tyndall
25 July 1853
From a typescript in RI MS T TS, volume 12, pp.4134

Royal Institution, | 25 July, 1853.
My dear Tyndall,
I have received your letter without date[1], so do not know whether you have been expecting to hear from me; I rather waited to learn your address, but by Dr Bence Jones' advice intend to send this to our friend Du Bois-Reymond's house. I really have nothing to say, but to convey my kindest remembrances and wishes to your and my friends at Berlin, and to set your mind free in every point however slight. I think your proposed course here will be excellent, and know very well that I shall enjoy it, and believe that all others will do so with me. I thought I saw Du Bois-Reymond's face in the Railway the other day, on the Northern road, but it must have been a mistake. Give my heartiest remembrances to him. Also

to M.M. Magnus, Rose, and such other friends of mine as you may meet. In thought I send my homage and respects to Humboldt, (and through him even to the King[2], who has honoured me with the Order of Merit[3]), but I refrain from troubling him. Though very much his junior in years, I feel the burden of formalities, and so think I ought to spare him and such as him. Sometimes I fear I may carry this too far, and that it may assume the appearance of indifference on my part; but I hope that will not be the case and that I am favourably interpreted, especially in the case of the man whom I most deeply venerate.

I was with Dr. Percy on Saturday night[4], who tells me they have been building an enormous geyser at Woolwich - not on purpose, however, but accidentally. The fact is that they have been building a new boiler which, as far as I understand it, consists of two tall concentric iron cylinders, having water between them and fire in the centre. Percy compares it to the geyser, and thinks of it with some degree of apprehension.

I wish I could transport myself suddenly to Berlin, and be with you in some of the Laboratories and Workshops. I should luxuriate in some of the manufactories of philosophical instruments, but - but such wishes remain long after the ability or fitness to satisfy them has passed away. A single day of such would set my head a ringing, and I should have to run away.

Ever, my dear Tyndall, | Yours very truly, | M. Faraday

1. Letter 2698.
2. Frederick William IV.
3. See Humboldt to Faraday, 18 August 1842, letter 1420, volume 3 and Bunsen to Faraday, 21 August 1842, letter 1421, volume 3.
4. That is 21 July 1853.

Letter 2707
John Tyndall to Faraday
28 July 1853
From a typescript in RI MS T TS, volume 12, pp.4023-4

July 28th.

Dear Professor Faraday

DuBois starts for England tomorrow and will carry this to Hull[1]. He came to me the evening before last with a countenance shining with pleasure, 'You would not guess what I have got' he exclaimed. I paused - one hypothesis alone seemed opened to me - "You have got a letter from Ambleside" I replied. 'No' said he 'I have got a letter for you' - He opened his pocket book and handed me a letter bearing a superscription well known to both of us - the letter in fact was yours[2].

Would that my wish and the wish of many able, honest and hospitable men added to my own could be effected in transporting you at a safe speed to Berlin. From one end of Germany to the other there is but one feeling towards you - a feeling which I believe it has never before been the lot of a man of science to excite: it will be a pity if you cannot afford time for a single visit to a country which I believe above all others pays you a noble reverence - You have visited France and Italy - establish the equilibrium of things by a visit to Germany. Whenever you make up your mind I offer my services as a guide. Mrs Faraday may rest assured that I will take great care of you - my greatest difficulty would be to preserve you from the loyalty of your clansmen, but even this I would undertake to manage.

I mentioned in my last letter[3] that I intended to try Trevelyan's[4] experiment with a new metallic body. I was led to choose quartz from the extraordinary conductive power which my last experiment proved it to possess. With the first crystal I had cut a permanent oscillation was obtained but no tone. With a second crystal cut somewhat differently I have obtained a distinct tone, sufficiently loud I think to be heard throughout the theatre when everything is perfectly quiet. I have failed to obtain vibrations with a piece of glass cut similarly to the quartz, this might have been anticipated. On the whole I think the matter might be worked up into a suitable Friday evening's lecture[5].

I cannot resist the temptation to send you a note which I received from a little friend of mine just before leaving England - a beautiful child-like little boy about 10 years old. He was with me at Queenwood. His cousin Elma and cousin Grace are strangers to me but he tells me all about them with the most perfect good faith that it will interest me. I should not venture to send you the note had not somebody told me that you could afford to play with little boys.

Believe me dear Prof Faraday | Most truly Yours | John Tyndall Universitäts Strasse 4 | Berlin.

1. For the meeting of the British Association.
2. That is letter 2706.
3. Letter 2704.
4. Arthur Trevelyan (1802–1878, B6). Scientific writer.
5. Tyndall (1854a), Friday Evening Discourse of 27 January 1854.

Letter 2708
Lovell Augustus Reeve[1] to Faraday
29 July 1853
From the original in RI MS Conybeare Album, f.46

[RI embossed crest] | July 29

My dear Sir,
I looked in to ask if you would kindly favour [me] with an opinion, or hint, as to what might be said as a matter of comment on the enclosed communication.

In last Saturday's Lit. Gazette I wrote an article on Electric Gas[2] from material supplied to me by Mr Robert Hunt, with the view of exposing what he termed a 'great sham'.

Just at the eleventh hour while making up my paper this week for press, I have a letter from the Electric Gas Company accompanied by a certificate from Mr Holmes[3] to the effect that this prepared water *is* without doubt converted by the magneto-electric machine into a non-explosive quietly burning Gas[4].

I ought to insert this communication - and yet I do not like to insert it without comment. Can you help me in this emergency? Mr. Hunt is in Cornwall, and Dr Playfair is also out of town.

I will come again at 4 o'clk on the chance of finding that this has reached you -
& much oblige | dear Sir | Yours faithfully | Lovell Reeve

Address: Professor Faraday

1. Lovell Augustus Reeve (1814–1865, DNB). Editor of the *Lit.Gaz.*, 1850–1856.
2. "Electric Gas", *Lit.Gaz.*, 23 July 1853, p.722.
3. Frederick Hale Holmes. Professor of Chemistry at the Royal Panopticon of Science (*Lit.Gaz.*, 23 July 1853, p.722) and one of the pioneers of electric light. See James (1997), 294.
4. This was published in *Lit.Gaz.*, 30 July 1853, p.745.

Letter 2709
Faraday to George Towler
30 July 1853
From the original in WIHM MS FALF

Mr Faraday presents his compliments to Mr. Towler & hasten to say that he cannot spare an hour or indeed any portion of time for discussion. If he once entered on such a course there would be many probably a hundred that he would have to meet before Mr. Towler's turn would come round[.]

Mr Towler speaks of non magnetic iron & a small needle. There is no such thing as non magnetic iron upon the surface of the earth nor can there be. Every piece of iron is a magnet by induction from the earth, a state which it takes whether the needle is there or not[.]
Royal Institution | 30 July 1853

Letter 2710
Faraday to John Barlow
1 August 1853
From the original in RI MS F1 D17

Royal Institution | 1 August 1853

My dear Barlow

I only learned on Friday last[1] that you had left a post office reference to Geneva or I should have written to you before[.] I am so accustomed to communicate with you that even when there is no other occasion than kind feeling I do not like to give it up. In real truth the kind feeling is after all the most important of any. I am hoping to hear from you some day soon and to hear also that you are better. I should like you to be so well as to be unconscious of it. Tell me soon how you both are and that you are enjoying the scenery & the circumstances: give us a delightful account but let it be a true one. We most sincerely hope & wish that both these points may coincide in one. We have had a fortnights trip into N Wales with constant wet weather - so after living in the hotels for a while we came back not having seen much of the country but we are pretty well[.] Nothing to brag about: nothing to complain of.

To day Mr. Vincent begins his holiday and the work folk go into the Library &c[2]. He wished me to offer his respects & to say that there was nothing particular to mention. Anderson also begins his holiday today - the Porters have had theirs. I wished Anderson to take his whilst I was here: he is very well & I expect will enjoy himself - i.e if the rainy weather ceases. We have had thus far a very rainy season & it still continues.

The works in the corner of the Hall are now in progress; but here again the rain teazes us sadly & much retards their progress but I think they will be a great comfort when they are finished[.] Mr Wright[3] of the Clarendon was somewhat frightened when he saw them & called whilst I was away. He saw Mr. Vincent & sent his surveyor[.] I suppose he had forgotten that he was upon our wall & not we upon theirs.

As to our painters it is a sad thing that drunkenness should take them away one after another. I have taken the man who succeeded Mr Newsham[4] for the present jobs of the season. He is I understand quite sober, he knows the house, and he works for Mr. Ellis[5] & others gaining their approbation: We shall see how things turn out[.]

I am glad to tell you that Percy will lecture here next season[6]. He has chosen the Metals for his course and I have no doubt that he will make it very interesting to our audience. In the point of character (which you know is often apart from the attraction) they are sure to do us good. The Museum of Economic Geology &c is no longer the Museum having changed its name into some other long phrase as Metropolitan School of &c &c[7] but I hear that it may perhaps change its name again as the point does not appear decided. I fancy the College of Chemistry is by this time identified with it but as you know I am an exceedingly bad Newsmonger & shall make all sorts of mistakes[.]

You would laugh when you heard of our Cab revolution - only think of a strike for three days[8]. It must have been a petty annoyance spread over a very large extent of population & I have no doubt produced extreme irritation with a great number of persons which yet was often extremely ludicrous. We expect much good will result from it in relation to cab conduct[.]

Whether this will find you at Geneva or whether it will have to follow you elsewhere I do not know. If you see De la Rive give my kindest remembrances to him & also to Marcet if in your company[.] We want to know something of you and Mrs. Barlow, for nothing had come to our ears or eyes in any way until I heard on Friday last that you were at Geneva & at the same time through two or three reporters, that Miss Grant had heard of you - the report being pretty good - I hope it is so in truth. My wife & I often think of where you will be & what you may be doing and we hope that Mrs. Barlow will think this part of the letter is as much to her as to you and that it is the bearer of many kindly remembrances founded upon a long continued course of affectionate intercommunication[.]

Ever My dear Barlow | Yours most truly | M. Faraday

Address: Revd. John Barlow MA | &c &c &c | Poste Restante | à Geneve

1. That is 29 July 1853.
2. See RI MM, 18 April 1853, **11**: 6; 6 June 1853, **11**: 19; 7 November 1853, **11**: 32.
3. Unidentified.
4. Richard Newsham. Plumber of 8 Little Stanhope Street, Mayfair. POD.
5. Unidentified.
6. Percy eventually withdrew his offer. See RI MM, 21 November 1853, **11**: 35.
7. The name was changed to the Metropolitan School of Science Applied to Mining and the Arts. On these changes see Bentley (1970).
8. Cab drivers went on strike on 27 July 1853 over a reduction in their fares. See *Ann.Reg.*, 1853, **95**: 91–2.

Letter 2711
Faraday to John Tyndall
1 August 1853
From a typescript in RI MS T TS, volume 12, pp.4135

Royal Institution, | 1 August, 1853.

My dear Tyndall,

Our letters have crossed[1], and so I must teaze you again for a little moment, to say that I am sure you will do right in purchasing the apparatus you think of; and also I am sure you will make heat most interesting to us. Your letter seems to imply that you are a little discouraged by the infinity of objects about you: but you are taking a rest now, and there is no reason why your mind should not in its quiescent state take the hue of every subject that comes near you; even though it do yield up that of the going, for that of the coming, subject. When you brace up your mind and settle with its undivided powers upon one subject, there is no fear that it will be any way deficient, or short of its ordinary vigorous tone. My kindest remembrances to all our joint friends.

Believe me to be | Ever truly yours, | M. Faraday.

1. That is letters 2704 and 2706.

Letter 2712
Philip Lucas[1] to Faraday
1 August 1853
From the original in IEE MS SC 2

Sir

Among the innumerable letters that you have doubtless received about table turning, the following plan of testing its truth may not have been suggested to you.

On the axis of the table above the legs, let another table rotate which shall extend about eighteen inches beyond the upper and placed at a sufficient distance from the top to enable the experimenters to stand upon it so as to place their fingers on the first table: it is evident that any movement of an involuntary or voluntary nature, will cause the lower table to rotate as it forms a portion of the upper whereas if it rotates from other causes, independent of other material, the whole will revolve. I annex a small drawing to delineate my meaning

and remain | Your obedient Servant | Philip Lucas Junr
Mr Bailey's[2] | Dr White's Grove | Hampstead | 1 Aug 1853

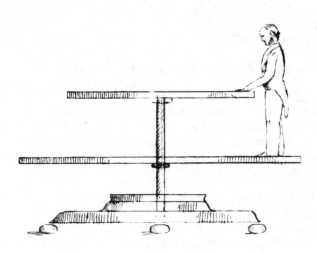

1. Unidentified.
2. POD gives Charles Bailey of Grove House, Hampstead Road. Otherwise unidentified.

Letter 2713
Lovell Augustus Reeve[1] to Faraday[2]
1 August 1853
From the original in RI MS F1 K36
The Literary Gazette | Office | 5, Henrietta St. | Covent Garden | London
| Established 1817 | Augt 1, 53
My dear Sir,
 Thinking you may like to see what has been said in Paris about the Electric Gas, I enclose a No, just received, of the Cosmos for your perusal. See the first article[2][.]
 I am, dear Sir, | Yours faithfully | Lovell Reeve

1. Lovell Augustus Reeve (1814–1865, DNB). Editor of the *Lit.Gaz.*, 1850–1856.
2. Recipient established on the basis that the writer and subject are the same as letter 2708.
3. "Grant Nouvelle. - Gaz Electrique", *Cosmos*, 1853, **3**: 197–9 which was mostly a translation of "Electric Gas", *Lit.Gaz.*, 23 July 1853, p.722.

Letter 2714
Faraday to William Charles Henry[1]
2 August 1853
From the original in HLHU b MS Am 1631 (130)
Royal Institution | 2 Aug 1853
My dear Dr. Henry

I have only three letters of Dalton[2] which I send you[3]: but let me have them again for they belong to a Portrait book[4]. In old time I was accustomed to destroy letters so that many do not remain with me of anybodys[.]

I do not know that I am unorthodox as respect the atomic hypothesis. I believe in matter & its atoms as freely as most people at least I think so. As to the little solid particles which are by some supposed to exist independent of the forces of matter and which in different substances are imagined to have different amounts of these forces associated with or conferred upon them (and which even in the same substance when in the solid liquid & gaseous state are supposed to have like different proportions of these powers) as I cannot form any idea of them apart from the forces so I neither admit nor deny them. They do not afford me the least help in my endeavour to form an idea of a particle of matter[.] On the contrary they greatly embarras[s] me for after taking account of all the properties of matter and allowing in any considerations for them then these nuclei remain on the mind & I cannot tell what to do with them. The notion of a solid nucleus without properties is a natural figure or stepping stone to the mind at its first entrance on the consideration of natural phenomena; but when it has become instructed the like notion of a solid nucleus apart from the repulsion which gives our only notions of solidity or the gravity which gives our notion of weight is to me too difficult for comprehension and so the notion becomes to me hypothetical & what is more very clumsy hypothesis[.] At that point then I reserve my mind as I feel bound to do in hundreds of other cases in natural knowledge.

I have published nothing on the matter save the old Speculation[5][.]

With many thanks for your kind invitation I am My dear Dr. Henry | Yours truly | M. Faraday

1. William Charles Henry (1804–1892, Farrar et al. (1977)). Physician and chemist.
2. John Dalton (1766–1844, DSB). Chemical philosopher who lived in Manchester. Developed a version of the atomic theory of matter.
3. Dalton to Faraday, 29 July 1840, 3 September 1840, 11 November 1840, letters 1302, 1311, 1325, volume 2.
4. That is RI MS F1 H. Henry wanted these letters for Henry (1854), but he did not make use of them.
5. Faraday (1844a).

Letter 2715
Faraday to William Charles Henry[1]
9 August 1853
From the original in RI MS F1 D18

R Institution | 9 August 1853

My dear Sir

I have not the slightest objection only as the letter was written carelessly & in a hurry it may convey my ideas very imperfectly[2][.] I cannot find a spare copy of my Speculation[3] (printed in Phil Mag for the first six months of 1844) or I would send it[.] If you are led to look at it then for *"mere* philosopher" near the top of the second page read *"wise* philosopher"[4][.]

I am My dear Sir | Most truly yours | M. Faraday
Dr. Henry | &c &c &c

1. William Charles Henry (1804–1892, Farrar et al. (1977)). Physician and chemist.
2. That is giving Henry permission to publish the second paragraph of letter 2714 in Henry (1854), 132–3.
3. Faraday (1844a).
4. *Ibid.*, 137.

Letter 2716
Faraday to Pierre Antoine Favre[1]
9 August 1853
From the original in the possession of Herbert Pratt

London | Royal Institution | 9 August 1853

Sir

I have received your very kind letter and also your most important work[2]. I want words to express the deep sense I entertain of your kindness and consideration for one who is not unfrequently discouraged because he cannot do all he desires to do and can have no future hope of imitating the industry and depth of research which characterise such a work as yours. Nevertheless the good will of such men as yourself is to me a strong motive to perseverence as far as it is permitted to my strength. But where physical strength and *memory* fail there the expectation & hopes cannot well be strengthened. If you have the opportunity pray present my most sincere respects and congratulations to M. Silbermann[3] and believe me to be with the highest esteem

Your Most faithful Servant | M. Faraday
à Monsieur | Monsieur P.A. Favre | &c &c &c

Address: A Monsieur | Monsieur P.A. Favre | &c &c &c | 11 Rue d'Enfer | à Paris

1. Pierre Antoine Favre (1813–1880, DSB). Head of the chemistry laboratory at the Central School of Arts and Manufactures, Paris.
2. Possibly Favre (1853).
3. Johann Theobald Silbermann (1806–1865, P2, 3). Scientific instrument maker in Paris.

Letter 2717
John Barlow to Faraday
10 and 11 August 1853[1]
From the original in IEE MS SC 2

Lake of Geneva | Aug 10h

My dear Faraday,
How are you all? I trust nothing of a harassing or distressing nature shortened your visit at Beaumaris. I saw your name in the newspaper among those of the Lord Mayor's[2] guests on the 14 of July but I trust that the list (as in the case of Queen's Balls) was made from those who were invited, not from those who were seen at the entertainment[3]. ... Our journey has been very prosperous hitherto and I earnestly hope that the weather has been as fine in England as that with which we have been favoured.

At Paris we were joined by Charles Herries[4] (the nephew & brother of your friends) and with him we have since been travelling - our route has been Bale (viâ new railroad) Geneva - by the Münster Thal, Neufchatel & Rolle. Then to that lovely village of St. Gervais (near Sallanches) & Chamouni, where we spent a fortnight - I mean dividing the time between St Gervais & Chamouni. At Chamouni I met a man whose name I could not find out, though I ought to know it familiarly[5]. He is a Fellow of Trinity College Dublin. He had Brougham's[6] paper on light[7] referred to him[8]. He was one of the first in making magnetic observations after Gauss's fashion[9]. He knows you personally, and he asked me whether certain observations of Kreil?[10] at Prague on the moon's influence on the magnet[11], had not made you revise your views? (meaning of course your papers of 1850–51[12]). He is a light-haired man, about 40, & has a wife[13], a luxury allowed to Irish Fellows....

When I was last at Geneva, De la Rive was at Vichy. He is due at home now. Mr. Drummond[14] ought also to be at Geneva by this time. I hope to send you news of both. I dined at Malagny with M. Marcet - a distinguished-looking young Spanish lady was on a visit to his daughter. This girl is a Pizarro, and there is a curious story connected with her.... Her uncle fell desperately in love with a beautiful Spanish lady. *Her* friends objected, because they felt assured that her excessive beauty would ensure for her a still more eligible match. *His* friends were equally adverse, because they had heard the girl's mother ill spoken of. He yielded to their representations. He promised to give up the marriage, & only stipulated for their assent to his visiting his love for the last time to bid her farewell. She was then at a country house high in the mountains. She received him with bitter reproaches, and avowed her disbelief in his attachment. He challenged her to put him to the test. She said "If you really love me you will not refuse to take off your coat & swim across the ice-cold pond yonder".

"Excited as he was, he plunged in" caught a pleurisy from which he never recovered but died within a twelvemonth. The Heroine of this curious romance *is the present Empress of the French*[15].

Geneva Aug. 11th, 4.00 P.M.

A thousand thanks in Mrs. Barlow's name and my own for your letter[16], which was cordially welcomed by both of us. My wife had one of those obscure attacks (which sometimes harrass her,) while we were in the Münster Thal. It gave way to a mustard poultice but so did the cuticle also. There must be something particularly malignant in the *sinapic* of that district.

We have indeed enjoyed the glorious scenery which has daily been presented to our view in such rich abundance. From Martigny we went to Villeneuve by Monthey (instead of Bex) and the prospect including a bridge over the Rhine was one of the most striking I ever witnessed. Snow mountains - an amphitheatre of near rocks, not too near, a foaming river, and most luxuriant foliage. This richness of verdure is of course the result of previous rains: but it is very brilliant.

I earnestly hope that the sunshine is on its way to England....

You are sure to succeed in what you do for the Institution:- I quite agree with you that Percy is a great prize. If your health and inclination enable you to give the Christmas Lectures[17], I shall be very sanguine for next year. I hear that a Mr. Hawkins[18], who is restoring those apocryphal animals the Megatherium &c for the Crystal Palace[19] is thinking of writing to propose to give us an Evening next year[20]. Do you know anything about him? If you meet any of the Paleontologists & happen to think of it, inquire. But we have many sterling men who will help us. To say nothing of yourself & Tyndall[.] Owen has promised[21], so has Grove after his fashion[22]. If you write to Airy you will poke him I am sure[23]. In short we are sure to get on.

Mrs. Barlow keeps urging me so constantly to remember her most earnestly to you and Mrs. Faraday & Miss Jenny if she is with you that I

fear I am going over a thrice uttered effusion (instead of a thrice told tale) but it is genuine[.]
  Ever your attached | John Barlow
Dr Faraday &c
I will write again & tell you where to direct to us. I hope we shall be at Munich in the beginning of next month.

1. Dated on the basis that the second part of this letter is the reply to letter 2710 and also on the reference to Faraday's trip to Wales and the Friday Evening Discourses for the ensuing year.
2. Thomas Challis (1794–1874, B1). Lord Mayor of London, 1852–1853.
3. See *Morning Chronicle*, 15 July 1853, p.4, col. f, for an account of the conversazione at the Mansion House the previous day. There is no mention of Faraday in this account.
4. Charles John Herries (1815–1883, DNB). An Excise Commissioner, 1842–1856.
5. The following sentences almost certainly identify him as Humphrey Lloyd.
6. Henry Peter, Lord Brougham (1778–1868, DNB). Whig politician and writer on optics.
7. Brougham (1850).
8. This is confirmed by the register of papers in RS MS MM 14.43.
9. On this see Morrell and Thackray (1981), 524.
10. Karl Kreil (1789–1862, P1, 2). Professor of Astronomy at Prague.
11. Kreil (1852, 1853). On Lloyd's interest in this see Lloyd, H. (1853).
12. Faraday (1851d, e), ERE26 and 27.
13. Dorothea Lloyd, née Bulwer. Married Lloyd in 1840, see under his DNB entry.
14. Unidentified.
15. Eugénie (1826–1920, DBF). Empress of France, 1853–1870. This seems to be one of the many canards that were then circulating about her. See Ridley (1979), 171.
16. Letter 2710.
17. Faraday delivered six Christmas lectures on "Voltaic Electricity". His notes are in RI MS F4 J16.
18. Benjamin Waterhouse Hawkins (1807–1889, B5). Sculptor and anatomist.
19. On this see Rudwick (1992), 140–8.
20. He did not deliver one.
21. Owen (1854), Friday Evening Discourse of 10 February 1854.
22. Grove (1854), Friday Evening Discourse of 3 February 1854.
23. See letter 2760. Airy did not give a Friday Evening Discourse in 1854.

Letter 2718
Heinrich Gustav Magnus to Faraday
10 August 1853
From the original in RI MS F1 I89a

Berlin 10 Aug 53.
Dear Sir
  Allow me to introduce to you Prof Helmholtz[1] from Königsberg i/P[2] who has published the interesting papers "On the velocity of the Nervous Principle["][3], on Complementary Colours[4] and many others[.]
  Yours | Very sincere | G. Magnus
Dr. Faraday.

1. Hermann von Helmholtz (1821–1894, DSB). Professor of Physiology at Königsberg, 1849–1855.
2. "in Prussia".
3. Helmholtz (1850).
4. Helmholtz (1852).

Letter 2719
Anton Schrötter[1] to Faraday
10 August 1853
From the original in RS MS 241, f.112

An | Das Ehrenmitglied, P.F. | Herrn Michael Faraday | in | London
    Die kaiserliche Akademie der Wissenschaften hat in ihrer Gesammtsitzung am 26. Mai 1852 (:zu welcher auch die auswärtigen Mitglieder einberufen waren:) den Beschluss gefasst, nebst den bisher üblichen Notificationsschreiben, gleich allen anderen Akademien, ihren Mitgliedern auch Diplome auszustellen[2]. In Folge dessen gebe ich mir die Ehre das für Euer Hochwohlgeboren ausgestellte Diplom hiermit zu übersenden.
    A. Schrötter | General-Secretär
Wien, den 10. August 1853.

TRANSLATION
To | The honorary member, P.F. | Mr. Michael Faraday | in | London
    The Imperial Academy of Sciences has passed at its General Meeting on 26 May 1852 (to which external members had been called as well) the resolution to issue diplomas for their members as well, besides the customary letters of notifications, as all other academies do[2]. Therefore I honour myself to send the diploma issued to you, honorable sir, with this.
    A. Schrötter | General Secretary
Vienna, 10 August 1853.

1. Anton Schrötter (1802–1875, ADB). Austrian chemist and Secretary of the Imperial Academy of Sciences in Vienna from 1850.
2. See Hammer-Purgstall and Ettinghausen to Faraday, 26 February 1848, letter 2058, volume 3.

Letter 2720
Jacob Herbert to Faraday
17 August 1853
From the original in GL MS 30108/1/59

Trinity House, London, | 17 August 1853.

Dear Sir,

I am instructed to transmit to you the accompanying Extracts from Two Letters of Mr. William Wilkins, and from a Report of a Committee who have recently visited Cromer Light House[1], on the subject of the ventilation of the Lantern thereof; and to request you will favor the Elder Brethren by communicating your opinion upon the suggestions, for it's improvement, contained in the said Extracts, and what objections (if any) exist to their being carried into effect.

I am | Dear Sir, | Your very faithful Servant | J. Herbert
M. Faraday Esq. D.C.L. | &c &c &c

1. These are in GL MS 30108/1/59.

Letter 2721
George Robert Waterhouse to Faraday
18 August 1853
From the original in IEE MS SC 2

British Museum | Aug. 18 / 53

My dear Sir

For some days past I have thought, each day, that I should be able to get out and call upon you, my object being to tell you that I sent in your note backed, as strongly as possible, - neither the note nor the observations accompanying it, however, came before the Trustees - it was explained to me that a similar application (one of a very pressing nature) had quite recently been brought before the Trustees - that the officer who had charge of the object wished to be taken out of the Museum offered to take it himself & not to let it out of his sight & to bring it back again - it would not do - the Trustees sayed they felt themselves bound to abide to the rules which had been made. Under these circumstances I was begged to withdraw your letter & my notes relating to it, as by so doing I should save the Trustees from making a refusal which would be very painful to them[1].

I am most sorry for all this!
Believe me | faithfully yours | Geo R. Waterhouse
Prof M. Faraday

1. Faraday wanted to borrow a large silica crystal from the British Museum and wrote to Waterhouse on 7 August 1853 for permission to borrow it. Faraday, *Diary*, August 1853, **6**: 13060–1.

Letter 2722
Faraday to the Trustees of the British Museum
20 August 1853
From the original in BM CA

Royal Institution | 20 August 1853

My Lords & Gentlemen

I am engaged in the investigation of a great object in natural science, namely the relation of light to Electricity and Magnetism. I have advanced so far as to prove the influence and power of the two latter forces over a ray (Philosophical Transactions 1846 p1[1]), and now hope to reverse the order and evolve or disturb these forces by the action of light. For these researches I need the use of a peculiar crystal of Silica. I have sought for such and have obtained some specimens; but they are too small to allow much hopes of success. In the British Museum I have found one, which for its clearness, size, position of the plagiedral planes, and other circumstances, is eminently fitted for the research, and I cannot find such another. Under these circumstances I take the liberty of making application for the loan of this crystal, for the service of science. It is about $13\frac{1}{2}$ inches long, $4\frac{1}{2}$ in diameter and is well known to Mr. Waterhouse and the attendants[2]. I shall not, in the slightest degree, injure or even affect it; my only object being to pass a ray from the Sun through it whilst it is surrounded by a helix and in relation to a Galvanometer. I need hardly say that I will take the utmost care of it: my willingness at all times to assist the Museum authorities in the *preservation* of the objects under their care, when they think that I can, by my advice, aid them in such matters, will I hope give assurance in that respect[3]. I cannot tell for how long I may want it, for the experiments have to wait upon the Sun. If I could transport the apparatus to the British Museum I would propose that course; but the Galvanometer is an especial instrument from Berlin[4], and requires fixing with the care of an astronomical instrument[.] I therefore hope that the Trustees will permit me the use of this crystal in the Royal Institution. I would express my own deep thanks for such a favour, but, that I feel it would be unmeet for me to offer private feelings or desires in such a case; and as I work for the pure good & advancement of science, I have no doubt that the Trustees will do all that lies within their power to aid me. If by the use of the crystal an affirmative result were to be obtained, it would give the specimen a value far beyond any it could possess as a simple mineralogical illustration[5][.]

I have the honour to be | with profound respect | My Lords and Gentlemen | Your Very humble & faithful Servant | M. Faraday
To | The Trustees of | The British Museum

Endorsement: Acknowledged 23 Augt

1. Faraday (1846a), ERE19.
2. See letter 2721.
3. See, for example, Faraday to Hawkins, 24 March 1845 and 7 August 1845, letters 1700 and 1766, volume 2, and also letter 2466.
4. Faraday, *Diary*, 3 September 1853, 6: 13074.
5. The Minutes of the Trustees note that this request was approved on 10 September 1853.

Letter 2723
Faraday to Henry Ellis
20 August 1853
From the original copy in IEE MS SC 2

Royal Institution | 20 Aug 1853

My dear Sir Henry,

I beg, through you, as the proper channel to make the enclosed application to the Trustees of the British Museum[1][.] I have applied to Mr Waterhouse for advice how to proceed but I suppose I have been informal for he does not encourage me[2]. The Trustees certainly ought to have the power, under sufficient precautions, to grant such a request as mine; for the British Museum is especially for the advancement of science. If they have not, I presume some department of government has; but I think it can hardly be needful that I should make such application, or that I should move such bodies as the Royal Society or the British Association to make such application, to a Secretary of State for a purpose so simple &, as it seems to me, so fit. Will you do me the favour to aid my object and to let me know the result of my application. I am anxious if possible to make my experiments before the Sun loses its power otherwise, they will have to run on into next Year[.]

I am My dear Sir Henry | Your obliged & faithful servant, | M. Faraday
Sir Henry Ellis | &c &c &c

1. Letter 2722.
2. See letter 2721.

Letter 2724
James Braid[1] to Faraday
22 August 1853
From the original in IEE MS SC 2

Burlington House | Oxford Street | Manchester 22 Augt 1853

Sir,
Although I have not the honor of your personal acquaintance I have long been familiar with your important contributions to chemical & physical science; and it is no small gratification to me to have had my views of the nature & cause of Table Turning confirmed & so ably ellucidated by your ingenious & conclusive physical tests[2]. The influence of dominant expectant ideas not only on the muscular system, but on every function of the body, had long been a favourite study of mine, & therefore enabled me to publish a scientific explanation of "Table-moving" long before I witnessed a single experiment of the sort; and, when attending the Conversazione of the Manchester Athenaeum, to propose a test with a circle of brass wire, which was at once conclusive even to those who would not have been convinced otherwise, that it could not be Electricity which was the cause, & moreover, that when muscular contact & an opportunity of applying an conscious muscular action were removed, by the wire lying loosely on the table, that no motion of the table could be induced. I beg your acceptance of a letter published by me in self defence against some unfair attacks made upon me by "D.T." and also a paper lately published by me on "Hypnotic Therapeutics"[3] in which you will perceive my mode of accounting for various physiological influences & cures which may be realised through *mental impressions changing physical action*, thus producing effects *subjectively*, which the mesmerists attributed to *objective* influence of some magnetic or odylic influence passing from the body of the operator to the patient - just as the mesmerists wished to explain "Table Turning" as the result of an objective influence. In the appendix to my "Hypnotic Therapeutics" you will observe I have criticised Dr Elliotsons[4] Dr Ashburners[5] & the Revd Mr Sandbys[6] & Townshend's[7] Mesmeric residuum force theory, &, with the aid of your physical tests I suspect they will not move far from the point to which I have fixed them.

Dr Carpenter[8], to whose article you refer[9], is indeed a most lucid writer on every department of physiology & psychology, & was thus able at once to take up my views of hypnotic & mesmeric phenomena, in illustrating which he has done ample justice to my labours & researches in this curious & interesting field of inquiry[.]

I have the honor to be | Sir | Your obedient servant | James Braid
Prof Faraday.

1. James Braid (c1795–1860, DNB). Physician in Manchester.
2. See letter 2691 and "Professor Faraday on Table-Moving", *Athenaeum*, 2 July 1853, pp.801–3.
3. Braid [1853].
4. John Elliotson (1791–1868, DNB). Mesmerist. Formerly Professor of Medicine at University College London, 1831–1838.
5. John Ashburner (1793–1878, B1). Physician and mesmerist.
6. George Sandby (d.1880, age 82, GRO, CCD). Vicar of Flixton, 1843–1860.
7. Chauncey Hare Townshend (1798–1868, DNB). Poet and mesmerist.
8. William Benjamin Carpenter (1813–1885, DSB). Professor of Forensic Medicine at University College London, 1845–1856.
9. Carpenter (1852), Friday Evening Discourse of 12 March 1852.

Letter 2725
Faraday to Charles Manby
27 August 1853
From the original in WIHM MS FALF

Royal Institution | 27 Aug 1853

My dear Manby

In haste to introduce to you M. Wage[n]mann[1] who wishes to become a member of the Civil Engineers - I send you at the same time a Parafine candle of which he is the author & I think so well of these candles that I sent one to Mr Walker some time ago as a very interesting object. Think of them being made at not merely a marketable but a cheap cost. What shall we come to - for here the coal mine beats the bee.

Ever Truly Yours | M. Faraday

1. Paul Wagenmann of Bonn who took out patent number 1853/2958 on the manufacture of liquid hydro-carbons and parafine.

Letter 2726
Faraday to Jacob Herbert
29 August 1853
From the original copy in GL MS 30108/1/59

Royal Institution | 29 Aug. 1853

My dear Sir

In consequence of your letter regarding the Cromer lighthouse[1] and our conversation in your room I proceeded to the place and was there examining into the circumstances & condition both by day & night on the 23rd & 24th instant[2]: & will now give you briefly the results, which are included chiefly under two heads the *ventilation* & the *warming* of the lanthorn.

The ventilation appears to have been much cared for & presents rather a complicated system. There are eight circular ventilators in different places in the stone work of the lower part of the lanthorn. They are moderate in size & the perforations in the surface plate are such that each offers an *air way* of $5\frac{1}{2}$ square inches. They are good as far as they go but are few in number & small in size; supposing the lanthorn depended chiefly on them. Then there are *twenty* small ventilators, opening, one under each lower pane of glass & directly in front of it; each has an air way of $18\frac{2}{3}$ of a square inch. These as they come into action under the influence of the wind are very useful on the windward side of the lanthorn. Further there are Eleven tubes above which being partly open to the outer air at their origin above the glass, then go upwards near the enclosed roof for a distance of $4\frac{1}{2}$ feet[.] They are 3 inches in diameter. I suppose it is imagined that streams of fresh air entering by them will tend to propell the bad air out of the top at the cowl. They are of no use where they are for only the two or three to windward will ever have air entering by them; that air will have no more ejecting force over the bad air than if it entered below; & entering above it is lost as fresh air to the lighthouse. What little effect they have is more hurtful than good.

A most important adjunct to the ventilation exists in the Watch room windows. These are three in number being on the North East & West sides. They are 45 inches wide each, and the sashes open from above. The watch room is a very dry room. The tower itself also is dry but no air enters the lanthorn or watch room from below. Some of the brethren directed the keepers attention to the use of these windows in aiding ventilation; and since then the difficulty which occurred on special occasions of extreme cold or closeness has been removed: they have but to open the windward window more or less to keep the glass perfectly clear & the lanthorn in a right condition as to ventilation[.]

As I am about to propose to take the Eleven upper tubes down on the first convenient opportunity I will here enter into a brief explanation of the action of the many ventilators in this lanthorn: in fact the principles & effects concern all lanthorns being general in their nature. It might have been thought that *thirty nine* ventilators were enough for one lanthorn; but of this number, when all were open, only nine passed the fresh air inwards. I examined the currents through them again & again both at day & night time: the results were constantly the same. Of the eight larger ventilators only the two to windward admitted constant streams of air: the other six had varying currents i.e in & out but chiefly outwards. Of the twenty small ventilators only the four to windward gave constant entering currents, the two or three to leeward passed the air in & out irregularly; the 13 or 14 others gave a constant & strong current outwards[.] This is the natural effect of the wind acting on their hoods it is the same as that which I have found on former occasions at the Needles & elsewhere and is due to the principle which acts so beneficially in the wind guard. Of the Eleven upper ventilators air entered at the three to windward, but the other eight passed more air outwards than inwards[.] If it be now considered that the joint air passage of the 2 larger and 4 smaller ventilators giving access to air in the lower part of the lanthorn is only 18 square inches it will be understood how much additional aid is gained by the use of the watch room windows. If the window to windward be opened only one inch it offers a passage to air having an area of 45 square inches and as an equal aperture is opened between the upper & the lower sashes by the same act, 90 square inches of sectional area or of air passage is there, at once obtained. This amounts to five times that of all the ventilators which supply fresh air to the lamps & glass at the same time[.]

There appears to be plenty of air way out at the Cowl, provided air be let in below as described. The ventilation of the lamps is also perfect according to the description of the keepers & the action at the time that I saw it. Before the lamp tubes were put up, there was a continual condensation on the roof and droppings from the ribs. To remedy the dropping evil, a collecting gutter was put up against the ribs and a pipe from it conducted the condensed water into a vessel placed beneath[.] The keepers say that now there is no condensation on the roof nor do they ever have to place the vessel into which the water formerly ran. If this be really so on the most cold & trying nights then the gutter might as well be taken away with the eleven pipes before referred to, on any convenient opportunity[.]

Proceeding to the *warming* of the lanthorn; I find that the keepers complain of its temperature in cold weather. They state that the oil in the lower lamps then freezes: that they cannot make those lamps burn properly either by turning up the cottons or otherwise; that the oil will not freely descend to the cotton; & that on such occasions the quantity burnt in a lower lamp has not been more than two thirds of that burnt in an upper lamp for the same time. I can well understand this to be the case especially with the Rape oil[.] After the lamps are lighted the temperature is very different indeed in the upper & the lower parts of this fine high lanthorn. I found it so, and they say the difference is quarter in the cold weather. To counteract this effect there is a stove in the watch room. It has a jacket, from which the hot air passes by a pipe $6\frac{1}{2}$ feet long & only 3 inches in diameter into the lanthorn at one place in the floor. The chimney of the stove which in the watch room is 6 inches in diameter passes through the floor into the lanthorn & is then continued upwards as a copper pipe 10 inches in diameter & $13\frac{1}{2}$ feet long & then goes out through the roof. This stove is insufficient in time of need. They have had to make such large fires in it as seem to burn it out, and now when the fire is lighted the smoke enters into the air chamber & so into the lanthorn. They have had occasion to heat the stove so highly that the part of its chimney in the lanthorn has been dull red hot. The stove must then have done duty far more by the hot chimney in the lanthorn & by the warming of the air in the watch room than by the small supply of hot air which its pipe of 3 inches in diameter could convey: but such a kind of action forces the stove too far, & by burning it out soon causes its derangement.

I am fully persuaded that the presence of a simple stove in the lanthorn, without air jacket and with a horizontal flue, will be far more effectual than the present stove and burn far less fuel. The place is well fitted for an arrangement like that at the St. Catherines light where the horizontal part of the flue runs under the gallery[.] At Cromer the platform is 18 inches wide. There is a corner under the gallery at the head of the stairs which would do exceedingly well for the stove; the distance from it to the present place of the chimney is in one direction under the gallery 26 feet & in the other direction about 11 feet[.] I would recommend that the present upright chimney be replaced by one terminating above with a wind guard as at St. Catherines (See report of 29 June 1853[3].); that the horizontal part of the chimney kept at about an inch from the wall & from the platform be continued from the stove to this upright part in *both* directions; & that near the entry into the upright part there be throttle valves that either the one length or the other, or both may be used according to circumstances. In the vertical part of the chimney, the upper lengths should be put *into* & not *over* the lower length, as described in the report just referred to; & there should be a vessel termination to collect any condensation in this part[.]

The keepers at this house are remarkable for their practical sagacity: they have by observation & experience discovered several points in the working of the lights & lighthouse which are fully justified by principle & well illustrate it. They in this respect form a very striking contrast to some other establishments which I have visited[.] The lighthouse both by day & night presents an object well worthy of approbation[.]

The flag staff N has been set I suspect by a magnetic needle without the card; it is 20° or more west of the North star[4][.]

I am | My dear Sir | Ever Your faithful Servant | M. Faraday
Jacob Herbert Esq | &c &c &c

1. Letter 2720.
2. Faraday's notes of his visit are in GL MS 30108/1/59.
3. Letter 2692.
4. This letter was read to Trinity House Court, 6 September 1853, GL MS 30004/25, pp. 276–7. The Light Committee was instructed to put Faraday's suggestions into effect.

Letter 2727
Faraday to Jacob Herbert
3 September 1853
From the original copy in GL MS 30108/1/58

Royal Institution | 3 Septr. 1853

My dear Sir

Mr. Phillips[1] (from Wilkins & Co) has just been to me to report on the results of the changes made at St. Catherines. The ventilators are now fixed but he states that only those to leeward admit air into the lanthorn (with the wind at west by south), whilst all the others pass air outward, the doors to the tower being shut[.] There is certainly much that is peculiar to this lighthouse perhaps from its position under the high hill & cliffs of the neighbouring land, but I suspect that the walled in condition of the gallery has a part in the result he describes, affecting the hoods & outer apertures of the ventilators. The windguard which has replaced the cowl is according to his account effectual; the cowl he describes as having frequently revolved 5 or 6 turns together[.]

He reports, upon his own knowledge, that the stove will not burn properly when the tower is shut off from the lanthorn. This stove was cracked about a twelvemonth ago, & the keepers made no complaint of this kind to me when I was recently there. Mr Phillips has applied to me to know what I would advise. I could not take upon me to authorize him in matters not referred to in my former note[2] but told him I should write to you and I advise, now that his men are there, that under the peculiar conditions of this lighthouse the stove should be taken into the watch room below which would give 8 feet additional hot vertical draught to the chimney and that the chimney (the new part) be enclosed by an air pipe 9 inches in diameter open throughout. Elsewhere as at the South Foreland the arrangement now existing at the St. Catherines has been found quite effectual but here the air seems to drop down upon the lanthorn & reverse many of the usual results. He will apply to you on Monday[3] for instruction or authority[4][.]

I am | My dear Sir | Yours Very faithfully | M. Faraday

Jacob Herbert Esqr | &c &c &c | Secretary

1. Unidentified.
2. Letter 2692.
3. That is 5 September 1853.
4. This letter was read to Trinity House Court, 6 September 1853, GL MS 30004/25, pp.276–7. It was referred to the Light Committee.

Letter 2728
Henry Ellis to Faraday
12 September 1853
From the original in IEE MS SC 2

British Museum | Sept. 12th 1853

My dear Dr Faraday

Your letter[1] respecting the Crystal was laid before the Trustees at their meeting on Saturday last[2] and I have great pleasure in telling you, that impressed by the importance of the object which you have in view, the Trustees have given instructions to Mr Waterhouse to deliver the Crystal to you with the injunction that it be returned to the Museum Collection as soon as possible.

Faithfully Yours | Henry Ellis | Pr. Lib.
Michael Faraday Esq D.C.L.

1. Letter 2722.
2. That is 10 September 1853.

Letter 2729
Faraday to George Robert Waterhouse
14 September 1853
From the original in Natural History Museum, Palaeontology Archives, Keepers Letters

Royal Institution | 14 Sept 1853

My dear Sir

Will you do me the favour to let the bearer Mr Anderson have the crystal which the Trustees have permitted me the use of[1][.]

Ever Truly Yours | M. Faraday
G. Waterhouse Esq | &c &c &c

1. See letter 2728.

Letter 2730
Faraday to Edward Vivian[1]
14 September 1853
From the original in Torquay Natural History Society MS

Royal Institution | 14 Septr. 1853

My dear Sir

I feel it very difficult to give an opinion on your case not having the opportunity of seeing the instruments so as to judge of *how* they were constructed and at the same time of asking questions of you or the maker: I will not restrain myself however from making suggestions but remember I do so with insufficient data[.]

I do not see how the mere difference in size of the air spaces *a* & *b* can *cause* the change you mention especially as you say the alteration comes to an end after a certain time as six months[.] Is not the cause a real difference in the character of the atmosphere in the two air spaces? and is not the gradual change a consequence of the slow equilibration of these different atmosphere by their gradual transmission through the separating column of spirit? the effect going on until they are alike & then ceasing. I observe that the bulbs a.a.a. in which the air expands are all of them those which are *last* finished in the construction of the instrument and it seems to me that when the bulbs were finally sealed the air within them mingled as it would be with alcohol vapour has been affected & changed in part by the heat applied. Hot glass in a mixed atmosphere of air & alcohol vapour will form carbonic acid, acetic acid, & other products & in this way the air *a* may at the first be diffused to the air *b*. If for instance by any such change the air *a* is partly deprived of a portion of its oxygen then oxygen will slowly pass from *b* to *a* until both are alike again[.]

I only suggest this change in the oxygen & this kind of change by heat as illustrating the kind of difference which I suspect exists in the new instrument. The difference may be caused some other way. For instance the air may be introduced by some particular process - the mouth or lungs may be used or other means & in these the cause of the first difference may be[.] I have not even the means of guessing not knowing the particular method of construction pursued by the maker but I think you will find a difference of this kind at the root of the matter[.]

Before concluding let me ask why you say the difficulty is insuperable? The instruments at last come to a settled condition & if you do not graduate until then I suppose all would be right.

I am very pleased to renew our acquaintance in any manner and am My dear Sir | Very Truly Yours | M. Faraday
Edwd Vivian Esq | &c &c &c

Will the following point help you. Glass which has been heated is *said* to return to its final volume when cooled only very slowly and after some weeks or months. I think Bellani[2] in this way accounted partly for certain changes in thermometers which went on even for years[3]. Can any such effect occur with you | MF

1. Edward Vivian (1808–1893, B3). Partner in the Torquay Bank and meteorologist.
2. Angelo Bellani (1776–1852, DSB). Italian scientific instrument maker.
3. See "Variation of Thermometers", *Quart.J.Sci.*, 1823, **15**: 369–70.

Letter 2731
Jacob Herbert to Faraday
14 September 1853
From the original in GL MS 30108/2/60d

Trinity House, London, | 14th September 1853.
My dear Sir,

The Board is about to cause a Bell, of not less than one Ton, and not exceeding two Tons in weight, to be fixed at the South Stack Light House, near Holyhead,- and I am directed to say, that it would be satisfactory to the Elder Brethren, to be favored with your views, as to the form of Belfry which may be best adapted to receive it;- with reference to the great object of the propulsion of it's sound, in the direction in which it is required to be heard, that is to say,- seaward.

In their recent visit to Ireland, the Elder Brethren observed, that, at some of the Light Houses in that Country, which are provided with Bells, a sort of Dome was erected over the Bell; and the communication of your opinion, as to the benefit which such an adjunct may afford in the conveyance of the Sound will be acceptable.

It is probable, that the use of Bells may become more general than at present, at the Light Houses belonging to this Corporation,- and it is desirable, therefore, that the manner of their Suspension should be carefully considered.

In making this communication, allow me to draw your attention to your Letter of the 24 March last[1], having reference to this subject.

I remain, | My dear Sir, | Your very faithful Servant | J. Herbert
M. Faraday Esq. | &c &c &c

Address: M. Faraday Esq. | &c &c &c | Royal Institution | Albemarle Street

1. Letter 2660.

Letter 2732
William Edwards Staite[1] to Faraday
21 September 1853
From the original in GL MS 30108/2/64a.1

Liscard Vale, | near *Liverpool*, | 21st Sep. 1853.
Dear Sir,

It is now upwards of three years since I had the honour of exhibiting to you at the Baker St Bazaar[2], my Automatic Magnetic System for regulating the *Electric Light*[3],- since which time, however, I have greatly simplified the apparatus of my *Lamps*, rendering Clock-work altogether unnecessary. I also employ a new form of *Battery*, in which Alloyed Lead is used, as the positive metal, in a dilute solution of Nitrous Acid. The products form a valuable pigment, & the value nearly covers the cost of materials. I feel sure you take sufficient interest in my success to be pleased when I state that after some months trial of the Light in Liverpool by the Dock Committee, in a Tower erected on purpose to test its value as a Beacon-light on the river Mersey it has been pronounced as quite successful: the light is steady, Constant & Continuous for many hours without the necessity of attention or manipulation: in short, it burns from Sunset to Sunrise, with the same Certainty as an oil lamp, & at a cost of six pence per hour per 1000 Candles - *the Contract price*.

The Dock Comm[itte]e are now arranging for its immediate application to their Sandon Graving Docks: & they wish each Dry Dock to be furnished with two lights, one at each end, so as to neutralize each others' shadows, to enable repairs of Ships to be carried on after dark. I have taken the liberty of writing to you, to solicit the favour of your advice as to the mode I propose of reflecting the Light, so as to Confine it within prescribed limits & at the same time to diminish the glare without loss of luminosity. I shall be truly grateful for any hint or suggestion you may kindly offer on this point, & I shall of course consider any such communication as strictly confidential.

The enclosed sketch is what I think would answer the purpose very well, & the cost of such an arrangement (using white-wash'd wood, or wood painted white) would be trifling.

Hoping you will pardon the liberty I have taken, & with every sentiment of esteem & respect,

I beg to remain, | Dear Sir, | Your's faithfully | W.E. Staite
To | Professor Faraday, | &c. &c. &c. | Royal Institution.

1. William Edwards Staite (1809–1854, Fahie (1902), 375). Pioneer of electric lighting.
2. Staite's light was displayed there on 7–10 December 1848. See IEE MS SC 71/1, introduction, p.23.
3. For details of Staite's light see Fahie (1902).

Letter 2733
Faraday to Jacob Herbert
22 September 1853
From the original copy in GL MS 30108/2/60d

Royal Institution | 22 Septr. 1853

My dear Sir,

I have been in the country for some time & hence the delay in my reply to your letter of the 14th[1] instant in which you ask me for an opinion on the best form of belfry for a bell at the South Stack Lighthouse. This I find it difficult if not impossible to give you for I have had no experience nor any opportunity for observation in such matters up to this time. I do not know the particular locality now under consideration & have not visited any lighthouse or signal bell anywhere: nor have I yet had occasion to seek access to that experience which has been already obtained by others[.] I can therefore do little more than refer back to the vague generalities of my former letter (24 March 1853[2])[.]

Still I venture to utter a surmise or two. I think it probable that the bell should be covered over, but I should not expect that a dome would be the best form but rather a flat surface and I think that the distance of this surface would be influential & perhaps importantly so. The surface under the bell probably ought to be regular - flat or perhaps even conical. If the sound has to do duty only for a part of the horizon as for instance 180° then probably a wall-back on the other or unimportant side might be very valuable, but I am really unwilling to offer further notions without communication with those who have had experience in these matters. In any site where a bell is to be placed I conclude that the form of the ground and also of the near buildings will have to be considered in reference to their acoustic influence[3][.]

I am | My dear Sir | Your faithful Servant | M. Faraday
Jacob Herbert Esq | &c &c &c

---

1. Letter 2731.
2. Letter 2660.
3. This letter was read to Trinity House Court, 27 September 1853, GL MS 30004/25, p.286. It was referred to the Light Committee.

Letter 2734
Harriet Jane Moore to Faraday
23 September 1853[1]
From the original in RI MS F1 I160

9 Carlisle Parade | Hastings | Sept 23rd

My dear Mr Faraday

I was much obliged by your kind Lady's note. Pray tell her that I intend to have her autograph in her present, as so I consider it to be, since it is the work of own brother. She says that you watch the spider weaving their webs, I have often been much interested in observing their workmanship, & ingenuity in repairing any damage in their fabric. I wish you & Mrs. Faraday were here at this cheerful agreeable place, as it seems to me; I set off from Putney on Tuesday[2] morning alone, being very anxious to secure lodgings to my sister[3] before the Equinoxial gales set in; and by dint of hard work, secured these very pleasant rooms, where she joined me yesterday, and less fatigued by our journey than I had feared. Our rooms almost overhang the sea, which is to me a splendid object; on Tuesday the sunset was glorious & the moon shone upon the placid waters so beautifully that I thought how much you would have admired it. The instant [name illegible] came into this our sitting room, she exclaimed 'Mrs Faraday's carpet'. I had not noticed that it was the same as that in Warwick Cottage, so we have something in common. I am in some hopes that the mild sea breezes may invigorate my poor sister. We are very comfortable, & very friendly, and I shall stay till our friends arrive next week, & should

remain longer but that I do not like to leave Julia on duty too long. Give my kindest regards to dear Mrs. Faraday & believe me
Ever your's very truly | Harriet Moore

1. Dated on the basis that letter 2740 is the reply.
2. That is 20 September 1853.
3. Julia Moore (d.1904, age 100, GRO). Sister of Harriet Jane Moore.

Letter 2735
Christian Friedrich Schoenbein to Faraday
24 September 1853
From the original in UB MS NS 408
My dear Faraday,
Some weeks ago I returned from the journey I had undertaken to Bavaria, Austria &c. during our mid-summer-holidays and I can assure you that it was a very pleasant one. The first stay I made at Munich where I remained no less than 10 days finding that town highly pleasing and interesting both for the men and the things, I chanced to meet and see there. I think you would relish it as much as I did and if you should have any mind to cross the water once more, I strongly recommend you taking a trip to the Capital of Bavaria. The number of exquisite objects of painting, sculpture, architecture &c. accumulated there, is very great indeed and placed so closely together that you may see and enjoy them with perfect ease and comfort. Of course I met Liebig at Munich whom I knew before little more than by sight, but within the first five minutes we had found out the footing upon which both of us could move comfortably enough. You will laugh when I tell you that Liebig asked me to deliver a lecture before a very large audience in his stead and Mr. Schoenbein though reluctantly yielded to that strange demand. The subject treated was that queer thing called "Ozone" which ten or twelve years ago as you are perhaps aware, was declared by a Countryman of Your's and pupil of Liebig's to be a "nonens". Nothing was easier to me than proving its corporeal existence and our friend Liebig, in spite of the unfriendly feelings he formerly entertained towards my poor child, has now taken it into his favor and seems even to have fallen deeply in love with the creature. He has therefore repeatedly entreated me to write a sort of biography of my progeny and give and account of its education and the accomplishments it has acquired under my tuition during the last decennium. I do not know yet whether I shall comply with his wishes being not very fond of copying myself over and over again.

My trip on the Danube down from Ratisbonne to Vienna proved highly delightful to me, though I experienced the mishap of losing my pocket-book and along with it my passport, no joke to a traveller who was about to enter the austrian Empiry. No unpleasant results however issued from that adventure. The scenery down the river merits to be called beautiful; now and then the Danube is forced to make its way through very deep and narrow ravines the top of the hills being covered with ruined castles, churches, convents, country seats &c. and their declivities richly wooded, another time you enjoy a beautiful and extensive view on the Alps of the Tyrol, Salzburg, Styria &c.

Vienna itself is a fine and a noble town full of interesting objects of Science and the arts and its inhabitants have become proverbial for their good nature. There is therefore no wonder that I enjoyed there very agreeable days. In going home I passed through Prag, Dresden, Leipzic, Frankfort &c seeing little more of those cities than their steeples and towers, for having stayed out too long I was obliged to return to Bale as quickly as possible. Mrs. Schoenbein and the girls have during my absence been living in the hills according to our usual style of passing the midsummer holidays. My eldest daughter has been absent from home these last 5 months and lives very happy on the beautiful lake of Geneva at a little place called Rolle. She has almost grown up into womanhood, is very like her mother, only a little taller and upon the whole a good-natured and dutiful child. I think you would like her. Our friend de la Rive was kind enough to invite her to pass the approaching season of the vintage at his country seat near Geneva.

Now having talked so much about myself and my family it is time to ask you how you and your amiable Lady are doing. I hope well, in spite of the oriental and other affairs of the world[1]. I should feel over happy if it fell to my lot to see you once more and to accomplish my wishes I see no other means than your coming over to us.

Mrs. Schoenbein joins me in her kindest regards to Mrs. Faraday and I beg you to believe me for ever
Your's | most truly | C.F. Schoenbein
Bâle Septbr. 24, 1853.

Address: Doctor Michael Faraday | &c &c &c | Royal Institution | London

1. A reference to the events leading up to the Anglo-French war against Russia. See Lambert, A.D. (1990), 48–51 for the specifics.

Letter 2736
Faraday to Henry Ellis
28 September 1853
From the original in BM CA

Royal Institution | 28 Septr. 1853

My dear Sir Henry
    I shall this day, personally, return the crystal to Mr. Waterhouse[1] and beg you will have the goodness to express my sincere thanks to the Trustees of the British Museum for the favour granted me. I have optically examined the crystal, and find it just what I wanted; but from the delays which occurred, so much of the sunny weather has passed by that I have little hope now of any fit for my purpose this year. If however between this time and next Summer I am encouraged by results with other crystals I may probably make application then for a second loan of the specimen[.]
    I am | My dear Sir Henry | Your Very Obliged Servant | M. Faraday
Sir Henry Ellis | &c &c &c

1. See letters 2721, 2722, 2723, 2728 and 2729.

Letter 2737
Faraday to Lyon Playfair
1 October 1853
From the original in IC MS LP253

R Institution | 1 Octr. 1853

My dear Playfair
    I am greatly obliged by your card - but living out of town just now have not power use it today[.]
    Ever Truly Yours | M. Faraday

Letter 2738
Jacob Herbert to Faraday
6 October 1853
From the original in GL MS 30108/2/61

Trinity House, London | 6th October 1853

Sir,
    Mr. Walker having submitted Drawings and Specifications for a new Lantern for the Low Light at the Spurn Point, and it being deem'd

advisable that previously to the Works being enter'd upon the Board should have the Benefit of your Opinion as to the best Method of Ventilation to be adopted therein,- I am directed to acquaint you therewith and that Mr. Walker has been requested to communicate with you on the Subject.
    I am | Sir | Your most humble Servant, | J. Herbert
M. Faraday Esq D.C.L. FRS. | &c &c &c | 21 Albemarle Street

Letter 2739
Emmanuel Arago[1] et al[2]. to Faraday
6 October 1853
From the printed original in RI MS F1 H56
M
    Monsieur Emmanuel Arago, Monsieur Alfred Arago[3], Madame Emmanuel Arago, Mademoiselle Jeanne Arago, Monsieur[4] et Madame Jacques Arago, Monsieur[5] et Madame Victor Arago, Monsieur Joseph Arago[6], Monsieur Mathieu[7] et Madame Mathieu née Arago[8], Monsieur Etienne Arago[9];
    Monsieur et Madame Taponier et leurs Fils, Monsieur et Madame Conte, Monsieur Antonin Arago[10] et ses Enfants, Monsieur[11] et Madame[12] Gaston de Vilar et leurs Fils, Monsieur et Madame Frachon et leur Fille, Messieurs Victor, Roger et Emmanuel Arago, Monsieur[13] et Madame[14] Laugier et leurs Fils, Monsieur Charles Mathieu[15]:
    Ont l'honneur de vous faire part de la perte douloureuse qu'ils viennent de faire en la personne de M. François Arago, leur Père, Beau-Père, Grand-Père, Frère, Beau-Frère, Oncle et Grand-Oncle, décédé à l'Observatoire de Paris, le 2 octobre 1853, à l'âge de 67 ans.
Paris, le 6 octobre 1853.

Address: Monsieur Faraday | de l'Institut de France | Londres

TRANSLATION
Sir,
    Mr Emmanuel Arago, Mr Alfred Arago[3], Mrs Emmanuel Arago, Miss Jeanne Arago, Mr[4] and Mrs Jacques Arago, Mr[5] and Mrs Victor Arago, Mr Joseph Arago[6], Mr Mathieu[7] and Mrs Mathieu née Arago[8], Mr Etienne Arago[9];
    Mr and Mrs Taponier and their sons, Mr and Mrs Conte, Mr Antonin Arago[10] and his children, Mr[11] and Mrs[12] Gaston de Vilar and their sons,

Plate 10. Dominique François Jean Arago.

Mr and Mrs Frachon and their daughter, Messrs Victor, Roger and Emmanuel Arago, Mr[13] and Mrs[14] Laugier and their sons, Mr Charles Mathieu[15]:

Have the honour of informing you of the painful loss that they have just suffered in the person of M. François Arago, their Father, Father-in-Law, Grandfather, Brother, Brother-in-Law, Uncle and Great Uncle, who died at the Observatory in Paris, on 2 October 1853, at the age of 67.
Paris, 6 October 1853.

1. François Victor Emmanuel Arago (1812–1896, DBF). Lawyer and politician; son of Arago.
2. Those who do not have notes have not been further identified and are not indexed.
3. Alfred Arago (1816–1892, DBF). Painter; son of Arago.
4. Jacques Etienne Victor Arago (1790–1855, DBF). Writer and explorer; brother of Arago.
5. Victor Arago (1792–1867, Toulotte (1993), 296). Brother of Arago.
6. Joseph Arago (1796–1860, Toulotte (1993), 296). Brother of Arago.
7. Claude-Louis Mathieu (1783–1875, Robert et al. (1889–91), 4: 309). Astronomer and politician; brother in law of Arago.
8. Marguerite Mathieu, née Arago (1798–1859, Toulotte (1993), 296). Sister of Arago.
9. Etienne Vincent Arago (1802–1892, DBF). Politician and writer; brother of Arago.
10. Antonin Arago. Son of Jacques Etienne Victor Arago. Toulotte (1993), 296.
11. Gaston de Vilar (b.1818, Capeille (1910), 661). Married into Arago family.
12. Marie Vilar, née Arago. Married Gaston de Vilar. Capeille (1910), 661.
13. Auguste Ernest Paul Laugier (1812–1872, Vapereau (1870), 1067, (1893), 931). Astronomer; married Arago's neice.
14. Lucie Laugier, née Mathieu (1822–1900, Toulotte (1993), 296). Arago's neice.
15. Charles Mathieu (1828–1889, Toulotte (1993), 296). Arago's nephew.

Letter 2740
Faraday to Harriet Jane Moore
15 October 1853
From Gladstone (1874), 50

Royal Institution, 15th Oct., 1853.

My dear Miss Moore,

The summer is going away, and I never (but for one day) had any hopes of profiting by your kind offer of the roof of your house in Clarges Street. What a feeble summer it has been as regards sunlight! I have made a good many preliminary experiments at home, but they do not encourage me in the direction towards which I was looking[1]. All is misty and dull, both the physical and the mental prospect. But I have ever found that the experimental philosopher has great need of patience, that he may not be downcast by interposing obstacles, and perseverance, that he may either overcome them, or open out a new path to the bourn he desires to reach. So perhaps next summer I may think of your housetop again. Many thanks for your kind letter[2] and all your kindnesses uswards. My wife had your note yesterday, and I enjoyed the violets, which for a time I appropriated.

With kindest remembrances and thoughts to all with you and her at Hastings.

I am, my dear Friend, | Very faithfully yours, | M. Faraday

1. Letter 2736.
2. Letter 2734.

Letter 2741
Wenceslas Bojer[1] to Faraday
15 October 1853
From the original in RS MS 241, f.135

Museum, Royal College | October 15th 1853.
To | Michael Faraday Esq F.R.S.
Sir

I have the pleasure to inform you that at a meeting of the Council of the Royal Society of Arts & Sciences of Mauritius, you were unanimously elected an Honorary member of the Society, and further that the Council have entrusted to me the agreeable task of preparing your "Diploma" which I have the honor to forward to you with the Council's earnest hope that you will kindly accept the title as the highest mark of respect the Council can bestow upon a man so eminent in science as you are, and of gratitude for the service you have rendered by your great discoveries to physical & chemical science.

The Diploma will be delivered to you by my worthy friend & colleague Mr. James Morris[2] of London.

I have the honor to be | Sir | Your most obedient Servant | W. Bojer V.P.

1. Wenceslas Bojer (1795–1856, DMB). Naturalist in Mauritius and Vice-President of the Royal Society of Arts and Sciences of Mauritius.
2. James Morris (1810–1869, DMB). Professor of Classics at the Royal College, Mauritius, 1845–1849 and Secretary, Royal Society of Arts and Sciences of Mauritius. Then a Mauritian representative in London.

Letter 2742
George Biddell Airy to Faraday
17 October 1853
From the original press copy in RGO6 / 468, f.179
Royal Observatory Greenwich 1853 Octr. 17
My dear Sir
In speaking about the inductions among the parallel wires to and from Liverpool, I failed in conveying to you my conjectural reasons for supposing that there would be no induction[1]. I *intended* to express what I have diagrammatized on the following page (pray pity the mental struggles of a smatterer).

In case 1, the wire furnished with battery would produce a certain induction in the wire near it.

In case 2, I *supposed* that an induction of the opposite kind would be produced.

Therefore in case 3, I *imagined* that the effects of the two inductions would neutralise each other.

And case 4 (which is the case of wires to send from Liverpool) *appeared* to me to be, in the course of its currents, the same thing as Case 3: and thus I supposed that there would be no induction.

How many of these steps are erroneous?

Yours very truly | G.B. Airy
Professor Faraday

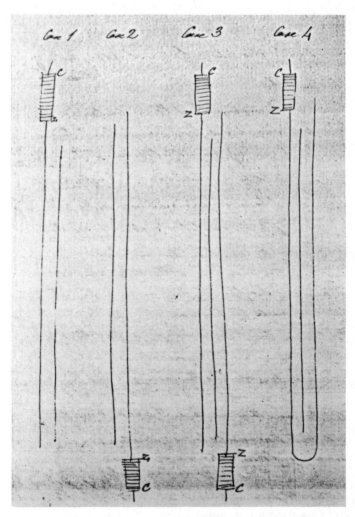

1. This refers to a set of experiments that Faraday, Airy and others had seen performed on 15 October 1853 at the Lothbury Wharf Office of the Electric Telegraph Company. The experiments were on long distance telegraphy and displayed the phenomena of telegraphic retardation. Faraday's notes are in Faraday, *Diary*, 15 October 1853, 7: pp.401–8. For a discussion of this work see Hunt (1991).

Letter 2743
Faraday to George Biddell Airy
19 October 1853
From the original in RGO6 / 468, f.182

R Institution | 19 Octr. 1853

My dear Sir

From what you said about the notes last Saturday night[1] I send you mine to look at if you like. Because of my very bad memory I am obliged to make them long. I must ask you to return them but make any abstract or copy that you like. I have even sent the notes I made the week before[2] at the Wharf[3] on the same subject.

In regard to your note[4] I send back the diagram with the wires marked a and b. Then for simplicitys sake let us suppose that the action[s] are perfect & permanent i.e that the whole of a shall in each case be equally electrified - the insulation being *perfect*. Then no induction would take place in any of the cases if the wires b were also perfectly insulated. It is only when being uninsulated they can assume the contrary state that the induction occurs. (I neglect that which theoretically would occur across the *thickness* of the wire for it is as nothing)[.]

But if b was in each case uninsulated or touching the ground then induction would occur but the amount of induction would be twice as much on it in the cases 3 & 4 as those in 1. & 2. the battery being of the same power.

I hope I have caught your meaning & that any case which may occur to your mind will be included in the explanation I have endeavoured to give[.]

Ever Truly Yours | M. Faraday
The | Astronomer Royal | &c &c

---

1. See Faraday, *Diary*, 15 October 1853, 7: pp.401–8 which noted Airy's presence at the experiments. See note 1, letter 2742.
2. Faraday, *Diary*, 4 October 1853, 7: pp.393–401.
3. That is Lothbury Wharf where the experiments were conducted.
4. Letter 2742.

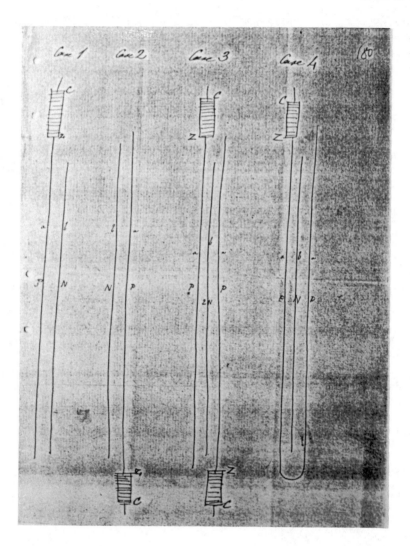

Letter 2744
Faraday to Edward Hawkins[1]
20 October 1853
From the original in BM DWAA MS Correspondence 1826–1860, volume 5, 1723

Royal Institution | 20 Octr. 1853

My dear Hawkins

I should not think it advisable to place glass, which, by its exfoliation is manifesting its tendency to decompose, in a damp room[.]

Ever Truly Yours | M. Faraday

1. Edward Hawkins (1780–1867, DNB). Keeper of Antiquities at the British Museum, 1826–1860.

Letter 2745
Faraday to Edward Meryon[1]
20 October 1853
From the original in FACLM H MS c1

New Road | Hammersmith | 20 Octr. 1853

Dear Sir

I think you must refer to a little apparatus which by the position of a series of little ivory balls shews the progress of undulations & is intended to illustrate the wave theory of light. If so the one you saw probably belonged to the Revd. Robt. Walker[2] of Oxford. There are others about town. I think that Newman 122 Regent St our instrument maker has one, if not he knows where they are. The apparatus has no reference to the water lines of ships that I am aware of[.]

Ever faithfully Yours | M. Faraday

Dr. Meryon | &c &c &c

1. Edward Meryon (1809–1880, B2). Physician.
2. Robert Walker (1801–1865, B3). Professor of Experimental Philosophy at the University of Oxford, 1839–1865.

Letter 2746
George Biddell Airy to Faraday
24 October 1853
From the original press copy in RGO6 / 468, f.187

Royal Observatory Greenwich | 1853 October 24

My dear Sir

With many thanks I return your notes on Staiths[1] and Clark's experiments[2]. I have made no extracts, for your resumé aided by my ocular sight (and giving me some facts of the early experiments which I did not know) has given me a tolerably clear mental view of the whole.

My ideas of galvanism and induction were never of the clearest, and now they are more disturbed than ever.
I am, my dear Sir, | Yours very truly | G.B. Airy
Professor Faraday

1. An error for Samuel Statham (d.1864, age 58, GRO, gutta percha manufacturer) who conducted the experiments. See Faraday, *Diary,* 4 October 1853, 7: p.393.
2. See letter 2743.

Letter 2747
George Biddell Airy to Faraday
25 October 1853
From the original in IEE MS SC 2

Royal Observatory | Greenwich | 1853 October 25
My dear Sir
You know that in all matters magnetical, meteorological, and chemical, I consider myself & Co. here as mere machines, fit to act up to other people's ideas, but having no ideas of our own.
In this consideration I take leave to ask you -
What is the value of Schönbein's Ozonometer?
Is it worth while for us to observe with it here?
If you say that it is, I shall put it in hand for daily observation and registry.
I am, my dear Sir, | Yours very truly | G.B. Airy
Professor Faraday

Letter 2748
G.G. Wilson[1] to Faraday
26 October 1853
From the original in IEE MS SC 2

217 Strand | 26th Oct 1853
Dear Sir
I am personally unknown to you and the only introduction I can offer to you is this. I have just seen a gentleman of the name of Fitzgerald[2] who called here in the way of business and who tells me that he had the pleasure of meeting you at a party last evening when you were speaking upon the subject of table turning, and that your opinion was totally opposed to the possibility of anything like Spiritual or Satanic Agency in the matter. Some most curious and startling facts have occurred at my place of residence in connexion with this much vexed subject - much

vexed I beg to repeat notwithstanding your letter to the Times[3] which every one has read or at least ought to have read.

I am not a showman, but simply a bankers' clerk and therefore will not run about to exhibit the thing but if in the interest of science you would take the trouble to travel as far as Ilford in Essex where I live I fully believe that you would go away with a very different opinion upon the subject from that which you hold at present.

Should you think it worth your while to take any notice of this communication

My address is | Mr G.G. Wilson | at Messrs Strahan & Co[4] | 217 Strand

and if you drop a line I will either arrange to accompany you to my home in the evening (which may be any you like to appoint after tomorrow) or will furnish you with the exact address - as we live in a new neighbourhood and are rather difficult to find.

I am Dear Sir | Yours very respectfully | G.G. Wilson
Mr Faraday Esq | &c &c &c

P.S. | In order to make it quite clear to you that I have and can have no interest in misinforming or attempting to deceive you - I add in plain English - that no money or present of any kind would be accepted from any one however pleased and satisfied with the demonstration he might be.

1. Unidentified.
2. Unidentified.
3. Letter 2691.
4. Strahan, Paul, Paul and Bates, bankers of 217 Strand. POD.

Letter 2749
Faraday to George Biddell Airy
27 October 1853
From the original in RGO6 / 468, f.188-9

Royal Institution | 27 Octr. 1853

My dear Sir

I am very glad you are about to observe for ozone[1]. - I think it may, and probably will, become a matter of great importance in relation to the atmosphere & its natural actions. Schonbein's Ozonometer is the best thing we have at present; but the subject is very likely to open out.

Referring to your previous note[2]; May I say that perhaps your mind has been for a moment embrarrassed [sic] by thoughts, mingling the conditions of dynamic & static induction; which differ very greatly from each other. In our recent observations together[3], though the source of Electricity was the Voltaic pile, still the induction we looked after was pure *static* induction; with the one exception the phenomenon which you noticed, namely the *advance* of the needle on breaking battery contact.

I have thought much on Induction and series xi, xii & xiii[4] of my Exp Researches (which I think you have) are on the subject. I will not ask you to go through them but perhaps paragraphs 1175–8 and 1295–306[5] might suggest a clearing thought to you in relation to the wire results. You will see at Par 1333[6] that I have *in principle* anticipated the effect as *to time* of conduction by the wire consequent upon induction exerted by neighbouring matter. I had quite forgotten the anticipation[.]

The case of dynamic induction you will find in Paragraphs 1048 &c[7].

Ever My dear Sir | Yours | M. Faraday
Professor Airy | &c &c &c

1. See letter 2747.
2. Letter 2746.
3. Faraday, *Diary*, 15 October 1853, 7: pp.401–8.
4. Faraday (1838a, b, c), ERE11, 12 and 13.
5. Faraday (1838a), ERE11, 1175–8, 1295–1306.
6. Faraday (1838b), ERE12, 1333.
7. Faraday (1835), ERE9, 1048 *et seq*.

Letter 2750
Faraday to Charles Giles Bridle Daubeny[1]
27 October 1853
From the original in Magdalen College, Oxford, MS 400

Royal Institution | 27 Octr. 1853

My dear Daubeny

I have read with delight, and thank you heartily for, your inquiry. What should we do unless some of those who have a right to bear witness for the true interests of the present age, were not to speak out. You have, as one of these, performed an important duty well. I am greatly flattered to see that my name is thought by you useful as illustrating any part of your argument, and the kind appreciation which you make of one part of my peculiarities at p 18[2] opens my thoughts to you:- for it is as true as it is kind; and that which the world attributes to address, or to religion, or to some other queer cause, is just what you have stated it to be. When I first had the opportunity at the Royal Institution of pursuing science, I longed for much but hoped for little; yet I resolved, as you have said, to withdraw from Society that I might at least have time to learn. In this manner I gained *time* and I saved or rather avoided *expense*. As soon as circumstances enabled me, I withdrew from Professional business, by which I gained *more time*, being doubly a gainer in that point; whilst the absence of income & the absence of expenditure neutralized each other, and so left me, if I may say so, a free man. It was a great delight to me to find that, though I thus ran counter to the customs of life, I was still able to secure to myself the kind feelings & friendship of yourself and some others; which make up to me all of society that I desire[.]

Ever My dear Daubeny | Yours | M. Faraday

1. Charles Giles Bridle Daubeny (1795–1867, DSB). Professor of Chemistry at Oxford University, 1822–1855.
2. Daubeny (1853), 18 referred to "that rigid rule of exclusion from society which has enabled Faraday to carry out his great investigations".

Letter 2751
Josiah Latimer Clark to Faraday
31 October 1853
From the original in IEE MS SC 2

The Electric Telegraph Company | (Incorporated 1846) | Engineer's Office, | 448, West Strand, | London, 31 Octr. 1853.

Dear Sir,

I return you the notes you were so kind as to lend me, with many thanks for their perusal. I have taken the liberty to enclose a list of corrections on some points on which you were not rightly informed, and a diagram of the sending apparatus used in some of the Experiments[1].

I find we get a return charge quite sensible to the tongue from a coil of 100 yards of gutta percha wire covered with lead. I even reduced the Experiment so far as to receive a perceptible charge from 10 feet of percha wire in a tumbler of Acidulated Water, and lastly from a Leyden Jar. I have no doubt a Leyden battery would give a very perceptible charge.

I am determined to get evidences (if possible) of the disturbance of one wire by another, we ought to see it when [the] circuit is *broken* despite of imperfect insulation. I am making some careful experiments to demonstrate this last link of identity between Galvanic & frictional Electricity of which I will send you a full account & you can then if you wish see them repeated. If you would like a coil or two of wire to Experiment upon either lead covered or plain I have no doubt our people will be glad to send them to you. We are always anxious to further the objects of science in any way. I am anxious to try some delicate Electrometer Experiments about which I shall perhaps have an opportunity of speaking to you.

I remain | Yours very faithfully | Lat. Clark
M. Faraday Esq | Royal Institution

1. See Faraday, *Diary*, 4 and 15 October 1853, 7: pp.393–408 and note 1, letter 2472.

Letter 2752
Henry Allen¹ to Faraday
2 November 1853
From the original in IEE MS SC 2

Brighton | Novr 2 | 1853

Sir

When your letter on "Table turning" was first published², it staggered my preconceived opinions on the probable cause of the Phenomenon.

Your name, and the energy with which you applied yourself to solve the new problem demanded respect and consideration: Subsequent experience however, reduced me to this dilemma, that I must either succumb entirely to the 'argumentum ad hominem', and lay aside the evidence of my senses or search for a solution, on some other principle.

Repeated trials, made in concert with persons beyond suspicion of trickery, resulted in convincing me that the agency of muscularity would be utterly powerless to effect the movements & biddings of the several tables that have walked, & run, & writhed under my manipulation.

Granting your Theory to be reasonable, that muscular pressure, involuntarily, but insensibly energetic, could after an allotted period, first move, and then slowly, and then more rapidly, turn a large table, it is nevertheless inconceivable that an individual *willing*, and (it may or may not be) communicating his wish to his coadjutor, should, suddenly, in an instant, render the table motionless (while rapidly revolving[)] or under the same circumstances, cause it to revolve with immediate and equal velocity in the *opposite* direction, being conscious that the difficulty of reversing the motive power, will be in proportion to the rapidity of the revolution in the opposite direction.

The reveries of some monomaniacal clergymen and others, who have created and invoked a satanic agency are truly pitiful, so unhallowed a use of Table-turning has been made by them that I will pass them by.

Though you may not think me superstitious you will probably give me credit for being credulous, for believing, as I do, that by a simple act of volition on my part, a table has been made to indicate by particular & unmistakable movements, certain ideas then fixed in my mind. The aberrations in the Correctness of the dumb answers being the *exception*, not the *rule*, the result of my repeated experiences on this point, I will withhold, but you should question the sanity of my views in other ways on the question before us, let me therefore simply communicate to you what occurred in my Drawing room a day or two since, no doubt closely approximating to the marvellous.

A small three legged mahogany Table, had, under the repeated manipulations of a pupil of mine, my son, and myself, uniformly failed to discover any signs of motion. I began to suspect that my young friend, from his peculiar temperament, might be the Cause, and instead of him, I substituted a young lady in one quarter of an hour the table moved off and rapidly, its motion, rapidity & revolution, being entirely under the Control of our will. Our manual pressure amounted to simple contact; our volition, under this Contact causing it to raise *one* leg, then another, then *two* at once, slowly and measuredly, at last the mere touch of any *one* of us was found sufficient to produce the same results, as those produced by a Union of Contact and that almost instantaneously.

And now I come to the experiment which more especially seems to negative the hypothesis of the agency of muscular action, and confirm that of some occult power in the individual extracted by contact with the inanimate object, combined with strong volition, dependent more or less on the temperament of the individual.

After we had ascertained that the table was under our almost instantaneous control, my son, 15 years of age, & myself went down to the Dining-room and laid our hands gently on the sensible horizon of a large globe fixed in a mahogany pedestal it moved off rapidly in a few seconds, and was equally obedient with the table to our will, raising *one* or *two* legs in accordance with our volition; a large mahogany chair was next made to traverse the room; a large Japanned tea-tray with Cups, Saucers, Metal tea pot, &cc, as readily and quickly yielded to our touch: Desks, Tea caddys, in fact, every moveable Article we could think of candlesticks, China, plates & Glass-Dishes circumambulated, when they came in contact with our fingers[.] Indeed we felt ourselves so charged with this occult power, that hardly any thing could resist us, and for the last evening or two, the table, which the first time, required a quarter of an hour to start it, now requires a minute or two only.

Pray excuse me for troubling you but a strong desire on scientific grounds, to know whether & how all this can be reconciled with your Theory induces me to hope that to will give the question a little consideration and kindly put me in possession of the result of your investigation.

I have the honor to be | Sir | Yours faithfully | Henry Allen | Vicar of Patcham & | Chaplain to the Troops | Brighton

1. Henry Allen (d.1865, age 57, GRO, CCD). Vicar of Patcham, 1843–1865.
2. Letter 2691.

Letter 2753
Benjamin Collins Brodie to Faraday
4 November 1853[1]
From the original in IEE MS SC 2

13 Albert Road | Regent's Park | Nov. 4

My dear Sir,

We are anxious at the Chemical Society to throw a little more life and interest into our meetings. The enclosed Paper explains the plan by which it is hoped to effect this and also the nature of the request which I have to make to you, which is whether you will kindly consent to give one of the discourses mentioned in it[2]. I will with your permission, take an early opportunity of calling upon you to explain a little more fully what we desire.

I am, | very truly yours | B.C. Brodie

Professor Faraday

1. Dated on the basis of the reference to discourses at the Chemical Society.
2. The idea of discourses at the Chemical Society was approved at the June 1853 meeting of its Council and reported at the 1854 Annual General Meeting. See *J.Chem.Soc.*, 1854, 7: 158–9. Faraday did not deliver such a discourse.

Letter 2754
Faraday to John Stevens Henslow
5 November 1853
From the original in ULC Add MS 8177

Royal Institution | 5 Novr. 1853

My dear Henslow

I send you first a bottle containing Sodium. If it is turbid (the liquid) when you get it, let it rest quietly & the liquor will become clear & you will see some of the globules of sodium as metals very well I think. *Do not open the bottle* or the preparation will be spoiled, it is 30 years or more old[.]

As for potassium I have been in the habit in lecture[s] of placing a moderately clean piece about the size of a pea between two thick glass plates

then pressing the plates together with a little lateral motion added till the potassium is spread out as large as a shilling and holding the plates together with clips of bent copper. Somewhere the potassium will shew itself metallic & keep so under the glass for some hours[.]

As to the wires, there is some fine copper drawn to $\frac{1}{350}$ of inch - some fine platina drawn to $\frac{1}{216}$ of inch - 2 reels of Platina in silver the platina being $\frac{1}{1000}$ and $\frac{1}{2000}$ of inch[.] If you hold the ends of these in the *side* of a candle flame you may melt down the silver & shew the platina as the specimens will shew you. The platina is then best shewn to a company by letting it glow as an ignited body in the flame only the $\frac{1}{2000}$ will melt in the candle unless care be taken[.]

There are also two cards containing like specimens of platina respectively the $\frac{1}{10000}$ $\frac{1}{20000}$ $\frac{1}{30000}$ of an inch in diameter. Here you want a glass to see them[.]

I have put in some pieces of Gilt silver wire & silver copper wire but here the covering metal is *thick*. I cannot get at any others[.]

I must ask you to return the *Sodium & the *fine wires on cards & *reel. I am sorry I cannot leave them with you[.]

Ever My dear Henslow | Yours Truly | M. Faraday
I send this note by Post & the other things as a packet by the rail | MF

Letter 2755
Lyon Playfair to Faraday
10 November 1853
From the original in RI MS Conybeare Album, f.13
Department of Science and Art, | Marlborough House, Pall Mall, London.
| 10th day of Nov 1853
My dear Sir
I am late an hour beyond my promise but this has been owing to official visitors who I was compelled to see[.]
Yours Sincerely | Lyon Playfair
Prof Faraday

Letter 2756
Faraday to John Stevens Henslow
14 November 1853
From the original in Sutro Library MS Crocker 11

Royal Institution | 14 Novr 1853

My dear Henslow
    I send you specimens of

Platina $\frac{1}{216}$ of inch in diameter
Copper $\frac{1}{350}$ ——
Platina in silver - the platina $\frac{1}{1000}$
D° —— $\frac{1}{2000}$

and in [three words illegible] in, or on a card where the platina is only $\frac{1}{5000}$ of an inch in diameter[.]
    The latter piece is difficult to [four words illegible] is to place it on the surface of some nitric acid in a glass, the silver dissolves & the platina is left floating but of such fine tenuity that it is almost impossible to move it without breaking it. I think these were prepared by Brockedon[1] by means his jewelled holes[2][.]
    Ever My dear Henslow | Yours | M. Faraday

Endorsement: Royal Institution | 14 Novr. 1853

1. William Brockedon (1785–1854, DNB). Painter, author and inventor.
2. See Brockedon's Friday Evening Discourse on wire drawing, 11 May 1827, *Quart.J.Sci.*, 1827, **23**: 462–4.

Letter 2757
John Stevens Henslow to Faraday
21 November 1853
From the original in RI MS F1 K20

Hitcham | Hadleigh Suffolk | 21 Nov. 1853

My dear Faraday,
    I was induced to delay my return till Saturday[1] that I might go to Lady Lyell's[2] on Friday evening where I hoped to see several old friends & was not disappointed. I have opened your box of interesting wires[3] & shall endeavour to make an illustration by crossing them over a piece

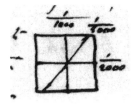

of glass & placing a lens before the centre, which will show the relative thickness better than if they were parallel. Another with hair, spiders web & the copper, may be placed beside it, & a mem. *below* that platina *has been* wire-drawn to $\frac{1}{30,000}$ inch. I think if I fasten down about $\frac{1}{4}$ of inch over the glass, & then apply the nitric acid I need not afterwards, interfere with the finest, to risk the breaking [of] it. Can you (without trouble) tell me the relative bulk of water & that assumed by the oxygen & Hydrogen resulting from its decomposition (at ordinary temperatures)? I want some such illustration as this

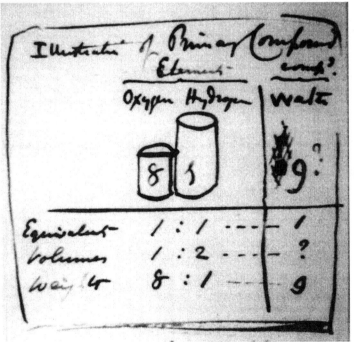

One might have spheres instead of cylinders (as 1:2) for the volumes, but the evidence to the eye not so palpable. Don't trouble yourself to reply if you cannot off hand - I will try to find it out, if it is recorded in any of the few works on chemistry I possess.

Ever sincerely yours | J.S. Henslow

1. That is 19 November 1853.
2. Mary Elizabeth Lyell, née Horner (1808–1873, Burkhardt et al. (1985–94), 4: 652). Conchologist who married Lyell on 12 July 1832.
3. See letters 2754 and 2756.

Letter 2758
Faraday to John Stevens Henslow
22 November 1853
From the original in ULC Add MS 8177

R Institution | 22 Novr. 1853

My dear Henslow

	grains
1 cubic inch of water weighs	252.458
— oxygen	0.3438
— hydrogen	0.021483
— 2 vols hy & 1 vol oxygen	0.128922

so $\dfrac{252.458}{0.128922}$ = about 1958.

so 1958 vols of mixed O & H make 1 vol of water and therefore your middle line of 1 : 2 gives or requires $\frac{1}{653}$ in the water column. The 9 as equivalent for water is right i.e the 9 which you have? opposite

is right[1][.]

Ever Truly Yours | M. Faraday
Revd. J.S. Henslow | &c &c &c

1. See letter 2757.

Letter 2759
David Brewster to Faraday
24 November 1853
From the original in IEE MS SC 2
My Dear Mr. Faraday,
   I was much obliged to you for your kind attention to my request respecting the subscription for the statue to M. Arago[1]; and I was gratified to find that we had thought of the same sum.
   I have examined the interesting specimen of *Tourmaline* which you requested Dr. Herapath[2] to send me[3]. I have sent him my Report by the same Post which carries this; it expressed my wish that he would shew it to you. It is *certainly* all *Tourmaline*, without Quartz or Feldspar.
   I am, | My Dear Mr Faraday, | Ever most Truly yours | D. Brewster
St. Leonard's College | St. Andrews | Novr 24th 1853
P.S. Will you kindly answer the following Queries.
1. What crystals change colour by simply pricking them, and where has any notice of them been published? I think *Iodide of Mercury* is one, & I think Mr Talbot has somewhere described the experiment[4].
2.   A Lady in whom I am interested has a Red mark of some extent on her cheek. Would it be safe to apply white lead, or zinc paint to hide it? Or do you [know] of any process for this purpose?

1. According to *Athenaeum*, 29 October 1853, p.1293 both Brewster and Faraday were members of the committee to erect a statue of Arago.
2. William Bird Herapath (1820–1868, DSB). Bristol physician and chemist.
3. Herapath (1853).
4. Talbot (1836), 1–3 and (1842).

Letter 2760
Faraday to George Biddell Airy
9 December 1853
From the original in RGO6 / 405, f.151
                                                         Royal Institution | 9 Decr 1853
My dear Sir
   Mr Barlow has been speaking to me as if he had hopes that we might hear you here some F.E. this season[1][.] Shall I encourage him or not? You cannot for a moment suppose (by my question) that I am not as *earnest* & *more so* than he is to hear you but I would not that you should be troubled for us beyond that degree which consists with your willingness and is in a manner its own reward by the pleasure which it gives you to give pleasure & delight to others[.]

With kindest remembrances to Mrs. & Miss Airy I am
My dear Sir | Ever Truly Yours | M. Faraday
The | Astronomer Royal | &c &c &c

1. Airy did not deliver a Friday Evening Discourse during 1854.

Letter 2761
Faraday to William Robert Grove
9 December 1853
From the original in RI MS G F29

R Institution | 9 Decr. 1853

My dear Grove

Though I am exceedingly deaf yet I was very sorry to hear that you had been in this house & I had not seen you. At present I may not go out but if you are here again & I out of bed do let me see you. I have told the porters to watch for you.

Yours Ever | M. Faraday

Letter 2762
Macedonio Melloni to Faraday
12 December 1853
From the original in IEE MS SC 2

Moretta di Portici près Naples | ce 12 Decembre 1853.
Mon illustre ami!

J'ai consigé, il y a quelques jours, au Consul piémontais de Naples un exemplaire de mon second mémoire sur le magnétisme des roches[1] avec priere de vous le faire parvenir le plutôt possible. Cependant comme je sais, par expérience, la lenteur de ces sortes de transmissions, je crois convenable de vous écrire directement par la poste afin de vous informer d'avance des principaux resultats contenus dans ce travail.... Je ne sais si je me trompe, mais il me semble que quelques-uns d'entre eux ne sont pas tout-à-fait indignes de fixer l'attention de la Societé Royale.

Ma position dans ce pays est toujours la même.... je ne conserve plus que ma place d'académicien et les restes d'un patrimoine sérieusement compromis par la passion fatale qui m'a poussé dans le tourbillon scientifique.... Encore dois-je la possibilité de continuer, tant bien que mal, ma vie d'étudiant à la protection des Ambassadeurs prussiens ou pour mieux dire à l'influence de l'illustre savant, que vous avez déjà déviné, sur

Plate 11. Macedonio Melloni.

le Roi de Prusse² leur seigneur et maître! D'autre part Arago de glorieuse mémoire ne pouvant plus agir comme autrefois en ma faveur auprès des puissances politiques à l'ordre du jour a néammoins continué à me prodiguer les trésors de sa plus vive amitié jusqu'aux derniers moments de sa précieuse existence.... Et vous même, cher et illustre confrère, ne m'avez-vous pas comblé de bontés toutes les fois qu'il vous a été possible de le faire?... N'ai-je pas encore entre mes mains cette lettre si affectueuse que vous voulûtes bien m'écrire pour me consoler de mes dernières mésaventures?... Ces marques d'estime et d'interêt de la part d'hommes tels que Humboldt, Arago et Faraday sont pour moi le plus haut dégre de bonheur auquel puissent atteindre les veritables *amoureux de la Science* et tant qu'elles se maintiendront actives et florissantes à mon égard, les vicissitudes malheureuses de la vie ne parviendront jamais à abattre le courage de

votre très-dévoué et très reconnt. Servit. et ami | Macédoine Melloni

Address: Monsieur | Monsieur Michel Faraday | de la Société Royale de Londres, de | l'Institut de France des Académies | des sciences de Berlin; de Copenhague &c &c | Londres

TRANSLATION
Moretti di Portici near Naples | this 12 December 1853. My illustrious friend,

I gave, a few days ago, to the Piedmontese Consul in Naples a copy of my second memoir on the magnetism of rocks¹ with a request that he should convey it to you at the earliest opportunity. However as I know, from experience, the slowness of this sort of transmission, I think it best to write to you directly by post in order to forewarn you of the principal results contained in this work ... I do not know if I am mistaken, but it seems to me that some of these are not altogether unworthy of drawing to the attention of the Royal Society.

My position in this country is still the same ... All I possess, is my work and the remains of an inheritance which is seriously reduced by the fatal passion which has drawn me into the turbulent world of science ... I still owe the possibility of continuing this life of learning, for better or worse, to the protection of the Prussian ambassadors or rather, as you may have guessed, to the influence of that illustrious savant, the King of Prussia², their lord and master. On the other hand, Arago, of glorious memory, unable to act as before in my favour with the political powers of the day, nevertheless continued to lavish on me the treasures of the keenest friendship right up to the last moments of his precious existence.... And you too, dear and illustrious colleague, have you not showered me

with kindness at every opportunity? ... Do I still not have in my hands that deeply affectionate letter that you kindly sent me to console me after my last misfortunes? ... These testimonies of esteem and interest from men such as Humboldt, Arago and Faraday are for me the highest degree of happiness to which *lovers of Science* can aspire and so long as they remain active and flourishing towards me, the unhappy vicissitudes of life will never succeed in breaking the spirit of
    your very devoted and most grateful Servant and friend | Macedoine Melloni.

1. Both his first and second memoir are in Melloni (1853).
2. Frederick William IV.

Letter 2763
Faraday to John Gibson Macvicar[1]
13 December 1853
From the original in NLS MS 7178, f.61

Royal Institution | 13 Decr. 1853

Dear Sir

I am much obliged to you for your paper having received it at the same time with your letter[.] I cannot however give you my opinion for the older I grow the more reserved I become in drawing conclusions. Even in the simplest case a conclusion should be drawn only after consideration of all the data connected with the question and these are constantly so numerous, that my mind does not willingly engage itself in the labour necessary except in especial investigations of my own.

I do not know how your view would agree with the researches of Regnault who in experiments of many hours (and I think days) continuance and conducted apparently in the most careful manner found *no difference* when dogs & other animals were made to live & breathe in atmospheres consisting of Oxygen and Hydrogen only all the Nitrogen being removed & replaced by hydrogen. His results were published in a quarto volume a few years since[2][.]

I am My dear Sir | Very Truly Yours | M. Faraday
Revd. Dr MacVicar | &c &c &c

1. John Gibson Macvicar (1800–1884, DNB). Minister at Moffat from 1853 and scientific writer.
2. Faraday seems to be confusing Regnault (1847) with Regnault and Reiset (1849).

Letter 2764
Faraday to John Hope Shaw[1]
19 December 1853
From the original in SELJ MS 8/1

Royal Institution | 19 Decr. 1853

Sir
    I deeply regret that I cannot accede to your request: but the progress of time & the state of my head are entirely against it. I have under medical advice been obliged to restrain my exertions for years past to the Royal Institution where I have been for 40 years. I have there lessened my lecturing duty from year to year and though now announced to deliver six at Christmas[2], it was up to Yesterday doubtful (& is so still) whether I shall be able because of an attack in the throat to which I am liable accompanied for the first time with extreme deafness.

    I am sorry to give you such an answer but it has been accepted as sufficient for several years past by my nearest & highest friends, and it is known that I speak no where out of the Royal Institution: not even at our meetings of the British Association[.]

    I have the honor to be | Sir | Your Obedient humble Servant | M. Faraday

Jno Hope Shaw Esq | &c &c &c

1. John Hope Shaw (1792–1864, Taylor, R.V. (1865), 520–3). Leeds solicitor and seven times President of the Leeds Philosophical and Literary Society, including 1854–1855. Clark (1924), 228.
2. Faraday delivered six Christmas lectures on "Voltaic Electricity". His notes are in RI MS F4 J16.

Letter 2765
Faraday to Thomas Twining
19 December 1853
From the original in the possession of Günther Gerisch

Royal Institution | 19 Decr. 1853

My dear Sir
    I have just read your Letters[1] and now make it an opportunity of thanking you for them. Though I am not familiar with the department of practical knowledge, all important as it is, to which they relate yet I can respect & reverence it in others & if you succeed in getting that done for other districts which you have effected for Nassau[2]: the resulting body of evidence must have extreme value to those who engage in the amelioration of the condition of man working for his daily bread.

I could not help thinking whilst I read the letters how much healthy, mental, pleasure must have been added to that portion of enjoyment which you would desire from the beauty of the country itself. I hope you enjoyed yourselves thoroughly[.]

My wife joins me in kindest remembrances to you & Mrs. Twining[3][.]

Ever My dear Sir | Very Truly Yours | M. Faraday
T. Twining Jur. Esq | &c &c &c

1. Twining (1853).
2. A German Duchy.
3. Victorine Twining, née von Hagen (1830–1889, Pearce (1988), 2). Married Thomas Twining in 1850.

Letter 2766
George Biddell Airy to Faraday
19 December 1853
From the original press copy in RGO6 / 405, f.153

Royal Observatory Greenwich | 1853 Decr. 19

My dear Sir

I have kept your letter of the 9th[1] by me, partly because I have been very busy and a little while absent, partly because I wished to try whether I could rake up any thing that would be likely to suit you.

To this moment I have not. And on the whole I should be glad to feel myself free from the tie of a lecture engagement. I have a good deal of unusual official work laid out for the spring[2].

But if any thing occurs to me as a fitting subject, I will give you notice and will be prepared to consult your wish[3]. The mere trouble of giving a lecture is a very inconsiderable thing.

I am, my dear Sir, | Yours very truly | G.B. Airy
Professor Faraday | &c &c &c

1. Letter 2760.
2. This would appear to concern the publication of observations made at the Cape of Good Hope. Airy, W. (1896), 219.
3. To give a Friday Evening Discourse.

Letter 2767
George Biddell Airy to Faraday
19 December 1853
From the original press copy in RGO6 / 468, f.195

1853 Decr. 19

The time occupied by the passage of a galvanic pulse from Greenwich to Brussels (about 170 miles under earth and water, and 100 miles in air), is between $\frac{1}{9}$ and $\frac{1}{10}$ of a second of time[1].

GBA
Professor Faraday

1. See A.B.G. [i.e. Airy], "Telegraphic Longitude of Brussels", *Athenaeum*, 14 January 1854, pp.54–5.

Letter 2768
Warren De La Rue to Faraday
22 December 1853
From the original in IEE MS SC 2
[De La Rue's Letterhead] | 7 St Mary's Road, Canonbury | December 22nd 1853

Dear Mr Faraday

Permit me to say that I am much flattered that so much of your attention has been bestowed on my drawing of Saturn[1] as to have called forth the conjectures respecting the relative positions of that planet's rings which you are so good as to communicate to me.

The engraving represents the shadow of the planet on the rings correctly for a given epoch, but in other respects except for the position of the two satellites, it must be regarded as a summary of many observations:- for the moments of fine or even fair definition, on the most favorable night, are of too short duration to admit of a complete drawing being made, and I found it generally better to confine my attention to the recording of some one phenomenon only.

The inference I drew from the configuration of the shadow of the planet on the rings, coincides with your own, namely, that the middle ring is in a plane less elevated than the outer ring:- the outline, moreover, of the nebulous ring shows that *it* is more elevated than the two others, and on several occasions of fine definition I have had a distinct impression of its overlaying the middle ring as I have depicted.

I have repeatedly remarked that the outer and middle rings and the division between them were of different breadths at opposite ends of the same diameter, showing that their centres of gravity as well as their centres of rotation must be eccentric. This joined to their planes not being coincident must produce a complexity in their mutual perturbations and their action on the planet very interesting to the Saturnian mathematicians.

With respect to the probable section of the rings I would remark that on Oct. 16th of this year I observed this appearance

the shadow on the ring B being convex toward the planet - indicating *its* section to be thus

there was a faint penumbra visible beyond the dark shadow which was not itself so dark as the division between A & B. The night was as fine as any I have observed on. The shadow on the ring B was just visible on the Eastern side of the ball - but not on the ring A on that side.

One subject of interest with respect to Saturn is that the distance of the inner edge of the ring B from the planet's centre is less than the drawings of the older observers would appear to indicate was the case in their day, and hence M. Otto Struve[2] who has studied the subject[3] suggests that the ring may be gradually (or rather I should say rapidly) collapsing. He consequently felt desirous to have Huyghens[4] object glass mounted in order to ascertain if the appearance was due to the telescope and in accordance with his request the Royal Society have as you know determined on so doing[5]. My own belief is that the older astronomers did not clearly observe the darker portions of the ring B and hence that its edge appeared to them more distant from the planet than it really was.

I enclose a few diagrams which I employed myself and furnished to other observers, and which, from being drawn out in accordance with the data given in the Nautical Almanac, so far as it goes, will show by comparison with my drawing how much I have had to change the relative diameters of the rings to agree with my impressions, which I would remark are confirmed in the main by M. Otto Struve's micrometric measurements.

I must apologize for the length of this note which I fear will have tired your patience.

Very truly Yours | Warren De la Rue
Michael Faraday Esq | &c &c &c

1. On this see *Month.Not.Roy.Ast.Soc.*, 1854, **14**: 134.
2. Otto Wilhelm Struve (1819–1905, DSB). Vice-Director of Pulkovo Observatory, 1848–1862.
3. Struve (1851).
4. Christiaan Huygens (1629–1695, DSB). Dutch natural philosopher. On this work see Van Helden (1980).
5. See RS CM, 26 May 1853, **2**: 252–3 and 30 November 1853, **2**: 271. The latter noted the award by the Royal Society of a grant to De La Rue of £250.

Letter 2769
William Crawford Williamson[1] to Faraday
23 December 1853
From the original in IEE MS SC 2

Owens College | Manchester Dec. 23 / 53

Dear Sir

Dr Cumming[2] of London has just been lecturing here and as I understand from several parties, he made a public statement *that you had withdrawn your letter explanatory of Table turning as being insufficient to account for the phenomena*[3].

Since we have been in the habit of quoting your apparently conclusive experiments, in confirmation of our own notions respecting this absurdity, such a statement coming from such a quarter, seemed to require, either contradiction or substantiation; and I have taken the liberty of addressing you in a direct manner, in preference to adopting any more round-about mode of ascertaining the truth; fully convinced that your philosopher spirit will acquit me of any impertiency in doing so.

The advocates & practitioners of this extraordinary delusion are still so numerous, and the mischief they are likely to do, judging by what has occurred in America, so fearful, that it behoves every conscientious man to lift up his voice against them in a solemn manner. The publication of your letter effectually silenced them in this neighbourhood - but the statement alleged to have been made by Dr Cumming will go far to recussitate what we had hoped was now a caput mortuum[4].

I must apologise for troubling you with this matter. I have not had the opportunity of seeing my colleague, Principal Scott[5] - this day or two, or I would have asked him to undertake the enquiry; being I believe a personal acquaintance, &, consequently, one from whom it would have come with greater propriety.

I remain Dear Sir | Very Truly Yours | William C. Williamson
To Professor Faraday

1. William Crawford Williamson (1816–1895, DNB). Professor of Natural History, Anatomy and Physiology at Owens College, Manchester, 1851–1892.
2. John Cumming (1807–1881, DNB). Minister at the Scottish National Church in Covent Garden, 1832–1879.
3. Cumming was advertised as delivering a lecture entitled "The signs of the times" at the Scotch Church, Manchester on 19 December 1853, *Manchester Guardian*, 17 December 1853, p.1, col. a. For a text of the a lecture of the same title, delivered in London on 13 December 1853, see Cumming (1854). The reference to Faraday is on pp.179–80.
4. "A dead head". That is the residuum left by a process of chemical analysis.
5. Alexander John Scott (1805–1866, DNB). Principal of Owen's College, Manchester, 1851–1857.

Letter 2770
Faraday to William Crawford Williamson[1]
27 December 1853
From [Williamson] (1870), 292

Royal Institution, 27th December, 1853.

Dear Sir, - That I have withdrawn my letter on table-turning is a very fit assertion for the table-turners to make, and quite in keeping with the whole subject, inasmuch as it is utterly untrue. Do me the favour to contradict the assertion everywhere. At first I considered that the matter might be gently dealt with, as a mistake generally; but now I consider that it is simply contemptible.

1. William Crawford Williamson (1816–1895, DNB). Professor of Natural History, Anatomy and Physiology at Owens College, Manchester, 1851–1892. Recipient, and author of the letter, established on the basis that this is the reply to letter 2769.

Letter 2771
Ellis N. Field[1] to Faraday
28 December 1853
From the original in GL MS 30108/2/61

Light house Cromer | Dec 28th / 53

Sir,

In answer to yours, I send you the best information I can. The stove answers well, with all winds, by attending to the ventilation *below* the lower lamps burn as well as the upper[.] We are all quite well.

Yours Truly | Ellis N. Field

1. Keeper of the lighthouse at Cromer. Otherwise unidentified.

Letter 2772
Alexandre-Edmond Becquerel to Faraday
30 December 1853
From the original in IEE MS SC 2
Monsieur et illustre physicien

il y a bien longtemps que je voulais vous envoyer quelques épreuves de spectres solaires faites sur la préparation chimique que j'ai obtenu il y a plusieures années, (ann. de physique et de chimie 3eme serie tome 25 page 447[1].) afin que vous puissiez juger de quelle manière les différentes couleurs de la lumière peuvent se reproduire sur une substance chimiquement impressionnable, et même se conserver pourvu que les impréssions soient conservées a l'abri de la lumière du jour.

je n'ai malheureusement que quelques épreuves assez médiocres, et qui sont même en partie decomposées, par suite des experiences que j'ai faites avec elles; cependant j'ai voulu profiter du depart de Monsieur Odling[2] pour londres pour vous les envoyer, car cela me procure l'honneur d'entrer en correspondance avec un des plus illustres physiciens de l'europe. comme actuellement je me remets a travailler de nouveau cette question, je vous enverrai cet été de nouvelles épreuves, car elles sont beaucoup plus belles au moment où on les forme, que lorsqu'on les a faites depuis un certain temps; en effet, chaque fois qu'on les observe à la lumière, les teintes s'alterent un peu.

je vous enverrai également, lorsque la saison permettra de faire les experiences, des images faites à la Chambre obscure et colorées d'elles mêmes, afin de vous montrer que l'on peut peindre avec la lumière, ainsi que je l'ai fait voir il y a bientot six ans. il me restait bien quelques peintures que j'aurais pu vous expédier, mais bien plus alterées que les images des spectres que je vous adresse, et c'est pour ce motif que j'ai mieux aimé attendre une nouvelle occasion.

je regarde donc le problème de *la possibilité de peindre avec la lumière,* comme scientifiquement resolu. je ne dis pas pratiquement, car les images sont assez longues a obtenir, et une fois obtenues, elles ne sont fixes qu'autant qu'on les conserve à l'obscurité. la substance chimiquement sensible, qui a reçu des teintes diverses, n'a pas perdu toute impressionnabilité pour cela, et change encore quand on l'expose a l'action d'une lumière quelconque. je n'ai pas encore pu fixer les teintes de façon a ce qu'elles ne changent plus même a la lumière; je ne sais si cela est possible mais je cherche toujours a resoudre la question.

pour voir les images que je vous adresse il est nécessaire de les regarder dans une piéce bien eclairée, mais de ne les considérer que peu de temps, et de les enfermer bien vite dans une boite pour les laisser à l'obscurité jusqu'à ce que on veuille les voir de nouveau. on peut les regarder aussi a la lampe. à chaque fois qu'on les observe, elles s'alterent de plus en plus; car telles qu'elles sont, elles sont deja bien alterées et bien moins belles que lorsque je les ai faits il y a 5 ans.

je termine cette lettre, Monsieur, en me felicitant que ce petit envoi m'ait donné l'occasion d'avoir l'honneur de vous écrire, et de vous prier d'agréer l'assurance de la plus haute consideration d'un de vos admirateurs
Edmond Becquerel
30 decembre 1853 | Paris 57 rue Cuvier. au jardin des plantes.

TRANSLATION
Sir and Illustrious physicist

I have for a long time wanted to send you some prints of the solar spectra made on a chemical preparation which I obtained several years ago (*Annales de physique et de chimie* 3rd series, volume 25, page 447[1]), so that you may judge in what way the different colours of light can be reproduced on a substance on which an impression can be made chemically, and even be preserved, provided that the impressions are kept hidden from the light of day.

I have unfortunately only a few prints of mediocre quality, and which have to a certain extent decomposed, because of the experiments I have carried out with them; however, I wanted to take advantage of the visit to London of Mr Odling[2] to send them to you, since that procures for me the honour of entering into correspondence with one of the most illustrious physicists in Europe. Since I intend shortly to work on this question again, I will send you new prints this summer, for they are much more beautiful at the moment they are made than those made a while ago; in fact, each time one observes them in the light, the tints change a little.

I shall also send you, when the season allows the experiments to take place, naturally coloured images made in the Camera Obscura, in order to show you that one can paint with light, as I showed six years ago. I still have some pictures that I could send you, but they are much more altered than the images of the spectra that I am sending, and that is why I have preferred to wait for another occasion.

I regard therefore the problem of *the possibility of painting with light* as scientifically proven. I am not saying practically, since it takes a long time to obtain the images, and once obtained, they are not fixed unless they are kept in darkness. The chemically sensitive substance, which received the various tints, has not lost all its impressionability for that and still changes when it is exposed to any kind of light. I have not been able to fix the colours in such a way that they no longer change in the light; I do not know if it is possible, but I am constantly searching to remedy this situation.

To see the images that I am sending you, it is necessary to look at them in a well lit room, but to look at them only for a short time, and to put them back quickly in a box in order to leave them in darkness until they are to be viewed again. You can also look at them under a lamp. Each time they are observed, they alter a little more; for such as they are, they are already very different and far less beautiful than when I made them five years ago.

I end this letter, Sir, congratulating myself that this little parcel has given me the opportunity to write to you, and I ask you to accept the assurance of the highest consideration of one of your admirers | Edmond Becquerel

30 December 1853 | Paris 57 rue Cuvier. at the Jardin des Plantes.

1. Becquerel (1849b).
2. William Odling (1829–1921, DSB). Chemist at Guy's Hospital.

Letter 2773
Faraday to Charles W. Woolnough[1]
2 January 1854
From a lithographic copy at the front of Woolnough (1881)

Royal Institution | 2 January 1854

Sir

I beg to thank you for your kindness in sending me a copy of your very practical work on Marbling[2]. I feel much interest in the subject because of its associations with my early occupation of book-binding and also because of the very beautiful principles of natural philosophy which it involves[.] The Marbled cloth is very good indeed[.] I suppose it is not done by a continuous process but in sheets[.] Indeed I see that is the case at p 72[3][.]

I am Sir | Your Very Obliged Servant | M. Faraday

C.W. Woolnough Esq

1. Unidentified.
2. Woolnough (1853).
3. Ibid., 72.

Letter 2774
Faraday to Henry Ellis
5 January 1854
From the original in BL add MS 65155, f.111

Royal Institution | 5 Jany 1854

Many thanks My dear Sir Henry for your very kind note[.]
    Ever faithfully Yours | M. Faraday
Sir Henry Ellis | &c &c &c

Letter 2775
Michel Eugène Chevreul[1] to Faraday
9 January 1854
From the original in IEE MS SC 2

9 de Janvier 1854

Monsieur et cher confrère,
    Je vous écris en courant pour vous demander où je puis trouver ce que vous avez écrit ou professé sur les *tables tournantes*[2]. Je me trouve rengagé à revenir sur ce sujet par une ancienne lettre *sur le mouvement d'un pendule formé d'un corps doux suspendu à un fil que lon tient à la main*. cette lettre fut imprimée dans la revue des deux mondes en 1833[3] - autant que jai pu le savoir par ce qu'on ma dit, votre opinion se rattacherait à la mienne. Or à la veille de publier un ouvrage sur la *baguette dévination le pendule* et les *tables*[4] je ne puis omettre de parler de vous[5], ayez donc la complaisance de me repondre le plutot possible et croyez, monsieur et cher confrere, a Vous mes sentiments de profonde estime et d'attachement
    E Chevreul

TRANSLATION

9 January 1854

Sir and Dear Colleague,
    I am writing to you quickly to ask you where I can find where you have written or expounded on *table turning*[2]. I find myself engaged in returning to this subject through an old letter *on the movement of a pendulum formed from a soft material suspended from a thread that one holds in one's hand*. This letter was printed in the *Revue des deux mondes* in 1833[3]. In as much as I understand from what I have been told, your opinion is similar to mine. Now on the eve of publishing a work on *divining rods, pendulums, and tables*[4], I cannot omit to speak of you[5], please therefore have the kindness

to reply as soon as possible and believe, Sir and dear colleague, my sentiments of profound esteem and attachment for you

E. Chevreul.

1. Michel Eugène Chevreul (1786–1889, DSB). Director of dyeing at the Gobelins tapestry works and Professor of Chemistry at the Muséum d'Histoire Naturelle.
2. Letter 2691 and "Professor Faraday on Table-Moving"; *Athenaeum*, 2 July 1853, pp.801–3.
3. Chevreul (1833).
4. Chevreul (1854b).
5. *Ibid.*, 219–22.

Letter 2776
Faraday to Mary Fox
12 January 1854[1]
From the original in the possession of George W. Platzman

Brighton | 12 Jany 1853 [sic]

My dear Lady Fox

We are here for rest - but both of us are confined to the house with very heavy colds. We shall not return from here except just in time to prepare for & give the Friday Evening (the 20th)[2] so that I do not think I can make any engagement before Monday the 23rd. Then if you will give your friend a note (you did not mention his name) he can call on me but pray do not give yourself the least personal trouble in the matter. Trusting that you & Sir Charles are well - which we really are not

I am | My dear Lady Fox | Yours Very Truly | M. Faraday

1. Dated on the basis of the reference to the Friday Evening Discourse.
2. Faraday (1854a), Friday Evening Discourse of 20 January 1854.

Letter 2777
Faraday to John Peter Gassiot
17 January 1854
From the original in RI MS F1 N1/48

Three long bands of fine oiled silk crape (each being effectively a plate of fine dry oil - impervious & insulating.) respectively 108 inches long and $6\frac{1}{2}$ inches wide[.]

Two similar bands of tin foil each 106 or 107 inches long and only 6 inches wide[.]

These made into a common band of Silk ———
foil ———
silk ———
foil ———
silk ———
so that the oiled silk shall overlap the edges of the tin foil[.]
The compound band is then bent backwards and forwards thus

so as to make a packet $6\frac{1}{2}$ inches wide & 18 inches long which lies in the bottom of the box and yet is equivalent to 702 square inches of doubly coated surface or a square of 25 inches in the side - the induction being through a thickness no more than that of *one oiled silk*. Then by proper easy fittings one end of one of these plates is connected with the primary current wire on one side of the *breaker* & the other plate with the same current wire at the other side of the breaker - hence the good effect.

Ever dear Gassiot Truly yours | M. Faraday
17 Jany 1854

Endorsement: Grove PM. Vol 4 501 | 4 series[1]
Ruhmkorffs small coil
30 metres long
2 millime diameter
200 convolutions of the wire
2503 metres length
$\frac{1}{4}$ millim diameters
10,000 convolutions
Primary wire $2\frac{1}{2}$ mil. thick = .08 Inch
       40 metres long about 1560 feet
       300 turns
Secondary $\frac{1}{3}$ mil. thick   = .01 inch
       70,000 metres 227,500 yds above 116 miles
       30,000 turns
Millimetre = .03957 Inch
Metre = 39.37100 Kater[2] PT 1818 | 39.370079[3]

1. Grove (1852b), 501 which described Rühmkorff's coil. See Gassiot (1854) for his interest.
2. Henry Kater (1777–1835, DSB). Captain in the Royal Engineers and man of science. Treasurer of the Royal Society, 1827–1830.
3. Kater (1818), 109. Gassiot inserted an extra zero into this number.

Letter 2778
Faraday to Jacob Herbert
18 January 1854
From the original copy in GL MS 30108/2/61

Royal Institution | 18th Jany 1854

My dear Sir

I have a letter from you of the date of 6th Octr 1853[1], yet to answer: I have not forgotten nor neglected it. I have had the opportunity of conversation with Mr Walker several times. I have also written to Cromer lighthouse to enquire what the effect of the changes lately made there according to my instructions had been during the recent very severe winter weather; when strong wind & cold had to be encountered and regulated:- The answer[2] is as follows & most satisfactory. "Sir - In answer to yours I send you the best information I can. The stove answers well with all winds by attending to the ventilation below: the lower lamps burn as well as the upper. We are all quite well. Yours &c Ellis N. Field[3]. Dec. 28, 1853".

The result of this & other experience & a consideration of the principles concerned induce me to submit to the Trinity Board the following system for adoption in a lighthouse in the next possible case: I strongly anticipate such good results as may justify the recommendation of the plan (with more or less modification according to peculiar localities or circumstances) as a pattern in future cases. The watch chamber to be dry. The lanthorn & watch chamber to be cut off from the tower by a door as usual; and those for warming & ventilation to be considered virtually as one chamber. The ventilation as to entering air to be effected by the use of four windows in the watch chamber like those I have described in the Cromer report (29 Aug 1853)[4] any one of which (according to the wind) will when open an inch or two be sufficient for the purpose. This air to be heated by a stove standing (unless there be special reasons) in the middle of the watch room. The stove not to be jacketted or double; or, if it be, both the top & bottom part to be removed:- The cross braces on which the lamp &c stands, support the ceiling of the guard room & the floor of the lanthorn about 6 inches apart; the space thus opened to be made hot air passage; by opening the centre of the ceiling of the Guard room for a diameter of 5 feet covering the space with a light grate or a screen of trellis wire work, and opening also the floor of the lanthorn all round for a distance of two feet or perhaps 18 inches from the wall. The guard room & the lanthorn to be separated by a door on the stairs for the sake of pure caution which may however not be finally needed. I expect that the air entering by the window open to windward in the guard room will be effectually warmed by the stove, will freely pass upward, & entering the aperture above will be diffused all round the lanthorn under the platform & glass - and do its proper work perfectly. I expect that a minimum quantity of coals & of air will be required to produce the desired state;- that the warm air will ascend freely & generally into the lanthorn;- that none but warm air will ascend;- that the air will be in a minimum degree of dampness in respect of that taken from the tower or other parts;- and that all the heat generated by the stove will be conveyed into the lanthorn. The chimney of the stove should be of copper as usual; should pass through the ceiling & floor to any convenient side of the lanthorn according to the nature & position of the latter; but should be within the lanthorn in its ascent, and if manageable on the coldest side. I do not expect that the usual ventilators in the sides of the lanthorn will be required; but I think they ought not to be dismissed in the first instance - or until the efficacy & sufficiency of the proposed plan has been fully proved.

Mr Walker does not see any difficulty in the construction; he now has the proposition before him for consideration and with a view to the preparation of drawings[5][.]

I am | My dear Sir | Ever Truly Yours | M. Faraday
Jacob Herbert Esq | &c &c.

1. Letter 2738.
2. Letter 2771.
3. Keeper of the lighthouse at Cromer. Otherwise unidentified.
4. Letter 2726.
5. This letter was read to the Trinity House By Board, 24 January 1854, GL MS 30010/38, p.395. It was referred to the Light Committee.

Letter 2779
William Bird Herapath[1] to Faraday
18 January 1854
From the original in RI MS

Jany 18 1854 | 32 Old Market St | Bristol

My dear Sir,

I really must apologise for having kept your most interesting specimen of Tourmaline so long but my reason for doing so was that I might have an opportunity of exhibiting it to our Bristol Microscopical Society. I had to give a paper there on Wednesday last[2] & therefore took Tourmaline for my subject[3].

It is evident that the proper name for the coloured portion of this crystal is Rubelite - a variety of Tourmaline - it owes its colour to Borate of Manganese containing nearly 5 per cent of the oxide of manganese and about 4 per cent of Boracic acid - some Lithia also.

There are also transparent and colourless Tourmalines found but it is invariably the case that these colourless crystals lose their overtly absorption properties, but still retain their doubly Refractive powers.

The colourless base of Tourmaline consists of Silicate of Alumnia sometimes also potassa - soda; Lime & Lithia - always some Boracic acid - Prot or Per Oxide of Iron is added to the coloured crystals - which in our case gives a deep Black or a Brown colour - and in the other a light bottle green - manganese also occurs in the Brown and Red varieties - the Blue or neutral tinted crystals "Indicolite" are coloured by Maganese & Iron - a good deal of Lithia.

The transparent portion has been well described by Sir David Brewster[4] but he has omitted the existence of Double refractive properties which it undoubtedly possesses as well as slight absorbent power where he has indicated.

There is a specimen of Rubelite in the Bristol Phil. Museum which closely resembles yours, but it is in the original crystal and perfectly uncut. This is a very interesting specimen as it shows that the prism

of this outline in section and consists of plates of coloured & colourless strata arranged at right angles to the prismatic axis and consequently when an optician had prepared a section of it for experiments on polarisation by cutting parallel to its length it would appear as a striated plate alternatively white and pink or purple and whitish pink. The coloured portions polarise and absorb as Tourmaline does - the colourless are only Doubly refractive.

Thanking you for your great kindness & trusting that the crystal will arrive perfectly safe

Permit me to remain | My Dear Sir | Yours very truly obliged | W. Bird Herapath

Profr Faraday

P.S I enclose Sir David Brewsters opinion which is very different from my present one as you perceive[5][.]

1. William Bird Herapath (1820–1868, DSB). Bristol physician and chemist.
2. That is 11 January 1854.
3. On his interest in the topic see Herapath (1853).
4. Brewster (1853).
5. See letter 2759.

Letter 2780
Faraday to Julius Plücker
27 January 1854
From the original in NRCC ISTI

Royal Institution | 27 January 1854

My dear Plucker

I am very tired - rather poorly - and want something to cheer me - so I have resolved to write to you. For it seems to me very long since I have seen your handwriting[1] and the last time we met was for a very short time. It begins to come back upon me that I saw you on your way to Hull[2] & that you shewed me some beautiful tables & curves of the progression of magnetic and diamagnetic force and I am most desirous of seeing them in an English dress that I may slowly & deliberately embue my mind with them. I hope you reached home safely[.] I could not be with you at Hull. Every year I feel less able to encounter the quick hurry and excitement of vigorous & active spirits. One of our philosophic physicians Sir Henry Holland has in one of his medical essays dealt with the question of *time* in regard to the operation of thought[3][.] He shews that the time often becomes sensible:- and in advancing years especially - often increases to a very considerable amount. I feel persuaded by my own experience and by many observations of others that he is right:- for with me the time necessary to apprehend an idea is very sensible and when it becomes necessary to take up many dissimilar ideas in *quick* succession then the necessity of a certain amount of time makes the operation a real labour. Such is the case at meetings like those of the B. association at Hull and I cannot bear them.

I have not been working much lately only a little upon metal wires covered with Gutta Percha. When these are immersed in water they form remarkable Leyden arrangements[.] Only think of a cylinder of Gutta Percha coated inside with copper & outside with Water and though the wire be but $\frac{1}{16}$ of an inch thick still 100 miles give an inside coating of about 8270 square feet and an outside coating of 33000 square feet. I have worked with 1500 miles of such wire at once and the results are exceedingly curious & interesting[.] The report of the experiments is now in the Printers hand and though I have not separate copies to send yet I trust you will soon get it in the proceedings of the Friday Evenings[4].

With kindest remembrances of the mutual pleasures we have had together

I am | My dear Plucker | Ever Truly Yours | M. Faraday

Address: Professor Plucker | &c &c &c &c | University | Bonn | on the *Rhine*

1. Letter 2634 was the last from Plücker.
2. For the 1853 meeting of the British Association.
3. Holland (1852), chapter 4.
4. Faraday (1854a), Friday Evening Discourse of 20 January 1854.

Letter 2781
Faraday to Christian Friedrich Schoenbein
27 January 1854
From the original in UB MS NS 409

Royal Institution | 27 January 1854

My dear friend
 Your letter of Octr. last[1] was well timed for it found me somewhat tired & out of health and by its happy affectionate feeling was quite a cheerer. I do not find that as my philosophical part wears out I at all diminish in my desire for the kindly sympathizing & brotherly feelings which have grown up with it. Your holiday trip must have been a delightful one but such things are for quasi young men. I have become a mere looker on[.] Still I and my wife do get a few short trips for instance to Wales or Norfolk or Brighton but as to crossing the Channel again I doubt it. I enjoy greatly the account of your meeting with Liebig and the Ozone affair: - it was very excellent & came off well for you[.] I like such an end to a controversy and I think you must feel that you have had a very refined revenge upon your too hasty and too positive opponents[.] Furthermore I think the Chronology of Ozone as you speak of it would be a very desirable thing[.]
 Your family account is very pleasant and I try to imagine Miss Schoenbein upon the model of what I remember of Madame Schoenbein when we were in Basle[2]:- but I have no doubt my idea is a great mistake. No matter it is very pleasant, and you must give our kindest remembrances to Madame Schoenbein. I do not suppose there is any body else at home who remembers me. It would be a delightful thing to accept your invitation & pop in:- but unless I can go by the telegraph line I am afraid that will not happen[.]
 By the bye I have lately been examining some very curious facts obtained with telegraph lines of which you will see a report in our proceedings in due time for I gave an account of them last Friday to our Members[3]. They cover copper wire with Gutta Percha here (for insulation in submarine & other cases) so perfectly that it remains beautifully insulated. I worked with 100 miles in coils immersed in the water of a canal yet with 360 pairs of plates the conduction through the gutta percha was able to deflect a delicate galvanometer only 5°. The copper wire is $\frac{1}{16}$ of an inch in thickness and the thickness of the Gutta Percha on it is about

$\frac{1}{10}$ of an inch - so that 100 miles gives a Leyden jar of which the inner coating (the copper wire) has a surface of 8272 square feet and the outer coating (the water at the G.P.) four times that amount or 33000 square feet. This wire took a charge from a Voltaic battery and could give back the electricity in a discharge having all the characters of a Voltaic current[.]

Furthermore such a wire when under ground or under water is so affected by the transition of dynamic into static electricity as to require a hundredfold the amount of tension for the transmission of an electric pulse as the same wire suspended in the air:- an effect of this kind is the interpretation of the extraordinary diversity in the expression of electric velocity given by different experimenters. But you will hear of all this in the report, when it comes out which will be soon.

Our librarian Mr. Vincent tells me that the Berichte der Verhandlungen der Naturforschenden Gesellschaft Basle, *Band 1 to 8* are not in our Library and he cannot get them here. He thinks your University distributes them to different bodies. If so is it possible for us to have that privilege? I ask you in all ignorance. But do not by any means let me be ignorantly intrusive.

Ever My dear Schoenbein | Yours | M. Faraday

Address: Dr. Schoenbein | &c &c &c | University | Basle | on the Rhine

1. Letter 2735.
2. In 1841. See Schoenbein to Faraday, 27 September 1841, letter 1364, volume 3.
3. Faraday (1854a), Friday Evening Discourse of 20 January 1854.

Letter 2782
Faraday to Arthur-Auguste De La Rive
28 January 1854
From the original in BPUG MS 2316, f.78–9

Royal Institution | 28 January 1854

My dear & kind friend

It seems a very long time since I wrote to or heard from you[1], but I have no doubt it has been my own fault. I often verify to myself the truth of an old school copy, "Procrastination is the thief of time"; and when I purpose to write it seems to me as if my thoughts now were hardly worth utterance to the men of persisting intellect & strength. But there are ties besides those of mere science and worldly relationship and I venture to think I have some such with you. These I can not easily relinquish for they grow dearer as other more temporal things dissolve away and though one cannot talk so often or so glibly about them because of their far more serious character still from time to time we may touch these cords and I shall think it a happiness whilst they respond and vibrate between us. Such ties exist but in few directions but they are worth all the rest[.]

I had a word from Schonbein a little while ago[2] and he called you to mind by speaking of his daughter who was I think then with you: and it called up afresh the thoughts of the place when very many years ago. I first saw it and your father[3] [in] 1814 or 5 but the remembrances of that time are very shadowy with me[4][.] Then came up the picture of the time when I and my wife were there with you[5] and your happy family and a strong thought of the kindness I have had from your house through two generations and now comes the contemplation of these generations passing away[.] Surely though we have both had trials & deep ones yet we have also had great mercies & goodness shewn us; above all the *great hope*. May the year that we have entered be full of peace to you, - and sweet pleasure among your children[.]

I have lately had a subject brought before me, in electricity, full of interest. My account of it is in the printers hands & when I receive it I will send you a copy by post[6][.] Briefly it is this. Copper telegraph wires are here covered perfectly with Gutta Percha so that hundreds of miles may be immersed in water and yet a very small discharge *through* the gutta percha occur when a very intense voltaic battery (300 or 400 pr of plates) is connected with it - 100 miles of such wire in water with the two exposed extremities insulated can be charged by one pole of a Voltaic battery and after separation from the battery for 5 or 10 minutes will give a shock or a current to the body or a galvanometer - or fire gunpowder or effect other electric actions either static or dynamic[.] The 100 mile[s] is in fact an immense Leyden jar & because the copper is $\frac{1}{16}$ of an inch in diameter and the gutta percha $\frac{4}{10}$ of an inch thick or $\frac{4}{16}$ of an inch extreme diameter the surface of the copper or inner lining of the jar is equal to 8270 square feet & the outer coating or water surface equal to 33000 square feet. But besides this fact of a charge given, kept, and then employed, such a wire in water has its power of conveying electricity wonderfully affected - not its final power for that is the same - for that is the same for an equal length of the wire in air or in water but its power in respect of brief currents or waves of electric force even to the extent of making the time occupied in the transmission vary as 100 to 1 or more. In a few days you shall have the account. I do not know whether I have told you yet of the pleasure I had in your Vol. I[7], but I long for Vol. II[8][.] Many thanks for all your kindness in it & on every occasion[.]

Ever My dear De la Rive | Yours M. Faraday

Address: a Monsieur | Monsieur Aug de la Rive | &c &c &c &c | Geneva | Switzerland

1. Letter 2610 was the last from De La Rive.
2. Letter 2735.
3. Charles-Gaspard De La Rive (1770–1834, DSB). Swiss chemist.
4. In the summer of 1814. See Bence Jones (1870a), **1**: 253–4.
5. In 1835. See Faraday to Magrath, 19 July 1835, letter 807, volume 2.
6. Faraday (1854a), Friday Evening Discourse of 20 January 1854.
7. De La Rive (1853–8), **1**.
8. De La Rive (1853–8), **2**.

Letter 2783
John Stevens Henslow to Faraday
28 January 1854
From the original in RI MS Conybeare Album, f.39

St Albans | 28 Jr 1854

My dear Faraday,
I have had the enclosed in my pocket book for the last 4 or 5 weeks - hoping to leave it as I passed thro' town on a visit to Darwin[1] on a former trip, & again at the present time[2] - but I have not had an opportunity - so pack it off by post. I have altered the label for I found I had written it wrong from having transferred the writing of it from right to left without *inverting* the order - forgetting that the threads crossed. You have therefore my hair & the spiders thread no longer figuring as overthin & overthick. I hope & trust the attack you mentioned in the last[3] has proved transitory. I am vastly better & can walk to advantage & no longer to a disadvantage. It seems my attack was what is now called neuralgic - & a hint that I ought to be more cautious. The Doctors have prescribed 3 excellent doses, viz. Warmth, recreation, & good living! I am getting on with the Museum Notices - & gave them[4] a lecture on crystallography last Tuesday[5][.]
Ever Yours truly & | Sincerely | J.S. Henslow

1. Charles Robert Darwin (1809–1882, DSB). Naturalist who lived at Down in Kent.
2. See Darwin to Lyell, 18 February 1854, Burkhardt et al. (1985–94): **5**: 173–5.
3. Not found.
4. That is the Ipswich Museum.
5. That is 24 January 1854.

Letter 2784
Faraday to Pierre Antoine Favre[1]
1 February 1854
From the original in Pierpont Morgan Library, Dannie and Hettie Heineman Collection, Science no. 80

London | Royal Institution | 1 February 1854

My dear Sir

I hope you will excuse my letter though written in English:- for I cannot resist the pleasure of saying how much I have enjoyed both your letter and your Thesis[2]: both of which, if I may make free to say so, bear the impress of the sound philosopher and the hard working careful experimentalist. Both have been very interesting to me and the latter especially and the more so because it not only traces the heat function of the forces of matter round the circuit in a most definite and decided manner but because it justifies the ideas one entertains more or less that all the functions of these forces are mutually related or convertible. It is very delightful when such researches as yours appear for they enable the philosopher to take a firm stand where before all was hesitation doubt and suspicion. I may illustrate that point by referring to the question whether it is the mere oxidation of the zinc or the sum of all the actions up to its combination with the acid which produces the final amount of current power and I think you have very happily settled the question by testing it through the *heat* concerned.

I am My dear Sir | With great regard your | Very faithful Servant | M. Faraday
a Monsieur | Monsieur P.A. Favre | &c &c &c &c

Address: a Monsieur | Monsieur P.A. Favre | &c &c &c | Rue St. Jacques No 36 | à Paris

1. Pierre Antoine Favre (1813–1880, DSB). Head of the chemistry laboratory at the Central School of Arts and Manufactures, Paris.
2. Favre (1853) was presented as a thesis at the University of Paris.

Letter 2785
Faraday to William Buchanan
4 February 1854
From the original in the possession of Joan Ferguson

Royal Institution | 4 Feby 1854

Dear friend & brother

We long to see your face and all friends here earnestly desire to see you:- conversation never brings up your name without that desire being manifest:- and it is now a great while since you were here[1]. Mr Boosey and myself were talking of it the other day and I found our thoughts jumped together. He is restrained in some degree by a circumstance which I dare say you are aware of namely the probability of a marriage in his house this summer[2] but we have no such thing before us. So come and see us. We should feel honored to have you on a visit to our place. We will make you as much at home in the Old place as we can and you know its conditions & inconveniences and you shall do exactly as you please:- and I hope by the time you are likely to come that my dear wife will be stronger and as able as she will be happy to attend to you. I write this early in the year that you may have time to review the probabilities and perhaps give us some little notion when the visit may come off:- You see we trust so freely in your love & willingness as to think it will come off:- but at whatever time so that we have an idea when: it will be pleasure to us and to all here[.]

Give our love to Mrs. William[3] and say we look to her as an advocate in our cause. I trust her health will be so far amended this year as to be able to spare you. Our love to Mr. Dixon[4] + & Mr. Leighton. I hope and believe that all our friends with you will think our minds are often on them - but though I fear to go on mentioning names I cannot omit Mrs. Anderson[5] & her daughters[.]

Ever My dear friend & brother | Your affectionate | M. Faraday
Wm. Buchanan Esq | &c &c &c
+ Mr Waterstone[6] of whom my wife has just reminded me[.]

1. See letter 2153.
2. Andrew Reid (1823–1896, Reid, C.L. (1914)), a Newcastle printer and nephew of Sarah Faraday, married, on 13 June 1854, Ellen Boosey who was then aged 21. GRO.
3. Elizabeth Buchanan, née Gregory. Wife of William Buchanan. See DNB under his entry.
4. Unidentified.
5. Elizabeth Anderson who moved to Edinburgh in 1850. See letter 2312.
6. Unidentified.

Letter 2786
Arthur-Auguste De La Rive to Faraday
6 February 1854[1]
From the original in IEE MS SC 2

Geneve | le 6 janvier [sic] 1854

Monsieur & très cher ami,

Je ne puis vous dire combien j'ai été sensible à votre bon souvenir. Votre lettre[2] m'a profondément touché. Il est si rare de voir des hommes de Science comme vous, mettre audessus de tout, les sentiments élevés de l'âme & les espérances éternelles, que rien ne peut faire autant de bien à un coeur affligé comme le mien, que des paroles comme celles que vous voulez bien m'adresser. Voilà bientôt quatre années que ma vie a été brisée & mon isolement m'est toujours plus pénible, non pas que je ne sois entouré d'enfants grands & petits aussi aimables que possible; mais il me manque cette compagne[3] confidante de toutes mes impressions, de mes soucis comme de mes jouissances, que rien ne peut remplacer sur cette terre. Il faut donc aller en avant avec des espérances d'un autre ordre; chaque année est un pas de plus vers ce moment où mes espérances, s'il plait à Dieu, seront réalisées; mais en attendant on a souvent des moments de découragement & de tristesse qui font trouver la vie bien longue & bien dure. Cependant la Providence me traite encore avec bien de la bonté. Ma fille[4] que est mariée à jeune homme[5] excellent plein de pieté, est entrée dans une famille pieuse & respectable à tous égards (la famille *Tronchin*). Elle a un petit garçon de neuf mois très prospère & très gentil. Mon fils[6] a épousé sa cousine[7], la petite fille de Madame Marcet, et il va bientôt être père[8]. J'ai encore deux charmantes filles bien jeunes, l'une de 16[9] l'autre de 9 ans[10]. Enfin mon fils cadet qui va bientôt avoir vingt ans[11], est entré l'année dernière comme élève à l'Ecole Polytechnique, après des examens assez brillants. Voilà mon histoire; je vous la fais sans craindre de vous ennuyer, parceque votre amitié pour moi, héritage précieux de mon excellent père[12], fait que je suis sur que vous voudrez bien y mettre quelque intéret.

J'aimerais bien vous voir & m'entretenir avec vous; mais je n'ai plus le courage d'aller en Angleterre; les bon amis que j'y ai perdus & surtout le souvenir de ma chère femme avec qui je devais y aller il y a quatre ans, me rendent ce voyage trop mélancolique pour que j'ose encore y penser. Mais vous qui n'avez pas les mèmes impressions, pourquoi ne viendriez-vous pas cet été passer quelques semaines chez moi à la campagne avec Madame Faraday? J'ai un appartement à vous offrir, ma nièce & belle-fille qui a eu le plaisir de vous voir à Londres, vous recevrait de son mieux. Vous meneriez une vie calme & tranquille comme cela convient à des hommes de notre âge & sérieux; vous vous feriez du bein dans notre bon air de la Suisse, & vous en feriez beaucoup à votre ami. Je n'ai aucun projet d'absence du 1er Mai au 1er Novembre 1854; ainsi vous pouvez choisir l'époque qui vous conviendrait le mieux dans cet intervalle de six mois.- Faites cela; le voyage est très rapide & très facile maintenant; je suis sur que Made Faraday s'en trouverait bien aussi.

Je vous remercie des détails intéressants que vous m'avez donnés dans votre lettre sur les expériences faites avec les fils télégraphiques. Ne croyez-vous pas que le retard observé dans la transmission de l'électricité quand le fil plonge dans l'eau vient précisement de ce qu'il joue le role d'une bouteille de Leyde qui exige un certain temps pour se charger; cela n'en est pas moins très curieux.

J'ai donné l'ordre qu'on vous envoyât mon premier volume sur l'Electricité qui vient de paraitre en français[13]. Je suis bien avancé dans le second[14], & j'attends une occasion pour en envoyer le manuscrit en Angleterre.- Pour en revenir aux succès, vous trouverez dans le volume *français* beaucoup de choses nouvelles & en particulier à la fin (§6 du chap. VIème de la 3ème partie), une théorie nouvelle du magnétisme & du diamagnétisme que je crois assez satisfaisante[15]; je n'entre pas dans plus de détails puisque vous pouvez lire le morceau, si cela vous intére⟨sse⟩.

Oserais-je vous adresser deux questions auxquelles vous pouvez seul me répondre:
1°) Avez-vous jamais réussi dans vos expériences à produire des courants d'induction dans d'autres liquides que le mercure - ou des métaux fondus, par example dans les solutions acides ou salines?
2°) Croyez-vous que les liquides puissent conduire une partie quelconque de l'électricité qu'ils transmettent sans éprouver de décomposition & avez-vous, depuis que vous avez traité ce sujet, été apellé à l'examiner de nouveau & à arrêter vos idées sur ce point important? Je suis, je l'avoue, très perplexe à cet égard, & je serais plutot disposé maintenant à admettre que dans quelques cas l'électricité peut traverser en partie les liquides électrolytiques sans les décomposer. Cependant je reconnais que c'est un point très difficile à constater à cause des effets secondaires qui jouent toujours un grand rôle dans ces phénomènes. Je mettrais un grand intéret à savoir où vous en êtes à cet égard.

J'espère que vous avez reçu dans le temps ma petite notice sur Arago[16] que je vous ai envoyée par Marcet; je l'ai faite en quelques heures & elle s'est bien ressentie de la précipitation que j'ai été, malgré moi, obligé d'apporter à cet hommage que je tenais à rendre promptement à la mémoire de notre ami.

Veuillez avoir la bonté de me rappeler au bon souvenir de Madame Faraday sur laquelle je compte pour vous engager à nous faire une visite l'été prochain, & agréez, très cher maitre & ami, l'assurance des sentiments de respect & d'affection que je vous porte
Votre dévoué | A. de la Rive

Address: Monsieur Faraday | Associé étranger de l'Institut | de France | Royal Institution | Albermarle Stt | Londres.
Postmark: 6 February 1854

TRANSLATION

Geneva 6th January [sic] 1854

Sir and very dear friend,

I cannot tell you how moved I was by your remembering me. Your letter[2] touched me profoundly. It is so rare to see a man of Science, such as yourself, put above all else the exalted sentiments of the soul and eternal hopes, and nothing can do as much good to an afflicted heart such as mine, than words such as the ones that you kindly addressed to me. Soon it will be nearly four years that my life was shattered and my isolation is still very painful, not that I am not surrounded by children big and small as kind as anyone could wish; but I miss that companion[3] who shared everything with me, my worries and my joys, whom nothing can replace on this earth. I must therefore go forward with hopes of another order; each year is a step further towards the moment where my hopes, please God, will be realised; but whilst I wait, I often have moments of discouragement and sadness which make life seem very long and very hard. Even though Providence still treats me with a lot of kindness. My daughter[4] who married an excellent young man[5], full of piety, has entered into a pious and respectable family in every way (the *Tronchin* family). She has a very healthy and very amiable little boy of nine months. My son[6] married his cousin[7], the granddaughter of Madam Marcet and he is soon to be a father[8]. I have two other charming daughters who are still very young, one is 16[9], the other 9 years old[10]. Finally my youngest son who is going to be twenty[11], enrolled as a student at the Ecole Polytechnique last year, after some rather brilliant exams. That is my history; I tell it to you without fear of boring you, because your friendship for me, which I inherited from my excellent father[12], makes me sure that you will be interested.

I should very much like to see you and to talk to you; but I do not have the strength to go to England; the good friends that I have lost there and above all the memory of my dear wife with whom I was supposed to go four years ago, make this journey far too melancholic for me to think of it yet. But you do not have the same impressions, why do you not come this summer to spend some weeks at my house in the country with Mrs Faraday? I have an apartment to offer you and my niece and daughter in law who had the pleasure of meeting you in London, would receive you as best she could. You would lead the calm and quiet life that befits serious men of our age; our good Swiss air would do you good and you would do a lot of good to your friend. I am not intending to be away between 1st May and 1st November 1854, thus you can choose the period that suits you best in this interval of six months. Do that; the journey is very quick and easy now; I am sure Mrs Faraday would enjoy it also.

I thank you for the interesting details that you have given me in your letter on the experiments conducted with telegraph wires. Do you not think that the delay observed in the transmission of electricity when the wires plunge into water comes precisely because it plays the same role as a Leyden jar which takes a certain amount of time to charge itself; that does not make it any less curious.

I have asked that my first volume on Electricity that has just appeared in French[13] be sent to you. I am well advanced with the second[14] and I am looking for an occasion to send the manuscript to England. To return to the successes, you will find in the *French* volume many new things and in particular at the end (§6 of the VIth chapter of the 3rd part), a new theory on magnetism and diamagnetism which I believe to be quite satisfying[15]; I am not going to go into detail, since you can read the passage if it interests you.

Dare I ask you two questions which only you can answer?
1st Have you ever succeeded in your experiments to produce induced currents in liquids other than mercury or molten metals, for example in acid or saline solutions?
2nd Do you think that liquids can conduct any part of the electricity they transmit without experiencing any decomposition and have you, since you dealt with this subject, ever thought of re-examining it and changing your ideas on this important point? I am, I admit it, very perplexed about this, and I would be rather disposed now to admit that in certain cases electricity can cross in part electrolytic liquids without decomposing them. However, I recognise that this is a very difficult point to prove because of secondary effects that always play a big role in these phenomena. I would be very interested to know what you think about this.

I hope that you received some time ago my little notice on Arago[16] which I sent you through Marcet; I wrote it in a few hours and you can clearly tell the haste with which, despite myself, I was obliged to write this homage which I wanted to render promptly to the memory of our friend.

Please be kind enough to remember me to Mrs Faraday on whom I am counting to persuade you to come and pay us a visit next summer, and accept, very dear master and friend, the assurance of the feelings of respect and affection that I bring you
your devoted | A. de la Rive

1. Dated on the basis of the postmark and that this is the reply to letter 2782.
2. Letter 2782.
3. Jeanne-Mathilde De La Rive.
4. Jeanne-Adèle Tronchin, née De La Rive (1829–1895, Choisy (1947), 51). Married 1852.
5. Louis-Nosky-Rémy Tronchin Officer in Swiss army. Choisy (1947), 51.
6. William De La Rive.
7. Cécile-Marie De La Rive, née De La Rive (1831–1893, Choisy (1947), 51). Married William De La Rive on 15 March 1852.
8. Sophie-Louisa De La Rive (1854–1902, Choisy (1947), 52).
9. Adélaïde-Eugénie-Augusta De La Rive (1838–1924, Choisy (1947), 51).
10. Françoise-Amélie-Alice De La Rive (1844–1914, Choisy (1947), 51).
11. Charles-Lucien De La Rive (1834–1924, Choisy (1947), 51). Swiss physician and writer.
12. Charles-Gaspard De La Rive (1770–1834, DSB). Swiss chemist.
13. De La Rive (1854–8), **1**.
14. De La Rive (1853–8), **2**.

15. De La Rive (1854-8), **1**, 557-79.
16. De La Rive (1853).

Letter 2787
Faraday to Mary Buckland
7 February 1854
From a typescript in RI MS

Royal Institution | 7 Feby 1854

My dear Mrs. Buckland

We have received your kind remembrance and thank you most earnestly for it. It is an odour which I very much delight in but my head will not bear it for long time together so I put them in a glass cover them over & take a sniff now & then. It is so with the Jasmine & the hyacinth:- and it is so with many things in this world[.]

My wife has been & is very much indisposed. She has very little strength for walking or any kind of exercise and at present from the addition of a prolonged & heavy cold is very deaf. She feels your kind expressions deeply[.] I wish she were able to take advantage of your hearty good will: but she sends you her very sincere thanks & earnest wishes for your health & happiness[.]

I am glad to hear a little now & then of the Dean and that the circumstances are known under which he feels quiet & in repose. Such being the case surely no one would wish to press in upon his attention & disturb him.

I remember that upon a former occasion you wished to know when Dr. Conolly spoke here. That only is the reason why I send you the enclosed paper in relation to next Friday week the 17th instant[1].

I have put bye your last letters so carefully that I cannot [find] them and my memory has let slip *with every thing else* your address, so that I must send this to the Cloisters. I hope they will forward it[.]

Ever My dear Mrs. Buckland | Your Faithful Servant | M. Faraday

Endorsement: Taken out of an envelope & forwarded | FTB[2]
Address: Mrs. Buckland | Deanery | Westminster

1. Conolly (1854), Friday Evening Discourse of 17 February 1854.
2. Francis Trevelyan Buckland (1826-1880, DNB). Army surgeon and naturalist.

Letter 2788
Joseph Dalton Hooker[1] to Faraday
8 February 1854
From the original in RI MS Conybeare Album, f.23

Kew | Feby 8th / 54

My dear Sir

I have to thank you for the very kind letter which you have written me, in your own name & that of the Managers of the Rl. Institution[2]. I assure you that I not only feel most deeply the warm interest expressed in myself, but the flattering opinion formed of my capabilities. It is with sincere regret that I definitely decline the proposals made in your letter, & not without the most earnest consideration of the whole matter. My reasons are very numerous, & I feel that several of them are conclusive in themselves.

In the first-place I have always felt very strongly (& often expressed myself so) that when officers in a public service undertake scientific duties that should engross their whole attention, & receive salaries for the same, it becomes their duty to *devote* themselves to the accomplishment of their tasks. I have been twice abroad, on Government employ, & have collected materials on both, that are not half worked out; & that will not be for years to come, & for the prosecution of which, I have received many years' salary & continue to receive it.

On my return from the Antarctic Voyage, the Treasury voted me £1000 to be spent on illustrating a Flora of the Southern ocean[3], & the Admiralty kept me on the full pay list (of Asst. surgeons) for my own salary;- that work is not completed nor can it be for 2 years to come. The materials of all succeeding voyages have been sent to me for the purpose of incorporation, & my pay was increased by the Admiralty two years ago mainly on this account. The Tasmanian Flora[4] is to form one part, & the colonists of that Island only last week, sent me the announcement that they had voted me £350 in the Legislative Council, which they beg of my acceptance, in the hope that I will allow no consideration to interfere with the speedy performance of the work, which has been announced for now 9 *years*. This little matter alone, for which I was wholly unprepared (I do not know a single member of the council & never had any communication with them) renders it my first duty to go on with that work.

My Indian collections stand next in order, these alone cannot be arranged & distributed (as I have engaged that they shall be) within three years more, & until they are gone over & catalogued I can make no satisfactory progress in the development of those laws that have governed the distribution of the Plants of the Asiatic continent upon whose investigation I have long been engaged, nor can I proceed with the "Flora Indica", upon which Dr Thomson[5] & I are at work[6], & which our fellow-Botanists consider the most important that can be undertaken.

There are personal considerations, at present pressing upon me, but there are others in prospect. Had I any time at my disposal, it should be given to my Father[7], who has neither the leisure nor the means to devote the necessary attention to his Library & Herbarium. This latter is the finest in Europe, it is maintained at considerable private cost, much more for my use than his own, it is absolutely essential to my daily work that it should be kept in scientific order, but owing to its rapid increase & the want of the means of providing a scientific curator, it is rapidly deteriorating in some respects.

Added to this my father's public duties are increasing with his years, & being allowed no assistant of any kind, it becomes the more incumbent upon me to allow of no consideration interfering with my duty to him. At present the whole charge of Library, Herbarium, & Scientific correspondence devolve entirely upon himself in a private capacity, though all maintained as essential to his public position; the correspondence for the Museum, Garden & aboretum besides the superintendence of the whole, equally devolves upon himself, & except myself he has no assistant in any one capacity. With his advancing age these duties very shortly become ostensibly too onerous for one man, they are so already in reality; & this constitutes the third reason I have for declining the invitation from the Royal Institution, (& your most kind offer of an application to Lord John Russell[8]) - that there is no possible prospect of my being able to continue my duties at the Rl. Institution, were I to commence them. It has been for some time the unanimous opinion of my (Botanical) friends that I should be attached to the Garden, applications to that effect have indeed been laid before the Government, now two years ago, when Lord John Russell[9] gave me the temporary salary of £400, for 3 years in lieu, as the means of enabling me to publish my Indian materials, & keeping me at Kew till circumstances should favor my permanent attachment to the Gardens. Had I the materials for lecturing prepared, the case would be somewhat altered, but having performed the duties of the Botanical Professor in Edinburgh University, I know from experience that lectures in this department of science beyond all others, require copious illustrations by diagrams. I had then 400 diagrams (of my Father's) which have since been parted with, together with my lectures, so that I should have to start afresh, with the certain prospect of not being able to continue long at my post. My distance from Town (where I have no other business) is another objection, for though we have a rail-road, I cannot make the journey without losing 3 hours in the transit.

I fear I have trespassed upon your patience with these details, but I most truly feel the responsibility of declining an invitation couched in such terms as that you forward, & backed by considerations involving the progress of Botanical Science (however little I should prove capable of forwarding it by the means at my disposal.)

I need hardly say that the position of Lecturer at the Royal Institution, is one of which I should be very proud, & should (as is very possible) the government desire me to lecture at Kew, in connection with any position they may appoint me to, I would feel it a privilege to be allowed to offer my services, for an occasional lecture, to the Managers of the Institution. As it is, the opinion of all my friends so entirely coincides with my own, as to the inexpediency of my undertaking any such duty, that I must definitely decline[10].

Believe me Ever | most respectfully yours | Jos. D. Hooker

1. Joseph Dalton Hooker (1817–1911, DSB). Botanist and Assistant Director of the Royal Botanic Gardens, Kew, 1855–1865.
2. See RI MM, 6 February 1854, **11**: 44 where Faraday was asked to invite Hooker to give a course of lectures after Easter. This reference also establishes the recipient of this letter.
3. Hooker (1844–7, 1853–5, 1855–60).
4. Hooker (1855–60).
5. Thomas Thomson (1817–1878, DNB). Scottish naturalist who worked mostly in India.
6. Hooker and Thomson (1855).
7. William Jackson Hooker (1785–1865, DSB). Botanist and Director of the Royal Botanic Gardens, Kew, 1841–1865.
8. Lord John Russell (1792–1878, DNB). Cabinet minister without portfolio, 1853–1854. Hooker as a government employee would have had to obtain permission for the appointment at the Royal Institution. On this and on the appointment generally see Huxley (1918), **1**: 377.
9. As Prime Minister.
10. Faraday reported this decision at RI MM, 20 February 1854, **11**: 46.

Letter 2789
Julius Plücker to Faraday
8 February 1854
From the original in IEE MS SC 2

Bonn 8th of February | 1854.

My dear Sir!

I have been very much enjoied to receive your last kind letter[1]; I'll bring against forward my bad English for answering it.

Three weeks ago I gave two copies of a rather extensive[2] paper to my bookseller, one of them addressed to you, the other one to Prof. Wheatstone. Unhappily just the day before a parcel had departed for London. The next departure will take place, I hope, very soon. To this paper on magnetic and diamagnetic induction the table with curves, you speak of in your last letter, is annexed. The general conclusions are enumerated p.51–56. The fundame[n]tel fact, represented by the curves, is deduced from observations carefully made by the large Electromagnet and a delicate balance. Take as unity of the magnetic attraction that corresponding to the unity of intensity of the current; then you may find experimentally the attraction (y) corresponding to any given intensity [(]x). In this way you will get for different substances very different curves, but following all the same general law. (First set of curves). To the intensity of the current I substituted that of the acting power of the Magnetic pole (second set) and finally to the attraction, w[h]ich undergo the different substances, the intensity of the magnetism induced by the pole in these substances (third set of curves).

I operated only on gaz oxigene, melted phosphorous and different powders. The conclusions I anticipated with regard to bars of soft iron and steel (acier trempé) have been since plainly confirmed by experiment. Iron is nearly double as strong magnetic then steel but the magnetism of steel increases more rapidly then [sic] that of soft iron does. I am engaged in a new series of experimental results, w[h]ich will, I hope, confirm all my theoretical views.

Going on at the same time in my experimental Essays relatif to magnetisme and electromagnetism. I have just finished for the "Annalen" the first part of a paper on *mixed vapours*[3].

I hapily returned home last September. Since that time I worked as much as I could, not so much as I would. I have every day two or three public lectures to give; after having given them I feel myself rather tired. That is very hard work, allso, as soon as the lectures are closed, I want to make up my mind and therefore I run during one month or two through the world. Being returned home I feel myself restored.

Pray, sir, present my compliments to Mad. Faraday[.] With my best wishes for your health very truly
    Yours | Plücker

Address: Professor Faraday | &c &c | Royal Institution | London

1. Letter 2780.
2. Plücker (1854a).
3. Plücker (1854b).

Letter 2790
Christian Friedrich Schoenbein to Faraday
10 February 1854
From the original in UB MS NS 410

Bâle Febr. 10th 1854.

My dear Faraday
    At last I have seen again some lines from the Master of the Royal Institution[1] and I can assure you that the mere sight of his handwriting

gave me infinite pleasure, as it yielded me a visible proof of his being still amongst the living, and able to handle the pen, for I will not conceal it from you that the long silence he kept this time, had already begun to cause feelings of uneasiness about the well-being of the dearest of my friends.

What you tell me of your late electrical experiments makes me very curious to learn the details of them, which I hope will soon be the case. It seems to me that we are as yet very far from having arrived at a standstill in electrical researches.

As to my little scientific doings I have continued to study the influence exerted by temperature upon the colors of substances and obtained some pretty results. You are perhaps aware that some time ago I tried to prove that a great number of oxycompounds being more or less colored at the common temperature, would turn colorless on being sufficiently cooled down each of such substances having its peculiar temperature at which its color entirely disappears. I think I have satisfactorily proved that even common Ink is in that case and you may easily convince yourself of the correctness of the statement. Color a weak solution of gallic acid by some drops of a dilute solution of perchloride of iron dark blue even to opaqueness; put the colored liquid into a frigorific mixture of muriatic acid and snow until frozen, and you will of course obtain a dark colored ice; cool it then down to about 40° below zero or somewhat less and you will have a colorless ice, which on increasing its temperature again will reassume its color before having arrived at its melting point. From some reasons I was led to conjecture that there must exist a series of bodies that exhibit the reverse behaving i.e. grow colored on their temperature being sufficiently lowered, and my conjectures proved to be correct. The coloring matter of a great number of red and blue flowers such as Dahlias, Roses &c. being associated to sulphurous acid, are at the common temperature nearly or entirely colorless; now aqueous solutions of those matters having been uncolored by aqueous sulphurous acid become beautifully and intensely recolored on being sufficiently cooled down to lose their color again on raising the temperature of the ice, and I must not omit to mention that the colorless state is reassumed before the melting of the ice.

I have particularly worked upon the coloring matter of a certain sort of dark brown Dahlia very common with us, which exhibits the change of color indicated in a most beautiful manner. On account of the easy mutability of that matter in its discolored state, I preserve it by the means of filtering paper, which I rub with the leaves of the flower and suffer it to dry. Such paper, of which I send you a little specimen, yields very easily the coloring matter to water coloring beautifully the latter. A fresh solution of that kind should always be employed on making the experiment and you will be successful, when you employ my paper for preparing the solution.

It is a fact worthy of remark that such a solution rendered colorless by SO^2 turns colored also by heating it to its boiling point.

In want of something better you might perhaps give the substance of my late researches on colors and the connexion with the chemical constitution of the matters exhibiting them in a friday Evening, for the effects are very striking[2]. Part of the results are described in the X volume of the proceedings of the Phil. Society of Bâle[3], part in a memoir published in the proceedings of the Academy of Vienna[4] which most likely will be republished in Liebigs Annals[5] and some, notably those above mentioned are not yet made known at all.

You are most likely aware that Dr. Baumert[6] has of late confirmed the results previously obtained by de la Rive, Marignac[7], Berzelius[8] and myself, as to the capability of the purest i.e. absolutely anhydrous Oxigen of being thrown into its ozonic state by the means of the electrical discharge and I am therefore inclined to think that we can no longer doubt of the important fact that oxigen may exist in two different states in an active and inactive one, in the ozonic condition and in the ordinary state.

Now such a fact cannot fail bearing upon a great number of chemical phenomena and I am just now drawing up a sort of memoir[9] in which I try to embody the Ideas and Views on Electrolysis, Thermolysis and Photolysis (sit venia verbis[10]) I have been carrying about in my head these many years, ideas so very strange and queer that they will meet with but very little favor.

To give you some notion about their singularity and heterodoxical character allow me to state some of them, but in doing so I must ask you the favor to consider them as mere Ideas and Views.

1. There are no other electrolytes (taken [sic] that term in the limited sense, you attach to it) than oxycompounds.
2. There are no compound Ions such as acids, and it is only the basic oxide of salts upon which the electrolysing power of the current is exerted.
3. The theory of Davy on the nature of Chlorine, Bromine, Iodine, the acids and salts is unfounded.
4. Electrolysation depends in the first place upon the capability of common oxigen to assume the ozonic state when put under the influence of electrical discharge and in the second place upon the power of the current to carry under given circumstances matters from the positive to the negative electrode i.e. in the direction of the current itself.
5. The transfer of the electrolytic fluid from the positive to the negative electrode as observed by Wiedemann[11] and others is closely connected with the travelling of the kation in the same direction.

6. The travelling of the anion, i.e. Oxigen is only apparent or relative being caused by the real travelling of the kation.
7. Chemical decomposition caused by electricity heat and light depends upon allotropic modifications of one or the other constituent part of the compounds decomposed.
8. Chemical synthesis caused by electricity, heat and light is closely connected with allotropic modifications of one or the other matter concerned in that chemical process.
9. The notions of chemical affinity such as they are entertained at present cannot be maintained any longer.

You see such assertions are bold enough, so bold indeed, that I am afraid even you, the boldest philosopher of our age, will shake your head; but I thi⟨nk⟩ there is no harm in going a little too far, truth will make its way in spite o⟨f it⟩ and if the feelings of our cook-like Chymists, who are brewing on and on their liquors and puddings without paying much attention to the conditions of the primary matters they are continually mixing together, should be roused even to wrath I would not only care very little about it but even take some pleasure in it, for I cannot deny that now and then I grow very angry about the narrow or little-mindedness of the generality of the tribe. Being now in a confessing

mood of mind, I will openly tell you that Davy's theoretical views are most particularly unpalatable to my scientific taste and I cannot help thinking that they have retarded rather than accelerated the progress of sound chemical science. As to some of his scientific doings they are certainly of a superior kind and nobody can value them more than I do. The heterodoxical memoir alluded to will not henceforth go forth to the world, for I shall try to work it out as well as I can[12]. In April next I think to fetch my eldest daughter[13] back again from the "Welchland"[14] to put the second[15] there. Your imagination gives you a correct idea of Miss Schoenbein, for she is really in many respects a second edition of her Mother. Our phil. Society will take great pleasure in sending you the whole series of their proceedings and in receiving, what your Institution is publishing. As the crossing of the channel and coming over to Switzerland is a matter of a couple of days I will not give up the pleasing hopes of seeing you and Mrs. Faraday once more with us in Bâle, where you have more friends and admirers than you are aware of.

Pray present my most humble compliments to your Lady and believe me
Your | most affectionate friend | C.F. Schoenbein

NB. Mrs. Schoenbein and the Children charge me to remember them kindly to you.
As I have something to send to Southampton you will receive my letter from that town.
P.S. In reading over the preceding lines I feel I have written a very bad english letter but I will not write another for fear of making it still worse. Being entirely out of the habit of speaking, writing and I may say even reading in your native tongue, I must necessarily lose the knowledge of it. And that you must take for my excuse. S.

Address: Doctor M. Faraday | &c &c &c | Royal Institution | London

1. Letter 2781.
2. Faraday gave an account of this to the General Monthly meeting of the Royal Institution. *Proc.Roy.Inst.*, 1854, **1**: 400.
3. Schoenbein (1852a).
4. Schoenbein (1853).
5. It was republished as Schoenbein (1854c).
6. Friedrich Moritz Baumert (1818–1865, P1, 3). Teacher of chemistry at the University of Breslau. For this work see Baumert (1853).
7. Jean Charles Galissard de Marignac (1817–1894, DSB). Professor of Chemistry at Geneva, 1841–1878. For this work see Marignac (1845).
8. See Berzelius, *Jahres-Bericht*, 1847, **26**: 58–64.
9. Schoenbein (1854b).
10. "if you will pardon the expression".
11. Wiedemann (1852).
12. Schoenbein (1854b).
13. Emilie Schoenbein.
14. That is French speaking Switzerland.
15. Wilhelmine Sophie Schoenbein.

Letter 2791
George Biddell Airy to Faraday
11 February 1854
From the original press copy in RGO6 / 468, f.204

Royal Observatory Greenwich | 1854 February 11

My dear Sir

I am much obliged by the copy of your lecture on the long-telegraph-wire-experiments[1]. I was burning for it, and probably in two days more should have inquired of you whether it was printed.

I like your notions on the entanglement of the induction and conduction, though I cannot yet place the mechanics of them in a clear form before my own mind.

I had returned (alone) before your lecture, and had dreamed of attending it. But a desperate trustee-meeting concerning a new arrangement of a Charity Trust left me hungry and weary two or three hours before the lecture, and I was glad to be quiet.

Yours very truly | G.B. Airy

Professor Faraday

1. Faraday (1854a), Friday Evening Discourse of 20 January 1854.

Letter 2792
George Biddell Airy to Faraday
17 February 1854
From the original in RI MS F3 G124

Royal Observatory Greenwich | 1854 Feb 17

My dear Sir

From time to time I look at your lecture of January 20[1] and reflect thereon.

Page 3 at the bottom[2], you refer very pointedly to the great extent of *surface* of the copper wire as producing the striking results, and you do not refer to any other character of the metal. But do you not suppose that the *longitudinal* extension had much to do with it:- or in plain words that it depended on the wire's being long and thin? Do you suppose that a set of copper sheets amounting in the aggregate to 8300 square feet would produce the same effect?

As the world is full of sheets of copper, I should think that this experiment might be tried without great expense, if desirable.

In your lecture you have not adverted to the peculiar character which the galvanic pulse seemed to have acquired at the third galvanometer, namely a double or treble throb lasting in the whole a full second of time. When I was making longitude-signals with Edinburgh we remarked something of the same kind, so that we were induced by what we then saw to extend our interval of certain signals from 2s to 3s in order to avoid confusion. There is always a difficulty in ascertaining the state of force at intervals shorter than or comparable with the time of natural vibration of the needle. Can you devise any thing which will exhibit it for shorter intervals? I conjecture it as quite possible that there is a nascent affection of the same kind at Galvanometer No. 2.

I am, my dear Sir, | Yours very truly | G.B. Airy

Professor Faraday

1. Faraday (1854a), Friday Evening Discourse of 20 January 1854.
2. That is *ibid.*, 348. The offprint of this paper was separately paginated. Airy's copy is in RGO6 / 468, f.209–14.

Letter 2793
Faraday to George Biddell Airy
18 February 1854
From the original in RGO6 / 468, f.207–8

Royal Institution | 18 Feby 1854

My dear Sir

With respect to all the phenomena described *up to A* page 4 of the accompanying report[1], length goes for *nothing*, except as affording surface, and therefore needed no other notice at bottom of page 3. Plates of copper & water would do just as well in association with Gutta Percha; but one would have to adjust the amount of surfaces: thus, with the wire one surface of the Gutta percha is 8300 square feet & the other 33000 square feet: if arranged as plates we must take the mean of these numbers or 20650 as the square feet of Gutta percha in plates $\frac{1}{10}$ of an inch thick, which has to be coated on one side with copper sheathing & on the other with water. So a hundred sheets of such gutta percha each 14 feet by 15 feet or thereabouts, arranged in any of the many ways by which their surfaces could be coated with metal & water, so as to form a Leyden arrangement, would answer the same purpose as the 100 miles of wire[.]

In the phenomena described *after* A, length goes for much, and is referred to pp 6, 8 10 &c as you will see by the marginal ink numbers 1, 2, 3, 4[2]:- and if you wish to see the reasoning more developed I must refer you back to the Philosophical Transactions for 1838 and especially to paragraphs 1328, *1330, 1331, 1334*[3] of the Experimental Research there printed.

In the latter results, described after A, where induction effects are combined with that effect due to *length* of wire which may be indifferently considered as either conduction or retardation, the induction effect is distributed along the wire; not being equal in every part as in the previous experiment, but diminishing in amount from the battery to the earth; but all such variations are evident at a glance & I need say no more about them[.]

As to the double or treble throbs, preparations for their observation with a view to the determination of their character would require great care. I saw nothing which struck my mind as indicating any thing which was not referable to momentum of the needles, disturbance of their polarity, - trembling contact, & other circumstances; all of which would have to be sought for & then cleared away, before I should be able to draw any conclusion in favour of a resolution of one pulse of power into several, by simple length of wire. The expansion of the time. i.e. the production of a slow action in the distance by a quick action at the battery end was clear enough, and is referred to page 9 fig 5[4]; briefly it is true, but I thought the point too evident to need many words[.]

Ever My dear Sir | Very Truly Yours | M. Faraday
The Astronomer Royal | &c &c &c

1. That is Faraday (1854a), 349. Friday Evening Discourse of 20 January 1854. See letter 2792. The page reference is to the separately paginated offprint of the paper. (Airy's copy is in RGO6 / 468, f.209–14).
2. As marked in Airy's copy.
3. Faraday (1838b), ERE12, 1328, 1330, 1331, 1334.
4. That is Faraday (1854a), 354. (See note 1).

Letter 2794
Faraday to George Biddell Airy
20 February 1854
From the original in RGO6 / 468, f.215-7

Royal Institution | 20 Feby, 1854

My dear Sir

It has occurred to me that perhaps the fact (described in p 2 of the Evening notice[1]) that many successive shocks could be obtained from one charge of the wire by quick tapping touches, has directed your mind to the condition of length[2]; for there is an effect of length of wire in that case, though almost insensible, as the results with 100 miles of wire in air described at p.3[3], shew. The effect is of this kind. With a given wire, length opposes resistance to conduction:- so when the static electricity, employed in sustaining the induction in the wire, is discharged by touching one end, the resistance of the length of wire has to be overcome and so *time* is required[.] If the wire is touched for discharge at both ends at once, which was the case in several of our experiments at the wharf[4], then the resistance is only one fourth; for the double wire may then be considered as a wire of double mass and only half length. Supposing the induction to be entirely accumulated in the 100 plates of Gutta percha referred to in my last letter[5], & that they were at one end of 50 miles of *air wire*, at the other end of which the electricity was to be discharged by successive taps, then I believe that the resistance would be the same as that with the 100 miles of water wire charged & discharged at *one* end only, & the effects the same. On the assumption that the conducting power of metal wires is directly as the sectional area & inversely as their length, a mile of copper wire of the $\frac{1}{112}$ of an inch in diameter should offer the same resistance; and a few feet of wet thread should produce the same result:- which from numerous familiar experiments with ordinarily charged Leyden batteries I have no doubt it would do.

But though such an effect in relation to time is due to length of wire, that cause is almost insensible here, for the experiments which you saw with Bain's[6] printing telegraph (p.9 of the notice[7]) shewed that with 750 miles of *air wire* or even 1500 miles the retardation was scarcely sensible; while with the *induction or underground wire* it was 1 and 2 seconds.

This long *time* I believe to be due to that conduction which every insulator shews more or less and which therefore follows upon every act of induction; as I have shewn in my old researches, see paragraphs 1323, 1324[8] & the other paragraphs there referred to. As soon as the 100 miles of water wire are charged inductively, the two electric states begin to travel through the gutta percha between the two surfaces; and hence the leaking of electricity which always occurs. There is as *true* a conduction through the Gutta percha of $\frac{1}{10}$ of an inch thick as through the copper wire 100 miles long; but the amount is so different that the conduction of Gutta percha is almost infinitely small when compared with that of copper. Nevertheless this act of conduction causes that the electricity leaves in part the surfaces of the wire & opposed water, & penetrates the gutta percha; so that the particles of Gutta percha next the wire become positively charged & those next the water negatively charged (the wire being first charged Positive). This charge occupies *time*; and at the wharf Mr Statham[9] was continually occupying time for contact with the battery, to charge, as he said, the wire fully. Then when the wire is discharged it requires the time again for the return of the Electricity from the Gutta percha and it is this time which enabled me to divide the charge into as many as 40 distinct portions[.]

My thoughts are so familiar with these considerations, which flow as natural consequences from principles long since published, that I am apt to consider them as of little importance and not worth pointing out. Your notes[10] make me think that perhaps I have been too brief and ought to have enlarged more upon the principles of insulation and conduction and the many beautiful conditions & effects that flow from them[.]

I am My dear Sir | Very Truly Yours | M. Faraday
Geo B. Airy Esq | &c &c &c

1. That is Faraday (1854a), 347. Friday Evening Discourse of 20 January 1854. The page reference is to the separately paginated offprint of the paper. (Airy's copy is in RGO6 / 468, f.209–14).
2. See letters 2791 and 2792.
3. That is Faraday (1854a), 348. (See note 1).
4. That is Lothbury Wharf. See note 1, letter 2742.
5. Letter 2793.
6. Alexander Bain (1810–1877, DNB). Telegraph engineer.
7. That is Faraday (1854a), 353–4. (See note 1).
8. Faraday (1838b), ERE12, 1323, 1324.

9. Samuel Statham (d.1864, age 58, GRO). Gutta percha manufacturer.
10. Letters 2791 and 2792.

Letter 2795
Thomas Sopwith[1] to Faraday
20 February 1854
From the original in RI MS Conybeare Album, f.39

[Athenaeum letterhead] | Athenaeum Club | Feb 20 / 54

My dear Sir

Will you if you conveniently can;- tell me what *works* or *designation* as to merit or *discoveries* in science &c. &c should be added to the name of W. Lassell[2] Astronomer Liverpool. He is nominated here by Lord Rosse & the usual designations not having been appended I write to another friend & yourself hoping it may be in your power to supply this information which is wanted before *Noon* tomorrow[3][.]

Yours most faithfully | Thos. Sopwith
M. Faraday Esq | &c &c &c

1. Thomas Sopwith (1803–1879, DNB). Mining engineer.
2. William Lassell (1799–1880, DSB). Liverpool astronomer.
3. Lassell was elected a member of the Athenaeum Club, under rule 2, in 1857. Waugh [1894], 85. Faraday had ceased to be a member. See letters 2187 and 2188.

Letter 2796
Faraday to Charles Babbage
24 February 1854
From the original in BL add MS 37195 f.452

R Institution | 24 Feby 1854

My dear Babbage

I think the best thing I can do is to send you my copy of Gmelin[1] Vol II and ask you to look at the upper half of page 62[2] - it contains more information than I could give you. I do not know whether you ever lend books if so I trust you get them back again in due time[.]

Ever Yours | M. Faraday

1. Leopold Gmelin (1788–1853, DSB). Professor of Chemistry at Heidelberg, 1817–1851.
2. Gmelin (1848–72), **2**: 62 provided references to data on the physical properties of water.

Letter 2797
Faraday to William Whewell
27 February 1854
From the original in TCC MS O.15.49, f.35

Royal Institution | 27 Feby. 1854

My dear Dr. Whewell
I send you a copy of a report of a F.E. here[1]:- it is an account of some remarkable results obtained by experiments with long insulated telegraph wires subject to induction & I think the phenomena & their causes will interest you.

I am told there is some hope that you will favour us with a lecture on Education[2]. Permit me to say how much it would gladden me to have your thoughts upon the higher points of this great subject brought before the audience in our lecture room. The social phenomena presented by the reception of table turning &c fully shew that for those who esteem themselves amongst the fully educated there is still an education of the mind required[.]

There is no one whom I should so much like to hear upon mental education as yourself[.]

Ever My dear Sir | Your faithful Servant | M. Faraday
Revd. Dr Whewell | &c &c &c

Our Managers meet this day week is it likely we may then hear of your assent?

1. Faraday (1854a), Friday Evening Discourse of 20 January 1854.
2. Whewell (1854). Delivered on 29 April 1854.

Letter 2798
Faraday to Edward William Brayley
28 February 1854
From the original in RI MS F1 D19

Royal Institution | 28 Feby 1854

My dear Sir
In signing certificates it is not my inclination but my rules which govern. I am sorry your case does not come in my rules but perhaps I am wrong & have forgotten and you may have a paper in the P. Trans. *then I sign at once*. In the absence of that qualification the other is some striking and philosophical work or operation like the construction of the Crystal Palace - or the first Submarine telegraph cable or some other object of equal interest & importance that I may be a judge of. I am very sorry that

I cannot do myself the pleasure of signing on the present occasion but I am obliged to deny myself almost weekly at this time of the year[1].
Ever Truly Yours | M. Faraday
E.W. Brayley Esq | &c &c &c

1. Faraday did not sign the certificate. RS MS Cert 9.333. Brayley was elected a Fellow of the Royal Society on 1 June 1854.

Letter 2799
Faraday to Arthur-Auguste De La Rive
1 March 1854
From the original in BPUG MS 2316, f.80–1

Royal Institution | 1 March 1854

My dear friend

Your kindness and invitation[1] moves our hearts to great thankfulness youwards: but they cannot roll back the years and give us the strength & ability of former times. We are both changed: my wife even more than I; for she is indeed very infirm in her limbs, nor have I much expectation that in that respect she will importantly improve:- but we are both very thankful for each others company and for the abundant blessing God has granted to us. I do not think it probable that either of us shall cross the sea this year or move a hundred miles from home; but we shall often during the Summer recall to mind your very pleasant invitation.

Your volume[2] & the new matter I shall look forward to with eagerness. My little report[3] I have no doubt you have received ere this, you will there perceive how much the induction you referred to in your letter has to do with the phenomena described[.]

Now in reference to your questions; the first whether I have ever obtained induction currents through liquids not being metals? - I have not worked on the subject since 1832. At that time I obtained *no current* with a tube of Sulphuric acid (Exp Res. 200[4]) but the current obtained in metals passed through liquids (Exp Res. 20[5]). I should not at all despair of obtaining the current by the use of Electromagnets and thick wire Galvanometers (3178)[6] but I have never obtained them[.]

With regard to the second question I have never seen any reason to withdraw from the opinion I formed in the year 1834 that water & such liquids could conduct a very feeble portion of electricity without suffering decomposition. I venture to refer you to the paragraphs in the Exp Researches namely 968 to 973, also 1017 and 1032[7]. I have never contested the point because having once advanced it I have not since found any reason to add or alter;- and I left it to make its way. You will find at the end of Par 984[8] reference to a point which has always had great weight with me. When electrolytes are solid as in the case of nitre or chloride of sodium at common temperatures or water at or below 0°F and when they according to all appearances *cannot* conduct as electrolytes, they still can conduct electricity of high tension; as is shown at par: 419 to 430[9]. If they have this power to such a considerable degree with electricity able to open the gold leaves, it is almost certain they have it to a certain degree with electricity of lower tension; and if the solid electrolytes have such power I cannot see any reason why their liquefaction should take it away. It would seem to me rather unphilosophical to admit it for the solid and then, without proof, to assume that it is absent in the liquid: for my part, I think the proof is all the contrary way. The power seems to be present in a very low degree but I think it is there. So much for that matter[.]

If I were in your company I should have a long chat with you about Palagis experiments[10][.] I cannot understand them as to any new principle that is involved in them; and if there be not a new principle I fear they are only mistakes; i.e imperfect forms of old results where the two developed forces are before hand present. I cannot conceive it possible that if a sphere (metallic) of 3 inches diameter be inside a metallic sp⟨here⟩ of 12 feet (or any other) diameter & touching its side, its mere removal into the centre of the large sphere or any other position in it will cause any electricity to appear.

Adieu My dear friend for the present[.]
Ever Affectionately Yours | M. Faraday
Profr. Aug de la Rive | &c &c &c &c

Address: a Monsieur | Monsieur Aug de la Rive | &c &c &c | Geneva | Switzerland

1. In letter 2786.
2. Faraday (1854a), Friday Evening Discourse of 20 January 1854.
4. Faraday (1832b), ERE2, 200.
5. Faraday (1832a), ERE1, 20.
6. Faraday (1852c), ERE29, 3178.
7. Faraday (1834), ERE8, 968–73, 1017, 1032.
8. *Ibid.*, 984.
9. Faraday (1833b), ERE4, 419–30.
10. Palagi (1854).

Letter 2800
Faraday[1] **to Arthur-Auguste De La Rive**
7 March 1854
From the original in BPUG MS 2316, f.82–3

Royal Institution | 7 March 1854

My dear Friend

Your question "whether I have ever succeeded in producing induction currents in other liquids than mercury or melted metals, as for instance in acid or saline solutions?"[2] has led me to make a few experiments on the subject[3], for though I believed in the possibility of such currents, I had never obtained affirmative results: I have now procured them, and send you a description of the method pursued. A powerful Electro magnet of the horseshoe form, was associated with a Grove's battery of 20 pairs of plates. The poles of the battery were upwards, their flat end faces being in the same horizontal plane; they are 3.5 inches square and about 6 inches apart. A cylindrical bar of soft iron 8 inches long and 1.7 in diameter was employed as a keeper or submagnet: the cylindrical form was adopted, first, because it best allowed of the formation of a fluid helix around it; and next, because when placed on the poles of the magnet and the battery connexions made and broken, the magnet and also the keeper rises and falls through much larger variations of power and far more rapidly than when a square or flat faced keeper is employed; for the latter if massive has, as you know, the power of sustaining the magnetic conditions of the magnet in a very great degree when the battery connexion is broken. A fluid helix was formed round this keeper having 12 convolutions, and a total length of 7 feet; the fluid was only 0.25 of an inch in diameter, the object being to obtain a certain amount of intensity in the current, by making the inductive excitement extend to all parts of that great length, rather than to produce a quantity current by largeness of diameter, i.e by a shorter mass of fluid.

The helix was easily constructed by the use of 8.5 feet of vulcanized caoutchouc tube, having an internal diameter of 0.25 & an external diameter of 0.5 of an inch: such a tube is sufficiently strong not to collapse when placed round the iron cylinder. The 12 convolutions occupied the interval of six inches, & two lengths of 9 inches each constituted the ends. This helix was easily and perfectly filled, by holding it with its axis perpendicular, dipping the lower end into the fluid to be used & withdrawing the air at the upper; then two long clean copper wires 0.25 of an inch in diameter, were introduced at the ends, and being thrust forward until they reached the helix, were made secure by ligaments, and thus formed conductors between the fluid helix and the Galvanometer. The whole was attached to a wooden frame so as to protect the helix from pressure or derangement when moved to & fro. The quantity of fluid contained in the helix was about 3 cubic inches in the length of 7 feet. The

Galvanometer was of wire 0.033 of an inch in diameter and 164 feet in length, occupying 310 convolutions: it was 18 feet from the magnet & connected with the helix by thick wires, dipping into cups of mercury. It was in the same horizontal plane with the magnetic poles & very little affected by direct action from the latter.

A solution formed by mixing 1 vol. of strong sulphuric acid & 3 of water was introduced into the helix tube, the iron keeper placed in the helix, & the whole adjusted on the magnetic poles in such a position, that the ends of the copper connectors in the tube were above the iron cylinder or keeper, and were advanced so far over it as to reach the perpendicular plane, passing through its axis: in this position the lines of magnetic force had no tendency to excite an induced current, through the metallic parts of the communication. The outer ends of the copper terminals were well connected together & the whole left for a time, so that any voltaic tendency due to the contact of the acid & copper might be diminished or exhausted: after that the copper ends were separated & the connexions with the Galvanometer so adjusted, that they could be in an instant either interrupted, or completed, or crossed, at the mercury cups. Being interrupted, the magnet was excited by the full force of the battery & thus the direct magnetic effect on the Galvanometer was observed; the helix had been so arranged, that any current induced in it should give a deflection in the contrary direction to that caused directly by the magnet; that the two effects might be the better separated. The battery was then disconnected & when the reverse action was over, the Galvanometer connexions were completed with the helix; this caused a deflection of only 2° due to a voltaic current generated by the action of the acid in the helix on the copper ends: it shewed that the connexion throughout was good, and being constant in power, caused a steady deflection, and was thus easily distinguished from the final result. Lastly the battery was thrown into action upon the magnet, and, immediately the galvanometer was deflected in one direction, & upon breaking battery contact, it was deflected in the other direction, so that by a few alternations, considerable swing could be imparted to the needles. They moved also in that particular manner, often observed with induced currents, as if urged by an impact or push at the moments when the magnet was excited or lowered in force; and the motion was in the *reverse* direction, to that produced by the mere direct action of the magnet. The effects were constant. When the communicating wires were crossed, they again occurred, giving reverse actions at the galvanometer. Further proof that they were due to currents induced in the fluid helix, was obtained, by arranging one turn of a copper wire round the iron core or keeper in the same direction as that of the fluid helix, and using one pair of plates to excite the magnet; the induced current caused in the copper wire was much stronger than that obtained with the fluid, but it was always in the same direction.

After these experiments with the highly conducting solution, the helix was removed, the dilute acid poured out, a stream of water sent through the helix for some time, distilled water then introduced and allowed to remain in it awhile, which being replaced by fresh distilled water, all things were restored to their places as before & thus a helix of pure water submitted to experiment. The direct action of the magnet was the same as in the first instance but there was no appearance of a voltaic current, when the galvanometer communications were completed; nor were there any signs of an induced current upon throwing the magnet into & out of action. Pure water is too bad a conductor to give any sensible effects with a Galvanometer & magnet of this sensibility & power.

I then dismissed the helix, but, placing the keeper on the magnetic poles, arranged a glass dish under it & filled the dish with the same acid solution as before; so that the liquid formed a horizontal disc 6 inches in diameter nearly, an inch deep & within 0.25 of an inch of the keeper; two long clean platinum plates dipped into this acid on each side of the keeper and parallel to it, and were at least five inches apart from each other; these were first connected together for a time, that any voltaic tendency might subside, and then arranged so as to be united with the galvanometer when requisite as before. Here the induced currents were obtained as in the first instance, but not with the same degree of strength. Their direction was compared with that of the current induced in a single copper wire passed between the fluid and the keeper, the magnet being then excited by one cell, & was found to be the same. However, here the possibility exists of the current being in part or altogether excited upon the portions of the wire conductors connected with the platinum plates; for as their ends tend to go beneath the keeper & so into the circuit of magnetic power formed by it and the magnet, they are subject to the lines of force in such a position, as to have the induced current formed in them; and the induced current can obtain power enough to go through the liquid, as I shewed in 1831[4]. But as the helix experiment is free from this objection, I do not doubt that a weak induced current occurred in the fluid in the dish also.

So I consider the excitement of induction currents in liquids not metallic as proved; and as far as I can judge, they are proportionate in strength to the conducting powers of the body in which they are generated. In the dilute sulphuric acid, they were of course stronger than they appeared by the deflection to be; because they had first to overcome the contrary deflection which the direct action of the magnet was able to produce: the sum of the two deflections in fact expressed the force of the induced current. Whether the conduction by virtue of which they occur is electrolytic in character or conduction proper I cannot say. The present phenomena do not aid to settle that question, because the induced current may exist by either one or the other process. I believe that conduction proper exists and that a very weak induction current may pass altogether by it, exciting for the time only a tendency to electrolysis, whilst a stronger current may pass partly by it & partly by full electrolytic action.

I am My dear friend | Ever most truly yours, | M. Faraday
Prof. | Aug de la Rive | &c &c &c &c

1. Apart from the signature, this letter is not in Faraday's hand.
2. In letter 2786.
3. Faraday, *Diary*, 4 March 1854, 6: 13119–45.
4. Faraday (1832a), ERE1, 20.

Letter 2801
John Conolly to Faraday
7 March 1854
From the original in RI MS Conybeare Album, f.39

Hanwell | March 7, 1854.
My dear Sir,

It appears that the Board of Management of the Asylum for Idiots has lately issued numerous letters requesting aid; of which, not having been able to attend regularly, I was not aware until your kind letter was forwarded to me with the enclosed acknowledgement. If I had known what was about to be done, I should scarcely have allowed you to be troubled with an application. Such letters are very properly sent to Lombard Street or the Stock Exchange; but are not, I think, justifiably sent to those whose modes of life are remote from the ways of the City.

Permit me, however, to add my very sincere thanks to those of the Board for your kind and liberal donation which will assuredly be applied to good purposes.

Believe me, my dear Sir, with the sincerest respect and regard,
Very faithfully Yours | J. Conolly
Profr. Faraday | &c &c &c

Letter 2802
Faraday to Arthur-Auguste De La Rive
8 March 1854
From the original in BPUG MS 2316, f.84

Royal Institution | 8 Mar 1854
My dear De la Rive

I send you the enclosed letter[1] in such shape that you may publish it if you think it worth while[.] It has been copied so as to be a little better in writing than if you had had the original. I wish I could have written it in French. As the experiments arose out of your question I send the matter to you first if you publish it in the Bibliotheque[2] then I shall afterwards give my rough copy to the Philosophical Magazine[3] as the translation from your Journal. If you should not find it expedient to print it, then I would alter the heading a little and send it to the Phil Mag as original. Do exactly as you like with it[.]

Ever My dear friend | Yours Affectionately | M. Faraday

Address: A Monsieur | Monsieur Aug. de la Rive | &c &c &c | Geneva | Switzerland

1. Letter 2800.
2. Faraday (1854c).
3. Faraday (1854d) which stated (p.265) that it was in the Bibliothèque.

Letter 2803
Lord Wrottesley to Faraday
8 March 1854
From the original in IEE MS SC 2

Wrottesley | 8 Mar 1854

Dear Sir

The Parliamentary Committee of the British Association appointed to watch over the interests of Science have considered that it would greatly assist them in the due performance of that duty if they were occasionally favored with opinions from distinguished cultivators of Science in whose judgment and discretion confidence may be securely reposed, of a kind calculated from the subject matter to which they relate, to afford valuable information as to the objects, to which the labours of the Committee might be most beneficially directed.

I have therefore to request that you will be so kind as to send me at your earliest convenience a reply to the following query,

Whether any and what measures could be adopted by the Government or the Legislature to improve the position of Science, or of the Cultivators of Science in this country?-[1]

I remain | Yours truly | Wrottesley | (Chairman)
Professor Faraday

1. For the report of the committee on this see Wrottesley (1855). For the background see Layton (1981), 188–92.

Letter 2804
Faraday to William Whewell
10 March 1854
From the original in TCC MS O.15.49, f.36

Royal Institution | 10 Mar 1854

My dear Dr. Whewell

I am tempted once more before you answer our application[1] to write for having got hold of the enclosed list of lectures and *men* I wished you to see how desirable it would that we should have a discourse either at *the beginning* or *the end*, general in its nature and shewing the idea of education as needful for all classes of men & minds - mental education which in a man of thought goes on within him from first to last[2]. No man could do this in my opinion as you would do it.

Are circumstances such as to enable you to do it with convenience & satisfaction to yourself[.]

Ever | Yours Faithfully | M. Faraday
Revd. Dr. Whewell | &c &c &c

Dr Daubeny[3]
1 On the importance of the study of chemistry as a means of education for all classes of the community[4].
Dr Tyndall
2 On the importance of the study of physics as a means of education for all classes[5].
Mr Paget[6]
3 On the importance of the study of Physiology as a means of education[7].
Dr Booth[8]
4 On the importance of mathematical studies as a means of education[9].
Dr Hodgson[10]
5 On the importance of the study of Social economy as a means of education[11].
Dr R. Latham[12]
6 On the importance of the study of language, classics, &c as a means of education[13].

Endorsed on the list in another hand: The Liniment does no seem to afford me any relief.

1. See letter 2797.
2. Whewell (1854). Delivered on 29 April 1854.
3. Charles Giles Bridle Daubeny (1795–1867, DSB). Professor of Chemistry at Oxford University, 1822–1855.
4. Daubeny (1854). Delivered on 20 May 1854.
5. Tyndall (1854b). Delivered on 27 May 1854.

6. James Paget (1814–1899, DNB1). Assistant Surgeon at St Bartholomew's Hospital, 1847–1861.
7. Paget (1854). Delivered on 3 June 1854.
8. James Booth (1806–1878, DNB). Vicar of St Anne's, Wandsworth, 1854–1859.
9. This lecture was not delivered.
10. William Ballantyne Hodgson (1815–1880, DNB). Political economist.
11. Hodgson (1854). Delivered on 10 June 1854.
12. Robert Gordon Latham (1812–1888, DNB). Director of the Ethnographical Department of the Crystal Palace, from 1852.
13. Latham (1854). Delivered on 13 May 1854.

Letter 2805
Faraday to Lord Wrottesley
10 March 1854
From the original copy in IEE MS SC 2

Royal Institution | 10 Mar 1854

My Lord

I feel unfit to give a deliberate opinion on the course it might be advisable for the Government to pursue if it were anxious to improve the position of Science and its cultivators in our country[1]. My course of life and the circumstances which make it a happy one for me are not those of persons who conform to the usages & habits of Society. Through the kindness of all from my Sovereign downwards I have that which supplies all my need and in respect of honors I have as a scientific man received from *foreign* countries and Sovereigns those which belonging to very limited & select classes surpass in my opinion any thing that it is in the power of my own to bestow.

I cannot say that I have not valued such distinctions on the contrary I esteem them very highly but I do not think I have ever worked for or sought after them[2][.] Even were such to be now created here the time is passed when these would possess any attraction for me and you will see therefore how unfit I am upon the strength of any personal motive or feeling to judge of what might be influential upon the minds of others. Nevertheless I will make one or two remarks which have often occurred to my mind.

Without thinking of the effect it might have upon distinguished men of Science or upon the minds of those who stimulated to exertion might become distinguished I do think that a Government should *for its own sake* honor the men who do honor & service to the country. I refer now to honors only not to beneficial rewards; of such honors I think there are none. Knighthoods & Baronetcies are sometimes conferred with such intentions but I think them utterly unfit for that purpose. Instead of conferring distinction they confound the man who is one of twenty or perhaps fifty with hundreds of others; they depress rather than exalt him, for they tend to lower the especial distinction of mind to the common places of society. An intelligent country ought to recognise scientific men among its people as a class. If honors are conferred upon eminence in any class as that of the law or the Army, they should be in this also[.] The aristocracy of the class should have other distinctions than those of lowly & highborn rich & poor yet they should be such as to be worthy of those whom the Sovereign & the country should delight to honor and being rendered very desirable & even enviable in the eyes of the Aristocracy by birth, should be unattainable except to that of Science[.] Thus much I think the Government & the country ought to do for their own sake & the good of Science, more than for the sake of the men who might be thought worthy of such distinction. The latter have attained to their fit place whether the community at large recognize it or not[.]

But besides that & as a matter of reward & encouragement to those who have not yet risen to great distinction I think the Government should in the very many cases which come before it having a relation to scientific knowledge employ men who pursue science, provided they are also men of business: this is perhaps now done to some extent but to nothing like the degree which is practicable with advantage to all parties[;] the right means cannot have occurred to a government which has not yet learnt to approach & distinguish the class as a whole[.] At the same time I am free to confess that I am unable to advise how that which I think should be, may come to pass. I believe I have written the expression of feelings rather than the conclusions of judgment, and I would wish Your Lordship to consider this letter as private rather than as one addressed to the Chairman of a Committee[.]

I have the honor to be | My Lord | Your Very faithful Servant | M. Faraday

Lord Wrottesley | &c &c &c

1. See letter 2803.
2. This sentence is quoted in Wrottesley (1855), liv.

Letter 2806
William Whewell to Faraday
12 March 1854
From the original in TCC MS O.15.49, f.65

Trin. Lodge, Cambridge, | Mar. 12, 1854

My dear Dr. Faraday

I do not know anything which could weigh so much with me, in making me wish to give a lecture at the R.I.[1], as your thinking it would be likely to be interesting and informative. I think I have a few thoughts "On the influence of the History of Science upon Intellectual Education"[2] which perhaps may not have occurred exactly in the same form to other persons, and which may serve to answer the speculations which may be delivered, on the influence of special branches of Science. If you think this likely to answer the purpose of the R.I. I should suppose such a lecture would come best at the beginning of the series; and in that case. I would prepare myself for such a lecture to be delivered, I suppose, soon after Easter. You will let me know whether this proposal appears to you worth following out - and believe me
Always truly yours | W. Whewell

1. See letter 2804.
2. Whewell (1854). Delivered on 29 April 1854.

Letter 2807
Arthur-Auguste De La Rive to Faraday
13 March 1854
From the original in IEE MS SC 2

Genève | le 13 mars 1854

Mon cher & digne ami,

J'ai reçu successivement vos excellentes lettres du 1er & du 8 mars[1], & je viens vous en adresser tous mes remerciements.- L'intéressant mémoire que vous m'avez envoyé m'est[2] arrivé juste à temps pour figurer dans le numéro de la *Bibl. Univ.* qui paraitra le 15[3]. Pour aller plus vite en besogne, nous nous en sommes partagés la traduction Marcet & moi, mais j'ai revu le tout. Je suis bien reconnaissant que vous ayez pensé à me donner la primeur de cet article & de la découverte importante qui y est renfermée. Je ne crois pas possible que le courant d'induction déterminé directement dans un liquide, comme il l'est dans votre expérience, puisse être d'une nature électrolytique; ne pourrait-on pas s'assurer de ce qu'il en est, en voyant si en ne laissant passer plusieurs fois de suite qu'un des deux courants d'induction, les électrodes sont ou non polarisés; ils ne doivent l'être que peu ou point si le courant n'est pas électrolytique.

Je vous enverrai immédiatement le n° de la *Bibl. Univ.* où votre article va paraître; permettez-moi de vous envoyer les précédents qui ont paru en *janvier* & *février* ainsi que ceux qui paraitront à l'avenir. Je serais bien heureux de les voir figurer dans votre bibliothèque.- Vous trouverez dans le même numéro de *mars* la traduction complète de vos belles recherches sur les fils télégraphiques qui m'ont enfin donné la solution de cette question relative à la soi-disante vitesse de l'Electricité, qui m'avait toujours embarassé[4].

Je regrette bien d'être obligé de renoncer au plaisir de vous voir cet été Madame Faraday & vous; j'espère, si Dieu le permet, aller dans un an ou 15 mois vous faire une visite en Angleterre; je remets toujours le moment de faire ce voyage qui me sera bien pénible à cause des souvenirs que j'ai dans ce pays. Mais la douceur que j'aurai de voir quelques amis que j'y ai encore sera une compensation à mes impressions pénibles.

J'espère que vous avez reçu mon premier volume Français[5] que je vous ai fait envoyer de Paris; je travaille avec vigueur au second[6] qui est bien difficile à faire dans l'état actuel de la science.

Veuillez avoir la bonté de me rapeller au bon souvenir de Madame Faraday & agréez, vous aussi, Monsieur & excellent ami, l'assurance des sentiments respectueux & affectueux
de Votre tout dévoué | A. de la Rive

TRANSLATION

Geneva | 13th March 1854

My dear and worthy friend,

I received successively your excellent letters of 1st and 8th March[1] and I write to address all my thanks to you. The interesting paper[2] that you sent me arrived just in time to appear in the number of the *Bibl. Univ.* which will be published on the 15th[3]. In order to get it done quicker, Marcet and I shared the translation, but I reviewed all of it. I am very grateful that you let me be the first to publish your paper and the important discovery that it contains. I do not believe it to be possible that an induced current determined directly in the liquid, as it is in your experiment, can be of an electrolytic nature; could one not assure oneself if it is or not, by seeing what would happen if, after repeatedly passing only one of the two currents of induction, the electrodes were polarised or not; they should be but little or not at all polarised if the current is not electrolytic.

I shall immediately send you the number of the *Bilbl. Univ* in which your article will appear; allow me to send you the previous ones which appeared in *January* & *February* as well as those that will appear in the future. I would be very happy to see them figure in your library. You will find in the *March* issue the complete translation of your beautiful research on telegraph wires which have finally given me the solution to the question relative to the so-called speed of Electricity, which always confused me[4].

I very much regret that I am obliged to renounce the pleasure of seeing Mrs Faraday and your good self this summer; I hope, God willing, to pay a visit to England in a year or 15 months; I am always putting back the moment of making this journey which will be painful for me because of the memories that are associated with that country. But the sweetness that I shall have of seeing some friends that I still have will be a compensation for my painful impressions.

I hope that you received my first French volume[5] that should have been sent to you from Paris; I am working vigorously on the second[6] which is very difficult to do bearing in mind the current state of science.

Please have the kindness to remember me to Mrs Faraday and accept, you too, Sir and excellent friend, the assurance of the respectful and affectionate sentiments

of your totally devoted | A. de la Rive.

1. Letters 2799 and 2802.
2. Letter 2800.
3. Faraday (1854c).
4. Faraday (1854b).
5. De La Rive (1854–8), 1.
6. De La Rive (1854–8), 2.

Letter 2808
Faraday to William Whewell
14 March 1854
From the original in TCC MS O.15.49, f.37

Royal Institution | 14 Mar 1854

My dear Dr. Whewell

Your letter[1] has been a very great gratification to our Committee and to myself:- they have asked me to convey our very sincere thanks to you for your great kindness. Saturday the 28th of April[2] is the first lecture of the series for which day I conclude we may now make the arrangement with the title you have given me "On the influence of the history of science upon intellectual education"[3][.]

As I sent you a list of names[4], I may as well tell you that I did not intend to join in the series, just because I do not feel competent; but our Managers wish it very much & I may not be able to resist the wish, in which case, in order that I may be safe, I think I must confine myself to something like personal experience or observation on mental education, and shall probably come the Saturday after you[5][.]

Ever Your faithful Servant | M. Faraday

1. Letter 2806.
2. That is Saturday 29 April 1854.

3. Whewell (1854). Delivered on 29 April 1854.
4. With letter 2804.
5. Faraday (1854). Delivered on 6 May 1854.

Letter 2809
Faraday to Jacob Herbert
14 March 1854
From the original copy in GL MS 30108/2/63

Royal Institution | 14 March 1854

My dear Sir

Having this day examined[1] one division of a Cata dioptric apparatus constructed by Mr. Chance of Birmingham and compared it with one of French construction which the Corporation possess mounted in the Corporation frame, I am of opinion that in the colour of the glass, the working of the various pieces & the fitting of the whole together the former is equal to the latter: and from the effect upon the screen I believe that one would not be distinguishable from the other when seen at Sea[2]:

I am my dear Sir | Your Very faithful Servant | M Faraday
Jacob Herbert Esq | &c &c &c

1. See Trinity House By Board, 28 February 1854, GL MS 30010/38, p.450 and 14 March 1854, p.465 for the arrangements for this examination.
2. This letter was read to Trinity House By Board, 21 March 1854, GL MS 30010/38, pp.475–6. It was ordered that a copy of this letter be sent to Chance, with a note that the Elder Brethren agreed with Faraday's view.

Letter 2810
Jacob Herbert to Faraday
14 March 1854
From the original in GL MS 30108/2/89

Trinity House, London, | 14th March 1854

Sir,

I am directed to forward to you the accompanying Samples of White Lead, from the parties whose names are marked thereon, who have sent in Tenders for the supply required for the Corporation's Service in the present year; and to request, that you will be pleased to analyze the same, and report the result, for the information of the Board[1].

I am | Sir, | Your most humble Servant | J. Herbert
M. Faraday Esq. D.C.L. | &c &c &c

1. Faraday's analyses are in GL MS 30108/2/89. His report, dated 21 March 1854, was read to the Trinity House Wardens Committee, 28 March 1854, GL MS 30025/23, pp.66–7 which formed the basis on which Trinity House placed their order for white lead.

Letter 2811
Lord Wrottesley to Faraday
15 March 1854
From the original in RI MS Conybeare Album, f.25

Wrottesley | 15 Mar 1854

Many thanks, My dear Sir, for your suggestions, which I duly received[1].

Yours truly | Wrottesley

1. Letter 2805.

Letter 2812
J. Laffitte[1] to Faraday
15 March 1854
From the original in IEE MS SC 2

3 Arabella rou [sic] Pimlico | London 15 mars 1854

Monsieur Faraday à Londres

Il y a environ 12 ans que j'eus l'honneur d'etre introduit auprès de vous par Monsieur Lubbock[2].

Depuis cette époque, ayant quitté Londres, je n'ai pu continuer d'avoir le plaisir de me présenter chez vous.

Maintenant un voyage d'affaires m'ayant appelé en Angleterre, je prends la liberté de soumettre à votre jugement eclairé, l'idée d'un nouveau moteur, pour lequel nous avons pris un brevet en france.

Pour ne pas abuser de vos moments, j'ai fait un résumé aussi succint que possible & si vous pensez que cette idée puisse présenter des résultats ecrites, je serai heureux de vous communiquer des détails plus circonstanciés & de solliciter vos conseils & votre patronage pour me diriger dans son application

Je suis Monsieur | avec un profond respect | Votre très dévoué | J Laffitte

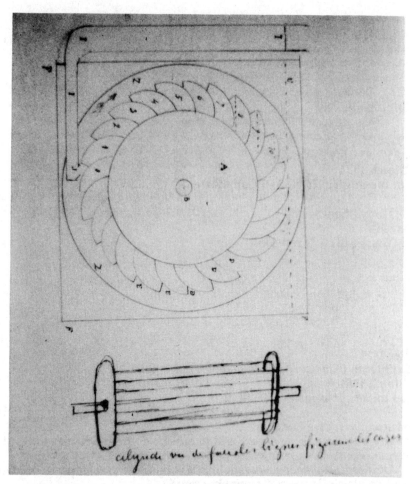

cylindre vu de face, les lignes figurent les cases

L'air introduit dans l'eau tendant toujours a s'élever, agit de toute sa force ascensionnelle contre les obstacles qu'il rencontre.
Voilà l'idée & voici son application.
A Un cylindre tournant sur un axe horizontal est garni de cases demi cylindriques dans toute sa longueur: il est renfermé dans une boite remplie d'eau.
I Au dessous de cette boite est introduit un tube dont l'ouverture est placée sous le cylindre & vis à vis la case n° 1[.] En supposant que cette case soit de la contenance d'un litre d'air elle eprouve la force d'un kilo & donne un mouvement de rotation au cylindre qui présente alors les autres cases, alternativement à l'orifice du tube. Par leur position sur le cylindre

10 de ces cases eprouvent l'effet de la force motrice augmentée par l'extension de l'air.

Ainsi après le 1er tour un tube d'air produit 15 K° de force. Le souffle d'un homme dans le tube I suffit pour produire ce resultat, qui peut etre augmenté 5 fois plus au moyen d'un levier & d'un excentrique & donner un pouvoir de 75 k°.

L'introduction de l'air par des soufflets à double vent placés dans les cotés de la boite & fonctionnants tous seuls après avoir été chargés en commencant [sic] est l'affaire de la mécanique & d'une application facile.

Ce que nous soumettons à vos lumières c'est l'introduction de l'air dans l'eau comme moteur.

Address: M. Faraday Esq | Royal Institution | Albermarle St 21 | London

TRANSLATION

3 Arabella Road, Pimlico | London 15 March 1854
Mr Faraday in London

About 12 years ago I had the honour of being introduced to you by Mr Lubbock[2].

Since that time, having left London, I was unable to continue to have the pleasure of meeting you.

Now that business affairs have brought me to England, I take the liberty of submitting to your enlightened judgement the idea of a new motor, for which we have taken out a patent in France.

In order not to waste your time, I have made as brief as possible a summary & if you think that this idea can present useful results, I will be happy to send you more details & to seek your advice & your patronage to direct me in its application.

I am Sir | with profound respect | Your most devoted | J Laffitte

Air introduced into water tends always to rise, and acts with all its ascending force against obstacles that it meets.

There is the idea & here is its application.

A A cylinder turning on a horizontal axis is surrounded by semi-cylindrical cases along its entire length: it is enclosed in a box filled with water.

I Below this box is introduced a tube of which the opening is placed below the cylinder & facing case n° 1. Supposing that this case has the capacity of 1 litre of air, it will sustain the force of one kilo & give a movement of rotation to the cylinder which will then, in turn, present other cases to the opening of the tube. By their position on the cylinder 10 of these cases will experience the effect of the force of movement increased by the extension of the air.

Thus after the first turn a tube of air can produce 15kg of force.

A mere human breath in tube I would be enough to produce this result, which can be increased 5 times more by means of a lever & an eccentric & give a power of 75k°.

The introduction of air through bellows with a double vent, placed in the corners of the box and working by themselves having been charged to start with, is a mechanical affair and can easily be sorted out.

What we submit to your judgement is the introduction of air into water as a motor.

1. Unidentified.
2. John William Lubbock (1803–1865, DSB). Banker, astronomer and Treasurer of the Royal Society, 1830–1835 and again 1838–1845.

Letter 2813
Macedonio Melloni to Faraday
21 March 1854
From the original in IEE MS SC 2

Naples | ce 21 mars 1854

Mon illustre ami!

Merci, mille fois merci, des paroles si bienveillantes que vous avez eu la bonté de m'écrire - elles ont soulagé mon coeur et rendu mon esprit susceptible d'apprécier toute l'importance de vos nouvelles et magnifiques découverts - Ainsi les physiciens ont voulu à toute force perseverer dans leur idée de mesurer la vitesse absolue de l'electricité pour chaque éspèce de conducteur métallique, lorsque vous leur aviez annoncé depuis longtems, que cette vitesse dependait de la tension du fluide transmis.....
Or le fait est venu confirmer les prévisions du génie, et la sceince professionelle se trouve justement punie de la sotte préférence accordée à ses propres routines.

Tout le monde doit bein convenir aujourd'hui que l'electricité se propage plus ou moins lentement de l'une à l'autre extremité d'un même fil télégraphique, selon qu'elle subit une induction laterale plus ou moins intense. Maintenant, si j'ai bien compris votre pensée, l'induction abaisserait le dégré de tension; et de cet abaissement résulterait l'altération observée de la vitesse du courant eléctrique.

Permettez, mon illustre ami, que je vous soumette une objection qui pourrait s'elever contre cette manière de voir. La cause du retard qu'eprouve le courant eléctrique, ne serait-elle pas analogue à ce qui se passe dans un conduit destiné à alimenter une série de citernes placées le long de sa course? Parce que le tems que l'eau emploie à parcourir toute la longueur du conduit est plus long lorsque les citernes sont vides que lorsqu'elles sont pleines ou fermées il n'en resulte pas pour cela que le liquide marche plue vite dans le second cas que dans le premier.

En me voyant avancer une espèce d'attaque contre votre théorie sur l'origine commune des forces d'induction et de transmission n'allez pas croire, je vous en prie, que je rejette absolument cette théorie si simple et si elegante, ni que je soutienne l'egalité de vitesse pour toute espèce d'eléctricité dynamique: car, je me sens, au contraire, très-incliné à admettre, et l'identité des causes qui produisent les phénomènes de la transmission et de l'induction, et l'inegale vitesse des courants eléctriques par suite de leurs divers dégrés de tension. Je crois même que cette dernière proposition pourrait se démontrer directement par l'expérience.

Supposons, par exemple, deux long fils métalliques egaux, isolés, et communiquant, par leurs extremités les plus éloignées, avec deux galvanomètres de même sensibilité, dont les bouts libres soient plongés dans le sol. Imaginons, en outre, deux piles ou eléctro-moteurs voltaiques à tensions très-differentes, mais produisant des déviations à peu près égales sur les deux galvanomètres susdits lorsqu'on les fait communiquer respectivement, par un de leurs poles, avec l'extremité la plus rapprochée des fils; tandis que l'autre pole est en communication avec la terre. Supposons enfin que, par un moyen quelconque, on puisse établir ou interrompre simultanément les communications des piles avec les fils.

Le mouvement successif des deux galvanomètres pendant l'expérience ne conduirait-il pas à la démonstration cherchée? Et ne verrait-on pas l'index magnétique correspondant au circuit de la pile composée d'un grand nombre de couples à petites surfaces sortir plus vite de sa position d'equilibre que celui qui appartient appartient [sic] au circuit de la pile formée d'un petit nombre de couples à grandes surfaces?

Je voudrais bein savoir votre opinion là dessus; et, dans le cas affirmatif, je serais vraiment enchanté de vous voir prendre les dispositions nécessaires pour mettre en oeuvre ce projet d'expériences. Je le serais d'autant plus, qu'en faisant quelques essais pour construire un eléctroscope très-sensible[1] j'ai acquis la conviction qu'un fait parfaitement analogue existe à l'égard de l'electricité statique.

On transmet à un metal isolé une forte charge eléctrique moyennant un mince et long conduit des ces matières à moitié conductrices que Volta appellait *semi coibenti*. On fait ensuite communiquer l'extremité libre du conduit avec le sol; et, malgré cela, le corps métallique entouré d'air sec et tranquille se conserve eléctrisé pendant des journées entières. Ainsi le principe eléctrique; qui avait d'abord parcouru le conduit, ne peut plus retrocéder par la meme voie lorsque sa tension a atteint une certaine limite d'abaissement.

Au moyen de cette propriété j'espère me procurer un eléctroscope analogue à celui de Bohnenberger[2], mais debarassé des imperfections qu'on lui a souvent reprochées. Cet instrument indiquera immediatement et nettement la nature de l'eléctricité explorée; et il aura en outre l'avantage de montrer, si je ne me trompe, que le principe eléctrique ne rayonne pas comme la lumière, et la chaleur, qu'il ne se déplace pas par influence de l'extremité antérieure à l'extrémité postérieure, des métaux isolés, et qu'il se propage réellement dans toute sorte de corps par une suite de polarisations moléculaires comme vous l'admettez depuis longtems contre l'avis opposé de la presque totalité des physiciens.

La construction de cet eléctroscope est assez avancée et j'espère vous en envoyer bientôt une description complete. Mais, de grace, repondez d'abord aux observations précédentes et veuillez bien me pardonner si, par des motifs déduits de ma position actuelle, je vous prie d'adresser votre reponse, ainsi que tout autre envoi de papiers, livres ou brochures, a Mr Flauti Secrétaire perpétuel de l'Académie Rle des Sciences de Naples - je vous serais même fort obligé si vous vouliez avoir la grande bonté de transmettre cet avis au Secrétariat de la Société Royale d'Edimbourg, lorsqu'il s'offrira une occasion de lui écrire - La maniere la plus sure, la plus prompte, et la plus économique de nous faire parvenir les envois c'est de les expédier *par Marseille et les paquebots de la Mediterranée*. Excusez ces details, recevez encore une fois mes plus vifs remerciments et croyez moi pour toujours votre tout dévoué admirateur et ami Macédoine Melloni
P.S. Nous ne recevons plus ici, en fait de journaux scientifiques, que les Archives des Sciences naturelles de Genève et les Annales de Chimie de Physique et d'histoire naturelle de Paris. Si, en lisant le Philosophical Magazine, les Annales de Poggendorf, ou autres feuilles periodiques, vous y trouviez des articles interessants en physique, ou s'il vous arrive de publier la moindre chose dans ces journaux ou ailleurs, vous accompliriez une oeuvre vraiment charitable en nous les envoyant, manuscrits ou imprimés par la voie indiquée tantôt-

Endorsed by Faraday: I expect the smaller intense pile would tell quickest but there are many points to include in the considerations. See Clarke.
Address: Monsieur | Monsieur M. Faraday de la Société | Royale de Londres, de l'Institut de France &c &c &c | Royal Institution, Albemarle Street | London.

TRANSLATION

Naples | this 21st March 1854

My illustrious friend!

Thank you, a thousand times thank you, for the most generous words that you had the kindness to write to me - they soothed my heart and made my spirit truly appreciate the real importance of your news and of your magnificent discoveries. Thus physicists have wanted at all costs to persevere with the idea of measuring the absolute speed of electricity for each type of metallic conductor, when you announced a long time ago, that this speed depended on the tension of the transmitted fluid ... Now fact has confirmed the predictions of genius, professional science finds itself justly embarrassed by foolishly choosing to be hidebound.

Everyone must agree today that electricity is propagated more or less slowly from one end to the other of a telegraphic wire, depending on if it is subject to a lateral induction of greater or lesser intensity. Now, if I have understood your thoughts correctly, induction would lower the degree of tension; and this lowering would result in the alteration observed in the speed of the electric current.

Permit me, my illustrious friend, to propose an objection that could be raised against such a view. The cause of the delay that is felt by the electric current, would it not be analogous to what happens in a pipe that is meant to fill a series of water tanks placed along its course? Because the time that the water takes to run the length of the pipe is longer when the tanks are empty than when they are full or shut, the result is that the liquid flows faster in the second case than in the first.

Seeing me advance a kind of attack against your theory on the common origin of the forces of induction and transmission, please do not believe, I beg you, that I am rejecting out of hand this theory which is so simple and so elegant, neither that I claim an equal speed for all types of dynamic electricity; for, on the contrary, I feel very inclined to accept both the identity of the causes which produce the phenomena of transmission and of induction and the unequal speed of electric currents arising from their different degrees of tension. I believe moreover that this last proposition could be demonstrated directly by experimentation.

Let us suppose, for example, two equal long metal wires, isolated and attached at the far ends to two equally sensitive galvanometers; the free ends [of the wires] being plunged into the ground. Let us imagine, moreover, two piles or voltaic electric generators of very different tensions, but which produce more or less equal deviations on the two galvanometers. The piles are put in contact by each of their poles, with the wires and with the earth respectively. Let us suppose finally that by whatever means, one can simultaneously restore or break the contact of the piles with the wires.

Would not the successive movement of the two galvanometers during the experiment not conduct one to the demonstration one is seeking? Would one not see the needle of the galvanometer attached to the pile composed of a great number of small surface pairs swing more than the needle of the galvanometer attached to the pile formed by a small number of pairs of greater surface?

I would very much like to know your opinion on this; and if you agree, I would be truly pleased to see you take the necessary steps to set up this series of experiments. I should be all the more pleased because in trying to build a very sensitive electroscope[1] I have become convinced that it is just the same for static electricity.

A strong electric charge is transmitted to an isolated metal with long, thin pipe made of those semi conducting materials that Volta called *semi-coibenti*. Then one puts the free end of the pipe in contact with the ground; and despite that, the metallic body surrounded by dry and still air, remains charged for days on end. Thus the electricity which had at first run through the pipe, cannot recede by the same route when its tension has reached a particularly low level.

Using this knowledge, I hope to procure an electroscope similar to that of Bohnenberger[2], but devoid of the imperfections that have often been levelled at it. This instrument will immediately and neatly indicate the nature of the explored electricity; and it will moreover, have the advantage of showing, if I am not mistaken, that electricity does not radiate like light and heat, that it is not displaced by the influence of the one end on the other in isolated metals, but that it is propagated truly in every sort of body by a series of molecular polarisations as you proposed a long time ago, against the view expressed by nearly all other physicists.

The construction of this electroscope is quite advanced and I hope soon to be able to send you a complete description. But, please, first reply to my previous observations and please excuse me if, for reasons connected with my current position, I ask you to address your reply, and any papers, books or brochures, to Mr Flauti, the permanent secretary of the Académie Royale des Sciences de Naples - I should also be most grateful if you could kindly pass this on to the Secretariat of the Royal Society of Edinburgh, when you have occasion to write to them. The safest, fastest and most economical method of sending correspondence to us is *through Marseilles and the steam ships of the Mediterranean*. Please excuse these details and please accept once again my warmest thanks and believe me to be for ever your most devoted admirer and friend

    Macedoine Melloni

P.S. We no longer receive any scientific journals except the *Archives des Sciences naturelles* of Geneva and the *Annales de Chimie de Physique et d'histoire naturelle* from Paris. If, when you read the *Philosophical Magazine*, Poggendorff's *Annalen* or other periodicals, you find any interesting articles on physics, or if you publish the smallest thing in these journals or elsewhere, you would do a great act of kindness in sending them to us, either in manuscript or printed form, by the method indicated above.

1. Melloni (1854b).

2. Gottlieb Christoph Bohnenberger (1732–1807, P1). Electrical experimenter in Germany. On his electroscope see Hackmann (1978), 25.

Letter 2814
Faraday to William Whewell
22 March 1854
From the original in TCC MS O.15.49, f.38

Royal Institution | 22 Mar 1854

My dear Dr. Whewell
    If you have no correction to make in the accompanying proof[1] do not trouble yourself to write[.]
    Ever faithfully Yours | M. Faraday

1. Probably of the lecture list for the series on education, dated April 1854. See the copy in RI MS GB 2: 84.

Letter 2815
Faraday to Frederick Oldfield Ward[1]
29 March 1854
From the original in JRULM MS 341 (56)

Royal Institution | 29 Mar 1854

My dear Sir
    Your facts are very important and when such like have been multiplied & closely examined we may be able to form some idea of the manner in which the poisonous particles are differentiated at present we can only suppose a manner[.] Nevertheless the facts if confirmed are equally important for if the presence of a body causes injury it is not necessary to wait until we know how it does so before we dismiss it[.]
    Ever My dear Sir | Yours Very Truly | M. Faraday
F.O. Ward Esq | &c &c &c

Endorsement: Reply to note about the "*Emerald green*" i.e the *arsenio-acetate of copper*, now used as a pigment - certainly with injury to workmen employing it - probably with danger to the occupants of rooms decorated with it.

1. Frederick Oldfield Ward (d.1877, age 60, GRO). Sometime scientific student at King's College, London. Hofmann (1875), 1138.

Letter 2816
Unidentified correspondent to Faraday
5 April 1854
From Faraday (1854f), 88

—— April 5, 1854.
Sir, - I am one of the clergymen of this parish, and have had the subject of table-turning brought under my notice by some of my young parishioners; I gave your solution of it as a sufficient answer to the mystery[1]. The reply was made, that you had since seen reason to alter your opinion. Would you have the politeness to inform me if you have done so? With many apologies for troubling you,
I am, your obedient servant, | ——

1. That is letter 2691.

Letter 2817
Faraday to James David Forbes
8 April 1854
From the original in SAU MS JDF 1854/56

Royal Institution | 8 April 1854
My dear Prof Forbes
I have just had a letter from Melloni[1] in which he says "je vous prie d'adresser votre reponse, ainsi que tout autre envoi de papiers, livres ou brochures, à Mr Flauti Secretaire perpetuel de l'Academie Rle des Sciences de Naples - Je vous serais meme fort obligé si vous vouliez avoir la grande bonté de transmettre cet avis au Secretariat de la Societe Royale d'Edinbourg, lorsqu'il s'offrira une occasion de lui écrire - La maniere la plus sure &c de nous faire parvenir les envois c'est de les expedier par Marseille et les paquebots de la Mediterranée"[.] With kind remembrances I am ever Very Truly Yours | M. Faraday

1. Letter 2813.

Letter 2818
Christian Friedrich Schoenbein to Faraday
9 April 1854
From the original in UB MS NS 411
My dear Faraday
These lines will be delivered to you by Mr Merian[1] of Bâle a former pupil and the son of a most intimate friend of mine the well known swiss geologist Peter Merian[2]. My young friend being an Engineer and going to England with the particular view of seeing your railways and establishments for manufacturing locomotives &c you would render him a great service by getting him introduced to some superintending railway engineers and manufacturers of locomotives. Mr. Merian is a very excellent man, distinguished mathematician, well versed in engineering, and in every respect highly respectable. You may therefore strongly and confidently recommend him to any of your friends and I need not say that by doing so you will lay me under very great obligations.

You have no doubt received my last letter[3] as well as a memoir of mine which I sent you through Mr. Gould[4] the Ornithologist and I have gratefully to acknowledge the receipt of your last paper on Electricity[5].

Its contents proved highly interesting to me and most particularly to that part of it which refers to the variations of the velocity of the current.

Having repeatedly been called upon by Mr. Liebig to draw up for his annals a paper embodying all the leading facts relative to Ozone I have at last complied with the wishes of my new friend and send you a copy of it[6]. From a note of Liebig's joined to my paper[7] you will perceive that the celebrated Chymist of Munich has taken a lively interest in the matter and in a letter, he wrote me a couple of days ago he expresses his conviction that the discovery of the ozonic Condition of Oxigen and the facts connected with that subject, will exert a great influence upon the future development of Chemical Science[8]. I have been of a similar opinion these many years.

My paper on the chemical effects produced by Electricity, Heat and Light[9], of which I talked to you in my last letter is going to be printed and as soon as finished, you shall have it, but I am sorry for you to say that it is written in my native tongue, being however not very voluminous you may easily get it translated for you. I should like very much indeed that you would take notice of its contents.

Mrs. Schoenbein and the girls are doing well and charge me with their best compliments to their friend at the Royal Institution. I join my kindest regards to Mrs. Faraday and am for ever
Your's | most truly | C.F. Schoenbein
Bâle April 9, 1854.
Dr. M. Faraday | &c &c &c

1. Rudolf Merian (1823–1872, NDB under Peter Merian). Swiss engineer.
2. Peter Merian (1795–1883, NDB). Swiss geologist and politician.
3. Letter 2790.
4. John Gould (1804–1881, DSB). Ornithologist.
5. Faraday (1854a), Friday Evening Discourse of 20 January 1854.
6. Schoenbein (1854a).
7. *Ibid.*, 258.
8. See Liebig to Schoenbein, 19 September 1853, in Kahlbaum and Thon (1900), 33–4, for a similar expression of Liebig's opinion.
9. Schoenbein (1854b).

Letter 2819
Sarah Faraday and Faraday to Caroline Deacon
12 and 13 April 1854
From the original in the possession of Elizabeth M. Milton

Brooke Lodge | Red Hill | Reigate | April 12th | 1854
My dear Caroline

I feel that you should have had a letter before this from me for there is a plaintive tone in what I hear to others that shews you want a little cheering & that you are still not so strong as we could wish, but the spring is often a time to feel rather good for nothing & the wind is rather easterly so we must hope as April goes out we shall be more like ourselves as they say - though I think it is like ourselves to feel weak & feeble pretty often[.] I often feel inclined to quote the Protoplast[1] & now something comes in to my mind but I have not time, we met the *authoress*[2] at Mr Barlows a bright gay fashionable looking lady of about 30 to 35 which has been a matter of astonishment to us ever since we heard it, we can hardly realize it, as the Americans say, she lamented my being so long indisposed & I ventured to say I had been making acquaintance with her these two months & gave her a squeeze of the hand which she returned warmly & now I do not suppose we shall say much more, for it would not be safe for me to begin to discuss the objectionable points. I should soon get out of my depth I do not think she is known as the author but hearing your uncle say something of the book, she wrote him a note thinking she ought to tell him. As I was saying we met at Mr Barlows being out for the first time in the evening since Christmas.

Now here we are near Reigate in comfortable lodgings[.] Red Hill is a quiet new place in itself & looks rather bare but the country walks are freely open & very pleasant all round about, but you know Reigate this is rather early & I cannot do much in the walking way but Jane & her Uncle are enjoying it much & are now out looking at the sunset & I am scribbling in the dark, at least I can hardly see it is $\frac{1}{2}$ past 7.

13th Again Jane & her uncle are looking after the sunset & I am talking to my dear Caroline; I do not think I told you what Miss Conacher[3] said of your portrait she was quite shocked & gave utterance to a lengthened O or two Os "Mrs. Deacon & she with such a sweet face" but still it gives me pleasure to have it. Now I must answer your letter at least a little piece of it which I never noticed for a week after I received it; with respect to Miss [name illegible] - Susans[4] friend has left & she has now no communication - do you keep up a correspondence. Your Uncle is busy with his Lecture on *Mental Culture*[5] a task which he does not much enjoy he feels so unequal to writing a lecture which is to be read & with out experiments or scarcely any, one too which will be keenly criticised - he feels his deficiencies & says it shall be the last which I agree to, but the part he took on the table turning subject seemed to make it necessary that he should be one in the coming course - indeed I suppose his remarks gave rise to the course. Margery will most likely be away at the time of your uncles lecture but I wanted to come here & hear it read before hand & help to criticise.

Tomorrow is Good Friday & we rather expect the three Buchanans to spend the day with us, we feel anxious to as our part in keeping those lads in a right course, feeling for their having no father[6] & being pleased to see how steadily & satisfactorily they have gone on so far, they were used to so much liberty in Edinbro' & had such fine walks about[.] Now poor Nathaniel[7] says he has never been in the country since he came to London except once with us at Hammersmith[8] & Kew[.] Hammersmith was not country I must say I had no temptation to walk out of our own garden but how I long for a little chaise to carry me over the ground but we can get nothing conveniently not even a Donkey with a side saddle, so we have sent to have a saddle from Reigate but I doubt my power of sitting upon a Donkey for a mile or two but we shall see[.]

My hearing is very indifferent but not so bad nearly as it has been & I keep up my flesh people say I look stout & well indeed I have been taking my food well with extra nourishment so it is no wonder if I am rather stout[.] I am glad you were not tempted to enter into long engagement with *the Lady*, at the time you saw her for it could not have been a satisfactory beginning.

You did send me the paper you thought you had forgot of Constances remarks & we were much amused[.] Mrs. Christian[9] speaks pleasantly of her I went one day to Notting Hill & saw her & the children. I am sorry to hear she is not well & uncertain when she will be able to travel she has had an interesting visit in many respects particularly with Respect to the Church which you will be glad to hear of from her[.] I do not wonder at her being the worse for such a visit, she must lead such a different life at home[.]

If your parcel is of consequence from Miss [name illegible][10] I could very well send Susan some time to fetch it, let me know whether you would like me to do so. I am a little anxious about Susan, her sisters husband Mr Clark[11] has emigrated to N America & his wife is to follow, I hope Susan may not be tempted to go after them. Mrs. Girdlers[12] daughter has some letters from Australia where she is married to a black man serving in the family she was in, a widower with a child but she writes very happily[.]

I enclose a Mare's tail have you seen it in this state. The flowers are abundant here of the kinds you mention I got into a little wood this morning it was sweet indeed. Jane desires her love with thanks for your letter. She has had a nice long letter from Frank[13] he talks of continuing in Paris a year longer[.]

I have not heard a word of my dear father[14] but I dare say you hear from Ellen[15][.] I feel for her poor girl on his advancing infirmities but sufficient for the day. With kind remembrance to Thomas & William[16] & the little one

I remain very affectionately your Aunt | S. Faraday

Dear Caroline

I have only to put my loving mark for Sarah has I dare say told you all. We are very comfortable & happy here. Sarah very lame but very cheerful. I send you a few violets. Love to Constance, Yourself your husband and others about you[.]

Your affectionate Unkle M. Faraday

1. [Baillie] (1853).
2. E.C.C. Baillie. Writer, otherwise unidentified.
3. Euphemia Conacher. A member of the London Sandemanian Church.
4. Unidentified.
5. Faraday (1854f). Delivered on 6 May 1854.
6. David Buchanan (1779–1848, DNB). Edinburgh journalist and Glasite.
7. Nathaniel Buchanan otherwise unidentified.
8. In October 1853. See letter 2745.
9. Unidentified.
10. The same name as before.
11. Unidentified.
12. Unidentified.
13. Frank Barnard.
14. Edward Barnard.
15. Ellen Barnard (1823–1899, GRO under Vincent). A niece of Sarah Faraday's.
16. Unidentified.

Letter 2820
Jacob Herbert to Faraday
13 April 1854
From the original in GL MS 30108/2/64a.4

Trinity House, London, | 13th April, 1854.

Letters, of which the enclosed are Copies[1], having been laid before the Board, I have been directed to acquaint you with their Contents, and to signify the request of the Elder Brethren that you will attend to any Communication which the Chevalier Wm Hähner[2], may make to you either personally or in writing, consequent upon the Board's having referred him to you on the subject of the Electric Light, which he represents he has "newly invented," - and which he considers applicable to the Illumination of Light Houses[3].

I enclose also a Copy of my Letter to the Chevalier[4], and beg to add the request of the Board to be favored with any observations you may think right to offer in relation to his project, after he has communicated with you thereon.

I am, | Sir, | Your most humble Servant | J. Herbert
M. Faraday Esq:

1. Buschek to Herbert, 5 April 1854, GL MS 30108/2/64a.2, Hähner to Herbert, 5 April 1854, GL MS 30108/2/64a.3.
2. Guillaume Hähner, otherwise unidentified.
3. See Trinity House By Board, 11 April 1854, GL MS 30010/38, p.493
4. Herbert to Hähner, 13 April 1854, GL MS 30108/2/64a.5.

Letter 2821
Faraday to Mason[1]
18 April 1854
From the original in SI D MS 554A

Mr Faraday's compts to Mr Mason[.] Sir Humphry Davy made the experiment Mr. Mason suggests & described it in the Phil Transactions with no useful results.
R.I. | 18 April 54

1. Unidentified.

Letter 2822
Jacob Herbert to Faraday
19 April 1854
From the original in GL MS 30108/2/88

Trinity House, London, | 19th April 1854.
Sir,

I am instructed to forward to you the accompanying Samples of Water which have been received from the Skerries Light House Establishment for the purpose of an analysis thereof being made: I also enclose herewith an Extract from a Letter of the Surgeon at Holyhead[1], who has attended the Keeper's family, together with a Copy of my Communication to the Superintendent for the Milford District thereupon[2], and of his reply, dated 14 Inst[3]:-; And I am to signify the Request of the Elder Brethren that you will favor them, after examining the water, by reporting your opinion as to it's wholesomeness for Drinking or Culinary purposes, - and whether the quality of it is such as is calculated to produce the effects adverted to in the Surgeon's Letter.

I am Sir, | Your most obedient Servant | J. Herbert
M. Faraday Esq. | &c &c &c

1. Walshew to Herbert, 13 March 1854, GL MS 30108/2/88, in which he referred to lead poisoning as the cause of the family's illnesses.
2. Herbert to Bailey, 22 March 1854, GL MS 30108/2/88.
3. Bailey to Herbert, 14 April 1854, GL MS 30108/2/88.

Letter 2823
George Biddell Airy to Faraday
22 April 1854
From the original press copy in RGO6 / 468, f.18–9

Royal Observatory Greenwich | 1854 April 22
My dear Sir

Simultaneously with the Ozone observations made here, we have had observations made at the Hospital Schools and on the declivity of Lewisham Hill (Dartmouth Terrace). You know the locality generally, the three are nearly but not quite in a straight line: the distance of the Hospital Schools about 1300 feet, that of the Lewisham station 1 mile. I enclose you the results. Also those at Bexley Heath.

The reason for my troubling you with these is that I am struck with their discordance. They seem to have no particular connexion, except that on the whole the morning numbers are larger than the evening numbers. In other respects they are so unlike that they seem to suggest - either that our modus operandi is wrong, or that the phaenomenon registered is something so very local as to be of no particular use.

Pray instruct me on all this.

I am, my dear Sir, | Yours very truly | G.B. Airy
Professor Faraday

Letter 2824
Faraday to George Biddell Airy
26 April 1854
From the original in RGO6 / 468, f.20-1

Royal Institution | 26 April 1854

My dear Sir

The impression produced on my ⟨m⟩ind by your reports[1] is first that observers have not yet learned what is requisite for a safe or a constant observation on their own part and next what circumstances about any given locality may affect the result. I do not know how far your observers have compared themselves with themselves or with others but without some proof of certainty in the results at one place it would be hardly worth while comparing them in different places. The following questions will illustrate my meaning[.]

Does an observer making three of four simultaneous observations in the same place obtain a like result by all?- if there is a difference what is the extent?

Does another observer obtain the like accordance or discordance with himself in the same place?- What is the state of accordance between him & the former observer?

Do observations made on two or four sides of the observatory by the same person agree together?- if not is the difference constant?

Do those made by *two observers or more* simultaneously in these different places agree for the same place & time?

If the results are satisfactory & the power of observing seems to be obtained then observations at places further apart would be required. I imagine a great difference is to be expected between a clear open space and the neighbourhood of a building and I conceive that no general results can be expected to agree well before the influence of all minor circumstances has been ascertained - the corrections in fact worked out[.]

Ever My dear Sir | Truly Yours | M. Faraday
G.B. Airy Esq | &c &c &c

1. See letter 2823.

Letter 2825
Faraday to Jacob Herbert
27 April 1854
From the original copy in GL MS 30108/2/88

Royal Institution | 27 April 1854

My dear Sir

I beg to report to you upon the two waters received from the Skerries[1]; in reference especially to the probability that ill health may have been the result of their use by the lighthouse keepers. The supposition is, that a poisoning, as by lead, has been produced by them: but there is not the slightest chance of such an effect from waters in the state of those received by me. The two specimens were distinguished as *tank* water and *Cottage* water. Both poured out perfectly clear from the bottles and neither contained a trace of lead. They contained small portions of saline matter: that labled *Tank* the least: the salts in it were a little sulphate of lime and a larger proportion of common salt, the latter probably derived from sea spray - both in wholesome proportions. The Cottage water contained the same substances with a little carbonate of lime and the common salt was in larger proportion than before still the water was good.

In both these bottles I could find at the bottom a very small portion of settled matter; and in that labled *Tank* the deposit contained a trace of lead. In that from the *Cottage* there was no signs of it. Whether the lead was in solution in the Tank water when put into the bottle, or whether it had been drawn up from a deposit in the tank by the pump I cannot say. Lead cisterns often contain such deposits & yet the clear water from them is perfectly good & wholesome. If at any given time turbid water had been drawn up from the *Tank* & used in that state then I could have supposed that it had something to do with the injurious effects described but I do not think the water as it has come to me could produce them.

On one occasion when I was at the Needles lighthouse I found the roof had been painted and that the weather & rain had carried off much of the carbonate of lead; making the roof most ugly and sadly injuring the water for the time for any domestic purpose[2]. Can any thing of this kind have happened at the Skerries and thus have injured the water for a time? If so the ill health would be referable to a given period.

It might perhaps be desirable to supply a filter containing a layer of charcoal or sand or both to the lighthouse[.] If sediment is ever pumped up with the water the filter would retain it & even lead in solution would be taken out. Supposing the water really & always right so that no filter was required still it might have a moral effect & reassure the keepers minds: but then it might also have the contrary effect & frighten them without cause[3][.]

I am | My dear Sir | Your Very faithful Servant | M. Faraday
Jacob Herbert Esq | Secretary | &c &c &c

1. See letter 2822.
2. This was at St Catherine's. See letter 2692.
3. This letter was read to Trinity House Court, 2 May 1854, GL MS 30004/26, pp.17–18. It was ordered that extracts of this letter be sent to the surgeon at Holyhead.

Letter 2826
Faraday to Mary Fox
28 April 1854
From the original in RI MS F1 D20

Royal Institution | 28 April 1854

My dear Lady Fox
    I feel myself very much honored by the exchange of knives. I shall have to apply yours to very rough work:- My wife is not strong[.] She sends her kindest remembrances. In respect of your kind proposal for Monday[1] I am afraid we must decline it for we both wish for rest & I have much work before me next week[2][.] With Sincere thanks
    I am Most truly Yours | M. Faraday

1. That is 1 May 1854.
2. That is for Faraday (1854f) delivered on 6 May 1854.

Letter 2827
Joseph Ketley[1] to Faraday
3 May 1854
From the original in IEE MS SC 2
To Professor Faraday

Providence New Chapel Hs. | Georgetown Demerara | 3rd May 1854
My dear Sir
    That which may prove to be of universal consequence in the application of Gutta Percha to works of importance, alone induces me to submit to your notice, the change which has taken place in the elastic properties of Gutta Percha which I purchased from the Company about five years ago. I enclose for your inspection a piece of the same, with their stamp upon it, in order that it might be identified as theirs:- And you will find to your amazement, as I found to my mortification, that it is more brittle than glass! Now if, in five years it has undergone this strange transformation, what might not be the dreadful consequences, if, trusting to its supposed durable qualities, and to its complete adoption as an effectual and never failing isolator, it should be found that in a few years it should become sufficiently brittle, in the deep, as to suffer the corroding sea water to obtain access to the wires? - and no one be able to account for the sudden failure? - I can but think that this is a matter of serious import. I would have sent this direct to the Company; but felt that it would be preferable first to submit it to your own consideration as being conversant

with the substance by having tested & analysed samples for scientific object[2].

I venture however to state to you my own observations. Two years ago I observed that the Gutta Percha I had by me presented the appearance of a sort of grey oxide, which at first I took to be the effect of a damp atmosphere but upon rubbing it between my thumb & finger found it to be somewhat more substantial than damp:- it was not damp, but dry - & from that moment I began to observe, inasmuch as I had understood that Gutta Percha would not easily decompose & hence was recommended for holding Pure water. After this I had occasion to observe that water (rain water, which here is unmingled with smoke, &c., & is our ordinary drink) standing a few weeks in a Gutta Percha Bucket so far decomposed its surface, as to produce a smooth sort of fungus & to present ropy streaks of what appeared like disintegrated Gutta Percha, in the water. Hence forth I took care never to use the Gutta Percha Carboy for water, nor to drink out of the Gutta Percha Cups which I had bought of the company for travelling purposes.

About 12 months ago, a gentleman brought me about 2 or 3 feet of Gutta Percha Rope or Cord about $\frac{1}{4}$ in diameter, which had been recommended for strength - but alas! it was as brittle as sealing wax! - I have kept a portion of it by me - and when a few days since I found how brittle the shoe soles had become I must needs try the said cord - when to my surprize I found it would scarcely bear touching!- I do not know whether the said cord was from Wharf Road or another establishment but I enclose a short piece for your convincing[.]

I cannot tell whether this change is to be attributed to *time* and *these latitudes*; or to another cause, which a little further on I will venture to suggest - but in either case I look on it with great concern for the probable future consequences, where it is unsuspectingly confided in, for great works.

Before I suggest the other probable cause I will state further, two things - first some of the Gutta Percha that I have, retains its tough, or elastic property *nearly* unchanged. I say *nearly*, because it will bend considerably, but not so elastically that I cannot break it:- which, my impression is, was not the case at first - I have the idea that on certain occasions I tried to break it five years ago and failed - some breaks now more easily than others - some after a scratch with the point of a knife - and some with the slightest touch - for it will not bend at all - has no elasticity whatever.

I cannot but think, my dear Sir, that this should be enquired into; and enquiries be made of a searching kind - because, I repeat it, - because, of possible Consequences.

Having said this much I will now mention to you; that after learning that the wonderful substance called Gutta Percha was the exudation sap of a tree, I set myself in earnest to search after trees yielding milky sap. I found some 3 or 4 or more: and began to think I could recommend a trade in that article from here. One tree in particular yields an abundance of thick jelly like milk every full moon when the sap rises to the extremities - not so at the dark moon. I collected a quantity, and after subjecting it to the bath, somewhat after the manner of the Company, I thought myself prepared to announce the fact - for it appeared tenacious - slightly elastic - and in all respects, or nearly so, seemed to answer the description - Until alas! when it had become thoroughly cold and dry, I found it just as brittle as I now find some of the pieces I bought of the company! Meanwhile I wrote to the Society to ask for a description of the Tree - which I received - but finding my research so disappointing I did not think it worth while to trouble either myself or them any further.

Now then, my dear Sir, for the lesson at least - the probability.

May not the company have been supplied with the brittle saps bearing a resemblance to the true Gutta Percha - somewhat similar to those yielded by our trees here? - May it not have occurred that some of these brittle products have been mixed up in large proportions with the real article & so occasioned the strange result? And might not some of the more brittle substances have been sometimes sent out with little, perhaps sometimes, none of the tenacious Gutta Percha, mingled with it? And in any supposition ought not this thing to be seriously enquired after? For if these spurious sorts have become articles of commerce with the company, then dreadful will be the results! - And if not, then the true article is not safe for great undertakings! - And at some future time - not distant, the telegraphy of the deep will become a terrible failure:- so I sh[oul]d fear.

Allow me however to mention another fact. Yesterday I took a piece of brittle Gutta Percha, to melt the edge of it before a glow of fire in order to rub it on the side of my slipper that had become worse:- and wishing to see whether it would break. To day on taking the same piece in my hand I find that the former elasticity is in some measure restored - but this will not continue[.]

In proof of it I will state another fact. Sometime ago I warmed and rolled into thin flat plates some of the Gutta Percha, with a view to fasten them on the soles of my slippers. I did not so use them - & now those thin pieces are even more brittle than thin amalgamated sheet zinc.

I have by me Gutta Percha tubing and also cord, which has apparently retained its strength except that it will break somewhat easier than I remember to have observed before.

I do not think I need apologize in thus troubling you, if at least I am right in my ideas of the importance of the question. I thank you much for your kind reply to my former query: And hope I may be favoured with your judgment on this matter meanwhile I remain My dear Sir

Yours faithfully | Joseph Ketley

Address: To Professor Faraday | &c &c &c

1. Joseph Ketley (1802–1875, *Congregational Year Book*, 1876, 348–50). Congregational missionary in British Guiana.
2. See Faraday (1848a).

Letter 2828
Christian Friedrich Schoenbein to Faraday
4 May 1854
From the original in UB MS NS 412
My dear Faraday
    Mr. Stehlin[1], Juris utriusque Doctor, of Bâle will perhaps take the liberty to call upon you to enquire after the address of Mr. Grove and in that case I beg you to be friendly to my young friend who is an excellent and uncommonly well informed man. Going to England with the intention of making himself acquainted with the law and courts of the country you may perhaps be able to favor the views of Dr. Stehlin by giving him an introductory line to some of your friends who happen to be a lawyer or otherwise connected with a court or a lawyer's inn.
    I am back again from the journey I made the other day to the lake of Geneva and thank God brought home my eldest daughter in perfect health. She has turned out a good girl, being highly affectionate to her parents and sisters. I think you would like her.
    Pray let me soon hear from you and believe me
    Your's | most truly | C.F. Schoenbein
Bâle Mai 4th 1854.
Mrs. and Miss Schoenbein join me in their kindest regards to Mrs. Faraday and Yourself.

1. Unidentified.

Letter 2829
Faraday to Charles Manby
8 May 1854
From the original in WIHM MS FALF

                                                    R Institution | 8 May 1854
My dear Manby

I do not know anything of our F.E. arrangements - all is under Mr. Vincent our Librarian but I will put your letter into his hands believing that if he can help you he will[.]

Yours Truly | M. Faraday

Letter 2830
George Biddell Airy to Faraday
12 May 1854
From the original press copy in RGO6 / 468, f.24–5

Royal Observatory, Greenwich | 1854 May 12

Dear Sir

Since receiving your letter about Ozone[1], I have carefully examined into the circumstances of observation. The observers are so closely connected, and their habits are so completely formed one upon another, that I do not conceive that there is any risk of discordance from personal peculiarity. But another fact has come out which may have something to do with it, and concerning which I wish to learn whether it is recognized (it is new to me).

The fact is, that the tint of the paper sometimes *goes off*. In early morning it is sometimes deeply coloured, and by 9 o'clock the colour has nearly vanished. If at such a time a strip of the paper is torn off early and dipped into water, it gives a full purple: while that which remains, if treated in the same way at a later hour, shews no colour or very little.

I cannot say with certainty that this diminution of tint is different at different places, but I am inclined to think that it is.

I am, my dear Sir, | Yours very truly | G.B. Airy
Professor Faraday

1. Letter 2824.

Letter 2831
Faraday to George Biddell Airy
13 May 1854
From the original in RGO6 / 468, f.26

Royal Institution | 13 May 1854

My dear Sir

It is some time since I observed with the test papers at Brighton[1] & I then met with effects like that you described[2]: but I concluded that they had been met or corrected in some way so as to make the results proportionate & useful. It must be useless to set about a series of observations whilst such an interference is left unattended to[.]

Ever My dear Sir | Truly Yours | M. Faraday
G.B. Airy Esq | &c &c &c

1. See letter 2356.
2. Letter 2830.

Letter 2832
Faraday to Christian Friedrich Schoenbein
15 May 1854
From the original in UB MS NS 413

Royal Institution | 15 May 1854.

My dear Schoenbein

Your letters[1] stimulate me by their energy and kindness to write but they also make me aware of my inability for I never read yours even for that purpose without feeling barren of matter and possessed of nothing enabling me to answer you in kind:- and then on the other hand I cannot take yours and think it over and so generate a fund of philosophy as you do for I am now far too slow a man for that[.] What is obtained tardily by a mind not so apt as it may have been is soon dropped again by a failing power of retention and so you must just accept the manifestation of old affection & feeling in any shape that it may take however imperfect. I received your paper and though a sealed book to me at present I have put it into the hands of Mr. Stokes whose researches on light I think I mentioned to you[2][.]

I made the experiments on the Dahlia colour which you sent me and they are very beautiful. Since then I have also made the experiment with ink and Carbonic acid (liquid) and succeeded there also to the extent you described. I had no reason to expect from what you said that dry ink would lose its colour but I tried the experiment & could not find that the carbonic acid bath had the power to do that. Many years ago I was engaged on the wonderful power that water had when it became ice of excluding other matters[3]. I could even break up compounds by cold thus if you prepare a thin glass test tube about the size of the thumb and a feather so much larger that when in the tube and twirled about it shall rapidly brush the sides: if you prepare some dilute sulphuric acid so weak that it will easily freeze at 0° Faht. & putting that into the tube with the feather you put all into a good freezing mixture of salt & snow:- if finally whilst the freezing goes on you rotate the feather continually & quickly so as to continually brush the interior surface of the ice formed clearing off all bubbles & washing that surface with the central liquid you may go on until a half or two thirds or more of the liquid is frozen & then pouring out the central liquid you will find it a concentrated solution of the acid. After that if you wash out the interior of the frozen mass with two or three distilled waters so as to remove all adhering acid & then warm the tube by the hand so as to bring out the piece of ice it upon melting, will give you pure water

not a trace of sulphuric acid remaining in it. The same was the case with common salt solution, Sul. Soda, Alcohol, &c &c and if I remember rightly even with some solid compounds of water. I think I recollect the breaking up of crystals of Sulphate of Soda by cold and I should like very much now to try the effect of a carbonic acid bath on crystals of Sulphate of copper. So it strikes me that in the effect of the cold on the colourless dahlia solution the reappearance of the colour may depend upon the separation of the Sulphurous acid from the solidifying water.

Your nine conclusions in the letter you last sent[4] me are very strong and will startle a good many but if the truth is with them I should not mind the amazement they will produce nor need you mind it either but the chemist, of which body I do not count myself one now a days, will want strong proof & be slow to convince. As to the electrical matters I referred to I expect you have received by post a printed account of what I there referred to[5].

I think some of my letters must have missed, you scold me so hard[6]. As I cannot remember what I have sent or said I am obliged to enter in a remembrancer the letters written or received and looking to it find the account thus: 1852 Decr. 8. S. to F[7] - Dec. 9. F. to S[8] - Dec 29. S to F[9]. 1853 July 24 S to F[10]. - July 25 F to S[11]. - Octr S to F[12] - 1854. Jany 27. F to S[13] - Feby 17. S to F[14]. - May 15. F. to S[15]. and considering that I have little or nothing to say & you are a young man in full vigour that is not so very bad an account so be gentle with your failing friend.

You say that in April you are to fetch a daughter from the "Welch land" &c. I had the foolish thought (perhaps), that you were coming to England & have been hoping to see you but I suppose mine was all [a] mistake for here is May. As for us we do not expect to move far from home now the imagination ramble and the desire also but the body is too heavy and earthly. Our kindest remembrance to Madame Schoenbein & to all

who remember us. Young folks cannot be expected to retain much idea of old ones after so long a while[.]
    Ever My dear friend | Affectionately Yours | M. Faraday

Address: Dr. Schoenbein | &c &c &c | University | Basle | on the Rhine

1. Letter 2790. Faraday had clearly not yet received letters 2818 and 2828. For the latter see letter 2898.
2. In letter 2604.
3. See Faraday, *Diary*, 16 April 1850, 5: 10844–52 and *Athenaeum*, 15 June 1850, pp.640–1 for an account of Faraday's Friday Evening Discourse of 7 June 1850 "On certain Conditions of Freezing Water".
4. Letter 2790.
5. Faraday (1854a), Friday Evening Discourse of 20 January 1854.
6. In letter 2790.
7. Probably letter 2578 since Faraday seems to date the letters from Schoenbein by date of receipt.
8. Letter 2604.
9. Letter 2607.
10. Letter 2699.
11. Letter 2705.
12. Letter 2735.
13. Letter 2781.
14. Letter 2790.
15. Letter 2832.

Letter 2833
Joseph Toynbee[1] to Faraday
16 May 1854
From the original in RI MS Conybeare Album, f.21

                                        18 Savile Row | May 16, 54.
My dear Sir,
    I shall be very glad to see Mrs Faraday and yourself tomorrow at 10 o'Clock.
    Yours very faithfully | Joseph Toynbee
M. Faraday Esq

1. Joseph Toynbee (1815–1866, DNB). Ear surgeon.

Letter 2834
Macedonio Melloni to Faraday
18 May 1854
From the original in IEE MS SC 2

Moretta di Portici pres Naples ce 18 May 1854.
Cher et illustre ami!

J'ai suspendu jusqu'à present ma reponse à votre lettre du 19 April[1] dans l'espoir de pouvoir vous envoyer, d'un jour à l'autre, la déscription complète de mon nouvel eléctroscope[2]. Mais la crainte que ce retard soit sinistrement interprêté m'oblige à vous ecrire ces lignes, afin de vous dire d'abord, que vos souffrances m'ont profondément affligé, et que je ne décesse de faire les voeux les plus ardents pour votre prompt retablissement dans l'état de parfaite santé - Je vous remercie ensuite bien vivement, mon excellent ami, de la peine que vous vous êtes donnée pour rectifier mes idées à l'égard de votre ingénieuse théorie sur les relations que vous établissez entre l'induction la conductibilité l'isolement et la capacité eléctrique des corps[3].... Ayez la bonté d'attendre quelques semaines, et peut être pourrais-je decider la question de savoir si dans les phénomènes de l'induction et de la conductibilité il y a simple polarité moléculaire ou transport réel de fluide eléctrique de l'une à l'autre extremité des conducteurs isolés, comme on l'admet encore dans la plupart des traités de physique... Vous comprenez bien que mes espérances de reussite sont essentiellement fondées sur l'appareil thermoscopique dont je vous parlais tantôt.... Sa construction avance et, malgré les nombreuses difficultés artistiques, et économiques qui semblaient s'accumuler exprès pour m'empêcher d'atteindre le but, je puis déjà vous annoncer *avec certitude* que cet instrument a complètement réussi - J'ai supprimé le conducteur imparfait dont je vous parlais dans ma lettre précédente[4]; mais son emploi primitif m'a conduit à une heureuse application d'un principe nouveau, ou plutôt à l'application d'une consequence jusqu'à present negligée des principes connus au moyen de laquelle mon appareil a acquis une sensibilité et une netteté d'indications vraiment étonnantes. Pour vous en donner une idée je dirai que, *dans son etat actuel d'imperfection*, il donne des deviations de 20 a 25° sur un cercle d'un decimetre de diametre par le contact de la lame zinc et cuivre soudée de Volta; et que ces déviations se maintiennent assez longtems pour permettre d'explorer tout à son aise, *jusque dans les tems les plus humides*, l'espèce d'electricité dont elles dérivent.-

Je crois qu'avec cet instrument, porté à son dernier dégré de perfection, on parviendra à faire rejeter définitivement, même par ses propres partisans, le principe de M. Palagi d'une prétendue eléctrisation des corps par le simple changement de distance[5]; principe qui n'est, selon moi, qu'une consequence des phénomènes d'induction dûs à l'éléctricité developpée par le frottement des corps mobile sur les tubes ou les tringles qui servent à le rapprocher du corps fixe.- Il me semble enfin que nous pourrons employer cet instrument avec succès pour voir s'il y a réellement de l'eléctricité dégagée pendant l'évaporation des liquides, dans l'acte de la végetation, et dans une foule d'autres questions analogues qui ont été, si je ne me trompe, fort mal étudiées dans ces derniers tems à cause de l'emploi impropre du galvanometre... Et, pour en citer un seul exemple, quel rapport y a-t-il entre les courants eléctriques *que l'on développe artificiellement* par le contact, direct ou indirect, des extremités du galvanometre, avec les feuilles et les racines d'une plante arrachée du sol, et le dégagement supposé de l'electricité statique dans l'etat naturel de cet être organique?....

Je finis en renouvellent mes voeux les plus ardents pour le rétablissement de votre précieuse santé et en me déclarant avec toute l'effusion de mon ame

votre très-affectionné et très reconnt | Serviteur et ami | Macédoine Melloni

Address: Monsieur | Monsieur Michel Faraday | de la Société Royale de Londres, de | l'Institut de France, de la Société Ita | lienne des Science &c &c &c | Royal Institution | Albemarle Street | Londres

TRANSLATION

Moretta di Portici near Naples this 18th May 1854
Dear and illustrious friend!

I have delayed with my reply to your letter of 19th April[1] in the hope of being able to send you, any day, the completed description of my new electroscope[2]. But the fear that this delay might be wrongly interpreted obliges me to write these lines, first of all to tell you that your sufferings have distressed me deeply, and that I have not ceased hoping most ardently for your prompt recovery to a state of perfect health - second, I thank you most sincerely, my excellent friend, for the trouble you have taken to rectify my ideas regarding your ingenious theory on the relationship that you have established between electrical induction, conductibility, insulation and capacity of various bodies[3].... Please have the kindness to wait a few weeks, and perhaps I will be able to settle the question and know if in phenomena of induction and conductibility there is a simple molecular polarity or a real movement of electric fluid from one end of an isolated conductor to the other, as is still accepted by most of the treatises on physics... You can well understand that my hopes of achieving this are essentially founded on the thermoscopic apparatus that I mentioned earlier.... Its construction is progressing and despite numerous technical and economic difficulties that seem to pile up expressly to stop me from achieving my goal, I can already announce *with certitude* that this instrument has completely succeeded - I have replaced the imperfect conductor that I mentioned in my previous letter[4]; but its primitive working led me to a happy application of a new principle, or rather to the application of a hitherto neglected consequence of the known principles by means of which my apparatus acquired a sensitivity and a clarity of indications that is truly surprising. To give you an idea I will say that *in its current imperfect state*, it gives deviations of 20 to 25° on a circle of a diameter of one decimetre by the contact of Volta's soldered zinc and copper needle; and that these deviations are maintained long enough to allow one to explore at leisure *even in the most humid weather*, the type of electricity from which they derive.

I believe that this instrument, brought to its last degree of perfection, will lead to the ultimate rejection, even by its supporters, of M. Palagi's principle of a seeming electralisation of bodies by simply changing the distance[5]; a principle that is, according to me, merely a consequence of the theories of the phenomena of induction due to electricity developed by the rubbing of mobile bodies on tubes or triangles which serve to connect it to the fixed body.- It seems to me finally that we can use this instrument successfully to see if there is really any electricity released during the evaporation of liquids, in the act of vegetation, and in a whole range of other analogous questions which have been, if I am not mistaken, very badly studied recently because of the improper use of the galvanometer... And to cite a single example, what relationship is there between electric currents *which are developed artificially* by contact, direct or indirect, of the ends of a galvanometer, with leaves and roots of a plant wrenched from the soil, and the supposed release of static electricity in its natural state of this organic matter?....

I end in renewing my most ardent vows for the reestablishment of your precious health, declaring myself with all the effusion of my soul

Your very affectionate and very grateful | Servant and friend | Macédoine Melloni

1. Melloni (1854b),
2. Not found.
3. Faraday (1854a), Friday Evening Discourse of 20 January 1854.
4. Letter 2813.
5. Palagi (1854).

Letter 2835
Faraday to Frederick Gye
22 May 1854
From the original in Burndy Library, Massachusetts Institute of Technology

Royal Institution I 22 May 1854

My dear Sir
    If there be an opera on Thursday night[1] and if perfectly convenient to yourself can you treat me with a feast of sweet sounds? I have such confidence in you as to trust fully that you will say *no* if my desire is at all out of place[2].
    Every Truly Yours I M. Faraday
Fredk Gye Esq I &c &c &c

1. That is 25 May 1854.
2. Faraday would have heard part of "Fidelio" by the German born composer Ludwig van Beethoven (1770–1827, GDMM) who worked mainly in Vienna and "The Barber of Savile" by the Italian composer Gioacchino Antonio Rossini (1792–1868, GDMM). See *Times*, 25 May 1854, p.8, col. a.

Letter 2836
Charles Robert Leslie[1] to Faraday
22 May 1854
From the original in IEE MS SC 2

2 Abercorn Place I St John's Wood I May 22nd 1854

Dear Sir,
    Availing myself of the enclosed introduction from Sir Benjn Brodie, I beg you will allow me to ask a question which I feel sure nobody can answer so well as yourself; otherwise I should not think it right to intrude, for a moment, on time so valuable as yours.
    I have thoughts of publishing some essays on painting[2], and I should be glad to know whether or not I am right in supposing that the azure of the sky is occasioned by particles of water in a more complete state of solution than the vapour of which the clouds are composed, the local colour of water being, as Professor Wheatstone tells me, blue.
    I have fancied also that air, unimpregnated with water, would be invisible; and that (supposing the absence of water from the atmosphere) the space in which we see the sun and the other heavenly bodies would appear utterly stark, while these luminaries would be much more intensely light than we see them, and without any appearance of halo or rays.-

These suppositions may betray, to you, great ignorance, - but I am sure you will know how to excuse it.- I have only to add that I am in no immediate want of a reply, but that I shall feel greatly obliged by one, when it may best suit your convenience so to favour me.

I am, | dear Sir, | yours obediently, | and with great respect | C.R. Leslie

Michael Faraday Esq | F.R.S. &c &c.

1. Charles Robert Leslie (1794–1859, DNB). Painter.
2. Leslie, C.R. (1855).

Letter 2837
Faraday to Charles Robert Leslie[1]
25 May 1854
From the original in RI MS F1 A24

Royal Institution | 25 May 1854

My dear Sir

It gives me great pleasure to think that I may in any way be useful to you as it [is] very agreeable to my thoughts to suppose I have any knowledge which can bear upon your high & intellectual pursuit:- leaving you to judge of the applicability of what I may say, I shall plunge at once into the middle of your letter[2]. We only know of two states of existence for the water in the atmosphere, one as clear transparent vapour and the other as the vesicles which form clouds, or what is commonly known as visible steam. We have no philosophical reason for supposing that in the first state it can produce the blue colour of the sky; for all experiment goes to shew, that, in that state, it is as transparent & colourless as the air itself. Neither, if we were to assume that the local colour of water is blue, and that in the state of transparent vapour this colour is retained, would that account for a blue sky; for supposing the whole of the atmosphere to its very summit, were retained at the high temperature of 80°F, and that it were saturated with aqueous vapour, still the quantity of water present if condensed into the liquid state would not make a layer of more than $13\frac{1}{2}$ inches in depth. But considering the rapid diminution of temperature upwards, and other circumstances affecting the quantity of water present as transparent vapour at any one time in the atmosphere, we cannot suppose there is ever more than one fourth of this amount, and I leave you to judge how utterly insufficient this Would be to produce the blue skies seen in this country, much less those of Italy and other parts of the world. Three inches in depth of the Rhine water at Geneva, which is as blue as any water I know of, if held in a glass vessel between the eye & the sky would give scarcely an appreciable effect of colour.

But the other state of "vesicles" appears to be sufficient to account for the blue colour, *not because of any colour they have in themselves,* but because of their optical effect on the rays of light passing through the atmosphere[.] My own knowledge does not render me competent to give an opinion on this matter; but I place confidence in the investigations of a high mathematician, M. Clausius[3], whose paper you will find (& I think read with interest) in Taylor's memoires: and to facilitate your access to it I send you my copy, which if you can return it in a week or two I shall be obliged. His paper begins at p 326 and you will see that by the time he arrives at the end p 331, he considers that the *blue colour* of the sky, as well as the *morning & evening red* are fully accounted for[4][.]

In reference to another part of your letter, which speaks of air without water being invisible, and therefore of the sun as intensely luminous and without halo or rays, and the space appearing black; there is reason to believe that such effects (except the rays) would occur. As to the rays, the irradiation of a very brilliant center of light does not depend upon the atmosphere, but upon effects produced in & by the parts of the eye; and would occur if the atmosphere could be entirely removed. On the other hand there is no reason to believe that the presence of water in the atmosphere, in its perfectly dissolved state, would produce any of the effects you refer to or change the appearances from those presented by perfectly dry air. It is more probable that the vesicles Clausius & others speak of, are the cause of the general diffusion of the light coming from the Sun to the earth, which takes place even in the clearest atmospheres[.]

Believe me to be | My dear Sir | Very faithfully Yours | M. Faraday
C.R. Leslie Esqre RA | &c &c &c

1. Charles Robert Leslie (1794–1859, DNB). Painter.
2. Letter 2836.
3. Rudolf Clausius (1822–1888, DSB). German physicist.
4. Clausius (1853).

Letter 2838
Joseph Denman[1] to Faraday
25 May 1854
From the original in IEE MS SC 2

H.M. Yacht Victoria & Albert | Portsmouth | May 25th
My dear Sir

The relative advantages of the Screw and of the Paddle wheel are just now being carefully investigated with the view of deciding which will be the most suitable for Her Majesty's New Yacht[2].

The question of the rolling motion of the ship as dependent upon either mode of propulsion is of course an object of prominent consideration.

His Royal Highness Prince Albert has desired me to communicate with you upon the subject, especially referring to an experiment he recently discussed with you[3], showing the tendency of a wheel having a violent rotary motion to maintain a vertical position[4].

The general impression is that Paddle wheel steamers roll less than screw steamers, and His Royal Highness suggests the enquiry, whether this may not be due to the rotary motion of the paddles under the same general principles.

The weight of the paddle wheels is considerable, but on the other hand the motion is slow; in the new yacht for instance of 2300 Tons, the paddle wheels would be 32 feet in diameter, and would make 30 Revolutions. Their weight would be about 70 Tons.

From these data will you be so kind as to state for His Royal Highness's information whether you think the revolutions of the Paddle wheels would have an appreciable effect in diminishing the rolling motion.

Should the screw be adopted, it would be 16 Feet in diameter, about 10 Tons in weight and would make about 60 revolutions. Do you think the action of the screw working at right angles to the keel of the ship would have any effect on the ship's motion? in increasing rolling? - or in diminishing pitching motion?

His Royal Highness desires me also to suggest whether by the adoption of a heavy kind of Fly wheel working in the centre of the ship in a line with the keel+, the rolling motion of ships might not, (on the principle before referred to) be greatly diminished, if not altogether prevented. I am my dear Sir

Faithfully Your's | Jos Denman
+ unconnected with either propeller.
Professor Faraday | &c &c &c

1. Joseph Denman (1810–1874, B1). Naval officer. Commanding officer of the royal yacht, 1853–1862.
2. Also called the Victoria & Albert. Paddle wheels were chosen for this yacht which was launched on 16 January 1855. Gavin (1932), 129–43, 279.
3. Presumably when Prince Albert chaired Whewell (1854) or Faraday (1854f) on 29 April and 6 May 1854 respectively.
4. See Faraday, *Diary*, 11, 13, 14, 17 March 1854, 6: 13146–13188 for Faraday's experiments on this topic.

Letter 2839
Faraday to Joseph Denman[1]
27 May 1854
From the original copy in IEE MS SC 3

Royal Institution | 27 May 1854

My dear Sir

Your letter is full of interest and I feel great delight that any conversation in which I had part should be connected with so just an application of the principles of natural philosophy as has been made by His Royal Highness Prince Albert in the cases of the paddle wheel & the propeller[2][.]

You will be aware from the communication of his Royal Highness that all practical result may be referred to the following facts. A disc when rotating, resists any force tending to alter its place so as to change the plane of its rotation far more than if the disc were not rotating: and the resistance is the greater as the body is heavier;- as the parts have greater velocity or momentum & therefore as they are further from the axis of rotation, and as the change of plane is greater. Now the force of the paddle wheels & their positions in relation to a steam ship are such that they cannot but affect its rolling; and their tendency will be to diminish it. You will understand that the endeavour is not to preserve any particular plane as regards the horizon, but *that* in which the disturbing force finds the rotating disc; so if a wave causes the vessel to roll, the revolving bodies will tend to resist this roll; as the vessel endeavours to recover itself the tendency will be to resist the recovery also; but on the whole, the rolling will be obstructed and diminished. I have always considered that paddle wheels resist and diminish rolling by the hold the descending side takes (like a hand in swimming) upon the water; but I have not the slightest doubt, now, that they will act also by the effect His Royal Highness has pointed out. What the proportion may be I cannot say; or to what extent the weight of 70 tons disposed in forms about 32 feet in diameter & revolving once in 2 seconds would affect a ship of 2300 tons. But I should expect it would be very appreciable, & should not be surprised if it may form a considerable part of any superiority which paddle wheels have over screws.

The screw you refer to, though it would revolve with twice the velocity of the paddle wheels has only half their diameter & a third of their weight; so that it would present much less resistance to change of plane than the latter. Besides this it is at the extremity of the vessel; and therefore perhaps 6 or 8 times as far from the horizontal transverse line about which the ship tends to revolve when pitching, as the paddle wheels are from the horizontal longitudinal line about which the ship tends to revolve when rolling;- for the short motions of the roll will be much more resisted than the long motion of the pitch, because the place of rotation in the first case, is more quickly changed. I do not think that the screw would tend to increase rolling otherwise than as it would replace the paddle wheels which tend to diminish it.

The suggestion of his Royal Highness in regard to a central fly wheel is highly philosophic & perfectly justified by natural principles[.] At the same time I cannot undertake to say what amount of effect it would produce in any given case. Still the experiment could be made so simply and progressively that I think any marine engineer could ascertain the point practically in a very few days.

Suppose a boat with a heavy disc or fly wheel fitted up in the middle this being attached by running bands to an axle & handles in the fore or aft part so that a man (or two men if needful) could get the fly into rapid rotation; the boat being of such size that a third person standing across or from side to side could by the action of his limbs sway her right & left. He might do this when the fly is still, and also when in quick motion. He would soon find the resistance to his efforts in the latter case, and then a judgment might be formed as to the result of *a larger experiment* & as to the application to a ship. If more convenient two fly wheels might be used one on each side of the boat and the gear & men be in the middle; but the first experiment ought to be made with a boat that can be easily & quickly rocked or the results will not be so instructive as they might be.

Though I have spoken thus far of a disc revolving in a vertical plane yet it is of course evident that a horizontal or any other plane may be selected provided that the axis of rotation is perpendicular to the length of the boat.

Supposing that a great disc or fly wheel were revolving in the inside of a vessel parallel to & in the same direction as the paddle wheels, and a wave were to affect the vessel, rolling her so as to depress the Starboard side, the resistance set up by the disc would not be direct, but would have an oblique result tending to turn the ships head to Starboard. Has any thing of this kind been distinguished by the man at the wheel? Probably he could not tell it from the effect due to immersion of the Starboard Paddle wheel:- in the boat experiment it ought to be sensible.

I am | My dear Sir | Very Truly Yours | M. Faraday
Captn. Jos Denman | &c &c &c

1. Joseph Denman (1810–1874, B1). Naval officer. Commanding officer of the royal yacht, 1853–1862.
2. See letter 2838.

Letter 2840
Faraday to Arthur-Auguste De La Rive
29 May 1854
From the original in BPUG MS 2316, f.69–70

Royal Institution | 29 May 1854

My dear friend

Though feeling weary & tired I cannot resist any longer conveying to you my sincere thanks (however feebly) for the gift of your work in French[1]. I have delayed doing so for some time hoping to be in better spirits but will delay no longer, for delighted as I have been in the reading of it, my treacherous memory begins to let loose that which I gained from it:- for when I read some of the summaries a second time, I am surprised to find them there & then slowly find that I had read them before[.] The power with which you hold the numerous parts of our great department of science in your mind is to me most astonishing & delightful and the accounts you give of the researches of the workers & especially those of Germany exceedingly valuable & interesting to me. May you long enjoy & use this great power for the good of us all. We shall long for the second volume but we must have patience for it is a great work that you are engaged in[.]

You sent me also the Numbers of the Bibliotheque for January February & March & there again your kindness to me is deeply manifested & with me is deeply felt: but do not trouble yourself to send me the succeeding numbers for I have the work here & see it with great interest for it is to me a channel for much matter that otherwise would escape me altogether. I wish I could send you matter oftener, but my wishes far outmeasure my ability. My portfolio contains many plans for work but I get tired with ordinary occupation & then my hands lie idle.

Your theoretical views from p. 557[2] and onwards have interested me very deeply and I am glad to place them in my mind by the side of those ideas which serve to aid discovery & development by suggesting analogies and crucial experiments and other forms of test for the views which arise in the mind as vague shadows however they may develope into brightness. I have always a great difficulty about hypotheses from the necessity one is under of holding them loosely & suspending the mental decision. I do not know whether I am right in concluding that your hypothesis supposes that there can only be a few atoms in each molecule and that these are arranged as a disc or at all events disc fashion i.e in the same plane it seems to me that if we consider a molecule in its three dimensions it will be necessary to consider the atoms as all having their axis in planes parallel to one only of these directions however numerous these atoms may be. I speak of course of those bodies which you consider as naturally magnetic p 571[3]. Perhaps when I get my head a little clearer I may be able to see more clearly the probable arrangements of many

atoms in one molecule. But for the present I must refrain from thinking about it[.]
Our united kindest remembrances
Ever My dear friend | Your faithful | M. Faraday
Proff Aug de la Rive | &c &c &c

Address: A Professor | August de la Rive | &c &c &c &c | à Geneva | Switzerland.

1. De La Rive (1854–8), 1.
2. *Ibid.*, 557–79.
3. *Ibid.*, 571.

Letter 2841
Faraday to Henry Ellis
30 May 1854
From the original in BL add MS 70843, f.121
Royal Institution | 30 May 1854
My dear Sir Henry
I am delighted at the least opportunity of shewing you my view of your continual kindness[.]
Ever Truly Yours | M. Faraday
Sir Henry Ellis | &c &c &c

Letter 2842
Faraday to John William Parker[1]
30 May 1854
From the original in Columbia University Library MS Coll Herter
R Institution | 30 May 1854
My dear Sir
I return you the revise[2]. You may do as you like about sending me the sheets when corrected. I think I should not make any further change but I am anxious that those made should be put in and that errors should not drop in in making the corrections. As the title page of each lecture will I suppose be cancelled in the volume[3] so the date may go there where you have placed it.
Lady Brodie[4] (14 Savile Row) asked me if the lecture were published separately to tell the bookseller to send her a dozen copies. I suppose it is in your way to supply her with them and will leave it in your hands[.]

Very Truly Yours | M. Faraday
J.W. Parker Esq | &c &c &c

1. John William Parker (1792–1870, DNB). Printer and publisher.
2. Faraday (1854f).
3. Anon (1855).
4. Anne Brodie, née Sellon (d.1861, age 64, *Gent.Mag.*, 1861, **11**: 218). Philanthropist and wife of Benjamin Collins Brodie whom she married in 1816 (see under his DNB entry).

Letter 2843
Josiah Latimer Clark to Faraday
31 May 1854
From Melloni (1854c), 32–3
COMPAGNIE DES TELEGRAPHES ELECTRIQUES | (fondée en 1846). | Bureau des Ingénieurs, 408 [sic], West-Strand. | Londres, 31 mai 1854.
M. Latimer Clark au professeur Faraday.
 J'ai fait quelques expériences sur les vitesses comparées des courants de différentes tensions, et je vous envoie les bandes de papier qui montrent les résultats. Je n'ai pas réussi à rendre égales les déviations du galvanomètre produites par les courants les plus intenses, c'est-à-dire ceux qui dérivent d'un grand nombre de petites lames, avec celles qui proviennent d'un petit nombre de lames de grand surface. Je fais allusion en cela à la forme d'expérience suggérée par M. Melloni[1]; mais je crois que les résultats l'intéresseront néanmoins.
 Les expériences ont été exécutées sur une longueur de 768 milles de fil métallique recouvert de gutta-percha, sur la ligne qui va de Londres à Manchester, et revient ici deux fois; avec nos piles ordinaires à sulfate de cuivre, dont les éléments ont trois pouces carrés, et avec des tensions qui ont varié de 31 couples à environ 16 fois ce nombre, soit 500 couples.

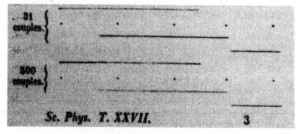

 Dans les bandes dont j'ai parlé, la ligne supérieure produite par un mécanisme local, indique le commencement de l'expérience et le temps durant lequel le courant était transmis.

La seconde ligne (de points) indique le temps en secondes et provient de la détente d'une petite roue touchée par un pendule, à chaque fois qu'il passe par le centre de l'arc d'oscillation.

La troisième ligne montre l'instant où le courant apparaît à l'extrémité que nous appelons *extrémité éloignée* (distant end) de la ligne de 768 milles de fil.

La quatrième ligne indique enfin le reste de la décharge de l'*extrémité approchée* (near end) du fil, que l'on plaçait en communication avec la terre, dès que l'on avait interrompu la communication avec la pile. Cette ligne n'a aucun intérêt pour le sujet de nos recherches actuelles. Or on voit, au moyen de la troisième ligne que, dans tous les cas, il s'est écoulé environ deux tiers de seconde avant que l'action devînt apparente à la distance de 768 milles, ce qui correspond à une vitesse d'environ 1000 milles par seconde. Cette vitesse est donc sensiblement égale, quelle que soit la tension du courant électrique.....

TRANSLATION
The Electric Telegraph Company | (Incorporated 1846) | Engineer's Office, | 448, West Strand, | London, 31 May 1854.
Mr Latimer Clark to Professor Faraday

I have done some experiments on the comparative speeds of currents of different tensions, and I am sending you the strips of paper which show the results. I have not managed to equalise the deviations of the galvanometer produced by the strongest currents, that is to say those which derive from a large number of small plates, with those which come from a small number of plates with larger surfaces. I allude to the kind of experiment proposed by Mr Melloni[1]; but I believe that the results will interest him nevertheless.

The experiments were carried out along a 768 mile length of metal wire covered with gutta percha, on the line which goes from London to Manchester and comes back here twice; with our ordinary copper sulphate piles, of which the elements are three square inches, and with tensions which varied between 31 pairs to about 16 times that amount, that is 500 pairs.

In the strips I spoke about, the top line produced by a local mechanism, indicates the beginning of the experiment and the time in which the current was transmitted.

The second line (the points) indicates the time in seconds and comes from the extension of a small wheel touched by a pendulum, each time it passed through the centre of the arc of oscillation.

The third line shows the moment at which the current appears at the end we call the distant end of the line of 768 miles of wire.

The fourth line indicates finally the remains of the discharge at the near end of the wire, which was connected to the earth, as soon as its communication with the wire was interrupted. This line has no interest for the subject of our current research. Now we see, by means of the third line that, in all of the cases, it took about two thirds of a second before the action became apparent at a distance of 768 miles, which corresponds to a speed of about 1000 miles per second. This speed is essentially equal, whatever the tension of the electric current.

1. In letter 2813.

Letter 2844
Guillaume Hähner[1] to Faraday
31 May 1854
From the original in GL MS 30108/2/64a.6
Monsieur le Professeur,

Ayant soumis à Monsieur Herbert, Secretaire de l'honorable Board of Trinity house les avantages de ma méthode d'Illumination des Phares, moyennant la lumière Electrique, et ayant été engagé par le susdit Board de le soumettre à Votre éxamen[2], j'ai l'honeur de Vous rémettre ci-joint la description du susdit procédé, accompagné d'un dessin de l'Appareil.

J'espère que cette description sera suffisante pour Vous mettre en état de porter un jugement sur ma méthode. J'ai fait construire un petit appareil dont le susdit dessin est une Copie exacte, et qui fonctionne parfaitement et serait presque suffisant pour un petit Phare. Il est à Votre disposition si Vous croyez en avoir besoin.

J'espère avoir l'honneur de Vous faire ma visite au mois de Juillet prochain; en attendant, si Vous aviez des éclaircissemens à me demander ou des Communications à me faire, veuillez me les diriger directement par lettre.

En tout cas, soit que l'invention Vous semble digne d'être prise en consideration, ou non, je compte, Monsieur, sur Votre entière discrétion, parcequ'il me serait d'un grand prejudice si l'invention fut rendu publique avant que j'ai pris les arrangements nécessaires.

Je saisis avec plaisir cette occasion pour Vous éxprimer la haute estime avec laquelle j'ai l'honeur d'être | Monsieur le Professeur | Votre très dévoué et obst. servr. | Guillme. Hähner | Consul de S.M. le Roi de Saxe[3]
Livourne en Toscane | ce 31 Mai 1854
A Monsieur le Professeur Faraday a Londres

Lighthouse Illumination by Electric Light

Till now no one has succeeded in using Electric light for lighthouses' illumination, because of the difficulty of obtaining the duration of the light; for it is well known how easily Electric light extinguishes, and what a care it affords to conserve it alighted for some time. This way of Illumination gives a light of great beauty, and is visible at a greater distance than any other artificial light.

Through the following apparatus this Illumination can be obtained in a very favorable way.

Two disks of coke, graphit, or some other similar stuff formed as toothed wheels are fixed strongly on two pegs or pins (pennies) of metal and turn on opposite sides with a very slow movement caused by a watch mechanism[.] In this way the disks move regularly and end their movement in a fixed time. - The two poles of the pile are in communication with the pegs of metal and in consequence with the two disks of coke.

In the movement of rotation the teeth of the disks find themselves in contact with each other, and in removing themselves slowly from the luminous arc which finishes suddenly when they are too far, but then two other points come in contact and renew the same phenom [sic] regularly. This way of illumination instead of being inconvenient presents the advantage of having a perfect eclipse, which is not to be obtained by the ordinary systems. The elements of the Pile could be plunged in sea water, which could serve as exciting liquid.

Explanation of the drawing:
a.b. Toothed Disks of coke
c.c.d. Wheels which move the said Disks of coke
e.e. Metallic bars which communicate with the pins or pegs, on which the disks of coke are fixed.
f.f. Wires or poles of the pile
g. Spirail [sic] which strains the wire
Wm Hähner

Endorsed by Faraday: Received June 6th 1854 | MF

TRANSLATION
Dear Professor,

Having submitted to Mr Herbert, Secretary of the Honourable Board of Trinity House the advantages of my method of lighthouse lighting using electric light, and having been asked by the said Board to submit it to your examination[2], I have the honour of enclosing the attached

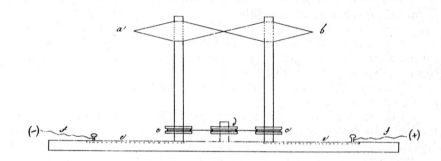

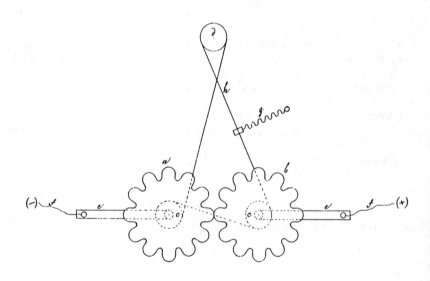

description of the said procedure, together with an illustration of the Instrument.

I hope that this description will be sufficient for you to form a judgement on my method. This is an exact copy of the apparatus that I have had built, which works perfectly and would be almost sufficient for a small Lighthouse. It is at your disposal if you think you might need it.

I hope to have the honour to visit you next July; in the meantime, if you would like any further information or would like to write to me, please do so directly by letter.

In any case, should the invention seem to you worthy of consideration or otherwise, I count, Sir, on your complete discretion, since it would prejudice me greatly if the invention became known before I had made the necessary arrangements.

I take this opportunity with pleasure to express to you the high esteem with which I have the honour of being | Professor | Your very devoted and obedient servant | Guillme Hähner | Consul of His Majesty the King of Saxony[3]
Livorno in Tuscany | this 31 May 1854
To Professor Faraday in London

1. Unidentified.
2. See letter 2820 and Trinity House By Board, 11 April 1854, GL MS 30010/38, p.493.
3. Frederick-Augustus II (1797–1854, NDB). King of Saxony, 1836–1854.

Letter 2845
Faraday to Thomas Twining
2 June 1854
From the original in RI MS F1 N1/24

Royal Institution | 2 June 1854

My dear Sir

We have just received a very beautiful mark of your kindness & my thanks for it came very well when I am about to claim more - but your earnest invitation makes us wish to accept it by coming to lunch with you at Twickenham[.] The days here are sadly embarrassed by one duty or another but on the supposition that next Tuesday[1] might suit and that we may perhaps have a little sun by the 6th of June I venture to ask whether we shall spend a few hours in your Garden on that day. But if you are going out or if on any account it will be inconvenient do not hesitate to say so. The Waterloo Station is I think the right one[.]

Ever My dear Sir | Yours Very Truly | M. Faraday
T. Twining Esq | &c &c &c

1. That is 7 June 1854.

Letter 2846
Faraday to Macedonio Melloni
2 June 1854
From Melloni (1854c), 31-2

Institution Royale, 2 juin 1854

Mon cher Melloni,
......M. Latimer Clark a fait l'expérience que vous avez demandée[1] et écrit un compte rendu des résultats; je vous envoie le tout ci-joint[2]. Il est très-difficile d'avoir les lignes complètement libres, pendant un certain intervalle de temps, en sorte que l'on a dû attendre les occasions favorables et opérer le mieux possible à plusieurs reprises, sans que j'aie pu assister aux expériences. Mais je crois que vous serez satisfait, et vous pouvez avoir pleine confiance dans l'exactitude de ses observations.
Votre affectionné | M. Faraday

TRANSLATION

Royal Institution, 2 June 1854

My dear Melloni,
Mr Latimer Clark has conducted the experiment that you asked[1] and has written an account of his results: I enclose a complete copy[2]. It is very difficult to have completely free lines, for a certain period of time, so that he had to wait for a favourable time and operate as best he could at several attempts, without my being able to be present at the experiments. But I believe you will be satisfied and you can have full confidence in the accuracy of his results.
Your affectionate | M. Faraday

1. In letter 2813. This suggests that Faraday had not yet received letter 2834 a view which is supported by the opening paragraph of letter 2870.
2. Letter 2843.

Letter 2847
Josiah Latimer Clark to Faraday
5 June 1854
From the original copy in IEE MS SC 22 / 16

448 West Strand | June 5th 1854

Dear Sir

"We have had Express (underground wire) needles demagnetised and underground Tunnel circuits fused during the last week".

The above is an extract from a letter from one of our Superintendents dated 1st June. It seems a very strange fact but I have no doubt of its truth. I attribute it to the inductive effect of lightning in the vicinity of our underground circuits. The London and North Western is the District referred to.

Yours very truly | (signed) Latimer Clark
Professor Faraday,

Letter 2848
Faraday to Josiah Latimer Clark
6 June 1854
From the original copy in IEE MS SC 22 / 16

Royal Institution | June 6th 1854

My dear Sir,

I am very greatly obliged by your most interesting note[1]. The fact is very interesting. You attribute it to the inductive action of lightning and I suppose lightning must have been the cause of it. Do you not think it possible that the lightning has acted by *shock on to the metal tubes* in the earth containing the wires; which would serve as lightning conductors to the flash, & that they may have induced at the moment on the wires within or may even have divided the flash with them.

I am, My dear Sir | Yours very truly | M. Faraday
Latimer Clark Esqre

1. Letter 2847.

Letter 2849
George Biddell Airy to Faraday
7 June 1854
From the original press copy in RGO6 / 468, f.228–9

Royal Observatory Greenwich | 1854 June 7

My dear Sir

Mr. Latimer Clark at my request has been so good as to make experiments on his long lines for the velocities of galvanic currents from batteries of different numbers of cells. This is of some importance to us longitudinarians, because we cannot always be sure that the number of cells at the opposite stations is the same, and our ultimate result would be erroneous to the amount of half the difference of the times of traverse corresponding to the two sets of battery cells.

The result, in a range from 62 to 500 cells, is very satisfactory, that there is no sensible difference of time.

I should not have thought this worth the trouble of your reading, only that pro tanto it seems to shew that there is no real difference depending on *intensity* in the application to signal communications. I do not mean that there is no difference in other effects.

The velocity in the subterraneous wires is low, somewhere about 1000 miles per second.

I am, my dear Sir, | Yours very truly | G.B. Airy
Professor Faraday | &c &c &c

Letter 2850
Faraday to George Biddell Airy
8 June 1854
From the original in RGO6 / 468, f.230

Royal Institution | 8 June 1854

My dear Sir

Many thanks for your note[1]. Curiously enough I wrote in March to Mr. L. Clarke about experiments as to any change in velocity of conduction dependant on the character of the current as to quantity & intensity; for Melloni wished for some results[2]. Mr. Clarke said he would make the experiments - and a week ago sent me all the results[3] which I sent off at once to Melloni at Naples[4][.] They are the same as those you refer to.

Did Mr. Clarke tell you the curious fact that some of the instruments connected with the *underground wires* have had their needles *reversed* & in some cases the *wire melted*[5]. I suspect that they have been struck through the earth (where near the surface) by lightning[6][.]

Ever My dear Sir | Yours Truly | M. Faraday

Geo B. Airy Esq | &c &c &c

1. Letter 2849.
2. Letter 2813.
3. Letter 2843.
4. Letter 2846.
5. See letter 2847.
6. See letter 2848.

Letter 2851
George Gabriel Stokes to Faraday
8 June 1854
From the original in RI MS Conybeare Album, f.7
Pembroke Coll. Cambridge | June 8th 1854
My dear Sir,
I much regret that I should have kept the pamphlet you sent me so long. I read it shortly after I got it, but then my lectures commenced, which for the time they lasted took up my attention; and as there were some points in it which I wished to refer to at leisure I kept it, not thinking how much time slipped away in the interval. I meant to [word illegible] one or two experiments in connexion with it, but of late I have had no sunlight. I hope that you have not in the mean time been wishing that you had it to refer to.
Believe me | Yours very truly | G.G. Stokes

Letter 2852
Edward William Cooke[1] to Faraday
8 June 1854
From the original in RI MS Conybeare Album, f.40
"The Ferns" | Kensington, | 8, June 1854
Dear Profr Faraday
According to promise I beg to remind you that the last meeting of the Kensington Conversazione takes place *this* evg Thursday at Campden House, 8 o'clk[2]. We trust you will be present, you will find much to admire in the way of art, & I think nature will not be *far behind* in attractiveness.
Your lecture which will - as usual, draw a very large audience to the theatre of the Royal Instn tomorrow evg[3] has induced me to send a hundred of my Venetian Photographs for the Library but pray use your own discretion as to admitting so many & reject any you please - my idea

was to form a continuous line but if you think a few gaps expedient Mr Vincent will attend to your wishes.

Yours very truly | E.W. Cooke
Profr Faraday F.R.S. &c &c &c

1. Edward William Cooke (1811–1880, B1). Painter.
2. On this exhibition of art see *Athenaeum*, 10 June 1854, p.720.
3. Faraday (1854e), Friday Evening Discourse of 9 June 1854.

Letter 2853
Faraday to Charles Grey
10 June 1854
From the original in RAW Vic add MS C/12

Royal Institution | 10 June 1854

Sir

Will you do me the favour to lay before His Royal Highness Prince Albert the accompanying copy of a lecture delivered in the Royal Institution. The great honor which His Royal Highness conferred on me by his presence at the time is that which encourages me to offer this testimony of duty and profound respect[.]

I have the honor to be | Sir | Your Very humble | faithful Servant | M. Faraday
The Honorable | Coll Grey | &c &c &c

1. Faraday (1854f), delivered on 6 May 1854.

Letter 2854
Faraday to Emma Maria Grove[1]
15 June 1854
From the original in RI MS G F30

Royal Institution | 15 June 1854

My dear Mrs. Grove

If we were to pop in on Monday Evening[2] about 7 or $\frac{1}{2}$ p 7 oclk in hopes of finding dinner over & you, my friend Grove, & the children (bless them) with nothing to do other than bear with our company is there any chance of our finding such to be the case?

Ever Most truly Your | faithful Servant | M. Faraday

1. Emma Maria Grove, née Powles (d.1879, age 68, GRO). Married Grove in 1837 see DNB under his entry.
2. That is 19 June 1854.

Letter 2855
James Lodge Mapple[1] to Faraday
17 June 1854
From the original in GL MS 30108/2/64a.7

June 17-54

Sir

In reply to yours I beg to state that my plan consists of 3 disks revolving on their centers two of the disks are of hard coke for the deflagration the other disk is of an insulating substance and is plac'd between the coke disks so as to keep them the proper distance apart which distance is adjustable the Coke disks are plac'd on uprights which spring together which springs press them against the insulating disk which is rotated and causes the Coke disks to rotate with it so as they are consum'd the spring'd uprights pressing them against the insulating disk keeps them the same distance apart of course they will get a little nearer as they get smaller but I think the difference will be very trifling and if any deposit takes place on either of the disks a spring scraper can be applied to remove it as the disks rotate I think the apparatus for the burning is complete and the constancy will depend on the battery but at present it needs trying[.]

Hoping you will excuse my lengthy description

I remain Sir | Your Obedient Servant | J.L. Mapple

14 Roslyn St. | Hampstead

To M. Faraday Esqr.

1. James Lodge Mapple. He took out a number of electrical patents during the 1840s and 1850s, but is otherwise unidentified.

Letter 2856
Faraday to John Welsh[1]
20 June 1854
From the original in PRO BJ1 / 9

R Institution | 20 June 1854

Dear Sir

I have tried to pay for the thermometer and have had so much trouble that I must apply to you for instructions[.] Shall I send you a check on Drummonds[2] or a Post office order? if the latter at what Post office shall it be payable. I made an order out for Mr. Ronalds[3] thinking he was Secretary at Kew & have had to go to Turnham Green & then send & write to the Chief Post office and even at this moment all is not right but I expect it will be in half an hour[.]

I am Dear Sir | Yours Very Truly | M. Faraday
John Welch [sic] Esq | &c &c &c

1. John Welsh (1824–1859, DNB). Superintendent of the Meteorological Observatory, Kew, 1852–1859.
2. Drummond. Bankers to the Royal Institution. See Bolitho and Peel (1985).
3. Francis Ronalds (1788–1873, DNB). Superintendent of the Meteorological Observatory, Kew, 1843–1852.

Letter 2857
Faraday to Stephen Henry Ward[1]
22 June 1854
From the original in WIHM MS FALF

Royal Institution | 22 June 1854

Many thanks My dear Mr. Ward for your kindness in sending me the account of the Wardian cases[2]. We are just starting for the Isle of Wight & shall take it with us. Best remembrances to all with you.

Ever Yours Truly | M. Faraday

1. Stephen Henry Ward (c1818–1880, B3). Physician and medical writer.
2. Ward, S.H. (1854), Friday Evening Discourse of 17 March 1854.

Letter 2858
John Tyndall to Faraday
25 June 1854
From a typescript in RI MS T TS, volume 12, p.4025

25th, June, 1854.

My dear Prof Faraday

I was very sorry to find on coming to the Institution on Monday last[1] that you had departed an hour before. I had been in town for two or three days previous, and once called at the Institution: But I had got wet in the country, sat in a draught on my way to town and done sundry other imprudent things: these brought on a cold and confined me to the house for two whole days. When I called at the Institution it was late and I was

ill, for otherwise I should have intruded my bodily presence upon you. I am now at work, and as usual sadly bewildered - I know nothing of magnetism - The experiments which every body seems to understand are those which puzzle me most - At least I find the accepted theories of magnetic action no refuge at present. Well patience is sure to bring something out of the bewilderment at least. Magnus informs me that he has sent you a paper on black sulphur[2] and that some specimens of the substance are also on the way to you. He obtains the black sulphur by melting the substance over and over again - perhaps 20 times - and cooling it very quickly each time. But I dare say you will be glad to forget science for the present so will not trouble you with it further.
    Kind remembrance to Mrs Faraday
    Believe me | Most sincerely Yours | John Tyndall

1. That is 19 June 1854. See letter 2854.
2. Magnus (1854).

Letter 2859
Faraday to John Tyndall
28 June 1854
From the original in APS MS

Ventnor | Isle of Wight | 28 June 1854

My dear Tyndall
    You see by the top of this letter how much habit prevails over me[1]. I have just read yours from thence[2] and yet I think myself there. However I have left its science in very good keeping & am glad to hear you are at Experiment: but how is the health? not well I fear. I wish you would get yourself strong first and work afterward:- as for the fruits I am sure they will be good, for though I despond for myself I do not for you. You are young I am old, you are fresh in thought I am exhausted in thoughts and little more than a thing of habits - but then our subjects are so glorious that to work at them rejoices & encourages the feeblest delights & enchants the strongest.
    I have not yet heard any thing from Magnus - thoughts of him always delight me - we shall look at his black sulphur together. I heard from Schoenbein the other day who tells me Liebig is full of ozone i.e of allotropic oxygen[3][.]
    Good bye for the present. We are moderately well - my wife weak.
    Ever My dear Tyndall | Yours Truly | M. Faraday

1. "Royal Institution" is crossed through above "Ventnor".

2. Letter 2858.
3. Letter 2818.

Letter 2860
Faraday to Henry Bence Jones
30 June 1854
From the original in RI MS F1 D21

South Cliff Cottage | Shanklin | Isle of Wight | 30 June 1854
My dear friend
    I do not know why I write to you but that you are a friend and are called specially to my mind by any indisposition coming over me or my wife. She has I suppose caught cold very easily and in order in part to avoid the consequences we have removed from Ventnor where we were to this place. But at Ventnor her ear plagued her & since we have been here (the second day now) a serious gathering occurred in it causing her much pain & trouble. Now I think it is going off after breaking - & she is easier but very deaf & very weak. She has brought no medicine or prescription with her hoping all from the air & place that was needful. I have an impression (and she has also) that the frequent syringing with warm water has made the ear susceptible to cold - perhaps the blistering has helped. I dare say all will come right again but the fact is I am somewhat of a coward on my wifes account and so I trouble you. If there is nothing to say do not think of writing. The place is very beautiful and the birds singing sweetly; and we are now on the top of the cliff. I am quite well and idling to perfection - it is the only thing that suits me. I forget where Lady Millicent[1] & the family are at present but I hope in such health as to cheer you. What a blessing health is? I hope you are enjoying it - and with it a cheerful active mind[.]
    Ever My dear friend | Yours faithfully | M. Faraday
Dr. B. Jones | &c &c &c

1. Millicent Bence Jones, née Acheson (d.1887, age 78, GRO). Married Bence Jones in 1842 (see under his DNB entry). She was lady by virtue of being the daughter of an Earl.

Letter 2861
John Tyndall to Faraday
30 June 1854
From a typescript in RI MS T TS, volume 12, pp.4026-8

June 30th, 1854

My dear Prof Faraday
It is just three quarters of an hour to bed time and I think you will not call me imprudent if, as a sedative to my thoughts, I employ this interval in writing to you. You have kindly asked after my health[1] - Though not robust it is still in a sound condition - My face is pale as usual, but it would be hasty to infer from this that there is any thing radically wrong with me. Dr Bence Jones once hinted to me that he thought I might be consumptive. I smiled inwardly at the time, for it brought to my mind a similar remark made by others who when they knew me better came round to the opposite opinion. The fact is, there is greater toughness in these attenuated muscles of mine than many give me credit for, and it has often been my lot in mere physical exertion to weary out men of far greater promise. I do intend however to give myself three weeks holiday. Francis and I have arranged a trip into Wales together and thence to Liverpool to the British Association. This will endow me with an amount of vigour sufficient to cope with the requirements of our next campaign. I have been for the last week endeavouring to decide a point or two in magnetism and have got myself into a labyrinth of difficulties; sometimes in the deepest intellectual darkness, relieved now and then by a gleam which cheers a man on to renewed effort. I often think that the qualities which go to constitute a good christian are those most essential to a man of science, and that above all things it is necessary for him to become as "a little child."[2] But how apt is a man to forget this docile spirit:- how apt to rise disaffected and unhappy from his task when he has failed to confirm some foregone conclusion on which he has set his heart:- And then again in the midst of all his discontent a still small voice[3] seems to reason with him and like the harp of David acting upon Saul drives away the evil spirit from his heart and makes him once more fresh and hopeful[4]. I have found myself, even recently, converted from a miserable, complaining, rebellious wretch, into a loyal and happy worker in less time than it has taken me to write this sentence. A thought has rifted and scattered the cloud of discontent as the wind disperses the mist upon the hills.

I wish I could hear you say that you were more than "moderately well" and that Mrs Faraday was more than moderately strong, and I do trust that the fresh breezes of the cozy little spot at which you now reside will increase the strength of both of you. Your last letter is to me in many respects more precious than 'much fine gold', but I have found difficulty in interpreting one phrase - ["]Though I despond as for myself" - I can only account for it on the supposition that nothing short of infinitude can satisfy the soul of man. When I look at the intellectual conquest which lies at your feet, and which, as I heard Verdet[5] say last autumn in Paris, stands unequalled since the time of Newton[6], the attitude which presents itself to my mind as the natural one is that of a warrior who, after the day's successful conflict, wipes his iron bow and looks with tranquil satisfaction upon the spoils of victory. With regard to my own feelings in the matter I can only say that the ejaculation of Baalam [sic] would express the asperation of my soul "Let me die the death of such a worker"![7]

They are still plying me with missives from the Society of Arts. Were I to undertake all they request of me my whole time would be scarcely sufficient for them. Dr Lotham[8] has written requesting me to give a lecture pointing out the connexion of physics with chemistry, on the one side, and with mathematics on the other. Of course I have been compelled to decline it. A Mr Michall[9] wrote requesting me to come and help him on Friday *and* Saturday to arrange some philosophical apparatus. He writes as if he thought he had nothing to do but to ask me; but notwithstanding this, lest they should consider my persistent refusals illnatured, I have consented to go down to them for a couple of hours tomorrow. A few days ago Dr Bence Jones came into the laboratory and told me that the time for voting the grants by the Royal Society had just arrived. I applied on Friday last for 50 or 100 pounds - On Wednesday evening I met Sabine at Col Yorke's[10] and he informed me that they had given me the 100 - It is very kind of them[11].

My candles are now burnt out and wishing you and Mrs Faraday good night, I end my scroll by subscribing myself

Most faithfully Yours | John Tyndall

1. In letter 2859.
2. See, for example, Matthew 18: 2-3.
3. 1 Kings 19: 12.
4. 1 Samuel 16: 23.
5. Marcel Emile Verdet (1824–1866, P2, 3). Professor of Physics at the Ecole Normale, Paris.
6. Tyndall, *Diary*, 5 September 1853, **5**: 251 notes this meeting with Verdet, but not the remark.
7. See Numbers 23:10.
8. Unidentified.
9. Unidentified.
10. Philip Yorke (1799–1874, *J.Chem.Soc.*, 1875, **28**: 1319). Officer in the Scots Fusilier Guards. Amateur man of science and sometime Manager of the Royal Institution.
11. See Tyndall, *Diary*, 2 July 1854, **5**: 354 and RS CM, 29 June 1854, **2**: 292.

Letter 2862
Macedonio Melloni to Faraday
1 July 1854
From the original in IEE MS SC 2

Moretta de Portici | ce 1er Juillet 1854

Cher et illustre ami

J'ai recu, il y a quelques jours, les documents relatifs aux expériences que vous avez eû la bonté de faire faire à ma requistion par l'ingenieur en chef de la compagnie anglaise des télégraphes eléctriques[1]: elles m'ont beaucoup interessé et je ne manquerai pas d'en rendre compte à mes collegues dans la premiere séance de notre académie des sciences... il se pourrait que l'exemple d'une société particulière si favorable aux recherches scientifiques induisit enfin le Ministre de l'instruction publique de Naples à fournir les petites ressources que l'Académie lui a demandées depuis longtems pour mettre en oeuvre mon projet d'expériences magnetiques autour du Vesuve..... Quoiqu'il arrive, recevez mes plus vifs remercîments pour cette nouvelle preuve de votre précieuse amitié et soyez bien convaincu qu'elle restera profondément gravée dans mon coeur!

Le nouveau modele de mon eléctrosope[2] sera bientôt fini et je ne manquerai pas de vous en envoyer de suite la description. En attendant je vais vous communiquer quelques resultats obtenus avec l'ancien modele, fort incomplet sans aucun doute, mais bien superieur, comme je vous le disais à tous les autres eléctroscopes, soit pour la sensibilité, soit pour la netteté des indications. Ces resultats me paraissent en général favorables à votre ingenieuse théorie de l'induction et de la conductibilité eléctrique; il y en a cependant quelques uns dont je ne puis encore me rendre compte et je vous serai bien obligé si vous vouliez m'éclairer là dessus. Voici les faits que je vais résumer le plus brièvement possible.

Il faut premettre [sic] d'abord qu'un corps est complètement garanti des actions d'attraction et de repulsion provenant d'une source eléctrique extérieure lorsqu'on l'abrite convenablement par un écran de métal mis en communication avec le sol. Cela posé, vous concevez qu'avec mon appareil, un écran percé, et un petit bâton de cire d'espagne eléctrisé je puis facilement étudier *la transmission* des milieux de differente nature pour *la force eléctrique rayonnante*. Or je trouve que cette force ne se propage jamais immédiatement dans les substances solides et liquides, comme quelques physiciens semblent l'admettre à l'égard du verre; mais toujours par la voie médiate, c'est-à-dire de couche en couche; ce qui rentre parfaitement dans vos idées. Cependant je ne comprends pas d'une maniere bien nette comment *la radiation eléctrique* parvenue à la surface du milieu se propage egalement dans tous les sens de sa masse, et surtout, comment *elle se renverse complètement dans certaines circonstances*

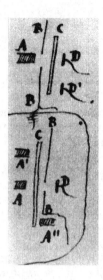

Soit A le corps eléctrisé, B l'ecran percé, C le milieu soumis au rayonnement eléctrique, D l'electroscope. L'action a lieu tout aussi bien lorsque l'eléctroscope est en face de l'ouverture que lorsqu'il se trouve placé en D'. De plus - Si l'on transporte la lame interposée C du côté de la source eléctrique, on observe encore le même effet en placant cette source en A, A' ou A".[3]

Dans ces differents cas ce n'est pas le seul mouvement de l'index eléctroscopique à l'état naturel que l'on obtient, mais aussi la qualité de l'electricité transmise, qui est toujours la même que celle de la source: en sorte que le petit baton de cire d'espagne porté en A, A' ou A" augmente toujours la divergence initiale de l'index eléctroscopique lorsque l'instrument a été d'abord eléctrisé dans le même sens que celui du corps inducteur et diminue toujours cette divergence dans le cas contraire.

Le resultats sont les mêmes, quelle que soit la proprieté isolante ou conductrice du milieu interposé et paraissent seulement un peu plus intenses pour les corps conducteurs, qui doivent être comme vous le pensez bien maintenus à l'état d'isolement. Ceux-ci peuvent même recevoir une disposition au moyen de laquelle on démontre que dans le phenomène de l'induction des substances conductrices il n'y a point ainsi qu'on le dit, de transport du fluide eléctrique de l'une à l'autre extremité du corps.

Je prends deux disques egaux, C', C" de fer blanc ou de laiton bien mince, je les reunis ensemble par un fil metallique; et, après avoir fixé l'un de ces disques, C' par exemple, contre l'ouverture centrale de mon ecran communiquant avec le sol, j'eléctrise positivement l'eléctroscope, j'approche le baton frotté de cire d'espagne d'abord en C', puis en C", et, j'observe à chaque fois le mouvement de l'index eléctroscopique. S'il est vrai, comme on l'admet dans tous les traités de physique, que dans un corps conducteur isolé soumis à l'induction, *le fluide homologue est repoussé jusqu'à l'extremité la plus eloignée*, on devrait evidemment avoir une diminution de la divergence eléctroscopique beaucoup plus grande dans le premier cas que dans le second[4]; ce qui n'est point; car on obtient un effet à peu près egal dans les deux positions du corps eléctrisé. Le transport supposé de l'une à l'autre extremité du corps métallique isolé n'existe donc pas[5].

On dirait même au premier abord, d'après l'ensemble des faits qui précèdent, que le phénomène de l'induction développe dans le corps induit une seule espèce d'electricité; ce qui serait en oppostion avec tout ce que nous savons sur cette branche de la science. Mais il est facile d'expliquer cette anomalie apparente en réflechissant que *le fluide de nom contraire à celui du corps inducteur ne peut étre que dissimulé*, et que, par consequent, il ne saurait exercer son action à travers l'epaisseur de la lame des disques ou des lames interposées, telles minces qu'elles soient; tandis que le fluide homologue est doué de tension et repandu en proportions plus ou moins grandes sur toutes les faces du conducteur, moins celle qui est en regard du corps eléctrisé.

Je me trompe beaucoup ou le theorème fondamental de l'induction eléctrique, tel qu'on le trouve ordinairement énoncé, devrait être modifié de maniere à ne pas confondre deux effets complètement distincts: l'état eléctrique pendant l'induction et après le contact et l'eloignement du corps inducteur. On connait parfaitement ce qui arrive dans ce dernier cas, et pas assez ce qui se passe dans le premier. En effet les moyens de l'analyse expérimentale sont alors soumis à une grave objection. Vous dites qu'un eléctroscope eléctrisé donne deux actions contraires lorsqu'on l'approche des deux extremités de votre cylindre métallique isolé soumis à l'induction. Mais l'instrument porté successivement à ces deux extremités est en des conditions toutes differentes: car il éprouve lui même une action inductive qui depend de sa proximité au corps inducteur - Abritez vos instruments derrière une lame métallique convenablement placée et mise en communication avec le sol; et vous arriverez comme moi, à la conclusion, *qu'une seule electricité est sensible pendant l'induc⟨tion⟩*, parcequ'elle se trouve à l'état libre, tandis que l'autre est dissimulée, et par consequent insensible. Ne serait-il pas plus convenable de presenter le theoreme de l'induction eléctrique sous la forme suivante?

Quand on approche un conducteur isolé d'un corps eléctrisé, le fluide naturel de ce conducteur est en partie décomposé: une certaine quantité du principe homologue est repoussée à l'état de tension sensible; la portion correspondante du principe contraire est attirée et dissimulée[6].

Dites-moi, de grace, votre opinion sur tout ce que je viens d'ecrire, et croyez moi pour toujours - votre très fervent admirateur et ami | Macédoine Melloni

P.S. Prenez garde de ne pas mettre une autre fois mon nom avant celui de Mr Flauti dans l'adresse de vos lettres..... autrement l'artifice ne servira à rien autre chose qu'à entraver davantage notre correspondance---

Address: Monsieur | Monsieur Michel Faraday | de la Société Royale de Londres de l'Institut | de France, de la Société Italienne des sciences &c &c &c | Londres

TRANSLATION

Moretta de Portici | this 1st July 1854

Dear and illustrious friend,

I received, a few days ago, the documents concerning the experiments that you kindly sent me through the engineer at the head of the English electric telegraph company[1]: they interested me greatly and I shall not fail to give an account of them to my colleagues at the first meeting of our académie des sciences... it could be that the example of a particular Society so favourable to scientific research will finally induce the Neapolitan Minister of Public Instruction to provide some small grants that the Académie has for a long time been requesting to set up my project of magnetic experiments around Vesuvius.....Whatever happens, please accept my warmest thanks for this new proof of your precious friendship and be convinced that it will remain deeply engraved upon my heart!

The new version of my electroscope[2] will soon be finished and I shall certainly send you a description. In the meantime I shall communicate to you a few results obtained with the old version, very incomplete without any doubt, but better as I was saying than any other electroscope, both for its sensitivity and for the clarity of its results. These results seem to me to favour, in general, your ingenious theory of induction and electric conductivity; there are, however, a few results which I do not understand and I would be most obliged to you if you could give your opinion on them. I give below the briefest possible summary.

It must first of all be stated that a body is completely guaranteed actions of attraction and repulsion coming from an exterior electric source when one screens it appropriately with a metal screen placed in communication with the ground. That having been said, you will appreciate that with my apparatus, a pierced screen, and a small stick of electrically charged Spanish wax, I can easily study *the transmission* in media of a different nature of *radiant electric force*. Now I find that this force never propagates itself immediately in solid or liquid substances, as some physicists seem to think as far as glass is concerned; but always by the mediate route, that is to say from layer to layer; which concurs entirely with your own ideas. However, I do not understand clearly how *electric radiation*, having come to the surface of a medium, propagates itself equally in every direction of its mass, and above all, how *it reverses itself completely in certain circumstances*.

Let A be the electrified body, B the pierced screen, C the medium submitted to electrical radiation, D the electroscope. The action takes place equally if the electroscope is opposite the opening or if it is placed in [position] D'. Moreover - if one puts the interposed screen C on the side of the electric source, one observes the same effect, be that source at A, A' or A" -[3]

In these different cases it is not only the movement of the electroscope indicator in its natural state which one obtains, but also the quality of the electricity that is transmitted, which is always the same as that of the source: so that the small stick of Spanish wax put at A, A' or A" increases always the initial divergence of the electroscopic index when the instrument is first electrified in the same direction as the induced body and this divergence is always diminished in the contrary case.

The results are the same whatever the isolating or conductive properties of the interposed medium and seem only a little more intense for conductive bodies, which must, as you say, be kept insulated. These can even receive a disposition by which one can show that in the phenomenon of the induction of conductive substances, there is not as has been stated, any transmission of electric fluid from one end of the body to the other. I take two equal discs C', C", of tin plate or very thin brass, I join them with a metal wire; and after having fixed one of these discs, C' for example, against the central opening of my screen that is connected to the floor, I electrify the electroscope positively, I bring the stick rubbed with Spanish wax first to C' then to C" and I observe each time a movement in the electroscope indicator. If it is true, as is said in all treatises on physics, that in an isolated conducting body submitted to induction *the equivalent fluid is pushed out to the furthest extremity*, one should evidently have a diminution of the electroscopic divergence much greater in the first case than in the second[4]; this is not the case; for one obtains a more or less equal effect in the two positions of the electrified body. The supposed movement from one end to the other of the isolated metallic body does not therefore exist[5].

One might even say, after all the preceding facts, that the phenomenon of induction developed in a body induces a single kind of electricity: which would be contrary to everything we know about this branch of science. But it is easy to explain this apparent anomaly if we reflect that *the fluid of the opposite name of the induced body can only be dissimulated* and that consequently it would not be able to exert its action through the thickness of the screen of the discs or screens that have been interposed, thin as they are; whilst the homologous fluid is endowed with tension and spread in greater or lesser proportion on all the sides of the conductor, except that which is facing the electrified body.

I wonder whether the fundamental theorem of electric induction, as it is normally worded, should not be modified in such a way as not to confuse two completely distinct effects: the electric state during induction and after the contact and the removal of the inducting body. What happens in the latter case is well known and not enough is known about the former. In fact the means of experimental analysis are then subjected to a grave objection. You say that an electrified electroscope gives two contrary actions when one approaches it with two ends of your isolated metal cylinder subjected to induction. But the instrument taken successively to these two extremities is in two totally different conditions: for it itself experiences an inductive action which depends on its proximity to the inductive body - Hide your instruments behind a suitably placed metal screen and communicating with the ground; and you will reach, as I did, the conclusion *that one kind of electricity is felt during induction*, since it is found in a free state, whilst the other is dissimulated, and consequently not felt. Would it not be better to present the theorem of electric induction in the following form?

When one approaches an isolated conductor of an electrified body, the natural fluid of this conductor is in part decomposed: a certain quantity of the homologous principal is pushed away to a state of measurable tension; a corresponding portion of the contrary principal is attracted and dissimulated[6].

Tell me, please, your opinion on what I have just written, and believe me for ever - your most fervent admirer and friend | Macedoine Melloni
P.S. Take care another time not to put my name before that of Mr Flauti in addressing your letters ..... otherwise the method will do nothing more than greatly hamper our correspondence ---

1. Josiah Latimer Clark. See letters 2843 and 2846.
2. Melloni (1854b).
3. Faraday wrote "yes" in the margin here.
4. Faraday wrote "no" here.
5. Faraday wrote "?" in the margin here.
6. Faraday wrote "yes" against this paragraph.

Letter 2863
Faraday to Henry Bence Jones
4 July 1854
From the original in RI MS F1 D22

South Cliff Cottage | Shanklin | Isle of Wight | 4 July 1854
My dear friend
Your letter was very kind and your words & advice of great value to us. The gathering in my wifes ear must have been very serious; occurring chiefly, or at first, in the left ear, - it extended or was joined in by the right ear; and though things are improving, & we believe steadily, yet much fluid or discharge comes from both. She has not been out of this house since we came into it and not much out of her room; but today appearances improve. She has taken one of the pills (on Monday[1]), has great faith in them, & takes another tonight:- Your thought even, however distant the possibility was, of coming here was very kind, but to think of such a thing would have grieved us very much. We hope you found Lady Millicent[2] well & all hearty & happy. We have a surgeon here & a good doctor & surgeon at Ventnor - but a word of yours on paper seems of far more value than a long examination & consideration by a stranger - and so I wrote - for the moral as well as the *physical* help;- and both came. The pain in the head was at first, & especially at nights, very sharp & great:- it still continues but is changed in character being more of a heavy dull throb but is very distressing at times. We have not poulticed the ears but syringed them with warm water & used the cotton as you said. I suppose we must expect these series of changes as the parts slowly recover their tone. My wife takes the quinine as you directed[.]
Ever My dear friend | Your grateful | M. Faraday
Dr. B. Jones | &c &c &c

1. That is 3 July 1854.
2. Millicent Bence Jones, née Acheson (d.1887, age 78, GRO). Married Bence Jones in 1842 (see under his DNB entry). She was lady by virtue of being the daughter of an Earl.

Letter 2864
Christian Friedrich Schoenbein to Faraday
4 July 1854
From the original in UB MS NS 414
My dear Faraday

Now-a-days people talk so much about the wonderful improvements of the ways of communication and intercourse being established between the different parts of the civilized world and to us it is a most difficult matter to send a little parcel from Bâle to London. Without that deplorable deficiency you had certainly received many weeks ago the paper enjoined[1], but I was forced to wait untill chance yielded me an opportunity to forward it to you. I should like very much you would read the memoir for it contains my views on the proximate cause not only of Electrolysis but also of what I have ventured to term Thermolysis, Photolysis, Electrosynthesis, Thermosynthesis and Photosynthesis, i.e. of chemical decompositions and compositions being effected by the agencies of electricity, light and heat. My leading idea is this, that the phenomena mentioned are due to allotropic modifications which the elementary bodies being concerned in those analytical and synthetical processes undergo, when placed under the influence of the agencies named.

HO is decomposed, because its O, on being put under the influence of the current happens to be transformed into $\overset{\circ}{O}$ (by which I mean ozonized Oxigen) which as such cannot form water with H. Oxide of Silver which I hold to be Ag$\overset{\circ}{O}$ is decomposed by heat, because this agency transforms $\overset{\circ}{O}$ into O, which cannot combine with Ag &c. &c. &c. Perhaps a friend of your's will take the trouble to translate the paper, for without reading the whole chain of my reasoning and arguments, I am afraid, you will not well understand the neological views of your friend. As to the electrosynthesis of oxigen and oxidable matters, I think I have been entirely successful in proving that it is due to the ozonisation of oxigen being effected by electrical discharge.

At this present moment I am busily engaged in researches on the desozonising influence being exerted by ponderable matters upon $\overset{\circ}{O}$ and the results already obtained leave, I think, no doubt that a number of substances enjoy conjointly with heat the power of transforming both free and latent $\overset{\circ}{O}$ into O, a fact which is interesting enough but by no means surprizing to me. Ozonized oxigen, by whatever means, electrical or chemical, it may have been generated on being put in contact with the peroxides of lead, manganese, silver, the oxides of mercury, the oxide of copper or silver and gold, the peroxide of iron &c is immediately brought back to its inactive state and the simplest way of showing this desozonizing action is as follows: Charge bottles with air being strongly ozonized by phosphorus, introduce some finely powdered peroxide of Silver, Lead, Manganese, Iron &c and shake the whole for half a minute or less and you will find that your Ozone is gone, no smell and action upon the test-paper being perceived any more. The substances just named being saturated with oxigen cannot, as oxidable matters do, take up Ozone and hence it seems to follow that in one case the disappearance of ozonized oxigen is due to its having been transformed into O, in the same way as this change of state is effected by heat.

Thenard's[2] peroxide of Hydrogen is to me $HO + \overset{\circ}{O}$ and you know well enough that the oxides, which according to my late experiments destroy the ozonized condition of oxigen, have also the power of decomposing $HO + \overset{\circ}{O}$ into HO and O.

Chlorate of potash is to my notion ozonized oxigen associated to muriate of potash, now this $\overset{\circ}{O}$ may speedily be transformed into O by the aforesaid oxides and peroxides and I find that peroxide of iron enjoys that power to a very remarkable degree, for $\frac{1}{1000}$ part of it only, being mixt with the melted salt will cause a lively disengagement of oxigen even at a temperature at which pure chlorate does not yet yield a trace of that gas. $\frac{1}{100}$ part of the peroxide named gives rise to such a violent elimination of oxigen as nearly to approach an explosion and produce an incandescence of the salt.

A small portion only of a large and intimate mixture of one part peroxide of iron and 50 parts of chlorate of potash being just heated to the point of fusion of the salt occasions such a rapid and complete decomposition of the latter that the whole mass quickly and spontaneously becomes incandescent without having time to fuse. The higher the degree of mechanical division given to the oxide employed the greater the desozonising or decomposing power of that matter. I entertain very little doubt that the same cause which acts in the peroxide of iron &c and determines the transformation of free $\overset{\circ}{O}$ into O also produces the same effect upon ozonized Oxigen being contained in the peroxide of Hydrogen Chlorate of potash &c; in other terms that the desozonisation of the oxigen of the oxy-compounds named and their decomposition are phenomena depending upon each other. It appears to me to be a very singular fact and therefore worthy of remark that the oxigen of all the oxides or peroxides which enjoy the power of desozonising free $\overset{\circ}{O}$ &c, exists either wholly or partly in the ozonized state itself. I hardly need add that what they call catalytic actions are to my opinion referable to allotropic phenomena. But of that more another time. From the preceding communications you will easily perceive that I cannot get out of the charmed circle drawn round me by that arch-conjurer called oxigen and I am afraid, so long as I can walk I shall move on that narrow ground.

I cannot conclude without expressing you my most grateful thanks for the kind letter[3], with which you favored me some weeks ago and I must tell it you over and over again that the mere sight of your hand writing gives me infinite pleasure and always conjures up the image of its author whom I revere and love more intensely than any other of my friends.

I read your remarks on the chemical effects produced by cold with the greatest interest; it is indeed a subject of research worth while to pay the greatest attention to and I very little doubt that your conjecture on the proximate cause of the recoloring of the Dahlia pigmentum is correct.

I must not omit to tell you that we have kept in readiness the numbers of the Phil. Society of Bale for the library of the Royal Institution these many months; but up to this present moment we have not yet found a convenient opportunity for sending them off and beg therefore not to be charged with carelessness.

Next month I shall take a trip to the eastern Cantons of Switzerland to attend a meeting of our Swiss Association and go perhaps for a week or so to Munich and Nuremberg. Mrs. Schoenbein intends to pass some time with their parents at Stuttgart and the girls who are at home will be placed on the heights of the Jura to inhale its bracing air, jump about like chamois on rocks and in dales, in woods and on meadows. They charge me to offer to yourself and Mrs. Faraday their kindest regards to which I join my own.

Believe me my dear Faraday | for ever | Your's C.F. Schoenbein
Bâle July 4. 1854.

1. Schoenbein (1854b).
2. Louis Jacques Thenard (1777–1857, DSB). Professor of Chemistry at Paris.
3. Letter 2832.

Letter 2865
Macedonio Melloni to Faraday
12 July 1854
From the original in IEE MS SC 2

Moretta de Portici près Naples 12 Juliet 1854
Cher et illustre ami
Dans ma dernière lettre[1] j'élevais quelques doutes à l'égard des consequences que l'on a cru pouvoir deduire jusqu'à present des expériences qui servent de base au theoreme fondamental de l'induction

électrostatique. Ces doutes ont passé dans mon esprit à l'état de certitude depuis qu'il m'a été permis de les soumettre à l'épreuve de l'analyse experimentale: et me voilà aujourd'hui bien convaincu que l'énonce du théoreme susdit doit être essentiellement modifié.

Veuillez, de grace, verifier les faits que je vais decrire, et si vous les trouvez exacts, comme je n'en doute point, ayez la bonté de les communiquer à la *Société Royale* et d'en faire inserer la traduction dans les *Transactions philosophiques*[2].

Lorsqu'on approche d'un corps électrisé A un conducteur isolé BC, le principe électrique contraire à celui de A se developpe en B, l'homologue en C. En effet si on place; d'après la méthode d'Aepinus[3] un corps metallique, isolé en contact avec l'une ou l'autre extremité du conducteur et si on l'approche ensuite d'un électroscope chargé d'une électricité connue, on obtient une action négative pour le contact B et positive pour le contact C lorsqu' A est électrisé positivement; et on a, au contraire, une action positive pour B et negative pour C dans le cas opposé où A est électrisé negativement.

Pour abréger l'expérience et la rendre peut être encore plus significative, il suffit d'avoir recours à la methode Wilke[4], qui consiste à composer le conducteur BC de deux pièces détachées que l'on separe, sans les toucher, sous l'influence électrique pour les eloigner ensuite de A et les presenter successivement à l'électroscope: car alors on trouve constamment les deux pieces électrisées en sens opposé, l'anterieure possedant toujours l'état électrique contraire à celui de A. Enfin, si on ne separe les deux pieces qu'apres l'éloignement de A on n'y observe plus aucune trace d'électricité, chacune d'elles se montrant alors à l'etat naturel: preuve qu'il n'y a eû pendant l'expérience aucune transfusion électrique de A en BC et que les phénomènes presentés par ce dernier corps proviennent uniquement de l'électricité naturelle de BC troublée dans son etat d'equilibre par la presence de A.

Le developpement des deux principes électriques dans un conducteur isolé par la simple action d'un corps électrisé placé à une certaine distance est donc un fait incontesté et incontestable.

Cependant les preuves experimentales que je viens de citer ne demontrent cette verité *qu'après l'action de A, et non pas pendant que cette action est en train de s'exercer, comme on l'admet dans tous les traités de physique*[.]

Vous pouvez vous convaincre, dit-on, de l'existence réelle des deux électricités en presence du corps inducteur, soit approchant successivement de B et de C le même électroscope électrisé, soit en suspendant le long de BC une série de pendules a fil de lin: car les signes électroscopiques sont contraires aux deux extremités du cylindre, et la pendule correspondante se mouvent en sens opposé lorsque vous en approchez un corps chargé d'une électricité connue.

Je reponds que ces expériences ne sont guère concluantes, puisque les appareils employés pour explorer l'etat électrique des deux bouts du cylindre sont soumis, eux mêmes, à l'influence de A et subissent en B une une [sic] perturbation électrique bien autrement intense que celle qu'ils eprouvent en C. Ne serait-il pas possible que le changement des actions attractives en repulsives, ou viceversa, derivât tout simplement de cette perturbation électrique *de l'analyseur* et non pas de la qualité differente des électricités qui dominent en B et en C?

Pour resoudre la question il faudrait donc trouver le moyen de soustraire les instruments à l'action perturbatrice du corps inducteur. Or ceci ne presente aucune difficulté. Prenez une lame métallique et fixez-la verticalement dans le voisinage du conducteur de la machine électrique, après l'avoir mise en communication avec le sol: Approchez du côté opposé une petite balle de moëlle de sureau suspendue à un long fil de lin: et vous pourrez tourner tant qu'il vous plaira le plateau de le machine, sans que le petit pendule dévie le moins du monde de la direction verticale. Les choses ne se passent pas tout-à-fait de même lorsque le pendule est isolé et électrisé; car alors celui ci eprouve une certaine tendance à se rapprocher de la lame; mais cette tendance dérive uniquement d'une *réaction* developpée par l'électricité du pendule, et n'a rien à faire avec la force électrique provenant de l'autre côté de l'autre côté de la lame; comme on peut s'en convaincre, soit en supprimant l'électricité du conducteur, soit en lui communiquant successivement les deux principes électriques: car dans l'un et l'autre cas l'inclinaison du pendule ne subit pas la moindre variation. Au reste l'attraction *de réaction* que la lame métalliique en communication avec le sol exerce sur le pendule électrisé diminue rapidement, comme toutes les forces de ce genre, lorsque la distance augmente; en sorte qu'elle devient sensiblement nulle à un fort petit éloignement de la lame.

Maintenant, si on tient d'une main un électroscope chargé d'une électricité connue et de l'autre une lame métallique et que l'on approche l'instrument tantôt de B et tantôt de C en le preservant soigneusement de l'influence de A au moyen de la lame maintenue à une certaine distance, on voit ces extremités du cylindre BC *exercer toutes les deux la même espece d'action électrique sur l'instrument,* C étant toutefois doué d'une action plus puissante que B.

Autrement: si on attache le long du cylindre BC la serie connue des pendules accouplés, et qu'on la soustrait à l'induction directe de A par des lames métalliques, convenablement placées, une baguette électrisée de verre, transportée successivement au dessus de chaque couple normalement à l'axe de BC et soigneusement abritée de l'action de A par une lame métallique qui communique avec le sol, *augmente ou diminue toutes les divergences des couples*, selon que A est électrisé positivement ou negativement. On peut même faire cette expérience d'une manière beaucoup plus frappante en disposant la baguette parallelement à l'axe du cylindre; car alors *toutes les divergences subissent en même temps la même phase d'augmentation ou de diminution*; ce qui dissipe *d'un seul coup de baguette*, les illusions que nous nous étions formées à l'égard des tensions électriques contraires developpés sur les parties anterieure et posterieure du corps soumis à l'induction.

En variant la forme de ce dernier corps on peut enfin rendre l'experience independante des ecrans qui servent à preserver les instruments d'analyse de l'action directe de A. Imaginons, en effet, que l'on ôte la partie cylindrique de BC moins une bande superieure assez forte pour soutenir les surfaces hémispheriques placées à ses extremités: supposons ces surfaces terminées interieurement par un plan muni d'un leger pendule à fil de lin. L'appareil étant isolé et fixé à une certaine distance de la machine électrique en activité, on voit les deux pendules s'ecarter simultanément des surfaces planes correspondantes; l'antérieur moins que le posterieur; mais *tous les deux en vertu de l électricité positive*, comme cela resulte evidemment de leur repulsion commune sous l'action électrique de la baguette de verre, portee successivement en B et en C. La même repulsion s'obtient lorsqu'on remplace l'hemisphere anterieur B par un disque très mince; ce qui prouve l'existence de l'électricité positive jusque tout près de la surface tournée vers A. Il va sans dire que si A est électrisé negativement, le sens électrique des apparences observées sen renverse, et que l'electricité negative est la seule sensible dans les diverses parties de l'appareil-

Ainsi *le cylindre BC soumis à l induction de A ne développe, a l'etat de tension apparente, que la seule électricité homologue à celle du corps inducteur. L'électricité contraire est completement dissimulée et ne devient sensible qu'après la separation et l'isolement des parties anterieurs de BC et la suppression de la force inductrice*[.]

On pourrait croire, au premier abord que l'existence de l'électricité homologue à celle du corps inducteur jusque dans la partie antérieure du corps induit est en contradiction formelle avec les expériences de Coulomb et des autres physicens qui ont trouvé cette parti électrisée en sens contraire[5]. Mais la contradiction n'est qu'apparente et s'explique naturellement par les deux phases opposées de tension insensible ou sensible que prend successivement *sur le plan d'épreuve* une des deux espèces d'électricité. En effet supposons, pour fixer les idées, que A soit positif et que le point anterieur du cylindre BC touché avec le plan d'epreuve possede une seule unité d'électricité sensible et quattre d'électricité dissimulée, qui dans ce cas sera negative. Au moment du contact, le plan d'épreuve sera électrisé positivement, puisque le seule unité electropositive possede l'etat de tension apparente. Mais lorsque ce plan, chargé de +1 d'électricité sensible et de −4 d'électricité dissimulée, s'éloigne de A pour subir l'essai de la balance de torsion, la dernière espece d'électricité acquiert, elle aussi, l'état de tension, neutralise la positive et reste en excès de trois unités. Se le point touché possedait trois unités d'électricité dissimulée et deux de sensible, le plan d'épreuve, positif pendant le contact de BC et la presence du corps A, accuserait sur la balance de torsion une electricité négative egale à une seule unité. Enfin, le plan d'épreuve serait, encore positif au moment du contact avec BC, mais ne donnerait plus à la balance de torsion aucun signe d'électricité apparente si le point touché possedait des proportions egales du principe électrique sensible et du principe électrique dissimulé.- Il est inutile de s'occuper des points placés au delà de cette limite, parceque on ne trouve plus alors dans les deux cas que le seule tension électropositive.

Tout se reduit, comme on le voit, à une lutte plus ou moins inégale des deux électricités qui donnent, tantôt un resultat et tantôt une [sic] autre, selon qu'elles se trouvent dans un état de developpement semblable on dissemblable.

Ainsi la dénomination de *point neutre*, adoptée par Coulomb pour signifier la partie du corps induit où les deux principes électriques possedent la même intensité, n'est pas, au fond, inexacte. Je crois cependant qu'elle doit être rejetée parcequ'elle tend à donner une idée fausse de la distribution de l'électricité sensible pendant le phénomène de l'induction: car alors le point en question ne se trouve pas à l'état naturel, et manifeste, au contraire, comme nous venons de le voir, une certaine tension électrique de même éspèce que celle du corps inducteur.

Il n'y a pas de doute que la principale cause de l'erreur où nous étions tous tombés jusqu'à ce jour n'ait été l'apparence trompeuse presentée par les pendules accouplés le long du cylindre métallique soumis à l'induction. En voyant les divergences de ces pendules plus fortes vers les deux bouts que dans la partie centrale du cylindre; et trouvant, d'autre côté, que les extremités de ce même cylindre donnaient des électricités differentes lorsqu'on les separait, à l'état d'isolement, dont l'action de la force inductive, on était naturellement porté à en deduire que les divergences extrémes n'étaient pas produites par le même principe.

Maintenant si vous me demandez la cause de cette singuliere disposition de l'electricité sensible dans le cylindre soumis à l'induction, je repondrai franchement que je ne saurais encore la formuler d'une maniere bien nette. Cependant l'explication qui me parait la plus plausible c'est que l'électricité homologue à celle du corps inducteur une fois developpée dans le corps induit, tend à s'y repandre d'apres les lois connues de la distribution électrique: et nous savons que dans un cylindre la tension est toujours moindre à la partie centrale qu'aux extremités. C'est vrai que l'électricité rencontre à l'extremité voisine du corps inducteur une force de repulsion plus puissante qu'à l'autre bout: aussi y a-t-il de ce côté un phénomène perturbateur que l'on supprime, je ne sais trop pourquoi, dans tous les traités de physique. Les doubles pendules s'inclinent vers A malgré l'électricité homologue dont ils sont pourvus: comme cela arrive toujours lorsqu'on met un corps mobile faiblement électrisé en presence d'un corps fixe doué d'une forte dose de la même éspèce d'électricité, et l'inclinaison des fils qui soutiennent les deux balles de sureau attachées à chaque couple dérivant de la même force attractive, produit naturellement entre les deux pendules une augmentation de divergence.

Mais en revenant à la nouvelle forme sous laquelle, je crois indispensable d'énoncer le théoreme fondamental de l'induction électrostatique, il est facile de voir qu'elle ne complique pas inutilement l'explication des faits qui en dependent: bien au contraire, elle tend à les présenter sous un point de vue unique et invariable, le seul qui soit reelement rationnel et conforme à l'observation.

Ainsi, par exemple, si les deux électricités induites se trouvaient contemporanément existantes à l'état de tension dans notre cylindre horizontal muni de pendules, comme on l'a supposé jusqu'à ce jour, elles devraient aussi exister dans le même état sur la partie métallique verticle et isolée d'un électroscope mis en presence d'un corps électrisé. Or, pourquoi en touchant la garniture superieure de l'appareil et en soustrayant ensuite l'instrument à l'action de la force inductrice, le trouvons nous électrisé en sens contraire? Evidemment parceque la seule électricité homologue était, sous l'action du corps inducteur douée de tension et mobile; tandis que l'autre était privée de tension et de mobilité. Dans le premier cas on faisait donc une supposition totalement differente de celle qu'il fallait adopter pour avoir l'explication du second. Cette contradiction manifeste n'exist plus dans le nouvel énoncé des phénomènes électriques développés par influence, où l'état different des deux électricités, que l'on imaginait pour se rendre compte de la charge inductive des électroscopes, est admis comme un fait démontré directement par l'expérience.

On pourrait citer aisément d'autres exemples analogues. On pourrait montrer surtout, comment l'énonciation exacte des états où se trouvent les deux principes électriques d'un corps isolé sous l'action de la force inductive permet de concevoir leur developpement sans avoir recours au transport de ces deux principes de l'une à l'autre extermité du corps induit.... Mais ce serait là une véritable termerité d'ecolier envers son maître.... Voila pourquoi je m'arrête tout court en me declarant comme toujours, bien sincerement

votre tout-devoué admirateur et ami | Macédoine Melloni

Address: Monsieur | Monsieur Michel Faraday | de la Société Royale des Londres, de | l'Institut de France, de la Societé Ita | lienne des Sciences, des Académies de Berlin, | Turin, Naples &c &c | Londres.

TRANSLATION

Moretta di Portici by Naples 12 July 1854

Dear and illustrious friend,

In my last letter[1] I raised some doubts regarding the consequences which have been taken to be deducible until now from the experiments which serve as the basis of the fundamental theorem of electrostatic induction. These doubts have passed in my mind to a state of certainty since I was allowed to submit them to the proof of experimental analysis: and here I am today quite convinced that the wording of the above theorem ought to be essentially modified.

If you would be so kind as to verify the facts that I am going to describe, and if you find them correct, as I have no doubt you will, please have the kindness to communicate them to the *Royal Society* and to make sure a translation is included in the *Philosophical Transactions*[2].

When one approaches a charged body A with an isolated conductor BC, the electric principle contrary to that of A is developed in B, homologous to C. In fact if one places, according to Aepinus'[3] method, an insulated metallic body in contact with one or other end of the conductor and if one then approaches an electroscope charged with a known electricity, one obtains a negative action for the contact B and a positive one for the contact C when A is charged positively; and one has, on the contrary, a positive action for B and a negative action for C in the opposite case where A is charged negatively.

To shorten the experiment and make it perhaps even more significant, it is sufficient to resort to Wilcke's[4] method, which consists of constructing the conductor BC from two detached pieces which are separated, without one touching them, under the influence of electricity, which moves them away from A and to present them in turn to the electroscope: for then one finds consistently that the two pieces are charged in opposite directions, the former always possessing the electric state contrary to A. Finally, if one does not separate the two pieces until after A is moved away, one can observe no trace of electricity, each showing themselves in their natural state: proof that during the experiment there had not been any electric transmission from A to BC and that the phenomena presented by this latter body came uniquely from the natural electricity of BC whose state of equilibrium was disturbed by the presence of A.

The development of the two electric principles in an isolated conductor by the simple action of a charged body placed at a certain distance is therefore an undisputed and indisputable fact.

However, the experimental proofs that I have just cited do not demonstrate this truth *until after the action of A, and not whilst this action is taking place, as is stated in all treatises on physics.*

One can be persuaded, it is said, of the real existence of the two electricities in the presence of a conducting body, by suspending along BC a series of pendulums of linen thread: for the electroscopic signs are contrary to the two extremities of the cylinder, and the corresponding pendulums move in the opposite direction when you approach it with a body charged with a known electricity.

In reply I would say that these experiments are not at all conclusive, since the instruments used to explore the electric state of the two ends of the cylinder are themselves subjected to the influence of A and undergo in B an electric disturbance which is of quite a different intensity than that which they experience in C. Would it not be possible that the change of attracting actions into repulsing ones, or vice versa, is derived quite simply from this electric disturbance *of the analyser* and not from the differing qualities of the electricities which are dominant in B and C?

In order to resolve this question, it is necessary, therefore, to find the means of subjecting the instruments to the disturbing action of the inducing body. Now this presents no difficulty. Take a metallic screen and fix it vertically, and in contact with the ground, close to the conductor of the electric machine: Approach from the other side with a small elderberry suspended on a long linen thread: and you can turn the plate of the machine as much as you like, without the little pendulum deviating from the vertical position at all. Things do not happen quite in the same way when the pendulum is insulated and charged; for then it experiences a certain tendency to come towards the screen; but this tendency stems uniquely from a *reaction* developed by the electricity of the pendulum, and has nothing to do with the electric force coming from the other side of the screen; as one can convince oneself, be it through suppressing the electricity of the conductor, or in communicating successively two electrical principles: for in the former case as in the latter the inclination of the pendulum is not subject to the merest variation. At least the attraction *of reaction* that the metallic screen which communicates with the ground exerts on the electrified pendulum, decreases rapidly, as do forces of this nature, when the distance increases; so that it becomes more or less nil at a very little distance from the screen.

Now, if one holds in one hand an electroscope charged with a known electricity and in the other a metallic screen which one brings towards the instrument sometimes from B and sometimes from C, taking care to keep it from the influence of A by means of the screen kept at a certain distance, one can see *both* ends of the cylinder BC *exert the same kind of electric action on the instrument*, despite C being capable of a stronger action than B.

In other words: if one attaches along a cylinder BC the known series of coupled pendulums, and if one subjects it to the direct induction of A by metal strips which have been appropriately placed, a charged glass rod, passed successively above each couple normally at the axis BC and carefully shielded from the action of A by a metal screen which communicates with the ground, *increases or diminishes all the differences of the couples*, depending on if A is charged positively or negatively. This experiment can be done in an even more striking way by placing the rod parallel to the axis of the cylinder; for then *all the differences undergo at the same time an increase or a decrease;* which dissipates in *one swipe of the rod* the impression that we had formed with regard to contrary electric tensions developed on the anterior and posterior parts of the body subjected to induction.

By varying the form of this latter body one can finally render the experiment independent of the screen which serves to preserve the instruments of analysis from the direct action of A. Let us imagine, in fact, that one takes away the cylindrical part of BC minus a top band strong enough to sustain the hemispherical surfaces placed at its ends: let us suppose that these surfaces end inside in a plane fitted with a light pendulum made of linen thread. The instrument being isolated and fixed at a certain distance from the electric machine that is active, one can see the two pendulums draw aside simultaneously from the corresponding surface planes; the anterior less than the posterior; but *both by virtue of the positive electricity*, as evidently results from their common repulsion under the action of the glass rod, taken successively to B and to C. The same repulsion is obtained when one replaces the anterior hemisphere B by a very thin disc; which proves the existence of positive electricity right up to the surface turned towards A. It goes without saying if A is charged negatively, the electric direction of the observed effects is reversed, and that negative electricity is only felt in different parts of the instrument.

Thus *the cylinder BC subjected to the induction of A develops, in the state of apparent tension, a single electricity homologous to that of the inducing body. Contrary electricity is completely dissimulated and becomes perceptible only after the separation and isolation of anterior parts of BC and the suppression of the inducing force.*

One could believe, at first sight, that the existence of electricity that is homologous to that of the inducing body right up to the anterior part of the induced body, contradicts completely the experiments of Coulomb and other physicists that have found this part to have the opposite charge[5]. But the contradiction is only an apparent one and can be explained naturally by the two opposing phases of imperceptible or perceptible tension that one of the two kinds of electricity takes in turn *on the experimental plane*. In fact, let us suppose, just to clarify our ideas, that A is positive and that the anterior part of cylinder BC touched by the experimental plane possesses a single unit of perceptible electricity and four [units] of dissimulated electricity, which in this case will be negative. At the moment of contact, the experimental plane will be charged positively, since only a single electropositive unit possesses a state of apparent tension. But when this plane charged with +1 [unit] of perceptible electricity and with –4 of dissimulated electricity is drawn away from A to undergo the test of the torsion balance, the latter type of electricity also acquires a state of tension, neutralises the positive and remains in excess by 3 units. If the point of contact possessed three units of dissimulated electricity and two of perceptible electricity, the experimental plane, positive during the contact with BC and the presence of the body A, would show on the torsion balance a negative electricity equal to one unit. Finally, the experimental plane would still be positive at the moment of the contact with BC, but would not give to the torsion balance any sign of apparent electricity if the touched point possessed equal proportions of the perceptible electric principle and of the dissimulated electric principle. - It is useless to occupy oneself with points placed beyond this limit, since one finds in the two cases only electropositive tension.

Everything is reduced, as you can see, to a more or less unequal battle between the two types of electricity, which give, sometimes one result and sometimes another, depending on if they find themselves in a state of similar or dissimilar development.

Thus the definition of *the neutral point*, adopted by Coulomb to mean that part of the induced body where the two electrical principles possess the same intensity, is not, in the final analysis, wrong. I believe however that it should be rejected because it tends to give a false idea of the distribution of perceptible electricity during the phenomenon of induction: for then the point in question is not found in its natural state and shows, on the contrary, as we have just seen a certain electrical tension of the same type as that of the inducing body.

There is no doubt that the primary cause of the mistake which we have all made up to now was the beguiling appearance presented by pendulums coupled along a metal cylinder subjected to induction. Seeing greater divergences towards the two ends than in the central part of the cylinder; and finding, on the other hand, that the ends of the same cylinder gave different electricities when they were separated to a state of insulation of which the action of the inductive force, one was naturally led to the conclusion that the divergences at the ends were not produced by the same principle.

Now if you were to ask me the cause of this singular disposition of perceptible electricity in a cylinder subjected to induction, I would reply, frankly, that I would not yet be able to formulate it in a clear way. However, the explanation that seems to me to be the most plausible is that electricity that is homologous to that of the inducing body, once developed in the induced body tends to spread according to the known laws of electrical distribution: and we know that in a cylinder the tension is always smaller in the central part than at the ends. It is true that electricity meets at the end closest to the inducing body, a repulsive force greater than at the other end: also there is on this side a perturbing phenomenon that has been suppressed, I am not quite sure why, in all the treatises on physics. The double pendulums incline towards A despite being charged with the same electricity, as if that always happens when one meets a mobile weakly charged body in the presence of a fixed body with a strong charge of the same electricity and the inclination of the wires which support the two balls of elderberry attach to each couple come from the same attractive force produce naturally between the two pendulums an increase in divergence.

But coming back to the new form under which I believe it indispensable to state the fundamental theorem of electrostatic induction, it is easy to see that it does not unnecessarily complicate the explanation of facts which depend on it. On the contrary it tends to present them under a unique and invariable point of view. The only one which is really rational and in accordance with observation.

Thus, for example, if the two induced electricities are at the same time in a state of tension in our horizontal cylinder equipped with pendulums, as has been supposed up to now they should also exist in the same state on the vertical and insulated metallic part of an electroscope put in the presence of a charged body. Now, why when we touch the upper casing of the apparatus and when we subject the instrument to the action of the inducing force, do we then find it charged in the opposite way? Evidently because only the same electricity was under the action of the inducing body was capable of tension and was mobile; whilst the other was incapable of tension and mobility. In the first case one made an assumption which was totally different to that which one should have adopted to have the explanation for the second. This manifest contradiction no longer exists in the new wording of the electric phenomena developed by influence where the different state of the two electricities which one had imagined in order to be aware of the inductive charge of electroscopes is admitted as a fact demonstrated directly by experiment.

One could easily cite other analogous examples. One could show above all how the exact enunciation of the states in which are found the two electric principles of an isolated body under the action of an inductive force permits to conceive their development without having to have recourse to the transport of the two principles from one end of the induced body to the other.... But that would be truly reckless of the pupil towards his master.... That is why I am stopping short and declaring myself as always most sincerely

your most devoted admirer and friend | Macedoine Melloni

1. Letter 2862.
2. See letter 2870.
3. Franz Ulrich Theodosius Aepinus (1724–1802, DSB). German electrical writer.
4. Johan Carl Wilcke (1732–1796, DSB). Electrical researcher.
5. Coulomb (1787).

Letter 2866
John William Ware Tyndale[1] to Faraday
13 July 1854
From the original in GL MS 30108/2/64.8

The Electric Power Light & Color Company, | 31, Pall Mall, | London July 13th 1854

Sir,

I beg to acknowledge the receipt of your letter, relative to the trial of the Electric Light at the Trinity House[2]; and I am requested by Dr. Watson to say that he will write to you in the course of a day or two, on the subject[3].

I am Sir | Your obedient Servant | J. W. Ware Tyndale | Secretary

M. Faraday Esq | &c &c

1. John William Ware Tyndale (1811–1897, B6). Barrister.
2. See Trinity House By Board, 20 June 1854, GL MS 30010/38, pp.601–2.
3. See letter 2868.

Letter 2867
Faraday to William Reynolds[1]
15 July 1854
From the original in SI D MS 554A

Royal Institution | 15 July 1854

My dear Sir

Your letter is doubly kind - kind on your part and kind on the part of Mr. Rathbone². I hasten to express my very earnest thanks to both for the unexpected proof of such good will towards me. I regret to say that I shall not be able to take advantage of the hospitality offered me but the uncertainty of my visit to Liverpool³ is very great and if as I desire I do get there it will only be for two or three days and it will be necessary for me to be in the town. My fear is that I shall not be there at all: for my dear wifes health must be my first care and there are some other influential points besides that which are against it. If there I hope that Mr & Mrs Rathbone⁴ will allow me to express my thanks personally for their very unexpected & great kindness[.]

Present my respects to Mrs. Reynolds⁵ & tell her I am greatly delighted to find that the dumpling nuts have turned up again[.]

I am My dear Sir | Your Very Obliged Servant | M. Faraday
W. Reynolds Esq | &c &c &c

1. William Reynolds (1803–1877, Greg (1905), 204). Liverpool merchant.
2. Richard Rathbone (1788–1860, Marriner (1961), 231). Liverpool merchant.
3. For the meeting of the British Association.
4. Hannah Mary Rathbone, née Reynolds (1798–1878, DNB). Writer and historian. Married 1817.
5. Hannah Mary Reynolds, née Rathbone (1791–1865, Greg (1905), 205). Married William Reynolds in 1831.

Letter 2868
Joseph John William Watson to Faraday
17 July 1854
From the original in GL MS 30108/2/64.10
Electric Power Light and Colour Company's | Works, | Frogmore Lane | Wandsworth | July 17 1854

Sir,

In answer to your enquiries of Mr Tyndale¹, the secretary to the Company, relatin[g] to the forthcoming trials of my invention at the Trinity House I wish to inform you that all apparatus, materials &c will be delivered at Tower Hill during this week and that everything will be in readiness for experiment by Saturday next (22nd) or Monday morning 9 oC.

In regard to the height of the lamp from the ground that will depend entirely on the position in which you intend to place it - the elevation of the lamp itself when arranged for burning is 4ft. 6in. but this is not 'en masse' as I measure from the base of the lamp to the top of the rod carrying the upper-electrode: this rod of course could project through a very small apperture in any illuminating apparatus. As, however, there may probably be some questions you would like to ask me before making your photometric arrangements I will do myself the pleasure of waiting on you at 10 ½ oC: on Saturday morning at the Institution when the plans for Monday can be decided on[2][.]

I am | Sir | Yours very faithfully &c obediently | Joseph J. W. Watson

M. Faraday Esq D.C.L | &c &c

1. John William Ware Tyndale (1811–1897, B6). Barrister.
2. There is a draft of this letter in GL MS 30108/2/64.9.

Letter 2869
François Napoleon Marie Moigno[1] to Faraday
17 July 1854[2]
From the original in RI MS Conybeare Album, f.75

Monsieur

en Venant à Londres je me faisais une fête de Vous voir et de Vous présenter mes humbles hommages. Venu dès le premier jour à l'institution royale j'ai appris avec une très grande douleur que la mauvaise Santé de Mme Faraday vous avait forcé de partir pour l'isle de Wight. Cà été pour moi et pour mon compagnon de voyage, Mr Duboscq[3], qui aurait fait avec tant de bonheur Devant Vous Ses grandes expériences d'optique avec la lumière électrique[4], un désappointement cruel.

Si par hazard Vous veniez à Londres dans le courant de cette Semaine, soyez assez bon, je vous en conjure pour nous prevenir de l' heure à la quelle nous pourrons vous remontrer. M. Duboscq donne une nouvelle Séance à polytechnique institution Vendredi[5] prochain à 7 heures du soir, devant un grand monde de Savans de Londres.

Je suis dans les Sentimens respectueses de la consideration la plus distinguée et dans l'espoir du prochain rétablissement de la Santé de Mme Faraday

Votre très humble Serviteur | l'abbé F. Moigno
17 juillet. | Hotel de Provence | Leicester Square.

TRANSLATION

Sir,

In coming to London, I was hoping to give myself the pleasure of seeing you and of presenting to you my humble homage. Having come on the first day to the Royal Institution, I learned with great sadness that Mrs

Faraday's poor health had forced you to leave for the Isle of Wight. This was for me and for my travelling companion, Mr Duboscq[3], who was hoping so much to have the pleasure of demonstrating to you his optical experiments with electric light[4], a cruel disappointment.

If by any chance you come to London in the course of this week, please be so good, I beg you, to let us know of a time when we can meet you. Mr Duboscq is giving a new lecture at the polytechnic institution next Friday[5] at 7 o'clock in the evening, before a large number of London savants.

I remain in the respectful sentiments of the most distinguished consideration and in the hope of the speedy recovery of the health of Mrs Faraday

Your most humble Servant | l'abbé F. Moigno
17 July. | Hotel de Provence | Leicester Square.

1. François Napoleon Marie Moigno (1804–1884, NBU). French scientific writer.
2. Dated on the basis of the reference to Sarah Faraday's health and their visit to the Isle of Wight. See letters 2859, 2860.
3. Jules Duboscq (1817–1886, P3). Scientific instrument maker in Paris.
4. For a description of Duboscq's apparatus see Anon (1854b), 94–7.
5. That is 21 July 1854. See *Athenaeum*, 15 July 1854, p.882.

Letter 2870[1]
Faraday to Macedonio Melloni
31 July 1854
From Guareschi (1909), 36–40

Royal Institutions, London, 31 july 1854.

My dear Melloni,

I have three letters of yours unanswered the two last of the dates of the 1st and the 12th instant[2]. I have been unable to answer them before, because the beginning of the month I was in the country, and I returned from it only to place myself under the surgeon's hands: and since then have been unable to write or work: I am now getting better but am still in his care.

When I received your last letter, I had no need to repeat your experiments; for they were all perfectly familiar to me, as far as I could judge by your description, and are necessary consequences of the theory of static induction, which I published in the Philosophical Transactions 17 years ago. Still I should have sent your letter to me to the Royal Society as you requested, if it had been open: but its sittings were closed on the 15 june, a month before your letter was written, so that I was unable to fulfil your desires in that respect.

Your difficulties present no difficulties to me, neither do I remember clearly the error or illusion into which we have all fallen and which you say is continued[3] in all the books. The fact is I have interpreted induction according to my own views in the last 18 years and have not carefully analyzed the words of recent treatises, but I do not remember that the statement of Pouillet[4] or De La Rive contradict my notions. As my views though given at length in the Philosoph. Trans. in the years 1837 and 1838 in the series XI, XII, XIII and XIV[5] of my Experimental Researches, have not been published either in French or Italian, they probably have never come under your notice: I endeavoured to send English copies of them at the time, through the Royal Society, but very likely they never reached you: so I am about to give you a brief summary of them, referring to the members[6] of the paragraphs in the Exp. Researches, that you may, if you desire it, look at the original matter.

But first let me remind you, that I do not as yet know the nature of your Electrometer and therefore may have misunderstood your statement altogether; though as regards the simple results with (fig. 1), shaded or not.[7] I do not think that likely, for they are precisely the same as my own old ones.

I profess to know nothing of the existence of either one or two electric fluids, or of the nature of the electric power, exerted either in a *P, or an N direction* (1298 note[8] 1667[9]), but I do not think that our ignorance of the essential nature of the electric force offers any difficulties in the consideration of the nature of induction, conduction, etc., provided we do not travel beyond *facts* and the *laws* which govern them.

In *induction*, it is only the *surface* of the conductor which is finally affected, and not its internal parts: and that whether the conductor be insulated or connected with the earth (1220, 1221, 1295, 1301[10]). If it be *uninsulated*, only that part of its surface is finally affected, on which lines of induction force, proceeding from the ex[c]ited or *inductric* body A, abut and terminate: if it be *insulated*, then the parts of the surface, from which an exact equivalents of new lines of force originate and proceed outwards from the compound system (fig. 2) towards neighbouring conductors, are affected also: the first surface is B, the second surface C and the neighbouring parts, and between them there is a part or zone of various from[11] according to circumstances, in the neutral or normal conductor.

The induction is limited by the induction *surfaces* (1231, 1297[12], 1361, 1372[13], 1483[14], etc.). The one which is primarily charged I have distinguished as the *inductric*, the other as the *inductors*.

The lines of induction force, used merely as representations of the disposition of the electric force, are described (1231, 1304[15], 1441, 1450[16]): they commence at one and terminate at the other, of the inducting surfaces. If the inductric body A be an ex[c]ited insulator, as a rod of shell lac, they then commence at the ex[c]ited particles.

Induction is not sustained *through* the body of the thinnest conductor. Theoretically, it occurs at the first instant: but conductors discharges its state within, and it is the surfaces only of the conductor which remains finally affected. An *uninsulated* gold leaf in a frame or ring, may have its opposite sides raised by induction to the highest *opposite or like* states, without the slightest interference of one side with the other.

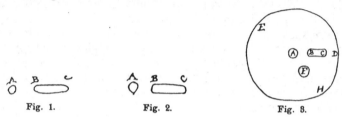

Fig. 1. Fig. 2. Fig. 3.

The lines of induction force across the dielectric or insulating medium may be curved (1215, 1219, 1221, 1224, 1230[17], 1374, 1449[18], 1614[19]. See also figures 7, 8, 9, 10, 11, 12). In experiments in open spaces they are almost universally curved.

Induction lines have a lateral relation (1224, 1295[20], 1449[21]) quite analogous to that of the lines of current force, when the piles of a voltaic battery are dipped apart from each other into a large mass of water or dilute solutions; and which have been so well illustrated by Nobili[22] in his metallochromic results[23].

When a charged sphere A (fig. 3) is in the center of a much larger sphere of conducting matter, the lines of force proceed as radii from A to every part of the outer sphere: the sum of force on the surface of A and the sum of contrary force on the inner surface of the surrounding sphere, are exactly equal to each other.

When an insulated conductor B, C, is introduced, then a certain amount of the lines of force from A terminate on B, produces an opposite state there, but an equal amount of force or of lines, originate about C and terminate on D; or in other words the lines of force which would have been passed across the space BC, if the conductor BC were away, have, through the conduction of the particles of BC, been replaced by the equivalents of contrary forces at the respective surfaces of B and C; at the same time, as the resistance or tension set up in induction (1368, 1370[24], etc.), is removed as regards the space BC, by the conduction; so, more electricity must induce from A towards B than in other directions, as towards E; and more inductions action is induced on D than elsewhere as at E. But though D differs thus from the other parts of the inner surface of the surrounding chamber, and equal to the amount of force existed[25] in *all* directions by A. I need not refer particularly to irregular and mixed cases, as the walls of a room (1434[26]) or more complicated results (1337[27], 1566[28],

1679^{29}, etc.): the principles are the same and the amount of action always definite.

If another conductor, either insulated or connected with the earth, be approached to the former conductor, then lines of induction force are transferred to it from the former induction bodies (1225^{30}, 1449^{31}). Thus, if F be such a body, then electric force or lines, which before proceeded towards B or towards H, are transferred to F; and if F be insulated, with the same development of contrary force as before. If F be *uninsulated* and large, the part behind it, at H, may even receive no charge, but be in the natural state: and the sum of power upon the inner surface of the envellope H, D, E will be less than the sum of contrary power on A, by so much as is disposed of on the surface of an uninsulated F.

I know *no* distinction between *free* and *dissimulated* electricity (1684^{32}). Both are cases of induction and change by induction. If the electricity between A and E is not dissimulated, neither is that between A and B, or between A and F. It might as well be said that the lightning which falls upon and kills a man is free to him, but is dissimulated to one who is a few miles distant: the difference is merely one of circumstance not of principle. Whether the inductric body induces upon me or upon another person, makes no difference in the action.

If the sum of power be 20, as much as 15 may be towards me and only 5 to him: but the land of action is alike in both cases and the *sum* of the power remains the same.

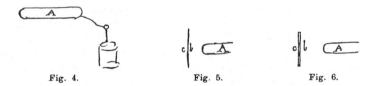

Fig. 4.   Fig. 5.   Fig. 6.

All *charge is induction*: all *induction is charge* (1177, 1178, 130^{33}, 1682^{34}). A short but imperfect summary is contained in the paragraphs from 1667 to 1684^{35}. You will find also in the Philosophical Magazine, 1843, vol.XXII, p. 200^{36}, in vol.II of the 8th Edition [sic] of my Experimental Researches, p. 279^{37}, some demonstrative experiments on static induction. I have been accustomed for years past to shew very many forms of experiments founded on these principles; I will describe one which seems to embrace many of the results you describe to me. A conductor A is sustained in a charged state by connexion with a (fig. 4) Leyden- jar: an insulated brass plate is brought near it, and then the surfaces examined by a Coulombs proof plane or carrier^{38} (which I hold to be unexceptionable when properly employed) and the surface of b seems to be contrary to A (fig. 5), whilst the surface of c is like that of a. The two plates close together and connected (fig. 6), replace the one plate a the surfaces b and c found like the former surfaces. Then the plates are opened about an inch being still connected by a brass rod on (fig. 7) which they hang, b and c are found as before; but within at d there is no electricity of either kind on either surface.

All the time there is electricity like that of A to be found at the edges of the plates, provided irregular lateral inductions be prevented. Then the plates are more opened out, still the electricity at (fig. 8) b and c is the same, but that at the edges of the near plate b, begins to lose its A character and either be neutral or assume that of b. Being more opened out, the electricity of b creeps round the edge of the plate (fig. 9), approaching nearer to the middle at the plate b is smaller, or the sign of A larger, or its power more intense; and the electricity of c also becomes more extended over the surface a round the edges of that plate, according to its size and position in relation to surrounding bodies. If an uninsulated conductor be brought towards any part of the compound system of plates a rod, as at 1, 2, 3, 4, it immediately renders these places electric like A, either causing or exalting their state, whilst it acquires through the induction, a state the contrary of A or like that of b. If ex[c]ited shell lac be brought near 1, 2, 3, 4 it immediately exerts its induction action upon the plates a, in addition to the action of A. All these and a thousand others are the simplest possible results of the theory.

You speak of screening the *"pendules accouple"* from the action of A whilst you examine them by an ex[c]ited electric (rod?). But in that case the pendules give no indication of the state of the part to which they are attached. They do not receive their final state by *conduction* from the part they are fixed to, but only by *induction* as a part of the conducting mass BC: if they are *expand* to the inductive force of A they will acquire the opposite state; if they are *perfectly screened* they will be neutral, or if they are so near so expand to surrounding conductors as to be in a position to *carry on* the forces, they will assume the c state, which is the same as the state of A.

Fig. 7. Fig. 8. Fig. 9.

I have often shewn my audience this condition of the pendulous balls[39], by placing the cylinder and its balls well insulated, in different positions as respects the *inductric ball* A: thus (fig. 10), when in this position, B and its balls will be in the reverse state to A, and c and its ball in the same state as A: but then (fig. 11) held it by an insulating handle thus and through B still acquires a state contrary to A, yet the balls attached to it by conducting matters shew the *other* state, or that of A and c.

Again, if the ball A and cylinder BC (fig. 12) retain their position, it is very easy to have the balls hanging to B, in the state of B, or by approaching an uninsulated ball or screen, either at 1 or 2 or 3, to make them assume the contrary state on that of C, or by adjustment of distance to be perfectly indifferent.

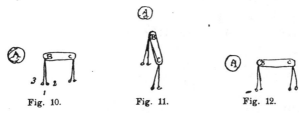

Fig. 10. Fig. 11. Fig. 12.

Trusting to the truth of the principle I have described I am accustomed to use wire gauze instead of a continuous metal plate, for screens and other apparatus, which I want my audience to see through: and I have plates like those described above constructed of such wire gauze. Through the openings are three or four times the diameter of the wire, yet no electricity of induction can pass through them, and a plate of such wire gauze is as impervious to conduction as a plate of metal.

I make a cylinder of such wire gauze part of the conductor of the machine, but the most delicate tests inside gave no indication of Electricity there. On the lecture table, I threw a net of common twine over my gold leaf electrometers connecting it well with the ground, and they are rendered perfectly safe from charge of the machine, which by induction would *destroy them at once* if not so guarded.

All of which, illustrates the powerful effect of screens in *static inductive action*. I have published no account of these things because they are simple consequences of my theory: but De La Rive who happened to see them once in the lecture room here, gave a brief account of them in the Geneva Journal[40].

And now, my dear friend, I will relieve you from a tiresome lecture. You speak of the over universal in books, and therefore I may say, that *Daniell* in his Introduction to Chemical Philosophy[41] adopts my views and therefore *as I suppose* is not in error.

That Harris I believe agrees with me; that as far as I am aware De La Rive does not put forth any error of the kind you refer to[42] in his recent work on Electricity[43] - or Pouillet - except that he speaks of dissimulated electricity[44] but by that means electricity of induction. But I will not tease you more.

Ever My dear Friend | Your affectionate servant | M. Faraday
A M | Macedoine Melloni

1. Faraday's notes for this letter are in IEE MS SC 3.
2. Letters 2834, 2862, 2865.
3. "contained" probably intended.
4. Claude-Servais-Mathias Pouillet (1790–1868, DSB). Professor of Physics in Paris.
5. Faraday (1838a, b, c, d), ERE11, 12, 13 and 14.
6. "numbers" probably intended.
7. "," probably intended.
8. Faraday (1838a), ERE11, 1298.
9. Faraday (1838d), ERE14, 1667.
10. Faraday (1838a), ERE11, 1220, 1221, 1295, 1301.
11. "form" probably intended.
12. Faraday (1838a), ERE11, 1297.
13. Faraday (1838b), ERE12, 1361, 1372.
14. Faraday (1838c), ERE13, 1483 where Faraday introduced the term inductric.
15. Faraday (1838a), ERE11, 1231, 1304.
16. Faraday (1838b), ERE12, 1441, 1450.
17. Faraday (1838a), ERE11, 1215, 1219, 1221, 1224, 1230.
18. Faraday (1838b), ERE12, 1374, 1449.
19. Faraday (1838c), ERE13, 1614.
20. Faraday (1838a), ERE11, 1224, 1295.
21. Faraday (1838b), ERE12, 1449.
22. Leopoldo Nobili (1784–1835, DSB). Professor of Physics in Florence.
23. Nobili (1830).
24. Faraday (1838b), ERE12, 1368, 1370.
25. "excited" probably intended.
26. Faraday (1838b), ERE12, 1434.
27. Faraday (1838b), ERE12, 1337.
28. Faraday (1838c), ERE13, 1566.
29. Faraday (1838d), ERE14, 1679.
30. Faraday (1838a), ERE11, 1225.
31. Faraday (1838b), ERE12, 1449.
32. Faraday (1838d), ERE14, 1684.
33. Faraday (1838a), ERE11, 1177, 1178. "130" was probably intended as another paragraph in series 11.
34. Faraday (1838d), ERE14, 1682.
35. Faraday (1838d), ERE14, 1667–84.
36. Faraday (1843).
37. Faraday (1844b), 279–84.
38. Coulomb (1787).
39. For example see Faraday's notes of his fourth lecture in a course of six on static electricity delivered on 30 April 1853. RI MS F4 J15, f.10–13.
40. See *Arch. Elec.*, 1843, **3**: 645–9.
41. Daniell (1839, 1843).
42. Faraday's inference from Melloni's remarks in letter 2865.
43. De La Rive (1853–8), **1**: 82–5.
44. Pouillet (1853), **1**: 486–9.

Letter 2871
Faraday to Thomas Byam Martin
7 August 1854
From the original in BL add MS 41370, f.333-6
[Royal Institution embossed letterhead] | Royal Institution | 7 Aug 1854
My dear Sir Byam
    I send you herewith the best answer I can make to your enquiries[1][.]
    Ever Your faithful Obedient Servant | M. Faraday

Observations &c.
                                                Royal Institution 7 August 1854.
    Very few of the questions are so put that I, in reference to their chemical or physical character, can give any consistent or distinct answer to them. The proposition is correct in theory, *i.e.* dense smoke will hide objects, and burning sulphur will yield fumes that are intolerable, and able to render men involved in them incapable of action, or even to kill them: but whether the proposition is *practicable* on the scale proposed and required, is a point so little illustrated by any experience, or by any facts that can be made to bear upon it, that for my own part I am unable to form a judgment. I have been on the crater of Vesuvius[2] and to leeward of the mouth; and have seen the vapours (which are very deleterious) pass up over my head and go off down the wind in a long and not rapidly expanding stream. I have, by changes in the wind, been involved in the vapours, and have managed with a handkerchief to the mouth and by running, to get out of their way. I should hesitate in concluding that ten or twenty vessels could give a body of smoke, the columns of which, at a mile to leeward, would coincide and form an impervious band to vision a mile broad; but I have no means of judging, for I know of no sufficient facts that can be of use as illustrations of the proposed applications.
    In reference to the burning of sulphur and formation of sulphurous acid, I may remark, that, as 400 tons of sulphur have been spoken of, perhaps the following considerations may help to give some general ideas, in the present state of the proposition, as to the probable effect of its fumes. If a ship charged with sulphur were burning in a current of air, a continuous stream of sulphurous acid fumes, mingled with air, would pass off from it. This stream, being heavier than air, would descend and move along over the surface of the water; and, I expect, would sink perpendicularly and expand laterally, so as to form a low broad stream. Its noxious height would probably soon be less than 15, or perhaps even 10, feet, (but I cannot pretend to more than a guess) and its width by degrees more and more. The water over which it would have to move, would tend continually to take part of the noxious vapour out of it. Now 400 tons of sulphur would require 400 tons of oxygen; and that it would find in about

1740 tons of air. Supposing that this product were mixed with ten times its bulk of unaltered air, it would make near upon 20000 tons of a very bad mixture; and one, which if a man were immersed in it for a short time, would cause death. Supposing that the 20000 tons of mixed deleterious air were converted into a regular stream, 30 feet high and 300 feet wide, then it would be about 6500 feet or a mile & a quarter long. Such is a representative result for 400 tons of sulphur, and hence an idea may be formed of the time during which with a given velocity of wind, the places involved in the stream may remain subject to its effects.

In respect of the seven questions[3], there is scarcely a point in them to which I am able to give an answer of any value.

As to 1^4.; I suspect much larger quantities of matter will be required than is supposed.- I do not imagine that if burnt in heaps coals would burn fast enough to give the smoke required.

2^5. The data are wanting.

3^6. I suspect the upper part of high buildings would frequently be free from the sulphurous vapours; and that sets of eddies of fresh air from above would occur behind.

4^7

5^8

6^9 The lateral extent at the distance of a mile very doubtful - would need proof[.]

7^{10}.

The proposition is, as I have said, correct in theory, but in its results must depend entirely on practical points. These are so untried and unknown, and there are so few general facts bearing on the subject, that I have the utmost difficulty in speaking at all to the matter. These circumstances must plead my excuse for the very meagre character of the only observations which I am at present able to offer. All I need add is, that if the project were known or anticipated, it would not be difficult for the attacked party to provide respirators, which would enable the men, in a very great degree or even altogether, to resist a temporary invasion of an atmosphere such as that described.

M. Faraday

Endorsement: Paper from Professor Faraday in answer to questions by the comtt[11].

1. This concerned the plan put forward by the retired Admiral Thomas Cochrane, 10th Earl of Dundonald (1775–1860, DNB) to attack Cronstadt using burning sulphur ships. On this proposal and its background see Lloyd, C. (1946).
2. In May 1814. See Bowers and Symons (1991), 107–10.
3. These "Questions sent to Mr Faraday" are noted in PRO ADM1 / 5632. The note to each of Faraday's answers (or lack thereof) gives the text of the question.

4. "1. Can it be shewn by any *proof*, the different requirements for success, the quantity of matter to be ignited the distance at which it will be of avail, & the amount of wind to render it effective; must these be nicely adjusted or each of them admit of considerable latitude?"
5. "2. What is the amount of effect in intensity anticipated on the individual, & if not totally destructive, to what period of time would it paralyse them."
6. "3. What proof is there that supposing the vapours to be intolerable along the surface over which the wind carries them, that under cover of parapets or by closing the windows or shutters, wh probably exist at the embrasures of the casements, this same vapour would penetrate & extend with sufficient intensity?"
7. "4. If the smoke is to conceal the ships from the viewing of the Batteries, how are the ships & smoke vessels themselves to approach, by, probably, *an intricate passage*, through the same smoke?"
8. "5. Where the batteries are dispersed as at Cronstadt, there must be separate smoke vessels, sufficient for each, & as their guns are in complicated lines & distances, mutually flanking each other, wd it not be a matter of difficulty to obtain a simultaneous effect on each, which wd be very necessary?"
9. "6. The extent laterally that wd be covered with effect by the vapour from each vessel would need proof."
10. "7. How are the smoke or vapor vessels to be brought into position with sufficient rapidity?
During this operation are they to lay to or anchor? or to move on? if the first how are our boats &c to pass through the smoke & if the last how are these vapour vessels themselves to direct their course."
11. This was the secret Admiralty committee established to consider Dundonald's plans. Its members were Martin, William Parker (1781–1866, DNB, Commander in Chief Devonport, 1854–1857), Maurice Frederick Fitzhardinge Berkeley (1788–1867, DNB, a Lord of the Admiralty, 1846–1857) and John Fox Burgoyne (1782–1871, DNB, Inspector General of Fortifications, 1845–1868). They submitted their report (which is in PRO ADM1 / 5632) to James Robert George Graham (1792–1861, DNB, First Lord of the Admiralty, 1852–1855) on 9 August 1854. This report, which, apart from the first sentence, quoted Faraday's report in full, recommended that Dundonald's scheme should not be put into effect. Graham accepted this recommendation.

Letter 2872
William Frederick Pollock to Faraday
7 August 1854
From the original in RI MS F1 I52a

59 Montague Square | 7th August 1854

My dear Faraday

I enclose a note which I have been requested to forward to you, by my friend F. Haywood[1] of Liverpool. He is one of the leading people in the place, and a most accomplished person. His house is a little out of Liverpool, and both my wife & myself have passed many very agreeable days there on various occasions - so that I really hope you may go there, if you are able to attend the coming meeting of the British Association.

With Mrs Pollock's regards to Mrs Faraday (in which I beg to join) and to yourself

Believe me | Yours very truly | W.F. Pollock

1. Francis Haywood (1796–1858, B5). Translator of German texts into English.

Letter 2873
Faraday to William Frederick Pollock
8 August 1854
From the original in the possession of Roy Deeley
[Royal Institution embossed letterhead] | Royal Institution | 8 Aug 1854
My dear Pollock
    You are very kind and so is Mr Haywood[1] to whom I have just written and I can hardly be sufficiently grateful for the present invitation and for others I have received from Liverpool[2]. I have told Mr Haywood that if I am able to see Liverpool it will only be for a day or two & that I must then go to a Hotel. I hardly know what I shall be able to do for what with my desire to work at this time of the year and the forbidding instructions of Dr B. Jones (under whose hands & Sir B. Brodies I have just been) I cannot tell what will happen[.] Many thanks for your great kindness[.] If I go to Liverpool I shall most surely see Mr Haywood.
    Our kindest regards to Mrs Pollock[.]
    Ever Truly Yours | M. Faraday
W.F. Pollock Esq | &c &c &c

Endorsed by Pollock: Relating to the meeting of the British Association at Liverpool in 1854, when Frank Haywood of Edge Lane Hall, wished to secure Faraday as one of his guests, and had begged me to convey his invitation to Faraday | W.F.P.

1. Francis Haywood (1796–1858, B5). Translator of German texts into English. See letter 2872.
2. See letter 2867.

Letter 2874
Fabian Carl Ottokar von Feilitzsch to Faraday
11 August 1854
From the original in IEE MS SC 2
Most honourable Sir!
Three years ago I had the honor to send you my researches about the physical distinction of magnetic and diamagnetic bodies[1]. By your kindness and under your protection they were admit[t]ed in the Philosophical Magazine[2] and consequently they were bespoken from several sides; thereby it was caused, that in the "Archives des sciences physiques et naturelles de Geneve" well founded doubts could by [sic] raised against my opinions[3]; I were obliged to assert their rightness, but on the other side I was convinced from the principle defended by me: "that the magnetism and the diamagnetism were only identical excitations of the matter", and it grew my task to maintain it by new researches. You have treated my first notices with so much indulgence, that I dare to hope, you, the creator of this new disciplin[e] will take it as a prove of my respects, that I put at your benevolence the results of my new treatise. On the other hand it is my duty vis-a-vis of you, as well as of all, who got knowledge of my first paper, to confess my error and to put some better thing in its place. If I had succeeded to gain your approbation, my highest wishes were accomplished. You will kindly allow me to send you the german treatise[4], as well as a short extract, that I have written in your language, trusting, that you will excuse its incorrectness.
I am, Sir, your | most devoted servent | Dr. von Feilitzsch | Professor at the university | of Greifswald
Greifswald in Prussia, | Aug. 11, 1854.

1. Letter 2350.
2. Feilitzsch (1851).
3. *Bibl.Univ.Arch.*, 1851, **16**: 50–1.
4. Feilitzsch (1854).

Letter 2875
Vincenzo Flauti to Faraday
12 August 1854
From the original in IEE MS SC 2

Reale Accademia delle Scienze | Società Reale Borbonica

Napoli il 12 agosto 1854

Sigr Cavaliere

Col dolore il più vivo partecipo a Lei, come già suo amico, la perdita che le scienze fisiche hanno fatto, nel giorno di jeri, dell'illustre cav Macedonio Melloni, vittima della feral malattia, che va terminando in Napoli individui e familglie intere[1].

Colgo questa dolente occasione per attestarle il mio profondo rispetto | il segrio perpo | V Flauti
All'insigne sigr M. Faraday | Cavr Membro della Società Reale di Londra | ed associato a tutte le principali Accademie | di Europa di America | Londra

Address: 21 Albemarle St | S. L'illustre Chev. M. Faraday | de la Societe Royale de Londres | &c &c &c | Londres

TRANSLATION
Reale Accademia delle Scienze | Società Reale Borbonica

Naples this 12 August 1854

Dear Sir,

It is with the deepest grief that I write to inform you, as a friend of his, of the loss that the Physical Sciences bore, yesterday, of the illustrious *cavaliere* Macedonio Melloni, a victim of the deadly illness in Naples that is wiping out individuals and entire families[1].

I take this painful opportunity to declare my profound respect | The Permanent Secretary | V Flauti
To the Illustrious Mr M Faraday | *Cavaliere*, Member of the Royal Society of London | and Associate of all the principal Academies of Europe and America | London

1. That is the cholera.

Letter 2876
Friedrich Wöhler to Faraday
12 August 1854
From the original in RS MS MC 5.176
Monsieur,
Je y a quelques jours j'ai eu la grande satisfaction de recevoir le diplome de membre de la Société Royale de Londres[1]. Ne connaissant ni la personne ni le nom du secrétaire de la Société Royale, j'ose m'adresser à Vous, Monsieur, en vous priant, de vouloir bien être l'interprète de mes sentimens de gratitude et de réconnaissance auprès de l'illustre Société pour l'honneur distingué, qu'elle m'a daignée de faire.

Veuillez agréer, Monsieur, l'éxpression de la haute considération, avec la quelle j'ai l'honneur d'être
Vôtre | tres devoué | Wöhler
Göttingen 12 Aug. 1854.
A | Monsieur Faraday

TRANSLATION
Sir,
A few days ago I had the great satisfaction of receiving the diploma of membership of the Royal Society of London[1]. Not knowing either the person nor the name of the Secretary of the Royal Society, I address myself to you, sir, and humbly ask you to be the interpreter of my feelings of gratitude and thanks to the illustrious Society for the distinguished honour that it has deigned to give me.

Please accept, Sir, the expression of the high consideration with which I have the honour of being | Your | very devoted | Wöhler
Göttingen 12 August 1854.
To | Monsieur Faraday

1. On this see RS CM, 29 June 1854, **2**: 290.

Letter 2877
Faraday to Charles Richard Weld
15 August 1854
From the original in Birmingham University Library MS L Adds 1083
[Royal Institution embossed letterhead] | R Institution | 15 Aug 1854
My dear Sir
    Herewith I send you a letter I have just received from Wohler[1][.]
    Ever Truly Yours | M. Faraday

- R. Weld Esq | &c &c &c

1. Letter 2876.

Letter 2878
Faraday report to Trinity House
15 August 1854
From the original in GL MS 30108/2/64.11
Report on Electric Light &c

15 Aug 1854 | R. Institution

I beg to report upon the result of the Experimental trials of the Electric light belonging to the Electric power & colour Company which have been recently carried on at the Trinity House[1][.] Dr. Watson has on the part of the Company done all that I desired, and the lamp has been in action at different times for the whole of five days. The trials sufficiently commend the light as most beautiful; and the lamp as one well fitted to bring the Electric forces into action. They also have shewn that the illumination can be sustained for 8 hours or any longer time if needed; and with an intensity much surpassing a Fresnel lamp. They have shewn also the facility of exchanging one lamp for another in less than a second of time:- the power of arranging an intermitting light, and the probable facility of placing 2, 3 or 4 centers of light in a space not more than an inch or an inch & a half wide:- but they have also revealed other circumstances, which would not otherwise have been known, and which I must take carefully into account. For much as I may desire that the Electric light, with its special advantages, may find ultimately its full application in light houses, I am bound before all other things to consider the security and the constancy of the service required, and the necessity that there should be allowed in places apart from society and often very difficult of access inasmuch as such are very frequently the localities of the most important lighthouses; where, if any where, the power of the Electric light would be required. The conclusion I have arrived at may be found at the end of this report; but I give the reasons at some length that they may stand as my justification[.]

The light is sustained by an arrangement of *Voltaic batteries*. The proposition of Dr. Watson, on the part of the company, is, that as the substances produced by the action of the battery are products of great commercial value, so the company will undertake the arrangement and care of the battery by their own servants, the conveyance of the acids, metals, & other materials, & the removal of the products from the lighthouses; undertaking to supply a constant light and charging a certain price for the light so supplied. Under such an arrangement the following points still remain to claim the attention of the Elder Brethren. A room or covered place would be required for the Voltaic battery, & for the stock of acids metals, cells, &c &c; also living rooms for the man or men in charge of the battery and a free supply of water. The battery employed at the Trinity house was composed of eight dozen pairs; each consisting of amalgamated Zinc & cast iron for the metals, and of strong Nitric acid & somewhat diluted sulphuric acid for the fluids. It occupied a space of 21 feet by 7 feet. Three carboys of oil of vitriol, 16 of Nitric acid, and 14 in the empty state to receive the diluted products, are necessary for a fortnights work of 12 hours to the day; & these with working room would require a space of 14 feet by 14 feet; so that a battery room equivalent to one of 18 or 20 feet square would be necessary even though the supply of materials &c were made regularly every fortnight. If the intervals were longer the place must be larger. The amount of habitation required for the battery men can be judged of from the data connected with the present keepers[.]

I am bound to state that, at the Trinity house, the battery action was accompanied by a serious inconvenience which can be provided against only by removing the building to a distance from the lighthouse[.] It consists in the emission of nitrous fumes, occurring always at the discharge of the battery when the light is extinguished, and often during the course of the general action. It is more liable to occur with acids that have been previously used; nevertheless such must be employed to the utmost for otherwise the required product would not be obtained, there would be great waste of materials and the commercial part of the plan would fail. The battery room requires an open air ventilation; and as it must be a fixture, would be objectionable on any side of a lighthouse, if very near to it. All the persons engaged in the experiments at the Trinity house suffered from these fumes: in the case of one of the workmen it caused spitting of blood, & I consequently interrupted the experiments and made them much shorter than was intended. The place was no doubt close not having a thorough draught, though opening by a large door to the air:- it was the place selected by Dr. Watson[.]

The staff employed about the lamp & battery consisted of three persons; one of superior intelligence having a good comprehension of the principles & practice of the lamp & battery, and two who were labourers of ordinary intelligence. The battery requires a peculiar kind of care[.] There are very numerous screw-clips about it which are conjoined & separated every time the battery is employed; if any one of these fail during the time of action, the Electric current is stopped & the light goes out. Occasionally one or another will become hot & then the power of the light sinks; these must be watched for and cooled with water. From time to time the zinc plates must be amalgamated; but the time is only to be ascertained by watching the action of the plates. About the time when the

Nitric acid becomes much charged with iron, a cell here & there will become hot and fume much, & at the same time it will oppose great resistance to the current and the light will diminish; then the connexion has to be made across the cell & the cell itself removed. These uncertain cells have to be watched for;- as many as eight occurred in one days work.

The existence of a battery staff separate from the keepers of the lighthouse appears to me a source of great risk in the constancy & character of the light. The light keepers would depend entirely upon the companys men for the supply and amount of light; for the power of producing & sustaining it would be out of their hands. On the other side the battery men would be obliged to attend to the interests of their employers; they must use the material supplied economically, and exhaust the Nitric acid to the full degree; which as I have stated tends to cause uncertain diminution of the light:- and in this way it is probable that conflicting interests might be involved and at the very same time the responsibility divided and rendered uncertain[.]

If it were proposed that the Trinity House should take the whole matter into their hands, then I need merely point out in the first place that the knowledge & intelligence required by the men in charge must be peculiar in its kind & much above that now needed, and that care & attention would be required to many more points than at present and would need to be unremitting:- and in the second that the Trinity House must become a manufacturing and Commercial body; in respect of which case I must observe that the light could not be obtained at any permissible price if the products of the action were discarded and thrown away; and if they were preserved to be sold for commercial purposes then in my opinion it is not likely that a price could be obtained for them which would be at all proportionate to the first expence of the materials the cost of the labour of using them & their conveyance to the lighthouses and back again.

Next in order to the battery are the *connexions* consisting of insulated copper wire leading to the lamp[.] These were (in the trial experiments) 250 feet long so that that distance has been practically proved[.] Their nature is simple but as everything depends upon their continuity it must be thoroughly understood by the keepers so that in case of accidents they may know how to search for failures & their places & rectify them[.]

The *lamp* is an arrangement of supports and adjustments intended to carry two insulated vertical cylinders or rods of hard carbon one over the other between the ends of which the light is produced. Though supplied by the company it would of necessity be under the charge of the light keepers and therefore I am bound to notice the liabilities & possible casualties. It is intricate as compared with the ordinary oil lamp, and the principles concerned in its construction and use are very different; being both more numerous & more refined. In the oil lamp the place of the light is constant, being determined by the fixed burner: in the Voltaic lamp it would change continually if left to itself, for as the lower carbon burns away the place of the light sinks & has to be raised very frequently by hand, or else continuously by clock work:- on the other hand as the upper carbon burns away it requires continual regulation in relation to the lower one; this is effected cleverly in the lamps by an electro magnetic arrangement beneath; but that involves the use of an electro magnet which is liable to change in force; of insulations of the carrying parts, and of mechanical arrangement of holders, levers, and screws which are sometimes liable to fail (failing from time to time in the actual experiments) and which require considerable & peculiar intelligence to understand & keep in order.

The lamp was left at various periods and at one time for 5 hours together; it was found at the end of that time burning brightly & had done so during the intervening time. Nevertheless it went out occasionally (and of course suddenly), sometimes by *contact* of the carbons and once by a *loose screw*. Dr. Watson proposes an Electro magnetic alarm which shall give notice if the light is extinguished. This can no doubt be applied but involves another object for intelligent care[.]

The Carbon cylinders, which are essential, are variable in their quality and the proportion of light varies much with them when the *same amount* of battery force is employed. The lower one burns away at the average rate of 7 inches in 12 hours and the upper one at the rate of 14 inches. They have to be replaced nightly or it may be twice in a night of more than 10 or 12 hours[.] They require careful centering & fixing in the apparatus. The Trinity house would be dependant upon others i.e the company, both for the supply and the quality.

I arrive now at the consideration of the *light* itself. It flickers much partly because the electric discharge is sometimes with flame & sometimes without, and partly because of variations in the places on the carbon through which the discharge occurs. This is so considerable that the light in a given direction will vary as 1 to 2 or even in higher rates as 1 to 4 or 5. When these flickerings are at short intervals as for instance less than a second then the effect is nearly lost upon an eye at a distance and an average degree of brightness is perceived. But there are other variations, which, occupying the time of three or four seconds or even a longer period, are more important. These depend chiefly upon the continually varying interval between the carbons, which having been adjusted by the Electromagnet, then goes on increasing to a certain extent, when it is again adjusted;- but also upon the quality of the carbons;- upon the general condition of the voltaic battery;- often upon the peculiar condition of a single cell;- and upon the actions of the men at the battery itself. They were occasionally of such degree and in such an interval of time as might I think in bad weather and under hasty observation lead a mariner to mistake a light intended to be fixed for a variable light.

The light of the two lamps used conjointly on the first day & of the single lamps employed on other days, urged in both cases by the force of 96 pairs of plates was always much above that of the great central three wicked oil lamp;- the average result was that it equalled 4.43 such oil lamps. As to the degree of variation caused by the circumstances already referred to, I found it very considerable, especially with a battery charged with second day acid, in which case individual cells were liable to be heated. On such occasions it has varied in the course of 10 minutes in the ratio of 4 to 1 nearly being at one time equal to 5.92 or almost 6 Fresnel lamps, and at another only to 1.57 or $1\frac{1}{2}$ of such standards. Over this variation the keeper of the light has no power & he cannot correct it by adjustment at the lamp.

The source of light is very compressed & intense; it is like a small sun instead of a diffuse & large lamp flame. Hence some supposed theoretical advantages but also some serious practical disadvantages. In the first place it cannot be counted upon as a center of light sending out rays of equal intensity in all directions. Such a result in a horizontal plane is indeed true, as regards the average light; but not in a vertical plane for then the light diminishes rapidly upwards and downwards from a plane of maximum intensity, because of the shadow cast in those directions by the carbon terminals. Another curious point is that the most intense light is not thrown in a horizontal direction. The electric force excavates the Positive carbon into a cup at its extremity whilst the negative carbon becomes conical at the end. As the positive carbon consumes most rapidly the upper is made to assume that condition:- hence the maximum light issues not in a horizontal plane but in a cone which is bent downward to such an extent that the maximum beam forms an angle of 17° to 20° with the horizontal line.

The compressed condition of the source of light causes it to throw great shadows from opaque objects. The lamp has two upright supporting pillars, about an inch in diameter and 5 or 6 inches from the light on opposite sides these throw a vertical shadow of 7 inches wide at 38 inches from the lamp, this subtending an angle of 10° or 11°:- they would of course eclipse the light to that amount to ships at sea. Dr. Watson proposes to alter the construction & to use flat bar supports instead of cylinders which would do away with much of this shadow. In like manner the upright bars of the window and other opaque objects in a lanthorn would cast shadows, and if of greater width than the light itself, the shadows would increase in width outwards & cause eclipses of the light. The reason why this does not happen sensibly or in the same degree with the Fresnel lamp is that there the light has a horizontal width of nearly 4 inches and so the absolute shadows vanish at a small distance. Dr. Watson proposes to have two or three electric lights arranged in the same horizontal plane; this would involve the use of two or three pairs of carbon conductors; the occurrence of shadows and loss of light from the extra carbons introduced; the necessity of a more complicated lamp;- and also the use of separate batteries for each light, for experience leads to the conviction that such a separate arrangement is required.

Finally the Electric light of one lamp was placed in the center of the catadioptric apparatus. The effect was very beautiful but inconvenient circumstances dependant upon the concentration of the light occurred. The apparatus is intended to gather in the rays through 120° or more in a vertical plane and cast them horizontally; and, as in the oil lamp arrangement, the rays are sent forth far more freely in directions above the horizontal plane than below it, (where the shadow of the burner interferes,) so the system of reflectors above the dioptric pannel or first class refractor is more important and more extensive than that below, & in the construction is attended to accordingly. But with the electric light the reverse of this is the case: it is the rays below which are most abundant, those above been cut off to a great extent by the cupped carbon as already described: and not merely so but even the maximum ray is depressed by 18° or 20°, so that much of the intention of the present catadioptric apparatus is rendered nugatory by the conductor of the Electric lamp. The reversion of the current would correct this in part, but that would involve a rapid consumption of the lower carbon; and it may be, give rise to other circumstances of which I have practically no knowledge; Dr. Watson having always made the upper carbon positive.

Neglecting the reflectors I will call attention to the action of the 1st class reflector which would remain the same in whatever way the electric battery was connected[.] The beams which proceed outwards from it are directed horizontally and as a general mass very well so but they are divided into horizontal planes which when they arrive at the screen produce bands, some luminous, some dark, and some most beautiful & intense in colour. The luminous and dark bands are produced by imperfections in the form of the separate prisms of glass constituting the pannel, or by the want of absolute perfection in their association together. Such defects are of no consequence when a flame $1\frac{1}{2}$ inches in height or more is in the focus; for then the rays from the various parts of the flame are superposed in the resulting beam, and at a moderate distance the differences in the bands almost disappear. But with a compressed light, like that of the Electric lamp, there is not this sort of correction. One shaden band which at a few inches from the refractor was only $\frac{1}{2}$ of an inch or less in breadth gradually opened out and was a dark band 6 inches wide on the screen at 46 feet distance; it had every appearance as if it would widen & hold its character more and more outwards, and I saw nothing that would indicate its disappearance at any distance. It would require the utmost perfection of workmanship to get rid of these bands.

The *coloured* bands are due to the difference in what is called the *dispersion* of the different rays of light. A single prism can & must produce them when the center of light is very small; and as the light is smaller & more powerful the colours are more intense. They were very bright upon the screen, and appeared as if they would travel out to a great distance still keeping their colour. With the oil flame of $1\frac{1}{2}$ inches high these coloured rays superpose & overlap each other; at the screen they are but small in effect & disappear at the distance of a hundred yards or so. Supposing the dioptric apparatus so perfectly worked & fitted as to produce no dark bands the coloured rays from the different prisms would overlap each other; but without the prisms were made achromatic (which they could not be because of the angles required) there is I think no probability of their disappearing when the E light is used[.]

So if the eye be carried from above downwards across the horizontal band of light & colours on the screen, it will pass from one colour to another; & from light to comparative darkness & then to light again. In like manner if a ship move from the horizon towards a light house with a similar lamp an observer on it will gradually see the light, sometimes of one colour & sometimes of another; but what is more important, he will see it *lighter* and *darker*; and I can easily conceive the possibility, that the mere lift of a large wave may raise the observer out of a strong ray into one of comparative weakness, or vice versa, & so give him the impression of an alternating or revolving light when it should appear fixed[.]

In order to correct this effect Dr. Watson proposed to place two lights, one above the other, that they might act the joint parts of the top & bottom of a lamp flame. This has not been tried. I conceive it would be very difficult to arrange them vertically one above the other; and if not vertical, it would require four or more lamps for one center, with separate batteries to each, & all the complications consequent thereon.

When the concentric lens was brought before the Electric light the effect was exceedingly beautiful and the illumination on the screen most vivid. The colours were less because here the rays from different parts of the lens were frequently overlaid & neutralized each other[.] Here again the influence of the central condition of the source was seen; for the illuminated space on the screen was not larger than the lens & it is probable that even at a great distance the divergence would be very little & the beam very narrow. The flame of the Fresnel lamp has a horizontal width of about 3 inches[.] The electric light is probably not not [sic] more than the $\frac{1}{8}$ of an inch in width in the same direction so that the effectual divergence of a beam from the latter would probably be not more than a twentieth or a thirtieth part of one from the former. Hence, though very bright it would endure for so short a time in any one direction that its occurrence might be missed; what is of more importance is, that having a very small vertical width, accident or a want of correction in the height of the carbons might easily throw the beam altogether above or below the horizon, or the place of a ship on the sea.

Much, therefore, as I desire to see the Electric light made available in lighthouses, I cannot recommend its adoption under present circumstances. There is no human arrangement that requires more regularity and certainty of service than a lighthouse. It is trusted by the Mariner as if it were a law of nature; and as the Sun sets so he expects that, with the same certainty, the lights will appear. The means of meeting this expectation must be provided for the most isolated positions, where, if there be a derangement, there is no help to be had except that which is on the spot, and where, therefore, no chance of risk that can be avoided should be permitted to occur: and still further to explore such precautions it must be remembered that the most important lighthouses are often in the most out of the way localities[.] It would not be easy even to replace a failing Voltaic lamp by an ordinary lamp but would require much time. I think therefore the Electric light should be tried in its other applications that its conditions & liabilities may be eliminated, its construction perfected - and a school provided w[h]ere a body of men may be taught how to use it with certainty; before it is introduced into lighthouses whose present optical construction may not consist with it, and whose present attendants would as I believe be unable to fitly guard against or contend with its peculiar liabilities.

One other kind of doubt I am bound to place before the consideration of the Elder Brethren[.] The production of the light which the Electric company proposes to supply is based upon commercial considerations. A manufacturer requires nitrate of Iron which can be made at the manufactory as it is wanted and in quantity according to the demand. It is by the light scheme proposed to make it at 100 or more different & distant places, very often difficult of access, and in quantities not dependant upon demand in the manufactory but upon the light required in different seasons. For this the materials of which a large preparation and strong acids in glass are to be conveyed to these spots, men retained to use them there, & then the products removed and all for years together with the regularity of clockwork. How this is to be done remuneratively, except by an enormous charge for the light, I cannot imagine; and therefore, independently of all the precautions and contingencies which I have noted I should suppose that the Trinity house would not think of changing its system and the whole construction of its lighthouse illuminating arrangement or put itself in dependence upon a commercial company before it had had a few years assurance of the permanency and stability of the Manufacture on which it would be called upon to depend[2][.]

M. Faraday

Royal Institution | 14 [sic] August 1854.

1. For the arrangements see letters 2866 and 2868. Faraday's notes of these trials are in GL MS 30108/2/64.12. There are also detailed notes of the experiments, in another hand, in Untitled notebook in IEE MS SC 2.
2. This report was laid before Trinity House By Board, 22 August 1854, GL MS 30010/39, p.79. Consideration was deferred until Trinity House By Board, 29 August 1854, GL MS 30010/39, p.90 when it was resolved to inform Watson that Trinity House would not proceed with the electric light.

Letter 2879
Vincenzo Flauti to Faraday
18 August 1854
From the original in RS MS 241, f.136
Reale Accademia delle Scienze | Società Reale Borbonica
Napoli 18 Agosto 1854
Signore
Con qual piacere che può sentire un occhio professore, il quale desidera credere l'Accademia di Napoli, di cui è Segretario perpetuo, onorata dalla corrispondenza di un uomo di merito si distinto come il suo, e che tanto ha contribuito e contribuisce a' progressi della Scienza Fisico-Chimica, mi è dato finalmente parteciparle la di Lei nomina a Socio corrispondente, Sovranamente approvata come si leverà dal decreto che gliene acchiudo insieme al Diploma accademico e ad un esemplare dello Statuto | Il Segretario perpetuo | V Flauti
All'insigne fisico cavaliere | Michele Faraday Socio corrispondente | della nostra Accademia | in | Londra

S "M" il Re mio Augusto[1] Signore nel Consiglio ordinario di Stato di 13 andante si è degnata approvare la proposta fatta dalla Reale Accademia delle Scienze con la quale Ella vien nominata suo Socio Corrispondente estero

Nel Real Nome e con mio particolar piacere le partecipo questa Sovrana determinazione e le trasmetto estratto del correlativo Real Decreto - Napoli 19 Luglio 1854 J Scorpa[2].
Sig. Michele Faraday.
Ministero | Real Secretaria di Stato | degli Affari Ecclesiastici | e dell' Istruzione Pubblica. | 1 Ripartimento | 2 carico | No.1102

TRANSLATION
Reale Accademia delle Scienze | Società Reale Borbonica
Naples | 18 August 1854

Sir,
With feelings of the utmost pleasure experienced by a professorial eye, which desires to believe that the Accademia di Napoli, of which one is the Permanent Secretary, is honoured by the correspondence of a man of such distinguished merit as yourself, and who has contributed and continues to contribute to the progress of the Science of Physics and Chemistry, I have finally been given the pleasure of communicating to you your Nomination as a Foreign Correspondent, sovreignly approved as you will see from the decree which I enclose together with the academic Diploma and a copy of the Statute. | The Permanent Secretary | V Flauti To the distinguished physicist *cavaliere* | Michael Faraday, Foreign Correspondent | of our Academy | in | London

His Majesty the King[1], my Sovereign Lord, at the ordinary Council of State, on 13th last, deigned to approve the proposal made by the Reale Accademia delle Scienze, by which you have just been nominated its Foreign Correspondent

In the Name of the King and with particular pleasure, I communicate this Sovereign desire and I send the relevant extract of the Royal Decree - Naples 19 July 1854 J Scorpa[2].
Mr Michael Faraday
Ministry | of the Secretary of State | for Ecclesiastic Affairs | and Public Instruction. | 1 Distribution | 2 Enclosures | No. 1102

1. Ferdinand II (1810–1859, DBI). King of Naples, 1830–1859.
2. Unidentified.

Letter 2880
George Biddell Airy to Faraday
19 August 1854
From the original in IEE MS SC 2

Royal Observatory Greenwich | 1854 August 19

My dear Sir
I have some thought of trying pendulum experiment[s] in a mine, as I did (unsuccessfully) a good many years ago[1]. At that time I saw R.W. Fox[2], and he was possessed with the idea that the vibrations of a pendulum might be influenced by magnetic currents in the rocks about it[3].

I should be very much obliged if you will tell me *in the first place* whether you think that there can be any such attraction: and if so, of what kind. I may point out what would be the effects of different kinds of force (A) If the force were a steady vertical force, it would be injurious (B) If the force were always opposed to the motion it would not be injurious (C) If the force aided the motion in the approach to perpendicular position, and opposed it in the recess from perpendicular position, it would be injurious. *In the second place*, would such force be annihilated by destroying the insulation of the pendulum.

I regret to trouble you with what may be foolish questions:- but if there is any reality in the thing it may be important - and to whom else can I go for information?

I am, my dear Sir, | Yours very truly | G.B. Airy
Professor Faraday

---

1. See Chapman (1993) for these experiments conducted in 1826 and 1828, and also for Airy's 1854 experiments at South Harton Colliery. See also Airy, G.B. (1856).
2. Robert Were Fox (1789–1877, DNB). Man of science at Falmouth.
3. See Fox, R.W. (1830).

Letter 2881
Faraday to Vincenzo Flauti
Mid-August 1854[1]
From an extract in Rend.Soc.Reale Borbon.Accad.Sci., 1854, 3: 80

Con grandissimo dolore ho intesa la morte del cav. Melloni, inaspettatissime per me, e nel momento che egli impegnato in aggiugnere scoperte importanti alle scienze, nelle quali si aveva acquistato un gran nome.

...

Povero Melloni la sua memoria non ritornerà mai senza profondo dolore a tutti gli amatori delle scienze naturali. Vi prego di manifestare il mio vivismo cordoglio alla di lui famiglia.

TRANSLATION

It was with great sadness that I learned of the death of Mr. Melloni, which was so unexpected, coming, as it did, at a time when he was engaged in making important discoveries for science, in which he had acquired a great renown.

...

Poor Melloni - no lover of the natural sciences will be able to recall his memory without deep sadness. I ask you to convey my sincere sympathies to his family.

1. Dated on the basis that this letter is the reply to letter 2875. It was read to the Accademia on 25 August 1854.

Letter 2882
Thomas Byam Martin to Faraday
c20 August 1854[1]
From the original in BL add MS 41370, f.371
My dear Mr. Faraday
  I am ashamed to think how long I have neglected to thank you for the interesting paper you sent to my Committee at the Admiralty[2]; the fact is I was taken seriously ill at the moment when we brought our business to a close and have been made to keep my papers out of sight since that time[3].
  I not only beg to thank you for your observations in reply to our questions but permit me to ask if there is not some fee usual, and due, for your obliging services - I am ignorant in such matters, and beg you will tell me candidly what is customary[.]

1. Writer and date established on the basis of provenance and that letter 2887 is the reply.
2. See letter 2871 and note 11.
3. Martin died on 21 October 1854.

Letter 2883
Faraday to George Biddell Airy
21 August 1854
From the original in RGO6 / 212, f.329-30

Surbiton | 21 Aug 1854

My dear Sir
  Your letter[1] sets the thoughts loose upon many points, none of which may be of the least consequence but you must take them for what they are worth;- talking would have dealt with them better than writing. In the first place unless the effects, & therefore the causes, are supposed to be very local I should imagine they would be as evident upon the surface as in the interior or nearly so. As respects general electric currents in the earth, they ought perhaps to be more evident when they are all below the thing

affected, than when the latter is in the midst of them, for in the first case they would all act in one direction & in the second the parts on opposite sides would act in opposite directions.

If the pendulum has nothing magnetic about its moving parts then I do not see how earth-currents of electricity can affect it. If it were magnetic the currents might tend to turn it round an axis between the polar parts; but as such a force would act in the manner of a couple, I suppose it would not interfere with the *time* of vibration[.]

You speak of Magnetic currents, by which I conclude is intended electric currents. If the idea is to include the *ordinary force* of the earth, then a magnetic pendulum would be in different relation to it according as it swung in the plane of the magnetic meridian or at right angles to it, inasmuch as the pull would, in the first case, be all on one side of the swing; and in the other different in amount & direction & equally disposed on both sides:- but all that will occur to you at once; & that the effect will be the same as *on the surface*[.]

If the rocks surrounding the wire have any *fixed local* magnetic force, making the resultant of the earths magnetism in a given place, different to what it would be at the surface;- then I must leave you to judge what the effect of such a difference should be.

Let us now suppose the pendulum free from magnetism. If composed of non metallic materials then I have no more to say:- but if its bob be a mass of metal then the following considerations arise. The bob moves to & fro about a fixed point; and hence its upper and lower parts move through different spaces in the same time. In these latitudes, where the dip is very great, the consequence is, that moving *across* the earths lines of magnetic force, electric currents tend to be found; and because of the difference between the amount of lines intersected by the upper and lower parts of the bob, are really found, as I have shewn in my old researches[2], & often obtained since:- so that as the bob swings it will become, virtually, a very feeble electromagnet with a horizontal magnetic axis; i.e if the bob swings east & west in one direction it will have whilst moving, magnetic force in a north & south direction: as the bob returns in the contrary direction it will have a north & south magnetic force equal to the first amount but in the *contrary* direction. If swung in any other azimuth like results will occur, the vertical magnetic axis being at right angles to the plane of motion, provided the ball be solid & uniform[.]

I suppose this alternate magnetic condition (which must be very small in amount) will not interfere as such because of the reasons before given i.e because the effect will be that of a couple.

But the assumption of the state is accompanied by a resistance to the motion which generates it. It might possibly therefore have the effect of *obstructing* the motion:- whether that would only diminish the arc of vibration or whether it would affect the *time* also you can tell better than

I can. The effect would be the same upon the *surface* of the earth and I directed Sabines attention to it a long while ago[3].

And now that I speak of the surface I may point out another consequence of pendulum motion. If a pendulum were swung at the places where the dip was 90° whether in one azimuth or another would, magnetoelectrically, make no difference: but if swung where the dip is less than 90° it might make a difference which difference would be a maximum where the dip is 0° & is of the following kind. Suppose a pendulum swing at the magnetic equator and in the direction of the magnetic meridian it would intersect no lines of magnetic force, but travel to & fro along them; & no electric currents would be induced in it, no magneto-electric state obtain, & no resistance to motion, of the kind spoken off [sic], be produced:- but if it were vibrated in a plane perpendicular to the magnetic meridian currents from the lower to the upper edge or from the upper to the lower would tend to be produced; and because the two edges move with different velocity would probably be produced. Whether such currents would be sensible in their effects *there* I cannot say - [(]it is hardly probable that they would be *here* where the dip is so considerable), even to the pendulum test.

As all these results depend upon the earths force & its direction, I do not think it likely that any supposed change from time to time in the currents of the earth are likely to affect a magnet irregularly, provided they are so small in amount as not to affect the magnet needle. Whatever leaves that untouched would not, I think, affect the pendulum.

I am My dear Sir | Your most truly | M. Faraday
G.B. Airy Esq | &c &c &c
Of course I speak only of what is known[.] Unknown results may remain for us to discover[.] MF

Address: Geo. B. Airy Esq | &c &c &c | Royal Observatory | Greenwich

1. Letter 2880.
2. Faraday (1832b), ERE2, 180.
3. See letter 2354.

Letter 2884
Julius Plücker to Faraday
21 August 1854
From the original in IEE MS SC 2

Bonn 21st of August | 1854.

Dear Sir!

I take the liberty to present to you two new papers. One of them is from Poggendorff's Annalen[1]; the other one I was obliged to write being this year Dean of the faculty[2]. I thought it proper to explain at this occasion, what I think the present state of our knowledge of the magnetism of gazes and crystals. Belonging to this subject several assertions were attributed to me, I never made. I passed myself through different errors. Therefore I wished to precise my present meaning.

I join to this parcel other copies of the latin paper for the Reverend Secretary of Royal Institution[3], to Mr Grove and Prof Tyndall. Your porter may, without giving you any trouble, deliver them by occasion.

I set off instantly for visiting my friends at Vienna.

Yours | most truly | Plücker

1. Plücker (1854b).
2. Plücker (1854c).
3. John Barlow.

Letter 2885
William Alexander Baillie Hamilton[1] to Faraday
22 August 1854
From the original in IEE MS SC 2

Immediate

Admiralty | 22, Aug 1854.

Sir,

I have received and laid before my Lords Commissioners of the Admiralty your letter of the 19th Instant in which you state that others can no doubt be found able and willing to investigate the merits of an Invention for working the Engines of a ship by Gas, and I am desired by their Lordships to acquaint you that they look in vain for any one so thoroughly competent as yourself to undertake this service for them, and they know of no one whose opinion would be so satisfactory. Under these circumstances therefore, and as their Lordships do not anticipate that the subject would occupy much of your time and attention, my Lords still venture to hope that you will favour them with your opinion[2][.]

I am Sir | Your most obedient Servant | W.A.B. Hamilton
To Dr. Faraday FRS | Royal Institution

1. William Alexander Baillie Hamilton (1803–1881, B1). Second Secretary of the Admiralty, 1845–1855.
2. The digest of the Surveyor of the Navy, PRO ADM88 / 10, entry 9394 noted that Faraday replied on 23 August 1854 saying that he could discuss the gas engine that day.

Letter 2886
Jacob Herbert to Faraday
25 August 1854
From the original in GL MS 30108/2/65

Trinity House London | 25th August 1854

Sir,

A recent committee of Inspection having stated in their Report that "both the Towers of the Haisbro' Lights[1] shew exceeding dampness and repairs executed one year require in places renewing the next,- that at the High Light House as much as 3 Quarts of Water in cold weather has been collected in the Receiver attached to the Ventilating Tubes - and that it is thought that if the Keepers were removed to a Cottage Dwelling and the Towers made hollow, with a Circular Staircase the evil would in a measure be cured"- I am directed to communicate the same to you, and to request that you will favor the Corporation with your opinion as to the probable advantage of such a change[2].

I am | Sir | Your most humble Servant | J. Herbert
Professor Faraday | &c &c

1. In Norfolk.
2. See Trinity House Wardens Committee, 24 August 1854, GL MS 30025/23, pp.176–7 for this request to Faraday.

Letter 2887
Faraday to Thomas Byam Martin
26 August 1854
From the original in BL add MS 41370, f.372

[Royal Institution embossed letterhead] | 26 aug 1854

My dear Sir Byam

I have received you kind note[1] and am very sorry to hear of your illness. There is no charge for I am not Professional. Do not get me into more consultations than can be helped. I have just come away from another at Somerset House[2]. I shall always attend to any thing *you* desire of me but wish the Admiralty - to procure other aid when they can[.]

Ever My dear Sir Byam | Your faithful Servant | M. Faraday

1. Letter 2882.
2. Presumably a reference to the subject mentioned in letter 2885.

Letter 2888
Jacob Herbert to Faraday
26 August 1854
From the original in GL MS 30108/2/64.15

Trinity House | 26, Aug: 54.
My dear Sir,
    Be assured that it affords me much pleasure to convey to you the enclosed Resolution of the Board on Tuesday last[1].
    I beg you to believe me always, with great respect & esteem,
    My dear Sir, | Very faithfully Yours | J. Herbert
M: Faraday Esq

1. Trinity House By Board, 22 August 1854, GL MS 30010/39, p.79 gave best thanks to Faraday for his work on electric light.

Letter 2889
John Barlow to Faraday
27 August 1854
From the original in IEE MS SC 2

Dresden | Aug 27, 1854
My dear Faraday
    You must write to me once more, as soon as you can do so without suffering. I am sure that that letter to me must have been a painful effort - The fingers cannot do their work satisfactorily when the nerves from the neck are irritated by disease & violence... We are very anxious also to have another report of Mrs. Faraday. She could not have made progress during your painful illness. Then I should desire to know which of the Porters has been in trouble. One naturally thinks of Lacy, who, I fear is transmitting a dreadful constitution to his children... Of all the mysterious dispensations of Providence, the fertility of mad, consumptive, & scrophulous families is, to me, the most inexplicable. It is, to my mind, the most (apparently) exceptional arrangement in the system of this world... I hope that I need not tell you that if the immediate administration of a sovereign or two will remove any part of the trial, I beg that you will advance it for me.... Our journey thus far has been very prosperous. I forget whether I wrote to you from Salzburg or from Ischl. We enjoyed the last named place extremely. The scenery is lovely, and the residence there of the Emperor[1] & Empress[2] does not spoil it. I saw them, one day, walking home from the village church (where they go every day). She is prettier than the pictures of her: the expression of her countenance is very pleasing. I am told that she is very amiable & benevolent. From Ischl to Vienna over the lake of

Gmunden & by the fall of the Traun to Linz - from Linz by Danube-Steamer to Vienna - at Vienna six days, & then here by Prague, where we remained three, & where your prompt letter just caught me. You perhaps heard that the cholera broke out in Munich soon after we left it. I was told that six people were lying dead in one day at the large hotel (the Bavaria). May Liebig & Hofmann & all our friends have escaped!... My accounts from England represent this disease to be more under control now, than on any of its former visitations. Lord Jocelyn's[3] death seems to have been, in great measure, the result of his own imprudence: Lord Beaumont's[4] appears hardly to have been a case of cholera, and the reports which I hear from the Westminster Hospital certainly indicate a very small proportion of deaths.... I quite fancy Schlagentweit's face. He is just the man to be panic-struck: If I at all understand these brothers, neither of them would distinguish himself much if made to confront a danger of this kind.....

.......As far as I can understand the patois of the people, there seems every prospect of an abundant harvest, except in fruit, which is neither good nor plentiful...

The more I see of other countries, the less am I disposed to encourage that swaggering language which our countrymen indulge in when speaking about England...... Educated classes are, I imagine, much of the same calibre of morals every where: but I should guess that the educated bear a larger proportion to the uneducated all over central Europe than is the case in England.... Then, if one comes into details - of the English farmers, I have a very bad opinion - of the English (especially London) tradesmen I think very little better. The foreign shopkeeper perhaps takes advantage of you in his prices, but he does not send false bills, or parcels in which the goods you have paid for are not forthcoming. This has happened to me repeatedly from the most eminent of the (so called) respectable London Tradesmen. In England nothing would induce me to associate with what is called an "Evangelical clergyman" or with any one else who made a parade of religion, because I never knew any such who was not at heart an infidel, a debauchee, or a rogue or, at best, a tool of these characters. Now this does not seem to be the case in Roman Catholic Countries. The discipline of the confessional must restrain breaches of the moral law. Your Connection[5] does the same thing by different machinery. Therefore I am always disposed to think well of and to confide in a member of it....

...I beg your pardon for this outbreak... Mrs. Faraday is however partly to blame for it.. She said that "I should think your form of worship ridiculous". Now I cannot imagine any thing less possible to deride than the simplicity and earnestness of your ritual: and I am sure that it must pervade the daily life of those who are exercised by it...

And yet there is the same earnestness in those who are well disciplined by the Roman system. I attended High Mass at St. Stephen's

Cathedral in Vienna on the Birth-day of the Emperor. The ArchBishop[6], assisted by a conclave of Bishops, officiated. There was the full Roman Pagent - Robes, incense, lights, music - Diplomatists, & statesmen, and officers in splendid uniforms, assembled in the most picturesque groupes to give effect to the spectacle. This was at the high altar.. at a side altar, very near this dazzling ceremony, a low mass was going on - perhaps fifty people were kneeling there. Not one of these raised his eyes or seemed at all conscious of what was going on so near him. In short, Priest & people did just what they would have done had they been alone in the Church.- No Protestant engaged in any of his services, would, as I am persuaded, be capable of such concentration - and yet these were people the majority of whom were in the humblest ranks.

...Don't imagine that I am exalting R. Catholics further than the letter of my words expresses.- It is a religion which, it I know myself, I never could become attached to. Still I appreciate their accomplishment of their steadiness which it is so difficult to attain.

I will now release you... Do not think about all this effervescence: but tell me how you, Mrs. Faraday & Miss "Jenny" are... If I can say or do any thing for you in Berlin, tell me[.]

Ever yours | John Barlow

1. Franz Josef (1830–1916, OBL). Emperor of Austria, 1848–1916.
2. Elizabeth (1838–1898, OBL) Empress of Austria, 1854–1898.
3. Robert, Viscount Jocelyn (1816–1854, CP). Conservative MP for King's Lynn, 1842–1854.
4. Miles Thomas Stapleton, 8th Baron Beaumont (1804–1854, B1). Colonel Commandant of the 4th West York Militia, 1853–1854.
5. That is the Sandemanian Church.
6. Josef Othmar von Rauscher (1797–1875, OBL). Archbishop of Vienna, 1853–1875.

Letter 2890
Faraday to John Barlow
30 August 1854
From the original in RI MS F1 D23

Royal Institution | 30 Aug 1854

My dear Barlow

You do not say[1] where I am to address you or when so I write instantly that I may catch you at Dresden - though perhaps I shall decide to send it to Berlin as you ask whether you can do any thing for me there. Remember me in kindness to all friends there - there are three or four I think of writing to, but now that I am better the hot weather is so enervating that I am *lazier* than ever. We are getting on pretty well here. Lacy is better & his family & I shall not mention your kindness you can do what you like when you come back. Anderson had to lay up but is better indeed pretty well again. He had a boil or something in the neck that had to be opened. Miss Savage[2] has just gone for her holiday. Mr. Vincent will be here very soon & is very well. Tyndall is at work hard the Laboratory & appears to be quite well[.] We are now at Surbiton near Kingston but I am in town three or four days in the week at work. I am now very well. My wife, I hope improving; pretty well but very deaf & feeble in the limbs & head.

As to the world I know nothing of it here nor do I care much for it - if it will let me alone - but only think of the nuisance of being found out at Surbiton, & teased with visits; and invited by the *Mayor of Kingston*[3] to *dine* at the *Venison feast* (annual) - &c &c[4][.] It is all meant very kindly but such kindnesses are not in my way, and I feel it unkind in me to refuse them[.]

I received a letter from Naples last Saturday[5] which will grieve you[6][.] Melloni died suddenly that is after a short illness on the 11th I think of this month (August)[.] We had been corresponding vigorously on some scientific matter[7] & instead of a letter from him I received the notice of death from his friend, M. Flauti of the Academy of Sciences there[.]

I must pass by your observations on religion &c & indeed must conclude in the briefest manner with our kindest wishes to you & Mrs. Barlow[.]

Ever Yours | M. Faraday

Address: Revd John Barlow MA | &c &c &c | Poste Restante | Berlin

1. In letter 2889.
2. Sarah Savage (d.1865, age 57, GRO). Housekeeper of the Royal Institution, 1835–1865. (RI MM, 19 July 1835, **8**: 363–4 and 6 March 1865, **12**: 97).
3. Frederick Gould (1817–1900, B5). Amateur man of science and Mayor of Kingston, 1853–1854.
4. Held on 31 August 1854. See *Surrey Comet*, 26 August 1854, p.7, col. a.
5. That is 26 August 1854.
6. Letter 2875.
7. Letters 2762, 2813, 2834, 2846, 2862, 2865 and 2870.

Letter 2891
Faraday to Thomas Croxen Archer[1]
30 August 1854
From the original in JRULM Ryl. Eng. MS 376/646q

[Royal Institution embossed letterhead] | 30 Aug 1854
Sir

All things seem to promise me the pleasure of being in Liverpool from Wednesday the 20 Septr to Saturday 23rd[2] so I will delay no longer in replying to your letter of enquiry of the 12 instant[.]
   I am Sir | Your Very faithful Servant | M. Faraday
Thos. C. Archer Esq | &c &c &c

1. Thomas Croxen Archer (d.1885, age 68, GRO, B1). Professor in the Liverpool Institution.
2. For the meeting of the British Association.

Letter 2892
Faraday to Friedrich Wöhler
30 August 1854
From the original in Niedersächsische Staat- und Universitätsbibliothek Göttingen, Philos 182: Faraday
                                     Royal Institution | London | 30 August 1854
My dear Sir
   You wrote me a letter[1] desiring me to thank the Royal Society[2] on your part and I immediately took the necessary steps for doing so[3]. But I cannot consider our communications concluded by that. I desire most earnestly to use the occasion as an excuse for the expression, of a few good wishes on my part to you, of the high gratification & delight which I have had in successively acquiring a knowledge of your great contributions to natural science[4] and the extreme satisfaction I have had in seeing your honors grow. Though it is long since we met either personally or by letter[5] still your idea is ever present with me and that must be my excuse to you for the freedom with which I write this letter, the freedom indeed of an old friend. May you long enjoy health & happiness & that great distinction of producing fruit which makes your name delightful both to friends & strangers[.]
   Ever My dear Sir | Your Very faithful & | true admirer | M. Faraday
Profr Wöhler | &c &c &c

Address: Professor Wöhler | &c &c &c &c | University | Gottingen

1. Letter 2876.
2. For his election as a Foreign Member. RS CM, 29 June 1854, **2**: 290.
3. See letter 2877.
4. The word "knowledge" is crossed out immediately before "science".
5. See Faraday to Wöhler, 15 August 1835 and Faraday to Vivian, 17 August 1835, respectively letters 809 and 810, volume 2.

Letter 2893
Jacob Gisbert Samuel van Breda and Wilhelm Martin Logeman to Faraday
September 1854
From Breda and Logeman (1854), 465–9
Sir,
 The experiments on electro-dynamic induction in liquids which you have published in a letter to M. de la Rive[1], have excited our lively interest, not only because the phaenomenon appeared to us to be of importance in itself, but especially because it seemed likely to throw some light upon the manner in which electricity is propagated in liquids.
 Do liquids conduct exclusively by electrolysis, or do they also possess a proper conductibility, similar to that of metals? An experiment that we have made may perhaps assist in the solution of this question. It is well known that the conductibility of liquids increases with their temperature, whilst the opposite effect takes place with the metals, a fact which is easily explained if we suppose that liquids, in general oppose less resistance to decomposition in proportion as their temperature is raised. If this explanation be the true one, the next thing to be ascertained is, whether a liquid will also exhibit this increase of conductibility for a current so weak as to traverse it without producing any apparent chemical decomposition. If this were the case, it would appear probable that the decomposition nevertheless took place, and that it was by its intervention that the current passed through the liquid. We have endeavoured to solve this question by the following experiment. We passed the current of a small Daniell's element through a column of distilled water 24 centimetres in length, contained in a glass tube of about 15 millimetres in diameter; the electrodes were of platinum wire. One of these electrodes was connected with the zinc pole of the battery, the other with one end of the helix of a galvanometer of which the wire made 1800 coils, the other end of which communicated with the copper pole. The tube was immersed in a waterbath, the temperature of which could be raised by means of a spirit-lamp. When the water was at 59°F., the needle of the galvanometer deviated 4°. When the lamp was lighted, this deviation was seen to increase regularly. At a temperature of 152°.6 F., the deviation was 7°, and at 190°.4F. it was 11°. The increase of the conductibility of the liquid by heat was therefore proved, even when traversed by an excessively feeble current. Had any chemical decomposition of the water taken place during this experiment? Its direct result led to the belief that such was the case, but we were also fortified in this opinion by the following circumstances. When the liquid was cooled, the communications remaining untouched, the needle of the galvanometer no longer showed any appreciable deviation. When the direction of the current in the column of water was reversed, the needle immediately deviated 8° and returned insensibly, but in a short time to 4°,

at which point it remained stationary. It was consequently an effect of the polarization of the electrodes that we observed in this case, a polarization which opposed the current at the first moment of its passing, without, however, being able to annul it, but which annulled it completely when it had become stronger by the passage of the stronger current through the heated liquid.

But is this polarization the peculiar effect, and consequently the irrefragable proof, of chemical action? There are many experiments which render this opinion, if not absolutely certain, at least exceedingly probable. We may mention in particular those of Schönbein, who found that the effect continues when the electrodes which have served to introduce a current into a liquid are immersed in another liquid through which no current has been passed, and also that effects exactly similar to those of the plates polarized by the current may be obtained by putting one of them only in contact with a gas (such as hydrogen or chlorine) for a very short time, and afterwards immersing them in acidulated water*. Some physicists, however, still maintain the opposite opinion. They explain polarization by an accumulation of electricity of different natures, either in the electrodes themselves, or in the adjacent portions of the liquid; these two electricities in recombining by a conductor uniting the two electrodes, after the connexion between these and the electromotor has been broken, would give rise to the current in the opposite direction to that of the latter, which is always observed in such cases.

It appeared to us that your *beautiful experiment of electrodynamic induction in liquids* might furnish a means of submitting this opinion to an experimental test, by trying whether the electrodes, employed in that experiment to conduct the instantaneous current of the fluid helix to the galvanometer, are or are not polarized by this current. To obtain a decisive effect it was necessary to reproduce the phaenomenon with more intensity than when, as in your experiments, the question was merely to prove the phaenomenon itself. For this purpose we made use of a tube of vulcanized Indian rubber, of about 1 centimetre in internal diameter and 13 metres in length. We twisted it round the two branches of the large electro-magnet intended for experiments in diamagnetism, which, if we are not mistaken, has the same form and the same dimensions as your own; it is covered by a coil of copper wire 3 millimetres in diameter and 180 metres long. The tube was entirely filled with a mixture of 6 parts by volumes of water and 1 part of sulphuric acid. It was terminated at both ends by glass tubes of about 4 centimetres in length; into each of these passed a platinum wire of 1 millimetre in diameter, the portion of which immersed in the liquid was about 2.5 centimetres in length. All being thus arranged, the ends of the two platinum wires were connected with the galvanometer of 1800 coils which was placed at a distance of 10 metres. We had ascertained previously that at this distance the magnet did not exercise any sensible action upon the needles. The moment the two ends of the copper-wire coil of the electro-magnet were put in connexion with the poles of a Grove's battery of 60 large elements, arranged in a double series of 30, the needle of the galvanometer deviated suddenly about 40°, and returned, after oscillations which occupied between 1 and 2 minutes, to 0°. When the circuit of the pile was interrupted, the galvanometer deviated again about the same number of degrees, but in the opposite direction, returning again to 0° in the same manner. The needle returning in both cases to 0°, one would be tempted, at first sight, to think that there was no polarization of the electrodes. But the strong impulsion communicated to the very astatic system of the galvanometer by the induced current, causing the needles to oscillate during a considerable period as we have just stated, the circuit remaining always complete, it appeared possible that the polarization, if it existed, had already exhausted itself before the needles had arrived at a state of repose. To get rid of this difficulty we put the two electrodes in direct communication with each other by means of a copper wire of only 10 centimetres in length, although they still remained in connection with the galvanometer. The induced current produced when the circuit of the pile was established, then passed by this wire rather than by the infinitely longer wire of the galvanometer, and the needles remained at rest. But when this wire was removed after the establishment of the communication with the pile, *we saw the galvanometer deviate instantly in an opposite direction to the deviation produced by the induced current of the preceding experiment, and rest, after a few oscillations, at 10°. By replacing the wire, then interrupting the circuit and again removing the wire, we saw the galvanometer deviate in the contrary direction, and rest, in the same manner, at about 10° on the other side of the divided arc.* In both cases the deviation diminished regularly by little and little until it became 0°; we did not exactly measure the time which this occupied, but it appeared to us to be about 30 or 40 seconds.

These experiments were frequently repeated, and always with the same result. We need not say that we always took the precautions pointed out by you, to prevent the effect of an induction in one of the metallic conductors.

Polarization therefore takes place in the electrodes which serve, not only to convey a current into a liquid, but to carry out the current induced in the liquid itself. It appears to us that this fact directly contradicts the theory which attributes polarization to an accumulation of the two electricities upon or around the electrodes; for in the present case not only would such an accumulation be infinitely less probable than in ordinary cases of polarization, but if it existed, it would necessarily give rise to a current not in a direction opposed to that of the principal current, but in the same direction.

May we therefore regard all polarization as an effect of electrolytic decomposition, and consequently as an irrefragable proof of the existence of this decomposition? If this be true, we shall be led to regard the opinion of those who admit the possibility of the transmission of a current, or of a portion of a current through a decomposable fluid without the occurrence of any decomposition, as resting on very slight grounds. Whenever we have passed a current, however weak, through such a liquid, we have always observed an undoubted polarization of the electrodes.

We shall take the liberty to describe one other experiment, which is still more convincing in this respect than that described at the commencement of this communication. We immersed two plates of platinum, 6 centimetres in length and 5 centimetres in breadth, at a distance of about 1 centimetre from each other, in distilled water. One of these plates communicated with the ground by a metal wire; they had previously been carefully cleaned by heating to redness, and consequently, when put in communication with the galvanometer, did not produce any sensible deviation. But as soon as *a single spark* from a common electrical machine had been thrown upon the plate which did not communicate directly with the ground, and the communication with the galvanometer had been established, the needle deviated from 3° to 4° in one direction, and the same distance in the opposite direction when the current of the spark was passed through the water the other way. This deviation could be brought to 15° or more by throwing several sparks instead of one upon one of the plates or by connecting the plate for a very short time with the conductor of the machine during the movement if its plate.

We fear that the importance of this letter will not be proportional to its length; if, however, its contents should seem to you to be worthy of attention we shall be happy to see it published in any manner you think proper.

We remain, Sir, &c., | J.G.S. Van Breda. | W.M. Logeman.
Haarlem, September 1854.
*Poggendorff's *Annalen,* vol. xlvi. p.109, and vol. xlvii. p.101[2].

1. Letter 2800 published as Faraday (1854b).
2. Schoenbein (1839a, b).

Letter 2894
Joseph John William Watson to Faraday
5 September 1854
From the original in GL MS 30108/2/64.16

31 Pall Mall | Sepr. 5 1854

Dr Sir,

Altho' I fear that I am out of order in the request contained in this note yet the importance of the subject may perhaps claim some little indulgence on your part. I was not a little surprised at the result of the communication made to me by Mr Herbert respecting the late trials at the Trinity House of the Electric Light[1], since I had fully anticipated some trial of the light in a situation where its great intensity might be displayed in a better situation than in the cellar of the Trinity House. Doubtless, however, you had excellent reasons for not continuing the exhibition to that point; but, as any condemnation from so high an authority as yourself would operate most injuriously against the Electric Company which has for its end, I consider, matters of great national importance and is therefore deserving of public support may I ask the great favour of your pointing out to me the *paths* for improvement by which this light may be made available for practical purposes such as Light house illumination. I am aware that I am perhaps asking too much but the wording of Mr Herberts communication emboldens me to still hope to see the electric light employed by the Trinity Board.

Any communication I may receive from you I can assure you I should consider as perfectly confidential and trusting that I shall not be disappointed

I am | Dear Sir | Yours most faithfully & obediently | Joseph J.W. Watson

M. Faraday Esq DCL | &c &c

1. See letter 2878.

Letter 2895
Faraday to Joseph John William Watson
8 September 1854
From the original copy in GL MS 30108/2/64.16

Surbiton | 8 Septr 1854

My dear Sir

I have received your letter of the 5th instant[1] in which you ask me to point out the *paths* for improvement in respect of the Electric light, so as to make it available for lighthouses. I am not able to do this:- because, having for many years thought of the light in relation to lighthouses, my expectations of its application generally, have become less and less. Had I seen any promising paths, I should have pursued them myself. I do not mean that others may not discover such, but only, that I am unable to point them out. This I may observe, that, many difficulties & obstructions appear & are known to me, who am by position aware of the requisites of the lighthouse service, which may appear as nothing to those unacquainted with these matters.

I have no objection to your knowing the grounds upon which I have reported to the Trinity House:- but my report[2] belongs to the Trinity House and I have no right over it. I am not aware how far it would be consistent with the correctness & general policy, for the Trinity house to give you a sight of the whole or part of the report; but I am sure that Mr Herbert will do all he can if you apply to him. Perhaps he could let you see it in his office.

I can easily understand that you, who are not aware of the care the lighthouse service needs, might think of and hope for, an experiment in a lighthouse. But a lighthouse is the *last place* in which an experiment should be made:- and if the Electric light ever rises to its application there, it must be after having been carried through & educated by other applications, where its failure or its variation would be of little comparative consequence. Great intensity of light, though very important in a lighthouse, is not nearly so important as constancy & certainty of action; extending over periods, not merely of hours, but of months and years. I once favoured the trial of a very promising proposition in a lighthouse. It not only cost a large sum of money and occasioned great labour & anxiety, but after many months of exertion, on my own part & that of others the apparatus had to be removed, for objections, less in force, in my opinion, than those which at present stand against the Electric light. The light is beautiful; and if it should be perfected hereafter so as to be in all points fitted for lighthouses generally, it may become of "national importance":- but at present it is of great "national importance" that it should not be let into the lighthouses because of the extreme responsibility of the lighthouse system and the evils that would be occasioned by any derangement of its character & lowering of its action[.]

I am | My dear Sir | Very Truly Yours | M. Faraday
Dr. Joseph Watson | &c &c &c

1. Letter 2894.
2. Letter 2878.

Letter 2896
Faraday to Richard Rathbone[1]
8 September 1854
From the original in WIHM MS FALF

[Royal Institution embossed letter head] | 8 Septr. 1854
My dear Sir

Your kindness is greater than I could imagine and I earnestly thank you and Mrs. Rathbone[2] for the proofs of it: but I still feel on several accounts that I must remain in the Town of Liverpool[3]. In the first place I had a very kind invitation before I received yours by Dr. Reynolds[4] and that I declined on the grounds that I advanced to Dr. Reynolds in my reply[5]. In the next place I may pass only two & certainly not more than three nights at Liverpool and having some poor and distant relations there[6], I shall want all the hours I can find to do what I wish to do. So that I really am constrained to keep myself very much at liberty.

Again thanking you most earnestly for your kindness I can only say that I am

Your Very greatly Obliged Servant | M. Faraday

Richard Rathbone Esq | &c &c &c

1. Richard Rathbone (1788–1860, Marriner (1961), 231). Liverpool merchant.
2. Hannah Mary Rathbone, née Reynolds (1798–1878, DNB). Writer and historian. Married 1817.
3. For the meeting of the British Association.
4. William Reynolds (1803–1877, Greg (1905), 204). Liverpool merchant.
5. Letter 2867.
6. See Faraday to Lyon, 28 December 1846, letter 1941, volume 3.

Letter 2897
William Lassell[1] to Faraday
12 September 1854
From the original in RI MS Conybeare Album, f.40

Bradstones, Sandfield Park | nr Liverpool 12th Sep 1854

My dear Sir

May I take the liberty of saying that if it should be your intention to visit Liverpool on the occasion of the approaching meeting of the British Association, it will give Mrs. Lassell[2] & me much pleasure to receive you as our guest during the week.

I do not know whether you are personally acquainted with Mr. Nasmyth[3], but as he will be staying with us at that time I venture to think his society would not render your sojourn the less agreeable.

I remain | My dear Sir | very truly yours | Wm Lassell

To | Dr. Faraday | &c &c

1. William Lassell (1799–1880, DSB). Liverpool astronomer.
2. Maria Lassell, née King (c.1791–1882 Private communication from Allan Chapman). Married Lassell in 1827.
3. James Nasmyth (1808–1890, DSB). Engineer and astronomer in Manchester.

Letter 2898
John Couch Adams[1] to Faraday
12 September 1854
From the original in RI MS F1 I116

Pembroke College | Sept 12 1854

Dear Sir

Will you excuse the liberty I take in writing to introduce a young friend of mine, Mr Monk[2] of St John's College, whom I have known during his three years residence at Cambridge, as an Undergraduate. Mr Monk, I believe, wishes to consult you about some point of chemical analysis, & I am sure that any information you may give him will be gratefully appreciated.

Believe me | Dear Sir | Yours very truly | J.C. Adams
Profr Faraday

1. John Couch Adams (1819–1892, DSB). Mathematical astronomer and Fellow of St John's College, Cambridge.
2. William Monk (b.1826, AC). Undergraduate at St John's College, Cambridge, 1851–1855.

Letter 2899
Faraday to Christian Friedrich Schoenbein
15 September 1854
From the original in UB MS NS 415

Royal Institution 15 Septr. 1854

My dear Schoenbein

Just a few scattered words of kindness not philosophy for I have just been trying to think a little philosophy (magnetical) for a week or two & it has made my head ache, turned me sleepy in the day time as well as at nights, and instead of being a pleasure has for the present nauseated me[.] Now you know that is not natural to me for I believe nobody has found greater enjoyment in physical science than myself, but it is just weariness which soon comes on but I hope will soon go off by a little rest. However rest is not to be had yet for as I have not been to the British Association for some years I have promised to go next week to Liverpool and I know from experience that is not rest[.] I do not intend to stop more than three days. Though I date from the Institution I may say that we are 12 or 14 miles out of town getting some fresh air. We are often obliged to go out of town and that is the reason why I have not seen your friend Mr. Stehlin[1] whose letter[2] I had I think some time after that of the 4th July though dated before it.

The July letter[3] was a great delight both your kindness and your philosophy most acceptable and refreshing. I hope to get your paper[4] translated but there is a great deal of vis inertia our way & I cannot overcome it as I would wish to do. It is the more difficult for me to criticize it because I feel a good deal of it myself, and am known to withdraw from the labour & responsibilities of Scientific work - and this makes me very glad that you have got hold of Liebig for I hope he will aid in developing your Ozone views[.]

Much of your letter of the 4th of July I should like to have sent to the Philosophical magazine it was such a fine free, brief, comment on Ozone in many of its positions & I think might have helped to call the attention of chemists where an elaborate memoir might fail but I did not take the liberty. In fact I should not like to send all you write for if I were to put in some of your former remarks about the errors of the acid theories[5] - the nonsense of organic chemistry[6] &c we should both be extinguished or at least sent to Coventry[.]

I said we were in the country & I met lately here the Dr. Drew (that I believe is the name) who undertook to obtain Ozone observations for you in England. He spoke as if his correspondents were discouraged by the uncertainty of their results and indeed Airy also wrote to me to ask me if I was aware that test papers which would give after exposure a certain degree of indication of ozone lost much of their power in 2 or 3 hours after & then gave a less degree[7][.] Dr. Drew talked about these points but I said little & rather referred him to you to whom he said he was about to send some communications[.]

You give a happy account of your family[.] You are a happy man to have such a family, and you are happy in the temperament which fits you for the enjoyment of it. May God bless every member of it and yourself with a cheerful & relying spirit & love to each other[.] Remember us to them all[.]

Ever My dear friend | Affectionately Yours | M. Faraday
Dr Schoenbein | &c &c &c

Address: Dr Schoenbein | &c &c &c | University | Basle | on the Rhine

1. Unidentified.
2. That is letter 2828.
3. Letter 2864.
4. Schoenbein (1854b).
5. Letter 2790.
6. Letter 2578.
7. Letter 2830.

Letter 2900
Joseph John William Watson to Faraday
18 September 1854
From the original in GL MS 30108/2/64.16

31 Pall Mall | Sept. 18: 1854

My dear Sir,
My letters by desire were not forwarded to Paris whence I have just returned or I should have replied earlier to your exceedingly obliging letter of the 8th of this month[1]. I have not applied to Mr Herbert to see your report[2] since from your letter I fear that I might read something which might tend to dull the ardour which I possess to bring the Electric light into use as a practical thing.

I do yet hope that I may once more, at no very remote day, write and obtain your presence at some experiments which may be deemed by you quite successful and even set at rest the lighthouse question.

Meantime, I am endeavouring to place the light in the Harbour Service where something of its effects as beheld from a long distance may be seen and appreciated.

With many thanks for your extreme courtesy and kindness.

Believe me, | My dear Sir, | Very faithfully Yours | Joseph J.W. Watson
M. Faraday Esq D.C.L | &c &c &c

1. Letter 2895.
2. Letter 2878.

Letter 2901
Faraday to Julius Plücker
19 September 1854
From the original in NRCC ISTI

Royal Institution | 19 September 1854

My dear Sir
At the very time that I had sat down to write to you in sincere acknowledgement of your last kind letter[1] I received one of the 21st August[2] and the papers[3]. The latter I immediately distributed giving the one copy to Dr. Tyndall and sending the others to the houses of Mr. Barlow and Mr. Grove. You go on working earnestly and well a great pleasure to yourself an[d] an honor to your friends and country and it is quite cheering to an old man like me to see it in your letters and your labours. Though cut off from the German language yet by dint of perseverance amongst some of my friends I get hold of the thoughts in your papers as

well as of those in the papers of other worthies of your country but then my memory is weak and soon holds them but indistinctly and then I mourn a little for the labour of recovering all I want to know and of doing that again & again is more than health can bear[.] I well know that if the time is come for me to cease running in the race I should be most ungrateful to murmer much. I ought rather to rejoice that I can enjoy the pleasure of looking on at the fine exertions of others[.]

I have not been at the British Associations for some years so tomorrow I go off to Liverpool to be present for 2 or 3 days at the one now approaching. I hope you will be successful & happy at Vienna. You must need a holiday for when you talk of your labours three lectures a day & successful research you quite frighten me[.]

I am Your Very true friend | M. Faraday
Professor Plücker | &c &c &c &c

Address: Professor Plücker | &c &c &c | University | Bonn | on the Rhine

1. Letter 2884.
2. Letter 2789.
3. Plücker (1854b, c).

Letter 2902
Faraday to Josiah Latimer Clark
26 September 1854
From the original in New York Public Library
[Royal Institution embossed letterhead] | 26 Septr. 1854
My dear Sir
I called yesterday to congratulate you[1][.] I trust every day makes you more & more happy in your change[.] Though I have no proper right still give my sincerest hopes & wishes to Mrs. Clarke[2] as an offering of respect tow[ar]ds you & to her through you[.]

Poor Melloni is dead. He left this life in the middle of the correspondence I had with him about your experiments[3]. I sent him all the data you sent to me[4] & he published some of them in a brief form[5]. I was waiting for his further results but now I fear there is no cause to expect any more. Those he had in hand were imperfect & I fear he cannot have lived to finish them[.]

Now the results which shewed that the force of a weak battery passed with equal rapidity along the line as that of a strong battery are very interesting to me. I could not keep copies of those sent to Melloni for want of time but if you could let me have them I think I should like to consider them in relation to my views and send a note or brief paper to the Phil Mag upon that point[6][.] Can you favour me so far?

Ever My dear Sir | Yours faithfully | M. Faraday

Latimer Clarke Esq | &c &c &c

---

1. Upon his marriage on 12 September 1854, GRO. (The marriage was dissolved in 1861, *Times*, 2 July 1861, p.11, cols. c-d).
2. Margaret Helen Clark, née Preece (age 25 at time of marriage, GRO).
3. Letters 2813, 2846, 2862.
4. Letter 2843 sent with letter 2846.
5. Melloni (1854c, d).
6. Faraday (1855c).

Letter 2903
César Mansuète Despretz[1] to Faraday
26 September 1854
From the original in IEE MS SC 2

Mon cher Confrère.

Mr. Ruhmkorff le plus habile constructeur de Paris et peut être del'Europe des appareils fondés sur vos beaux traveaux del'Induction, se rend à Londres. il desire y vendre le brevet d'un telegraphe qui parait réunir beaucoup de qualités. Il montera son appareil, vous pourrez en juger par vous même s'il vous était possible de lui rendre service, dans cette affaire, vous rendriez service au fabricant le plus honnète et le plus delicat, qu'il y ait dans le monde.

J'ai été deux fois à Londres sans pouvoir avoir le bonheur de vous rencontrer. J'espère que je serai plus heureux à un troisième voyage. En attendant, recevez l'assurance des sentimens distingués avec lesquels je suis | votre devoué | Despretz | m. del'Institut.

Paris le 26 7bre 1854

Address: A Mons. Mr. | Faraday m. de la Ste | royale de Londres & | à l'Institution royale

---

TRANSLATION

My Dear Colleague,

Mr Ruhmkorff, the most able maker, in Paris and perhaps Europe, of instruments based on your beautiful work on Induction, is coming to London. He would like to sell the patent of a telegraph that seems to unite many qualities. He will set up his instrument and you will be able to judge it for yourself. If it were possible for you to be of any service to him in this matter, you would be helping the most honest and most sensitive manufacturer there is in the world.

I have been to London twice without having the good fortune of meeting you. I hope to be more fortunate on a third trip. In the meantime, please accept the assurance of the distinguished sentiments with which I am | your devoted | Despretz | member of the Institut.
Paris, 26th September 1854

1. César Mansuète Despretz (1792–1863, DBF). French chemist.

Letter 2904
Samuel Finlay Breese Morse[1] to Faraday
30 September 1854
From the original in IEE MS SC 2

New York, United States. | Pokeepsie, Septr. 30th 1854

Sir,
I have had the gratification of making the acquaintance of C.D. Archibald[2], Esq F.R.S. who has visited this country on business in connection with the great and important project of uniting Europe and America by a submarine Telegraph, a project which has occupied my mind with much interest since the year 1842.

In the Prospectus of the British company formed for carrying out this project, I perceive with the highest gratification among the illustrious names who are giving it their countenance, your own illustrious name as *Electrician* of the company[3]. In the Company which had been previously formed on this side of the water for the same purpose, I have had the honor to be elected to a similar office. The negotiations for a Union of the cis-Atlantic and trans-Atlantic companies, conducted through Mr. Archibald's agency on the part of the latter, and the Board of Directors in New York on the part of the former have resulted successfully and in view, therefore, of these relations I take the liberty of addressing you on the subject of the feasibility of a submarine communication across the Atlantic.

In the autumn of 1842 I laid down probably the *first Submarine Telegraph line, ever laid,* in the harbor of New York, connecting Governor's Island, with Castle garden at the Battery, a distance of about one mile. The wire conductor was of copper No. 18 wound with cotton twine, passed through a shellac varnish, and then through a resinous compound of tar and asphaltum. Although this line was destroyed in the midst of my experiments by being accidentally drawn up the anchor of a vessel and cut off, it was not until I had passed communications through it from station to station. I had previously experimented on 33 miles of wire, (insulated in the same manner, and wound on reels,) for the purpose of ascertaining the number of pairs of plates in the Galvanic battery, which might be necessary to the successful working of my Telegraph on long distances, and in the autumn of (1843) eighteen hundred & forty three, having at my command 160 miles of wire thus insulated, I repeated the experiments confirming the inferences I had drawn from them, to wit; that in the application of the Battery to the Telegraph "while the distance increased in an Arithmetical ratio, an addition to the series of Galvanic pairs of plates increased the magnetic power in a geometric ratio." Hence I felt assured that the voltaic current could be propelled with effective power to any distance, unless some new & unperceived obstacle growing out of a varied condition of the several parts should arise to prevent this result.

In my report to the Secretary of the Treasury of the results of these experiments on 33 miles of conductors, and dated August 10th 1843, I perceived their bearing on a suboceanic Telegraph, and thus alluded to it to the Secretary; "The practical inference from this law is that a telegraphic communication on my plan, may with certainty be established across the Atlantic! Startling as this may *now* seem, the time will *come when this project will be realized.*"

That time thus predicted eleven years since, seems now to be near, and the project seriously undertaken. It is, nevertheless, due to the distinguished personages and gentlemen on both sides of the Atlantic who have lent their names and their means to the enterprize, that every part of the process should be carefully reviewed and a plan of operation proposed that shall be cautious as well as energetic, and economical as well as ample for insuring the success of the work.

The experiments to which I have alluded as made in 1842 were the basis of the battery adaptation to my Telegraph from its first establishment. There are some 40,000 miles of Telegraphic conductors in this country. As yet the longest connected line is in length two thousand & fifty nine miles, (2059) from New York to New Orleans. This line from its being erected through dense forests and marshes at the extreme South has never been in a condition in which I could test an all important experiment bearing directly upon the practicability of transmitting an available current from two electro-motors, one at each extremity of the circuit. To test this point satisfactorily, whether the current may thus be transmitted effectively a distance of 2000 miles throug[h] a metal conductor, so that the possibility of a submarine Ocean Conductor from Newfoundland to Ireland may be pronounced a fact, all the relay magnets on such a line as well as the intermediate batteries should be removed, & the line connected in one continuous circuit of the same conducting capacity throughout. This I have never been able to try to the extent necessary to pronounce upon the result, owing to the fact that this line has hitherto not been reliable throughout the whole distance, at any one time. I am about, to try, however, a similar experiment on the New York, Albany, and Buffalo Telegraph line, a distance of 506 miles but having five wires throughout the whole distance, so that I can try distances of 506, 1012, 1518, 2024, and 2530 miles at pleasure.

This line I am in expectation of seeing in such a condition in the coming month of October that I may be able soon to give you my results.

In the meantime please inform me if in your profounder and more extended researches in Electricity you have any reason to doubt the practicability of transmitting a current of galvanic electricity through such a length and whether you apprehend any new conditions in a submerged wire well insulated, that would be likely to interfere with, and obstruct the process.

This point I conceive is the most important to be settled, of all the apprehended difficulties of a submarine Telegraph across the Ocean. This point once determined favorably, the other parts of the enterprize range themselves under the head of simple engineering. British and American seamen will speedily do the rest, for all that will remain, will depend on the skill with which a perfect and indestructible well insulated wire cable, is deposited uninjured in its Ocean bed.

A question, indeed, arises which has been the subject of some speculation with me, to wit; Whether there is not a new condition of things in the fact of so deep a submersion in the Ocean, that may develope some new and unlooked for influence and impediment in the condition of electricity in a conductor of such length. If I have been rightly informed, there has been noticed in Europe some difference in the character of the conduction in wires above ground and beneath the ground[4]. As we have no subterranean wires in this country I am unable to verify these rumors by experiment. I have been informed that in subterranean wires extended more that 300 miles the current when continued in a closed circuit for a second of time does not cease at the moment of opening the circuit with that promptness with which it does in conductors above ground, so that the current lingers, as in the case of the magnetism of the Electro magnet when the helices of the magnet are of great size, and requires an appreciable time to be discharged.

Have you met with any facts on this subject corroborative of the rumors reported to me?

But taking for granted a successful result of the experiment on the propulsion of a current to the required distance that is to say from Newfoundland to Ireland, I have proposed that the cable conductor be constructed in the following manner, to wit;

The conducting wires of the circuit I propose to be of the purest copper, each not less than one eighth ($\frac{1}{8}$th) of an inch in sectional diameter.

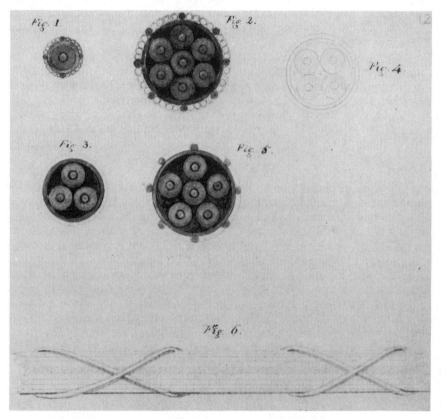

Each wire to be insulated to the thickness also of one eighth of an inch with gutta percha. If it should be decided by the company that in the first instance a *single conductor* shall be laid down, then as in Fig. 1. of the enclosed sheet of diagrams, representing a section of the cable, a thin tube of lead about one sixteenth of an inch ($\frac{1}{16}$th) in thickness is drawn over the wire conductor and its gutta percha covering, and then a series of strands of common iron wire and of hempen cord, or rope yarn of the same size, say four or five of the former, and the rest of the latter, are to be laid parallell with the interior conducting wire, on the exterior of the tube (Fig. 6.) and these are to be confined in place by two spiral cords wound in contrary directions and crossing each other around the cable at intervals say of nine or twelve inches.

If it is thought best to lay down, in the first instance, more than one conductor in the same package or fascis, then the number chosen may be three as in Fig. 3. or seven as in Fig. 2 these being the numbers most economically packed in a tube to form the fascis of conductors. Six wires as in Fig. 5, and four wires as in Fig. 4 do not pack in a tube economically.

On the propriety of such a mode of forming the cable I reason thus;

Supposing that more than one wire is adopted, the increased expense of preparing seven wires, over a less number is scarcely worthy of mention, when it is considered that the other expenses attending the enterprize, such as chartering of ships, laying out the cable &c. would be incurred nearly if not quite as largely for a smaller number of wires as for a larger number, especially too when, as an offset, a provision for an increased Telegraphic correspondence is thus made. I, therefore, propose the number *seven* as most convenient and economical for packing in a tube.

I propose the *leaden tube* for the sole purpose of protecting the gutta percha insulation from the action of sea water. This substance being of recent introduction in the arts has as yet not had its qualities for resisting, for a period of years, the action of salt water sufficiently tested, to warrant an unqualified recommendation of it without this outer protection of lead. Lead if I am rightly informed is not corroded beneath the salt water. On this point, however, I ask your opinion from your superior knowledge. But the thickness of the tube proposed is very small so small that without extensive aid it probably could not, in a length of only a few hundred feet, sustain its own weight. This feebleness of tenacity I compensate by the exterior series of strands of wire and hempen cord, and these are laid parallell to the interior conducting wire, in order to have the advantage of the full strength of the iron wires interspersed with the hempen cord. I depend on the *leaden tube* not for strength but only for protection of the insulation. I depend on the exterior series of strands of *iron wire* only for such *strength* to the cable as will fit it for being paid out without injury to the protecting tube, or interior conductor. The number of iron wires interspersed with the cord may of course be varied to suit the calculated weight of three or four miles of the cable suspended at a time in its reach from the ship to its Ocean bed. I propose *hempen cord* to fill the space between the iron wires to keep them in place as a cheap contrivance for that purpose, and I consider the whole exterior covering of the leaden tube doomed speedily to perish by the action of sea water, having performed its whole duty in simply imparting strength sufficient to lay the cable in its place.

In regard to the cable previous to its being laid down, and while in various stages of preparation, it will furnish opportunity to test step by step its fitness for the office it is to perform.

In view of the final deposite of the cable in its place, there are certain other points of importance to be determined. Such for example as the protection necessary for it near the shores in anchorage ground from the anchors of vessels and in deeper but still shallow water, from the icebergs in their spring visits from the Arctic Ocean. I have had my thoughts upon this part of the subject, but will wait a future opportunity to give the result of these thoughts, on the mode of avoiding the evils which might result from these and other disturbing causes.

In the month of July I took the liberty of sending to your address by my son in law Mr. Lind[5] and family who are now in Europe, two pamphlets containing some information respecting the Telegraph, which I hope you have received[.]

Accept Sir the assurance of | my sincere & profound respect | Your Obedient Servant | Sam. F.B. Morse
To Sir Michael Faraday, | F.R.S. &c &c | London.

1. Samuel Finlay Breese Morse (1791–1872, DAB). American artist and inventor.
2. Charles Dickson Archibald (1802–1868, B1). Writer on colonial topics.
3. This assertion, the source of which has not been located, was repeated in Morse (1914), 2: 343.
4. Faraday (1854g).
5. Otherwise unidentified.

Letter 2905
James Walker to Faraday
2 October 1854
From the original in RI MS F1 I65a
My dear Mr Faraday

As the Casquets will be No. 1 of your plan for improved ventilation I have said that the stoves sh[oul]d be of the size & form you approve, and am desired to consult with you respecting them. If you sh[oul]d be passing this way, to or from elsewhere will you kindly look in. I shall be at home tomorrow morning & forenoon; but Mr Cooper[1] will be in the way all Wednesday[2] also. It is desirable to have the stoves sent over[3][.]

I am | My dear Sir | very sincerely | J. Walker
Gt Geo St | 2 Oct | 1854
Professor Faraday

1. James Cooper (1817–1862, Min.Proc.Inst.Civ.Eng., 1863, 22: 624–5). Civil engineer and a partner of Walker's from 1851.
2. That is 4 October 1854.
3. See Trinity House By Board, 21 Faraday 1854, GL MS 30010/38, p.440 for approval to implement Faraday's ventilation plan at the Casquets.

Letter 2906
Michel Eugène Chevreul[1] to Faraday
6 October 1854
From the original in IEE MS SC 2

MINISTERE | de la Maison | DE L'EMPEREUR. | MANUFACTURE |
impériale | DES GOBELINS | Paris le 6 d'octobre 1854
Monsieur et honorable confrère,

Je profite de l'occasion qui m'est offerte par Mr Barnard votre neveu pour vous prier d'accepter 1° un exemplaire de l'ouvrage que je viens de publier sur la baguette divinatoire, le pendule explorateur et les tables tournantes[2]. 2° le rapport que j'ai fait au Jury francais sur l'exposition au palais de cristal des produits des gobelins de Beaurais et de la Saroincerie[3]; 3° des memoires de chimie appliquée à la teinture[4]. Je vous prierai d'offrir en mon nom des exemplaires des mêmes ouvrges à la Société Royale de Londres comme un hommage de la profonde estime pour les travaux d'un de ses membres étrangers.

Si vous prenez la peine de lire le premier ouvrage vous verrez que nos opinions sont bien près de l'identité.

J'ai eu beaucoup de plaisir a faire la connaissance de Mr votre neveu. Je lui trouve toutes les qualités possibles pour honorer sa carrière: et je ne doute pas de ses succès. Je peux être à la veille de publier mon ouvrage sur un moyen rationnel et experimental de définer les couleurs. J'ai trouvé un artiste[5] qui vient de reproduire heureusement mes cercles chromtiques par la chromalithographie[6] de sorte qu'il me sera facile de me faire entendre dorénavant.

cest Mr votre neveu qui m'a appris que lon avait traduit mon ouvrage sur le contraste en anglais[7]; je m'en felicite je vous assure.

Recevez, Monsieur et honorable confrère, l'expression de mes sentiments les plus affectueux | E Chevreul

TRANSLATION
MINISTERE | de la Maison | DE L'EMPEREUR. | MANUFACTURE |
impériale | DES GOBELINS | Paris 6 October 1854
Sir and honourable colleague,

I am taking advantage of the opportunity offered to me by Mr Barnard, your nephew, to ask you to accept 1st a copy of a work that I have just published on the divining rod, the exploratory pendulum and table turning[2]; 2nd the report which I gave to the French Jury on the exhibition at the Crystal Palace on the gobelin products of Beaurais and la Saroincerie[3]; 3rd some papers on the applied chemistry of tinting[4]. I would ask you to offer in my name copies of the same works to the Royal Society in London, as homage of the profound esteem for its work of one of its foreign members.

If you find the time to read the first work, you will find that our opinions are all but identical.

I have very much enjoyed making the acquaintance of your nephew. I find in him all the qualities possible to further his career; I have no doubt as to his success. I may be on the eve of publishing my work on a rational and experimental way of defining colours. I have found an artist[5] who has happily just reproduced my chromatical circles by chromolithography[6], in such a way that it will be easy for me to study it from now on.

It was your nephew who told me that my work on contrast had been translated into English[7]; I am very flattered by this, I can assure you.

Please accept, Sir and honourable colleague, the expression of my most affectionate sentiments | E Chevreul

1. Michel Eugène Chevreul (1786–1889, DSB). Director of dyeing at the Gobelins tapestry works and Professor of Chemistry at the Muséum d'Histoire Naturelle.
2. Chevreul (1854b).
3. Chevreul (1854c).
4. Chevreul (1854d).
5. René Henri Digeon. Parisian engraver. BDPSDG.
6. Chevreul (1855).
7. Chevreul (1839) was translated into English as Chevreul (1854a).

Letter 2907
Faraday to Jacob Herbert
11 October 1854
From the original copy in GL MS 30108/2/65
Report &c

Royal Institution | 11 October 1854

My dear Sir

Since Friday[1] last I have been to Haisboro[2], according to your instructions[3], & visited both the lights. I found nothing wrong in the lanthorns, nor any complaints from the Keepers respecting them. There is no sweating on the glass, for it is always perfectly dry internally, & the ventilation perfect. You are aware that the ventilating tubes are applied to the lamps, & the testimony of the keepers, is, that there *was* condensation on the glass before their application, but that there *is* none now. On a long cold night, sometimes between 3 & 4 pints of water are condensed in the collecting ball at the upper lighthouse; but none at the lower, where it is all carried out at the cowl by the ventilation.

The keepers, who have in part to live in the towers, complain of great dampness there; and there are sufficient proofs that the complaints are well founded. This part of the subject I leave to the far superior judgement of Messrs. Walker & Cooper[4]. But I may remark that the source of dampness appears to me to be from above, and is probably of the following nature. The towers are 60 years old or more and are of brick; they are painted externally & partly internally, but fissures & cracks occur in this covering, & there are also fissures between the large window frames which are numerous & the surrounding brick work. When a driving rain beats against & runs down the tower, it is in part drawn into these fissures; & sometimes the keepers have to put pails to catch what is driven through at the windows, though they are not made to open as casements. Such water as thus soaks into the brick work cannot evaporate outwards, because of the paint; & has to be removed entirely by internal evaporation; & as the rooms are closed, & fires are not kept up in the fire places as in old times (as the keeper of his own knowledge tells me was the case) so they are now very damp.

The division of the towers into rooms cuts off, as it were, the open connexion between the tower & the lanthorn, and it supplies as I think good illustrations of the value of that principle, which I have recently commended to the Trinity House[5]. The lanthorns are good, but the towers are, as to dampness, bad; & we have each exhibiting its state when separated from the other[6].

I am My dear Sir | Your Very faithful Servant | M. Faraday
Jacob Herbert Esq | Secretary &c &c &c

1. That is 6 October 1854. Faraday's notes of his visit are in GL MS 30108/2/65.
2. In Norfolk.
3. See letter 2886.
4. James Cooper (1817–1862, *Min.Proc.Inst.Civ.Eng.*, 1863, **22**: 624–5). Civil engineer and a partner of Walker's from 1851.
5. See letter 2778.
6. This letter was read to Trinity House By Board, 17 October 1854, GL MS 30010/39, p.132. It was referred to the Deputy Master, Warden and Light Committees.

Letter 2908
Faraday to Henry Bence Jones
12 October 1854
From the original in RI MS F1 D25

R Institution | 12 Octr. 1854

My dear friend

My hours here have not been such as gave me the choice of seeing you or I should have thanked you for past recent kindness & asked for more - for you gave my wife a prescription containing petrol &c to aid her on the occurrence of white spots in her mouth & concurrent derangement of stomach & I can not find it. Anderson tells me you are to be in the

Laboratory tomorrow morning. Can you remember enough to write it for her & he will send it to Surbiton. I trust very freely to your kindness and often feel some reproof in calling so freely upon it.

With Sincerest thanks | Ever Yours most truly | M. Faraday
Dr. B. Jones | &c &c &c

Letter 2909
Faraday to Charles Pooley[1]
14 October 1854
From the original in WIHM MS FALF

[Royal Institution embossed letterhead] | 14 Octr. 1854
Dear Sir
I have received your kind note & the book and thank you very much for both. I am glad the lecture[2] obtains your approbation[.] I have sufficient communications of the opposite nature to make such as yours very acceptable[.]

I am Sir | Your Very faithful Servant | M. Faraday
Chas. Pooley Esq | &c &c &c

1. Unidentified.
2. Faraday (1854f).

Letter 2910
Faraday to Thomas Byam Martin
14 October 1854
From the original in BL add MS 41370, f.374

[RI letterhead] Royal Institution | 14 Octr 1854
My dear Sir Byam
I have no doubt you have thought of the matter but I cannot resist referring to the combustion of I think above 2000 tons of Sulphur in the middle of a crowded town like Newcastle & *as regards a certain application*[1], the little comparative evil it has done[2][.]

I am | My dear Sir Byam | Your faithful Servant | M. Faraday

1. See letter 2871.
2. A reference to a destructive fire in Newcastle which started on 6 October 1854 and destroyed many chemical factories. See *Ann.Reg.*, 1854, **96**: 170–4.

Letter 2911
Faraday to Henry Bence Jones
14 October 1854
From the original in RI MS F1 D26

[Royal Institution embossed letterhead] | Saturday | 14 Octr 1854
My dear Dr. Jones
    Your Very kind note arrived this morning & was received with many thanks. Do not on any account trouble yourself to come out we shall be home Thursday Morning[1]. By virtue of your note I called on Rhumkorf this morning & saw him[.] I like him very much indeed[.] I must call on Gassiot next week & hope I shall find him better. I suppose it is bile[.]
    Ever Your obliged friend | M. Faraday

1. That is 19 October 1854.

Letter 2912
Faraday to Henry Stevens[1]
19 October 1854
From the original in UCLA UL Henry Stevens collection #801, Box 39, folder 6

Royal Institution | 19 Octr. 1854
Sir
    I beg to acknowledge the packages with many thanks and to return according to the instructions the enclosed signed paper[.]
    I am Sir | Your Very Obedient Servant | M. Faraday
H. Stevens Esq | &c &c &c

1. Henry Stevens (1819–1886, DAB). American born bookdealer who worked in London from 1845.

Letter 2913
Faraday to Charles Grey
20 October 1854
From the original in Kunstsammlungen der Veste Coburg, Autographensammlungen, Inv. IV, 997

Royal Institution | 20 Octr. 1854
My dear Sir
    There is a Gentleman in town at present from Paris, M. Leon Foucault[1] who has some most beautiful apparatus & experiments

connected with a matter which I know has interested his Royal Highness Prince Albert very much namely the fixity in space of a rotating disc[2]. I have just seen the results & gone into the proofs they give of the Rotation of the earth independent of astronomical considerations[3]. I leave you to judge whether it is expedient to place the matter under His Royal Highness's observation[4]. M. Foucault is a perfect gentleman[.] His apparatus is portable. If the Prince should desire to see the results M. Foucault would require only a steady table & perhaps a previous half hour to take his apparatus out of the boxes. The chief experiments which are several & very interesting could be made & explained briefly in twenty minutes or even less but their relation to the earth & its phenomena easily extend to a longer time with those interested in the matter. M. Foucault is residing at present with his friend Dr. Mussy[5] 14 Clifford Street Bond St[.]

 I have the honor to be | My dear Sir | Your Very humble Servant | M. Faraday
The Honorable | Genl. Grey | &c &c &c

1. Jean Bernard Léon Foucault (1819–1868, DSB). French physicist.
2. See letters 2838 and 2839.
3. Foucault (1851).
4. See "Court circular", *Times*, 27 October 1854, p. 6, col. f which noted Foucault's demonstration before Prince Albert the previous day.
5. Probably Henri Guéneau de Mussy (1814–1892, B2). French physician who lived in England between 1848 and 1872.

Letter 2914
Arthur-Auguste De La Rive to Faraday
27 October 1854
From the original in IEE MS SC 2
Genève | le 27 8bre 1854
J'ai eu la visite de mon jeune ami Mr Verdet[1] qui a passé huit jours chez moi à la campagne; il était bien reconnaissant d'une lettre qu'il avait reçu de vous; c'est un jeune physicien bien distingué & qui ira loin.- Oserais-je vous prier de faire parvenir la lettre ci incluse a Mr Tyndall dont j'ignore l'addresse.
Monsieur & très cher ami,
 Je n'ai pas encore répondu à votre si bonne & excellente lettre du 29 mai[2] dont j'ai été bien touché comme je le suis toujours de tous les témoignages d'amitié que vous me donnez. A mesure que l'on avance en âge & qu'on voit les rangs de ses amis s'éclaircir autour de soi, on a d'autant plus besoin des marques d'affection de ceux qui vous restent, & quand surtout on a été frappé comme je l'ai été de manière à n'avoir plus

qu'une moitié de vie, on est encore plus sensible aux consolations d'une amitié aussi sympathique & aussi sérieuse que la votre. Personne ne comprend mieux que vous que ce n'est plus dans ce monde que je dois chercher le bonheur & que mes pensées doivent s'élever plus haut.

J'ai su que vous aviez été peu bien cet été; j'espère que le repos vous aura fait du bien & que vous êtes maintenant tout-à-fait remis. Donnez moi de vos nouvelles dès que cela vous sera possible.

Vous me parlez dans votre lettre del hypothèse par laquelle je cherche à expliquer le magnétisme des corps & vous me faites l'objection que mes molécules devraient dans cette hypothèse, avoir la forme de disques. Cela semble en effet résulter de la conception que j'ai mise en avant. Cependant je ne crois pas que cette conséquence soit rigoureusement nécessaire.

Je distingue l'*atome chimique* de la *molécule intégrante* celle-ci étant formée par un groupe plus ou moins considérable d'atomes chimiques. J'admets que dans les corps qui sous le même volume renferment le plus grand nombre d'atomes chimiques, les atomes sont beaucoup plus rapprochées les uns des autres dans la molécule intégrante d'où nait un courant électrique pour chaque groupe, par l'effet de leur polarité. Or que la molécule intégrante soit sphérique, cubique, octaédrique ou rhomboëdrique, rien n'empêche les atomes de s'arranger de manière à former autour de la molécule des ceintures de courants tous parallèles les uns aux autres & dirigés dans le même sens. Le disque serait dans ce cas la section équatoriale d'une molecule sphérique. Quand le fer n'est pas aimanté, les molécules intégrantes se disposent naturellement de façon que l'action naturelle de tous leurs courants soit neutralisée, ce qui constitue le cas d'équilibre. Mais dès qu'une source extérieure telle qu'un aimant ou un courant vient à agir, alors tous les courants moléculaires s'orientent, & le corps est aimanté.

Quant au diamagnétisme, je crois qu'il est dû à une action inductrice moléculaire du même genre que l'induction que vous avez découverte, mais avec cette différence qu'elle persiste tant que la cause qui la produit est présente. Dès lors les courants moléculaires doivent être dans le corps induit dirigés en sens contraire des courants inducteurs, ce qui explique la répulsion exercée sur les corps diamagnétiques. Les dernières recherches de Matteucci & celles de Tyndall sur la polarité des corps diamagnétiques sont favorables à cette hypothèse[3]. Je n'ai pas la place dans une lettre de développer suffisamment mon explication; mais si vous me le permettez, je pourrai le faire une autre fois. Je suis convaincu qu'on peut ramener à la même cause les phénomènes du diamagnétisme & ceux de l'induction.

J'ai été très occupé pendant cet été de mon 2d volume[4] qui est presqu' entièrement achevé; j'espère que vous en serez content, c'est toute mon ambition, car il n'y a aucune opinion à laquelle je tienne autant qu'à la votre

Votre tout dévoué & affectionné | A. de la Rive

Address: Prof Faraday | &c &c &c | Royal Institution | Albermarle Stt | Londres

TRANSLATION
Geneva | 27 October 1854
I had a visit from my young friend M. Verdet[1], who spent eight days with me in the country; he was very grateful for a letter which he received from you; this is a young most distinguished physicist who will go far. Dare I ask you to forward the enclosed letter to Mr Tyndall, whose address I do not know.

Sir and very dear friend,
    I have not yet responded to your most kind and excellent letter of 29 May[2], which touched me as do all the testimonies of friendship that you give me. As one advances in years & one sees the ranks of one's friends thinning out around one, one has all the more need of the marks of the affection of those who remain & particularly when one has been hit, as I have, in such a way as to have but half one's life left, one is even more sensitive to the consolations of a friendship as sympathetic and serious as yours. No one understands better than you that it is not in this world that I should search for happiness and that my thoughts should be lifted higher.

    I learned that you have not been well this summer; I hope that rest will have done you some good and that you have now completely recovered. Please send me your news as soon as you can.

    You speak in your letter of the hypothesis through which I seek to explain the magnetism of bodies and you object to the fact that my molecules, in this hypothesis, would have to have the form of discs. This seems in fact to result from the conception that I first proposed. However I do not believe that this consequence needs to be rigorously applied.

    I distinguish the *chemical atom* from the *integral molecule*, the latter being formed by a greater or smaller group of chemical atoms. I admit that in bodies which in the same volume enclose the greatest number of chemical atoms, the atoms are much closer together in the integral molecule from which is born an electric current for each group, by the effect of their polarity. Now even if the integral molecule is spherical, cuboid, octahedral or rhomboidal, nothing stops the atoms from arranging themselves in such a way around the molecule as to form belts of currents all parallel to each other and pointing in the same direction. The disc would in this case be the equatorial section of a spherical molecule. When iron is not magnetised, the integral molecules arrange themselves naturally in such a way as the natural action of all their currents is neutralised, which constitutes the state of balance. But as soon as an external source such as a magnet or a current begins to act, then all the molecular currents arrange themselves and the body is magnetised.

As to diamagnetism, I believe that it is due to an inductive molecular action of the same sort as the induction you have discovered, but with this difference that it persists as long as the cause which produced it is present. It follows that molecular currents must in the induced body be directed in the opposite direction than the inducing currents, which would explain the repulsion exerted on diamagnetic bodies. The latest research by Matteucci and by Tyndall on the polarity of diamagnetic bodies is favourable to this hypothesis[3]. I do not have the space in a letter to develop my explanation adequately, but if you were to allow me, I could do so another time. I am convinced that one can reduce to the same cause the phenomena of diamagnetism and those of induction.

I have been very occupied this summer with my second volume[4] which is almost completely finished. I hope that you will be pleased with it, that is my ambition, for there is no opinion which I hold in higher regard than yours.

Your totally devoted and affectionate | A. de la Rive

1. Marcel Emile Verdet (1824–1866, P2, 3). Professor of Physics at the Ecole Normale, Paris.
2. Letter 2840.
3. Matteucci (1853b). Tyndall (1854c).
4. De La Rive (1853–8), **2**.

Letter 2915
Faraday to John Tyndall
31 October 1854
From Breda and Logeman (1854), 465

Royal Institution, | Oct. 31, 1854.

My dear Tyndall,

I send the enclosed letter from MM. Van Breda and Logeman[1] to you as an Editor of the Philosophical Magazine. If you should judge it proper for insertion in that Journal, I shall be very happy to see it there[2], but will beg you to accompany it on my part with the observation that it is not so conclusive in proving the negative (a thing very difficult to do) as to move me at present from the reserved condition of mind which I have recently expressed in respect of this matter[3].

Ever yours truly, | M. Faraday

1. Letter 2893.

2. This letter was published in Breda and Logeman (1854), 465–9.
3. That is Faraday (1854c, d).

Letter 2916
David Brewster to Faraday
4 November 1854
From the original in RI MS Conybeare Album, f.19

My Dear Dr. Faraday,
    I hope you will excuse the trouble I now give you. I have been requested to add the recent discoveries in *Electricity* to an article which I compiled many years ago for the Encyclopaedia Britannica[1], & I would reckon it a particular favour if you could tell me where I can get an account of the most recent discoveries without the labour of consulting the Transactions of Societies and scientific Journals. Delarives Treatise is, I believe not yet finished, the first volume only being published[2].
    I presume I will find your own discoveries in the Phil. Transactions.
    I am, | My Dear Dr. Faraday | Ever most truly yours | D. Brewster
St Leonards Coll. | St. Andrews | Nov 4th 1854

1. Brewster (1842). Brewster (1855b) was the revised version.
2. De La Rive (1853–8), 1.

Letter 2917
Faraday to George Biddell Airy
7 November 1854
From the original in RGO6 / 405, f.154

Royal Institution | 7 Novr. 1854
My dear Sir
    May we hope for the pleasure of having you tell *here* some Friday Evening the manner & results generally of your surveying experiments in the coal mine?[1] You know what great delight it would give to me personally & to all here; I will not weaken your sense of our thankfulness by putting it into many words. With kindest remembrances to Mrs. & Miss Airy I am as ever
    Truly Yours | M. Faraday
G.B. Airy Esq | &c &c &c

1. Airy, G.B. (1855b), Friday Evening Discourse of 2 February 1855. See letter 2880 for details of this subject.

Letter 2918
Thomas Boosey et al. to Dundee Glasite Church[1]
7 November 1854
From the original copy in the possession of Joan Ferguson
The Church of Jesus Christ at London
To the Church of Jesus Christ sojourning at Dundee
Dearly beloved Brethren

Your affectionate & faithful letter to us dated the 31st ulto[2] relating to what you justly termed a grievous error which has crept in amongst us, was read to us on Sabbath day last[3], along with the scriptures on which your judgement on the subject had been founded; and we feel called upon after a careful examination of those Scriptures, & as we hope a sober examination of ourselves how we have been holding them, cordially to acknowledge that we have indeed been grievously in error in supposing that there were any Scriptures in the Old or New Testament which in any way countenanced the light manner in which we have held the Commandment to all believers in Jesus Christ, to abstain from blood & from things strangled[4], with the evident reference that commandment has to the blood of the one great atonement, without the shedding of which there could have been no remission of sins.

The toleration which many have allowed themselves in eating what is called game, which they thought had been shot & afterwards purified or freed from the blood, must have originated & been continued by a want of the divine fear, or as you say a trembling at the Word of God, and an ignorance of that Word by which we must be judged at the last days; for we now quite agree with you in considering those Scriptures which were thought to sanction this toleration, viz the Romans 14 ch.1 to 4 verses[5] & 1 Timothy 4 ch 4[6] verse do not at all apply to the prohibition to eat of blood or things strangled, which commandment remains of the same force & application to all believers as it ever did. What can we then say or do under this conviction of the want of the Divine Fear, and ignorance of the Divine word of him who said, if you love me keep my commandments[7] but cast ourselves entirely upon his mercy and say with the leper of Old unclean, unclean[8], if the Lord thou wilt thou canst make us clean, and abundantly pardon us; for the blood of Jesus Christ (which we have great reason to fear we have counted as a Common thing) is still able to cleanse us from all sin.

Hoping dear Brethren you will forgive the trouble we have occasioned you for his sake by whose Obedience alone many are made righteous, and who can have compassion on the ignorant and those out of the way, and lead the blind by a way that they have not known - This letter has been read & approved of by the church who heartily concur in, expressing their love & thankfulness to you for your faithfulness & goodly

jealously on their behalf and it is signed in their name this 7th day of November 1854 by the Elders Deacons & Brethren subjoined
Thomas Boosey
Stephen Leighton
Benjm Vincent
George Whitelaw
John Leighton *Deacon*
Thomas Barker
Alexander Macomie
Edward Reid
William Vincent
M. Faraday
D.W. Martin
Robt Sims[9]

1. For the background and context of this letter see Cantor (1991a), 68–70.
2. A copy of this letter is also in this file.
3. That is 5 November 1854.
4. Acts 15: 29.
5. Romans 14: 1–4.
6. 1 Timothy 4: 4.
7. John 14: 15.
8. Leveticus 13: 45.
9. Robert Sims (d.1864, age 81, GRO). Sandemanian and plate glass merchant.

Letter 2919
George Biddell Airy to Faraday
8 November 1854
From the original press copy in RGO6 / 405, f.155–6

Royal Observatory Greenwich | 1854 November 8

My dear Sir

Most happy shall I be to give such a lecture as you propose[1] - if, after weighing the following considerations you will still accept it.

1. Local circumstances, which you can easily imagine, compelled me to give a lecture at South Shields[2] (the town to which the mine is near). So that a lecture at the Institution may now seem to the attendants thereof to be simply a second-hand or rechauffeé, and accordingly may not be acceptable.

2. I am getting on with the calculations as fast as I can: till they are done I do not even know whether the experiments are worth any thing: but before the time of lecturing I hope to know all about it. Suppose they should then prove good for nothing, what should you say to it?[3]

I am, my dear Sir, | Yours very truly | G.B. Airy

Professor Faraday

1. In letter 2917.
2. Airy, G.B. (1855a).
3. Airy, G.B. (1855b), Friday Evening Discourse of 2 February 1855.

Letter 2920
Faraday to George Biddell Airy
10 November 1854
From the original in RGO6 / 405, f.157
Royal Institution | 10 Novr. 1854
My dear Sir
Very hearty thanks for your kindness[1]. Whatever the results of the special investigation I long to hear your thoughts on the whole subject[2]. I think they said in America that even the difference of high & low water on a sea coast would be sensible or had been sensibly shewn I cannot recollect which. By sinking into the earth you have far more than that to handle[.] Besides I long to know whether the earths electricity or magneto-electrical action shews any signs of life in your results[3][.]

I have not seen Mr. Barlow since I received your letter but I am sure he would say the Season is, as to date, all before you. No doubt he would be glad to know that he might arrange the dates for other[s.] Let it be early if it be the same to you. I suppose we shall begin the end of January[.]

Ever My dear Sir | Yours Very Truly | M. Faraday
Geo B. Airy Esq | &c &c &c

1. See letter 2919.
2. Airy, G.B. (1855b), Friday Evening Discourse of 2 February 1855.
3. See letter 2880.

Letter 2921
Faraday to John Tyndall
11 November 1854
From a typescript in RI MS T TS, volume 12, p.4137
R. Institution, | 11 Nov. 1854.
Many thanks, my dear Tyndall, for your kind letter which I have just received. I was anxious about you, thinking you might be confined at home by a little indisposition (as you would call it) and writing, and should probably have called today, in the evening. Now I shall rest,

knowing how it is, and I hope you will enjoy the weather and the quietness and the time of work and the time of play, finding them all ministrants to your health and contented happiness. Here we jog on, and I have just undertaken the Juvenile Lectures at Christmas, thinking them the easiest thing for me to do[1]. Reading Matteucci[2] carefully, and also an abstracted translation of Van Rees' paper[3], is my weighty work; and because of the call it makes on memory, I have now and then to lay them down and cease to the morrow. I think they encourage me to write another paper on lines of force, polarity &c[4], for I was hardly prepared to find such strong support in the papers of Van Rees and Thomson[5] for the lines as correct representants of the power and its direction, and many old arguments are renewed in my mind by these papers. But we shall see how the maggot bites presently, and as I fancy I have gained so much by waiting, I may perhaps wait a little longer.

Ever, my dear Tyndall, | Yours truly, | M. Faraday

1. Faraday gave six Christmas lectures on the chemistry of combustion. See RI MS GB 2: 85.
2. Matteucci (1854) was cited in Faraday (1855b), [ERE29b], 3308.
3. Rees (1853) was cited in Faraday (1855b), [ERE29b], 3302.
4. Faraday (1855b), [ERE29b].
5. Thomson, W. (1854) which was a reprint, with additions, of Thomson, W. (1846). Faraday (1855b), [ERE29b], 3302 cited the later version.

Letter 2922
Faraday to Edward Sabine
30 November 1854
From a copy in Bate to Sheepshanks, 18 January 1855, RGO6 / 356, f.149–50

Mr: Bate himself (Mr B Bate as above)[1] gave me to understand that his sisters were not poor[2] or rather that his father left large sums of money to the survivors - but that would make no difference to me in anything I could do with propriety. But I cannot undertake to certify to accounts upon the evidence of private notes & papers regarding time &c of which at the time I had no knowledge. All I know is that Mr Bate was employed & as far as I am aware has never been paid[3].

(signed) M.F. 30 Novr: 1854.

1. The parentheses were inserted by Bate into the copy.
2. The following note was inserted by Bate at this point: "N.B. this remark was occasioned by an explanation which I felt called up to make to Dr. Faraday, being surprized to find him under an impression that my father's estate was involved in bankruptcy | B. Bate."
3. The letter refers to Bate's continued pursuit for payment for the work done by his father (Robert Brettel Bate (1782–1847, DNBmp), scientific instrument maker in the Poultry, London)

for the Royal Society Excise Committee, of which Faraday had been a member, in the 1830s. See letter 2301, McConnell (1993), 54–6, RS CM 15 June 1854, 2: 288–9 and 26 October 1854, 2: 294.

Letter 2923
Edward Sabine to Faraday
4 December 1854
From a copy in Bate to Sheepshanks, 18 January 1855, RGO6 / 356, f.149–50

Royal Society, | Somerset House, | Decr 4th 1854

Dear Faraday,

I return Mr: Bate's letters as the affair is now at an end[1]. The Govt applied to the R.S. for an opinion upon Mr. Bate's claims as stated by himself to the Treasury. The R.S. submitted the subject to a Committee consisting of Mr: Tite[2], Mr Grove and the Treasurer[3]; the two former being men of business accustomed to consider such matters - & their report was sent to the Treasury. As regards the Royal Society therefore the subject is closed. What reply the Treasury have made to Mr. Bate I have not heard[.]

Sincerely your's, | (signed) Edward Sabine.

1. See letter 2922 and note 3.
2. William Tite (1798–1873, DNB). Architect.
3. That is Sabine.

Letter 2924
Faraday to George Biddell Airy
8 December 1854
From the original in RGO6 / 405, f.160

R Institution | 8 Decr. 1854

My dear Sir

Mr Barlow has put me down for the F. Evg of Jany. 19th. i.e the first Evening[1])[.] You I think consent to favour us on the third ie Feby 2nd[2]. Having undertaken the Juvenile lectures at Christmas extending to the 9th. January[3] I feel I should be glad to be later in my Friday. Do not think I wish to draw on your good nature & so cause any inconvenience but if it should happen to be nearly the same to you, there are so many that would feel a delight in hearing you open our season that I thought I would put the case before you[.]

Ever Yours faithfully | M. Faraday

Geo B. Airy Esq | &c &c &c

1. Faraday (1855a), Friday Evening Discourse of 19 January 1855.
2. Faraday gave six Christmas lectures on the chemistry of combustion. See RI MS GB 2: 85.
3. Airy, G.B. (1855b), Friday Evening Discourse of 2 February 1855.

Letter 2925
George Biddell Airy to Faraday
11 December 1854
From the original press copy in RGO6 / 405, f.161–2

Royal Observatory Greenwich | 1854 December 11

My dear Sir

The circumstances which determined my selection of the beginning of February[1] (at least my preference of a time not earlier than that to an earlier time) is this - that the last few days of December and a portion of January are the only time when I can have a slight remission of labour and a little country life. Great things are guided by little ones, all the world over; and these moments of mine are ultimately determined by my children's school holidays.

It would upset my family arrangements very much, to engage to be in London on a definite day in January: although without doubt I shall have to come up on some days yet undefined.

I would do much to accommodate my movements to your wishes, but I am confident that the price which, in this instance, must be paid for it, is greater than you yourself would desire.

I am, my dear Sir, | Yours most truly | G.B. Airy
Professor Faraday

1. Airy, G.B. (1855b), Friday Evening Discourse of 2 February 1855. See letter 2924.

Letter 2926
Faraday to George Biddell Airy
12 December 1854
From the original in RGO6 / 405, f.163

Royal Institution | 12 Decr. 1854

My dear Sir

Your note[1] was very kind[.] I would not on any account that your family party should fail in one point of the happiness it promises[.] If I had had the least notion of such an event my note should never have come

near you[.] Kindest remembrances to Mrs. Airy & those I know of the company.
  Ever Yours faithfully | M. Faraday
G.B. Airy Esq | &c &c &c

1. Letter 2925.

Letter 2927
George Gabriel Stokes to Faraday¹
23 December 1854
From the original in IEE MS SC 2
                                              49 New Bond St. | Dec 23 / 54
My dear Sir,
  I am going away for Christmas, and do not expect to return till the 2d or 3d of Jany. If your apparatus should be ready in the mean time pray do not wait for me for you of course will be anxious to know the result[.]
  Yours very truly | G.G. Stokes

1. Recipient established on the basis of provenance.

Letter 2928
Faraday to 7th Earl of Shaftesbury¹
26 December 1854
From the original in SI D MS 554A
                                              Royal Institution | 26 Decr. 1854
My dear Lord
  Allow me to send a remembrance of former kindnesses which if no use you can put into the fire²[.]
  Ever My dear Lord | Your faithful Servant | M. Faraday
The | Earl of Shaftesbury | &c &c &c

1. Anthony Ashley Cooper, 7th Earl of Shaftesbury (1801–1885, DNB). Social reformer.
2. That is tickets for Faraday's six Christmas lectures on the chemistry of combustion. See RI MS GB 2: 85.

Letter 2929
7th Earl of Shaftesbury[1] to Faraday
29 December 1854
From the original in IEE MS SC 2

St. Giles House | Dec: 29, 1854

My dear Mr Faraday

I am much gratified by your kindness in sending me an admission to your Xmas lectures[2].

I sincerely wish that my Children were in town to hear them; but I shall hope to come & hear some of them myself.

Yours Truly | Shaftesbury

M. Faraday Esq

1. Anthony Ashley Cooper, 7th Earl of Shaftesbury (1801–1885, DNB). Social reformer.
2. See letter 2928. Faraday gave six Christmas lectures on the chemistry of combustion. See RI MS GB 2: 85.

Letter 2930
Faraday to Arthur Young[1]
3 January 1855
From the original in WIHM MS FALF

Royal Institution | 3 January 1855

My dear friend

I am not Professional and do not undertake analysis for any one:- not even for the Government. Further I am so little in the professional world that I do not know whom to recommend your friends to. There are plenty of professional chemists but I do not know of any who would or could undertake to investigate the qualities of an unknown bark[.]

Ever Truly Yours | M. Faraday

A. Young Esqr | &c &c &c

1. Arthur Young (1816–1888, Cantor (1991a), 302). Average adjuster.

Letter 2931
Thomas Boosey et al. to Edinburgh Glasite Church
3, 4, 8 and 9 January 1855
From the original in the possession of Joan Ferguson
The Church of Christ in London
To the Church of Christ at Edinburgh

London January 3rd 1855
Very dear Brethren

In the month of November last we received a letter from the Church at Dundee pointing out a grievous error existing among us with respect to the observance of the Divine Command to abstain from eating blood, namely, that we ate (or tolerated the eating) of the flesh of animals called "Game" when their blood had not been let out at the time it would flow - the time of death, and had endeavoured to satisfy our consciences by washing and cleansing the flesh after it was cold, to the best of our ability[1]. This cleansing the Church of Dundee considered it impossible, and unlawful to do - flesh with the blood being forbidden by God to be used as food and there being no direction for its purification. After we trust, sober consideration of the Scriptures relating to the subject, and committing ourselves into the hands of Him who is able to make wise unto Salvation[2], we came to the unanimous conclusion, that we had held the command to abstain from eating blood in a lax manner, that we had much reason to judge ourselves, and that we ought to be thankful for the admonition. We do not consider by this that we are in any way brought under the Levitical Law, or that we are prohibited from eating the flesh of any animal whatever, if it has been bled at the time of its death: but we feel that we are delivered from a state of doubt and hesitation.

We think it right to inform you that some of the friends here have always been of the Same Mind as our Dundee Brethren and thus we have been in the condemnation of not walking by the same rule and minding the same thing[3][.]

This matter and the Scriptures relating to it, have been also laid before our brethren at Newcastle, Old Buckenham, and Chesterfield and they are heartily joined with us in the same mind and the same judgement upon it.

We are dear Brethren, called upon to address *You* upon this painful subject, which has so long agitated the minds of the brethren, and which now threaten to cause a breach in the visible body of Christ; and this we hope to do in the spirit of Meekness and fear, yet of faithfulness and love. We have been informed of the manner in which the Church of Dundee has hitherto dealt with you in this matter; and we have heard the painful issue of the second stage of the discussion which took place on the 22nd December, from our two Elders Stephen Leighton and Benjamin Vincent, who were present on that occasion, having been called on by the Church of Dundee to accompany four of them, and thus being "the one or two more" required by the Scriptural direction[4]. We were much grieved to learn that you would not even hear our two Elders and two of the Brethren from Dundee, and that you opposed the doctrine and exhortation set before you by the two Elders from Dundee which we were told were clearly Scriptural and expressed in the words of truth & Soberness, from both the Old & New Testament[.] There remained, then, no other course to be pursued but to proceed to the last stage of the discipline as directed by our Lord Matt 17–18 "If he shall neglect to hear them" (the first offended and the one or two more) "tell it to the Church"[5]. This was done to *us* by our two Elders above mentioned who were the witnesses of what took place at Your meeting with the Dundee Brethren, and it was with much sorrow that we learnt from them that you maintain "That the flesh of any animal not strangled or suffocated, although the blood be not drawn from it at the time of death, may be made lawful food for Christians by having the blood afterwards cleansed from it["], which you say can be done with ease and certainty at any time between the death of the animal and the time of eating[.]

To this we cannot agree. Many of us feel persuaded from our own experience and competent Authority that it is almost if not quite impossible to effect the above mentioned cleansing: but whether this be the case or not, we regard such flesh as "flesh with the life thereof which is the blood thereof"[6] and therefore a forbidden thing by Gods command to Noah and all mankind - A Command which was given *before* the Levitical law, which was enforced by that law, and confirmed under the Gospel dispensation by the decree which went out from the Holy Ghost, and the Apostles and Church at Jerusalem[.] We consider therefore your views and practice in this matter to be contrary to the Divine Word, as given in Acts 15–28 & 29[7] compared with Gen. 9–3 4[8] Levit 17.13.14[9] & Deuty 12 23 24 25[10] and Ezek 33 24 25[11] which we commend once more to your serious consideration: for it is written "To the Law and to the Testimony: if they speak not according to the word it is because there is no light in them"[12] and again ["]All Scripture is given by inspiration of God, and is profitable for doctrine for reproof, for correction & inspiration in righteousness, that the man of God may be thoroughly furnished unto all good works"[13][.]

We are thus constrained, with much sorrow of heart, yet in hope and love to sustain the offence against you, commending you to God, who is able by his word and spirit to deliver you from your great error and delusion, and open your ear to discipline[.] In conclusion dear Brethren we cannot help reminding you that Christs disciples are kept in obedience to his commandments by his fear being put in their hearts and by their love to him whose blood was shed for the remission of their sins, and thus they find his commands not grievous[.]

It surely does not become us in this our day to weaken the effect of any one of those commands but it become us each to examine ourselves whether we are obeying them, not as a means of exalting our own righteousness, but as evidence of our faith & love[.]

We cannot help adding here, that we have observed with much fear, the apparent absence of fear for yourselves, and love for the Brethren at

Dundee in your correspondence and in the mode of conducting the discussion when they met you on Dec 22nd[.]

We feel that we have good reason to conclude that in these dealings with you and us, they have been actuated by that charity which rejoiceth not in iniquity but rejoiceth *in* the truth[14]. Surely it becomes us all to consider what spirit we are of; and to hearken [sic] to the voice of Him who decides that he will search Jerusalem with candles & punish the men who are settled on their lees[15], and who says I know thy works "Behold I come quickly["][16]. Our attention has been directed to the following Scriptures and we would commend them to you for sober consideration also Matt 28. 19 & 20[17] II Peter 3. 1.2[18] Mark 8 38[19] Heb 4. 12.13[20] II Corr 10. 4.5.6[21] Lamen 3. 31 to 42[22][.]

That through the long suffering mercy & goodness of God, by his word you may be led to repentance; and so brought back again to Jesus the Shepherd & Bishop of Souls, is very dear Brethren the earnest Prayers of the whole church & signed on its behalf by the following

Jany 3rd 1855
Thomas Boosey   Stephen Leighton   Elders
Benjamin Vincent   George Whitelaw

John Leighton   Alex Macomie   Deacons
Thomas Barker   E.K. Reid

William Vincent   M. Faraday
D.W. Martin   Anthony Lorimer[23]

Jany 4th Signed in the presence & on behalf of all the Brethren sojourning at Old Buckenham

Thomas Loveday[24]
William Fisher[25]
George Smith[26]

Jany 8 Signed in the presence & on behalf of the brethren sojourning at Chesterfield

John Oxley[27]

Jany 9 The Brethren & Sisters in Newcastle are all of one mind in considering the Churches of Dundee and London faithful in sustaining the Offence against the Church of Edinburgh & agree with the doctrine

W. Paradise
David Reid
Thomas Proctor[28]

1. See letter 2918 and Cantor (1991a), 68–70.
2. 2 Timothy 3: 15.

3. Philippians 3: 16.
4. Matthew 18: 16.
5. Matthew 18: 17.
6. Genesis 9: 4.
7. Acts 15: 28–9.
8. Genesis 9: 3–4.
9. Leviticus 17: 13–14.
10. Deuteronomy 12: 23–5.
11. Ezekiel 33: 24–5.
12. Isaiah 8: 20.
13. 2 Timothy 3: 16.
14. 1 Corinthians 13: 6.
15. Zepaniah 1: 12.
16. Revelation 3: 11.
17. Matthew 28: 19–20.
18. 2 Peter 3: 1–2.
19. Mark 8: 38.
20. Hebrews 4: 12–13.
21. 2 Corinthians 10: 4–6.
22. Lamentations 3: 31–42.
23. Anthony Lorimer. Sandemanian bookbinder. Cantor (1991a), 301.
24. Thomas Loveday. Foundryman and Sandemanian in Old Buckenham. (GRO certificate of son's birth).
25. William Fisher (d.1857, age 52, GRO). Cordwainer and Sandemanian in Old Buckenham.
26. Unidentified.
27. Unidentified.
28. Unidentified.

Letter 2932
Faraday to John Tyndall, Edward Frankland, John Barlow and Henry Bence Jones
16 January 1855
From the original in Newcastle University Library MS Album 38

Various philosophical notes of experimental investigation on foolscap paper, paged in series & partly bound in five volumes[1],- a quarto MS. book of Philosophical notes[2],- a second larger quarto of similar notes[3],- some of my printed papers collected in two bound volumes & illustrated by letters &c, - the one quarto[4] the other 8vo[5], and a bound copy of Davy's chemical elements[6], being a copy of that which, whilst abroad in 1814–5, he prepared for a second edition, (which was never published);- *these I offer for the Library of the Royal Institution*, if the Managers should think them worth a place:- if not to remain at the disposal of my executors[.]

Whatever my wife may think fit to give, from among my personal things, to relatives or friends as remembrances, will I hope be accepted as from both of us. We were always as one, and whatever she may do is what I should do.

I desire that, at my death, all cases in which I have become security should be put an end to, that they may not be continued over my wife. I

only remember one, namely, for Mr. Geo Whitelaw to Messrs Spottiswoods[7].
M Faraday.
16 January 1855
Dr. Tyndall | Dr. Frankland | Mr Barlow | Dr. Bence Jones

1. RI MS F2 C-G. That is Faraday, *Diary*, 1828–1832, **1**, pp. 327–430 to 21 January 1850, **5**: 10739.
2. RI MS F2A. That is Faraday, *Diary*, 1820–1823, **1**, pp. 1–117.
3. RI MS F2B. That is Faraday, *Diary*, 1823–1833, **1**, pp. 121–323.
4. RI MS F3B.
5. RI MS F3A.
6. Davy (1821). This is now RI MS HD 29.
7. That is Eyre and Spottiswoode, printers to the Queen.

Letter 2933
August Wilhelm Hofmann to Faraday
18 January 1855
From the original in RI MS Conybeare Album, f.15

R. College of Chemistry | Jan 18, 55.

My dear Mr Faraday,
May I take the liberty of carrying off your new apparatus for showing the products of the operation?
Yours ever sincerely | A.W. Hofmann
Professor Faraday

Letter 2934
Charles Fox to Faraday
27 January 1855
From the original in RI MS Conybeare Album, f.40

Spring Gardens | Jany 27/55

My dear Sir
Allow me to introduce to you Mr. Guido Weichold[1], who desires the use of a powerful magnet for an experimental purpose the nature of which he will explain to you.
He has been Tutor to my boys and is a most deserving young man. If you can put him in the way of getting what he wants I shall be gratified.
Ever yours faithfully | C. Fox
Michael Faraday Esqr F.R.S.

1. Unidentified.

Letter 2935
Thomas Boosey et al. to Edinburgh Glasite Church
31 January and 4 February 1855
From the original in the possession of Joan Ferguson

London Jan 31st 1855

The Church in London to those whom they hitherto corresponded as Brethren at Edinburgh

We duly received your Letter in reply to ours of the 3d inst[1][.] It was read to us on Sabbath the 21st inst after the lovefeast, & was kept for a week's consideration, & on that day week was read to us again & our minds taken upon it. We are grieved to find that you still maintain views & practices with respect to the covenant to abstain from eating blood, which we consider contrary to the Word of God; and that you still justify the manner in which your Elders have considered this Discipline, misinterpreting the Scriptural directions. We are thus constrained to conclude that you are not hearing the Church, or walking in the fear of the Lord with respect to His commandments, & that you are not following after that Charity which rejoiceth not in iniquity, but rejoiceth with the truth[2]. From these considerations, with deep sorrow of heart, and we trust with fear for ourselves, we feel compelled to withdraw from your communion in obedience to the Divine command, Matth. 18, 17[3], Rom. 16.17[4] - 2 Cor. 6.14[5] &c 1 Tim 6.5[6] - 2 John 9–11[7], praying that He who alone is able by His Word & Spirit to convince of sin may grant you repentance to the acknowledging of the truth.

Thomas Boosey
Stephen Leighton
Benjamin Vincent
George Whitelaw
Alex. Macomie
Thos. Barker
George Raith[9]
Sam Deacon[10]

E.K. Reid
W. Vincent
M. Faraday
D.W. Martin
Anthony Lorimer[8]

PS. The Elders of the Church of London have received a letter from Mr Oxley[11] stating that the brethren at Chesterfield are heartily joined with the Churches of Dundee & London in withdrawing from the Church at Edinburgh. Jany 31, 1855

Signed in the presence, & witness of, the Brethren at Old Buckenham

Thomas Loveday[12]
William Fisher[13]
George Smith[14]

The Preceding letter has been read to the brethren here at Newcastle on which they are of one mind with the Church at London

Signed Feby 4, 1855

William Paradise
David Reid
Thomas Proctor[15]
William Stark[16]
William Proctor[17].

1. Letter 2931.
2. 1 Corinthians 13: 6.
3. Matthew 18: 17.
4. Romans 16: 17.
5. 2 Corinthians 6: 14.
6. 1 Timothy 6: 5.
7. John 9: 11.
8. Anthony Lorimer. Sandemanian bookbinder. Cantor (1991a), 301.
9. George Raith (1804–1860, Cantor (1991a), 301). Counting house clerk.
10. Samuel Deacon (1790–1861, Cantor (1991a), 300). Newsagent.
11. Unidentified.
12. Thomas Loveday. Foundryman and Sandemanian in Old Buckenham. (GRO certificate of son's birth).
13. William Fisher (d.1857, age 52, GRO). Cordwainer and Sandemanian in Old Buckenham.
14. Unidentified.
15. Unidentified.
16. Unidentified.
17. Unidentified.

Letter 2936
Joseph Antione Ferdinand Plateau to Faraday
6 February 1855
From the original in IEE MS SC 2

Gand, 6 février 1855.

Mon Cher Monsieur Faraday.

Il y a a bien long-temps que je n'ai eu l'honneur de causer un moment avec vous; Je saisis donc avec joie l'occasion que m'offre l'envoi de la troisième livraison de ma petite *Physique*[1], livraison que vous recevrez sous bandes par la poste. Cette livraison termine la physique des corps pondérables; la première moitié seulement est de moi; la seconde, qui se compose de l'Acoustique, est de M. Quetelet. J'ai joint à l'exemplaire qui vous est destiné, un second exemplaire que je vous prie de vouloir bien remettre de ma part à M. Wheatstone. Dites lui, je vous prie, que la froideur qui semble exister entre lui et moi me fait grande peine. Je l'ai, à la verité, importuné un peu à propos de la reproduction non réalisée de ma *deuxième série*[2] dans les *Scientific Memoirs*[3], et aussi à propos d'une analyse de ce travail, que je désirais faire insérer dans un Journal anglais; mais je vous ai également importuné[4], et pourtant vous n'avez pas cessé de me témoigner de l'amitié; tachez, je vous en prie, de le ramener à de meilleurs sentiments envers moi.

J'ai lu avec admiration le compte rendu de vos belles expériences sur les effets de l'induction latérale dans un long fil métallique recouvert d'une substance isolante et plongé dans l'eau⁵. Que nous preparez-vous encore ? Il y a quelque temps que vous n'avez rien publié à ma connaissance, et je m'attends à l'apparition de quelque nouveau prodige.

Quant à moi, débarassé maintenant de l'encyclopédie populaire, j'ai repris la suite de mon travail sur les masses liquides sans pesanteur, et deux nouvelles séries⁶ sont fort avancées. Je regrette bien vivement que ce sujet soit si éloigné de ceux qui vous occupent: car il en résulte que vous ne pouvez y prendre un grand intérêt; cependant vous serez convaincu, je l'espère, que c'est une mine féconde en résultats.

Agréez, mon Cher Monsieur Faraday, l'assurance de tous mes sentiments de respectueuse affection.
Jh. Plateau
Place de Casino, 22.

TRANSLATION

Gent, 6th February 1855.

My Dear Mr Faraday.

It has been a long time since I have had the honour of talking for a moment with you; I seize therefore with joy the opportunity that is offered to me by the dispatching of the third part of my little *Physique*[1], a part that you will receive through the post. This part ends the physics of ponderable matter; I am the sole author of the first half; the second, which is on Acoustics, was written by M. Quetelet. I have added to the copy that is intended for you, a second copy that I would ask you kindly to give on my behalf to Mr Wheatstone. Tell him, I beg you, that the coldness that seems to exist between him and myself, gives me great pain. I have, to tell the truth, inconvenienced him a little, in connection with the non publication of my *second series*[2] in the *Scientific Memoirs*[3], and also in connection with an analysis of this work, which I wanted to have inserted in an English Journal; but I have equally inconvenienced you[4], and nevertheless you have not ceased to give testimonies of your friendship; please try, I ask you, to bring him towards better feelings towards me.

I read with admiration the account of your beautiful experiments on the effects of lateral induction in a long metal wire covered in an insulating substance and plunged into water[5]. What more have you got in store for us? It is some time since you published anything to my knowledge, and I await the appearance of some new marvel.

As for me, relieved of the popular encyclopaedia, I have taken up again my work on liquid masses with no weight, and two new series[6] are well under way. I deeply regret that this subject is so far from those you are occupied with: for the result is that you cannot take such great an interest; however, you will be convinced, I hope, that it is a mine prolific in results.

Please accept, My Dear Mr Faraday, the assurance of all my sentiments of respectful affection. | Jh Plateau
Place de Casino, 22.

1. Plateau and Quetelet (1851–5), part 3 published as part of the *Encyclopedie Populaire*.
2. Plateau (1849).
3. Plateau (1852).
4. See letter 2490.
5. Faraday (1854a), Friday Evening Discourse of 20 January 1854.
6. Plateau (1857, 1859).

Letter 2937
George Biddell Airy to Faraday
7 February 1855
From the original press copy in RGO6 / 470, f.57–8

Royal Observatory Greenwich | 1855 February 7

My dear Sir

We have observed on two occasions (under circumstances which appear to be general) an odd galvanic phenomenon, on which I wish to consult you.

For dropping the Time Signal Ball at Deal[1], there is a wire which, for about one minute of time before the ball-drop, is made continuous from Greenwich to Deal. At Deal it has connexion with the earth, but at Greenwich it has no connexion, as it is waiting to be joined to the battery at the proper instant. There are galvanometers at London and Tonbridge: and the ball-apparatus at Deal may be considered as a galvanometer of a delicate kind, for a very trifling current there drops the ball. The general direction of the line is very nearly west to east: deviating perhaps 30° from the magnetic west to east

Now the phenomenon is this. As soon as the contacts are completed so as to make the line continuous (not at Greenwich), the galvanometers at London and Tonbridge are deflected considerably (30°), but there is no sign of current at Deal.

Can you explain this? Will it always be so?

The direction of deflection is the same as that produced by a current from the copper pole of a Greenwich battery.

As soon as the Greenwich battery contact is completed from copper pole, the deflection is normal at London and Tonbridge and there is abundant current at Deal.

Almost immediately after the ball drop, Greenwich is connected with earth, and then the deflexions at London and Tonbridge cease entirely.

My assistant, now at Deal, finds that a battery nearly equal in strength to the dilute sulphuric acid battery, may be made with sea water. Can we rely on this?

Possibly Mr. Barlow will shew you some criticism of mine on a paper of yours[2].

I am, my dear Sir, | Yours very truly | G. B. Airy
Professor Faraday

1. On this see Chapman (1998), especially p.48.
2. See Airy to Barlow, 7 February 1855, Bence Jones (1870a), 2: 352–4 which was critical of Faraday (1855a), Friday Evening Discourse of 19 January 1855.

Letter 2938
Faraday to George Biddell Airy
9 February 1855
From the original in RGO6 / 470, f.59–60

Royal Institution | 9 Feby. 1855
My dear Sir

I should not like to draw a conclusion from the phenomena you describe[1] except upon more numerous and personal observations. Mr Latimer Clarke has told me that he has evidence of currents produced in underground wires referable (he thinks) to atmospheric inductions upon the surface & substance of the earth at different localities & unless yours be a *constant* phenomena it may be of that kind.

As to the sea water battery - you must not rely upon it before you have tried it for some time. A change of fluids (which for a time includes the condition of fresh fluids & fresh surfaces) may answer for a short time & yet the new fluid may not be satisfactory in the long run[.]

I have not seen your letter to Mr Barlow[2] yet but I have been ill & confined to my room[.] I dare say he will shew it to me in due time. In the meantime I send you a paper from the Phil Mag[3]. The speculative part I have no more opinion of than I have of the many speculations that float about (and must float) in mens minds but the experimental part contains many nuts which at present are hard to crack.

Ever My dear Sir | Yours Truly | M. Faraday

G.B. Airy Esqr | &c &c &c

1. In letter 2937.
2. See note 2, letter 2937.
3. Faraday (1855b), [ERE29b].

Letter 2939
William Whewell to Faraday
12 February 1855
From the original in TCC MS O.15.49, f.70

Trinity Lodge, Cambridge, | Feb. 12 1855

My Dear Dr Faraday

I have received papers containing your speculations on magnetism and especially the paper in the Phil. Mag. on Magnetic Philosophy[1]. I have read them with great interest, as I always read your speculations; but they require more time and leisure before I can fully possess myself of your views. At this imperfect stage of thought, and at the risk of proposing difficulties which your former papers have solved, will you allow me to make a remark on your notions as there given.

Your lines of magnetic force whether or not they contain the true theory, are an admirable way of exhibiting the facts; and I have always ascribed the success with which you have unravelled so many very complex phenomena to your starting from these lines. I do not say that they do not contain the true theory, or come nearest to it; for I think we are arriving at a point when the other two theories which you mention[2] not only cannot explain but cannot express the facts. And if the lines of physical force come to be the only way of expressing the laws of phenomena they *must* be accepted at least till they are resolved into something else[.] Now what I have to say is this: I do not think that the facts of magnetism alone, even those in your new paper, are the strongest examples or any examples of *this* peculiar advantage of the physical lines of force. So long as we confine ourselves to magnetism alone (paramagnetism) all the facts can be explained by the existence of two fluids. All your facts of chambers in which the magnetic force vanishes flow easily from that theory: for a self-repelling fluid is necessarily concentrated at edges and points: and the theory of two mutually attracting self-repelling fluids includes, so far as I see, all the facts by which you reduce the universal duality of the forces. But when instead of confining ourselves to one kind of polarity magnetism, we take in the related polarities, electric or voltaic currents, then your lines of force become the only way, so far as I see, of exhibiting the facts. I can make nothing of the other theories in that case, and I have not seen any attempt to apply them coherently. And the same is the case with diamagnetism. I do not see that either of the other theories enables us to explain the most obvious facts. Now what I have to say is this; the advantage of the physical lines of force theory thus residing in its application to diamagnetism and voltaic currents in their relation to magnetism, it would be a great boon to the ordinary thinker if you could explain it more fully in these relations. Your application of it to voltaic currents always makes me wonder at the clearness and readiness with which you conceive the relations of space, but I fear is not intelligible to ordinary readers. It might be made so by figures, diagrams, of the physical lines copiously used, and exhibiting many kinds of examples of the application of your views. And the same is the case with the application of the lines of force in diamagnetic phenomena. We - ordinary readers - would like to see the lines of force drawn in such cases as you have given in page 10 of this last paper[3], and in many other cases. It is probable that you have explained this matter is some of your previous papers; but it has I conceive a special bearing upon your present attempts to show the advantages of the lines of physical force. Your theorem (p.33) "pointing in one direction or another is a differential action due to the convergence or divergence of lines of force"[4] &c is a very curious proposition; and seems to me, or rather is, so far as I see, the only way of exhibiting the facts; but it wants much development to make it intelligible to us; and I want you to give it this development, by diagrams, copious and various, as well as by experiment. Excuse my liberty; I want to have all possible light thrown upon us from your abundant internal light[.]

Yours always truly | W. Whewell

1. Faraday (1855b), [ERE29b]. The other papers might have included Faraday (1854e, 1855a), Friday Evening Discourses of 6 June 1854 and 19 January 1855 respectively.
2. Faraday (1855b), [ERE29b], 3301 mentioned the aether theory and that of two magnetic fluids.
3. That is Faraday (1855b), [ERE29b], 3315.
4. Faraday (1855b), [ERE29b], 3361.

Letter 2940
Jacob Herbert to Faraday
14 February 1855
From the original in GL MS 30108/2/89

Trinity House London | 14 Feby 1855.

Sir,
 I am directed to forward to you the accompanying Samples of White Lead, from the parties whose names are marked thereon, who have sent Tenders for the supply required for the Corporations service, and the request that you will be pleased to analyze the same, and report the result for the Board's information.

 I am, | Sir, | Your most humble Servant | J. Herbert

M. Faraday Esq | &c &c &c

1. Faraday's analyses are in GL MS 30108/2/89. His report, dated 23 February 1855, was read at Trinity House Wardens Committee, 27 February 1855, GL MS 30025/23, pp.275–6 on which basis they awarded the contract for the supply of white lead.

Letter 2941
George Wilson to Faraday
17 February 1855
From the original in IEE MS SC 2

Elm Cottage: Edinburgh | Feb: 17: 1855

Michael Faraday Esq
Dear Sir
 Will you do me the honour to accept a copy of the enclosed lines on the late Edward Forbes[1]. He so often exchanged expressions of esteem and admiration for you, that the thought of him seems naturally to bring you up, and I feel as if I owed it to you to send you these inadequate verses.

 I have engaged to write his life. If from your important engagements you could spare time enough, to indicate by a line or two, how he deported himself, and was received as a Lecturer at the Royal Institution I should feel deeply indebted[2]; but I know too well how many and important are the demands on your time, to wonder if you decline to accede to my request.

 I remain | Yours very truly | George Wilson

1. Edward Forbes (1815–1854, DSB). Palaeontologist at the Geological Survey. A printed copy of the poem is in IEE MS SC 2.
2. Wilson and Geikie (1861) which contains several references to Forbes and the Royal Institution.

Letter 2942[1]
Faraday to William Whewell
23 February 1855
From the original in TCC MS O.15.49, f.39

Royal Institution | 23 Feby. 1855

My dear Dr. Whewell

Your letter[2] was very acceptable to me for it gives me courage, and I am heartily thankful to you for it. I have given many figures of lines of force at different times, and briefly refer you to them. Thus in the Paper on Magnetic conduction (2797)[3] there are figures at 2807. 2821. 2831. 2874. 2877[4]. 2972. 2993[5]. The paper on lines of Magnetic force (3099)[6] has a plate full of figures in reference to their delineation (3234)[7] drawn from nature; and the paper on Physical lines (3243)[8] also has a plate. Still I think that, as you suggest, figures more numerous still would be very useful. I have, however, been in some degree deterred from pressing these matters too hard, because I wanted to see how far that which has been advanced might be accepted or justified; and also because I wanted to trace more clearly to myself the origination & development of the lines of force round a wire carrying a current; through & about a helix - without or with an iron core; and amongst wires & helices in juxtaposition. Above all I want to obtain some *clear* idea of the coercitive force, & how it is that an electro helix, having but weak powers itself as a magnet, can raise up (or arrange) such a powerful system of lines when an iron core is introduced. I have other matters too in hand, regarding magnecrystallic action, which I hope to develop soon & think may turn out well. Your recommendation, however, shall not be forgotten & I shall probably soon begin to collect cases for illustration.

I conclude I am right in believing that if diamagnets & diamagnetism had been known to us before we knew any thing at all of Paramagnets & paramagnetism, the theory of two magnetic fluids would have applied to it, but could not then have included paramagnetism. It is this idea which makes me earnest in speaking of chambers of little or no action & places of weak magnetic action; for though the old theory of two fluids can account for them, they are not the less important to me who do not believe in that theory. I see in them proofs that the dualities must be related externally to the magnet; & so they come in as necessary consequences of the principles both of paramagnetic & diamagnetic action; but I hope to make all this clearer by degrees; & my hopes are greatly strengthened by the growing admission that the lines of magnetic force represent at present fairly the facts of magnetism. Into what they may ultimately resolve themselves, or to what they may lead I am sure I cannot say; but if I can only convert the theory of magnetic fluids, & that of electric currents, into two stools, the fall to the ground between them may be more useful than either of them as a seat in a wrong place.

Letter 2943

      I am | My dear Dr. Whewell | Yours faithfully | M Faraday

1. This letter is black-edged due to the death of Edward Barnard on 4 January 1855.
2. Letter 2939.
3. Faraday (1851d), ERE26.
4. Faraday (1851d), ERE26, 2807, 2821, 2831, 2874, 2877.
5. Faraday (1851e), ERE27, 2972, 2993.
6. Faraday (1852b), ERE28, 3099.
7. Faraday (1852c), ERE29, 3234.
8. Faraday (1852d), [ERE29a].

Letter 2943
Christian Friedrich Schoenbein to Faraday
27 February 1855
From the original in UB MS NS 416

My dear Faraday,

    From the very long silence I have kept, you will draw all sorts of conclusions, but I am quite sure, that none of them proves to be correct, for the simple reason that even Mr. Schoenbein himself cannot account for his taciturnity. I have been neither unwell, nor low-spirited nor overbusy, nor any thing else that could have prevented me from breaking it sooner, and least of all, I have forgotten my dear and amiable friend at the Royal Institution.

    But if I have not written to - I have written, at least, about you and in telling you so much I have revealed to you an author's secret which I beg you however to keep as yet to yourself[.] The matter stands thus: I have been composing a book these last six months, certainly not a scientific one, for doing such a thing suits, as you well know, neither the taste nor the powers of your friend; it is a sort of "quodlibet" or as the musical term runs a "pot pourri" i.e. a most variegated motley of things[1]. You recollect perhaps the trip I made to Munich and Vienna some time ago[2], and its having turned out so very pleasant induced me to try my graphic powers with the view of making Mrs Schoenbein and the girls, as it were, partners of my journey.

    Wives and Children are very partial judges of the litterary productions of their husbands and fathers and you will therefore not be surprized when I tell you that my excellent helpmate and young ladies made no exception to the rule. They found, indeed, every thing I had written and read to them so very excellent that they started one day the idea of having my scribbling published. However great my dislike to bookmaking is and how little I care for gaining laurels in the line of authorship, I at last yielded to the entreaties of my darlings, that is to say, promised to try what

I could do in the matter. And, indeed, I have finished the work and a copy legibly and nicely written out lies in my desk, but when it will go to the printer and published, that is a thing, which I cannot tell.

You will laugh, when I inform you that in spite of the embryonic state of my spiritual child, I have already baptized and given it the name "Glosses on Men and Things by an elderly Man". This title has, as you see, elasticity enough and I will not conceal it from you that I have made full use of its vagueness, having thronged all sorts of reflections and queer ideas into the opusculum.

On account of its motley character I should like you could read that strange composition, but it being written in german, I am afraid its contents will never come to your knowledge.

It is, however, time to return to yourself and tell you in what manner I have written about you. In the above mentioned book there is a little chapter bearing the title "Fachsmaenner"[3], gallicé "Spécialités" and anglicé perhaps - but I am unable to translate the word into your language, I mean to denote by that term Men devoting their whole life and mind to one object. By no means admiring what they call universal geniuses and being convinced that it is the "spécialités" to whom we owe every real progress in science, arts &c. I have, with a view of proving the correctness of my opinion, drawn up four slight sketches of such "Fachsmaenner", of Berzelius, von Buch[4], Cuvier[5] and of, of, but be it spoken out, of Faraday. I hope you will not tax me with indiscretion for having taken that liberty and believe that in doing so your friend has been actuated by the best motives.

As to science I have of late done nothing at all and do not recollect to have passed a Winter so inactively and lazily as the last. When spring calls forth again the dormant powers of the earth I hope I shall then feel too its congenial influence and be stirred into action, for there is matter enough to work upon and of laborers there are not too many[.]

My colle[a]gue Professor Wiedemann an excellent philosopher has (partly on my instigations) taken up Electrolysis again[6], that fundamental phenomenon, I used to call the true copula of Chymistry and natural philosophy and obtained some results that seem to speak very much in favor of my heretic opinion, according to which in all the oxysalts the electrolysing power of the current is solely and exclusively exerted upon their basic oxides and that there is no such thing as an oxy-compound Ion.

I proposed Mr. Wiedemann to electrolize salts containing the same base and acid in different proportions and see whether by the same current different or equal quantities of metal be eliminated from such salts. If my notion should happen to be correct, it is manifest that under the circumstances mentioned equal quantities ought to be eliminated. The salts as yet carefully electrolyzed are the mono- and tribasic acetates of lead and Mr. Wiedemann has ascertained that on electrolysing them by the same current they yielded equal quantities of lead. I may add that in those experiments my colle[a]gue uses as a sort of voltameter a solution of nitrate of silver i.e. the weight of metal being eliminated from that salt as the s<tan>dard measure of the amount of the electrolysing power of the current employed. Now upon one equivalent of silver Mr. W. obtained one equiv. of lead both from the neutral and tribasic acetate. Hence it seems to follow that the current has nothing to do with the acid, in other terms that the latter is no Anion. In my late paper "on the chemical effects of Electricity, Heat and Light"[7] I have circumstancially developed my notions on the Electrolysis of the Oxy-salts and you have perhaps taken notice of them.

I entertain no doubt you have spent the winter in high scientific spirits and performed some exploits in spite of the warlike mood of the public mind[8], which by the bye I do not relish at all and am inclined to consider as madness. I hardly need tell you how happy I should feel if you would favor me soon with your good news and not requite silence by silence. All my family are doing well and charge me with their best compliments to you and Mrs. Faraday, to which I join my own. Believe me my dear Faraday for ever

Your's most truly | C.F. Schoenbein

Bâle feb. 27, 1855.

Address: Doctor Michael Faraday | &c &c &c | Royal Institution | London

1. [Schoenbein] (1855).
2. See letter 2735.
3. [Schoenbein] (1855), 278–84.
4. Christian Leopold Buch (1774–1853, DSB). German geologist.
5. Georges Cuvier (1769–1832, DSB). French naturalist.
6. See Wiedemann (1852, 1856).
7. Schoenbein (1854b).
8. A reference to the Anglo-French war against Russia.

Letter 2944
Faraday to John Barlow
28 February 1855
From Bence Jones (1870a), 2: 355–6

Royal Institution: February 28, 1855.

My dear Barlow, - I return you Airy's second note[1]. I think he must be involved in some mystery about my views and papers; at all events, his notes mystify me. In the first, he splits the question into (a) action inversely as the square of the distance, and (b) metaphysics. What the first has to do with my consideration, I cannot make out. I do not deny the law of action referred to in all like cases; nor is there any difference as to the mathematical results (at least, if I understand Thomson[2] and Van Rees[3]), whether he takes the results according to my view or that of the French mathematicians. Why, then, talk about the inverse square of the distance? I had to warn my audience against the sound of this law and its supposed opposition on my Friday evening[4], and Airy's note shows that the warning was needful. I suppose all magneticians who admit differences in what is called magnetic saturation in different bodies, will also admit that there may be and are cases in which the law of the inverse square of the distance may not apply to magnetic action; but such cases are entirely out of the present consideration.

As to the metaphysical question, as it is called. If the admitted theory of gravitation will not permit us to suppose a new body brought into space, so that we may contemplate its effects, I think it must be but a poor theory; but I do not want a new body for my speculations, for, as I have said in the Friday evening paper, the motions of either planet or comet in an ellipse is sufficient base for the strict philosophical reasoning[5]; and if the theory will not permit us to ask a question about the conservation of force, then I think it must be very weak in its legs. The matter in the second note is quite in accordance with my views *as far as it goes*, only there is at the end of it a question which arises, and remains unanswered: When the attractive forces of the earth and moon in respect of each other diminish, what becomes of them, i.e. of the portions which disappear?[6]

Ever, my dear Barlow, yours truly, | M. Faraday.

1. Airy to Barlow, 26 February 1855, Bence Jones (1870a), **2**: 354–5. See also letters 2937 and 2938.
2. Thomson, W. (1854).
3. Rees (1853).
4. This warning was omitted from Faraday (1855a), Friday Evening Discourse of 19 January 1855, but a caution against theories of magnetism is in his lecture notes, RI MS F4 G46.
5. Faraday (1855a), 12–13.
6. Airy to Barlow, 3 March 1855, Bence Jones (1870a), **2**: 356–7 noted that Barlow had sent this letter to Airy "which I return to you, but without comment".

Letter 2945
Alessandro Palagi to Faraday
1 March 1855
From the original in IEE MS SC 2

Osservatorio Astronomico | Bologna 1° Maggio 1855
Signore

Datore di questa mia vi farà, o Signore, il mio Cugino Sigr. Dr Giulio Bassi[1], giovane distinto e appartenente ad una delle più vecchie famiglie della nostra Città. Parte Egli, insieme ad un Suo fratello, per la Francia e per l'Inghilterra a fine di diporto e d'istruzione; ed io colgo cosi favorevole occasione per procurargli l'onore della personale vostra conoscenza, che avrà per fortuna singolarissima.

Egli mi farà cortese di presentarvi, o Signore, il noto Opuscolo *Sulla distribuzione delle correnti elettriche nei conduttori*[2] ed io mi chiamerò sommamente avventurato, se Voi vi degnerete volgere il vostro sguardo su questo Scritto. Attendo dalla vostra benignità e dalla vostra sapienza amorevoli conforti e saggi consigli.

Piaciavi, o Signore, di conservarmi nella preziosa vostra grazia, mentre me pregio dell'onore di rassegnarmi con ossequio e con riverenza | Di V. S. Illustrissima | Dev-mo ob-mo Servo | Alessandro Palagi Monsieur | Monsieur Michel Faraday | &c. &c. &c. | London

Address: Monsieur | Monsieur Michel Faraday | &c. &c. &c. | Royal Institution | Albemarle Street | London

TRANSLATION

Astronomical Observatory | Bologna 1st March 1855

Sir,

The person who will give this letter to you, Sir, is my Cousin, Sigr. Dr. Giulio Bassi[1], a distinguished young man who belongs to one of the oldest families of our City. He is leaving, together with one of his brothers, for France and England for entertainment and education; and I am thus seizing this happy opportunity to procure for him the honour of your personal acquaintance, which will be for him fortune indeed.

He will do me the courtesy of presenting to you, Sir, the noted little work *On the distribution of electric currents in conductors*[2] and I should call myself exceedingly happy if you were to deign to glance at this work. I expect from your goodness and knowledge, kind comforts and wise advice.

Please, Sir, keep me in your precious grace, while I pride myself on the honour of signing myself with homage and reverence | of you, Illustrious Sir, | the most devoted and most obedient Servant | Alessandro Palagi

1. Unidentified.
2. Timoteo et al. (1855).

Letter 2946
Friedrich Wilhelm Heinrich Alexander von Humboldt to Faraday
3 March 1855
From the original in RI MS F1 I90b

Dans le moment où je lis, mon cher et illustre Confrere, avec le plus vif interet Vos importants et belles recherches sur la *Philosophie magnetique*[1], je trouve l'occasion de Vous renouveller par l'organe du Fils d'un de mes plus intimes amis[2], l'hommage de mon affectueuse admiration. Le porteur de ces lignes Mr Schönlein, fils du Premier Medecin du Roi[3], se prepare par des voyages dans l'Europe volcanique et non volcanique (je ne prends pas le mot dans le sens moral et politique) pour une Expedition Scientifique qu'il compte faire dans les regions tropicales du Nouveau Continent, à ses propres frais. Il a fait de fortes études de Géologie, de Physique et de la partie astronomique necessairs pour la determination des *positions* à Göttingen et à Berlin. Daignez, je Vous en supplie, lui montrer quelque interet et le recevoir avec cette bienveillance que Vous avez montré à tant de jeunes gens que le *Vieillard des Forêts de l'Orinoque* a osé Vous adresser.

Votre t.h.t.ob. et | tout *devoué* | serviteur | Al Humboldt
Berlin, | 3 mars | 1855

Address: To | Michael Faraday Esq, F.R.S. | Ord. Bor, Pour le Mérite Eg | London | 21 Albemarle Street | Al Humbolt by care | of Mr Schönlein | From Berlin

TRANSLATION

At the moment I read, my dear and illustrious Colleague, with the greatest interest of your important and beautiful research on the *magnetic philosophy*[1], I find the occasion to renew through the services of the son of one of my closest friends[2], the homage of my affectionate admiration. The person who is carrying these lines, Mr Schönlein, son of the First Doctor of the King[3], is preparing through expeditions into volcanic and non-volcanic Europe (and I do not use *volcanic* in a political or moral sense), for a Scientific Expedition which he is counting on making to the tropical regions of the New Continent, at his own expense. He has made a serious study of Geology, Physics and that part of Astronomy necessary for determining *positions*, at Göttingen and in Berlin. Please show him, I humbly ask you, some interest and receive him with the same kindness that you have shown to so many young people that the *Old Man of the Orinoco Forests* has dared to send you.

Your most humble, most obedient and | totally *devoted* servant | Al Humboldt
Berlin | 3 March | 1855

1. Faraday (1855b), [ERE29b].
2. Johann Lucas Schönlein (1793–1864, DSB). Professor of Medicine in Berlin, 1840–1859.
3. Frederick William IV.

Letter 2947
Isambard Kingdom Brunel to Faraday
6 March 1855[1]
From the original in IEE MS SC 2

Tuesday | March 7 [sic]

My dear Sir

I want to test the efficiency of ventilation of a room built as an experimental or specimen wardroom for hospitals to be sent out ready made to the East[2].

The room is 70 ft long 21 ft wide and a good height - it is ventilated by mechanical means *driving* air rising up through the floor at six places - along the centre and escaping through the roof - the air is driven in by a fan placed outside the room.

Can I by some easy means fill this room with a *very visible* smoke gas or *dusty* air (perhaps by driving it in with the fan -) - and then watch the displacement of this visible atmosphere by the fresh air? or could you kindly suggest some better mode or tell me how to carry out this one - it is important that no difference of temperature should assist or impede the ventilation as the object is to imitate what will have to be done in a hot climate where the air injected will probably be very little cooler than that of the chamber[.]

I know that I need not say anything to induce you to assist anybody in doing a usefull thing but nevertheless I will tell you that this is a case peculiarly claiming assistance. I am engaged in designing and constructing with all dispatch a *large quantity* of hospital accommodation to be sent out and erected near the seat of war - while the materials are being prepared this sample ward is made and will be erected by Thursday next[3] at Paddington for the inspection of the War Minister and the medical authorities and I want on that day or on Friday to test the ventilation. I will call on you at any time you may name today or Wednesday to take some instruction if you will be kind enough to give your thoughts to the subject[.]

Yours faithfully | I.K. Brunel

Endorsed by Faraday: Wednesday $\frac{1}{2}$ p3 or 4 o clk

1. Dated on the basis that letter 2948 was the reply and that 6 March 1855 was a Tuesday.
2. That is the prefabricated hospital designed by Brunel for use by the Army at Renkioi on the Dardanelles. On this and its display at Paddington see Brunel (1870), 461–73 and Noble (1938), 203–5. See also Rolt (1957), 292–8 and Toppin (1985–6).
3. That is 8 March 1855.

Letter 2948[1]
Faraday to Isambard Kingdom Brunel
6 March 1855
From the original in BrUL MS

R Institution | Tuesday | 6 Mar

My dear Brunel

I am engaged in very [word illegible] expts. which I cannot leave for a moment[2]. Tomorrow I could see you after the Levee[3]. Say about $\frac{1}{2} p$ 3 or 4 oclk, at 5 oclk I must be out again[.]

Ever Truly Yours | M. Faraday

Endorsement: P[erson] Farraday | D[ate]. 6, March 1855

1. This letter is black-edged. See note 1, letter 2942.
2. See Faraday, *Diary*, 6 March 1855, **6**: 13607–26.
3. See letter 2947. Faraday was present at the levée, but not Brunel. PRO LC6 / 14.

Letter 2949
Paolo Volpicelli to Faraday
6 March 1855
From the original in IEE MS SC 2
Illustre Sig Professore

Roma il 6 Maggio 1855

Ho ricevuto per mezzo dell'ottimo Signore Roberto Abbott[1] la sua pregievolissima del 9 Novembre 1854; e debbo ringraziare sommamente la S.V. Chiarissima, per avermi procurata la conoscenza di questo signore inglese, pieno di amabilità e d'istruzione. Avrei voluto fare molto per lui; ma egli ha ben poco profittato di me: questo è il solo difetto del carissimo Sig Abbott, il quale ha visitato con molta intelligenza tutto quello che vi ha in Roma, e ne' suoi intorni d'interessante, come appunto sogliono praticare gl'inglesi; a preferenza di ogni altra nazione. Io spero che il sig Abbott si ricorderà di me, e che vorrà comandarmi nell'avvenire.

Spero altresi, illustre signor professore, che vorrà ella onorarmi più presso onorarmi [sic] de' suoi comandi; e che quando qualche suo conoscente si porterà in Roma, vorrà diriggermelo, affinchè io possa avere la piacevole soddisfazione, di fare qualche piccola cosa per lei, e pe' suoi amici.

Gradisca, la prego, i sentimenti dell'altissima stima, del profondo ossequio, e della più sincera amicizia, coi quali ho l'onore profferirmi | Di V.S. Chiarissima | l'um[ilissimo] ed obb[ientissimo] | Servo Paolo Volpicelli
(al chiarissimo Sig Prof Cavr Faraday)

TRANSLATION
Illustrious Professor,

Rome, 6 March 1855

I have received through the excellent Mr Robert Abbott[1] your most kind letter of 9 November 1854; and I must thank you most kindly for having procured for me the acquaintance of this most amiable and most educated English gentleman. I would have liked to do a lot for him, but he took very little advantage of me: this was the only defect in dear Mr Abbott, who visited everything that there is of interest in and around Rome, with the sort of intelligence that the English, more than any other nation, know how. I hope that Mr Abbott will remember me and that he will want to take advantage of me in the future.

I also hope, illustrious professor, that you will wish to honour me again, honour me with your commands; and that when some acquaintance of yours comes to Rome, that you will send him to me, so that I can have the pleasurable satisfaction of doing some little thing for you, and for your friends.

Please accept, I beg you, the sentiments of the highest esteem, of the deepest homage, and of the sincerest friendship, with which I have the honour of being | of you, Dear Sir, the most humble and most obedient | Servant | Paolo Volpicelli

---

1. Possibly Arthur Robert Abbott (1833–1892, DQB). Teacher who was the son of Faraday's old friend Benjamin Abbott (1793–1870, DQB).

Letter 2950
James Scott Bowerbank[1] to Faraday
12 March 1855
From the original in RI MS F1 K6

3 Highbury Grove | Mar. 12, 1855

My dear Sir

I hope you will allow us to have the pleasure of adding your name to the list before it goes to press. If so will you favour me with a line to that effect by the close of this week[.]

I remain | My dear Sir | Yours most truly | J.S. Bowerbank
M. Faraday Esq D.C.L., FRS, &c &c

1. James Scott Bowerbank (1797–1877, DNB). Geologist.

Letter 2951[1]
Faraday to James Scott Bowerbank[2]
14 March 1855[3]
From the original in SI D MS 554A

R Institution | 14 Mar 1854 [sic]

My dear Sir

I would rather not be one of your Club at its first foundation. My reason is that I am continually objecting to proposals to sit &c and urging (what is true) that I dislike putting myself forward in shape or aiding in doing so. If I were to help to found the Club I should be justly chargeable with inconsistency on this point; so I hope you will do without me for the present[.]

Ever Truly Yours | M. Faraday
J.S. Bowerbank Esq | &c &c &c

1. This letter is black-edged. See note 1, letter 2942.
2. James Scott Bowerbank (1797–1877, DNB). Geologist.
3. Dated on the basis that this is the reply to letter 2950 and also that it is black-edged. See letters 2942, 2948 and 2962.

Letter 2952
George Biddell Airy to Faraday
15 March 1855
From the original press copy in RGO6 / 470, f.61
Royal Observatory Greenwich | 1855 March 15
My dear Sir
You may perhaps remember that I troubled you a little while ago about some galvanic currents that disturbed us[1]. - Upon mapping the wires and studying the connexions, we found that there was a connexion with a Battery not duly considered: and upon examining the Battery we suspected imperfect insulation. So I have mounted the Battery upon potted-meat-pots surmounted by inverted saucers, and all the strange currents have ceased. And so ends that Great Mystery.
That imperfect insulation is an odd thing. I wonder why the power of the battery did not go off in full force.
I am, my dear Sir, | Yours very truly | G.B. Airy
Professor Faraday

1. Letter 2937.

Letter 2953
Faraday to George Biddell Airy
16 March 1855
From the original in RGO6 / 470, f.62
Royal Institution | 16 Mar 1855
My dear Sir
Many thanks for your note[1][.] I am glad you have found out the mystery and that Mrs. Airy['s] kitchen apparatus has turned to account. It always interests me when some deep difficulty is aided by the application of common things for then principle shines forth. Are your troughs made of Gutta Percha? I suppose not[.]
Mr Walker[2] has been telling me of your clock at the London Bridge station, and I intend to go some day very soon & look at it. I suppose your activity has made Le Verrier[3] active also at the Paris observatory. The announcements of changes there seem at least to look like it[.]
Ever My dear Sir | Yours Very Truly | M. Faraday
G.B. Airy Esqr | &c &c &c

1. Letter 2952.
2. Charles Vincent Walker (1812–1882, DNB). Electrician. See Chapman (1998), 43–4.
3. Urbain Jean Joseph Leverrier (1811–1877, DSB). Director of the Paris Observatory, 1854–1870.

Letter 2954
Heinrich Buff[1] to Faraday
16 March 1855
From the original in IEE MS SC 2

Hoch verehrter Herr!

Vor zwei Jahren habe ich einige Versuche über die Electrolyse der Silber: und Kupfer: Lösungen, des reinen Wassers und der verdünnten Schwefelsäure bekannt gemacht, durch welche die Proportionalität der Stromstärke mit der Zersetzung, innerhalb sehr weiter Gränzen Bestätigung erhielt[2]. Dieselbe Frage ist seitdem wieder von verschiedenen Physikern und in verschiedenem Sinne discutirt worden. Eine allgemeine Geltung des electrolytischen Gesetzes wurde insbesondere hinsichtlich der Wasserzersetzung bestritten. Hierdurch veranlasst, habe ich mich ebenfalls wieder mit diesem Gegenstande beschäftigt. Es ist mir gelungen, mit Hülfe Wollaston'scher[3] Spitzen[4] die Zersetzung des Wassers und der wässrigen Lösungen durch electrische Ströme sichtbar zu machen, welche, wenn sie ganz zur Electrolyse verwendet werden, doch nicht mehr als 2,14 C.C. Wasserstoff im Laufe eines ganzen Jahres, d.h. stündlich den vierten Theil eines Cubick-Millimeters zu liefern vermögen. Diese Angaben sind durch die denselben zu Grunde liegenden Messungen, welche ich der Öffentlichkeit übergeben habe, wie ich glaube genügend gerechtfertigt. Sie dürften daher wohl geeignet sein die Annahme zu widerlegen, dass das Wasser einen in Betracht kommenden Theil des galvanischen Stroms, nach Art der Metalle zu leiten, im Stande sei.

Wenn man freilich das Verhalten der gemeinen Electricität in Betracht nimmt, wenn man bedenkt, wie geringe Mengen dieser Electricität erfordert werden um sehr bedeutende Spannungs-Effecte hervorzubringen, so bleibt immer noch die Möglichkeit, dass ein, allerdings sehr kleiner und vielleicht durch die Magnetnadel gar nicht messbarer, aber immerhin ein Theil des electrischen Fluidums durch das Wasser in ähnlicher Weise wie durch Metalldrähte geleitet werde.

Hinsichtlich dieses Punktes ist es mir nun kürzlich geglückt einige zum Theil neue Erfahrungen zu sammeln, welche ich mir erlaube Ihrer Beachtung vorzulegen; in der Hoffnung dass sie auch in Ihren Augen dazu beitragen werden, den Umfang der Geltung jenes wichtigen Gesetzes, durch dessen Entdeckung Sie zur Verständniss der chemischen Wirkungen des electrischen Stroms die einzige ganz allgemeine und zugleich die festeste Grundlage gegeben haben, um ein Bedeutendes zu erweitern.

Wenn die beiden Conductoren der Electrisi[e]rmaschine durch einen Multiplicatordraht in ununterbrochne Metallverbindung gesetzt werden, so wird bekanntlich der Übergang beider Electricitäten zu einander so vollständig vermittelt, dass durch Annäherung der Hand an den einen oder andern der Conductoren, ja selbst durch Berührung, Electricität in wahrnehmbarer Menge nicht abgeleitet werden kann. D.h. die Stellung der Galvanometernadel wird dadurch nicht merklich verändert. Der durch den Draht laufende Strom hat also ganz die Beschaffenheit eines galvanischen Stroms angenommen.

Mittelst einer Scheibenmaschine von 32 Pariser Zoll Durchmesser der Glasscheibe konnte ich bei dieser Art Schliessung Ströme erzeugen, welche die Nadel des von mir benutzten Galvanometers (einer Tangentenbussole) um 20 bis 24° ablenkten. Galvanische Ströme von dieser Stärke wirken auf das Wasser ganz dem electrolytischen Gesetze entsprechend und zwar erhält man auf 20° Ablenkung stündlich fast genau 0,01 C.C. Wasserstoffgas. Erfahrung und Rechnung zeigen sich so weit in befriedigender Übereinstimmung. Eine gleich kräftige Wasserzersetzung glaubte ich daher von der Einwirkung der gemeinen Electricität erwarten zu dürfen. Auch schien diese Annahme gerechtfertigt zu werden, als destillirtes Wasser in den Schliessungsbogen der Conductoren eingeschlossen wurde und man zu den Electroden Platinspitzen nahm. Vor dem Wasserestoffpole erhob sich, genau so wie unter der Einwirkung einer galvanischen Kette, bei gleicher Stromkraft eine ununterbrochne Säule äusserst feiner Gasbläschen, während die von dem Sauerstoffpole aufsteigende Gaslinie unverkennbar weniger massenhaft war, auch die Folge der einzelnen Bläschen viel deutlicher erkennen liess. Erscheinungen von demselben Charakter konnten übrigens auch durch schwächere Reibungsströme hervorgerufen werden und blieben selbst dann noch wahrnehmbar als nach allmäliger Abnutzung und Bestäubung der Reibzeuge die Nadel kaum noch einen Strom anzeigte.

In denselben Schliessungsbogen wurden nach Wiederherstellung der Maschine zugleich mit dem Wasser, verdünnte Schwefelsäure und Lösungen von Glaubersalz und Kupfervitriol eingeschaltet; und zwar so, dass der Übergang von der einen Flüssigkeit zur andern durch Wollaston'sche Spitzen stattfand. In welcher Ordnung nun diese Flüssigkeiten aufeinander folgen mochten; die drei erst genannten verhielten sich stets in ganz gleicher Weise, und so wie vorher für das Wasser beschrieben wurde. In der Kupferlösung entwickelte sich Gas nur an der Sauerstoff-Electrode. Nachdem die Einwirkung zwei Stunden gedauert hatte konnte man aber auch den Kupferabsatz an der negativen Electrode sehr deutlich erkennen. Dieses Verhalten der Kupferlösung haben Sie in den Experimental-Untersuchungen namentlich hervorgehoben und als chemisch electrische Zersetzung erkannt. Somit ist wohl kaum zu bezweifeln, dass die gleichzeitigen Vorgänge in den andern Flüssigkeiten ebenfalls electrolytisch waren.

Gleichwohl habe ich nicht unterlassen die Gase selbst auf ihre Natur zu prüfen. Zu diesem Zwecke diente ein enges, nur 1Linie weites Glasrohr, an dessen oberem [sic] Ende ein Platindraht in der Art eingeschmolzen war dass er etwa drei Linien weit frei in das Innere eindrang. Dieses Rohr mit reinem luftfreiem Wasser gefüllt, wurde mit dem unteren offnen Ende in ein Glas mit Wasser eingetaucht. Dann liess man reines Wasserstoffgas in den oberen Raum treten, bis die Flüssigkeit 4 oder 5 Linien unter das Ende des Platindrahts herabgesunken war. So entstand ein kleiner eudiometrischer Apparat in welchem, wie ich mich durch Vorversuche überzeugte, die geringste messbare Menge Sauerstoff angezeigt wurde, indem sie eine verhältnissmässige Menge des Wasserstoffs unter dem Einflusse eines durch den Platindraht geleiteten Funkenstroms verschwinden machte.

In diesem Eudiometer-Rohr wurde nun abwechselnd das von der negativen und das von der positiven Platinspitze aufsteigende Gas gesammelt, und jedes besonders auf seine Beschaffenheit geprüft. So überzeugte ich mich durch wiederholte Versuche, dass an der einen Spitze nur Wasserstoff, an der andern nur Sauerstoff entbunden wurde. Die einzelnen Versuche erforderten ein[e] lange Zeit fortgesetztes Drehen der Scheibe, zum Theil aus dem Grunde, weil die Wirksamkeit meiner Electrisirmaschine nicht lange ungeschwächt aushielt. Um z.B. 2 Linien Wasserstoffgas zu sammlen bedurfte es 4 Stunden Arbeit.

Das Resultat blieb in der Qualität und, soweit man aus der für die Gewinnung des Gases erforderlichen Zeit ein Utrheil ziehen durfte, auch in der Quantität ganz gleich, wenn man die vorher gut leitende Verbindung der beiden Conductoren der Maschine durch eine kurze Luftschicht unterbrach. Erst als nach allmäliger Vergrösserung der Unterbrechungsstelle die Platinspitzen zu leuchten anfingen, tratt [sic] auch im Verhältniss der Gasentwicklung eine auffallendere Veränderung ein. Die Gasmenge vermehrte sich, um so bedeutender, je tiefer, bei verstärkten Schlägen die von der Spitze ausströmende züngelnde Flamme in das Wasser eindrang; dabei wurden die Gasblasen mit Gewalt nach allen Richtungen umhergeschleudert. Allerdings erhielt man jetzt an beiden Drahtspitzen ein Gemenge von Sauerstoff und Wasserstoff; allein die eudiometrische Probe zeigte, sobald nur der Versuch lange genug fortgesetzt worden war, entschieden auf der positiven Seite einen Überschuss von sauerstoff, auf der negativen einen Überschuss von Wasserstoff. Es ist hieraus wohl kaum eine andere Folgerung zu ziehen, als dass der Wollaston'sche Versuch ein zusammengesetztes Phänomen darstellt, wobei die stets vorhandene Electrolyse von einer bald mehr, bald weniger starken durch Erhitzung bewirkten Zersetzung begleitet ist.

Jene chemisch electrische Zersetzung bleibt selbst dann nicht aus, wenn auch nur eine einzige Wollaston'sche Spitze, und ausser dieser keine andere Electrode in das Wasser taucht. Ich habe diesen sonderbaren Versuch, dessen Sie in der dritten Reihe Ihrer Experimental-Untersuchungen[5] Erwähnung thun, in folgender Weise wiederholt. Ein zugespitzter Platindraht wurde aus einiger Entfernung gegen die Oberfläche des Wasserbeckens gerichtet, in welchem die Wollastonsche Spitze unter dem Eudiometer-Rohr so aufgestellt war, dass das sich entbindende Gas gesammelt und dann geprüft werden konnte. War nun der zugespitzte Draht mit dem positiven Conductor, die Platinspitze mit dem negativen in leitender Verbindung, so entwickelte sich an der letzteren reines Wasserstoffgas; im umgekehrten Falle erhielt man Sauerstoffgas. Der andere Bestandtheil des Wassers musste sich folglich an der die Wasseroberfläche berührenden, durch den zugespitzten Draht electrisirten Luftschicht ausgeschieden haben. Die Richtigkeit dieses Schlusses wird durch den folgenden Versuch noch direkter bewiesen.- Der zugespitzte Draht wurde in ein zweites Eudiometer-Rohr eingeschmolzen, welches von gleicher Weite mit dem früheren war, und wie dieses theilweise mit Wasser und darüber soweit mit Wasserstoff gefüllt wurde, dass das Ende des Drahts 5–6 Linien von der Wasserfläche entfernt stand. Als nun das aus dem Glasrohr hervortretende Drahtende mit dem negativen Conductor verbunden wurde, die Wollastonsche Spitze aber mit dem positiven, so vermehrte sich die Gasmenge im zweiten Eudiometer-Rohr. Sie nahm dagegen ab, sobald man umgekehrt verfuhr, und das Drahtende zu dem positiven Conductor leitete. Bei dem letzten dieser Versuche war gleichzeitig das von der Wollaston'schen Spitze (die diesmal mit dem negativen Conductor verbunden war) aufsteigende Gas aufgefangen worden. Dabei zeigte sich nun, dass die Gasmenge in diesem Rohr ungefähr in demselben Verhältnisse zunahm, als sie sich in dem andern vermindert hatte. Gase wenn sie, wie in diesen Versuchen durch eine verstärkte electrische Spannung die Fähigkeit angenommen haben, das electrische Fluidum überzuführen, können also ähnlich den metallischen Oberflächen die Rolle einer Electrode übernehmen[6].

Genehmigen Sir schliesslich verehrter Herr die Versicherung der ausgezeigneten [sic] Hochachtung womit ich die Ehre
habe zu zeichnen. | H. Buff
Giessen am 16ten März | 1855.

TRANSLATION from the original in IEE MS SC 2
Letter addressed to Prof Faraday | by H. Buff | Professor of Physics in the University of | Giessen
Dear Sir,

Two years ago I published some researches upon the electrolysis of silver and Copper solutions, of pure water and dilute sulphuric acid, from which the dependence of the amount of decomposition upon the strength of the current received Confirmation within very wide limits[2]. This same question has since been discussed by different physicists and from different points of view. The general application of the electrical law has however been disputed particularly with respect to the decomposition of water. Induced by this circumstance I have been again occupied in the examination of this subject. I have succeeded by the aid of Wollaston's[3] points[4] in rendering evident the decomposition of water and of aqueous solutions by the electric current, which if entirely used for electrolysis does not liberate more than 2.14 Cubic Centimetres of hydrogen in the course of a whole year, i.e. the fourth part of a cubic millimetre per hour. These statements are I think sufficiently justified by those which I have published, upon which they are founded. They seem therefore well calculated to refute the assumption that water is capable of conducting an appreciable portion of the galvanic current like metal.

On considering the deportment of common electricity and reflecting how small a quantity of this electricity is required in order to produce very considerable effects of tension, the mind is impressed with the possibility that a certain portion of the electric fluid small indeed and perhaps no longer measurable by a magnetic needle, is conducted by water, in the same way as by metallic wires.

With regard to this point I have recently succeeded in collecting some new facts, which I may be permitted to bring under your notice: in the hope that they may in your eyes also, contribute to the further confirmation of that important law, by the discovery of which you have laid, at once the most general and the firmest foundation for the comprehension of the chemical action of the electric current.

If the two conductors of an electrical machine be placed in unbroken metallic union by a galvanometer wire, it is well known, that the passage of the two Electricities from one to the other is so completely effected, that even by touching either of the conductors with the hand, electricity is not carried off to any appreciable extent, i.e. the position of the needle of the galvanometer does not become perceptibly changed. The current passing through the wire has therefore essentially acquired the nature of a galvanic current.

By the use of a plate machine of 32 parisian inches in diameter, and by closing the circuit in this way I was enabled to produce a current which deflected the needle of the galvanometer which I employed (a tangential needle) 20° to 24°. Galvanic currents of this strength act upon water in a manner quite in accordance with the law of electrolysis, and in fact by a current deflecting the needle 20°, very nearly 0.01 of a cubic centimetre of hydrogen is collected per hour. Experiment and theory thus agree perfectly in this respect. I thought therefore that I might reasonably expect an equally powerful decomposition of water by common electricity. And this opinion appeared to be justified, for on interposing distilled water as a part of the circuit of the conductors, and using platinum points as the Electrodes, there arose from the negative pole (just as by the action of a galvanic current of equal strength) an unbroken series of bubbles of gas, whilst the stream of gas rising from the positive pole was unmistakably less coherent and the succession of the individual bubbles much more distinctly recognisable. Effects of the same character could be produced also by weaker frictional currents and remained perceptible even when the needle scarcely indicated the passage of a current, after the gradual wearing of the rubber and its becoming covered with dust. Into the same closed circuit after the restoration of the machine dilute sulphuric acid, and solutions of glauber's salts and of sulphate of Copper, were introduced with the coating in such a manner that the passage from the one fluid to the other took place through the Wollaston's points.

Now, in whatever order these solutions followed each other, the three above mentioned always comported themselves in a perfectly similar way, and as has been described in the case of water. In the copper solution gas was only disengaged at the positive electrode, but after the action had been continued for two hours the deposition of copper on the negative Electrode could be very distinctly perceived. This deportment of copper solutions you have specially brought forward in your "experimental researches" and recognised as an electro-chemical decomposition. Hence it can scarcely be doubted that the simultaneous actions in the other fluids were likewise electrolytic.

I have however, not omitted to examine the nature of the gases themselves. For this purpose I employed a narrow glass tube (only $\frac{1}{10}$th of an inch in diameter) into the upper end of which a platinum wire was sealed in such a way that it projected within the tube to the extent of about a quarter of an inch. This tube was filled with pure water free from air, and inverted in a glass of water. Then pure hydrogen gas was allowed to rise to the upper end, until the fluid had sunk about half an inch below the end of the platinum wire. Thus was obtained a small eudiometrical apparatus in which, as I had convinced myself by previous trials, the smallest measureable quantity of oxygen could be indicated since it would produce an appreciable diminution of the hydrogen under the influence of an Electric spark passed between the platinum wires.

In this eudiometric tube the gases disengaged from the positive and negative platinum poles were now collected. In this way I convinced myself by repeated experiments that from the one pole only hydrogen, and from the other only oxygen was disengaged. Each individual experiment required the turning of the plate to be continued for a very long time, partly because my electrical machine did not retain its state of activity for a long time undiminished. In order e.g. to collect $\frac{1}{5}$th of an inch of hydrogen gas four hours work was necessary.

The result remained the same as to quality, and as far as an opinion could be formed from the small quantity of gas disengaged in the necessary time also as respects quantity, whether the conductors of the machine be joined as in the preceding experiments by a good conducting union, or interrupted by a short stratum of air: only on gradually increasing the interrupting space the platinum points began to become incandescent and a remarkable change took place in the disengagement of gas. The quantity of gas increased in the same measure as the lambent flame emanating from the points, during the augmented changes penetrated deeper into the water; and the bubbles of gas were tumultuously disengaged in all directions. A mixture of oxygen and hydrogen was now collected from the points of both wires, but the eudiometrical examination indicated, as soon as the experiment had been continued long enough, that at the positive pole an excess of oxygen, and on the negative an excess of hydrogen was liberated. There is scarcely any other inference that can be drawn from this observation than that Wollaston's experiment is a mixed phenomenon, that the regular electrolysis which is always going on is accompanied by a more or less powerful decomposition produced by the heating of the electrodes.

This electro-chemical decomposition does not cease even when only a single Wollaston's point and no other electrode is immersed into the water. I have repeated this curious experiment, which you have mentioned in the third volume [sic] of your 'Experimental Researches'[5], in the following way. A pointed platinum wire was arranged at a short distance from the surface of a vessel of water in which the Wollaston's point was so placed under the eudiometer tube that the gas disengaged from it could be collected and subsequently examined. The pointed wire was now placed in contact with the positive conductor, and the platinum point with the negative, when pure hydrogen was liberated upon the latter, whilst under the reversed circumstances oxygen gas was then liberated.

The other constituent of the water must have been drawn off by the pointed wire into the electrified atmosphere, between it and the surface of the water. The correctness of this conclusion is proved by the following experiment. The pointed wire was fused into a second eudiometer tube of the same width as the former, and like it, partly filled with water, and above it with just so much hydrogen that the end of the wire stood at about half an inch distant from the surface of the water. Now when the end of the wire projecting from the glass tube was connected with the negative conductor, and the Wollaston's point with the positive, the volume of gas increased in the second eudiometer tube. On the other hand a diminution was observed when the circumstances were reversed, and the end of the wire united with the positive conductor. In the latter of these experiments the gas rising simultaneously from the Wollaston's point (which in this instance was connected with the negative Conductor) was simultaneously collected. It was then observed that the quantity of gas in this tube had increased, in the same proportion, as it had diminished in the other. Gases, when by an increased electrical tension (as in these experiments) they have acquired the power of conducting the electric fluid, thus become capable of playing the part of electrodes like metallic surfaces[6].

Believe me | dear Sir | Yours very sincerely | H. Buff
Giessen March 16, 1855

1. Heinrich Buff (1805–1878, NDB). Professor of Physics at Giessen from 1838.
2. Buff (1853).
3. William Hyde Wollaston (1766–1828, DSB). Man of science.
4. That is wire sealed into glass tubes. See Wollaston (1801), 430–1.
5. This would appear to be a reference to Faraday (1833a), ERE3, 327.
6. Faraday discussed this letter in Faraday (1855e), Friday Evening Discourse of 25 May 1855, p.124.

Letter 2955
Peter Theophilus Riess to Faraday
20 March 1855
From the original in IEE MS SC 2

My dear Sir

The interesting experiments of Mr. Latimer Clark, published under the aegis of Your great name in the last number of the Philos. magazine[1], together with the consequences drawn by the late Melloni, incline me to the following remarks. Melloni's conclusion, the velocity of propagation of the electricity to be independent of the electric density, appears to me not justified; on the contrary I incline to the view, that the velocity of electricity is proportional to its density. I believe, I have experiment to corroborate the result deduced by me from inquiries made sixteen years ago on the production of heat in the circuit of the Leyden battery. Then I maintained that the time of discharge of the battery is directly proportionate to the quantity of accumulated electricity, and inversely proportionate to its density. I have represented the time of discharge of the battery in §.436 of my treatise on frictional electricity by the relation $b(1/b + V)q/y^2$. The letter b and V depend of the nature of the metallic conductor, q is the quantity of electricity accumulated in the battery, y its density. It is evident by this expression, that, if the quantity of electricity 1 is brought in one jar and the quantity 2 in two jars, if the discharge of the first jar through a given conductor results in the time 1, the two jars will be discharged in the time 2; that on the contrary the time of discharge will remain 1, if the quantity 2 is accumulated in one jar. In the last case q as well as y have been duplicated, and the value of the fraction q/y is not altered. This case, I believe, furnishes an explanation of Clark's experiments. If we make the supposition, very plausible considering the great velocity of electricity, that the chemical action in Bains[3] telegraph will commence only after the transmission of the whole quantity of electricity evolved at each instant on each pole of the voltaic battery, the telegraph in Clarks experiments began to mark, when, by using the smaller battery, a certain quantity with certain density had passed the telegraph, and, by using the larger battery, when an nfold quantity with partly the nfold density had passed the telegraph. The law above announced, though found for the Leyden battery and hypothetically referred to the voltaic battery, informs that the time of discharge of the battery is the same in both cases, and it follows that the time for the progress of the discharge from the nearer to the further end of the long wire must be the same both for the smaller and the larger battery.

You remark, my dear Sir, that Melloni shortly before his death, was about to publish facts, by which he hoped to demonstrate clearly the errors of conclusions, arrived at by Coulomb, Poisson and others respecting electrostatic induction[4]. It is to be regretted, that he has really accomplished his intention. He sent on that subject a letter (dated 12 July 1854) to Baron Humboldt, and Humboldt communicated it to the Academy of Berlin, where I was acquainted with it[5]. Melloni undertakes to prove, that an inducteous body, during the presence of the inductric, receives only electricity of the same kind with the latter, and that the electricity of the opposite kind is only evolved after the remotion of the inductric body. Just the same view has been pronounced long ago in Germany, but has been left as wholly untenable and as produced only by misinterpreted experiments.

Melloni has communicated these inquiries also to Mr. Regnault, and You will find them published in the Comptes rendus de l'academie de Paris vol. 39 p.177[6][.]

I have the honour to remain with the greatest veneration | Yours P. Riess
Berlin 20 March 1855. | Spandau Strasse

Address: Professor Michael Faraday | member of all Academies of Science | London | (royal Institution).

1. Faraday (1855c).
2. Riess (1853), 1: 411–2.
3. Alexander Bain (1810–1877, DNB). Telegraph engineer.
4. Faraday (1855c), 163.
5. See Bericht.Akad.Wiss.Berlin, 1854, p.431.
6. Melloni (1854a).

Letter 2956
Jacob Herbert to Faraday
22 March 1855
From the original in GL MS 30108/2/89

Trinity House, | 22nd March 1855.

My dear Sir,

Herewith I send you three Samples of the White Lead, delivered by the Contractor, which have been taken at hazard from different kegs; And which I have to request you will analyze, and favor the Elder Brethren by reporting the result, at your convenience[1].

I am | My dear Sir, | Your very faithful Servant | J. Herbert
M. Faraday Esq | &c &c &c

Endorsed by Faraday: See Communication of date 14 Feby. 1855[2]

1. Faraday's analyses are in GL MS 30108/2/89 with a note that he reported them on 27 March 1855.
2. Letter 2940.

Letter 2957
Faraday to Edwin Atherstone[1]
26 March 1855
From the original in Somerset Record Office, DD/SAS G/3016

Royal Institution | 26 March 1855

My dear Sir

I thank you for your ticket - I cannot be there - but if I can find any one worthy of hearing you I shall take the liberty of giving your ticket to him[.]

Yours Very Truly | M. Faraday
Edwin Atherstone Esq | &c &c &c

Address: Edwin Atherstone Esq | &c &c &c | 17 Edward Street | Portman Square

1. Edwin Atherstone (1788–1872, DNB). Writer and poet.

Letter 2958
Jacob Herbert to Faraday
28 March 1855
From the original in GL MS 30108/2/89

Trinity House, London | 28th March 1855

My dear Sir,
The Elder Brethren having it in contemplation to try the prepared Zinc experimentally by painting two or three of the Light Houses this year with that preparation in lieu of White Lead, I am directed to transmit to you the accompanying Sample of Zinc Paint, in the state for mixing, which has been submitted as that which it is proposed to use, in the trial referred to; and to signify their request that you will favor them by analyzing the same and reporting the result, at your convenience[1].

I remain | My dear Sir | Your very faithful Servant | J. Herbert
M. Faraday Esq. D.C.L

1. Faraday's analysis is in GL MS 30108/2/89.

Letter 2959
Alessandro Palagi to Faraday
28 March 1855
From the original in IEE MS SC 2

Signore
Con questo corso di posta riceverete, o Signore, un Opuscolo che ha per titolo *Sulla distribuzione delle correnti elettriche nei conduttori*[1]. Questo lavoro, fatto in compagnia del Padre Bertelli[2], mio concittadino ed amico, ci è sembrato degno di vedere la luce e ci è parso ancora che in se contenga dei germi profittevoli alla Scienza della elettricità. Noi però attendiamo il giudizio imparziale e sapiente degli uomini dotti; siccome Voi siete intimamente persuasi che, quand'anche trovaste questi nostri studii privi d'interesse e di ne'un valore, non ce ne dareste biasimo, apprezzando in noi, non fosse altro, il buon volere.

E già un anno ch'io ho terminato, o Signore, una Serie di nuovi esperimenti eseguiti con macchine di mia invenzione, co' quali mi pare di provare fisicamente e logicamente che le tensioni elettriche, che i corpi isolati acquistano nel muoversi, non sono dipendenti e generate da confricazione dell'aria contro corpi solidi; non da elettricità ad'influenza o propria dell'atmosfera; non dal passare i corpi co'loro moti per i strati atmosferici diversamente e a vario grado elettrizzati; non da elettricità animale o di attrito comunque sviluppata o dalle macchine stasse, o dalle resti o dai movimenti dello sperimentatore: ma derivare sibbene quelle tensioni elettriche dall'azione scambievole dei corpi nell'avvicinarsi, o nell'allontanarsi reciprocamente.

Queste mie sperienze non le ho neanche pubblicate per la molta opposizione che mi viene fatta dai Professori dell'Università dai quali si sostiene che la elettricità atmosferica non appartiene alla Meteorologia, ma alla Fisica; epperò che io voglio entrare nell'altrui messe. Quasi che le Scienze fossero privativa di tale o tal'altra cattedra! E non vale il dire che un tal genere di elettricità, giusto perchè atmosferica, appartiene direttamente alla Meteorologia; che di tale elettricità trattano *ex professo* tutti i Meteorologi; che tutti i Fisici nei loro Trattati pongono quanto risguarda all'elettricità atmosferica nel *trattato* speciale di Meteorologia, e cosi non vale l'addurre qualsiasi altra ragione; chè dai Professori dall'Università, approfittando dell'ignoranza dei Superiori, si cerca ad ogni costo, e contro ancora il proprio convincimento, di recarmi noja; d'ingenerarmi imbarazzi; di rendermi avversi gli altrui Spiriti e di tenermi sempre invilito ed oppresso.

Nel corso di questa opposizione ingrata e di questa guerra maligna fui tentato più volte di pubblicare a mio onore la lettera dottissima, che aveste la bontà, o Signore, d'inviarmi il 7 Nov. 1853; ma poi me ne astenni per timore di recarvi dispiacere in farlo senza il vostro assenso. La rinomanza, che meritamente voi godete, o Signore, qui da noi, non solo, ma per tutto il mondo, farebbe stata a mia difesa un baluardo inespugnabile; il consiglio, che Voi mi date, o Signore, di proseguire in quegli studii avrebbe convinto e i miei Superiori, e miei concittadini di non trattarsi già di ameni e futili studii, ma sibbene di studii seri ed importanti.

Perdonate, o Signore, se vi ho scritto troppo a lungo e preoccupato troppo di me medesimo; la vostra bontà e il vostro sapere mi han fatto animo; sento troppo il bisogno di essere da Voi, o Signore, protetto ed animato!

Abbiatemi pertanto nella vostra grazia, o Signore, e credetemi pieno di altissima stima e profonda venerazione | Di Voi, illustre Signore | Devmo obb-mo Servo | Allessandro Palagi
Bologna 28 Marzo 1855
Al Chiarissimo Signore | Sigr. Michele Faraday | Membro della Società Reale di | *Londra*

TRANSLATION
Sir,

Through the post you will receive, Sir, a little work which is entitled *On the distribution of electric currents in conductors*[1]. This work, written in the company of Padre Bertelli[2], my fellow citizen and friend, seemed to us worthy of seeing the light of day, and seemed to us to contain a few germs of benefit to the Science of Electricity. We, however, await the impartial and wise judgement of gifted men, such as yourself; and we are intimately convinced that even if you found our private studies of little interest and value, you would never censure us, appreciating at least our good will.

It is already a year since I completed, Sir, a series of new experiments, carried out on machines of my own invention, which seemed to prove physically and logically that the electrical tensions which isolated bodies acquire in moving, are not dependent or generated by the friction of air against solid bodies; nor by electricity influenced by or of the atmosphere, nor by the movement of bodies through atmospheric strata diversly electrified or electrified to differing degrees; nor from animal electricity or friction from any source; nor from the machines themselves, or from the resting or movement of the experimenter: but that these electrical tensions derive from the mutual action of bodies in drawing near and being drawn away reciprocally.

These experiments of mine have not yet been published because of the great opposition of the Professors of the University who maintain that atmospheric electricity does not belong to Meteorology, but to Physics; and that I am treading on other people's territory. Almost as if the Sciences were a private thing belonging to this or that professorial chair! And it is not worth saying that this type of electricity belongs directly to Meteorology; that all Meteorologists deal *ex professo* with this type of electricity; that all Physicists in their treatises put anything regarding atmospheric electricity in a special *treatise* on Meteorology and therefore there needs to be no other reason; but the Professors of the University, profiting from the ignorance of their Superiors, find at any cost, and against their own conviction, ways of causing me trouble, of causing me difficulties, of turning others against me and of keeping me dejected and oppressed.

In the course of this disagreeable opposition and this maligning war I was tempted on several occasions, for my own honour, to publish the very learned letter that you had the kindness to send me, Sir, on 7 Nov. 1853, but I abstained for fear of displeasing you in doing so without your consent. The respect that you deservedly command, Sir, not just here among us, but in the whole world, would have been an unassailable bulwark in my defence; the advice that you give me, Sir, on how I should continue in these studies will carry weight with my superiors and will persuade my fellow citizens that it is not a question of amusing and futile studies, but studies that are serious and important.

Please excuse, Sir, that I have written at such length and have been preoccupied with myself; your kindness and your knowledge have given me courage; I very much feel the need, Sir, for your protection and friendship!

Please keep me in the meantime in your grace, Sir, and believe me to be full of the highest esteem and profound veneration, illustrious Sir, Your most devoted and most obedient Servant | Alessandro Palagi
Bologna, *28* March *1855*
To Mr | Mr Michael Faraday | Member of the Royal Society of | *London*

1. Timoteo et al. (1855).
2. Bertelli Barnabita. Otherwise unidentified.

Letter 2960
Faraday to William MacKenzie[1]
30 March 1855
From the original in the possession of Dennis Embleton

Royal Institution | 30 Mar 1855

Sir

I hasten to acknowledge your letter & the compliment which it conveys but that which you propose does not fall in with the tenor of my occupations or inclinations[.] I have never felt inclined to proceed further in authorship than is needful to write my own papers at least not for many years past[.]

I am Sir | Your Obedient Servant | M. Faraday
W. Mackenzie Esqr | &c &c &c

1. William MacKenzie. Publisher in Glasgow. Otherwise unidentified.

Letter 2961
Faraday to Henry Deacon
31 March 1855
From the original in Southwark Local Library Deeds 5585

Royal Institution | 31 Mar 1855

My dear Sir

I received your letter & since then your papers from Mrs. Deacon[1] & am very much obliged to you. It looks well so well that my wife has it in reading trusting to understand it. Next week we go into the country & I intend taking it as one of the things to read up. I trust you are getting on well in the manufactory matters.

Ever Truly Yours | M. Faraday
H. Deacon Esqr | &c &c &c

1. Emma Deacon, née Wade. See DNBmp under Henry Deacon.

Letter 2962[1]
Faraday to Harriet Fellows[2]
2 April 1855
From the original in WIHM MS 5634

Mr Faraday present his compliments & very sincere thanks to Lady Fellows for the Cocoons & silk for which he is very grateful. Tomorrow he will take them to Hastings & there occupy himself by winding the silk off & thinking of its uses[3]. They promise exceedingly well for his purpose.
Royal Institution | 2 April 1855

1. This letter is black-edged. See note 1, letter 2942.
2. Harriet Fellows, olim Knight (d.1874, age 65, GRO). Married the traveller and archaeologist Charles Fellows (1799–1860, DNB) in 1848. See under his DNB entry.
3. See Faraday, *Diary*, 28 and 29 March 1855, 6: 13728–31 where Faraday noted the problems he was experiencing with using silk in his torsion experiments.

Letter 2963
Lyon Playfair to Faraday
4 April 1855
From the original in IEE MS SC 2
Confidential

Marlboro' House. | 4th April / 55.
My dear Sir,
The Royal Commission for the Exhibition of 1851 have advised the Board of Trade to nominate you as one of the 40 jurors to represent England at the French Exhibition[1].

Although no remuneration is attached to the office, travelling expenses, & personal expenses at the rate of £1. per diem will be given.

The period for the action of the juries is not decided but I think it will be from the middle of June to the first week in July.

The jury to which it is proposed to attach you is that of Philosophical Instruments or Chemistry. As the list of English jurors is so small it is necessary to ascertain before hand whether they are able & willing to act.

Would you kindly, therefore, state to me whether your nomination to the office would be agreeable to yourself. The answer should arrive not later than Monday[2].

Your's truly | Lyon Playfair
Prof. Faraday FRS.

1. See Minutes of the 1851 Commission, 3 April 1855, IC MS.
2. That is 9 April 1855. Faraday was not one of the jurors.

Letter 2964
Faraday to Christian Friedrich Schoenbein
6 April 1855
From the original in UB MS NS 417

Hastings | 6 April 1855

My dear friend
    I have brought your letter[1] here, that I might answer its great kindness at some time when I could remember quietly all the pleasure I have had since the time I first knew you. I say remember it *all*, but that I cannot do; for as a fresh incident creeps dimly into view I lose sight of the old ones, and I cannot tell how many are forgotten altogether. But think kindly of your old friend;- you know it is not willingly but of natural necessity that his impressions fade away. I cannot tell what sort of a portrait you have made of me,- all I can say is, that whatever it may be I doubt whether I should be able to remember it:- indeed I may say I know I should not, for I have just been under the Sculptors hands[2], and I look at the Clay, & I look at the marble, and I look in the glass, & the more I look the less I know about the matter & the more uncertain I become[.] But it is of no great consequence lable the marble & it will do just as well as if it were like. The imperishable marble of your book will surely flatter[3].
    You describe your state as a very happy one; healthy, idle, & comfortable. Is it indeed so? or are you laying up thoughts which are to spring out into a rich harvest of intellectual produce? I cannot imagine you a *do-nothing* as I am; Your very idleness must be activity. As for your book, it makes me mad to think I shall lose it. There was the other[3], which the Athenaeum[4] or some other periodical reviewed in German, but we never saw it in English - I often lent it to others & heard expressions of their enjoyment, & sometimes had snatches out of it; but to me it was a shut book. How often have I desired to learn German; but head ache & giddiness have stopped it.
    I feel as if I had pretty well worked out my stock of original matter, & have power to do little more than reconsider the old thoughts. I sent you by post a notice of a Friday Evening here[5], and would have sent you a paper from the Philosophical Magazine[6]; - but I am afraid of our post, i.e, I am afraid that unawares I may put my friends to much useless expense. I receive almost daily now, papers & journals, which coming by post are charged to me two, three, & four shillings, until I absolutely cannot afford it; and fearing that with equal innocency I may be causing my friends

inconvenience I have abstained: However I hope that a friend of mine, Mr Twining, will in the course of a month or two put the paper I speak of in your way. You will therein perceive that I am as strong as ever in the matter of lines of magnetic force & a magnetic medium; and what is more I think that men are beginning to look more closely to the matter than they have done heretofore, and find it a more serious affair than they expected. My own convictions & expectations increase continually; *that* you will say is because I become more & more familiar with the idea. It may be so & in some measure *must* be so; but I always tried to be very critical on myself before I gave any body else the opportunity, and even now I think I could say much stronger things against my notions than any body else has. Still the old views are so utterly untenable *as a whole*, that I am clear they must be wrong; whatever is right.

I had forgotten that Wiedemann was in Basle give my kindest remembrances to him. I think I received a paper on electrolysis from him, but out here cannot remember & cannot refer[7]. Our sincerest remembrances also to Mrs. Schoenbein & the favourable family critics. I can just imagine them, hearing you read your M.S., & flattering you up, & then giving you a sly mischievous mental poke in the ribs, &c. They cannot think better of you than I do.

Ever My dear Schoenbein | Your attached friend, | M. Faraday

Address: Dr Schoenbein | &c &c &c | University | Basle | on the Rhine

1. Letter 2943.
2. It is not clear what to what this refers. The sculptor Matthew Noble (1818–1876, DNB) had made two marble busts of Faraday in 1853 (now in the Royal Society) and 1854 (now in the Royal Institution (plate 1)). See also Gladstone (1874), 79 and Margery Ann Reid's Diary, October 1854, RI MS F 13 B, pp. 27–9.
3. [Schoenbein] (1855).
4. Schoenbein (1842).
5. "Extracts from the Travelling Diary of a German Naturalist", *Athenaeum*, 22 July 1843, pp.664–6, 29 July 1843, pp.690–1.
6. Faraday (1855a), Friday Evening Discourse of 19 January 1855.
7. Faraday (1855b), [ERE29b].

Letter 2965
Faraday to Arthur-Auguste De La Rive
7 April 1855
From the original in Uppsala University Handskriftsavdelningen Erik Waller's Collection of Autographs

Royal Institution | 7 April 1855

My dear friend

I must just write you a letter though I have nothing to say; i.e nothing philosophical; but I hope to feel with you that when philosophy has faded away, the friend remains. Do not think that I cannot & do not rejoice in reading and understanding all that your vigorous mind produces, but for myself, I feel I have little or nothing to return; and though when my sluggish mind is moved, I can think determinately & write decidedly, yet being once written I fall back into quietude, and leave what has been said almost uncared for or unthought of; & so it is that I do not teaze you in letters with much of my philosophic opinions.

I am afraid too of the Post, for though I send you now & then a report of a Friday Evening meeting, being assured that they will go without charge to you, yet as to Papers from the Philosophical Magazine I am in the greatest uncertainty. I receive daily papers from abroad which are charged two, three or four shillings, and am absolutely obliged to refuse many:- and when they tell me at the post office here that such a paper can go to the continent at so much, I am in fear of some mistake, &, that inadvertently I run the risk of taxing my friends beyond their patience. However I have sent you, by a friend, a paper from the Philosophical Magazine[1], which perhaps you have seen already; & so I will say no more about it:- except that Mr. Twining (the friend) is an excellent Gentleman as you will find if he personally comes in your way[.]

But of other matter:- I hope & desire that you should enjoy good health & spirits and that your work will be a cheerer to you;- and further, that as you turn from it to graver thoughts & back again, *both* should minister peace & contentment to your mind. What a world this is! How the whole surface of the earth seems about to be covered with the results of evil passions. - Ambition - contest - inhumanity - selfishness. - Thou shalt love thy neighbour as thyself[2]; - how the extreme reverse of this shapes its self into the forms of Honor Patriotism: Glory, Loyalty, Reverence, &c &c[3]. Happy for us that there is a power who overrules all this to his own good ends, and who will one day make manifest the truth & cause the Light to shine out of the darkness[4].

I shall hope soon to hear of the volume of the work[5],- & shall rejoice to know that you are yourself in good strength; &, whilst in the flesh, still working on your way.- I am very well, but as before, continually failing in memory; but since I have given up lecturing & the occupations which require memory, I have been very well & cheerful, & free from giddiness.- My dear wife also, though infirm, is in good mind & we go on our way rejoicing in each others company.

Ever My dear friend | Yours faithfully | M. Faraday
M | Auguste de la Rive | &c &c &c

Address: A Monsieur | Monsieur De la Rive | &c &c &c | à Geneva

Postmark: Hastings

1. Faraday (1855b), [ERE29b].
2. Matthew 19: 19.
3. A reference to the Anglo-French war against Russia.
4. 2 Corinthians 4: 6.
5. De La Rive (1853–8), **2**.

Letter 2966
Faraday to Peter Theophilus Riess
7 April 1855
From the original in SPK DD

Royal Institution | London | 7 April 1855

My dear Sir

It was a very great pleasure to me to receive your kind letter[1]; - and written in such English as made me ashamed of my ignorance of the German language. I never cease to regret the latter circumstance; for I am aware of the great stores of knowledge in that language which would then be open to me in relation to my especial pursuits, and which some how or other the system of publication in our country almost entirely shuts out from me. I have several times within the last 15 years set about acquiring it, but a result over which I have no power, namely, a gradually failing memory, has on these occasions made the labour of head so great, that I have been obliged to refrain from such an endeavour, as also from many others. You gave me your book some time back[2]. I looked at it eagerly; but both by its language & its mathematical developments (for the use of symbols requires memory) it was shut out from me; and so I placed it in our library, where I am very glad to find it is of great use to others[.]

Your observations upon Mr. Clarkes experiments (for M. Melloni) are very interesting to me & I cannot doubt that you are right. I can see no difference of an essential kind, between the current produced by a Leyden Jar & that of a Voltaic battery, and your experiments & conclusions appear to me to be fully applicable to the case and perfectly justified. I think Melloni could have been but little acquainted with the great body of facts belonging to electricity; Static & dynamic;- perhaps he had only begun to enter upon the subject, and was caught, as all men are, by first appearances. I had several letters from him[3]; & in relation to his conclusions on induction, I had written him a very long letter[4] (in reply to a like long one from him[5]) against his views & statements. I heard from M. Flauti, the Secretary of the Academy at Naples, that it was received after his death, & had been read at the Academy[6] (which was not however my intention). It was my hope that it would have led him to revise his conclusions before he published them. From what you tell me I conclude that he had already published them[7]; & I am sorry for it.

Your letter to me for which I thank you very much indeed, makes me think that you approve of the correction which I put into the Philosophical Magazine[8], of Mellonis erroneous representation of Mr. L. Clarkes last results. As they were represented in the Italian journal[9], the diagram could only confuse the mind & give conflicting ideas.

I am My dear Sir | With Very Great Respect | Your Obliged & faithful Servant | M Faraday
Professor Riess | &c &c &c

Address: Professor Riess | &c &c &c | Spandau Strasse | Berlin
Postmark: Hastings

1. Letter 2955.
2. Riess (1853). See letter 2668.
3. Letters 2813, 2834, 2862, 2865.
4. Letter 2870.
5. Letter 2865.
6. See *Rend.Soc.Reale Borbon.Accad.Sci.*, 1854, 3: 93–4.
7. Melloni (1854a).
8. Faraday (1855c).
9. Melloni (1854e).

Letter 2967
Faraday to Alessandro Palagi
12 April 1855
From the original in Biblioteca Comunale dell'Archiginnasio Bologna, Collezione autografi Pallolti XII.699

Royal Institution | London | 12 April 1855
Sir
I hasten to acknowledge your letter of the 28th of last month[1] & the more so because I have waited for the paper[2] of which you speak in it & have not received it. Our Post office is very uncertain about printed papers which come by it and frequently charge so high a price through some little circumstance neglected that I am unable to take the papers in[.] For many came to me and almost daily & the orders which I have been obliged to give have I am afraid caused yours to be returned. However no paper from you can have arrived since your letter because I should have paid special attention to it.

I am very sorry that your former remarks should be a source of annoyance to you. There is no doubt that a principle so new & so bold as that you put forth would require to be supported by undeniable evidence but in the working of it out for the development of the natural truth there should have been no cause of offence any where. People may differ in their belief even as to natural things & yet bear with each other[.] I differ in many things just now from those about me & yet we all go on very harmoniously together. At the same time one must endeavour to criticise ones own views very strictly so that we may give them up if wrong or if right establish them more & more by facts. I hope all these troubles will pass away - & I shall look for your new facts in the Italian Journals or in the Bibliotheque de Geneve where I trust I may find it[.]

 I am Sir | Your Very Obliged | Servant | M. Faraday
Il Signore | Signore Palagi | &c &c &c

Address: Il Signore | Signore Palagi | &c &c &c | Bologna | Italy
Postmark: Hastings

1. Letter 2959.
2. Timoteo et al. (1855).

Letter 2968
Lord Wrottesley to Faraday
13 April 1855
From the original in IEE MS SC 2

               Wrottesley | 13 Apr / 55
My dear Sir
 If you have formed any decided opinions as to the giving away the Medals this year or as to supplying the vacancies in the List of Foreign Members, I should be much obliged to you to favour me with them before the meeting of the Council on Thursday next the 19th[1][.] There are 2 Royal Medals to be given besides the Copley.
  Yours very truly | Wrottesley
P.S | I shall go to Town on Wedy next the 18th to 1 Albemarle St
Professor Faraday

1. The meeting was postponed until 21 April 1855 (RS CM 21 April 1855, 2: 318). There is no indication as to what suggestions, if any, were made by Faraday.

Letter 2969
Arthur-Auguste De La Rive to Faraday
15 April 1855
From the original in IEE MS SC 2

Genève le 15 avril 1855.
Monsieur & très cher ami,

  Je suis profondément touché de la bonne pensée qui vous a mis la plume à la main pour m'écrire quelques mots d'amitié qui m'ont été encore plus doux & plus agréables que ne m'aurait été l'annonce d'une grande découverte[1]. A mesure que l'on devient vieux (& j'ai 53 ans) on sent toujours plus le prix des sentiments élevés & Chrétiens & la valeur d'amis comme vous avec lesquels on est en si pleine sympathie sur des points aussi essentiels.- J'ai passé de mauvais moments ce printemps. - Voilà cinq ans bientôt que j'ai perdu une amie[2] comme on n'en a qu'une & je sens chaque année un vide & un isolement plus grands quoique je sois entouré de nombreux & charmants enfants & même de quatre petits enfants. Ce sont d'immenses objects d'interet, mais ce ne sont pas des seconds vous-mêmes. Refléchissez que je n'ai ni soeur, ni belle soeur, en un mot aucune femme de ma génération avec qui je puisse parler de mes enfants, à qui je puisse communiquer mes impressions; je voudrais souvent être plus vieux.

  Ma santé physique s'est beaucoup améliorée; je suis très bien partout maintenant. Aussi, vais-je en profiter pour aller passer un mois à Londres afin d'achever de corriger les épreuves de mon second volume[3] qui ne me parviennent ici que très irrégulièrement. Je serai à Londres du 10 ou 15 Mai & j'irai reprendre ma chambre de *Suffolk Place* (Haymarket) dans la maison où était mon bon ami J.L. Prevost[4] & qui est occupé maintenant par son neveu Alexandre[5] qui m'a offert l'hospitalité de la manière la plus aimable. Vous n'avez pas d'idée combien je serai heureux de vous voir, de retrouver ce regard bienveillant, cette amitié si douce dont vous m'avez toujours gratifié.

  J'ai eu beaucoup de peine à finir mon second volume en le chemin qu'avaient fait plusieures questions depuis que je l'avais rédigé; il m'a fallu changer complètement certaines parties telles par example que la thermo-électricité; & bien des points de l'Electro-chimie ont du subir des modifications. Néamoins je suis assez satisfait des résultats auxquels je suis parvenu, & j'espère que vous le serez aussi. J'ai cherché à éclaircir bien des points par la voie expérimentale, & j'ai eu sous ce rapport un auxiliare précieux dans l'un de mes élèves Mr Soret[6] qui, après avoir passé quatre ans dans le laboratoire de Mr Regnault, est établi maintenant à Genève où il travaille très bien & beaucoup.

  J'ai bien reçu & lu avec beaucoup d'intérêt tout ce que vous m'avez envoyé & que vous avez publié sur le magnétisme & le diamagnétisme. Je persiste à croire à la polarité diamagnétique & je vous transmettrai mes raisons en même temps que je vous demanderai de vouloir bien faire quelques expériences avec des appareils que j'apporterai, & qui me paraissent pouvoir être de nature à éclaircir la question. Je n'ai point voulu parler de vos dernières recherches dans la *Bibl. Univ.* avant de m'en être entretenu avec vous.- Au fond la grande différence entre vous & moi, c'est que je suis tout *moléculaire* & que vous êtes tout *force*.- Je crois à un principe *passif* aussi bien qu'à un principe *actif*.- Mais cette discussion nous ménerait trop loin.

  Au reste il y a deux points sur lesquels nous nous entendrons toujours; le premier c'est que les faits, & les faits bien observés doivent dominer toutes les théories; le Second c'est qu'au-dessus des pauvres lois de la nature dont nous ne saisissons que des lambeaux, il existe un Etre qui dirige tout le monde matérial aussi bien que le monde moral, par sa providence d'une manière continue & générale.- Quand on est d'accord là dessus ainsi que sur tout ce qui touche aux sentiments du coeur, on peut bien différer sur quelques points de détail, sans danger pour l'amitié.-

  Mes souvenirs respectueux & affectueux à Madame Faraday; j'espère que sa santé est bonne ainsi que la vôtre. A bientôt & croyez, Monsieur & cher ami, aux sentiments de profonde & ancienne affection

de votre dévoué ami | A. de la Rive

TRANSLATION

Geneva 15 April 1855

Sir and very dear friend,

  I am profoundly touched by the thought which made you put a pen in your hand to write me some words of friendship, which were sweeter to me and gave me more pleasure than if you had announced a great discovery[1]. As one gets older (and I am 53), one always appreciates uplifted & Christian sentiments and the value of friends like you with whom one feels such an affinity on all the essential things. - I had some bad times last spring - It is nearly five years since I lost my unique friend[2] and as the years go by I feel a greater emptiness and isolation, even though I am surrounded by numerous and charming children and even by four grand children. I take an immense interest in them but they cannot replace you. If you think that I have neither sister, nor sister-in-law, in a word no woman of my generation, with whom to speak of my children, with whom to share my impressions; I often wish I were older.

  My physical health has greatly improved; I am very well at the moment. So I am taking advantage of this to spend a month in London in order to complete the correction of the proofs of my second volume[3], which only arrive here sporadically. I will be in London from 10 or 15 May & I am going to take up once again my room in *Suffolk Place* (Haymarket) in the house where my good friend J.L. Prevost[4] lived, & which is now occupied by his nephew Alexandre[5] who has offered me hospitality in the most amiable way. You have no idea how happy I shall be to see you, to find again that kind look, that gentle friendship with which you have always gratified me.

I have had a lot of trouble in finishing my second volume as since drafting it, several questions have been posed on the way; I have had to change several sections completely, such as that on thermo-electricity; and many points on Electro-chemistry have had to undergo modifications. Nevertheless, I am satisfied with the results I have reached and I hope you will be too. I have tried to enlighten many points by experimentation and I have had in this regard a precious assistant in one of my pupils, Mr Soret[6], who having spent four years in Mr Regnault's laboratory, is now established in Geneva where he is working well and a great deal.

I received and read with a great deal of interest everything you sent me and which you published on magnetism and diamagnetism. I still believe in the diamagnetic polarity and I will explain my reasons at the same time as I shall ask you if you would mind doing some experiments with some instruments that I shall bring and which seem to me to be able to enlighten this question. I have not wanted to speak of your latest experiments in the *Bibl. Univ.* before speaking to you - Deep down the great difference between you and me is that I am all *molecular* and you are all *force*. - I believe in a *passive* principle as well as an *active* principle. But this discussion would take us too far.

Besides, there are two points on which we shall always agree; the first is that facts and facts which have been well observed should dominate all theories; the second is that above the poor laws of nature of which we grasp but scraps, there exists a Being who guides the material world as he does the moral world, by his providence in a continual and general way. When one is in agreement on that as well as on everything that concerns the feelings of the heart, one can differ on some details, without any danger to friendship. -

My respectful and affectionate regards to Madame Faraday; I hope that her health, as well as yours, is good. I hope to see you shortly. Please believe, Sir and dear friend, the sentiments of profound and longstanding affection

of your devoted friend | A de la Rive

1. Letter 2965.
2. Jeanne-Mathilde De La Rive.
3. De La Rive (1853–8), 2.
4. Jean-Louis Prevost (1790–1850, DSB). Swiss physiologist.
5. Alexandre Pierre Prevost (1821–1873, P2, 3). Swiss physiologist.
6. Jacques Louis Soret (1827–1890, P2, 3). Swiss physicist.

Letter 2970
Faraday to William MacKenzie[1]
16 April 1855
From the original in the possession of Dennis Embleton

Royal Institution | 16 April 1855

Sir

I beg to acknowledge your letter of the 14th but am unable to assent to it. If I write at all about Electricity it would be in my own name - rather largely & not without much thought & intellectual labour but I have no intention of the kind at present[2][.]

I am Sir | Your Very Obedient Servant | M. Faraday

W. Mackenzie Esqr | &c &c &c

1. William MacKenzie. Publisher in Glasgow. Otherwise unidentified.
2. See letter 2960.

Letter 2971
William Cox to Faraday
16 April 1855
From the original in IEE MS SC 2

Cox's Hotel, | 55, Jermyn Street, | St. James. | April 16th 1855

Sir

I have staying here with me Mr Home[1] who is a medium for Spiritual demonstrating - & shall be very happy to give you the opportunity to show tables & chairs moving & other phenomena much more extraordinary, without *any Person* being *near*.

Yours respectfully | W. Cox

- Faraday Esqr

1. Daniel Douglas Home (1833–1886, DNB). Spiritualist medium.

Letter 2972
Faraday to William Cox
16 April 1855
From the original in IEE MS SC 2
    Mr Faraday is much obliged to Mr. Cox[1] but he will not trouble him - Mr Faraday has lost too much time about such matters already[.]
Royal Institution | 16 April

1. See letter 2971.

Letter 2973
William Cox to Faraday
c16 April 1855
From the original in IEE MS SC 2
Sir
    You are wrong in not seeing me[1] - I *have facts* which are at yr service *now* - after to day they will belong to others[.]
    respectfully yours W Cox

1. See letter 2972.

Letter 2974
John Tyndall to Faraday
18 April 1855
From the original in RI MS Conybeare Album, f.3
    18th April 1855 | [Royal Institution embossed letterhead]
My dear Mr Faraday
    You can hardly estimate how I prize your gift[1]. My highest ambition at the present moment is to work until I shall be able to present you with a volume worthy of your acceptance. If the head holds out I hope to accomplish this - if not it will be my sweetest consolation to reflect that you have not thought me unworthy of your kindness[.]
    Ever yours affectionately | John Tyndall

1. That is Faraday (1855d).

Letter 2975
Jacob Herbert to Faraday
19 April 1855
From the original in GL MS 30108/2/88

Trinity House | 19th April 1855.

My dear Sir,
  I send you herewith two Sample Bottles of Water which the Elder Brethren will be glad if you will examine, and acquaint me, whether they contain any, and if so what, description of impurities.
  I remain | My dear Sir, | Your very faithful Servant | J. Herbert
M. Faraday Esq | &c &c &c

Letter 2976
Faraday to Jacob Herbert
21 April 1855
From the original copy in GL MS 30108/2/88

Royal Institution | 21 April 1855

My dear Sir
  The waters are both very bad[1]. That from the *House tap* contains lead in solution, and there was a little deposit of lead at the bottom of the bottle. The water contains also Sea salt. I do not know where it comes from, but it is so free from the solutions usually occurring in common waters, & so peculiar in respect of the common salt & lead, that I think it must be *rain water* gathered upon a *leaden roof*,- within reach of the *spray* of the sea. Water so gathered will always be dangerous: for the purity of the water in respect of earthy salts & the presence of the marine salt both tend to make it act upon lead.
  The other specimen labled *Brack tank water* is very bad indeed. When poured out it was milky from the suspension of *hydrate* & *carbonate of lead* in it; & there was much dirty deposit at the bottom, which was chiefly carbonate of lead. This deposit is probably lead which has formerly been in solution:- when it has once been deposited & has settled at the bottom of the tank it does no harm to the water, unless it be stirred up & mixed with it, though it is always a sign that the water has at one time or other been poisoned by it. But *that* which is diffused through the water & *that* which is dissolved, are both dangerous. I do not know what relation this water has to the former, but from the likeness of character in many respects, suppose it is the former water collected in a brack tank. The change to an opalescent state & the gradual deposition of carbonate of lead accord with such a supposition[.]
  I am | My dear Sir | Yours Very truly | M. Faraday

Jacob Herbert Esqr | &c &c &c

1. See letter 2975.

Letter 2977
Alfred Swaine Taylor to Faraday
24 April 1855
From the original in IEE MS SC 2

15 St James' Terrace | Regents Park | April 24, 1855

My dear Faraday

I have great pleasure in replying to your question from authentic documents in my possession[1][.]

There was no nitrate of potash on any part of the premises.

There were 128 tons of Nitrate of soda, and there were 2800 tons of sulphur[.] Of these quantities there were in the seat or focus of explosion 47 tons of sulphur and 45 tons of nitrate of soda. These quantities were in a strong vault in the basement,- the sulphur being exposed and occupying the lower part of the vault:- and on a tarpaulin, placed on the sulphur, were piled the bags containing the nitrate of soda. The capacity of the vault was such that these articles (which alone were in this vault) - reached to within a foot of the ceiling[.]

There was a large quantity of sulphur amounting to some hundreds of tons stored between the entrance to this vault and the Woollen factory in which the fire commenced. This probably ignited, as it was in contact with the Woollen-factory-wall, and thus communicated like a train to the sulphur on the lower floor of the vault[.]

The quantity of sulphur consumed was enormous. It flowed out in rivers of blue fire from the two ends of the building, and made its way into the Tyne, realizing the description of Phlegethon[2] given by the poets.

Any documents in my possession with a plan of the building and an official table of the whole contents at the time of the conflagration are at your service.

I send for your perusal my Report to Lord Palmerston[3] which you can return to me on Friday evening next[4][.] Believe me

Dear Faraday | Your's most truly | Alfred S. Taylor

M. Faraday Esq

1. This relates to a destructive fire in Newcastle which started on 6 October 1854 and destroyed many chemical factories. See *Ann.Reg.*, 1854, **96**: 170–4. For Taylor's report see p.174. For Faraday's earlier interest in the fire see letter 2910.
2. A river of liquid fire in Hades.

3. Henry John Temple, 3rd Viscount Palmerston (1784–1865, DNB). Prime Minister, 1855–1858.
4. That is 27 April 1855.

Letter 2978
Nathaniel Barnaby[1] to Faraday
26 April 1855
From the original in IEE MS SC 2
Department of the | Surveyor of ye Navy. | Whitehall. | April 26th 1855.
Sir,
  With reference to the experiments made by you and Sir Humphry Davy in 1822–24 on the preservation of Copper Sheathing[2], I wish to take the liberty of troubling you with a question.
  The Iron and Zinc Protectors seem to have failed from two causes first. The conversion of the protecting metals into oxides, and the destruction of their influence, appear to have taken place before the Vessel could be docked to have the worn ones replaced, and Secondly. The rusty surfaces of the Protectors were nuclei for the deposition of earthy particles, and the formation of a bed for sea-weed, & shell-fish.
  It has struck me that both these evils could be avoided, if the protectors were placed in a water-tight cistern inside the Ship, at about the height of the water line, so that a communication could be kept up between the inside of the Box, and the sea water without, by means of a tube:- one or more of the Bolts passing thro' the bottom forming the connexion between the sheathing & the enclosed protectors - as shewn in the sketch.
  If you see any objection to this, I shall esteem it a great favour if you will tell me in a few lines addressed to me here - otherwise I hope to be able to get the Admiralty to make some Experiments on it.
  I have the honor to be Sir, | Very respectfully | Your obedient Servant | Nathl. Barnaby
Professor Faraday.

  A. A vessel coppered inside and having its internal surface connected with the Bolt B, & $\therefore$ with the metal Sheathing, by means of a slip of Copper passing over the upper edge & down the outside.

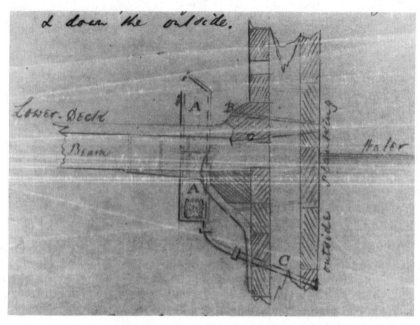

C - the tube which keeps up communication between the water in the cistern A & the sea

Address: M. Faraday Esqre F.R.S. F.G.S. &c &c | 21 Albemarle Street | London.

1. Nathaniel Barnaby (1829–1915, DNB3). Official of the naval construction department of the Admiralty.
2. See James (1992a).

Letter 2979
Faraday to Nathaniel Barnaby[1]
28 April 1855
From the original copy in IEE MS SC 2

R Institution | 28 April 1855

Sir

The proposition is of no value[2]. The protectors would preserve the inside of the box but would be utterly useless as regards the ships sheathing[.]

I am | Your Obedient Servant | M. Faraday
Nath Barnaby Esqr

1. Nathaniel Barnaby (1829–1915, DNB3). Official of the naval construction department of the Admiralty.
2. See letter 2978.

Letter 2980
William Allen Miller[1] to Faraday
28 April 1855
From the original in RI MS F1 I104a

King's Coll. London | April 28, 1855

Dear Dr. Faraday,
    I have just received your kind and valued present of the third volume of your Experimental Researches in Electricity[2]. Pray accept my best thanks for it. It will take its place by the side [of] your original proof sheets which once belonged to our excellent friend Daniell and which are thus doubly valuable to me.
    Believe me | my dear sir | very sincerely yours | Wm Allen Miller
Dr Faraday | &c &c &c

1. William Allen Miller (1817–1870, DSB). Professor of Chemistry at King's College, London, 1845–1870.
2. Faraday (1855d).

Letter 2981
William Whewell to Faraday
7 May 1855
From the original in TCC MS O.15.49, f.66

Trin. Lodge, Cambridge | May 7, 1855

Dear Faraday
    As you have kindly given me the First and the Third volumes of your collected "Experimental Researches"[1] I have no doubt you will be willing to give me the second volume[2]. Perhaps you imagine you have done so, but I believe you would be mistaken in such an imagination as I should not have forgotten it. I had rather have it as your gift than buy it at the shop.
    Believe me | Yours very truly | W. Whewell

1. Faraday (1839b, 1855d).
2. Faraday (1844b).

Letter 2982
Faraday to William Whewell
9 May 1855
From the original in TCC MS O.15.49, f.40

Royal Institution | 9 May 1855

My dear Dr Whewell
    Your note is a great pleasure to me[1]. I cannot think how I may have forgotten you in regard to Vol II[2], but I thank you for giving me an opportunity of repairing the omission.
    Ever Your faithful Servant | M. Faraday
    The volume will probably overtake the note for it shall go by rail today | MF

1. Letter 2981.
2. Faraday (1844b).

Letter 2983
William Thomas Brande to Faraday
19 May 1855
From the original in RI MS Conybeare Album, f.3

Tunbridge Wells | 19 May 1855

Dear Faraday
    Many thanks for your prompt attention to my request. It has enabled me to gratify the curiosity of an intelligent friend here, and who, being a great Microscopist, said nothing about the smallness of the sample. When sodium can be had for half a crown a pound, I presume Aluminium may be produced for five shillings so that perhaps at no very distant period, it may be better known - and the next generation may see roofs covered with it[1].
    If your wife has not forgotten the existence of such a person, remember me kindly to her and believe me
    My dear Faraday | always & very truly yours | Wm Thos Brande.
    If you will be good enough to leave *the 8vo Volume*[2] in the Hall, I will call or send for it in the course of next week - in the mean time accept my best thanks for your kind remembrance of your old Colleague.

1. A reference to the recent development by which aluminium could be obtained easily and its consequent comparative cheapness. On this see McConnell (1989).
2. Faraday (1855d).

Letter 2984
William Snow Harris to Faraday
20 May 1855
From the original in IEE MS SC 2

Plymouth 20 May 1855

My dear Faraday
    You have been always so generous & kind to me upon Philosophical Subjects, that I feel ashamed at offering any large apology for this further intrusion on your valuable moments. I am however desirous to satisfy my mind upon one or two points in which I am now deeply interested, will you then permit me to submit the following points to you.
    First - let me ask whether you consider the following expt. as a fair illustration of the evolution of Electricity during Chemical action
    - Expt:- put some coarse grains of the impure Zinc of commerce into a glass bottle A pour on them dilute sulphuric acid the water will decompose and hydrogen will escape at h. If during the effervescence a gold Leaf Electrometer be applied to the glass vessel A its leaves will diverge freely.

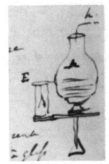

Well then here is clearly a development of common Electricity during chemical action.
    - Now if I clearly understand you in your views of current force - in the pile - Pure zinc or amalgamated zinc is not acted on in this way the water will not be decomposed although there may arise a large Electrical Tension between the metal and the fluid - directly a metal such as copper is put into the liquid & made to touch the Zinc then Chemical Action ensues and we have a current.

Now I want to satisfy my mind as to what takes place when we use Iron or impure zinc here we have at once the same result as is brought about by the introduction of the copper plate - in the impure zinc I suppose it is the presence of other metals which brings about the result - but how of Iron - say *pure* Iron will not filings of pure Iron cause chemical action in dilute acid without the presence of another metal?

With respect to what is called the "Contact Theory" I have carefully read through the subject and I am obliged to conclude - that it is perfectly *untenable* - the Phenomena of the Pile in all their generality upon Voltas simple view of the source of power are quite impossible.

Finally I would ask - whether it may not appear to you upon further reflection, that some confusion & misapprehension arises in the application of the terms positive & negative to the metals exhibiting Electrical disturbance after *contact* - say I bring a plate of *Zinc* to touch a plate of *Copper* and I find that Electricity has passed or is supposed to have passed from the *Copper* upon the Zinc. Surely in this case the Copper should be called the positive metal & the Zinc the negative - whereas Zinc is called *positive* now it is a cardinal point in your philosophy that the surface *from* which Electricity flows is to be considered positive - your *anode* for example is opposite the *positive* Electrode - positive because the current flows from it - reciprocally for the Cathode[.]

In ordinary Electricity we consider the prime Conductor positive in the Glass Machine because it gives off Electricity & the conductor of the rubber negative because it takes up Electricity from the Prime Conductor. Now this is precisely what by the Contact Theory the Copper does in relation to the Zinc it appears to me, that the views of pos & neg as originally expressed in the contact experiments should be reversed. I remain My dear Faraday

Most truly & sincerely yours | W. Snow Harris

I suppose in the Paper in the Trans for 1801 by Wollaston[1] page 427[2] - In which he says "If a piece of zinc and a piece of Silver have each an extremity immersed in dilute acid &c &c &c - the Zinc is dissolved &c .["] he does not mean *pure* zinc but the Zinc of commerce[.]

1. William Hyde Wollaston (1766–1828, DSB). Man of science.
2. Wollaston (1801), 427.

Letter 2985
Christian Friedrich Schoenbein to Faraday
26 May 1855
From the original in UB MS NS 418

My dear Faraday,

How could I employ the leisure hour of a fine May morning better and more agreeably than by devoting it to an epistolary conversation with my dear Friend Faraday, whom I besides owe an answer to his last amiable letter[1] and to-day, let me talk a little of Science.

As you cannot avert your mind from the contemplation of that mysterious agency called Magnetism, I am unable to let Oxigen out of sight and of late I have been actively working again on that curious subject not, I think, without some little success. You know that these many years I have entertained the notion according to which not only free but also Oxigen being chemically associated to some matter or other is capable of existing in two different conditions: in the common or inactive and the exalted or ozonic state and to distinguish by signs those different conditions from one another I have given to ozonised oxigen the symbol $\overset{\circ}{O}$ denoting the inactive O by its usual sign = O. Considering the peroxides of hydrogen, nitrogen (hyponitric acid), Barium, Manganese, Lead &c as compounds containing both sorts of oxigen, I have given them the formula $HO + \overset{\circ}{O}$, $NO2 + 2\overset{\circ}{O}$, $BaO + \overset{\circ}{O}$, $MnO + \overset{\circ}{O}$, $PbO + \overset{\circ}{O}$ &c and, as you are well aware, made these last six years many experiments with the view of separating from the oxycompounds mentioned and other similar ones their ozonised oxigen without obtaining however satisfactory results.

Some time ago Mr. Houzeau[2] communicated to the french academy a paper[3], in which he suggested ideas on the different states of the oxigen being contained in compounds being exactly the same which I for the first time ventured to express in Poggendorffs Annals seven or eight years ago and have since more fully developed in the publications of the Phil. Society of Bâle, notably so in the last number of the proceedings of that learned body[4]. The views recently put up by Mr. Houzeau are therefore rather old acquaintances of mine, but that Chymist has ascertained a novel fact, and as I consider it, a very interesting one. On adding peroxide of Barium to the monohydrate of sulphuric acid he obtained oxigen enjoying all the properties of Ozone. I have arrived at the same end, but in a somewhat different manner. You know, Silver being exposed to the action of ozonised oxigen at the common temperature is transformed into the peroxide of that metal and you will recollect that I sent you a small quantity of that compound, some years ago[5]. Now it is from this peroxide of Silver which I consider to be $Ag\overset{\circ}{O}^2$, that I succeeded to eliminate some Ozonised Oxigen.

On throwing the said peroxide into the monohydrate of sulphuric acid a most lively disengagement of a gaseous substance takes place conjointly with the formation of sulphate of Silver. The gas obtained in the manner indicated enjoys the following properties: its smell strongly resembles that of Ozone, but minute quantities being inhaled produce a sort of Asthma, as Ozone does; its electromotive power is strong; and like that of Ozone or chlorine, plates of Platinum or Gold becoming negatively polarised in the gas; it eliminates Jodine from the jodide of potassium and therefore turns instantaneously my test-paper dark blue, it rapidly transforms the yellow ferro-cyanuret of potassium even in its solid state into the red one; it suddenly oxidises sulphurous acid into sulphuric acid and sulphuret of lead into sulphate; it energetically and chlorinelike discharges the colors of organic matters such as Indigo, Litmus &c.; it colors blue the alkoholic solution of guajacum &c.

Now all these reactions being exactly those produced by Oxygen as modified by Electricity or phosphorus i.e Ozone I think, we may be allowed to conclude that the gas being disengaged out of the peroxide of Silver is, or contains, at least, the same principle i.e. Ozone.

Having but very minute quantities of that peroxide at my disposal, I, to my great regret, was forced to perform my experiments on a very small scale, but I had enough of the matter as to ascertain that the gas obtained was a mixture of $\dot{O}$ and O, in which the latter very much prevailed. Although there is no doubt to me that all the oxygen eliminated from the peroxide does in the moment of its being set free exist in the ozonic state, there are some obvious causes that account for the mixt nature of the gas i.e. for the transformation of $\dot{O}$ into O. One of them is the heat being disengaged at the points of contact between SO^3 and AgO^2 and the other the peroxide itself. As to the latter, you know perhaps that last year I ascertained the curious fact that a number of substances exert the same influence upon the ozonised oxygen as heat does i.e. destroy at the common temperature the ozonic condition of that oxygen without taking up a particle of it. The metallic peroxide enjoy that strange property in a very high degree and notably so that of silver, compounds as you see which to my opinion contain ozonized oxygen themselves. Now if a particle of peroxide of Silver not yet decomposed happens to come in contact with a particle of ozonised Oxygen being disengaged from another portion of the peroxide, that particle must become desozonised. There are perhaps some other causes unknown as yet that tend to change $\dot{O}$ into O in the case before us. I hope you still possess some of the peroxide of Silver I sent you some years ago and if so, you may even with that small quantity ascertain the correctness of my statements above made. In case you repeat my experiments I advise you to put a little peroxide into comparatively much Oil of Vitriol and do the thing at a low temperature from reasons that are obvious enough. To give you some visible proofs of the great chemical power of the oxigen having been eliminated from the peroxide of Silver by the means above indicated, I join three strips of paper one of them being impregnated with sulphuret of lead another with indigo solution, a third one with the coloring matter of litmus and you will perceive part of each of them to be bleached.

This was effected within a few instants by immersing a moistened end of the strip into the said oxigen.

From more than one reason I cannot help attaching some importance to the result of my experiments and believing that, if properly worked out and philosophically interpreted, it will lead to others of still greater consequence. And pray, let me reason and conjecture a little about it.

If it be allowed that the oxigen being contained in the peroxide of Silver exists in the ozonic condition and it being a fact that free $\overset{\circ}{O}$ is by heat transformed into O, does it not appear very likely that the same agency has the power of changing the $\overset{\circ}{O}$ of the peroxide into O and that this very change of condition is the proximate cause of the decomposition, which the peroxide undergoes when sufficiently heated? And if this conjecture should happen to be founded, are we not permitted to account in the same manner for the decomposition of all the other oxycompounds being effected by heat and yielding free inactive Oxygen? I am inclined to think that we are, and in a paper of mine printed last year I have given detailed reasons for entertaining such an idea. Now supposing my hypothesis to be true, I am afraid many of our present notions on the phenomena regarding chemical analysis, synthesis, affinities &c. cannot be maintained and must sooner or later be essentially modified. Stating that peroxide of Silver for instance consists of one Eq. of Silver and two Eq. of Oxigen and carbonic acid of one Eq. of Carbon and two Eq. of Oxigen is telling, if I may say so, but half the truth as regards the chemical constitution of the compounds named, for it implies the assertion that the oxigen being contained in those compounds is the same thing, an admission which according to my opinion cannot be allowed to be true, for $\overset{\circ}{O}$ is not O though the one may be transformed into the other.

But if the oxigen being chemically associated to other matters be capable of existing in different states and the bearings of the oxycompounds be so much influenced by the peculiar condition in which their Oxigen exists in them, are we not permitted to suspect that other elementary matters may also enjoy a similar capacity of assuming different conditions and be able to exist within compounds in those various states? May it not be presumed that the chemical behaviour of such compounds essentially depends upon the peculiar condition of their constituent parts? Is it to be believed that carbon exists in the oil of turpentine exactly in the same state as it does in charcoal, and is it not possible that the decomposition of all the organic substances is effected by heat, because this agency has the power of transforming carbon from one state into another? To condense these questions and others that easily suggest themselves into one, I ask, is it not very likely that, what they call "allotropism" acts a much more important and general part in Chymistry than it is thought of as yet? I for my part think it to be so.

Now no more of Science, theories and such like! We descend to daily life and my family. Being all of us highly in love with nature we are very fond of rambling in fields and woods, on hills and dales to admire the unfathomable riches of beauties being displayed there. May is called in German "Wonnemonat" which means month of joy, and well meriting that poetical denomination, it is of course a favorite of ours and we indulge during its reign as often as we can in our rambling propensities by taking trips in the neighbouring country. The Jura mountains are a particular point of attraction to us with their rich woodland, limpid rivulets, green valleys, bold rocks and fine views. I preface thus to tell you that some days ago on a fine morning a motley army consisting of big and small Children, male and female and old folks too were seen marching out of the old gates of Bâle tending their steps towards the "Gempenstollen" the highest and most prominent point of the Jura in our neighbourhood. Mr. Schoenbein well acquainted with all the recesses and by-ways round about us, and his family making up good part of the army was unanimously elected commander in Chief which important charge he accepted and filled it up to the best of his powers. The day turned out a glorious one nature exhibiting all her charms. By a great and gradually rising round about way leading through meadows covered with flowers, green fields, flowering orchards, beautiful beechwoods crowded with singing birds we reached after a four hours walk the summit of our favorite hill. A little fatigued the army desired to camp here and it was allowed to do so. The delicately green foliage of fine beeches and the crowns of stately firtree formed a splendid canopy and the mossy ground yielded soft resting places. Carrying our victuals with us the dinner was soon ready and I can assure you that we enjoyed our cold morsels infinitely better than we should have done, had we sat down at a sumptuous royal table. Our camp being placed upon the top of a gigantic projecting rock it commanded a most extensive and glorious view ∴ to the south at some distance, we saw the snowy heaventowering Alps of the Berner Oberland, nearer and to the west great part of the Vosges and Alsatia, to the north the Black Forest and Baden, nearest us the many valleys and summits of the Jura mountains. Being enchanted by that glorious sight we could not but most reluctantly break up our camp; but Mr. Schoenbein gave orders at last to march home again on a road however different from that we had come beautiful also beyond description. Having reached the foot of the hill the gypsy host was allowed to halt again for taking refreshments and by eight o'clock we approached the walls of the good town of Bâle, where the commander in chief discharged his troops not without having received before the thanks from high and low, old and young. I am sure you yourself and Mrs. Faraday would have highly relished the gypsy party but come over to us and we shall repeat it. Next midsummer we go to Langenbruck a village in some pass of the Jura intending to say there for a month. It would be a high treat to me and us all, if we could spend that time with you and Mrs. Faraday.

The gentleman who will deliver this letter to you is Mr. Schweitzer[6] of Bâle an old pupil of mine and whom I take the liberty to recommend most friendly to your kindness.

Mr. Wiedemann charges me with his best compliments to you, he is very actively occupied with electrolytical researches and has received some interesting results.

Should a parcel be directed to you under my address, pray take and keep it untill you find an occasional conveyance for Bâle. There is no hurry for it.

Excuse my immoderately long letter, let me soon have the favor of a letter and believe me

Your's | most truly | C.F. Schoenbein.
Bâle Mai 26th 1855.
Don't forget to remember me friendly | to Mrs. Faraday.

Address: Doctor Michael Faraday | &c &c &c | Royal Institution | Albemarle-Street | London.

1. Letter 2964.
2. Auguste Houzeau (1829–1911, DBF). French chemist.
3. Houzeau (1855).
4. Schoenbein (1854b).
5. Letter 2274.
6. Unidentified.

Letter 2986
John Peter Gassiot to Faraday
28 May 1855
From the original in IEE MS SC 2

Clapham Common | 28 May 1855

My Dear Faraday

During the progress of your lecture last Friday Evening[1] it occurred to me that the question you had raised as to conduction proper might be explained by some experiments I made a few years since & which are described in a paper published in the RS. transactions (PT 1844)[2] - it appears to my mind to explain that peculiar action which arises in a Voltaic Battery when it is thrown into a state of tension *before* actual

Electrolysis takes place, and that this is similar to the induced state of an Electrified Body previous to its receiving the actual discharge from the Electrical Machine[.]

I will briefly describe the Experiment - Charge in the usual manner (taking care to keep the outer portion of each cell dry) a number of cells or series of a Voltaic Battery sufficient to diverge the leaves of Gold leaf Electroscopes - one of which is attached to the + and the other to the - terminal, introduce into the circuit a delicate Galvanometer and two platinum wires the ends of which rest on a piece of bibulous paper saturated with a solution of Iodide of Potassium (as in original Experiment) - the Battery and entire apparatus being insulated - in this state the leaves of the Electroscopes will diverge one with + and the other with - Electricity. When all is thus arranged touch with a wire or with the finger - either terminal, the leaves of the electroscope will collapse, while those of the instrument attached to the other terminal will diverge with increased intensity - repeat this by touching first one and then the other terminal, the alternating and progressive effect shews some action must be passing through the entire Battery, but the needle of the Galvanometer (let the instrument be ever so delicate) is not deflected nor is there the slightest trace of Chemical action in the solution of Iodide of Potassium[.]

If as in my water Battery the terminals are brought sufficiently near to allow a spark to pass, or let the circuit be completed for a moment of time the deflection of the needle takes place & Iodine is evolved.

Have we not in this conduction proper through the battery without Chemical action[.]

Believe me | My Dear Faraday | Yours truly | John P. Gassiot

My Dear Faraday

I met Tyndall on Saturday[3] talking over with him the subject of yr lecture he appeared to me to think the Solution *might* be as I have represented - if you think th[is] worth inserting you may send it to the P.M., if I am wrong put it into *fire*[.]

Excuse this scrawl & the paper but I am off to Mark Lane[4] & have nothing else to write on[.]

Believe me | Yours | J.P. Gassiot

Monday Morning

Address: Dr. Faraday | Royal Institution | Albermarle Street

1. Faraday (1855e), Friday Evening Discourse of 25 May 1855.
2. Gassiot (1844).
3. See Tyndall, *Diary*, 26 May 1855, **6a**: 65.
4. Where Gassiot's office was located.

Letter 2987
Thomas Phillipps to Faraday
28 May 1855
From the original copy in Bod MS Phillipps-Robinson c.531, f.1-2

To Dr Faraday, Royal Institution DCL Athenaeum | 28 May 55

My dr Dr Faraday

I have followed yr advice & read the Reports in the Comptes Rendus upon Aluminium[1] but, alas! I must go to School again. The Nomenclature of Chymistry is so changed since I studied since I studied under my old friend Dr Kidd[2] of Oxford that I am at a loss to understand some of the new terms, & I suppose I must continue so, unless you will be my Professor & kindly teach me. Two terms in particular I cannot translate "*Chlorure* & *Chlorhydrique*". Are they connected with Oxymuriatic Acid? Is *Chalumcan* a Blowpipe, or a Common Smoking Pipe? I have sought in vain in this Library for a French *Technological* Dictionary.

Pardon my giving you so much trouble, but I will promise to make you amends as soon as I have turned all my Clay Mountains into Clay Silver or Aluminium.

Believe me | Very Truly Your's | T. Phillipps

1. Deville (1854a, b).
2. John Kidd (1775-1851, DNB). Regius Professor of Physic, University of Oxford, 1822-1851.

Letter 2988
Faraday to Thomas Phillipps
29 May 1855
From the original in Bod MS Phillipps-Robinson c.531, f.3-4

Royal Institution | 29 May 1855

My dear Sir Thomas,

There is perhaps no science that requires more closely following than Chemistry and you would have to read & think a good deal before you could possibly be on a level with the Chemistry of Aluminium[1]. However I can explain the terms you refer to. Oxymuriatic acid as it used to be called is now known to be an elementary body - is called chlorine & its compounds with metals are in French called Chlorures. "Chlorhydrique" is a compound of Chlorine & hydrogen the old Muriatic acid will pass by that name. "Chalumcan" is the blowpipe. I think you have mastered the subject pretty well if these are all your difficulties[.]

Ever Truly Yours | M. Faraday

1. See letter 2987.

Letter 2989
Faraday to Hugh Welch Diamond[1]
29 May 1855
From the original in the possession of C.J. Kershaw

Royal Institution | 29 May 1855

My dear Sir
    I am greatly obliged to you for the copy of Davy's letter[.] It is just his hand writing when I knew him which however was not until 9 years after in 1812. I think - It is really very interesting to see how thoroughly his thought & propositions anticipates the Museum as they seem to be carr[y]ing it out both in Jermyn St & the Society of Arts.
    Ever Truly Yours | M. Faraday
Hugh W. Diamond Esq | &c &c &c

Address: Hugh W. Diamond Esq | &c &c &c | Surrey County Lunatic Asylum | Near Wandsworth

1. Hugh Welch Diamond (1809–1886, DNB). Physician and photographer.

Letter 2990
Charles Lyell to Faraday
1 June 1855
From the original in RI MS Conybeare Album, f.9

53 Harley St. | Friday. Jun 1, | 1855.

Dear Faraday
    My friend the bearer of this Mr Wildy[1] is very desirous of attending Mr Tyndal's lecture tonight[2]. I told him he must not be surprised if he finds every ticket given away, but that he might take his chance of calling at the Institn. with this note from me & seeing if you could spare him a ticket. Excuse my giving you this trouble & send word vivâ voce that you have none left if that be the case[.]
    ever truly yours | Cha Lyell

1. Unidentified.
2. Tyndall (1855a), Friday Evening Discourse of 1 June 1855.

Letter 2991
Faraday to Frederick Gye
2 June 1855
From the original in SI D MS 554A

R Institution | 2 June 55

My dear Sir
    I hope you have not too freely anticipated Thursday Evening[1] - If you want the Stall for others who have more metallic thanks to give you than I have do not hesitate to let me know. In all cases I am your very much obliged
    M. Faraday
F. Gye Esqr | &c &c &c

1. That is 7 June 1855 when Faraday would have heard the "Huguenots" by Meyerbeer. *Times*, 7 June 1855, p.6, col. a.

Letter 2992
Thomas Phillipps to Faraday
2 June 1855
From the original in IEE MS SC 2

[Athenaeum letterhead] | 2 June | 55

My dear Dr Faraday
    Many thanks for yr translation[1], wch enables me to see that the actual production of the metal is not so costly & intricate as I supposed; for it appears by the Comptes Rendus[2] that one of the experimenters produced it by a Blow-pipe, in *globules*. To produce it in *Ingots* like Mr Deville[3], wd be both costly & intricate, & it *must* be produced in that form to make it *pay*. But perhaps a more simple method may be discovered by *yourself* before long.
    Mr Wohler told me the process which he used, but it seems he does not obtain it in Ingots, altho' the specimen which he sent to me might be called a *thin* Ingot[4].
    I am translating all the Passages in the Comptes Rendus relating to aluminium. Repeating my thanks for your kindness | believe me | Very truly your's | Thos Phillipps
    PS. I find a family of Fereday in Worcestershire; are you from *my* county?

1. Letter 2988.
2. Deville (1854a, b).
3. Henri Etienne Sainte-Claire Deville (1818–1881, DSB). Professor of Chemistry at Ecole Normale Supérieure, 1851–1880.

4. Wöhler to Phillipps, 10 April 1855, Bod MS Phillipps-Robinson d.50, f.73–5. On Wöhler's work see Weeks and Leicester (1968), 567–8.

Letter 2993
Faraday to Thomas Phillipps
4 June 1855
From the original in Bod MS Phillipps-Robinson c.531, f.5

R Institution | 4 June 1855

My dear Sir Thomas
   I do not know of any connexion between the Feredays of Worcestershire & our Family which is from Clapham in Yorkshire. I knew there were some Feredays Iron masters: &c but only by report[1][.]
   Ever Truly Yours | M Faraday

1. See letter 2992.

Letter 2994
Edward Sabine to Faraday
7 June 1855[1]
From the original in RI MS Conybeare Album, f.26

13 Ashley Place | June 7.

Dear Faraday,
   M. Plantamour[2] of Geneva who is here for a few days would be much pleased to be present at your lecture tomorrow night[3], and I have promised to ask for a ticket for him. I hope to accompany him myself, using for that purpose my General ticket of admission on friday evenings. I enclose two tickets which Mr Barlow was so kind as to send me for Tynda[l]l's lecture[4], (which I was unable to attend) - perhaps the date of these might be altered to Friday the 8th.
   Sincerely yours | Edward Sabine.
PS. Mr Walker[5] of Oxford comes up to Oxford to hear you on Friday, and will be at my house.

1. Dated of the basis of the references to Faraday's and Tyndall's Friday Evening Discourses.
2. Emile Plantamour (1815–1882, DHBS). Professor of Astronomy and Director of the Observatory at Geneva.
3. Faraday (1855f), Friday Evening Discourse of 8 June 1855.
4. Tyndall (1855a), Friday Evening Discourse of 1 June 1855.
5. Robert Walker (1801–1865, B3). Professor of Experimental Philosophy at the University of Oxford, 1839–1865.

Letter 2995
James Emerson Tennent[1] to Faraday
13 June 1855[2]
From the original in RI MS Conybeare Album, f.26

Board of Trade | June 13.

My dear Sir
I send you the broken fragment of the Chance mirror for inspection.
I hope [you] will remember to let Mr Walker & me see the specimen of aluminium[3][.]

It will give us great gratification if the accompanying card finds you unengaged; & that you will give us the pleasure of your company to dinner on the 29th.
Yours Faithfully | J. Emerson Tennent
M. Faraday Esq | DCL. &c

1. James Emerson Tennent (1804–1869, B3). Secretary of the Board of Trade, 1852–1867.
2. Dated on the basis that letter 2996 is the reply.
3. See McConnell (1989).

Letter 2996
Faraday to James Emerson Tennent[1]
13 June 1855
From the original in the possession of Y. Watanabe

Royal Institution | 13 June 1855

My dear Sir Emerson
You are very kind and I thank you sincerely for the invitation[2] but I must not accept it as I am known never to dine out in the circles of Society[.]
The fragment of mirror I will keep a few days & then return[.]
I have been puzzling my head ever since Saturday[3] trying to remember whom I had promised to send the Aluminium[4] to. I now send it - thanking you much for your kindness & hope you will excuse my forgetfulness
I am | Very Truly Yours | M Faraday

1. James Emerson Tennent (1804–1869, B3). Secretary of the Board of Trade, 1852–1867.
2. Letter 2995.
3. That is 9 June 1855.
4. See McConnell (1989).

Letter 2997
Jacob Herbert to Faraday
13 June 1855
From the original in GL MS 30108/2/88

Trinity House | 13th June 1855.

My dear Sir,
Herewith I send you Samples of Drinking Water (in two Bottles) used by the Keepers at the Beachy Head Light House, which have been sent up by direction of the Elder Brethren, who are desirous that you should analyze the same; and which I have to request you will do, and favor them by communicating to me the result.-

I remain | My dear Sir | Your very faithful Servant | J. Herbert
M. Faraday Esq.

Letter 2998
Faraday to Jacob Herbert
16 June 1855
From the original copy in GL MS 30108/2/88

Royal Institution | 16 June 1855

My dear Sir
The waters as they came to me in the bottles were not bad[1]. That labled *Assistant light keeper* was very good being soft, without colour, or taste; free from organic matter, and also from lead. It contains only small quantities of sulphate of lime & chloride of Sodium substances which might very properly be present. The water labled *Principal light keeper* was also soft & good in respect of taste, smell, & saline matters, but there was a little organic matter (I do not know what its source may be) which gave it a little colour and there was also a *small trace of lead*; very small but still distinct. There was no sediment of lead in either of the bottles[.]

Ever My dear Sir | Very truly Yours | M. Faraday
Jacob Herbert Esqr. | &c &c &c

1. See letter 2997.

Letter 2999
Arthur-Auguste De La Rive to Faraday
17 June 1855
From the original in IEE MS SC 2

Vichy le 17 juin 1855

Mon cher Monsieur & ami,

J'ai reçu votre bonne petite lettre[1] qui m'a bien touché comme tout ce qui me vient de vous. Vous avez été surpris de ma détermination; mais vous le serez moins quand vous connaitrez l'amie qui a bien voulu consentir à se réunir à moi pour que nous terminions ensemble les jours que la Providence daîgnera encore nous accorder sur cette terre.

Amie & contemporaine de ma chère première femme ainsi que de ma belle soeur Louisa Marcet[2], veuve de l'un de mes collègues & amis qu'elle a perdu il y a 16 ans, Madame Maurice Fatio était pour moi une de ces amies précieuses, si rares à trouver, soit à cause de la communauté des souvenirs qui nous unissaient, soit à cause de la tendre affection qu'elle avait pour mes enfants, soit surtout à cause de l'élévation de son caractère & de ses sentiments si profondément chrétiens. Il a fallu pour la décider à s'unir à moi qu'elle cherchât & vît dans le parti qu'elle a fini par prendre après deux mois d'hésitation & de scrupules, une direction providentielle, qu'elle y reconnût la volonté de Dieu. Or les circonstances qui ont accompagné cette époque d'incertitude ont été telles qu'il a semblé véritablement que notre union était voulue par Celui qui gouverne l'Univers. Ainsi c'est sous Son regard & sous Sa divine protection que nous nous sommes unis en implorant sur nous Sa bénédiction au nom de notre Seigneur Jesus-Christ.

Voilà, Monsieur & cher ami, les circonstances qui ont accompagné l'événement si sérieux & en même temps si doux qui vient de s'accomplir pour moi. Je suis convaincu que vous me pardonnerez ces détails que justifie l'amitié si précieuse que vous m'avez constamment témoignée & dont vous m'avez toujours donné tant de preuves. J'espère que cette amitié ne me fera jamais défaut & j'ose y compter.-

Je compte être à Londres vers le 8 ou le 10 juillet pour y terminer l'impression de mon second volume[3] qui a été retardée, j'ignore pourquoi. J'espère que vous me permettrez d'aller pendant ce temps passer quelques moments avec vous de temps à autres. Il y aurait même une ou deux expériences que j'aimerais bien faire avec vous, si cela vous est possible.

Je suis encore à Vichy jusqu'à la fin de Juin & je dois passer huit à dix jours à Paris avant d'aller à Londres pour terminer quelques affaires relatives à mon fils cadet[4] qui vient de sortir de la Ecole Polytechnique où il a achevé ses deux années d'une manière honorable, car il est sorti le 25ème sur une liste de 110. Mais j'ai préféré qu'il ne prît en France aucun service ni militaire, ni civil; & il va maintenant poursuivre ses études Scientifiques.-

Agréez, Monsieur & trés cher ami, l'assurance des sentiments les plus affectueux de votre tout dévoué | A. de la Rive

Address: Prof. Faraday F.R.S. | Associé de l'Institut de France &c | Royal Institution | Albemarle Stt | Londres.

TRANSLATION

Vichy 17 June 1855

My Dear Sir and Friend,

I received your good little letter[1] which touched me as does everything that comes from you. You were surprised by my decision; but you would be less so if you knew the friend who has consented to unite herself to me so that we may end in each other's company the days that Providence has accorded us on this earth.

A friend and contemporary of my dear first wife as well as of my sister-in-law, Louisa Marcet[2], widow of one of my colleagues and friends, whom she lost sixteen years ago, Madame Maurice Fatio was for me one of those precious friends, so rare to find, both because of the treasury of memories that united us and for the tender affection that she had for my children and above all for her upright character and her profoundly Christian sentiments. It took two months of hesitation and scruples, before she decided to unite herself to me, during which time she searched and lived the part that she ended by taking, a providential direction, in which she recognised the will of God. Now the circumstances which accompanied this period of uncertainty were such that it truly seemed that our union was willed by Him who governs the Universe. Thus it was under His gaze and under His divine protection that we were united, imploring Him to send us His blessing, in the name of our Lord, Jesus Christ.

There, Sir and dear friend, are the circumstances which accompanied this most serious and happy event that has happened to me. I am sure that you will forgive me these details that are justified by the precious friendship that you have always shown and of which you have always given me so many proofs. I hope that this friendship will never fail and I dare to count on it. -

I hope to be in London towards 8 or 10 July in order to finish the printing of my second volume[3] which has been delayed, I do not know why. I hope that you will allow me during this period to spend some time with you. There would even be one or two experiments which I would like to do with you, if it were possible.

I shall stay in Vichy to the end of June and I must spend eight or ten days in Paris before going to London to complete some business connected with my youngest son[4] who has just finished at the Ecole Polytechnique, where he completed his two years in an honourable fashion, coming 25th out of 110. But I preferred him not to take any military or civil position in France and he is now going to continue his Scientific studies.-

Please accept, Sir and very dear friend, the assurance of the most affectionate sentiments of your most devoted | A de la Rive

1. Not found.
2. Louisa De La Rive, née Marcet (1807–1834, Choisy (1947), 52). Daughter of Jane Marcet.
3. De La Rive (1853–8), **2**.
4. Charles-Lucien De La Rive (1834–1924, Choisy (1947), 51). Swiss physician and writer.

Letter 3000
John Tyndall to Faraday
1 July 1855
From a typescript in RI MS T TS, volume 12, pp.4029–33

Hotel Manchester | 1 Rue De Grammont | Paris, 1st, July.
My dear Mr Faraday

The sun shines straight into my bedroom this glorious Sunday morning, but the light he sends which beautifies the world and makes the heart of man joyful is accompanied by heat which dulls the brain and makes a man lazy. Two principles operate upon me at present, the old duality of the moral world - the love of ease and indulgence which would prompt me to stretch myself on my bed, and the sense of duty which urges me to write to you. For the present however the latter has triumphed, and I hope it will triumph to the end. The close of the day I saw you last saw me in Hampshire[1] where I remained for a week to get some writing done, and on Friday the 15th at midnight we loosed from the Docks at Southampton and steered for Jersey. I say '*we*' as I have a chemical companion along with me[2]. Our passage to Jersey was in the highest degree abominable, I never suffered so much in my life. The wind had been high during the day, but the evening was calm and pleasant, and I promised myself quite an agreeable passage. The sea however continued to swing furiously although the cause of its oscillation had ceased to act: we were tossed up and down and the consequences may be imagined by those who have had the felicity of experiencing the more exalted phases of sea sickness. All on board were ill - the strongest gave way and the stomachic perturbation was universal. We reached Jersey safely and were obliged to continue there until the following Tuesday as there was no boat previously to France. I say '*obliged*' for though Jersey might be a pleasant place for 3 days residence still with the exception of one day the weather was so bad that we were confined to our hotel. On Tuesday the 19th we set sail for St Malo and reached the place after a passage of four hours and a half. Here we gave our portmanteaus in charge to the Messageries who had them transmitted to Paris there to await our arrival, and with sufficient baggage to do for a foot journey of a fortnight we commenced our campaign through France. At St Malo we saw the tomb of Chateaubriand[3]. His bones are bedded in a rock on an island which when the tide recedes can be reached by footpassengers from the main land. When we were there the sun was shining on his tomb and the billows breaking and moaning against the rugged rocks beneath it. The place I believe had been one of the scenes of the poet's boyhood and at his own request he was buried here. From St Malo we walked to Dinan in Brittany - a place beautifully situated - The country round is richly wooded, while the contour of the land and the colouring of the rocks are beautiful. A magnificent viaduct has recently been thrown over a ravine near the town, which adds its picturesqueness to that of the scenery. Most of the stone we saw was a delapated granite, and it was manifestly stratified; at least its crystalline plates were so arranged as to lie flat as if they had been pressed into their position by the superincumbent weight. The rocks which presented this appearance were quite friable, and their aspect was perhaps calculated to throw some light upon the obscure question of slate cleavage. From Dinan we proceeded to a place called Dol. Thence to Avranches, whence we paid a visit to the celebrated Mont St Michel - a rock to attain which we had [to] traverse a league of flat strand which is covered at High Tides, and to wade here and there knee deep through water. On this wild rock a monastery was built ages ago, a most splendid piece of architecture which is now in a state of capital preservation. Its life was an oscillation between the tempests of war and the calm of religion - It is now a prison, and the splendid rooms, the Knights chamber and the refectory of the monks are converted into weaving rooms, where the prisoners work. I never saw a richer country than that which stretches from Dinan to Dol and from Dol to Avranches: Here and there you have views of vast extent and great beauty. The country is well wooded and through the foliage the rich green of the crops and the enclosures reveals itself. Apple trees are strewn every where and you meet cider at all turns[.] The people appear to like it much, indeed in many of the poorer cabarets you can get nothing but cider. To me it is not much pleasanter than vinegar. The villages at a distance with their brown and aged roofs look very picturesque, but they are vile within - In nothing is the difference between English and Continental life so strikingly exhibited as in the villages. Throughout the region which I have mentioned dirt is universal - Comparing the glorious country with the dwellings of the people, the words of Heber's[4] hymn were often in my mouth.

"Man alone is vile"[5]

This inattention to cleanliness is by no means the result of poverty. In dirty houses you often see solid pieces of furniture with old massive brass ornaments which are scrupulously polished while there are holes in the dirty earthen floor. The people do not appear to have that sense of household harmony which is everywhere present among the wives of England and the consequence is the presence of such incongruities as I have mentioned. As regard cleanliness comfort and good order the Hotel de Londres at Avranches is pleasant to think of. The way to the heart of a savage is said to be through his stomach and certainly the excellence of the fare in the hotel spoken of, combined with the moderation of the charge has given it a pleasant place in my memory. The scenery about Avranches is delightful. We always made a point of visiting the churches in each town. There is something soothing in the cool tranquil air of these places, and however one may dissent from their present uses, a solemn and religious feeling is natural in such places. I like them best empty. When the priest is present his theological mechanics ruins the earnestness and tranquility of the impressions otherwise attainable. You rarely find the churches without a stray worshipper - sometimes a dozen or more, almost all of them are women. Hence we might argue that they are either the chief saints or the chief sinners in France. Some of them you see with earnest countenances deeply engaged in their devotions, many, however mutter their prayers as mechanically as if they were repeating their multiplication table. They yawn and look unuterably vacant. At Avranches we had a solemn High Mass conducted for the soul of an inhabitant who had fallen before Sebastopol. A coffin and all adjuncts were there, but the Crimea held the body. This Romish faith must have had a wonderful, and I think on the whole a beneficial effect upon a half savage world. It seems vigorous now in France, but it is only seeming - its days are numbered, but with the instinct of an old man on the brink of the grave it still clings tenaciously to life. Women are its chief supporters at the present day and I suppose the reason is that they have more feeling and less intellect than men - I think I ought to ask you not to read this heretical utterance to Mrs Faraday or Miss Barnard. But I must hasten. Having passed through Caen, Havre, Rouen we reached Paris yesterday - I have as yet seen nobody but purpose calling upon some people to day. And now I will ask you to excuse the infliction of this badly written letter. With kind remembrances to Mrs Faraday and Miss Barnard
    Believe me most faithfully Yours | John Tyndall
I shall leave the hotel tomorrow. And if you should have any thing to say to me the address
    14 Rue du Cirque. Would be safest | J.T.

1. Tyndall, *Diary*, 9 June 1855, **6a**: 75.

2. Heinrich Debus (1824–1916, J.Chem.Soc., 1917, **111**: 325–31). Taught chemistry at Queenwood School, 1850–1867.
3. François René Chateaubriand (1768–1848, DBF). French statesman and writer.
4. Reginald Heber (1783–1826, DNB). Bishop of Calcutta, 1823–1826.
5. Heber (1827), 139.

Letter 3001
Myles Custance[1] to Faraday
5 July 1855
From the original in IEE MS SC 2

Conservative Club | St. James St. | July 5th 1855

Sir

Will you permit a stranger to intrude upon your politeness by requesting the favour of an answer to an enquiry which I wish to make as to the cause, or rather your opinion of the cause of the violent Gales of wind which prevailed for so long a time without any intermission until lately and which are still continued occasionally.

It occurred to me some time ago that the atmosphere must have undergone an extraordinary disturbance by the continued firing at Sebastopol. I attempted to make a rough calculation of the number of shots fired by the Artillery, Bombs, the explosion of ammunition waggons and mines together with the hundred of thousands of shots from small arms and also the great Guns on board the Ships continued daily for so long a time and very often during the greater part of the Night in addition. If you calculate the prodigious quantity of Gun powder thus expanded into gas a most prodigious disturbance of the atmosphere must of necessity I apprehend take place. I should not have intruded this letter upon your Notice had I not seen some time after this idea had struck me that a French Chymist whose name I cannot immediately recollect had declared himself to be of the same opinion[.]

It appears to me that the expansion of such a quantity of Gunpowder must be equivalent to adding an additional quantity of atmosphere to the Earth and that the addition of such a vast quantity must absorb a large amount of heat from the Natural atmosphere and thereby cause the wind to be so much colder than is usually the case at this period of the year. The "Times" Correspondent yesterday mentioned that in 2 days the Allies fired 23,960 shots from the Artillery alone independent of the number of shots fired from Sebastopol[2]. It would very much gratify me to have your opinion on the subject in order that I may be corrected if I am wrong in my opinion but if right that I may have the opportunity of waving your letter in the faces of my friends and thereby stop that ridicule and laughter which generally accompany the expression of my opinion[.]

I remain Sir | Yours very obediently | Myles Custance
M. Faraday Esq

1. Unidentified.
2. *Times*, 4 July 1855, p.9, col. f.

Letter 3002
Faraday to Charles Richard Weld
6 July 1855
From the original in RS MS RR 3.154

6 July 1855

My dear Sir
    I cannot doubt that any paper by Mr. Joule must be proper for the Transactions but I am unable to judge of the peculiar merits of the one you send me[1]. The object is (I conclude) to obtain data the best fitted to form the foundations of mathematical investigations into the nature of the electro-magnetic forces. Having no mathematical knowledge I am not competent to say whether the data here supplied are so direct in their consequences as to be thus fitted for that purpose. With my rough geometrical mode of looking at things I should have liked to know the number of spirals (which vary) in the different helices;- and the influence of the difference of helix diameter for the different bars:- also having a helix *constant* in diameter & length of wire, the difference caused by having the iron (of the same weight & length) in the form of a rod or a cylinder so as to be in one part or another of the space within the helix;- and other variations. But for ought that I know the mathematicians may not require these particularities but may be able with a given & constant length of wire to proceed at once from the data Mr. Joule gives. I have no doubt that the experiments are well & carefully made. I conclude that the iron was of like quality in all the cases & well annealed[.]
    Dr. Tyndall is abroad at present:- when he returns he will have the paper & your note[2][.]
    I am | Yours Very Truly | M. Faraday
C.R. Weld Esq

1. Joule (1856).
2. Tyndall's report on Joule's paper, dated 25 July 1855, is in RS MS RR 3.155.

Letter 3003
Faraday to the Editor of the Times
7 July 1855
From Times, 9 July 1855, p.8, col. f
To the Editor of the Times

Sir, - I traversed this day by steamboat the space between London and Hungerford Bridges between half-past 1 and 2 o'clock; it was low water, and I think the tide must have been near the turn. The appearance and smell of the water forced themselves at once on my attention. The whole of the river was an opaque pale brown fluid. In order to test the degree of opacity, I tore up some white cards, into pieces moistened them so as to make them sink easily below the surface, and then dropped some of these pieces into the water at every pier the boat came to; before they had sunk an inch below the surface they were indistinguishable, though the sun shone brightly at the time; and when the pieces fell edgeways the lower part was hidden from sight before the upper part was under water. This happened at St. Paul's-wharf, Blackfriars-bridge, Temple-wharf, Southwark-bridge, and Hungerford; and I have no doubt would have occurred further up and down the river. Near the bridges the feculence rolled up in clouds so dense that they were visible at the surface even in water of this kind.

The smell was very bad and common to the whole of the water; it was the same as that which now comes up from the gully holes in the streets; the whole river was for the time a real sewer. Having just returned from out the country air, I was, perhaps, more affected by it than others; but I do not think I could have gone on to Lambeth or Chelsea, and I was glad to enter the streets for an atmosphere which, except near the sinkholes, I found much sweeter than on the river.

I have thought it a duty to record these facts that they may be brought to the attention of those who exercise power or have responsibility in relation to the condition of our river; there is nothing figurative in the words I have employed or any approach to exaggeration; they are the simple truth. If there be sufficient authority to remove a putrescent pond from the neighbourhood of a few simple dwellings, surely the river which flows for so many miles through London ought not to be allowed to become a fermenting sewer. The condition in which I saw the Thames may perhaps be considered as exceptional, but it ought to be an impossible state, instead of which I fear it is rapidly becoming the general condition. If we neglect this subject, we cannot expect to do so with impunity; nor ought we to be surprised if ere many years are over, a hot season give us sad proof of the folly of our carelessness.

I am, Sir, your obedient servant, M. Faraday.
Royal Institution, July 7.

Plate 12. Faraday giving his card to Father Thames. Cartoon from *Punch*, 21 July 1855, **28**: 27. See letter 3003.

Letter 3004
John Tyndall to Faraday
10 July 1855
From a typescript in RI MS T TS, volume 12, pp.4034–6

14 Rue Du Cirque Paris. | 10th, July 1855.

My dear Mr Faraday

I repeated your question[1] "what remembrance is there of Arago?" yesterday to a friend, and the reply was that there is no remembrance. He left so many opponents behind him in the Academy that all unity of action is destroyed, and the sum collected was too small for what was proposed. I was at the Academy yesterday, and conveyed your sentiments of affection to Biot. He enquired very kindly after you - He was kind enough to place me in a good position, from which as he remarked I might see "les personages." At the end of a seat near me I saw a fine looking old man, and noticed that many of the members, on passing him squeezed him silently by the hand, while tears appeared in the eyes of some of them. The old man appeared to have been smitten by some calamity with which his friends sympathised - I learned afterwards that it was Thenard[2] that he had lost his son quite recently and had suffered a second heavy domestic affliction some time before. It was the first day of his appearance since his son died. Regnault was president and once he said that the *'parole'* was with M. Le Prince de Canino[3]- "Le Prince de Canino n'existe pas" responded the person addressed -- "Monsieur Bonaparte s'il vous plaît" I do not know whether it is that the title is abolished or that Lucien's republican tendencies caused him to disavow it. At the conclusion of the meeting I felt great pleasure in making the acquaintance of Prof DelaRive who is here with his wife. He will remain here for eight or nine days and then proceed to London. I had never seen him before but his kind frank countenance agreed with the image that I had previously formed of him. I find in Wartmann[4] an exceedingly agreeable companion: he and I walked together for several hours yesterday - we were at Versailles on Sunday with Wheatstone. From all I can see Wartmann is not only well instructed in science but possesses qualities of heart which are pleasant and refreshing to those who require something more than the mere culture of the brain. He gave me a curious picture of the strifes and heartburnings existing among the scientific men of Paris. I could not help comparing this life of ambition, this grasping at and feverish longing after the honours of the world, with the quiet of your existence, and I hope I derived profit and strength from the comparison. Arago appears to have been unfortunate in this respect, and from all I can gather regarding him he appears to have lacked that stability of soul, that reliance upon higher things than mere worldly renown, which I think ought to make the true philosopher. I read some time ago his life of Ampère[5] and I could not help thinking at the time that he was incapable of appreciating fully Ampère's character. The latter appeared to me to possess qualities in a high degree which in Arago were at least rudimentary. LeVerrier[6] I am told moves though the Academy as through a vacuum - he has no connexion with anybody. Thus do we find this august body split up into antagonisms which are scarcely worthy of ordinary illeducated citizens. Thenard is one of those whom all parties respect - the universal sympathy evidenced yesterday was a proof of this. He has in fact pursued science for the love of science, and has not used it merely as a stepping stone to worldly advancement.

While I write Wheatstone, Edmond Becquerel and other members of the section[7] are in the next room and Wheatstone's voice rings in my ears: He talks as if he were inspired being apparently carried quite beyond his own control. He evidently rates pretty highly those honours and marks of recognition and I am inclined to think that he pays in some measure the inevitable penalty, and often has an uneasy mind. Foucault[8] has set up his pendulum in the Exhibition: I saw him this morning for the first time: he has an arrangement of electromagnets underneath the pendulum by which its power is sustained, and the time of oscillation indefinitely prolonged. The fortnight I spent in Normandy[9] was of immense value to me since my railway campaignings I have not felt a greater infusion of physical energy than during that tour. I still continue very strong and hope to be able to keep my mind tranquil and consequently my health good while I remain at Paris. I shall endeavour to get back at the end of the month as indeed I long to be at my work again.

Thank you very much for communicating my request to Anderson - He sends me my letters regularly, and now with kindest remembrances and best wishes for the health and happiness of Mrs Faraday Miss Barnard and yourself

Believe me | most faithfully Yours | Tyndall

1. This was in a letter from Faraday to Tyndall that has not been found, but is referred to in Tyndall, *Diary*, 11 July 1855, **6a**: 148.
2. Louis Jacques Thenard (1777–1857, DSB). Professor of Chemistry at Paris.
3. Charles Lucien Jules Laurent Bonaparte, Prince de Canino (1803–1857, DSB). French zoologist.
4. Elie François Wartmann (1817–1886, P2, 3). Professor of Physics at Geneva from 1848.
5. Arago (1854).
6. Urbain Jean Joseph Leverrier (1811–1877, DSB). Director of the Paris Observatory, 1854–1870.
7. That is of the Paris exhibition. See Eve and Creasey (1945), 60.
8. Jean Bernard Léon Foucault (1819–1868, DSB). French physicist.
9. See letter 3000.

Letter 3005
Faraday to Cornelia Augusta Hewett Crosse[1]
12 July 1855
From Gladstone (1874), 80

July 12, 1855.

.... Believe that I sympathise with you most deeply[2], for I enjoy in my life-partner those things which you speak of as making you feel your loss so heavily.

It is the kindly domestic affections, the worthiness, the mutual aid in sorrow, the mutual joy in happiness that has existed, which makes the rupture of such a tie as yours so heavy to bear; and yet you would not wish it otherwise, for the remembrance of those things brings solace with the grief. I speak, thinking what my own trouble would be if I lost my partner; and I try to comfort you in the only way in which I think I could be comforted.

M. Faraday

1. Cornelia Augusta Hewett Crosse, née Berkeley (d. 1895, age 68, GRO). Married Andrew Crosse in 1850. See under his DNB entry.
2. On the death of Crosse on 6 July 1855.

Letter 3006
Faraday to Justus Liebig
17 July 1855
From the original in Bayerische Staatsbibliothek MS Liebigiana II.B. Faraday, M.

Royal Institution | London | 17 July 1855.

My dear Liebig

Now that I think of writing to you, it seems very long since I wrote last, and I seem as if I had left a pleasure unenjoyed:- but I have often thought of you, & had thoughts even of seeing you; though ever as the proposed time drew near, things before unthought of grew into realities, & the dreams which seemed sometimes as lively as realities, passed away;- & so it is with our life and so I suppose it ought to be. But the thoughts of you are pleasant; and my wife & I often think of the days at York[1], and then set too, to imagine what the years between then & now have done with you. I do not mean as to *progress, discovery,* and *fame* for that we know; but as to the personality of the *Man Liebig,* whose company and converse we enjoyed so much there, that it has left an enduring impression on my failing memory. We both desire to recall ourselves *by that time* to your kindliest remembrance of us.

And now I want to ask you to do me a favour if it lies within your power.- My nephew Mr. Frank Barnard, who desires to improve himself in Art as applied in Art manufacture, has been in Paris for some years & now purposes to visit Munich, that he may profit by its schools & Art treasures. I should be glad if you could give him the opportunity of a few minutes conversation, and if you know what he ought to do to gain admission to the places of study, to tell him. He is earnest to advance himself, but of course needs the information, which he can gain only by enquiry: If you can in this respect put him in the way it will be a great kindness to him & to me. I am quite sure he will not trouble you more than you will desire.

I hardly know who of my friends is in Munich with you:- for my memory fails me & I forget persons in relation to places. I know that I have many friends every where, & not as I am aware of one enemy or one who dislikes me. I ought to be grateful.

Ever My dear Liebig | Yours Most truly | M Faraday
Profr. Liebig | &c &c &c

Address: The | Baron Liebig | &c &c &c &c | Munich

1. At the meeting of the British Association in 1844. See Liebig to Faraday, 19 December 1844, letter 1660, volume 3.

Letter 3007
Lyon Playfair to Faraday
19 July 1855
From the original in RI MS F1 K34
[Athenaeum Letterhead] | 32 Ladbroke Square | Notting Hill | 19th July / 55

My dear Professor

The Scotchmen, myself being an atom of that Nation-loving race, are very desirous to make the Meeting of the [British] Association at Glasgow a successful one. The Meeting takes place on the 12th Septr. in a Town situated on a *moderately clean river*[1], the seat of badly smelling Chemical Manufactures, but within 2 hours of Loch Lomond where Rob Roy[2] became a hero, by remaining unhung as a thief, & within 3 hours of Loch Katrine where the fair lady of the Lake[3], her father & array of followers mysteriously inhabited an island on which two little Bushmen would scarcely be able to squat with our present vast ideas of space.

I hope all these Attractions will lead you to Glasgow where as the Clyde claims no relationship with the Thames, I can promise you a warm reception[.]

Yours Sincerely | Lyon Playfair
Prof. Faraday

1. A reference to letter 3003.
2. Robert MacGregor (1671-1734, DNB). Scottish outlaw and the subject of [Scott] (1818).
3. The subject of Scott (1810).

Letter 3008
Faraday to Lyon Playfair
20 July 1855
From the original in IC MS LP254

20 July 1855

My dear Playfair
  I should like much to see Glasgow this summer[1] - for I have many friends there - but I cannot tell as yet - though I fear it will not be. I am amused at one of your temptations the moderately clean river, for though it is so above bridge yet when I was there last[2] nothing could be worse than its state at the Broomielaw, amongst the boats. I should be very glad to think that that had been altered[.]
  Ever Yours Truly | M. Faraday

1. For the meeting of the British Association. See letter 3007.
2. Faraday was in Glasgow in 1849. See letter 2212.

Letter 3009
Faraday to Frederick Gye
20 July 1855
From the original in Museum of History of Science, Oxford, MS Museum 63

[Royal Institution embossed letterhead] | 20 July 1855
  Thanks my dear Sir for a very great treat, but if I had known before I wrote to you that it was to be the first night of Meyerbeers Opera I should not have moved you[.]
  Ever Your Obliged | M. Faraday
F. Gye Esqr | &c &c &c

1. Faraday heard the British premier of Meyerbeer's "L'Etoile du Nord". *Times*, 19 July 1855, p.8, col. a.

Letter 3010
Charles Lyell to Faraday
24 July 1855
From the original in RI MS F1 I58

53 Harley St | July 24, 1855.

My dear Faraday

I have received a letter marked *private*, from my friend Sir Edmund Head[1], now Governor General of Canada in which he says "could you get me from Faraday an answer to the following question - How far are *first rate* mathematical acquirements essential for conducting with advantage magnetic observations such as have been carried on at the Observatory at Toronto".

He says, he asks this question with a sincere desire to know what is necessary in order that the best may be done for the interests of science, & he should be glad if the answer could be got in the shape of a short note from you & one which he could use.

He has (intentionally I have no doubt) left me in the dark as to the particular bearing of the question, but I know him to be a man who will do what he can for the advancement of science.

Every most truly yours | Cha Lyell

1. Edmund Walker Head (1805–1868, DNB). Governor General of Canada, 1854–1861.

Letter 3011
Faraday to H. Duckworth[1]
25 July 1855
From the original in RI MS F1 N1/25

[Royal Institution embossed letterhead] | 25 July 1855

Sir

Each one has to judge for himself (in the case you speak of) from the evidence given. My conclusion is that the mite had nothing to do with the Galvanic action but got in by accident[.]

I am Sir | Your Obedient Servant | M. Faraday
H. Duckworth Esqr | &c &c &c

1. Unidentified.

Letter 3012
Arthur-Auguste De La Rive to Faraday
26 July 1855
From the original in IEE MS SC 2

Paris jeudi 26 juillet 1855

Monsieur & très cher ami,

    J'apprends par mon ami Prevost[1] que vous êtes encore à Londres & que vous ne quittez cette ville que lundi prochain. Monsieur Tyndall m'avait induit en erreur en me disant, il y a déjà 15 jours, que vous aviez déjà quitté Londres. Ce motif joint à d'autres avait fait que j'avais prolongé sans regret mon Séjour à Paris. Maintenant l'arrivée de mon frère[2] & de deux de mes enfants m'oblige de retarder mon départ jusqu'à lundi prochain 30 juillet. J'arriverai donc à Londres seulement ce jour là ou le lendemain matin mardi 31, désirant passer le dimanche tranquille & ne pas être en voyage ce jour là.

    Maintenant je viens vous demander si je ne pourrai pas aller un matin vous faire une visite à la campagne & si vous ne revenez point quelquefois à Londres de la campagne où vous serez.- Vous comprenez combien je tiens à vous serrer la main & à m'entretenir quelques instants avec vous.- Veuillez donc avoir la bonté de m'écrire deux mots chez MM. *Morris Prevost & c*[3] pour me donner vos instructions.

    Je compte rester à Londres 10 à 15 jours pour corriger les épreuves qui m'attendent[4], puis j'irai faire un tour en Ecosse avec ma femme & l'une de mes filles que j'emmène avec moi. Je repasserai donc de nouveau à Londres vers la fin d'Aout. J'espère donc bien avoir le plaisir de vous voir, d'autant plus que, comme vous le savez, je n'ai pas à Londres un plus excellent ami que vous.

    Agréez, Monsieur & cher ami, l'assurance de mes sentiments dévouées & les plus affectueux

    Auge. de la Rive

Address: Monsieur Faraday | Royal Institution | Albermarle Stt | Londres.

TRANSLATION

Paris, Thursday 26 July 1855

Sir and very dear friend,

    I learn from my friend Prevost[1] that you are still in London and that you are not leaving that city until next Monday. Mr Tyndall had misled me by saying, two weeks ago, that you had already left London. This reason linked to others made me prolong my stay in Paris with no regrets. Now the arrival of my brother[2] and two of my children obliges me to delay my departure until next Monday, 30 July. I shall arrive in London, therefore, only that day or the following morning, Tuesday 31, desiring to spend Sunday quietly and not be travelling that day.

Now I write to ask you if I could visit you one morning in the country and if you perhaps sometimes come back to London from the country where you will be staying. - You know how much I would like to shake hands with you and to spend some time with you. - Please therefore have the kindness to write a couple of words to me at Messrs. *Morris Prevost & co*[3] to give me your instructions.

I count on being in London 10 to 15 days to correct the proofs that are awaiting me[4] then I am going on a tour of Scotland with my wife and one of my daughters who will be travelling with me. I shall pass through London again towards the end of August. I therefore hope to have the pleasure of seeing you, all the more because, as you know, I have in London no more excellent friend than you.

Please accept, Sir and dear friend, the assurance of the devoted and most affectionate sentiments of
    Auge de la Rive

1. Alexandre Pierre Prevost (1821–1873, P2, 3). Swiss physiologist.
2. Eugène De La Rive (1804–1872, Choisy (1947), 52). Swiss lawyer and politician.
3. Merchants of 24A Gresham Street. POD.
4. Of De La Rive (1853–8), **2**.

Letter 3013
Faraday to Longman[1]
6 August 1855
From the original in IEE MS SC 3
            [Royal Institution embossed letterhead] | 6, August 1855
Gentlemen
    I hasten to express my thanks for your kindness in sending me a copy of Dr. Arnotts[2] work on Warming & Ventilation[3] which I trust to read shortly both with pleasure & profit[.]
    I am Gentlemen | Your Most Obedient Servant | M. Faraday
Messrs. Longman & Co

1. That is Longman, Brown, Green and Longmans, publishers in Paternoster Row. Wallis (1974), 40.
2. Neil Arnott (1788–1874, DNB). Scientific writer.
3. Arnott (1855).

Letter 3014
Peter Theophilus Riess to Faraday
9 August 1855
From Bence Jones (1870a), 2: 350–1

Berlin: August 9, 1855.
My dear Sir, - Returning from a journey in Silesia, I had yesterday the great pleasure to find, as a present from you, the third volume of the "Experimental Researches."[1] What a wonderful work these researches are in every respect! Incomparable for exhibiting the greatest progresses for which science ever was indebted to the genius of a single philosopher, highly instructive by indicating the means whereby the great results were found.

If Newton, not without reason, has been compared to a man who ascends to the top of a building by the help of a ladder, and cuts away most of the steps after he has done with them, it must be said that you have left to the follower, with scrupulous fidelity, the ladder in the same state as you have made use of it.

Accept my warmest thanks for your great kindness, to have laid in my hands the object of my continual study and admiration.

And believe me, dear Sir, ever to be yours most faithfully, | P. Reiss [sic]

1. Faraday (1855d).

Letter 3015
Faraday to Andrew Orr[1]
13 August 1855
From the original in WIHM MS FALF

[Royal Institution embossed letterhead] | 13 August 1855
My Lord
I grieve to think I may not be able to permit myself the delight of being at Glasgow at the meeting[2] but if when the time comes I find it possible I will assuredly[.] It lies not within my power to say at present but I thank Your Lordship & your companions in kindness for the favour of your letter[.]

I have the honour to be | My Lord | Your Very Obliged Servant | M. Faraday
The | Lord Prevost | of Glasgow | &c &c &c

1. Andrew Orr (1802–1874, B2). Lord Provost of Glasgow, 1854–1857.
2. Of the British Association.

Letter 3016
Benjamin Humphrey Smart[1] to Faraday
13 August 1855
From the original in IEE MS SC 2

[Athenaeum letterhead] | Athenaeum Aug. 13,1855

My dear Sir,

Though you have not a more hearty well-wisher than myself, and I securely count on your friendly feelings, yet as we have both something else to do than writing complementary letters, our correspondence is infrequent. For my part, I cannot recollect ever having taken up any pen to write to you, but when I had a favour to ask; and this indeed is the cause of my writing now.

Accompanying this, you will find a slip having on it the title, and the table of contents, of a very little work of mine[2] now going through the press. What I have to beg of you is this - that you would allow me to place the following inscription at its beginning in lieu of a preface:
To Michael Faraday Esq. | Hon. D.C.L.Oxf. F.R.S.; F.G.S. Prof. Chem. R.I. | &c &c &c
This Essay | though in a department of Philosophy distinct | from that in which his name stands illustrious, | yet being attempted to be carried out in the | inductive spirit which his example eminently recommends, | is, | with vivid recollections of early friendship and continued kindness, | affectionately inscribed.

If you grant me this favour, pray let me know whether I have placed you titles accurately; and believe me
My dear Sir | Sincerely and faithfully | Yours | B.H. Smart
To Michl Faraday Esqr. | &c &c &c | Royal Institution

1. Benjamin Humphrey Smart (1786–1872, DNB). Writer and teacher on elocution.
2. Smart (1855) dedicated as below to Faraday.

Letter 3017
Benjamin Humphrey Smart[1] to Faraday
16 August 1855
From the original in IEE MS SC 2

[Athenaeum letterhead] | Athenaeum, Augt. 16,1855.

My dear Sir,

You are quite right - I ought not to have asked your permission[2] - indeed that was not my meaning, nor do I understand you to have granted any; but without your *knowledge* I could not have felt justified in the step I wished to take.

I send the page - it will not be worked off for at least a week. Should you therefore see that any correction of any kind is desirable, pray mark it and return the proof:- otherwise you need not take the least further trouble, but believe me.

My dear Sir | Ever Sincerely Yours | B.H. Smart
Mich. Faraday Esq

1. Benjamin Humphrey Smart (1786–1872, DNB). Writer and teacher on elocution.
2. See letter 3016.

Letter 3018
Thomas Rawson Birks[1] to Faraday
24 August 1855
From the original in IEE MS SC 2
Dear Sir
Though I am personally a stranger to you, I venture to hope that you will forgive me for intruding upon your time by a few lines of inquiry, as my only object is to advance the science, to which you have contributed so many important discoveries.

I have now at intervals for nearly twenty years since my residence at Cambridge had an idea in my thoughts, which I conceive may furnish a key to unite several branches of physical science, & throw a fresh light on chemical & electrical science. I have just of late been using some intervals of leisure from other duties in unfolding it to myself more clearly, with some view to publication. It is one part of the theory that all chemical elements are really compounds of the first order, & that all their properties flow by mechanical laws originally from their atomic numbers & electric order. But this part of the theory is not ripe enough at present to trouble you with reference to it. My inquiry relates to the theory of electricity alone.

You are well aware that the theory of Poisson, simplified by Dr Whewell in the Enc. Metr.[2] explains all these phenomena by two electric fluids. From Delarive's work I infer that you incline to reject these fluids entirely[3], & to explain the chief phenomena by simple induction only. My own theory wd lead me to an intermediate view, that there is no distinct electric fluid, but that it depends on the increase or diminution of ethereal electricity at the surface of charged bodies. This wd differ from your view, if I am right in my impression, because yours wd exclude all action in vacuo; & from that of Coulomb & Poisson, because the influence, on their hypothesis, depends on distance only, & in mine, on inclination also.

Now I venture to trouble with the inquiry whether an experiment of this nature has been ever tried which wd, I think, decide between their view, & the modified one in question. Let two charged square discs, moveable on their centres, be placed opposite & near each other, & their repulsion carefully measured. Let them next be inclined on their axes to each other, say at an $\angle$ of 30[.] On the hypothesis of two fluids, & non conducting surfaces, the repulsion shd be increased, since the nearness of the upper halves will more than compensate for the recession of the lower halves. But if the action depends, as I conceive, on the size of the inclination, the force wd be diminished to nearly one half. Now I wish, if it be not troubling you too much, to inquire, whether such an experiment has been made by Sir Snow Harris or yourself, & with what result, of if not, whether you think it wd be worth the trial. Hoping you will excuse the liberty I have taken, I remain, with most sincere respect & esteem

Yours very truly | T.R. Birks | Late Fellow of Trin Coll. | Camb Kelshall Rectory | Royston | Augt 24th 1855

1. Thomas Rawson Birks (1810–1883, AC). Rector of Kelshall, 1844–1866.
2. [Whewell] (1845). On the authorship of this see Wilson, D.B. (1991), 243, 246.
3. De La Rive (1853–8), 1: 144–54.

Letter 3019
Faraday to Thomas Phinn[1]
27 August 1855
From Bence Jones (1870a), 2: 365

Royal Institution: August 27, 1855.

Sir, - I am sure that when the Lords Commissioners of the Admiralty look again at the enclosed printed advertising paper[2] which you have sent to me, and which I return herewith, they will see that it is not such a document as I can be expected to give an opinion upon. My Lords will do me the favour to remember that, as I have said on former occasions[3], though I am always willing to help the Government in important cases, and when it is thought that others cannot give satisfactory information, still I am not professional; and being engaged in deep philosophic research, am desirous of having my time and thoughts as little engaged by extraneous matters as possible.

I have the honour to be, Sir, your very obedient servant, | M. Faraday.

1. Thomas Phinn (1814–1866, B2). Second Secretary to the Admiralty, 1855–1857.
2. According to Bence Jones (1870a), **2**: 364 this was to do with Charles Crews's disinfecting powder and anti-miasma lamp (patent number 1855/732). See also Admiralty Digest, PRO ADM12 / 604, class 37.1.
3. For example, Faraday to Auckland, 29 July 1847, letter 2009, volume 3.

Letter 3020
John Tyndall to Faraday
27 August 1855
From a typescript in RI MS T TS, volume 12, pp.4037–41

Queenwood near Stockbridge | 27th Aug. 1855

My dear Mr. Faraday,

The thought of you wandered across my mind this morning before I rose, and I said in my heart I will commence the day by writing to him. I thought of writing to you after that terrible thunderstorm which visited this neighbourhood[1], but I heard that it stretched over London also, consequently I infer over Sydenham and that you yourself were a witness of its grandeur. I dont think I have ever seen lightning so incessant and vivid. The evening was beautiful. I watched the grouping of the clouds in the west after the sun had sunk; it was magnificent, and eastward the moon, almost full, hung her shield at intervals in the blue space between the masses of cloud - all was calm towards 8 o'clock and summer lightning, as I thought it, played fitfully on the south west. As the evening advanced however the lightning drew nearer and rumbling thunder became audible, and towards 10 o'clock the spectacle was grand. I wandered with a friend[2] to a summit in the neighbourhood and watched the flashes along the horizon, these were reflected from the clouds above and an intense brilliancy shivered at intervals over the heavens and the earth. The yellow light of the gas in the windows of the house seemed utterly extinguished by the superior glare of the lightning. The flashes and peals drew nearer and as a matter of precaution my companion and myself withdrew from the height and sought shelter in the house; here we stood in a porch watching the fitful but almost incessant glare, a wild gust of wind rushed in upon us carrying heavy rain along with it and compelled us to close the door. Louder rolled the cannonade and one crash broke above us which almost seemed intended to shiver the building to atoms. About 120 boys were in bed at the time; I thought of the little fellows and went upstairs to comfort them if necessary. The younger ones were greatly terrified and some of them were crying bitterly; I talked cheerily to them and when the vivid gleams broke in upon them exclaimed "beautiful!" By degrees they also began to consider the flashes as things of beauty and not of terror, but some of them were greatly surprised to find that when they shut their eyes and covered their heads with the bedclothes they still saw the flashes. I waited with them until I could assure them that all danger was past, until there was no cloud near to give us even a taste of the *returning stroke;* but long after this the discharge continued along the eastern horizon, and like quivering red hot bars the flashes sped up and down. Francis who was down here on Sunday[3] told me he noticed circular flashes, and was very confident of the reality of the fact: I myself noticed several times the clouds surrounded by a fiery rim intensely luminous, this I mentioned to Francis but he would not admit it as being the explanation of what he saw. You drew attention to the same source of error many years ago[4].

The people here have contrived to make me very comfortable in a little lodge which stands at the entrance to the college lawn and which serves on Sundays as a meeting house for the few quaker youths in the establishment. I have planted my apparatus. The work goes on slowly, for it is terribly difficult[.] The greatest care is required to render the facts safe and when they are safe one does not know what to make of them. My strength oscillates as usual: when I give up work for a day and swing my limbs over the breezy downs I return with an amount of force which seems inexhaustable, but two days thought produces a wonderful diminution of my energy. I write at present upon the legs of the vigour which I acquired on Sunday last in the New Forest. Francis, the chemist here[5], and I myself started early and wandered along bowery roads and through fine parks, until we found ourselves at noon knee deep in the fern of the forest[6]. Visited the stone which stands upon the spot where Rufus[7] was shot by Walter Tyrrell[8]. The people though they generally bear the character of barbarians were extremely civil and intelligent: there was an admirable precision in their directions when they showed us our way. The same quality was exhibited by the children. It was really remarkable to observe the exactitude with which they described the twists turnings and landmarks along the forest tracks. The day was glorious; the sun smote the foliage and seemed to drip as liquid gold into the shaddows beneath, the breezes made music amid the branches and as the rust cleared gradually away from the inner man our hearts responded to the sound. I had left home a decrepit wretch but as I advanced I found my vigour rush upon me. Passed through sunny little hamlets and by brown thatched cottages around which roses and runners and laurels clustered till finally we reached Lyndhurst; it was $2\frac{1}{2}$ o'clock, and having breakfasted early we put up at an inn and fortified ourselves with the wholesome country fare. Thence to the nearest station, thence to Southampton, then to Dunbridge from which we had a walk of four miles. The full harvest moon sailed above us through the cloudless sky, frosting the beech leaves with silver, and near her was the evening star. The air was clear and bracing and we finally crossed our respective thresholds leaving a day behind us to which memory can scarcely furnish a parallel.

As Francis sat beside me at dinner on Saturday in the same room with the boys; a question buzzed though the little host whether that was Prof. Faraday! Francis said to me at the time that the prospect of such a number of fine little fellows would be a pleasure to you, and the thought then and there occurred to me that supposing you wished to change your scene for a day you might possibly turn your thoughts and steps hitherward. One sunny day upon the chalk would do you good, one calm night spent away from the turmoil of city life. I can promise you a good bedroom with linen sweet and white. The place is also remarkable as furnishing evidence how nature deals with those who do violence to her mental laws, that here as in physics she is inexorable. This place was the Harmony Hall of the socialists; but the memory of them is clean gone from the place. There is a gas-house here where the hydrocarbon principle is applied and that wonderful Bughead cannel is distilled for the illumination of the establishment, there is a new and singular washing machine - there are the boys and the playgrounds which would be pleasant to your eyes: there is a most remarkable avenue of yew trees adjacent gloomy and glorious at the same time - there is another of elm trees which might have suggested the pillared aisles of our Cathedrals, and there will be gladness in all our hearts and a welcome in every eye should you set your foot within the establishment.

I had a letter some days ago from Frankland in which he urges a promise I made to him some time ago to spend three days with him in the lake district: this I intend to do, and we then go together to Glasgow to the meeting of the Association. I shall probably leave this place about the 6th. and spend a day in London, so as to gather a few things together that may possibly interest the physical section. I have had a letter from Duboscq[9] containing a list of apparatus for a course of lectures on optics concerning which I shall ask your counsel at a future day[10]. Now I think I have sufficiently afflicted you with this long and badly written scrawl. The bell has rung - it is 8 o'clock and the boys are moving to breakfast. Desiring my best wishes to Mrs. Faraday and Miss Barnard I say for the present goodbye.

affectionately yours | John Tyndall

Mr. Best[11] one of our members wrote to me when in town asking me to open certain proceedings of the Hants & Wilts educational union by a lecture - this I declined, and to day I have declined another invitation from the same gentleman and one also from Leeds. But lest he should think me unkind living so near Southampton I went over to the meeting on Wednesday last[12] - purposing simply to shew them the respect of my personal attendance they however induced me to say something - and indeed the proceedings were well conducted so that when called upon there seemed no reasonable objection to my saying a few words. I hear an account of the meeting has got into the papers so if you should see it what I have said will account for my appearance there[13].

1. See Tyndall, *Diary*, 23 August 1855, **6a**: 167–8.
2. Heinrich Debus (1824–1916, *J.Chem.Soc.*, 1917, **111**: 325–31). Taught chemistry at Queenwood School, 1850–1867.
3. That is 26 August 1855.
4. Faraday (1838c), ERE13, 1641.
5. That is Debus.
6. Tyndall, *Diary*, 26 August 1855, **6a**: 168–71.
7. William II (c1056–1100, DNB). King of England, 1087–1100.
8. Walter Tyrrell (f.1100, DNB under Tirel). Alleged assassin of William II.
9. Jules Duboscq (1817–1886, P3). Scientific instrument maker in Paris.
10. Duboscq to Tyndall, 22 August 1855, RI MS T.
11. Samuel Best (1802–1873, B1). Rector of Abbots-Anne, Andover, 1831–1873. Member of the Royal Institution from 1839.
12. Tyndall, *Diary*, 22 August 1855, **6a**: 164–5.
13. *The Hampshire Advertiser*, 25 August 1855, p.6, col. c noted Tyndall's presence at the meeting but did not record any of his remarks.

Letter 3021
Faraday to Caroline Deacon
2 September 1855
From the original in the possession of Elizabeth M. Milton

[RI embossed letterhead] | Sydenham | 2 Sept. 1855

My dear Caroline

I must write a word of love & affection to you - though as to an answer to your letter or reasons for this or that I will refer you to mine to Margery accompanying this - I rejoice you are altogether once more and in a place that I hope will do you good in health - give you delight in its beauties - and have many remembrances for the future. I should like to be with you but am best at home with my dear wife[.] Give my love to that great girl Constance & if she has forgotten me raise up some idea of me in her mind that will serve for present use and when you write home remember me kindly to your husband who if I recollect is somehow engaged now in chemical pursuits which I have left[.] Adieu my dear niece[.] May love joy & peace be with you.

Your Affectionate Unkle | M Faraday

Letter 3022
Faraday to John Tyndall
3 September 1855
From a typescript in RI MS T TS, volume 12, p.4138

ROYAL INSTITUTION OF GREAT BRITAIN. | 3 Sept. 1855.
My dear Tyndall,

Esteem my feelings as in the exact inverse proportion of my note, which is no fit answer to your very kind and pleasant letter[1]. But I have been a good deal lowered by a persistent attack of diarrhoea, and which, though I think it has been stopped, has left its marks in giddiness and feebleness of head. I once had very feeble hopes of being at Glasgow[2] with you all, but they are all gone, and I find once more what I am good for:- nothing. I thought if I waited longer before I answered you, I might lose you at Queenwood, and only write you a more stupid note than the present. Your picture of Queenwood excites many past remembrances, but I have no hopes of converting them into futures. We have all been poorly here - Wife, Jane and self - but we are all mending. Jane is away from us. My wife sends her kindest remembrances.

Ever yours, | M Faraday
Prof. Tyndall, | &c &c &c

1. Letter 3020.
2. For the meeting of the British Association.

Letter 3023
John Tyndall to Faraday
5 September 1855[1]
From a typescript in RI MS T TS, volume 12, p.4042

Royal Institution
Dear Mr Faraday

Your note reached me this morning an hour before I left Queenwood. I was glad and sorry at once on reading it. Most sincerely do I hope that you will soon be quite strong again. It would have been a delight to me had you resolved to go to Glasgow[2], now I am not quite sure that I act right in going and your presence there would have been a kind of quieter to my conscience. I know I could employ my time far better at home, but then I should be deemed unkind. I had just got hold of my work - just established that affinity between me and it which is always a work of some pain to me when I am obliged to quit it, and to enfeeble that affinity by other employment. However as I grow older I shall see more clearly how I ought to act, how far conform with the requirements of others and how

Letter 3025

far follow out my own notion of things - good bye for the present - say good bye to Mrs Faraday for me
    Most faithfully Yours | John Tyndall

1. Dated on the basis that this is the reply to letter 3023 and that Tyndall left Queenwood on this day. Tyndall, *Diary*, 5 September 1855, **6a**: 171.
2. For the meeting of the British Association.

Letter 3024
John Crichton[1] to Faraday
6 September 1855[2]
From the original in RI MS F1 I75

Dundee | 6 December [sic] 1855

Very Dear Brother
    Some of our Friends here say they have observed your name as one of the members of the British Association for the promotion of Science who are to hold a meeting at Glasgow on the 12th of this month and as Glasgow is only about three hours distant by Rail from this are anticipating that you may be induced to prolong your Journey to Dundee and so enjoy the happiness of being once more filled with your Company[.]
    If so ordered I need not say how happy I shall be by your making my House your Home whilst here and again in my old age enjoying the Liveliness of your Company along with as many of the Brethren as the Room will contain[.]
    With kindest Regard to you Mrs F and all enquiring friends I am
    Very Dear Brother | Affectionately Yours | Jno Crichton
Professor Faraday

1. John Crichton (1772–1860, B1). Dundee surgeon and member of the Glasite church there.
2. Dated on the basis of the reference to the Glasgow meeting of the British Association.

Letter 3025
James Clark to Faraday
8 September 1855
From the original in RI MS Conybeare Album, f.22

Birk Hall | Sep. 8h 1855

My Dear Faraday,

I have heard that it is your intention to be at the meeting of the British Association at Glasgow this month, & I called in Albemarle Street the night before I left London, to ascertain if such was your intention. My object was to ask you, if you did go to Glasgow, to come on for a few days, or as long as you can stay, to Birk Hall - we are only 40 miles from Aberdeen, & just in the mouth of the Highlands, where the air is most bracing, and will I doubt not do you much good & enable you to work with more than your usual energy during the winter. I have asked Liebig to come, & you might come together, - from Glasgow to Aberdeen is a short day. Sleep at Douglas Hotel & a coach, which reaches Aberdeen next morning at 10.45 o'clock will bring you to Ballater a little before 4 oclock[.] There I will meet you & Birkhall is only two miles further. If you go to Glasgow, I do hope you will come on, as I am persuaded a visit to our hills will be very beneficial to your health & give us great pleasure to see you[.]

Very faithfully yours | Ja Clark

Letter 3026
John Tyndall to Faraday
24 September 1855[1]
From a typescript in RI MS T TS, volume 12, pp.4043-6

Monday, 25th [sic], Sep, 1855,
My dear Mr Faraday

I have a bad pen and no knife fit to mend it, so I fear it will give your eyes some trouble to make out my blurred sentences. It is very still here at present - The sun set in golden clouds half an hour ago - The hive of life which this roof usually covers has been out nutting all day and is not yet returned, so I am left alone to make use of the tranquil evening hour in writing to you. In passing through London *en route* to Glasgow[2] I was glad to learn that your health was restored, on coming through on Saturday[3] I was equally glad to hear the good tidings confirmed. I thought of running down to Sydenham to see you and Mrs Faraday, but a second thought suggested that I might only disturb you - indeed the idea sometimes occurs to me that I give you a great deal too much trouble to read letters some of which contain almost nothing that could be expected to interest you, nevertheless some blind instinct causes me to write, and to hope that I do not pester you. Previous to going to Glasgow Frankland, myself and a third friend made four days journey through the lake district of Cumberland. Frankland himself has taken a small house in Windemere, has a boat on the lake and spends his time fishing and getting his thin cheeks tanned with the hue of health. We started one glorious morning,

Letter 3026

went to Coniston; thence through a beautiful valley to Langdale Pikes and put up at a farm house at the base of the mountain. The clouds gathered and rain fell during the night the wet weather prolonged itself into the following day, but taking advantage of a lighter hour we set out once more. We had to face a steep and rugged pass, and to clamber with great toil over the rough boulders strewed along the bed of a mountain torrent[.] The rain fell heavily all the time and drenched me through. We reached the summit, turned into a wrong track which cost us upwards of ten miles additional labour. We had to ascend a second pass almost as steep and rugged as the first, the rain still falling - sometimes our pathway was converted into a brook through which we trudged ankle deep in water. My companions were furnished with mackintoshes, I had a large horserug over my shoulders which drank in the rain like a sponge and became enormously heavy chiefly along its dependent edges which flapped wearily against my drenched limbs. Notwithstanding this I was the freshest of the party when the day was concluded. Having reached Wastdale we got shelter in a farm house - there were no public houses - and I was happy enough to secure a pair of trousers belonging to a corpulant mountaineer into which I crept while my own were drying. Next day was a repetition of the same. Crossing Blackstair to Buttermere the rain descended and drenched us once more. The exertion however kept a warm moisture on the body and although the toil was excessive I enjoyed it - it left no rust on the muscles. From Buttermere, which is a glorious place, we passed by Honister Crag to Borrowdale - saw Lodore, crossed Derwent Water in a boat and put up at Keswick. Thence to Penrith, thence to Glasgow where I was soon plunged amid the duties of secretary[4]. From this I was not able to extricate myself until the last journal was prepared. Stevelly[5] was ill and did not make his appearance. This rendered my duties very laborious. The meeting was very successful as regards numbers, but the papers brought forward were not remarkable. Indeed I left Glasgow with the opinion which I entertained on entering it, that I should have contributed more to the advancement of science by staying at home and doing my work than by going there. The longer I live the more I learn to value private in preference to associated effort. It is to the solitary worker that science has to look for its advancement rather than to the brilliant gatherings of a society however pleased that society may be with its own performance. I described the experiments which I made before you and De la Rive[6]. Thomson and myself had a short discussion afterwards, the memory of which was unpleasant to me for a day or two, but which I have now almost forgotten[7]. I have been told that the discussion was misrepresented in the newspapers, but this I do not know as I have never seen any report of it whatever. I thought at the time that a degree of partisanship was shown almost approaching to unfairness; but it is perfectly possible that I was a partisan myself and simply saw my own image in others. The whole matter seems infinitely small regarded from a proper point of view, so I will say no more about it.

Brewster read a paper on what he called the triple spectrum[8]. You know he imagines the spectrum to be composed of three distinct spectra, Red, yellow and blue his views have been criticised by Whewell[9] and others, and in the papers brought forward at Glasgow he makes a terrible assault upon his antagonists. It was most cleverly written, full of that kind of talent which excites your admiration without influencing your convictions. On Wednesday morning[10] I was weary enough and having completed my duties Frankland and myself proposed spending a day on Loch Lomond. We went, and while there learned that Loch Katrine, the scene of Scotts[11] poem of the Lady of the Lake[12] was within a few miles of us. Walked to Loch Katrine across the country. Crossed the lake in an elegant little screw steamer, and spent the night in the jaws of the Trosachs. I never saw a scene of deeper loveliness than that presented by the head of Loch Katrine. Next day we went to Sterling[13] and thence to Glasgow whence I started on Friday at 9 o clock and reached London at about 10 o clock the same evening.

I have got to work, but it will take a day or two to get thoroughly into it as I was before starting. I am trying whether the *total* magnetic intensity of bismuth is increased by compression. This I find to be the case, though I must still make some corroborative experiment. What stands in handbooks regarding the increase of specific gravity needs I think careful examination. With the power of compression, and the balance at my command here I have not been able to prove that any sensible change of specific gravity takes place. I shall however apply more delicate tests by and by - and now farewell for the present with kind wishes to Mrs Faraday and Miss Barnard

Believe me | most sincerely Yours | John Tyndall

1. Dated on the basis that 24 September 1855 was a Monday.
2. For the meeting of the British Association.
3. That is 22 September 1855.
4. Of the Mathematics and Physics Section. *Rep.Brit.Ass.*, 1855, p.xxvii.
5. John Stevelly (c1794–1868, Moody and Beckett (1959), **2**: 616). Professor of Natural Philosophy at Queen's College, Belfast, 1849–1867.
6. Tyndall (1855b). For these experiments see Tyndall, *Diary*, 14 August 1855, **6a**: 160–1.
7. See Tyndall, *Diary*, 15 September 1855, **6a**: 195–8 in which he noted his debate with Thomson and Whewell. *Morning Chronicle*,18 September 1855, p.3, col. a noted what it termed a "somewhat sharp discussion" over the polarity of bismuth, in which was contested what Faraday's current views on the subject were.
8. Brewster (1855a), but for a fuller account see *Athenaeum*, 6 October 1855, pp.1156–7.
9. See Whewell (1837), **2**: 360–1.
10. Tyndall, *Diary*, 19 September 1855, **6a**: 183–5.
11. Walter Scott (1771–1832, DNB). Scottish novelist.
12. Scott (1810).
13. Tyndall, *Diary*, 19 September 1855, **6a**: 185–7.

Letter 3027
Faraday to John Tyndall
6 October 1855
From a copy in untitled notebook in IEE MS SC 2

Sydenham | 6th Octr. 1855.

My dear Tyndall,

I was put into a very mixed mood by your last letter[1]; glad to hear from you, that you were out of the turmoil, had enjoyed the beauties of the Lakes and was happily at home again: but sorry for some annoyances which I saw you had met with at Glasgow. These great meetings, of which I think very well altogether, advance Science chiefly by bringing scientific men together and making them to know and be friends with each other; and I am sorry when that is not the effect in every part of their course. I know nothing except from what you tell me, for I have not yet looked at the reports of the proceedings: but let me as an old man, who ought by this time to have profited by experience, say, that when I was younger I often misinterpreted the intentions of people, and found they did not mean what at the time I supposed they meant and further, that as a general rule it was better to be a little dull of apprehension when phrases seemed to imply pique and quick in perception when on the contrary they seemed to convey kindly feeling. The real truth never fails ultimately to appear, and the opposing parties are, if wrong, sooner convinced when replied to forbearingly than when overwhelmed. All I mean to say is that it is better to be blind to the results of partizanship, and quick to see good will. One has more happiness in oneself, in endeavouring to follow the things that make for peace. You can hardly imagine how often I have been heated in private when opposed, as I have thought, unjustly and superciliously, and yet have striven and succeeded I hope in keeping down reply of the like kind: & I have I know never lost by it. I would not say all this to you if I did not esteem you as a true philosopher and friend.

I have not been altogether idle but I am of necessity very slow now. I have not read the journals, because, when able I have been at work and writing a paper[2]. The latter goes on slowly, but I think will be a useful contribution of facts; and may help to advance the logic of magnetism a little though not much. The secret of magnetic action is like a Sebastopol at least in this point that we have to attack it in every possible direction and make our approaches closer and closer on all the sides by which we can force access. My working is mainly with magnecrystals and the effects of heat on them.

You say it is still "here" but do not say where the "here" is; and my memory is so treacherous that now I have written this letter I shall not be able to send it until I go to London; but whenever you receive it believe me to be as ever

Yours Very Truly | M. Faraday

Profr. Tyndall | &c &c &c
My wifes best remembrances.

1. Letter 3026.
2. Faraday (1856), ERE30.

Letter 3028
Jacob Herbert to Faraday
11 October 1855
From the original in GL MS 30108/2/66

Trinity House, London, | 11th October, 1855.
Sir,
    Messrs. Chance, Brothers, of Birmingham, being about to supply this Corporation with a Catadioptric Apparatus for a fixed Light of the First Order, to illuminate 315° of the Horizon,- I am directed to acquaint you therewith and to transmit to you the enclosed Copy of a Letter from them[1] on that subject,- and, at the same time to signify the request of the Elder Brethren that you will, at such time as shall be perfectly convenient to yourself, visit the Factory of Messrs. Chance at Birmingham for the purpose of examining the said Apparatus, and that you will report your opinion thereof, for the Board's information[2].
    I am farther to request that you will communicate with Messrs. Chance agreeably to their wish, as expressed in the concluding paragraph of their Letter.-
    I am, | Sir, | Your most humble Servant | J. Herbert
Michael Faraday Esq: D.C.L., F.R.S.

1. Chance to Berthon, 8 October 1855, GL MS 30108/2/66.
2. See Trinity House By Board, 9 October 1855, GL MS 30010/39, p.548 for the decision to ask Faraday to make this visit.

Letter 3029
Faraday to Jacob Herbert
20 October 1855
From the original copy in GL MS 30108/2/66

Royal Institution | 20 Octr. 1855
Sir

In reference to your letter of the 11th instant[1], I beg to state that I proceeded to the Glass works at Birmingham on Tuesday last[2] for the purpose of examining Messrs. Chances Catadioptric apparatus[.] It was well fitted up so as to give me every opportunity of looking at it. I found the colour of the glass very good indeed as good as the pannels approved of on former occasions. Some of the ribs were striated in parts but it is impossible to avoid this altogether without enormous expence & as far as my memory serves the apparatus is not in that respect beneath those I have seen from or in France. The fitting & work of the glass & frames is good. I took a pannel down & examined it by the suns rays and found the focal distances of the several parts perfectly adjusted for the lamp distance[.]

At night when the lamp was burning I examined it from steps & a stage placed at the distance of 57 yards so as to command it at different levels and by revolving the apparatus on all sides & found the action as it ought to be. Then retreating to the distance of 225 yards, I examined it again & still had every reason to be satisfied with the apparatus as being equal in all points to the French Apparatus. I therefore beg to express my approval of it[3][.]

I am Sir | Your Very humble Servant | M. Faraday
Jacob Herbert Esqr | &c &c &c

1. Letter 3028.
2. That is 16 October 1855.
3. This letter was read to Trinity House By Board, 23 October 1855, GL MS 30010/39, p.574. It was referred to the Light Committee with the instruction to proceed with Chance's apparatus.

Letter 3030
Faraday to John Barlow
26 October 1855
From the original in RI MS F1 D27

RI. | 26 Octr. 1855

My dear Barlow
 Did I lend you a strip of Aluminium[1] or no[.] I let three or four persons have it & not remembering who[,] have lost it now that I want it[.]
 Ever Yours | M Faraday

1. On aluminium see McConnell (1989).

Letter 3031
Lyon Playfair to Faraday
30 October 1855
From the original in IEE MS SC 2

Ladbroke Square | 30th Oct / 55

My dear Professor

The Abbé Moigno[1] requested me to bring over for you the accompanying box of Lucifers. Their peculiarity consists in the red Phosphorous being mixed with the Emery on the Sand paper, thus rendering two conditions necessary for their ignition & thus diminishing the danger ordinarily attendant on them[.]

Sincerely Yours | Lyon Playfair

1. François Napoleon Marie Moigno (1804–1884, NBU). French scientific writer.

Letter 3032
Faraday to Lyon Playfair
30 October 1855
From the original in IC MS LP255

Royal Institution | 30 October 1855

My dear Playfair

I was very sorry to miss you this morning but inexorable time stole you away and I know so well the value of it to an occupied man that I cannot but allow to others the power I am often obliged to reserve myself. Many thanks for the box which is very curious & apparently good. I had seen them before - Do you think they will come into use?[1]

I was much struck by an observation Liebig made to me the other day when he was here & we were talking over just such a box. "But your match makers have not the phosphorus disease in the jaw" said he "it only occurs in Germany" and when I asked him his reason for that he seemed to give it in the bad ventilation & closeness of the German shops combined with the presence of phosphorous vapour[.]

Ever Your Obliged | M. Faraday
Lyon Playfair Esqr | &c &c &c

1. See letter 3031.

Previous Publication of Letters

This notes where the text of the letters in this volume have been previously published. It does not repeat the citation to letters which have only been located in published works. Nor does it note quotations of parts of letters used in critical studies of Faraday. Apart from Bence Jones (1870b), where there are substantial changes from Bence Jones (1870a), (particularly in volume one) differences between different editions of texts are not noted.

2146 Duveen (1949). Williams et al. (1971), **1**: 513.
2147 Williams et al. (1971), **1**: 513.
2152 Williams et al. (1971), **2**: 541–3.
2164 Faraday, *Diary*, 5, pp.196–8. Williams et al. (1971), **2**: 543–5.
2174 Williams et al. (1971), **2**: 545–6.
2183 Plücker (1849b). Williams et al. (1971), **2**: 546–8.
2184 Williams et al. (1971), **2**: 548–50.
2185 Williams et al. (1971), **2**: 550.
2189 Williams et al. (1971), **2**: 551.
2190 Williams et al. (1971), **2**: 551.
2191 Williams et al. (1971), **2**: 552.
2193 Spargo (1992), 47.
2197 Williams et al. (1971), **2**: 553.
2198 Williams et al. (1971), **2**: 553.
2199 Williams et al. (1971), **2**: 554.
2200 Williams et al. (1971), **2**: 555.
2205 Williams et al. (1971), **2**: 557.
2208 Williams et al. (1971), **2**: 558.
2210 Williams et al. (1971), **2**: 558–9.
2211 Thompson, S.P. (1910), **1**: 216–8. Williams et al. (1971), **2**: 559–61.
2214 Williams et al. (1971), **2**: 561–2.
2217 Williams et al. (1971), **2**: 562–3.
2219 Bence Jones (1870b), **2**: 244–5.
2221 Bence Jones (1870b), **2**: 245.
2225 Williams et al. (1971), **2**: 563.
2226 Williams et al. (1971), **2**: 564.
2229 Bence Jones (1870b), **2**: 245–6.
2232 Williams et al. (1971), **2**: 564–5.

2236   Williams et al. (1971), **2**: 565–6.
2237   Williams et al. (1971), **2**: 566–8.
2239   Williams et al. (1971), **2**: 569.
2241   Williams et al. (1971), **2**: 570.
2243   Bence Jones (1870a), **2**: 251–2, (1870b), **2**: 246–7. Williams et al. (1971), **2**: 571.
2245   Bence Jones (1870a), **2**: 247, (1870b), **2**: 242. Harding (1920), **2**: 325–6. Williams et al. (1971), **2**: 571–3.
2246   Williams et al. (1971), **2**: 573–4.
2248   Bence Jones (1870b), **2**: 251–2
2249   Williams et al. (1971), **2**: 574–6.
2250   Williams et al. (1971), **2**: 577–8.
2253   Williams et al. (1971), **2**: 541.
2255   Schettino (1994), 422–5.
2256   Bence Jones (1870a), **2**: 272–3, (1870b), **2**: 267–8. Williams et al. (1971), **2**: 579.
2263   Pelseneer (1936), 448. Williams et al. (1971), **2**: 579–80.
2268   Bence Jones (1870a), **2**: 273–4, (1870b), **2**: 268–9. Harding (1920), **2**: 327–8. Williams et al. (1971), **2**: 581.
2273   Williams et al. (1971), **2**: 581–2.
2274   Kahlbaum and Darbishire (1899), 184–5.
2276   Williams et al. (1971), **2**: 582–3.
2279   Bence Jones (1870a), **2**: 274–5, (1870b), **2**: 269–70
2280   Bence Jones (1870a), **2**: 275, (1870b), **2**: 270.
2283   Bence Jones (1870b), **2**: 228.
2287   Bence Jones (1870a), **2**: 262–3, (1870b), **2**: 257–8. Kahlbaum and Darbishire (1899), 185–6.
2291   Randell (1924), 127–8. Williams et al. (1971), **2**: 583. Storey et al. (1988), 105–6.
2292   Randell (1924), 129–30. Williams et al. (1971), **2**: 583–4. Storey et al. (1988), 108–9.
2305   Williams et al. (1971), **2**: 584–5.
2309   Williams et al. (1971), **2**: 585–6.
2310   Williams et al. (1971), **2**: 586.
2311   Todhunter (1876), **2**: 363–4. Williams et al. (1971), **2**: 587–8.
2316   Williams et al. (1971), **2**: 588.
2317   Williams et al. (1971), **2**: 589.
2318   Bence Jones (1870b), **2**: 251. Thompson, S.P. (1898), 207.
2320   Williams et al. (1971), **2**: 590.
2322   Williams et al. (1971), **2**: 591.
2323   Bence Jones (1870a), **2**: 268–9, (1870b), **2**: 263–4.
2325   Williams et al. (1971), **2**: 591–2.
2327   Williams et al. (1971), **2**: 593–5.
2328   Williams et al. (1971), **2**: 592–3.

2330 Williams et al. (1971), **2**: 595–6.
2332 Williams et al. (1971), **2**: 597.
2342 Bence Jones (1870a), **2**: 257, (1870b), **2**: 252.
2343 Bence Jones (1870a), **2**: 263–5, (1870b), **2**: 258–60. Part in Thompson, S.P. (1898), 206. Kahlbaum and Darbishire (1899), 186–8.
2344 Bence Jones (1870a), **2**: 275–6, (1870b), **2**: 270–1. Williams et al. (1971), **2**: 597.
2346 Williams et al. (1971), **2**: 598.
2347 Pelseneer (1936), 440–50. Williams et al. (1971), **2**: 598–9.
2348 Kahlbaum and Darbishire (1899), 189.
2350 Feilitzsch (1851). Williams et al. (1971), **2**: 599–603.
2351 Williams et al. (1971), **2**: 604–5.
2353 Bence Jones (1870a), **2**: 266–7, (1870b), **2**: 261–2. Kahlbaum and Darbishire (1899), 189–91.
2355 Randell (1924), 130–1. Storey et al. (1988), 230.
2356 Bence Jones (1870a), **2**: 267–8, (1870b), **2**: 262–3. Kahlbaum and Darbishire (1899), 192.
2357 Williams et al. (1971), **2**: 605–6.
2362 Williams et al. (1971), **2**: 606–7.
2364 Bence Jones (1870a), **2**: 269–72, (1870b), **2**: 264–7. Williams et al. (1971), **2**: 608–9.
2367 Williams et al. (1971), **2**: 609–10.
2368 Williams et al. (1971), **2**: 576–7.
2373 Bence Jones (1870a), **2**: 292–4, (1870b), **2**: 287–9. Williams et al. (1971), **2**: 610–11.
2377 Williams et al. (1971), **2**: 611–12.
2378 Bence Jones (1870a), **2**: 257–61, (1870b), **2**: 252–6. Williams et al. (1971), **2**: 612–14.
2388 Bence Jones (1870a), **2**: 286–7, (1870b), **2**: 281–2. Kahlbaum and Darbishire (1899), 193–5.
2391 Williams et al. (1971), **2**: 614–15.
2396 Quetelet (1851a). Williams et al. (1971), **2**: 615–18.
2398 Williams et al. (1971), **2**: 619.
2406 Williams et al. (1971), **2**: 620.
2409 Williams et al. (1971), **2**: 620–2.
2411 Bence Jones (1870a), **2**: 294–6, (1870b), **2**: 289–91. Williams et al. (1971), **2**: 623.
2412 Pelseneer (1936), 451. Williams et al. (1971), **2**: 624.
2413 Bence Jones (1870a), **2**: 287–8, (1870b), **2**: 282–3. Kahlbaum and Darbishire (1899), 195–6.
2415 Williams et al. (1971), **2**: 624–9.
2416 Williams et al. (1971), **2**: 629–30. Reingold and Rothenberg (1972–99), **8**: 179–80.
2428 Williams et al. (1971), **2**: 630–2.

2429 Williams et al. (1971), **2**: 632.
2430 Williams et al. (1971), **2**: 632–4. Reingold and Rothenberg (1972–99), **8**: 188–90.
2432 Williams et al. (1971), **2**: 634–5.
2435 Williams et al. (1971), **2**: 635–6.
2437 Williams et al. (1971), **2**: 636–7. Bron and Condax (1998), 16.
2438 Bron and Condax (1998), 18–19.
2439 Williams et al. (1971), **2**: 637–8.
2441 Kahlbaum and Darbishire (1899), 196–7.
2444 Williams et al. (1971), **2**: 639–40.
2448 Williams et al. (1971), **2**: 640. Reingold and Rothenberg (1972–99), **8**: 203.
2450 Williams et al. (1971), **2**: 640–1.
2452 Bence Jones (1870a), **2**: 296–7, (1870b), **2**: 291–2. Williams et al. (1971), **2**: 641–2.
2453 Bence Jones (1870a), **2**: 288–9, (1870b), **2**: 283–4. Kahlbaum and Darbishire (1899), 198–9.
2457 Williams et al. (1971), **2**: 642.
2460 Williams et al. (1971), **2**: 643.
2463 Williams et al. (1971), **2**: 643.
2474 Williams et al. (1971), **2**: 880–1.
2478 Bence Jones (1870b), **2**: 284–7. Williams et al. (1971), **2**: 643–5.
2479 Williams et al. (1971), **2**: 645.
2480 Williams et al. (1971), **2**: 646.
2481 Williams et al. (1971), **2**: 646–7.
2482 Kahlbaum and Darbishire (1899), 199–200.
2487 Williams et al. (1971), **2**: 648.
2489 Williams et al. (1971), **2**: 648–9.
2490 Williams et al. (1971), **2**: 649–50.
2494 Williams et al. (1971), **2**: 650–1.
2496 Williams et al. (1971), **2**: 651–2.
2497 Williams et al. (1971), **2**: 652–3.
2498 Williams et al. (1971), **2**: 653.
2502 Williams et al. (1971), **2**: 653–4.
2506 Tee (1998), 96.
2512 Williams et al. (1971), **2**: 654.
2515 Williams et al. (1971), **2**: 655.
2525 Bence Jones (1870a), **2**: 311–13, (1870b), **2**: 306–8.
2526 Kahlbaum and Darbishire (1899), 200–2.
2531 Williams et al. (1971), **2**: 655–6.
2534 Kahlbaum and Darbishire (1899), 202–4.
2535 Williams et al. (1971), **2**: 656–7.
2536 Williams et al. (1971), **2**: 657.
2537 Williams et al. (1971), **2**: 658–9.

2538   Williams et al. (1971), **2**: 659–60.
2546   Thompson, S.P. (1910), **1**: 214–6. Williams et al. (1971), **2**: 555–7.
2554   Williams et al. (1971), **2**: 661–3.
2555   Williams et al. (1971), **2**: 663.
2556   Williams et al. (1971), **2**: 663–4.
2562   Kahlbaum and Darbishire (1899), 204.
2563   Williams et al. (1971), **2**: 664–5.
2571   Williams et al. (1971), **2**: 764–6.
2574   Williams et al. (1971), **2**: 665–7.
2576   Williams et al. (1971), **2**: 668.
2577   Bence Jones (1870a), **2**: 314–16, (1870b), **2**: 309–11. Williams et al. (1971), **2**: 668–9.
2578   Kahlbaum and Darbishire (1899), 205–7.
2586   Williams et al. (1971), **2**: 669–71.
2587   Williams et al. (1971), **2**: 671.
2591   Williams et al. (1971), **2**: 671–2.
2592   Williams et al. (1971), **2**: 672–3.
2595   Williams et al. (1971), **2**: 673–4.
2604   Bence Jones (1870a), **2**: 297–8, (1870b), **2**: 292–3. Kahlbaum and Darbishire (1899), 207–10.
2607   Kahlbaum and Darbishire (1899), 210–12.
2608   Williams et al. (1971), **2**: 674.
2610   Bence Jones (1870a), **2**: 316–17, (1870b), **2**: 311–21. Williams et al. (1971), **2**: 675–6.
2612   Part in Williams et al. (1971), **2**: 686.
2613   Williams et al. (1971), **2**: 676–7.
2616   Williams et al. (1971), **2**: 677–8.
2618   Williams et al. (1971), **2**: 678–9.
2628   Williams et al. (1971), **2**: 680–1.
2629   Williams et al. (1971), **2**: 681–2.
2631   Williams et al. (1971), **2**: 682–3.
2633   Williams et al. (1971), **2**: 660–1.
2634   Williams et al. (1971), **2**: 683–4.
2637   Williams et al. (1971), **2**: 685.
2646   *Proc.Roy.Inst.*, 1853, **1**: 275. *Phil.Mag.*, 1853, **5**: 465–6. Williams et al. (1971), **2**: 686–7.
2647   Bence Jones (1870a), **2**: 319–21, (1870b), **2**: 314–16.
2648   Williams et al. (1971), **2**: 688–9.
2651   Williams et al. (1971), **2**: 689–90.
2652   Bence Jones (1870b), **2**: 317.
2691   *Ill.Lond.News*, 2 July 1853, **22**: 530. Williams et al. (1971), **2**: 690–2.
2699   Kahlbaum and Darbishire (1899), 212–14.
2702   Williams et al. (1971), **2**: 692.
2703   Bence Jones (1870a), **2**: 317–19, (1870b), **2**: 312–14.

2705 Bence Jones (1870a), **2**: 307–8, (1870b), **2**: 302–3. Kahlbaum and Darbishire (1899), 214–16.
2710 Williams et al. (1971), **2**: 693–4.
2714 Part in Henry (1854), 132–3.
2715 Williams et al. (1971), **2**: 694–5.
2717 Williams et al. (1971), **2**: 751–3.
2722 Bence Jones (1870a), **2**: 301–2, (1870b), **2**: 296–7.
2723 Bence Jones (1870a), **2**: 303, (1870b), **2**: 298.
2728 Bence Jones (1870a), **2**: 303, (1870b), **2**: 298.
2735 Kahlbaum and Darbishire (1899), 216–19.
2736 Bence Jones (1870a), **2**: 304, (1870b), **2**: 299.
2743 Williams et al. (1971), **2**: 695.
2747 Williams et al. (1971), **2**: 696.
2749 Williams et al. (1971), **2**: 696–7.
2751 Williams et al. (1971), **2**: 697.
2759 Williams et al. (1971), **2**: 698.
2762 Williams et al. (1971), **2**: 698–9. Schettino (1994),457–60.
2764 Williams et al. (1971), **2**: 699–700.
2768 Williams et al. (1971), **2**: 700–1.
2772 Williams et al. (1971), **2**: 702–3.
2777 Williams et al. (1971), **2**: 705–6.
2780 Williams et al. (1971), **2**: 706–7.
2781 Kahlbaum and Darbishire (1899), 219–21.
2782 *Bibl.Univ.Arch.*, 1854, **25**: 169–70. Bence Jones (1870a), **2**: 322–4, (1870b), **2**: 317–19. Williams et al. (1971), **2**: 707–8.
2784 Williams et al. (1971), **2**: 708–9.
2786 Williams et al. (1971), **2**: 703–5.
2788 Williams et al. (1971), **2**: 710–13.
2789 Williams et al. (1971), **2**: 709–10.
2790 Kahlbaum and Darbishire (1899), 221–6.
2792 Williams et al. (1971), **2**: 713.
2793 Williams et al. (1971), **2**: 714–15.
2794 Williams et al. (1971), **2**: 715–16.
2797 Williams et al. (1971), **2**: 716–17.
2798 Williams et al. (1971), **2**: 717.
2799 Bence Jones (1870a), **2**: 327–9, (1870b), **2**: 322–4. Williams et al. (1971), **2**: 717–19.
2800 Faraday (1854c, d). Bence Jones (1870a), **2**: 330–5, (1870b), **2**: 325–30. Williams et al. (1971), **2**: 719–22.
2802 Bence Jones (1870a), **2**: 335–6, (1870b), **2**: 330–1. Williams et al. (1971), **2**: 722.
2803 Williams et al. (1971), **2**: 722–3.
2804 Williams et al. (1971), **2**: 723–24.

2805 Bence Jones (1870a), **2**: 336–9, (1870b), **2**: 331–4. Gladstone (1874), 111–13. Williams et al. (1971), **2**: 724–5.
2806 Williams et al. (1971), **2**: 726.
2807 Williams et al. (1971), **2**: 733–4.
2808 Williams et al. (1971), **2**: 726.
2813 Williams et al. (1971), **2**: 727–9. Schettino (1994),468–71.
2818 Kahlbaum and Darbishire (1899), 226–8.
2823 Williams et al. (1971), **2**: 729.
2824 Williams et al. (1971), **2**: 730.
2828 Kahlbaum and Darbishire (1899), 228–9.
2830 Williams et al. (1971), **2**: 732–3.
2832 Bence Jones (1870a), **2**: 339–41, (1870b), **2**: 334–6. Kahlbaum and Darbishire (1899), 229–32.
2834 Williams et al. (1971), **2**: 734–5. Schettino (1994), 472–4.
2837 Williams et al. (1971), **2**: 735–6.
2839 Bence Jones (1870a), **2**: 341–4, (1870b), **2**: 336–9.
2840 Bence Jones (1870a), **2**: 344–6, (1870b), **2**: 339–41. Williams et al. (1971), **2**: 737–8.
2846 Guareschi (1909), 40. Schettino (1994), 474.
2850 Williams et al. (1971), **2**: 738.
2860 Williams et al. (1971), **2**: 738–9.
2862 Williams et al. (1971), **2**: 739–43. Schettino (1994), 475–8.
2863 Williams et al. (1971), **2**: 743.
2864 Kahlbaum and Darbishire (1899), 232–6.
2865 Williams et al. (1971), **2**: 744–8. Schettino (1994), 486–92
2870 Schettino (1994), 500–6.
2871 Williams et al. (1971), **2**: 749–51.
2874 Williams et al. (1971), **2**: 753–4.
2880 Williams et al. (1971), **2**: 754–5.
2882 Williams et al. (1971), **2**: 755–6.
2884 Williams et al. (1971), **2**: 756–7.
2887 Williams et al. (1971), **2**: 757.
2889 Williams et al. (1971), **2**: 757–9.
2890 Williams et al. (1971), **2**: 760.
2893 Williams et al. (1971), **2**: 761–4.
2899 Bence Jones (1870a), **2**: 346–7, (1870b), **2**: 341–2. Kahlbaum and Darbishire (1899), 236–8.
2901 Williams et al. (1971), **2**: 766.
2902 Williams et al. (1971), **2**: 767.
2910 Williams et al. (1971), **2**: 767.
2914 Williams et al. (1971), **2**: 768–9.
2915 Williams et al. (1971), **2**: 769.
2921 Bence Jones (1870a), **2**: 347–8, (1870b), **2**: 342–3.
2924 Williams et al. (1971), **2**: 770.

2925 Williams et al. (1971), **2**: 770.
2936 Williams et al. (1971), **2**: 774–5.
2938 Williams et al. (1971), **2**: 775–6.
2939 Williams et al. (1971), **2**: 776–7.
2942 Williams et al. (1971), **2**: 777–8.
2943 Kahlbaum and Darbishire (1899), 238–42.
2944 Bence Jones (1870a), **2**: 355–6, (1870b), **2**: 350–1. Williams et al. (1971), **2**: 779.
2952 Williams et al. (1971), **2**: 782.
2953 Williams et al. (1971), **2**: 782.
2954 Williams et al. (1971), **2**: 783–9.
2964 Bence Jones (1870a), **2**: 360–2, (1870b), **2**: 355–7. Kahlbaum and Darbishire (1899), 242–4.
2965 Williams et al. (1971), **2**: 790–1.
2966 Williams et al. (1971), **2**: 791–2.
2969 Williams et al. (1971), **2**: 792–3.
2971 Bence Jones (1870a), **2**: 362, (1870b), **2**: 357.
2972 Bence Jones (1870a), **2**: 362, (1870b), **2**: 357.
2973 Bence Jones (1870a), **2**: 363, (1870b), **2**: 358.
2977 Williams et al. (1971), **2**: 794.
2978 Appleyard (1931), 133–4.
2979 Appleyard (1931), 134.
2983 Williams et al. (1971), **2**: 795.
2984 Williams et al. (1971), **2**: 795–7.
2985 Kahlbaum and Darbishire (1899), 245–51.
2986 Part in Williams et al. (1971), **2**: 797–8.
2992 Williams et al. (1971), **2**: 798.
2999 Williams et al. (1971), **2**: 799–800.
3002 Williams et al. (1971), **2**: 800.
3003 Bence Jones (1870a), **2**: 363–4, (1870b), **2**: 358–9. Williams et al. (1971), **2**: 801.
3006 Hartmann (1939), 381. Williams et al. (1971), **2**: 802.
3007 Williams et al. (1971), **2**: 749.
3012 Williams et al. (1971), **2**: 803.
3014 Bence Jones (1870b), **2**: 345–6.
3019 Bence Jones (1870b), **2**: 360.
3024 Williams et al. (1971), **2**: 817.
3027 Williams et al. (1971), **2**: 803–4.
3032 Reid, T.W. (1899), 165–6.

Bibliography

ADIE, Richard (1851): "On the Connection between the Colour and the Magnetic Properties of Bodies", *Edinb.New Phil.J.*, **51**: 44–8.

AIRY, George Biddell (1848–9): "Substance of the Lecture ... on the large Reflecting Telescopes of the Earl of Rosse and Mr. Lassell", *Month.Not.Roy.Ast.Soc.*, **9**: 110–21.

—— (1849–50): "On the Method of observing and recording Transits, lately introduced in America; and on some other connected subjects", *Month.Not.Roy.Ast.Soc.*, **10**: 26–34.

—— (1851): "On the Total Solar Eclipse of 1851, July 28", *Proc.Roy.Inst.*, **1**: 62–8.

—— (1853): "On the results of recent calculations on the Eclipse of Thales and Eclipses connected with it", *Proc.Roy.Inst.*, **1**: 243–50.

—— (1855a): *Lecture on the Pendulum-Experiments at Harton Pit, Delivered in the Central Hall, South Shields, October 24, 1854*, London.

—— (1855b): "On the Pendulum-experiments lately made in the Harton Colliery, for ascertaining the mean Density of the Earth", *Proc.Roy.Inst.*, **2**: 17–22.

—— (1856): "Account of Pendulum Experiments undertaken in the Harton Colliery, for the purpose of Determining the Mean Density of the Earth", *Phil.Trans.*, **146**: 297–342.

AIRY, Wilfrid (1896): *Autobiography of Sir George Biddell Airy*, Cambridge.

ALVEY, Norman (1990): *Education by Election: Reed's School, Clapton and Watford*, St Albans.

AMPÈRE, André-Marie (1823): "Mémoire sur la théorie mathématique des phénomènes électro-dynamiques uniquement déduite de l'expérience", *Mém.Acad.Sci.*, **6**: 175–388.

ANDREWS, Thomas (1852): "On a Method of Obtaining a Perfect Vacuum in the Receiver of an Air-pump", *Phil.Mag.*, **3**: 104–8.

ANON (1757): *The Plate-Glass-Book*, London.

—— (1835): "Shewing how the Tories and the Whigs extend their patronage to Science and Literature", *Fraser's Mag.*, **12**: 703–8.

—— (1851): *Great Exhibition of the Works of Industry of All Nations, 1851. Official Descriptive and Illustrated Catalogue*, 3 volumes, London.

—— (1852a): *Exhibition of the Works of Industry of All Nations, 1851. Reports by the Juries on the Subjects in the thirty classes into which the Exhibition was divided*, 2 volumes, London.

—— (1852b): *Lectures on the Results of the Great Exhibition of 1851 delivered before the Society of Arts, Manufactures, and Commerce*, London.

—— (1854a): *Lectures delivered before the Young Men's Christian Association, in Exeter Hall, from November 1853, to February 1854*, London.

—— (1854b): *The Illustrated Hand Book of the Royal Panopticon of Science and Art*, London.

—— (1855): *Lectures on Education delivered at the Royal Institution of Great Britain*, London.

—— (1940): *The Record of the Royal Society of London*, 4th edition, London.

APPLEYARD, Rollo (1931): *A Tribute to Michael Faraday*, London.

ARAGO, Dominique François Jean (1854): "Ampère" in *Oeuvres complètes*, Paris, **2**: 1–116.

ARNETT, John Andrews (1837): *An Inquiry into the Nature and Form of the Books of the Ancients; with a history of the art of bookbinding, from the times of the Greeks and Romans to the present day*, London.

ARNOTT, Neil (1855) *On the Smokeless Fire-Place, Chimney-Valves, and other means, old and new, of obtaining healthful warmth and ventilation*, London.

ARROW, Frederick (1868): *The Corporation of Trinity House of Deptford Strond: A Memoir of its Origin, History, & Functions*, London.

BABBAGE, Charles (1832): *On the Economy of Machinery and Manufactures*, 2nd edition, London.

—— (1851): *The Exposition of 1851; or, Views of the Industry, the Science and the Government, of England*, London.

—— (1864): *Passages from the Life of a Philosopher*, London.

BACHE, Alexander Dallas (1839): *Report on Education in Europe, to the Trustees of the Girard College for Orphans*, Philadelphia.

BADDELEY, Paul Frederick Henry (1850a): "On the Dust-storms of India", *Phil.Mag.*, **37**: 155–8.

—— (1850b): "On the Dust-storms of India", *J.Asiatic Soc.Bengal*, **19**: 390–4.

—— (1852): "On Dust Whirlwinds and Cyclones", *J.Asiatic Soc.Bengal*, **21**: 140–7, 264–9, 333–6.

[BAILLIE, E.C.C.] (1853): *The Protoplast. A series of papers*, 2 volumes, London.

BARLOW, Peter (1823): *An essay on magnetic attractions*, 2nd edition, London.

BAUMERT, Friedrich Moritz (1853): "Ueber eine neue Oxydationsstufe des Wasserstoffs und ihr Verhältniss zum Ozon", *Pogg.Ann.*, **89**: 38–55.

BECQUEREL, Alexandre-Edmond (1847): "Note sur la phosphorescence produite par insolation", *Comptes Rendus*, **25**: 632–3.

—— (1849a): "Recherches relatives à l'action du magnétisme sur tous les corps", *Comptes Rendus*, **28**: 623–7.

—— (1849b): "De l'image photochromatique du spectre solaire, et des images colorées obtenues à la chambre obscure", *Ann.Chim.*, **25**: 447–74.

—— (1850a): "De l'action du magnétisme sur tous les corps", *Ann.Chim.*, **28**: 283–350.

—— (1850b): "De l'action du magnétisme sur tous les corps (deuxième Mémoire)", *Comptes Rendus*, **31**: 198–201.

—— and FREMY, Edmond (1852): "Recherches electrochimiques sur les propriétés des corps électrisés", *Ann.Chim.*, **35**: 62–105.

BENCE JONES, Henry (1852): *On animal electricity: being an abstract of the discoveries of Emil Du Bois-Reymond*, London.

—— (1870a): *The Life and Letters of Faraday*, 2 volumes, London.

—— (1870b): *The Life and Letters of Faraday*, 2nd edition, 2 volumes, London.

BENTLEY, Jonathan (1970): "The Chemical Department of the Royal School of Mines. Its Origins and Development under A.W. Hofmann", *Ambix*, **17**: 153–81.

BERGEMANN, Carl Wilhelm (1851): "Beiträge zur Kenntniss eines neuen metallischen Körpers", *Pogg.Ann.*, **82**: 561–85.

BERON, Pierre (1850): "Les Causes du Magnétisme terrestre prouvées", *Proc.Roy.Soc.*, **5**: 978–9.

BERZELIUS, Jöns Jacob (1819): "Researches on a new Mineral Body found in the Sulphur extracted from Pyrites at Fahlun", *Ann.Phil.*, **14**: 97–106.

BIDLINGMAIER, Rolf (1989): "Leben und Wirken Christian Friedrich Schönbeins", pp.33–57 *Christian Friedrich Schönbein 150 Jahre Entdeckung des Ozons*, Metzingen.

BIOT, Jean-Baptiste (1821): "Sur l'Aimantation imprimée aux métaux par l'électricité en mouvement", *J.Sav.*, 221–35.

—— (1830): "Magnetism", *Edinburgh Encyclopaedia*, **13**: 246–78.

BOATO, Giovanni and MORO, Natalia (1994): "Bancalari's Role in Faraday's Discovery of Diamagnetism and the Successive Progress in the Understanding of Magnetic Properties of Matter", *Ann.Sci.*, **51**: 391–412.

BOIS-REYMOND, Emil Heinrich du (1848–9): *Untersuchungen über thierische Elektricität*, 2 volumes, Berlin.

—— (1850a): "Réponse à la réclamation de priorité de M. Matteucci", *Comptes Rendus*, **30**: 512–5.

—— (1850b): "Seconde réponse à la réclamation de priorité de M. Matteucci", *Comptes Rendus*, **30**: 563–6.

—— (1850c): "Troisième réponse à M. Matteucci", *Comptes Rendus*, **31**: 91–5.

BOLITHO, Hector and PEEL, Derek (1985): *The Drummonds of Charing Cross*, London.

BOS, H.J.M., RUDWICK, M.J.S., SNELDERS, H.A.M. and VISSER, R.P.W. (editors) (1980): *Studies on Christiaan Huygens: Invited Papers from the Symposium on the Life and Work of Christiaan Huygens, Amsterdam, 22–25 August 1979*, Lisse.
BOUCHERIE, Auguste (1840): "Mémoire sur la conservation des bois", *Ann.Chim.*, **74**: 113–57.
BOUTIGNY, Pierre Hippolyte (1849): "Quelques faits relatifs à l'état sphéroïdal des corps. Epreuve du feu. Homme incombustible", *Comptes Rendus*, **28**: 593–7.
BOWERS, Brian and SYMONS, Lenore (1991): *Curiosity Perfectly Satisfyed: Faraday's travels in Europe 1813–1815*, London.
BRAID, James [1853]: *Hypnotic Therapeutics, illustrated by cases. With an appendix on table-moving and spirit-rapping*, Edinburgh.
BRANDE, William Thomas (1848): *A Manual of Chemistry*, 6th edition, 2 volumes, London.
BREDA, Jacob Gisbert Samuel van and LOGEMAN, Wilhelm Martin (1854): "On the Conductibility of Liquids for Electricity", *Phil.Mag.*, **8**: 465–9.
BREWSTER, David (1842): "Electricity", *Encylopedia Britanica*, 7th edition, **8**: 565–663.
—— (1853): "On the Optical Phaenomena and Crystallization of Tourmaline, Titanium, and Quartz, within Mica, Amethyst, and Topaz", *Phil.Mag.*, **6**: 265–72.
—— (1855a): "On the Triple Spectrum", *Rep.Brit.Ass.*, 7–9.
—— (1855b): "Electricity", *Encylopedia Britanica*, 8th edition, **8**: 523–627.
BROCK, William H. (1997): *Justus von Liebig: The Chemical Gatekeeper*, Cambridge.
—— and MEADOWS, A.J. (1984): *The Lamp of Learning: Taylor & Francis and the Development of Science Publishing*, London.
BRON, Pierre and CONDAX, Philip, L. (1998): *The Photographic Flash: A concise illustrated history*, Allschwil.
BROUGHAM, Henry Peter (1850): "Experiments and Observations upon the Properties of Light", *Phil.Trans.*, **140**: 235–58.
BROWN, Kenneth L. (1976): *People of Salé: Tradition and change in a Moroccan city 1830–1930*, Manchester.
BROWNE, Horace Baker (1946), *Chapters of Whitby History, 1823–1946. The story of Whitby Literary and Philosophical Society and of Whitby Museum*, Hull.
BRUNEL, Isambard (1870): *The Life of Isambard Kingdom Brunel, Civil Engineer*, London.
BUFF, Heinrich (1853): "Ueber das electrolytische Gesetz" *Ann.Chem. Pharm.*, **85**: 1–15, **88**: 117–24.
BUNSEN, Frances (1868): *A Memoir of Baron Bunsen*, 2 volumes, London.

BURGESS, Geoffrey Harold Orchard (1967): *The Curious World of Frank Buckland*, London.
BURKHARDT, Frederick et al. (1985-94): *The Correspondence of Charles Darwin*, 9 volumes, Cambridge.
BUTTMANN, Günther (1970): *The Shadow of the Telescope: A Biography of John Herschel*, New York.
CANTON, John (1768): "An easy Method of making a Phosphorous that will imbibe and emit Light, like the Bolognian Stone; with Experiments and Observations", *Phil.Trans.*, **58**: 337–44.
CANTOR, Geoffrey (1989): "Why was Faraday excluded from the Sandemanians in 1844?", *Brit.J.Hist.Sci.*, **22**: 433–7.
—— (1991a): *Michael Faraday: Sandemanian and Scientist. A Study of Science and Religion in the Nineteenth Century*, London.
—— (1991b): "Educating the Judgment: Faraday as a Lecturer", *Bull. Hist.Chem.*, **11**: 28–36.
CAPEILLE, J. (1910): *Dictionnaire de Biographies Roussillonnaises*, Perpignan.
CARLYLE, Thomas (1841): *On Heroes, Hero-Worship, & the Heroic in History*, London.
CARPENTER, William Benjamin (1852): "On the Influence of Suggestion in Modifying and directing Muscular Movement, independently of Volition", *Proc.Roy.Inst.*, **1**: 147–53.
CAWOOD, John (1977): "Terrestrial Magnetism and the Development of International Collaboration in the Early Nineteenth Century", *Ann.Sci.*, **34**: 551–87.
CAWOOD, John (1979): "The Magnetic Crusade: Science and Politics in Early Victorian Britain", *ISIS*, **70**: 493–518.
CHAPLIN, William Robert [1950]: *The Corporation of Trinity House of Deptford Stroud from the year 1660*, London.
CHAPMAN, Allan (1993): "The Pit and the Pendulum: G.B. Airy and the Determination of Gravity", *Antiquarian Horology*, **21**: 70–8.
—— (1998): "Standard Time for All: The Electric Telegraph, Airy, and the Greenwich Time Service" in James (1998), 40–59.
CHATIN, Gaspard-Adolphe (1851): "Présence de l'iode dans l'air", *L'Institut*, **19**: 145–6.
CHEVREUL, Michel Eugène (1833): "Lettre a M. Ampère sur une classe particulière de mouvemens musculaires", *Rev.Deux Mondes*, **2**: 249–57.
—— (1839): *De la loi du contraste simultanè des couleurs*, Paris.
—— (1854a): *The Principles of Harmony and Contrast of Colours, and their applications to the arts*, London.
—— (1854b): *De la baguette divinatoire, du pendule dit explorateur et des tables tournantes, au point de vue de l'histoire, de la critique et de la méthode expérimentale*, Paris.

—— (1854c): *Rapport sur les tapisseries et les tapis des manufactures nationales, fait a la commission Française du jury internationale de l'exposition universelle de Londres*, Paris.
—— (1854d): "Recherches chimiques sur la teinture", *Mém.Acad.Sci.*, **24**: 407–547.
—— (1855): *Cercles chromatiques*, Paris.
CHOISY, Albert (1947): *Généalogies Genevoises*, Geneva.
CHRISTIE, Samuel Hunter (1826): "On magnetic influence in the solar rays", *Phil.Trans.*, **116**: 219–39.
CLARK, E. Kitson (1924): *The History of 100 Years of Life of the Leeds Philosophical and Literary Society*, Leeds.
CLAUSIUS, Rudolf (1853): "On the Blue Colour of the Sky and the Morning and Evening Red", *Taylor Sci.Mem.*, 326–31.
CLIFTON, Gloria (1995): *Directory of British Scientific Instrument Makers 1550–1851*, London.
COLERIDGE, Samuel Taylor (1835): *Specimens of the Table Talk of the late Samuel Taylor Coleridge*, London.
CONOLLY, John (1854): "On the Characters of Insanity", *Proc.Roy.Inst.*, **1**: 375–81.
[COOKE, George Wingrove] (1853): "Modern Miracles. Spirit Rapping and Table Turning", *New Quart.Rev.*, **2**: 297–316.
CONNOR, R.D. (1987): *The Weights and Measures of England*, London.
COULOMB, Charles Augustin de (1787): "Cinquième Mémoire sur l'Electricité et le Magnetisme", *Mém.Acad.Sci.*, 421–67.
—— (1789): "Septième Mémoire sur l'Electricité et le Magnetisme", *Mém.Acad.Sci.*, 455–505.
COWELL, Frank Richard (1975): *The Athenaeum: Club and Social Life in London 1824–1974*, London.
CRAWFORD, D.G. (1930): *Roll of the Indian Medical Service 1615–1930*, London.
CRAWFORD, Elspeth (1985): "Learning from Experience", in Gooding and James (1985), 211–27.
CROOK, J. Mordaunt and PORT, M.H. (1973): *The History of the King's Works Volume VI 1782–1851*, London.
CROOKES, William (1871): *Psychic Force and Modern Spiritualism: A Reply to the "Quarterly Review" and Other Critics*, London.
CROSLAND, Maurice (1992): *Science under Control: The French Academy of Sciences 1795–1914*, Cambridge.
CUMMING, John (1854): "The Signs of the Times" in Anon (1854a), 157–96.
CURWEN, E. Cecil (1940): *The Journal of Gideon Mantell Surgeon and Geologist*, London.
DANIELL, John Frederic (1839): *An Introduction to the Study of Chemical Philosophy*, London.

—— (1843): *An Introduction to the Study of Chemical Philosophy*, 2nd edition, London.
DAUBENY, Charles Giles Bridle (1853): *Can Physical Science obtain a home in an English University? An Inquiry suggested by some remarks contained in a late number of the Quarterly Review*, Oxford.
—— Charles Giles Bridle (1854): *On the Importance of the Study of Chemistry as a branch of Education for all Classes*, London.
DAVY, Humphry (1812): *Elements of Chemical Philosophy*, London.
DE LA RIVE, Arthur-August (1839): "Note sur la seconde coloration du Mont-Blanc", *Bibl.Univ.*, **23**: 383–91.
—— (1853): "François Arago", *Bibl.Univ.*, **24**: 264–76.
—— (1853–8): *A Treatise on Electricity, in Theory and Practice*, 3 volumes, London.
—— (1854–8): *Traité d'Electricité théorique et appliquée*, 3 volumes, Paris.
DEVILLE, Henri Etienne Sainte-Claire (1854a): "De l'aluminium et de ses combinaisons chimiques", *Comptes Rendus.*, **38**: 279–81.
—— (1854b): "Note sur deux procédés de préparation de l'aluminium et sur une nouvelle forme du silicium", *Comptes Rendus.*, **39**: 321–6.
DICKENS, Charles John Huffam (1850): *The Personal History of David Copperfield*, London.
DIXON, Robert Vickers (1849): *A Treatise on Heat*, Dublin.
DOUGLAS, George and RAMSAY, George Dalhousie (1908): *The Panmure Papers*, 2 volumes, London.
DRAPER, John William (1844): "On Tithonized Chlorine", *Phil.Mag.*, **25**: 1–10.
—— (1845): "Account of a remarkable difference between the Rays of Incandescent Lime and those emitted by an Electric Spark", *Phil.Mag.*, **27**: 435–7.
DRUMMOND, Thomas (1830): "On the Illumination of Light-houses", *Phil.Trans.*, **120**: 383–98.
DUNS, John (1883): "William Stevenson", *Hist.Berwicks.Natural.Club.*, **10**: 289–99.
DUVEEN, Denis (1949): "Michael Faraday on Honors", *J.Chem.Ed.*, **26**: 441–2.
EBEL, Otto (1913): *Women Composers: A Biographical Handbook of Women's Work in Music*, 3rd edition, New York.
EVANS, David S., DEEMING, Terence J., EVANS, Betty Hall and GOLDFARB, Stephen (1969): *Herschel at the Cape: Diaries and Correspondence of Sir John Herschel, 1834–1838*, Austin.
EVE, A.S. and CREASEY, C.H. (1945): *Life and Work of John Tyndall*, London.
FAHIE, J.J. (1902): "Staite and Petrie's Electric Light - 1846–1853", *Elec.Eng.*, **30**: 297–301, 337–40, 374–6.

FALLON, John P. (1992): *Marks of London Goldsmiths and Silversmiths 1837–1914*, London.
FARADAY, Michael (1831): "On Mr. Trevellyan's recent Experiments on the Production of Sound during the Conduction of Heat", *J.Roy.Inst.*, **2**: 119–22.
—— (1832a): "Experimental Researches in Electricity. On the Induction of Electric Currents. On the Evolution of Electricity from Magnetism. On a new Electrical Condition of Matter. On Arago's Magnetic Phenomena", *Phil.Trans.*, **122**: 125–62.
—— (1832b): "The Bakerian Lecture. Experimental Researches in Electricity. - Second Series. Terrestrial Magneto-electric Induction. Force and Direction of Magneto-electric Induction generally", *Phil.Trans.*, **122**: 163–94.
—— (1833a): "Experimental Researches in Electricity. - Third Series. Identity of Electricities derived from different sources. Relation by measure of common and voltaic Electricity", *Phil.Trans.*, **123**: 23–54.
—— (1833b): "Experimental Researches in Electricity. - Fourth Series. On a new Law of Electric Conduction. On Conducting Power Generally", *Phil.Trans.*, **123**: 507–22.
—— (1834): "Experimental Researches in Electricity. - Eighth Series. On the Electricity of the Voltaic Pile; its source, quantity, intensity, and general characters", *Phil.Trans.*, **124**: 425–70.
—— (1835): "Experimental Researches in Electricity. - Ninth Series. On the influence by induction of an Electric Current on itself:- and on the inductive action of Electric Currents generally", *Phil.Trans.*, **125**: 41–56.
—— (1838a): "Experimental Researches in Electricity. - Eleventh Series. On Induction", *Phil.Trans.*, **128**: 1–40.
—— (1838b): "Experimental Researches in Electricity. - Twelfth Series. On Induction (continued)", *Phil.Trans.*, **128**: 83–123.
—— (1838c): "Experimental Researches in Electricity. - Thirteenth Series. On Induction (continued). Nature of the electric current", *Phil.Trans.*, **128**: 125–68.
—— (1838d): "Experimental Researches in Electricity. - Fourteenth Series. Nature of the electric force or forces. Relation of the electric and magnetic forces. Note on electric excitation", *Phil.Trans.*, **128**: 265–82.
—— (1839a): "Experimental Researches in Electricity. - Fifteenth Series. Notice of the character and direction of the electric force of the Gymnotus", *Phil.Trans.*, **129**: 1–12.
—— (1839b): *Experimental Researches in Electricity*, London.
—— (1843): "On Static Electrical Inductive Action", *Phil.Mag.*, **22**: 200–4.
—— (1844a): "A speculation touching Electric Conduction and the Nature of Matter", *Phil.Mag.*, **24**: 136–44.
—— (1844b): *Experimental Researches in Electricity*, volume 2, London.

—— (1846a): "Experimental Researches in Electricity. - Nineteenth Series. On the magnetization of light and the illumination of magnetic lines of force", *Phil.Trans.*, **136**: 1–20.

—— (1846b): "Experimental Researches in Electricity. - Twentieth Series. On new magnetic actions, and on the magnetic condition of all matter", *Phil.Trans.*, **136**: 21–40.

—— (1846c): "Experimental Researches in Electricity. - Twenty-first Series. On new magnetic actions, and on the magnetic condition of all matter - continued", *Phil.Trans.*, **136**: 41–62.

—— (1846d): "Mémoires sur de nouvelles actions magnétiques et sur l'état magnétique de toute la matière", *Bibl.Univ.Arch.*, **2**: 42–55, 145–64.

—— (1847a): "Ein und zwanzigste Reihe von Experimental-Untersuchungen über Elektricität", *Pogg.Ann.*, **70**: 24–59.

—— (1847b): "On the Diamagnetic conditions of Flame and Gases", *Phil.Mag.*, **31**: 401–21.

—— (1848a): "On the Use of Gutta Percha in Electrical Insulation", *Phil.Mag.*, **32**: 165–7.

—— (1848b): "Ueber die diamagnetischen Eigenschaften der Flamme und der Gase", *Pogg.Ann.*, **73**: 256–86.

—— (1849a): "Experimental Researches in Electricity. - Twenty-second Series. On the crystalline polarity of bismuth and other bodies, and on its relation to the magnetic form of force", *Phil.Trans.*, **139**: 1–18.

—— (1849b): "Experimental Researches in Electricity. - Twenty-second Series (continued). On the crystalline polarity of bismuth and other bodies, and on its relation to the magnetic and electric form of force (continued)", *Phil.Trans.*, **139**: 19–41.

—— (1850): "Experimental Researches in Electricity. - Twenty-third Series. On the polar or other condition of matter", *Phil.Trans.*, **140**: 171–88.

—— (1851a): "On the Magnetic Characters and Relations of Oxygen and Nitrogen", *Proc.Roy.Inst.*, **1**: 1–3.

—— (1851b): "Experimental Researches in Electricity. - Twenty-fourth Series. On the possible relation of Gravity to Electricity", *Phil.Trans.*, **141**: 1–6.

—— (1851c): "Experimental Researches in Electricity. - Twenty-fifth Series. On the magnetic and diamagnetic condition of bodies", *Phil.Trans.*, **141**: 7–28.

—— (1851d): "Experimental Researches in Electricity. - Twenty-sixth Series. Magnetic conducting power. Atmospheric magnetism", *Phil.Trans.*, **141**: 29–84.

—— (1851e): "Experimental Researches in Electricity. - Twenty-seventh Series. On Atmospheric magnetism - continued", *Phil.Trans.*, **141**: 85–122.

—— (1851f): "On Atmospheric Magnetism", *Proc.Roy.Inst.*, **1**: 56–60.

—— (1851g): "On Schönbein's Ozone", *Proc.Roy.Inst.*, **1**: 94–7.

—— (1852a): "On the Lines of Magnetic Force", *Proc.Roy.Inst.*, **1**: 105–8.
—— (1852b): "Experimental Researches in Electricity. - Twenty-eighth Series. On Lines of Magnetic Force; their definite character; and their distribution within a Magnet and through Space", *Phil.Trans.*, **142**: 25–56.
—— (1852c): "Experimental Researches in Electricity. - Twenty-ninth Series. On the employment of the Induced Magneto-electric Current as a test and measure of Magnetic Forces", *Phil.Trans.*, **142**: 137–59.
—— (1852d): "On the Physical Character of the Lines of Magnetic Force", *Phil.Mag.*, **3**: 401–28.
—— (1852e): "On the Physical Lines of Magnetic Force", *Proc.Roy.Inst.*, **1**: 216–20.
—— (1853a): "Observations on the Magnetic Force", *Proc.Roy.Inst.*, **1**: 229–38.
—— (1853b): *The Subject Matter of a Course of Six Lectures on the Non-Metallic Elements*, London.
—— (1853c): "MM. Boussingault, Frémy, Becquerel, &c. on Oxygen", *Proc.Roy.Inst.*, **1**: 337–9.
—— (1854a): "On Electric Induction - Associated cases of current and static effects", *Proc.Roy.Inst.*, **1**: 345–55.
—— (1854b): "De l'induction électrique et de l'association des états statique et dynamique de l'électricité", *Bibl.Univ.Arch.*, **25**: 209–28.
—— (1854c): "Sur le développement des courants induits dans les liquides", *Bibl.Univ.Arch.*, **25**: 267–74.
—— (1854d): "On Electro-dynamic Induction in Liquids", *Phil.Mag.*, **7**: 265–8.
—— (1854e): "On Magnetic Hypotheses", *Proc.Roy.Inst.*, **1**: 457–9.
—— (1854f): *Observations on Mental Education*, London.
—— (1854g): "On Subterraneous Electro-telegraph Wires", *Phil.Mag.*, **7**: 396–8.
—— (1855a): "On some points of Magnetic Philosophy", *Proc.Roy.Inst.*, **2**: 6–13.
—— (1855b): "On some Points of Magnetic Philosophy", *Phil.Mag.*, **9**: 81–113.
—— (1855c): "Further Observations on associated cases, in Electric Induction, of Current and Static Effects", *Phil.Mag.*, **9**: 161–5.
—— (1855d): *Experimental Researches in Electricity*, volume 3, London.
—— (1855e): "On Electric Conduction", *Proc.Roy.Inst.*, **2**: 123–32.
—— (1855f): "On Ruhmkorff's Induction Apparatus", *Proc.Roy.Inst.*, **2**: 139–42.
—— (1856): "Experimental Researches in Electricity. - Thirtieth Series. Constancy of differential magnecrystallic force in different media. Action of heat on magnecrystals. Effect of heat upon the absolute magnetic force of bodies", *Phil.Trans.* **146**: 159–80.

—— (1861): *A Course of Six Lectures on the Chemical History of a Candle; to which is added a Lecture on Platinum*, London.
FARRAR, W.V., FARRAR, Kathleen R. and SCOTT, E.L. (1977): "The Henrys of Manchester. Part 6. William Charles Henry: The Magnesia Factory", *Ambix*, **24**: 1–26.
FAVRE, Pierre Antoine (1853): *Recherches thermochimiques sur les composés formés en proportions multiples [et] Recherches thermiques sur les courants hydro-électriques*, Paris.
FEILITZSCH, Fabian Carl Ottokar von (1851): "On the Physical Distinction of Magnetic and Diamagnetic Bodies", *Phil.Mag.*, **1**: 46–51.
—— (1854): "Erklärung der diamagnetischen Wirkungsweise durch die Ampère'sche Theorie", *Pogg.Ann.*, **92**: 366–401, 536–76.
FISCH, Menachem and SCHAFFER, Simon (editors) (1991): *William Whewell: A Composite Portrait*, Oxford.
FOLKARD, Henry Coleman (1901): *The Sailing Boat*, 5th edition, London.
FORBES, Edward (1849): "Figures and Descriptions illustrative of British organic remains", *Mem.Geol.Survey*, Decade 1.
FORBES, James David (1833): "Experimental Researches regarding certain Vibrations which take place between Metallic Masses having different Temperatures", *Trans.Roy.Soc.Edinb.*, **12**: 429–61.
—— (1834): "Experimental Researches regarding certain Vibrations which take place between Metallic Masses having different Temperatures", *Phil.Mag.*, **4**: 15–28, 182–94.
FORGAN, Sophie (1977): *The Royal Institution of Great Britain, 1840–1873*, University of London (Westfield College) PhD thesis.
FOUCAULT, Jean Bernard Léon (1851): "Démonstration physique du mouvement de rotation de la terre au moyen du pendule", *Comptes Rendus*, **32**: 135–8.
FOWNES, George (1850): *A Manual of Elementary Chemistry, Theoretical and Practical*, 3rd edition, London.
FOX, Francis (1904): *River, Road, and Rail: Some Engineering Reminiscences*, London.
FOX, Robert Were (1830): "On the electro-magnetic properties of metalliferous veins in the mines of Cornwall", *Phil.Trans.*, **120**: 399–414.
FRANKLIN, Benjamin (1751–4): *Experiments and observations on electricity, made at Philadelphia in America*, London.
FRESNEL, Augustin Jean (1827): "Mémoire sur la Double Réfraction", *Mém.Acad.Sci.*, **7**: 45–176.
FULTON, John F. and THOMSON, Elizabeth H. (1968): *Benjamin Silliman 1779–1864 Pathfinder in American Science*, New York.
GASSIOT, John Peter (1844): "A description of an extensive series of the Water Battery; with an Account of some Experiments made in order to test the relation of the Electrical and the Chemical Actions which take

place before and after completion of the Voltaic Circuit", *Phil.Trans.*, **134**: 39–52.

—— (1854): "On some Experiments made with Ruhmkorff's Induction Coil", *Phil.Mag.*, **7**: 97–9.

GAUSS, Carl Friedrich (1841): "General Theory of Terrestrial Magnetism", *Taylor Sci.Mem.*, **2**: 184–251.

GAUTIER, Alfred (1852): "Notice sur quelques recherches récentes, astronomiques et physiques, relatives aux apparences que présente le corps du soleil", *Bibl.Univ.Arch.*, **20**: 177–207, 265–82.

GAVIN, Charles Murray (1932): *Royal Yachts*, London.

GERNSHEIM, Helmut and GERNSHEIM, Alison (1955): *The History of Photography from the earliest use of the camera obscura in the eleventh century up to 1914*, Oxford.

GILBERT, William (1600): *De Magnete, magneticisque corporibus, et de magno magnete tellure; Physiologia nova, plurimis & argumentis, & experimentis demonstrata*, London.

GLADSTONE, John Hall (1874): *Michael Faraday*, 3rd edition, London.

GLAESER, Ernest (1878): *Biographie Nationale des Contemporains*, Paris.

GMELIN, Leopold (1848–72): *Hand-book of Chemistry*, 19 volumes, London.

GOODING, David (1978): "Conceptual and experimental bases of Faraday's denial of electrostatic action at a distance", *Stud.Hist.Phil.Sci.*, **9**: 117–49.

—— (1982): "A Convergence of Opinion on the Divergence of Lines: Faraday and Thomson's Discussion of Diamagnetism", *Notes Rec.Roy.Soc.Lond.*, **36**: 243–59.

—— (1985): "'In Nature's School': Faraday as an Experimentalist" in Gooding and James (1985), 105–35.

—— (1990): *Experiment and the Making of Meaning: Human Agency in Scientific Observation and Experiment*, Dordrecht.

—— (1991): "Michael Faraday's Apprenticeship: Science as a Spiritual Path", in Ravindra (1991), 389–405.

—— and JAMES, Frank A.J.L. (editors) (1985): *Faraday Rediscovered: Essays on the Life and Work of Michael Faraday, 1791–1867*, London.

—— PINCH, Trevor and SCHAFFER, Simon (editors) (1989), *The uses of experiment: Studies in the natural sciences*, Cambridge.

GRAHAM, Thomas (1842): *Elements of Chemistry; including the applications of the science in the arts*, London.

—— (1849): "On the Motion of Gases. Part II", *Phil.Trans.*, **139**: 349–91.

GREENAWAY, Frank, BERMAN, Morris, FORGAN, Sophie and CHILTON, Donovan (editors) (1971–6): *Archives of the Royal Institution, Minutes of the Managers' Meetings, 1799–1903*, 15 volumes, bound in 7, London.

GREG, Emily, (1905): *Reynolds-Rathbone Diaries and Letters 1753–1839*, [Edinburgh].
GREGORY, William (1845): *Outlines of Chemistry, for the use of students*, 2 volumes, London.
GRIMWADE, Arthur G. (1982): *London Goldsmiths, 1697–1837: Their Marks and Lives*, 2nd edition, London.
GROVE, William Robert (1849): "On the Effect of surrounding Media on Voltaic Ignition", *Phil.Trans.*, **139**: 49–59.
—— (1852a): "On the Electro-Chemical Polarity of Gases", *Phil.Trans.*, **142**: 87–101.
—— (1852b): "On the Electro-chemical Polarity of Gases", *Phil.Mag.*, 4: 498–515.
—— (1854): "On the Transmission of Electricity by Flame and Gases", *Proc.Roy.Inst.*, **1**: 359–62.
GUARESCHI, Icilio (1909): "Nuove notizie storiche sulla vita e sulle opere di Macedonio Melloni", *Mem.Accad.Sci.Torino*, **59**: 1–59.
GULL, William Withey (1851): "On some points in the Physiology of Voluntary Movement", *Proc.Roy.Inst.*, **1**: 37–41.
GURNEY, Goldsworthy, (1823): "Oxy-hydrogen Blow-pipe", *Trans.Soc.Arts*, **41**: 70–7.
HACKMANN, Willem D. (1978): "Eighteenth Century Electrostatic Measuring Devices", *Ann.Ist.Mus.Stor.Sci*, 3^2: 3–58.
HAECKER, Paul Wolfgang (1842): "Versuche über das Tragvermögen hufeisenfömiger Magnete und über die Schwingungsdauer geradliniger Magnetstäbe", *Pogg.Ann.*, **57**: 321–45.
HAIGHT, Gordon S. (1954): *The George Eliot Letters. Volume 1, 1836–1851*, New Haven.
HALDAT DU LYS, Charles-Nicolas-Alexandre (1845): "Recherches sur l'universalité de la force magnétique", *Mém.Soc.Roy.Sci.Nancy*, 155–84.
HARDING, Marius Christian (1920): *Correspondance de H.C. Orsted avec Divers Savants*, 2 volumes, Copenhagen.
HARRIS, William Snow (1831): "On the Transient Magnetic State of which various Substances are susceptible", *Phil.Trans.*, **121**: 67–90.
—— (1834a): "On some Elementary Laws of Electricity", *Phil.Trans.*, **124**: 213–45.
—— (1834b): "On the Investigation of Magnetic Intensity by the Oscillations of the Horizontal Needle", *Trans.Roy.Soc.Edinb.*, **13**: 1–24.
—— (1851): *Rudimentary Electricity: being a concise exposition of the general principles of electrical science, and the purposes to which it has been applied*, 2nd edition, London.
HARTMANN, Ludwig (1939): "Michael Faraday und Justus Liebig: Ein unbekannter Briefwechsel", *Sudhoffs Archiv Gesch.Med.Naturwiss.*, **32**: 371–98.

HAUKSBEE, Francis (1712): "An Account of Experiments concerning the Proportion of the Power of the Load-stone at different Distances", *Phil.Trans.*, **27**: 506–11.

HEARNSHAW, F.J.C. (1929): *The Centenary History of King's College London 1828–1928*, London.

HEBER, Reginald (1827): *Hymns written and adapted to the Weekly Church Service of the Year*, London.

HEDLEY, William Percy and HUDLESTON, Chistophe Roy [1964]: *Cookson of Penrith, Cumberland and Newcastle upon Tyne*, [Kendal].

HELMHOLTZ, Hermann (1850): "Ueber die Fortpflanzungsgeschwindigkeit der Nervenreizung", *Pogg.Ann.*, **79**: 329–30.

—— (1852): "Ueber die Theorie der zusammengesetzten Farben", *Pogg.Ann.*, **87**: 45–66.

HENRY, William Charles (1854): *Memoirs of the Life and Scientific Researches of John Dalton*, London.

HERAPATH, William Bird (1853): "On the Manufacture of large available Crystals of Sulphate of Iodo-quinine (Herapathite) for Optical Purposes as Artificial Tourmalines", *Phil.Mag.*, **6**: 346–51.

HERSCHEL, Frederick William (1800): "On the Power of penetrating into Space by Telescopes; with a comparative Determination of the Extent of that Power in natural Vision, and in Telescopes of various Sizes and Constructions; illustrated by select Observations", *Phil.Trans.*, **90**: 49–85.

—— (1801): "Observations tending to investigate the Nature of the Sun, in order to find the Causes or Symptoms of its variable Emission of Light and Heat; with Remarks on the Use that may possibly be drawn from Solar Observations", *Phil.Trans.*, **91**: 265–318.

HERSCHEL, John Frederick William (1845a): " Ἀμόρφωτα, No. I. On a Case of Superficial Colour presented by a homogeneous liquid internally colourless", *Phil.Trans.*, **135**: 143–5.

—— (1845b): " Ἀμόρφωτα, No. II. On the Epipolic Dispersion of Light, being a Supplement to a paper entitled "On a Case of Superficial Colour presented by a homogeneous liquid internally colourless"", *Phil.Trans.*, **135**: 147–53.

—— (1849): *Outlines of Astronomy*, 2nd edition, London.

HERTZ, Emanuel (1931): *Abraham Lincoln: A New Portrait*, 2 volumes, New York.

HEYDON, Peter N. and KELLEY, Philip (1974): *Elizabeth Barrett Browning's Letters to Mrs. David Ogilvy 1849–1861*, London.

[HICKSON, William Edward] (1851): "Electro-Biology", *Westm.Rev.*, **55**: 312–28.

HOFMANN, August Wilhelm (1875): "The Faraday Lecture. The Life-work of Liebig in Experimental and Philosophic Chemistry; with Allusions to his Influence on the Development of the Collateral Sciences, and of the Useful Arts", *J.Chem.Soc.*, **28**: 1065–140.

HODGSON, William Ballantyne (1854): *On the Importance of the Study of Economic Science as a branch of Education for all Classes*, London.
HOLLAND, Henry (1852): *Chapters on Mental Physiology*, London.
HOOKER, Joseph Dalton (1844–7): *Flora Antarctica*, 2 volumes, London.
—— (1853–5): *Flora Novae-Zelandiae*, 2 volumes, London.
—— (1855–60): *Flora Tasmaniae*, 2 volumes, London.
—— and THOMSON, Thomas (1855): *Flora Indica: being a systematic account of the plants of British India, together with observations on the structure and affinities of their natural orders and genera*, London.
HOSKING, William (1851): "On Ventilation by the Parlour Fire", *Proc. Roy.Inst.*, **1**: 76–83.
HOUZEAU, Jean Auguste (1855): "Recherches sur l'oxygène à l'état naissant", *Comptes Rendus*, **40**: 947–50.
HUDSON, Derek (1949): *Martin Tupper: His Rise and Fall*, London.
HUMBOLDT, Friedrich Wilhelm Heinrich Alexander von (1846–58): *Cosmos: Sketch of a Physical Description of the Universe*, 4 volumes, London.
HUNT, Bruce J. (1991): "Michael Faraday, Cable Telegraphy and the Rise of Field Theory", *Hist.Tech.*, **13**: 1–19.
HUTTON, Charles (1795): *A Mathematical and Philosophical Dictionary*, 2 volumes, London.
HUXLEY, Leonard (1918): *Life and Letters of Sir Joseph Dalton Hooker*, 2 Volumes, London.
JACOBI, Moritz Hermann and LENZ, Heinrich Friedrich Emil (1839): "Ueber die Anziehung der Elektromagnete", *Pogg.Ann.*, **47**: 401–18.
JAMES, Frank A.J.L. (1983a): "The Conservation of Energy, Theories of Absorption and Resonating Molecules, 1851–1854: G.G. Stokes, A.J. Ångström and W. Thomson", *Notes Rec.Roy.Soc.Lond.*, **38**: 79–107.
—— (1983b): "The Study of Spark Spectra 1835–1859", *Ambix*, **30**: 137–62.
—— (1985): "'The Optical Mode of Investigation': Light and Matter in Faraday's Natural Philosophy" in Gooding and James (1985), 137–61.
—— (1991): "The Military Context of Chemistry: The Case of Michael Faraday", *Bull.Hist.Chem.*, **11**: 36–40.
—— (1992a): "Davy in the Dockyard: Humphry Davy, the Royal Society and the Electro-chemical Protection of the Copper Sheeting of His Majesty's Ships in the mid 1820s", *Physis*, **29**: 205–25.
—— (1992b): "Michael Faraday, The City Philosophical Society and the Society of Arts", *Roy.Soc.Arts J.*, **140**: 192–199.
—— (1997): "Faraday in the pits, Faraday at sea: the role of the Royal Institution in changing the practice of science and technology in nineteenth-century Britain", *Proc.Roy.Inst.*, **68**: 277–301.
—— (editor) (1998): *Semaphores to Short Waves: Proceedings of a Conference on the Technology and Impact of Early Telecommunications held at the Royal Society for the encouragement of Arts, Manufactures and Commerce on*

Monday 29 July 1996, organised by The British Society for the History of Science, The Newcomen Society and the RSA, London.

JAMIN, Jules (1848) "Mémoire sur la couleur métaux", *Ann.Chim.*, **22**: 311–27.

JOULE, James Prescott (1847): "On the Effects of Magnetism upon the Dimensions of Iron and Steel Bars", *Phil.Mag.*, **30**: 76–87, 225–41.

—— (1850): "On the Mechanical Equivalent of Heat", *Phil.Trans.*, **140**: 61–82.

—— (1856): "Introductory Research on the Induction of Magnetism by Electrical Currents", *Phil.Trans.*, **146**: 287–95.

[JOYCE, Arthur J.] (1849): "The Progress of Mechanical Invention", *Edinb.Rev.*, **89**: 47–83.

JUNGNICKEL, Christa and McCORMMACH, Russell (1986): *Intellectual Mastery of Nature. Theoretical Physics from Ohm to Einstein. Volume 1. The Torch of Mathematics 1800–1870*, Chicago.

KAHLBAUM, Georg Wilhelm August and DARBISHIRE, Francis V. (1899): *The Letters of Faraday and Schoenbein, 1836–1862*, Basle and London.

—— and THON, Eduard (1900) *Justus von Liebig und Christian Friedrich Schönbein. Briefwechsel, 1853–1868*, Leipzig.

KATER, Henry (1818): "On the length of the French Mètre estimated in parts of the English standard", **108**: 103–9.

KENYON, Frederic G. (1897): *The Letters of Elizabeth Barrett Browning*, 2 volumes, London.

KNOBLAUCH, Karl Hermann and TYNDALL, John (1850): "Ueber das Verhalten krystallisirter Körper zwischen den Polen eines Magnetes", *Pogg.Ann.*, **79**: 233–41; **81**: 481–99.

KOENIGSBERGER, Leo (1906): *Herman von Helmholtz*, Oxford.

KREIL, Karl (1852): "Einfluss des Mondes auf die magnetische Declination", *Denk.Akad.Wiss.Vienna*, **3**: 1–47.

—— (1853): "Einfluss des Mondes auf die Horizontale Componente der magnetischen Erdkraft", *Denk.Akad.Wiss.Vienna*, **5**: 35–90.

LAMBERT, Andrew D. (1990): *The Crimean War: British grand strategy, 1853–56*, Manchester.

LAMBERT, Johann Heinrich (1766a): "Analyse de quelques expériences faites sur l'aiman", *Mém.Acad.Berlin*, **22**: 22–48.

—— (1766b): "Sur la courbure du courant magnétique", *Mém.Acad.Berlin*, **22**: 49–77.

LAMONT, Johann von (1852): "On the Ten-year Period which exhibits itself in the Diurnal Motion of the Magnetic Needle", *Phil.Mag.*, **3**: 428–35.

LATHAM, Robert Gordon (1854): *On the Importance of the Study of Language as a branch of Education for all Classes*, London.

LAYTON, David (1981): "The Schooling of Science in England, 1854–1939", in MacLeod and Collins (1981), 188–210.
LESLIE, Charles Robert (1855): *A Hand-book for Young Painters*, London.
LESLIE, John (1821): *Geometrical Analysis, and Geometry of curve lines*, Edinburgh.
LIEBIG, Justus (1856): "Ueber Versilberung und Vergoldung von Glas", *Ann.Chem.Pharm.*, **98**: 132–9.
LLOYD, Christopher (1946): "Dundonald's Crimean War Plans", *Mariner's Mirror*, **32**: 147–54.
LLOYD, Humphrey (1853): "On the influence of the moon upon the position of the freely-suspended horizontal magnet", *Proc.Roy.Irish Acad.*, **5**: 383–7, 434–40.
LYELL, Charles (1852): "On the Blackheath Pebble-bed, and on Certain Phaenomena in the Geology of the Neighbourhood of London", *Proc.Roy.Inst.*, **1**: 164–7.
—— (1881): *Life, Letters and Journals of Sir Charles Lyell*, 2 volumes, London.
McCONNELL, Anita (1989): "Aluminium and its Alloys for Scientific Instruments, 1855–1900", *Ann.Sci.*, **46**: 611–20.
—— (1993): *R.B. Bate of the Poultry: The Life and Times of a Scientific Instrument Maker*, Pershore.
McKIE, D. and DE BEER, G.R. (1951–2): "Newton's Apple", *Notes Rec.Roy.Soc.Lond.*, **9**: 46–54, 333–5.
MacLEOD, Roy and COLLINS, Peter (editors) (1981): *The Parliament of Science: The British Association for the Advancement of Science 1831–1981*, Northwood.
MAGNUS, Heinrich Gustav (1854): "Ueber rothen und schwarzen Schwefel", *Pogg.Ann.*, **92**: 308–23.
MANTELL, Gideon Algernon (1850): "Notice of the discovery by Mr. Walter Mantell in the Middle Island of New Zealand, of a living specimen of the Notornis, a bird of the Rail family, allied to the Brachypteryx, and hitherto unknown to naturalists except in a fossil state", *Proc.Zoo.Soc.*, **18**: 209–12.
—— (1852): "On the Structure of the Iguanodon, and on the Fauna and Flora of the Wealden Formation", *Proc.Roy.Inst.*, **1**: 141–6.
MARIGNAC, Jean Charles Galissard de (1845): "Sur la production et la nature de l'ozone", *Comptes Rendus*, **20**: 808–11.
MARKHAM, R.A.D. (1990): *A Rhino in High Street: Ipswich Museum - the early years*, Ipswich.
MARRINER, Sheila (1961): *Rathbones of Liverpool 1845–73*, Liverpool.
MARTIN, John (1849a): *Thames and Metropolis Improvement Plan*, London.
—— (1849b): *Plan for ventilating Coal Mines*, London.
MARTIN, Thomas Commerford and WETZLER, Joseph (1887): *The Electric Motor and its Applications*, New York.

MARTINEAU, Harriet (1849–50): *The History of England during the Thirty Years' Peace: 1816–1846*, 2 volumes, London.
—— (1858): *Pictorial History of England during the Thirty Years' Peace: 1816–1846*, New edition, London.
MASSON, Antione-Philibert (1845): "Etudes de photométrie électrique. 1er et 2e Mémoires", *Ann.Chim.*, **14**: 129–95.
—— (1850): "Etudes de photométrie électrique. 3e Mémoire", *Ann.Chim.*, **30**: 5–55.
—— (1851): "Etudes de photométrie électrique. 4e et 5e Mémoires", *Ann.Chim.*, **31**: 295–326.
MATTEUCCI, Carlo (1850a): "Electro-Physiological Researches. - Eighth Series", *Phil.Trans.*, **140**: 287–96.
—— (1850b): "Réclamation de priorité à l'occasion des communications récentes de M. Du Bois-Reymond, sur des recherches d'électricité", *Comptes Rendus*, **30**: 479–80.
—— (1850c): "Réponse aux deux dernières Lettres de M. du Bois-Reymond, insérées dans les nos 17 et 18 des Comptes rendus de l'Académie, et en général à toutes les observations faites par le même auteur sur quelques-unes de mes recherches d'électrophysiologie", *Comptes Rendus*, **30**: 699–707.
—— (1853a): *Lettre ... a Mr. H. Bence Jones F.R.S. &. &. Editeur d'une brochure intitulée On Animal Electricity ou extrait de découvertes de Mr. Du Bois-Reymond*, Florence.
—— (1853b): "Recherches expérimentales sur le magnétisme de rotation et sur la polarité diamagnétique", *Comptes Rendus*, **37**: 303–6.
—— (1854): *Cours spécial sur l'induction, le magnétisme de rotation, le diamagnétisme, et sur les relations entre la force magnétique et les actions moléculaires*, Paris.
MAURY, Matthew Fontaine (1851): *Investigations of the Winds and Currents of the Sea*, Washington.
MELLONI, Macedonio (1850): *La thermochrôse, ou la coloration calorifique démontrée par un grand nombre d'expériences, et considérée sous ses divers rapports avec la science de la chaleur rayonnante*, Naples.
—— (1853): "Ricerche intorno al magnetismo delle Rocce", *Mem. Accad.Sci.Naples*, **1**: 121–64.
—— (1854a): "Recherches sur l'induction électrostatique", *Comptes Rendus*, **39**: 177–83.
—— (1854b): "Nouvel électroscope", *Comptes Rendus*, **39**: 1113–7.
—— (1854c): "Sur l'égalité de vitesse que prennent les courants électriques de tensions différentes dans le même conducteur métallique" *Bibl.Univ. Arch.*, **27**: 30–7.
—— (1854d): "Sopra alcuni fenomeni di elettricismo statico e dinamico, recentemente osservati da Faraday, ne'conduttorri de'telegrafi sotterranei e sottomarini", *Rend.Soc.Reale Borbon.Accad.Sci.*, **3**: 30–8.

—— (1854e): "Sull'eguaglianza di velocità che le correnti elettriche di varia tensione assumono nello stesso conduttore metallico", *Ann.Sci.Mat.Fis.*, **5**: 319–25.

MELSENS, Louis Henri Fréderic (1849): "Nouveau procédé pour l'extraction du sucre de la canne et de la betterave", *Ann.Chim.*, **27**: 273–310.

MERRYWEATHER, George (1851): *An Essay Explanatory of the Tempest Prognosticator in the Building of the Great Exhibition for the Works of Industry of all Nations*, London.

MOODY, Theodore William and BECKETT, James Camlin (1959): *Queen's, Belfast 1845–1949: The History of a University*, 2 volumes, Belfast.

MORRELL, Jack B. and THACKRAY, Arnold (1981): *Gentlemen of Science: Early Years of the British Association for the Advancement of Science*, Oxford.

MORRIS, Peter J.T., RUSSELL, Colin A. and SMITH, John Graham (1988): *Archives of the British Chemical Industry, 1750–1914: A Handlist*, Stanford in the Vale.

MORSE, Edward Lind (1914): *Samuel F.B. Morse His Letters and Journals*, 2 volumes, Boston.

MORSON, Anthony F.P. (1997): *Operative Chymist*, Amsterdam.

MUNK, William (1878): *The Roll of the Royal College of Physicians of London*, 2nd edition, 3 volumes, London.

MUSSCHENBROEK, Petrus van (1725): "De Viribus Magneticis", *Phil. Trans.*, **33**: 370–8.

NEWTON, Isaac (1726): *Philosophiae naturalis principia mathematica*, 3rd edition, London.

NOBILI, Leopoldo (1830): "Mémoire sur les couleurs en général, et en particulier sur une nouvelle échelle chromatique déduite de la métallochromie a l'usage des sciences et des arts", *Bibl.Univ.*, **44**: 337–64, **45**: 35–59.

NOBLE, Celia Brunel (1938): *The Brunels Father and Son*, London.

OERSTED Hans Christian (1820): "Experimenta circa effectum Conflictus electrici in Acum magneticam" *Schweigger J.Chem.Phys.*, **19**: 275–81.

—— (1848): "Précis d'une série d'expériences sur le diamagnétisme", *Ann.Chim.*, **24**: 424–35.

OURSEL, Noémie Noire (1886): *Nouvelle Biographie Normande*, 4 volumes, Paris.

OWEN, Richard (1851): "On Metamorphosis and Metagensis", *Proc.Roy.Inst.*, **1**: 9–16.

—— (1854): "On the Structure and Homologies of Teeth", *Proc.Roy.Inst.*, **1**: 365–74.

PAGET, James (1854): *On the Importance of the Study of Physiology as a Branch of Education for all Classes*, London.

PALAGI, Alessandro (1854): "Sur les variations électriques que subissent les corps lorsqu'ils s'éloignent ou se rapprochent les uns des autres", *Bibl.Univ.Arch.*, **25**: 372–80.

PASTEUR, Louis (1850): "Recherches sur les propriétés spécifiques des deux acides qui composent l'acide racémique", *Ann.Chim.*, **28**: 56–99.

PEARCE, Brian L. (1988): *Thomas Twining of Twickenham: His work, his Museum, and The Perryn House Estate*, Twickenham.

PELLATT, Apsley (1849): *Curiosities of Glass Making: with details of the processes and production of ancient and modern ornamental Glass Manufacture*, London.

PELSENEER, J. (1936): "Notes on some unpublished letters from Faraday to Quetelet", *Ann.Sci.*, **1**: 447–52.

PELTIER, Ferdinand Athanase (1847): *Notice sur la vie et les travaux scientifiques de J. C. A. Peltier*, Paris.

PELTIER, Jean Charles Athanase (1842): "Recherches sur la cause des phénomènes électriques de l'atmosphère, et sur les moyens d'en recueillir la manifestation", *Ann.Chim.*, **4**: 385–433.

PERKINS, Jacob (1826): "On the progressive compression of water by high degrees of force, with some trials of its effects on other fluids", *Phil.Trans.*, **116**: 541–7.

PERRY, George (1811): *Conchology, or the natural history of shells: containing a new arrangement of the genera and species*, London.

PHILLIPS, John (1847): "On the Aurora Borealis of October 24th, 1847; as seen at York", *Proc.Yorks.Phil.Soc.*, **1**: 70–1.

PIDDINGTON, Henry (1848): *The Sailor's Horn-Book for the Law of Storms: being a practical exposition of the theory of the Law of Storms, and its uses to mariners of all classes in all parts of the world, shewn by Transparent Storm Cards and Useful Lessons*, London.

—— (1851): *The Sailor's Horn-Book for the Law of Storms: being a practical exposition of the theory of the Law of Storms, and its uses to mariners of all classes in all parts of the world, shewn by Transparent Storm Cards and Useful Lessons*, 2nd edition, London.

PLATEAU, Joseph Antoine Ferdinand (1843): "Mémoire sur les phénomènes que présente une masse liquide libre et soustraite a l'action de la pesanteur", *Mém.Acad.Sci.Bruxelles*, **16**. [Separately paginated].

—— (1844): "On the Phaenomena presented by a free Liquid Mass withdrawn from the Action of Gravity", *Taylor Sci.Mem.*, **4**: 16–43.

—— (1849): "Recherches expérimentales et théoriques sur les figures d'équilibre d'une masse liquide sans pesanteur", *Mém.Acad.Sci. Bruxelles*, **23**. [Separately paginated].

—— (1852): "Experimental and Theoretical Researches on the Figures of Equilibrium of a Liquid Mass withdrawn from the Action of Gravity", *Taylor Sci.Mem.*, **5**: 584–712.

—— (1857): "Recherches expérimentales et théoriques sur les figures d'équilibre d'une masse liquide sans pesanteur", *Mém.Acad.Sci. Bruxelles*, **30**. [Separately paginated].

—— (1859): "Recherches expérimentales et théoriques sur les figures d'équilibre d'une masse liquide sans pesanteur", *Mém.Acad.Sci. Bruxelles*, **31**. [Separately paginated].

—— and QUETELET, Lambert-Adolphe-Jacques (1851–5): *Physique*, 3 parts, Brussels.

PLÜCKER, Julius (1847): "Ueber das Verhältiss zwischen Magnetismus und Diamagnetismus", *Pogg.Ann.*, **72**: 343–50.

—— (1848a): "Ueber das Verhalten des abgekühlten Glases zwischen den Magnetpolen", *Pogg.Ann.*, **75**: 108–10.

—— (1848b): "Ueber das Gesetz, nach welchem der Magnetismus und Diamagnetismus von der Temperatur abhängig ist", *Pogg.Ann.*, **75**: 177–89.

—— (1848c): "Ueber die verschiedene Zunahme der magnetischen Anziehung und diamagnetischen Abstossung bei zunehmender Kraft des Elektromagneten", *Pogg.Ann.*, **75**: 413–9.

—— (1849a): "Ueber die neue Wirkung des Magnets auf einige Krystalle, die eine vorherrschende Spaltungs-Fläche besitzen. Einfluss des Magnetismus auf Krystall-Bildung", *Pogg.Ann.*, **76**: 576–86.

—— (1849b): "On the Magnetic Relations of the Positive and Negative Optic Axes of Crystals", *Phil.Mag.*, **34**: 450–2.

—— (1849c): *Praemissa enumeratione novorum phaenomenorum recentissime a se in doctrina de magnetismo inventorum*, Bonn.

—— (1849d): "Ueber die Fessel'sche Wellenmaschine, den neueren Boutigny'schen Versuch und das Ergebniss fortgesetzter Beobachtungen in Betreff des Verhaltens krystallisirter Substanzen gegen Magnetismus", *Pogg.Ann.*, **78**: 421–31.

—— (1849e): "On the Repulsion of the Optic Axes of Crystals by the Poles of a Magnet", *Taylor Sci.Mem.*, **5**: 353–75.

—— (1849f): "On the Relation of Magnetism to Diamagnetism", *Taylor Sci.Mem.*, **5**: 376–82.

—— (1851a): "Ueber das magnetische Verhalten der Gase", *Pogg.Ann.*, **83**: 87–108.

—— (1851b): "Numerische Vergleichung des Magnetismus des Sauerstoffgases und des Magnetismus des Eisens", *Pogg.Ann.*, **83**: 108–14.

—— (1851c): "Ueber die magnetische Polarität und die Coërcitiv-Kraft der Gase", *Pogg.Ann.*, **83**: 299–302.

—— (1852a): "Ueber die Theorie des Diamagnetismus, die Erklärung des Ueberganges magnetischen Verhaltens in diamagnetisches und mathematische Begründung der bei Krystallen beobachteten Erscheinungen", *Pogg.Ann.*, **86**: 1–34.

—— (1852b): "Ueber die Reciprocität der elektro-magnetischen und magneto-elektrischen Erscheinungen", *Pogg.Ann.*, **87**: 352–86.
—— (1854a): "Ueber das Gesetz der Induction bei paramagnetischen und diamagnetischen Substanzen", *Pogg.Ann.*, **91**: 1–56.
—— (1854b): "Untersuchungen über Dämpfe und Dampfgemenge", *Pogg.Ann.*, **92**: 193–220.
—— (1854c): *Commentatio de crystallorum et gazorum conditione magnetica qualis hodie intelligitur*, Bonn.
—— and BEER, August (1850–1): "Ueber die magnetischen Axen der Krystalle und ihre Beziehung zur Krystallform und zu den optischen Axen", *Pogg.Ann.*, **81**: 115–62, **82**: 42–74.
—— and GEISSLER, Johann Heinrich Wilhelm (1852): "Studien über Thermometrie und verwandte Gegenstände", *Pogg.Ann.*, **86**: 238–79.
POGGENDORFF, Johann Christian (1848): "Ueber die diamagnetische Polarität", *Pogg.Ann.*, **73**: 475–9.
POISSON, Siméon-Denis (1811): "Mémoire Sur la Distribution de l'Electricité à la surface des Corps conducteurs", *Mém.Inst.*, 1–92, 163–274.
POST, Robert C. (1976): *Physics, Patents, and Politics: A Biography of Charles Grafton Page*, New York.
POTTER, Richard (1830): "An Account of Experiments to determine the quantity of Light reflected by Plane Metallic Specula under different Angles of Incidence. With a Description of the Photometer made use of", *Edinb.J.Sci.*, **3**: 278–88.
POUILLET, Claude Servais Mathias (1853): *Eléments de Physique Expérimentale et de Météorologie*, 6th edition, 2 volumes, Paris.
PROUT, William (1834): *Chemistry Meteorology and the Function of Digestion Considered with reference to Natural Theology*, London.
QUETELET, Lambert-Adolphe-Jacques (1849): *Sur le Climat de la Belgique. Troisème Partie. De l'Electricité de l'air*, Brussels.
—— (1850): "Sur l'électricité atmosphérique", *Bull.Acad.Sci.Bruxelles*, **17**: 3–13.
—— (1851a): "On Atmospheric Electricity, especially in 1849", *Phil.Mag.*, **1**: 329–32.
—— (1851b): *Sur le Climat de la Belgique. Quatrième Partie. Pressions et ondes atmosphériques*, Brussels.
RANDELL, Wilfrid L. (1924): *Michael Faraday (1791–1867)*, London.
RAVINDRA, Ravi (editor) (1991): *Science and Spirit*, New York.
RAYMOND, Meredith B. and SULLIVAN, Mary Rose (1983): *The Letters of Elizabeth Barrett Browning to Mary Russell Mitford, 1836–1854. Volume 3*, Waco.
REALF, Richard (1852): *Guesses at the Beautiful. Poems*, Brighton.
REES, Richard van (1846): "Over de verdeeling van het magnetismus in staalmagneten en electromagneten". *Verhandl.Nederlandsche Inst.Weten.*, **12**: 94–118.

—— (1853): "Ueber die Faraday'sche Theorie der magnetischen Kraftlinien", *Pogg.Ann.*, **90**: 415–36.

REGNAULT, Henri Victor (1847): "Relation des expériences ... pour déterminer les principales lois et les données numériques qui entrent dans le calcul des machines a vapeur", *Mém.Acad.Sci.*, **21**: 3–767.

—— and REISET, J. (1849): "Recherches chimiques sur la respiration des animaux des diverses classes", *Ann.Chim.*, **26**: 299–519.

REICH, Ferdinand (1849): "On the Repulsive Action of the Pole of a Magnet upon Non-magnetic Bodies", *Phil.Mag.*, **34**: 127–30.

REID, Christian Leopold (1914): *Pedigree of the Family of Ker ... [and] Ker-Reid*, Newcastle.

REID, Thomas Wemyss (1899): *Memoirs and Correspondence of Lyon Playfair*, London.

REINGOLD, Nathan and ROTHENBERG, Marc, et al. (1972–99): *The Papers of Joseph Henry*, 8 volumes, Washington.

RIDLEY, Jasper (1979): *Napoleon III and Eugénie*, London.

RIESS, Peter Theophilus (1846): "On the Incandescence and Fusion of Metallic Wires by Electricity" *Taylor Sci.Mem.*, **4**: 432–75.

—— (1852): "Sur la décharge de la batterie de Franklin", *Bibl.Univ.Arch.*, **19**: 177–95.

—— (1853): *Die Lehre von der Reibungselektricität*, 2 volumes, Berlin.

ROBERT, Adolphe, BOURLOTON, Edgar and COUGNY, Gaston, (1889–91): *Dictionnaire des Parlementaires Français*, 5 volumes, Paris.

ROBERTS-JONES Philippe (1974): "Madou et Quetelet", *Bull.Classe Beaux-Arts Acad.Roy.Belg.*, **66**: 200–4.

ROBISON, John (1822): *A System of Mechanical Philosophy*, 4 volumes, Edinburgh.

ROGET, Peter Mark (1831): "On the Geometric Properties of the Magnetic Curve, with an account of an Instrument for its Mechanical Description", *J.Roy.Inst.*, **1**: 311–9.

—— (1832): *Magnetism*, in volume 2 of *Natural Philosophy* in the Library of Useful Knowledge, London.

ROLT, Lionel Thomas Caswall (1957): *Isambard Kingdom Brunel*, London.

ROWCROFT, Charles (1850): *Evadne, or an Empire in its Fall*, 3 volumes, London.

ROYLE, John Forbes (1847): *A Manual of Materia Medica and Therapeutics*, London.

RUDWICK, Martin J.S. (1992): *Scenes from Deep Time: Early Pictorial Representations of the Prehistoric World*, Chicago.

RUPKE, Nicolaas (1994): *Richard Owen: Victorian Naturalist*, New Haven.

RUSSELL, John Scott (1852): "On Wave-line Ships and Yachts", *Proc. Roy.Inst.*, **1**: 115–9.

SABINE, Edward (1845–57): *Observations made at the Magnetical and Meteorological Observatory at Toronto in Canada*, 3 volumes, London.

—— (1850–3): *Observations made at the Magnetical and Meteorological Observatory at Hobarton, in Van Diemen Island*, 3 volumes, London.
—— (1851a): "On Periodical Laws discoverable in the mean effects of the larger Magnetic Disturbances", *Phil.Trans.*, **141**: 123–39.
—— (1851b): "On the Annual Variation of the Magnetic Declination, at different periods of the Day", *Phil.Trans.*, **141**: 635–41.
—— (1852): "On Periodical Laws discoverable in the mean effects of the larger Magnetic Disturbances - No.II", *Phil.Trans.*, **142**: 103–24.
SAUSSURE, Horace-Bénedict de (1779–96): *Voyages dans les Alpes*, 4 volumes, Neuchatel.
SAVART, Félix (1833): "Mémoire sur la Constitution des Veines liquides lancées par des orifices circulaires en mince paroi", *Ann.Chim.*, **53**: 337–86.
SCHETTINO, Edvige (1994): *Macedonio Melloni Carteggio (1819–1854)*, Florence.
SCHISCHKOFF, Georgi (1971): *Peter Beron (1798–1871): Forscherdrang aus dem Glauben an die geschichtliche Sendung der Slawen*, Meisenheim.
SCHLAGINTWEIT, Adolph and SCHLAGINTWEIT, Hermann Rudolph Alfred (1850): *Untersuchungen über die physicalische Geographie der Alpen in ihren Beziehungen zu den Phaenomenen der Gletscher, zur Geologie, Meteorologie und Pflanzengeographie*, Leipzig.
SCHOENBEIN, Christian Friedrich (1839a): "Beobachtungen über die elektrische Polarisation fester und flüssiger Leiter", *Pogg.Ann.*, **46**: 109–27.
—— (1839b): "Neue Beobachtungen über die Volta'sche Polarisation fester und flüssiger Leiter", *Pogg.Ann.*, **47**: 101–23.
[—] (1842): *Mittheilungen aus dem Reisetagebuche eines deutschen Naturforschers*, Basle.
—— (1849a): "Ueber die chemische Theorie der Volta'shen Säule", *Pogg.Ann.*, **78**: 289–306.
—— (1849b): *Denkschrift über das Ozon*, Basel.
—— (1850a): "De la théorie chimique de la pile Voltaïque", *Bibl.Univ.Arch.*, **13**: 192–212.
—— (1850b): "Ueber den Einfluss des Lichts auf die chemische Thätigkeit des Sauerstoffs", *Verhandl.Schweiz.Naturforsch.Gesell.*, 44–51.
—— (1850c): *Uber den Einfluss des Sonnenlichtes auf die chemische Thätigkeit des Sauerstoffs und den Ursprung der Wolkenelektrizität und des Gewitters*, Basle.
—— (1851a): "On a peculiar Property of Ether and some Essential Oils", *J.Chem.Soc.*, **4**: 133–43.
—— (1851b): "On some secondary physiological effects produced by atmospheric electricity", *Trans.Med.Chir.Soc.*, **34**: 205–220.

—— (1852a): "Ueber die Beziehungen des Sauerstoffes zur Electricität, zum Magnetismus und zum Lichte", *Bericht Verhandl.Naturforsch. Gesell.Basel*, **10**: 50–80.
—— (1852b): "Ueber die Natur und den Namen des Ozons", *J.Prak.Chem.*, **56**: 343–9.
—— (1852c): "Ueber die quantitative Bestimmung des Ozons", *J.Prak. Chem.*, **56**: 349–53.
—— (1852d): "Ueber die mittelbare Bleichkraft des Quecksilbers", *J.Prak.Chem.*, **56**: 353–4.
—— (1852e): "Ueber die mittelbare Bleichkraft des Stibäthyls", *J.Prak. Chem.*, **56**: 354.
—— (1852f): "Ueber den Einfluss einiger Salze auf die chemische Thätigkeit des gewöhnlichen Sauerstoffgeses", *J.Prak.Chem.*, **56**: 354–7.
—— (1852g): "Ueber Eisenoxydsalze", *J.Prak.Chem.*, **56**: 357–9.
—— (1852h): "Ueber das Verhalten der schwefligen Säure einigen Jodmetallen", *J.Prak.Chem.*, **56**: 359.
—— (1853): "Ueber Farbenveränderungen", *Sitzungsber.Math.Naturwiss. Classe Kaiserl.Akad.Wissen.*, **11**: 464–91.
—— (1854a): "Ueber verschiedene Zustände des Sauerstoffes" *Ann.Chem. Pharm.*, **89**: 257–300.
—— (1854b): "Ueber die chemischen Wirkungen der Electricität, der Wärme und des Lichtes", *Verhandl.Naturforsch.Gesell.Basel*, **1**: 18–67.
—— (1854c): "Ueber Farbenveränderungen", *J.Prak.Chem.*, **61**: 193–224.
[—] (1855): *Menschen und Dinge. Mittheilungen aus dem Reisetagebuche eines deutschen Naturforschers*, Stuttgart and Hamburg.
SCHWABE, Samuel Heinrich (1844): "Sonnen-Beobachtungen im Jahre 1843", *Ast.Nach.*, **21**: 233–6.
SCOFFERN, J.C. and HIGGINS, W.M. (1853): *The Victoria Gold Valuer's Ready Reckoner*, London and Melbourne.
SCOTT, Walter (1810): *The Lady of the Lake. A poem in six cantos*, Edinburgh.
[—] (1818): *Rob Roy*, 3 volumes, Edinburgh.
SECORD, James A. (1989): "Extraordinary experiment: Electricity and the creation of life in Victorian England" in Gooding et al. (1989), 337–83.
SHAIRP, John Campbell, TAIT, Peter Guthrie and ADAMS-REILLY, A. (1873): *Life and Letters of James David Forbes*, London.
SMART, Benjamin Humphrey (1855): *Thought and Language: An essay having in view the revival, correction, and exclusive establishment of Locke's Philosophy*, London.
SMITH, Charles C. (1891): "Memoir of Col. Thomas Aspinwall", *Proc. Mass.Hist.Soc.*, **7**: 32–8.
SMITH, Crosbie and WISE, M. Norton (1989): *Energy and Empire: A biographical study of Lord Kelvin*, Cambridge.

SMITH, Denis (1998): "James Walker (1781–1862): Civil Engineer", *Trans.Newcomen Soc.*, **69**: 23–55.

SPARGO, Peter E. (1992): "Faraday, Joule and the Mechanical Equivalent of Heat", *Trans.Roy.Soc.S.Afr.*, **48**: 47–53.

STALLYBRASS, Oliver (1967): "How Faraday "Produced Living Animalculae": Andrew Crosse and the Story of a Myth", *Proc.Roy.Inst.*, **41**: 597–619.

STEINLE, Friedrich (1996): "Work, Finish, Publish? The Formation of the Second Series of Faraday's Experimental Researches in Electricity", *Physis*, **33**: 141–220.

STEVENSON, Alan (1850): *A rudimentary treatise on the history, construction, and illumination of Lighthouses*, 2 parts, London.

STEVENSON, William (1853): "Abstract of Observations on the Aurora, Cirri, &c. made at Dunse", *Phil.Mag.*, **6**: 20–46.

STOKES, George Gabriel (1852): "On the Change of Refrangibility of Light", *Phil.Trans.*, **142**: 463–562.

—— (1853): "On the Change of Refrangibility of Light, and the exhibition thereby of the Chemical Rays", *Proc.Roy.Inst.*, **1**: 259–64.

STOREY, Graham, TILLOTSON, Kathleen and BURGIS, Nina (1988): *The Letters of Charles Dickens, Volume Six, 1850–1852*, Oxford.

STORY-MASKELYNE, Mervyn Herbert Nevil (1851): "On the Connexion of Chemical Forces with the Polarization of Light", *Proc.Roy.Inst.*, **1**: 45–9.

STRUVE, Otto (1851): "Sur les dimensions des anneaux de Saturne", *Mém.Acad.Sci.St.Petersburg*, **7**: 439–76.

SYMONS, George James (1882): *Lightning Rod Conference*, London.

TALBOT, William Henry Fox (1836): "Facts relating to Optical Science. No. III", *Phil.Mag.*, **9**: 1–4.

—— (1842): "On the Iodide of Mercury", *Phil.Mag.*, **21**: 336–7.

—— (1852): "On the Production of Instantaneous Photographic Images", *Phil.Mag.*, **3**: 73–7.

TARBE DE St-HARDOUIN, F.P.H. (1884): *Notices Biographiques sur les Ingénieurs des Ponts et Chaussées*, Paris.

TAYLOR, Brook (1721): "An Account of some Experiments relating to Magnetism", *Phil.Trans.*, **31**: 204–8.

TAYLOR, J. Leaky (1988): *The Society for the Relief of Widows and Orphans of Medical Men: A History of the first 200 years 1788–1988*, London.

TAYLOR, Jeremy (1660): *Ductor Dubitantium or the Rule of Conscience in all her generall measures; Serving as a great Instrument for the determination of Cases of Conscience*, 2 volumes, London.

TAYLOR, Richard Vickerman (1865) *The Biographia Leodiensis; or, Biographical Sketches of the Worthies of Leeds and Neighbourhood, from the Norman Conquest to the present time*, London.

TEE, Garry J. (1998): "Relics of Davy and Faraday in New Zealand" *Notes Rec.Roy.Soc.Lond.*, **52**: 93–102.
THOMPSON, D. (1955): "Queenwood College, Hampshire. A Mid-19th Century Experiment in Science Teaching", *Ann.Sci.*, **11**: 246–54.
THOMPSON, Silvanus P. (1898): *Michael Faraday, His Life and Work*, London.
—— (1910): *The Life of William Thomson Baron Kelvin of Largs*, 2 volumes, London.
THOMSON, James (1849): "Theoretical Considerations on the Effect of Pressure in Lowering the Freezing Point of Water", *Trans.Roy.Soc.Edinb.*, **16**: 575–80.
THOMSON, William (1846): "On the mathematical theory of electricity in equilibrium", *Camb.Dubl.Math.J.*, **1**: 75–95.
—— (1849): "An Account of Carnot's Theory of the Motive Power of Heat; with Numerical Results deduced from Regnault's Experiments on Steam", *Trans.Roy.Soc.Edinb.*, **16**: 541–74.
—— (1851): "On the Theory of Magnetic Induction in Crystalline and Non-crystalline Substances", *Phil.Mag.*, **1**: 177–86.
—— (1854): "On the Mathematical Theory of Electricity in Equilibrium", *Phil.Mag.*, **8**: 42–62.
TIMOTEO, D., BARNABITA, Bertelli and PALAGI, Alessandro (1855): *Sulla distribuzione delle correnti elettriche nei conduttori esperienze*, Bologna.
TODHUNTER, Isaac (1876): *William Whewell*, 2 volumes, London.
TOPPIN, David (1985–6): "The British Hospital at Renkioi", *Roy.Eng.J.*, **99**: 225–36, **100**: 39–50.
TOULOTTE, Muriel (1993): *Etienne Arago 1802–1892 une vie, un siècle*, Perpignan.
TREVELYAN, Arthur (1831): "Notice regarding some Experiments on the Vibration of Heated Metals", *Trans.Roy.Soc.Edinb.*, **12**: 137–46.
—— (1835): "Further Notice of the Vibration of Heated Metals; with the Description of a new and convenient Apparatus for experimenting with", *Phil.Mag.*, **6**: 85–6.
TURNBULL, H.W. (1961): *The Correspondence of Isaac Newton, Volume III, 1688–1694*, Cambridge.
TWINING, Thomas (1853): *Letters on the Condition of the Working Classes of Nassau, being a report on their intellectual and technical training, their earnings and household economy, and the institutions established for their benefit*, London.
TWYMAN, Alan (1988): *In Search of the Mysterious Doctor Weekes (A Fragment of Sandwich History)*, Sandwich.
TYNDALL, John (1851a): "On the Laws of Magnetism", *Phil.Mag.*, **1**: 265–95.

—— (1851b): "On Diamagnetism and Magnecrystallic Action", *Rep.Brit. Ass.*, 15–18.
—— (1851c): "On Diamagnetism and Magnecrystallic Action", *Phil.Mag.*, **2**: 165–88.
—— (1853): "On the influence of Material Aggregation upon the manifestations of Force", *Proc.Roy.Inst.*, **1**: 254–9.
—— (1854a): "On the Vibration and Tones produced by the Contact of Bodies having different Temperature", *Proc.Roy.Inst.*, **1**: 356–9.
—— (1854b): *On the Importance of the Study of Physics as a branch of Education for all Classes*, London.
—— (1854c): "De la polarité diamagnétique", *Bibl.Univ.Arch.*, **27**: 215–23.
—— (1855a): "On the Currents of the Leyden Battery", *Proc.Roy.Inst.*, **2**: 132–5.
—— (1855b): "Experimental Demonstration of the Polarity of Diamagnetic Bodies", *Rep.Brit.Ass.*, 22–3.
—— (1868): "On Faraday as a Discoverer", *Proc.Roy.Inst.*, **5**: 199–272.
—— (1870): *Faraday as a Discoverer*, New edition, London.
—— (1871): "Science and Spirits" in John Tyndall, *Fragments of Science*, London, 427–35.
—— (1879): "Science and the 'Spirits'" in John Tyndall, *Fragments of Science*, 6th edition, 2 volumes, London, **1**: 496–504.
VAN HELDEN, Albert (1980): "Huygens and the astronomers" in Bos et al. (1980), 147–65.
VAPEREAU, Louis Gustave (1870): *Dictionnaire universel des contemporains*, 4th edition, Paris.
—— (1880): *Dictionnaire universel des contemporains*, 5th edition, Paris.
—— (1893): *Dictionnaire universel des contemporains*, 6th edition, Paris.
VAUPEL, Elisabeth C. (1991): "Justus von Liebig und die Glasversilberung" *Prax.Naturwiss.Chem.*, (15 July), 22–29.
WALLIS, Philip (1974): *At the Sign of the Ship: Notes on the House of Longman, 1724–1974*, Harlow.
WARD, Nathaniel Bagshaw (1842): *On the Growth of Plants in Closely Glazed Cases.* London.
WARD, Stephen Henry (1854): "On the Growth of Plants in Closely-glazed Cases", *Proc.Roy.Inst.*, **1**: 407–12.
WARREN, Samuel (1853): *The Intellectual and Moral Development of the Present Age*, 2nd edition, London.
WATSON, Joseph John William (1853): *A Few Remarks on the Present State and Prospects of Electrical Illumination*, London.
WAUGH, Francis Gledstanes [1894]: *Members of the Athenaeum Club from its foundation*, np.
WEBER, Wilhelm Eduard (1848): "Ueber die Erregung und Wirkung des Diamagnetismus nach den Gesetzen inducirter Ströme", *Pogg.Ann.*, **73**: 241–56.

—— (1849): "On the Excitation and Action of Diamagnetism according to the Laws of Induced Currents", *Taylor Sci.Mem.*, **5**: 477–88.
WEEKS, Mary Elvira and LEICESTER, Henry M. (1968): *Discovery of the Elements*, 7th edition, Easton.
WESTFALL, Richard S. (1980): *Never at Rest: A Biography of Isaac Newton*, Cambridge.
WHEWELL, William (1835): "Report on the Recent Progress and Present Condition of the Mathematical Theories of Electricity, Magnetism and Heat", *Rep.Brit.Ass.*, 1–34.
—— (1837): *History of the Inductive Sciences, From the Earliest to the Present Times*, 3 volumes, London.
—— (1840): *The Philosophy of the Inductive Sciences, founded upon their History*, 2 volumes, London.
[—] (1845): "Theory of Electricity", *Encyclopedia Metropolitana*, **4**: 140–70.
—— (1854): *On the Influence of the History of Science upon Intellectual Education*, London.
WHISTON, William, (1719): *The Longitude and Latitude Found by the Inclinatory or Dipping Needle*, London.
WHITE, Francis (1847): *General Directory of the Town and County of Newcastle-upon-Tyne*, Sheffield.
WIEDEMANN, Gustav Heinrich (1852): "Ueber die Strömung von Flüssigkeiten vom positiven zum negativen Pol der geschlossenen galvanischen Säule", *Bericht Verhandl.Akad.Wiss.Berlin*, 151–6.
—— (1856): "Ueber die Bewegung der Flüssigkeiten im Kreise der geschlossenen galvanischen Säule und ihre Beziehungen zur Elektrolyse", *Pogg.Ann.*, **99**: 177–233.
WILLIAMS, Frances Leigh (1963): *Matthew Fontaine Maury Scientist of the Sea*, New Brunswick.
WILLIAMS, L. Pearce, FITZGERALD, Rosemary and STALLYBRASS, Oliver (1971): *The Selected Correspondence of Michael Faraday*, 2 volumes, Cambridge.
[WILLIAMSON, William Crawford] (1870): ["Michael Faraday"], *Lond. Quart.Rev.*, **34**: 265–95.
WILSON, David B. (1991): "Convergence: Metaphysical Pleasure Versus Physical Constraint" in Fisch and Schaffer (1991), 233–54.
WILSON, George (1848): "On the Action of the Dry Gases on Organic Colouring Matters, and its relation to the Theory of Bleaching", *Trans.Roy.Soc.Edinb.*, **16**: 475–95.
—— and GEIKIE, Archibald (1861): *Memoir of Edward Forbes*, Cambridge.
WOLF, Johann Rudolf (1852a): "Sonnenflecken-Beobachtungen in der ersten Hälfte des Jahres 1852; Entdeckung des Zusammenhanges zwischen den Declinationsvariationen der Magnetnadel und den Sonnenflecken", *Mitt.naturforsch.Gesell.Bern*, 179–84.

—— (1852b): "Liaison entre les taches du Soleil et les variations en déclinaison de l'aiguille aimantée", *Comptes Rendus*, **35**: 364.

—— (1852c): *Neue Untersuchungen über die Periode der Sonnenflecken und ihre Bedeutung*, Bern.

—— (1852d): "Sur le retour périodique de minimums des taches solaires; concordance entre ces périodes et les variations de déclinaison magnétique", *Comptes Rendus*, **35**: 704–5.

WOLLASTON, William Hyde (1801): "Experiments on the chemical Production and Agency of Electricity", *Phil.Trans.*, **91**: 427–34.

WOOD, Christopher (1995) *Dictionary of British Art Volume IV: Victorian Painters*, 3rd edition, Woodbridge.

WOOLNOUGH, Charles W. (1853): *The art of marbling, as applied to book-edges and paper*, London

—— (1881): *The Whole Art of Marbling as Applied to Paper Book-edges etc*, London.

WROTTESLEY, John (1855): "Report of the Parliamentary Committee of the British Association to the Meeting at Glasgow in September 1855", *Rep.Brit.Ass.*, xlvii–lxiii.

Index

This indexes the letter and notes, but not the introductory material apart from the biographical register where entries are indicated by a *. Numbers refer to letters and not to pages.

Individuals whose name changed during the period covered by this volume (either by marriage or by obtaining a new title) are indexed under the name used in 1855 with cross references where necessary. Where names in the index are the same, the eldest is given first. Books and articles (using short titles) are indexed only under the author. The phrase "writes to" is reserved to index only the writers and recipients of letters published in this volume.

Abbott, Arthur Robert: 2949.
Abbott, Benjamin: 2949.
*Abel, Frederick Augustus: 2478, 2502.
    Faraday writes to: 2519.
Absorption bands: 2616.
Académie de Nancy: 2444.
Académie des Sciences: 2235, 2246, 2985, 3004.
    Comptes Rendus: 2613, 2647, 2955, 2987, 2992.
    Elections to:
        Becquerel: 2363, 2364.
    Mémoires: 2654.
Académie Royal des Science des Naples: 2813, 2817, 2862, 2879, 2890, 2966.
Accademia Pontificia de' Nuovi Lincei: 2345.
*Acland, Henry Wentworth:
    Faraday writes to: 2331.
    Writes to Faraday: 2330, 2332.
Adams, John Couch: Writes to Faraday: 2898.
Adelaide Gallery: 2235.
Adie, Richard: Writes to Faraday: 2382.

Admiralty:
    Disinfectant: 3019.
    Proposed attack on Cronstadt: 2871, 2882, 2885, 2887.
    Protection of copper sheeting: 2978, 2979.
Aepinus, Franz Ulrich Theodosius: 2865.
*Aikin, Arthur: .
    Faraday writes to: 2159.
    Writes to Faraday: 2160, 2171.
Aikin, Lucy: 2171.
Air pump: 2327.
*Airy, Elizabeth: 2463, 2575, 2760, 2926.
Airy, Elizabeth: 2487.
*Airy, George Biddell: 2351, 2944.
    Faraday writes to: 2223, 2259, 2463, 2485, 2538, 2575, 2642, 2743, 2749, 2760, 2793, 2794, 2824, 2831, 2850, 2883, 2917, 2920, 2924, 2926, 2938, 2953.
    Friday Evening Discourses: 2223, 2227, 2257, 2259, 2410, 2485, 2487, 2642, 2717, 2760, 2766, 2917, 2919, 2920, 2924, 2925, 2926.
    Magnetism: 2535, 2536, 2538, 2575.
    Ozone: 2747, 2749, 2823, 2824, 2830, 2831, 2899.
    Pendulum experiments: 2880, 2883, 2917, 2919, 2920.
    Telegraphic experiments: 2742, 2743, 2746, 2749, 2767, 2791, 2792, 2793, 2794, 2849, 2850, 2937, 2938, 2952, 2953.
    Writes to Faraday: 2227, 2257, 2460, 2465, 2487, 2535, 2742, 2746, 2747, 2766, 2767, 2791, 2792, 2823, 2830, 2849, 2880, 2919, 2925, 2937, 2952.
*Airy, Richarda: 2223, 2259, 2463, 2485, 2487, 2575, 2642, 2760, 2917, 2926, 2953.
Albert Edward, Prince of Wales: 2531.
*Albert Francis Charles Augustus Emanuel, Prince:
    Attends lecture at Royal Institution: 2156, 2157, 2853.
    Foucault: 2913.
    Great Exhibition: 2405, 2472.
    Invites Faraday to Windsor: 2342.
    Royal yacht: 2838, 2839.
    Writes to Faraday: 2472.
Alfred, Prince: 2531.
Allen, Henry: Writes to Faraday: 2752.
Allen, John: Writes to Faraday: 2675.
Alps: 2313.
American Association: 2430.
*Ampère, André-Marie: 2237, 2350, 2497, 3004.
Anderson: 2663.
Anderson, Elizabeth: 2312, 2785.

*Anderson, Charles: 2206, 2207, 2215, 2450, 2548, 2637, 2672, 2710, 2729, 2890, 2908, 3004.
*Andrews, Jane Hardie: 2511, 2514, 2515, 2556, 2563, 2569.
*Andrews, Thomas:
    Faraday writes to: 2514, 2515, 2556, 2563, 2568.
    Sarah Faraday writes to: 2569.
    Writes to Faraday: 2511, 2697.
Annales de Chimie: 2813.
Annual Register: 2277, 2279, 2280.
Antarctic expedition: 2788.
Antonine column: 2300.
Arago, Alfred: Writes to Faraday: 2739.
Arago, Antonin: Writes to Faraday: 2739.
*Arago, Dominique François Jean: 2199, 2255, 2640, 2762, 2786, 3004.
    Death of: 2739.
    Health of: 2428, 2577, 2589.
    Statue to: 2759.
Arago, Etienne Vincent: Writes to Faraday: 2739.
Arago, François Victor Emmanuel: Writes to Faraday: 2739.
Arago, Jacques Etienne Victor: Writes to Faraday: 2739.
Arago, Joseph: Writes to Faraday: 2739.
Arago, Victor: Writes to Faraday: 2739.
Archer, Thomas Croxen: Faraday writes to: 2891.
Archibald, Charles Dickson: 2904.
    Faraday writes to: 2358.
Arctic expedition: 2665.
Argand lamp: 2234, 2580, 2594, 2616, 2617.
Arnett, John Andrews: 2187, 2188.
Arnott, Neil: 3013.
Ashburner, John: 2724.
Askin, Charles: 2483.
Aspinwall, Thomas: 2566.
Athenaeum: 2435, 2451, 2453, 2636, 2646, 2691, 2964.
Athenaeum Club:
    Elections to: 2151, 2795.
    Faraday resigns from membership: 2479, 2480.
    Library illumination: 2682.
Atherstone, Edwin: Faraday writes to: 2957.
Atkinson, Charles Caleb: Faraday writes to: 2194, 2196.
Atmospheric waves: 2455.
Atomism: 2914, 2969.
Aurora: 2351, 2352, 2357, 2359, 2586, 2591, 2592, 2646, 2648.
*Babbage, Charles: 2197.
    Faraday writes to: 2429, 2796.

Bache, Alexander Dallas: 2426, 2430.
*Baddeley, Paul Frederick Henry: Writes to Faraday: 2281, 2339, 2399, 2431, 2484.
Bailey, Charles: 2712.
Baillie, E.C.C.: 2819.
Baillière, Hippolyte: 2364.
Baillière, Jean-Baptiste-Marie: 2444.
Baily, Francis: 2301.
Bain, Alexander: 2794, 2955.
Baker Street Bazaar: 2732.
Bancalari, Michele Alberto: 2343.
Barchard, William:
    Faraday writes to: 2167.
    Writes to Faraday: 2168.
*Barker, Thomas:
    Writes to Dundee Glasite Church: 2918.
    Writes to Edinburgh Glasite Church: 2931, 2935.
Barkly, Henry: 2541.
*Barlow, Cecilia Anne: 2215, 2315, 2464, 2571, 2710, 2717, 2890.
    Faraday writes to: 2582.
*Barlow, John: 2442, 2492, 2819, 2884, 2901.
    Airy's criticisms of Faraday: 2937, 2938, 2944.
    Faraday writes to: 2176, 2178, 2215, 2298, 2315, 2410, 2414, 2467, 2542, 2602, 2710, 2890, 2932, 2944, 3030.
    Royal Institution: 2215, 2255, 2289, 2365, 2430, 2710, 2889, 2890.
        Brande's retirement: 2524, 2525, 2529.
        Friday Evening Discourses: 2717.
        By:
            Airy: 2223, 2760, 2920, 2924.
            Buckland: 2253.
            Russell: 2467.
            Sabine: 2596.
            Stokes: 2612.
            Story-Maskelyne: 2376.
            Whewell: 2225, 2226.
        Regulations: 2380, 2410, 2414.
        Tickets for: 2150, 2400, 2447, 2994.
        Lecture courses: 2176, 2178, 2717.
            Brodie and: 2207, 2209, 2210, 2215, 2242, 2243.
        Members elected: 2361.
    Proceedings: 2389.
    Visits:
        Austria: 2889.
        France: 2571, 2717.

    Germany: 2889.
    Sussex: 2464.
    Switzerland: 2571, 2710, 2717.
    Writes to Faraday: 2571, 2717, 2889.
    Writes to Sarah Faraday: 2464.
Barlow, Peter: 2379.
Barnabita, Bertelli: 2959.
Barnaby, Nathaniel:
    Faraday writes to: 2979.
    Writes to Faraday: 2978.
*Barnard, Edward: 2219, 2284, 2309, 2703, 2819.
    Death of: 2942.
Barnard, Edward: 2703.
Barnard, Ellen: 2819.
*Barnard, Frank: 2577, 2906, 3006.
    Faraday writes to: 2589.
    In Paris: 2819.
*Barnard, Jane:
    Mentioned by Barlow: 2717, 2889.
    Mentioned by Faraday: 2219, 2589, 3022.
    Mentioned by Sarah Faraday: 2819.
    Mentioned by Tyndall: 3000, 3004, 3020, 3026.
Barnard, William: Death of: 2147.
Barometer: 2486.
Barry, Charles: Writes to Faraday: 2304.
Barry, Charles: Faraday writes to: 2150.
Basle:
    Museum: 2441, 2453.
    Philosophical Society of: 2781, 2790, 2864, 2985.
Bassi, Giulio: 2945.
Batavian Society of Experimental Philosophy, Rotterdam: 2461.
*Bate, Bartholomew: 2301, 2922, 2923.
Bate, Robert Brettel: 2301, 2922.
Baumert, Friedrich Moritz: 2790.
Baxter, Thomas: 2312.
Beale, Lionel Smith: 2645.
Beaufort, Francis: 2547.
    Faraday writes to: 2545.
Beaumont, Miles Thomas Stapleton, 8th Baron: 2889.
Becker, Carl Ludwig Christian: 2558.
*Becquerel, Alexandre-Edmond: 2199, 3004.
    Magnetism of oxygen: 2363, 2364, 2372, 2373, 2378, 2439, 2444.
    Ozone: 2705.
    Phosphorus: 2616.

Photography: 2772.
Writes to Faraday: 2772.
*Becquerel, Antoine-César: 2199, 2444.
    Faraday writes to: 2364, 2373.
    Writes to Faraday: 2363, 2372.
Beethoven, Ludwig van: 2835.
Belfour, Edmund: Faraday writes to: 2158, 2491.
Belgium: 2264.
    Climate: 2263, 2264, 2455.
*Bell, Thomas: 2586.
    Faraday writes to: 2377, 2423, 2489, 2512.
Bellani, Angelo: 2730.
*Bence Jones, Henry: 2571, 2706, 2861.
    Animal electricity: 2547, 2640, 2647.
    Faraday writes to: 2285, 2471, 2547, 2860, 2863, 2908, 2911, 2932.
    Gymnotus: 2545, 2547.
    Medical advice:
        To Faraday: 2873.
        To Sarah Faraday: 2471, 2860, 2863, 2908.
    Medical Chirurgical Society: 2388, 2482, 2523, 2534.
    Portrait of Faraday: 2504, 2523.
    Writes to Faraday: 2541.
Bence Jones, Millicent: 2860, 2863.
Bergemann, Carl Wilhelm: 2439, 2449.
Berkeley, Maurice Frederick Fitzhardinge: 2871.
Bern Philosophical Society: 2554, 2571.
Beron, Peter: 2282.
Berriedale, James Sinclair, Lord: 2374.
*Berzelius, Jöns Jacob: 2435, 2699, 2790, 2943.
Best, Samuel: 3020.
Biblical references:
    Acts 15: 28–9: 2918, 2931.
    Corinthians:
        1: 5: 5–6: 2336.
        1: 13: 6: 2931, 2935.
        2: 2: 6–11: 2336.
        2: 4: 6: 2965.
        2: 6: 14: 2935.
        2: 10: 4–6: 2931.
    Deuteronomy 12: 23–5: 2931.
    Ecclesiastes 1: 2–3: 2378.
    Ephesians 3: 21: 2432.
    Ezekiel 33: 24–5: 2931.
    Fallen nature: 2647.

Genesis 9: 3–4: 2931.
God: 2965.
Gospel: 2610.
Hebrews:
    4: 12–13.
    18: 12: 2462.
Isaiah:
    8: 20: 2931.
    44: 20: 2336.
Jesus Christ: 2286.
John:
    9: 11: 2935.
    14: 15: 2918.
Kings:
    1: 19: 12: 2861.
Lamentations: 3: 31–42: 2931.
Leveticus:
    13: 45: 2918.
    17: 13–14: 2931.
Luke 16: 19–31: 2462.
Mark 8: 38: 2931.
Matthew:
    5: 13: 2607.
    18: 2–3: 2861.
    18: 16: 2931.
    18: 17: 2336, 2931, 2935.
    19: 19: 2965.
    26: 41: 2462.
    28: 19–20: 2931.
Numbers 23: 10: 2861.
Peter:
    2: 3: 1–2: 2931.
Philippians: 3: 16: 2931.
Proverbs 3: 5: 2336.
Revelation:
    1: 12: 2336.
    3: 11: 2931.
    16: 13: 2703.
Romans:
    14: 1–4: 2918.
    16: 17: 2935.
Samuel:
    1: 16: 23: 2861.
Timothy:

1: 3: 15: 2336.
1: 4: 1: 2703.
1: 4: 4: 2918.
1: 6: 5: 2935.
2: 3: 15: 2931.
2: 3: 16: 2931.
  Zepaniah 1: 12: 2931.
Bibliothèque Universelle: 2388, 2802, 2807, 2813, 2840, 2870, 2874, 2967, 2969.
Bigsby, Mrs: 2548.
*Biot, Jean-Baptiste: 2415, 2634, 3004.
Birks, Thomas Rawson: Writes to Faraday: 3018.
Bischof, Carl Gustav Christoph: 2152.
Blaikley, Alexander: 2557.
Bleaching: 2212.
Blunt, George William: Faraday writes to: 2426.
Board of Ordnance: 2500, 2501.
Board of Trade: 2963.
Bohnenberger, Gottlieb Christoph: 2813.
Boileau, John Peter: 2503.
*Bois-Reymond, Emil Heinrich du: 2571, 2706, 2707.
    Animal electricity: 2332, 2427, 2704.
        Bence Jones: 2579, 2647.
        Controversy with Matteucci: 2640, 2647.
        Experiments at Royal Institution: 2520, 2521, 2522.
        Untersuchungen über thierische Elektricität: 2232, 2256.
    Faraday writes to: 2256.
    Writes to Faraday: 2232.
Bojer, Wenceslas: Writes to Faraday: 2741.
Bolley, Alexander Pompius: 2388, 2413.
Bonaparte, Charles Lucien Jules Laurent, Prince de Canino: 3004.
Bonn, University of: 2152, 2214, 2237, 2884.
*Boosey, Thomas: 2312, 2785.
    Writes to Dundee Glasite Church: 2918.
    Writes to Edinburgh Glasite Church: 2931, 2935.
Booth, James: 2804.
Bossange, H.: Writes to Faraday: 2255.
Boucherie, Auguste: 2184.
Boutigny, Pierre Hippolyte: 2184, 2237.
Bowerbank, James Scott:
    Faraday writes to: 2951.
    Writes to Faraday: 2950.
Bowie, Henry: Faraday writes to: 2303.
Boyd, Alexander: 2703.

Boyd, Mrs J.: 2703.
Boyd, Mary: 2703.
Braid, James: Writes to Faraday: 2724.
*Brande, William Thomas: 2151, 2392, 2595.
    Brodie's position: 2206, 2207, 2210, 2242.
    Faraday writes to: 2525.
    Offers to deliver Faraday's lectures: 2178.
    Resignation from Royal Institution: 2516, 2524, 2525, 2527, 2529, 2532.
        Honorary Professor of Chemistry: 2525, 2527.
    Writes to Faraday: 2524, 2527, 2529, 2983.
Brandis, Dietrich: 2328, 2346.
*Brayley, Edward William: Faraday writes to: 2389, 2499, 2651, 2798.
Breadalbane, John Campbell, 2nd Marquis of: Writes to Faraday: 2306.
*Breda, Jacob Gisbert Samuel van: 2915.
    Faraday writes to: 2297, 2398.
    Writes to Faraday: 2296, 2893.
*Brewster, David: 2648, 2779, 3026.
    Writes to Faraday: 2759, 2916.
Bristol:
    Microscopical Society: 2779.
    Philosophical Museum: 2779.
British Association: 2723, 2764.
    1844 Meeting, York: 3006.
    1847 Meeting, Oxford: 2556.
    1849 Meeting, Birmingham: 2214, 2218, 2221.
        Faraday attends:
            Speaks at: 2219.
            Vice President of: 2219.
    1851 Meeting, Ipswich: 2427, 2439.
        Dumas speaks at: 2474.
        Faraday attends: 2442, 2449, 2453.
        Tyndall speaks at: 2454, 2463.
    1852 Meeting, Belfast: 2511, 2514, 2553, 2556, 2563, 2568, 2569, 2592, 2637.
    1853 Meeting, Hull: 2707, 2780.
    1854 Meeting, Liverpool: 2861.
        Faraday attends: 2901.
        Faraday invited to: 2867, 2872, 2873, 2897.
            Accepts: 2891, 2896, 2899.
    1855 Meeting, Glasgow: 3007, 3008, 3015, 3020, 3022, 3023, 3024, 3025, 3027.
        Tyndall at: 3026.
    Parliamentary Committee: 2803, 2805, 2811.

British Museum: 2482.
    Faraday advises: 2466, 2722, 2744.
    Faraday borrows silica crystal from: 2721, 2722, 2723, 2728, 2729, 2736.
        Applies pressure to do so: 2723.
    Faraday writes to: 2722.
    Seeks Faraday's advice: 2446.
Britton, John: Faraday writes to: 2338.
Brockedon, William: 2392, 2756.
Brodie, Anne: 2842.
    Faraday writes to: 2581.
*Brodie, Benjamin Collins: 2836.
    Medical advice to Faraday: 2556, 2581, 2873.
*Brodie, Benjamin Collins: 2215, 2247, 2376, 2670.
    Faraday writes to: 2207, 2210, 2243.
    Writes to Faraday: 2206, 2209, 2242, 2753.
Brougham, Henry Peter, Lord: 2717.
*Brunel, Isambard Kingdom:
    Faraday writes to: 2182, 2597, 2948.
    Writes to Faraday: 2181, 2947.
Brush, George Jarvis: 2400.
Brussels: Académie Royale de: 2323, 2396.
Buch, Christian Leopold: 2943.
Buchanan (2): 2819.
Buchanan, Charlotte: 2153, 2450.
Buchanan, David: 2153, 2819.
Buchanan, David: 2284, 2286.
Buchanan, Elizabeth: 2153, 2785.
*Buchanan, George: 2153, 2276, 2450.
Buchanan, Nathaniel: 2819.
*Buchanan, William:
    Daughter: 2153.
    Faraday writes to: 2153, 2284, 2286, 2336, 2340, 2785.
    Sarah Faraday writes to: 2335, 2337.
Buckland, Francis Trevelyan: 2787.
*Buckland, Mary: 2147, 2253.
    Faraday writes to: 2689, 2787.
*Buckland, William: 2253, 2787.
    Faraday writes to: 2147.
Buff, Heinrich: Writes to Faraday: 2954.
Bunsen, Christian Karl Josias: 2328, 2675.
Bunsen, Henry George: 2675.
Bunsen, Mary Louisa: 2675.
Bunsen, Robert Wilhelm Eberhard: 2699, 2705.

Burckhardt: 2441.
*Burdett Coutts, Angela Georgina: Faraday writes to: 2154, 2299, 2307.
Burges Alfred: 2628.
Burgoyne, John Fox: 2871.
Butler, George:
    Faraday writes to: 2501.
    Writes to Faraday: 2500.
Calcutta Medical Board: 2281.
Cambridge, Adolphus Frederick, Duke of: 2306.
Cambridge:
    Philosophical Society: 2169.
    Peterhouse: 2169.
    St John's College: 2898.
Canton, John: 2616.
Capillarity: 2164, 2236.
Carlyle, Thomas: 2427.
Carnot, Nicholas Léonard Sadi: 2251, 2254.
Caroline see Deacon, Caroline
Carpenter, William Benjamin: 2679, 2724.
Catalysis: 2864.
Challis, Thomas: 2717.
    Faraday writes to: 2676.
Champ, Mary Ann: 2549.
*Chance, James Timmins: 2244.
    Lighthouse lenses: 2643, 2809, 3028, 3029.
    Mirror: 2690, 2995, 2996.
Charteris, Francis Richard: 2684.
Chateaubriand, François René: 3000.
Chater: 2244.
Chatin, Gaspard-Adolphe: 2435.
Chemical analysis: 2265, 2324, 2597.
    For Board of Ordnance: 2500, 2501.
    For Markland: 2671.
    For Select Committee on National Gallery: 2684, 2685.
    For Trinity House:
        Oil: 2528, 2539, 2543.
        Water: 2230, 2231, 2385, 2386, 2533, 2822, 2825, 2975, 2976, 2997, 2998.
        White lead: 2288, 2319, 2507, 2513, 2518, 2533, 2655, 2661, 2692, 2810, 2940, 2956, 2958.
        Zinc paint: 2958.
    For Ward: 2815.
    For Way: 2650, 2658.
Chemicals:

Acetate of zinc: 2382.
Aluminium: 2983, 2987, 2988, 2992, 2995, 2996, 3030.
Bromine: 2366, 2367.
Calcareous spar: 2308, 2333, 2344.
Carbonic acid: 2147.
Charcoal: 2439.
Chlorine: 2353, 2366.
Collodion: 2162, 2664.
Elements: 2362, 2367.
Gold: 2573.
Iodide of mercury: 2759.
Iron:
    Molten: 2184, 2237.
Nickel: 2483.
Nitrous gas:
    Optical properties: 2607.
Oxygen: 2526, 2578, 2985.
Oxygen compounds:
    Colour: 2578, 2790.
Sodium: 2983.
Sulphuret of lead: 2348.
Tourmaline: 2759, 2779.
Chemical Society: 2265, 2388.
    Discourses: 2753.
Chemistry:
    Organic: 2578, 2604, 2899.
    State of: 2578, 2790.
Chevreul, Michel Eugène: Writes to Faraday: 2775, 2906.
Cholera: 2199, 2315, 2875, 2889.
Christian, Mrs: 2819.
*Christie, Samuel Hunter: 2331, 2341, 2415, 2592.
Chuard: 2443.
Church, Walter S.: 2400.
City Philosophical Society: 2187, 2188.
Clark: 2819.
Clark, Edwin: 2374.
*Clark, James: Writes to Faraday: 2156, 2157, 3025.
*Clark, Josiah Latimer:
    Faraday writes to: 2848, 2902.
    Telegraphic experiments: 2746, 2751, 2813, 2843, 2846, 2849, 2850, 2902, 2938, 2955, 2966.
        Effect of lightning: 2847, 2848, 2850.
        Melloni on: 2862.
    Writes to Faraday: 2751, 2843, 2847.

Clark, Margaret Helen: 2902.
*Clarke, Charles Mansfield: 2525, 2527, 2529.
    Faraday writes to: 2505.
    Writes to Faraday: 2516.
Clarke, William: 2312.
Clausius, Rudolf: 2837.
Clouds: 2648, 2649, 2659, 2667.
Clowes, William: 2173.
Clutterbuck, Henry: Writes to Faraday: 2572.
Cochrane, Thomas, 10th Earl of Dundonald: 2871.
Cold, chemical effects of: 2832, 2864.
Colding, Ludvig August: 2245, 2268.
Coleridge, Samuel Taylor: 2701.
Colour: 2906.
Conacher, Euphemia: 2819.
*Conolly, John: 2787.
    Writes to Faraday: 2161, 2801.
Constance see Deacon, Constance
Consumptive sanatorium: 2473.
Cooke, Edward William: Writes to Faraday: 2852.
Cooke, George Wingrove: Writes to Faraday: 2674.
Cookson, Isaac: 2244.
Cookson, William Isaac: 2588, 2593.
Cooper, James: 2905, 2907.
Copper, protection of: 2978, 2979.
Cosmos: 2713.
*Coulomb, Charles Augustin: 2415, 2444, 2865, 2870, 2955, 3018.
Coutts see Burdett Coutts
*Cowper, Edward: 2279, 2280, 2392.
    Writes to Faraday: 2277.
*Cox, William:
    Faraday writes to: 2972.
    Writes to Faraday: 2971, 2973.
Cramp, George: Faraday writes to: 2321.
Crews, Charles: 3019.
Crichton: 2254.
Crichton, John: Writes to Faraday: 3024.
*Crosse, Andrew:
    Acarus: 2277, 2280.
    Death of: 3005.
Crosse, Cornelia Augusta Hewett: Faraday writes to: 3005.
Cumming, John: 2769.
Cunningham: 2666.
Custance, Myles: Writes to Faraday: 3001.

Cuvier, Georges: 2943.
Cyclones: 2339, 2648.
Daily News: Faraday writes to: 2349.
Dalton, John: 2634, 2714.
Dance, William: 2187.
Daniell, Edmund Robert: 2219.
*Daniell, John Frederic: 2870, 2893, 2980.
    Hygrometer: 2281.
Danish Polytechnic School: 2245.
Darker, William Hill: 2376.
Darwin, Charles Robert: 2783.
Daubeny, Charles Giles Bridle: 2804.
    Faraday writes to: 2750.
Davis, Richard Hayton: 2683.
*Davy, Humphry: 2821, 2989.
    Chemical theories: 2790.
    Continental Tour: 2188.
    Elements of Chemical Philosophy: 2932.
    Preservation of copper: 2978.
    Royal Institution:
        Lectures: 2187.
        Resignation from: 2525.
*Deacon, Caroline: 2284, 2407, 2450, 2503, 2531.
    Faraday writes to: 2462, 2551, 2703, 2819, 3021.
    Faradays help financially: 2703.
    Sarah Faraday writes to: 2819.
*Deacon, Constance: 2407, 2462, 2503, 2551, 2703, 2819, 3021.
    Faraday writes to: 2531.
Deacon, Emma: 2961.
*Deacon, Henry: 2309.
    Faraday writes to: 2265, 2961.
Deacon, Samuel: Writes to Edinburgh Glasite Church: 2935.
*Deacon, Thomas John Fuller: 2450, 2462, 2531, 2551, 2703, 2819, 3021.
    Faraday seeks work for: 2503.
    Faraday writes to: 2407, 2503.
    Mental state of: 2284.
Debus, Heinrich: 3000, 3020.
*De La Beche, Henry Thomas: 2442.
    Faraday writes to: 2202, 2203, 2418.
    Writes to Faraday: 2384.
De La Pryme, Charles: Faraday writes to: 2509.
De La Rive, Adélaïde-Eugénie-Augusta: 2786.
*De La Rive, Arthur-Auguste: 2183, 2200, 2358, 2571, 2710, 2717, 2735, 2870, 2893, 3026.

Children: 2782.
Experiment of: 2199, 2208.
Faraday's recollections of visits to: 2782.
Faraday writes to: 2208, 2378, 2432, 2577, 2782, 2799, 2800, 2802, 2840, 2965.
Illness of: 2428, 2432.
In London: 2185, 2191, 2199.
Ozone: 2699, 2790.
Second marriage: 2999, 3004.
Treatise on Electricity: 2375, 2428, 2610, 2627, 2651, 2782, 2786, 2799, 2807, 2840, 2914, 2916, 2965, 2969, 2999, 3012, 3018.
Writes to Faraday: 2199, 2375, 2428, 2610, 2786, 2807, 2914, 2969, 2999, 3012.
De La Rive, Cécile-Marie: 2610, 2786.
De La Rive, Charles-Gaspard: 2782, 2786.
De La Rive, Charles-Lucien: 2786, 2999.
De La Rive, Eugène: 3012.
De La Rive, Françoise-Amélie-Alice: 2786.
*De La Rive, Jeanne-Mathilde: 2199, 2786, 2969, 2999.
Death of: 2378, 2432.
De La Rive, Louisa: 2999.
*De La Rive, Louise-Victoire-Marie: 2999, 3004, 3012.
De La Rive, Sophie-Louisa: 2786.
*De La Rive, William: 2378, 2577, 2610, 2786.
Faraday writes to: 2627.
*De La Rue, Warren: 2392.
Writes to Faraday: 2205, 2269, 2273, 2481, 2768.
Dence, Charlotte: 2161.
Denman, Joseph:
Faraday writes to: 2839.
Writes to Faraday: 2838.
Department of Science and Art: 2695, 2755.
Descartes, René du Perron: 2633.
Despretz, César Mansuète: Writes to Faraday: 2443, 2903.
Deville, Henri Etienne Sainte-Claire: 2992.
Devonshire, William George Spencer Cavendish, 6th Duke of: 2473.
Diamond, Hugh Welch: Faraday writes to: 2989.
*Dickens, Charles John Huffam: Writes to Faraday: 2291, 2292, 2295, 2355.
Digeon, René Henri: 2906.
Dilke, Charles Wentworth: Faraday writes to: 2275.
Dixon: 2785.
Dixon, Robert Vickers: 2654.
Dove, Heinrich Wilhelm: 2427.
*Draper, John William: 2353, 2616, 2617.

*Drew, John: 2699, 2705, 2899.
Drew, Richard: 2593.
Drummond: 2717.
Drummond, Thomas: 2597.
Drummond: 2856.
Dublin:
    Trinity College: 2717.
Duboscq, Jules: 2869, 3020.
Duckworth, H.: Faraday writes to: 3011.
Dudley caverns: 2219.
Duff, John: 2153.
*Dumas, Ernest-Charles-Jean-Baptiste: 2246, 2474, 2571.
    Faraday writes to: 2240.
    Illness of: 2200.
*Dumas, Hermenie: 2184, 2191, 2199, 2200, 2240, 2246, 2474, 2493.
*Dumas, Jean-Baptiste-André: 2198, 2199, 2240, 2442, 2571.
    Faraday writes to: 2191, 2200, 2246, 2474, 2493.
    Sons: 2246.
    Writes to Faraday: 2184, 2235.
Duncan, Captain: 2374.
Dundonald, Thomas Cochrane, 10th Earl of: 2871.
Dust storms: 2281, 2339, 2399, 2431, 2484.
Earth, rotation of: 2913.
Ecole Polytechnique: 2786, 2999.
Edinburgh Journal of Science: 2490.
Edinburgh New Philosophical Journal: 2382.
Edinburgh Review: 2146.
Edinburgh University: 2788.
Edmonson, George: 2451.
*Eichtal, Adolphe Seligman d': 2184, 2191, 2200.
Electric Gas Company: 2708.
Electric Power and Colour Company: 2687, 2866, 2868, 2878, 2894.
Electric Telegraph Company: 2742, 2751, 2843, 2862.
Electricity: 2818, 2945, 2959.
    Animal: 2256, 2332, 2427.
        Experiments at the Royal Institution: 2520, 2521, 2522.
        Eels: 2531, 2541, 2545, 2547, 2631.
    Atmospheric: 2263, 2264, 2330, 2331, 2343, 2352, 2353, 2396, 2397, 2412, 2484, 2558, 2959.
        Causes dust storms: 2281, 2339, 2431.
    Batteries: 2732, 2800, 2937, 2938, 2986.
        Chemical action: 2984.
    Cosmical implications: 2262, 2591.
    Earth: 2880, 2883, 2920.

Faraday loves subject: 2512.
Fluid: 2415, 2870, 3018.
Gas: 2708, 2713.
Life: 2277, 2280, 3011.
Light produced by: 2219.
    Duboscq: 2869.
    Hähner: 2820, 2844.
    Mapple: 2855.
    Oxford heliometer: 2259.
    Staite:
        Automatic regulation: 2732.
    Watson: 2599, 2606, 2608, 2609, 2614, 2619, 2687, 2866, 2868, 2888, 2894, 2895, 2900.
        Faraday's report on: 2878.
        Shown at Charing Cross: 2599.
Physiological effect: 2330.
Spark: 2437, 2438.
Specific inductive capacity: 2647.
Static:
    Electrometers: 2272, 2281, 2396, 2415, 2751, 2986.
        Melloni: 2813, 2834, 2862, 2865, 2870.
    Faraday cage: 2415, 2870.
    Induction: 2865, 2870.
    Relationship of charge: 2263.
    Unit jar: 2415.
Theory of the pile: 2388.
Thermo: 2257, 2969.
Telegraph: 2264, 2289, 2374, 2430, 2781, 2903.
    Bain: 2955.
    By balloon: 2431, 2484.
    Printing: 2794.
    Retardation: 2742, 2743, 2746, 2749, 2751, 2780, 2781, 2782, 2786, 2791, 2792, 2793, 2794, 2797, 2799, 2807, 2813, 2834, 2843, 2846, 2849, 2850, 2862, 2902, 2936, 2955, 2966.
    Submarine: 2798.
        Atlantic: 2904.
    Time from Greenwich to Brussels: 2767.
    Time signalling: 2937, 2938, 2952, 2953.
    Used to determine longitude: 2257, 2259, 2792, 2849.
    Used to provide meteorological information: 2667.
Vacuum: 2515.
Voltaic: 2601.
    Pile: 2274.
Electro-chemical action: 2647, 2969.

Electrolysis: 2790, 2799, 2800, 2807, 2864, 2893, 2943, 2954, 2964.
    Ions: 2790.
Induced currents in electrolytes: 2786.
Electro-magnetism: 2379, 2387, 2634, 3002.
Electro-magnetic engine: 2321, 2445.
Electro-magnetic induction: 2372, 2647.
    Induction coil: 2777, 2903.
Electro-magnetic rotations: 2647.
Electro-magnets: 2152.
Elias, Pieter: 2278.
    Writes to Faraday: 2383.
Elizabeth, Empress of Austria: 2889.
Elliotson, John: 2724.
Elliott, Edward: 2502.
    Faraday writes to: 2544.
Ellis: 2710.
*Ellis, Henry:
    Faraday writes to: 2446, 2723, 2736, 2774, 2841.
    Writes to Faraday: 2728.
Encyclopaedia Britannica: 2916.
Encyclopaedia Metropolitana: 3018.
Engraving: 2184, 2294.
Erdmann, Otto Linné: 2562.
Espie, Robert: Writes to Faraday: 2701.
Eugénie: 2717.
*Euler, Leonhard: 2441, 2453, 2482.
Evans and Askin: 2483.
Fabricius, Johannes: 2586, 2592.
Faraday, James: 2623.
*Faraday, James: 2266, 2284, 2312.
    Faraday writes to: 2283.
Faraday, Lucy Reid: 2283, 2284.
Faraday, Margaret: 2283.
Faraday, Margery Ann: Death of: 2283, 2284.
Faraday, Michael:
    Accustomed to destroy letters: 2714.
    Airy defers to: 2747.
    America: 2416.
    As civil engineer: 2257.
    As hypocrite: 2337.
    As lecturer: 2262.
    Autograph: 2469, 2581.
    Boldest philosopher of our age: 2790.
    Book binding: 2773.

Cannot be spared: 2607.
Declines to write articles: 2960, 2970.
Dedication of books to: 2640, 2647, 3016, 3017.
"Diamagnetic conditions of Flame and Gases": 2164, 2327, 2343, 2347, 2353, 2364, 2372, 2373, 2378, 2449.
Does not take pupils: 2387.
Early life: 2623.
"Experimental Researches in Electricity": 2245, 2499.
    Collected: 2387.
        Volume 1: 2981.
        Volume 2: 2981, 2982.
        Volume 3: 2974, 2980, 2981, 2983, 3014.
    Series 1: 2799, 2800.
    Series 2: 2799, 2883.
    Series 3: 2954.
    Series 4: 2799.
    Series 8: 2388, 2799.
    Series 9: 2749.
    Series 11: 2388, 2415, 2749, 2870.
    Series 12: 2388, 2749, 2793, 2794, 2870.
    Series 13: 2232, 2749, 2870, 3020.
    Series 14: 2870.
    Series 15: 2232.
    Series 19: 2232, 2651, 2722.
    Series 20: 2250.
    Series 21: 2217, 2249, 2250, 2350, 2364, 2372.
    Series 22: 2152, 2174, 2190, 2211, 2226, 2454.
        Copies sent: 2169, 2185, 2237.
    Series 23: 2217, 2246, 2287, 2326, 2350.
        Copies sent: 2289.
        Figure: 2320, 2322.
        Royal Society: 2250, 2268.
    Series 24: 2354, 2364.
        Copies sent: 2388, 2389, 2392, 2400, 2449, 2514.
        Reactions to: 2409, 2415, 2490.
    Series 25: 2317, 2318, 2326, 2368, 2411.
        Copies sent: 2388, 2389, 2392, 2400, 2413, 2432, 2449, 2514.
        Reactions to: 2394, 2409, 2415, 2427, 2428, 2439, 2443, 2490.
        Royal Society: 2327, 2331, 2343, 2347, 2349, 2364, 2378, 2379.
    Series 26: 2318, 2326, 2411, 2717, 2942.
        Copies sent: 2388, 2389, 2392, 2400, 2413, 2432, 2449, 2514.

Reactions to: 2394, 2409, 2415, 2427, 2428, 2439, 2443, 2490, 2536, 2615.
Royal Society: 2327, 2331, 2334, 2343, 2347, 2349, 2364, 2378, 2379.
Series 27: 2318, 2326, 2411, 2620, 2717, 2942.
   Copies sent: 2388, 2389, 2392, 2400, 2413, 2432, 2449, 2514.
   Reactions to: 2394, 2409, 2415, 2427, 2428, 2439, 2443, 2490, 2615.
   Royal Society: 2327, 2343, 2347, 2349, 2364, 2378, 2379.
Series 28: 2631, 2942.
   Copies sent: 2514, 2534.
   Reactions to: 2550, 2553.
   Royal Society: 2468, 2489.
Series 29: 2631, 2799, 2942.
   Copies sent: 2514, 2534.
   Reactions to: 2553.
   Royal Society: 2489.
[Series 29a]: 2494, 2631, 2942.
   Copies sent: 2534.
   Reactions to: 2536.
[Series 29b]: 2631, 2921.
   Copy sent: 2938.
   Reaction to: 2939.
Series 30: 3027.
Health of: 2153, 2176, 2177, 2178, 2182, 2212, 2261, 2412, 2426, 2434, 2459, 2538, 2626, 2651, 2781.
   Age: 2668, 2705, 2859.
   Cold: 2514, 2625, 2776.
   Cold shiverings: 2309.
   Comments of others on: 2214, 2255, 2439, 2456, 2464, 2572, 2606, 2610, 2634, 2653, 2783, 2889, 2914, 2969, 3026.
   Confusion: 2556.
   Deafness: 2761, 2764.
   Depression: 2563, 2569.
      Dull in spirits: 2577.
   Diarrhoea: 3022.
   Doctors: 2450, 2870, 2873.
   Faceache: 2218.
   Fatigue: 2620.
   Feebleness: 3022.
   Headache: 2208, 2239, 2315, 2322, 2326, 2448, 2449, 2568, 2605, 2620, 2764, 2787, 2899, 2964.
      Giddiness: 2316, 2449, 2452, 2556, 2964, 3022.
   Influenza: 2388.

Lazy: 2602.
May not go out: 2761.
Memory of: 2191, 2201, 2229, 2287, 2343, 2370, 2411, 2432, 2474, 2514, 2577, 2604, 2620, 2668, 2716, 2901, 2921, 2966.
    Examples: 2353, 2787.
    Treachery of: 2657, 2840.
Quinsy: 2452, 2453.
Slow: 2832, 2965.
Sore throat: 2388, 2450, 2453, 2764.
Stiff muscles: 2315, 2432, 2452.
Strength failing: 2716.
Teeth: 2309, 2315.
Tired: 2781.
Vacant mind: 2413.
Voice: 2309.
Weak: 2563.
Weariness: 2208, 2448, 2620, 2705.

Honours and titles:
    Knighthood ascribed to: 2146, 2484.
    Prussian Order of Merit: 2146, 2706.
Invitations to: 2419, 2693, 2995, 3024, 3025.
    Accepts: 2434, 2845.
        Royal Geographical Society Dinner: 2680
    Declines them: 2248, 2299, 2491, 2508, 2530, 2641, 2676, 2826, 2996.
    Kingston Venison Feast: 2890.
Languages:
    German: 2232, 2256, 2346, 2449, 2647, 2668, 2901, 2964.
        Has tried to learn: 2966.
Niece: 2653.
Nephew: 2690.
Not a chemist: 2832, 3021.
Origin of family: 2993.
Portrait collection: 2323, 2347, 2364, 2393, 2611, 2714.
Professional business: 2324, 2750, 2882, 2887, 2930, 3019.
Remembrancer: 2832.
Science:
    Research suspended: 2426.
Society: 2287, 2318, 2750.
"Speculation": 2714, 2715.
Testimonials:
    Declines giving: 2194, 2452.
        Tyndall aware of Faraday's views: 2454.
Views:

On atoms and molecules: 2705, 2714, 2840.
On beauty of nature: 2448.
    On sunsets: 2318.
On clubs: 2951.
On compliments: 2537.
On committees: 2221.
On criticism: 2967.
On dating theories: 2360.
On education: 2691, 2693, 2695, 2697, 2797, 2804.
On experience of life: 2509.
On experiment: 2411, 2859.
    Reward of: 2316.
    Seeing: 2352, 2632.
    Slowness of: 2538, 2575, 2740.
On force: 2784, 2944.
On friendship: 2782, 2965, 3006.
On head full of visions: 2310.
On honorary memberships: 2647.
On honours: 2805.
On humility: 2411.
On illness: 2200.
On mathematics: 2411.
    Geometrical mode: 3002.
On nature: 2604.
On organic chemistry: 2604, 2899.
On politics: 2208.
On powers of matter: 2604.
On President of the Royal Society: 2302.
On Providence: 2250.
On reservations in drawing conclusions: 2763.
On science:
    and government: 2805.
    as a republic: 2344.
        Tyndall on: 2379.
On scientific controversy: 2647, 2781, 3027.
On theories of magnetism: 2964.
On time: 2503, 2556, 2750, 2780.
On truth of nature: 2310.
On use of common things: 2953.
On work: 2411.
On writing: 2551.
Visits:
    Blackheath: 2462, 2466, 2467, 2468, 2471.

Brighton: 2248, 2261, 2275, 2353, 2354, 2356, 2358, 2600, 2602, 2604, 2605, 2607, 2776, 2781, 2831.
Country: 2346, 2733.
East Anglia: 2546.
    Cambridge: 2320.
    Lowestoft: 2548, 2549, 2551.
    Norfolk: 2781.
        Old Buckenham: 2297, 2298, 2299, 2320, 2548, 2549, 2551.
Filey: 2210.
Glasgow: 2212, 3008.
Hammersmith: 2745, 2819.
Hampstead: 2568, 2569.
Hastings: 2412, 2413, 2961, 2962, 2964, 2965, 2966, 2967.
Isle of Wight: 2857, 2860, 2863, 2869, 2870.
Leaving town: 2268, 2458.
Norwood: 2307, 2308, 2309, 2312, 2314, 2315, 2317, 2318.
Recollections of:
    Continental Tour, 1813–1815:
        Danube: 2705.
        Geneva: 2782.
        Munich: 2557.
        Vesuvius: 2871.
    France: 2215.
    Paris:
        Jardin des Plantes: 2200, 2240, 2246.
    Geneva: 2378, 2782.
    Oxford: 2689.
Redhill: 2819.
Returning home: 3003.
    Does not wish arrival time to be divulged: 2549.
Sea side: 2240.
Surbiton: 2883, 2890, 2895, 2899, 2908.
Sydenham: 3020, 3026, 3027.
The North: 2208, 2211, 2215, 2216, 2217.
Tynemouth: 2446, 2448, 2449, 2450, 2452, 2453, 2456, 2459.
Wales: 2696, 2703, 2710, 2717, 2781.
Wimbledon: 2171, 2175, 2176, 2182.
Working time: 2221, 2268, 2438.
Faraday, Robert: 2283.
*Faraday, Sarah: 2147.
    Barlow writes to: 2464.
    Brother: 2734.
    Cousin, death of: 2224.

Faraday mentions to:
    Airy: 2642.
    Andrews: 2514, 2556.
    Barlow: 2215, 2315, 2710, 2890, 2932.
    Bence Jones: 2471, 2860, 2863, 2908, 2932.
    Brodie: 2581.
    Buchanan: 2153, 2336, 2340, 2785.
    Buckland: 2689, 2787.
    Crosse: 3005.
    Deacon, C.: 2703, 3021.
    Deacon, H.: 2961
    De La Rive: 2378, 2782, 2799, 2965.
    Dumas: 2191.
    Fox: 2371, 2776, 2826.
    Frankland: 2932.
    Henry: 2416, 2448.
    Leighton: 2195.
    Liebig: 2557, 3006.
    Moore: 2318, 2740.
    Owen: 2381.
    Phillips: 2352.
    Plücker: 2449.
    Reynolds: 2867.
    Schoenbein: 2343, 2413, 2453, 2604, 2781.
    Twining: 2201, 2224, 2765.
    Tyndall: 2859, 2932, 3022, 3027.
    Vincent: 2312, 2450, 2548.
Faraday writes to: 2219.
Health of: 2471, 2689, 2785, 2819, 2826, 2859, 2867, 2869, 2889, 3022.
    Cold: 2776, 2787.
    Deafness: 2787, 2819, 2833, 2860, 2863, 2890.
    Infirm: 2799, 2819, 2890, 2965.
        Difficulty in walking: 2556, 2569.
        Rheumatism: 2556.
    Stomach: 2908.
Mentioned by:
    Aikin: 2171.
    Andrews: 2511, 2697.
    Barlow: 2571, 2717, 2889.
    Brande: 2983.
    Crichton: 3024.
    De La Rive: 2199, 2428, 2786, 2807, 2969.
    Dumas: 2184.
    Henry: 2430.

Latham: 2177.
Moigno: 2869.
Moore: 2314, 2456, 2734.
Phillips: 2351.
Pollock: 2872.
Plücker: 2634, 2789.
Quetelet: 2396, 2455.
Schoenbein: 2274, 2441, 2526, 2562, 2578, 2607, 2699, 2735, 2790, 2818, 2828, 2864, 2943, 2985.
Somerville: 2653.
Spence: 2419.
Tyndall: 2698, 2707, 2858, 2861, 3000, 3004, 3020, 3023, 3026.
Role in disposing of Faraday's property: 2932.
Writes to Andrews: 2569.
Writes to Buchanan: 2335, 2337.
Writes to Deacon: 2819.
Writes to unidentified correspondent: 2623.
Writes to Vincent: 2549.
Farnes, Francis Thomas: 2266.
Favre, Pierre Antoine: Faraday writes to: 2716, 2784.
*Feilitzsch, Fabian Carl Ottokar von: 2360.
Writes to Faraday: 2350, 2874.
Fellows, Charles: 2962.
Faraday writes to: 2662.
Fellows, Harriet: Faraday writes to: 2962.
Fenwick, John Edward Addison: 2267, 2270.
Fenwick, Katherine Somerset Wyttenbach: 2267.
Ferdinand II: 2255, 2879.
Fereday family: 2992, 2993.
Field, Ellis N.: 2778.
Writes to Faraday: 2771.
Fisher, William: Writes to Edinburgh Glasite Church: 2931, 2935.
Fitzgerald: 2748.
*Flauti, Vincenzo: 2813, 2817, 2862, 2890, 2966.
Faraday writes to: 2881.
Writes to Faraday: 2875, 2879.
Fleming, John: 2281, 2399.
Fluorescence: 2601, 2604, 2607, 2616, 2620, 2637, 2832.
Forbes, Edward: 2202, 2941.
*Forbes, James David: 2704.
Faraday writes to: 2233, 2817.
Force: 2232.
Foucault, Jean Bernard Léon: 2913, 3004.
Fowler, Charles: 2677.

Fowler, Charles: 2677.
Fownes, George: 2595.
    Manual: 2285.
*Fox, Charles: 2371, 2552, 2776.
    Writes to Faraday: 2934.
*Fox, Mary: Faraday writes to: 2371, 2776, 2826.
Fox, Robert Were: 2880.
France: 3000.
    Attack on Salé: 2500.
    Paris Exhibition: 2963, 3004.
    Politics of: 2184, 2191, 2199, 2589.
Francis, James Bicheno: 2181, 2182.
*Francis, William: 2152, 2333, 2451, 2861, 3020.
    Faraday writes to: 2636.
*Frankland, Edward: 2683, 3020, 3026.
    Faraday writes to: 2932.
    Fellowship of the Royal Society: 2621, 2622.
Franklin, Benjamin: 2415.
Franklin, John: 2665.
Franz Josef, Emperor of Austria: 2889.
Frederick-Augustus II: 2844.
*Frederick William IV: 2706, 2762, 2946.
Frémy, Edmond: 2705.
Fresnel, Augustin Jean: 2237.
Fresnel lamp: 2878.
Fresnel, Léonor: 2692.
Galvani, Luigi: 2232.
Galway College: 2550.
Gas:
    Anhydrous: 2276.
    Engine:
        For ships: 2885.
    Condensation of:
        Olifiant gas: 2197.
    Lighting: 2682.
*Gassiot, John Peter: 2665, 2911.
    Faraday writes to: 2777.
    Writes to Faraday: 2986.
*Gauss, Carl Friedrich: 2465, 2538, 2717.
Gautier, Jean Alfred: 2646.
Gay-Lussac, Joseph Louis: 2363.
Geissler, Johann Heinrich Wilhelm: 2634.
Geological Museum: 2415, 2418, 2710, 2989.
Geological Survey: 2202.

Gilbert, William: 2444.
Giles, Mrs: 2161.
Gilpin, William: Faraday writes to: 2641.
Girdler, Mrs: 2819.
Glaisher, James: 2648.
Glasgow:
    University of: 2169.
Glass: 2244, 2650, 2658, 2744.
    Heavy: 2205, 2233, 2289, 2305, 2418.
    Silvering: 2600.
Gmelin, Leopold: 2796.
Goethe, Johann Wolfgang von: 2638.
*Gordon, Alexander: 2234, 2656, 2660.
Gould, Frederick: 2890.
Gould, John: 2818.
Grace: 2661.
Graham, James Robert George: 2871.
*Graham, Thomas: 2511, 2568, 2595.
    Writes to Faraday: 2236, 2390.
Grant, Miss: 2677, 2679.
*Grant, Miss: 2464, 2571, 2710.
Grant, Robert Edmond: 2194.
Gravesend, fire at: 2312.
Gravity: 2415, 2631.
    And electricity: 2415.
Gray, John Edward: Faraday writes to: 2329.
Great Exhibition: 2261, 2388, 2416, 2429, 2430, 2439, 2443, 2444, 2447, 2448, 2467.
    Juries: 2405, 2432, 2455, 2906.
        Medal: 2472.
    Royal Commission: 2963.
    Sydenham: 2552, 2798.
        Apocryphal animals: 2717.
Gregory, William: 2362.
Greifswald, University of: 2350.
*Grey, Charles:
    Faraday writes to: 2853, 2913.
    Writes to Faraday: 2342.
Grove, Emma Maria: 2341.
    Faraday writes to: 2854.
Grove, George: Faraday writes to: 2517, 2552.
*Grove, William Robert: 2164, 2237, 2239, 2350, 2512, 2699, 2777, 2828, 2854, 2884, 2901, 2923.
    Cell: 2259, 2376, 2800.

Faraday writes to: 2170, 2247, 2341, 2520, 2761.
Friday Evening Discourse: 2253, 2717.
Grove House Asylum: 2161.
Gulf Stream: 2409.
Gull, William Withey: Writes to Faraday: 2579.
Guncotton: 2162.
Gurney, Goldsworthy: Lamp: 2160.
Gutta percha: 2484, 2751, 2780, 2781, 2782, 2793, 2794, 2827, 2843, 2953.
*Gye, Frederick: Faraday writes to: 2680, 2835, 2991, 3009.
Gye, Mrs Frederick: 2680.
Haarlem Society of Sciences: 2237, 2296, 2297, 2328, 2398.
Haecker, Paul Wolfgang: 2278, 2354.
Hähner, Guillaume: 2820.
    Writes to Faraday: 2844.
Haldat Du Lys, Charles-Nicolas-Alexandre: Writes to Faraday: 2444.
Hall, John: 2525.
Hallam, Henry: 2503.
Hamilton, William Alexander Baillie: Writes to Faraday: 2885.
Hampshire and Wiltshire Educational Union: 3020.
Hansen, Peter Andreas: 2302.
Hanwell Asylum: 2161, 2801.
Harcourt, William Vernon:
    Faraday writes to: 2369.
    Writes to Faraday: 2368.
Hardinge, Henry: 2502.
Hardinge, Richard: 2501.
Hare, Robert: 2437, 2438.
*Harris, William Snow: 2272, 2384, 2436, 2457, 2489, 2870, 3018.
    Writes to Faraday: 2415, 2984.
Hartz mountains: 2314.
Hastings, Thomas: 2501.
Hauksbee, Francis: 2415.
Hawes, Benjamin: 2200.
    Faraday writes to: 2198.
Hawkins, Benjamin Waterhouse: 2717.
Hawkins, Edward: Faraday writes to: 2466, 2744.
Haywood, Francis: 2872, 2873.
Head, Edmund Walker: 3010.
Heat, mechanical equivalent of: 2192.
Heber, Reginald: 3000.
Helmholtz, Hermann von: 2718.
Henderson, John: 2552.
Henry: 2274.
Henry, Harriet: 2430.

*Henry, Joseph:
    Faraday writes to: 2416, 2448.
    Writes to Faraday: 2430.
Henry, William Charles: Faraday writes to: 2714, 2715.
*Henslow, John Stevens:
    Faraday writes to: 2366, 2754, 2756, 2758.
    Writes to Faraday: 2362, 2367, 2757, 2783.
Herapath, William Bird: 2759.
    Writes to Faraday: 2779.
Herbert, George:
    Faraday writes to: 2600.
    Writes to Faraday: 2597.
*Herbert, Jacob: 2590, 2600, 2608, 2844, 2894, 2895, 2900.
    Faraday writes to: 2172, 2204, 2231, 2234, 2238, 2386, 2574, 2580, 2628, 2643, 2660, 2673, 2692, 2726, 2727, 2733, 2778, 2809, 2825, 2907, 2976, 2998, 3029.
    Writes to Faraday: 2163, 2220, 2228, 2230, 2288, 2385, 2507, 2513, 2518, 2528, 2533, 2539, 2543, 2561, 2565, 2566, 2570, 2599, 2606, 2655, 2656, 2661, 2681, 2687, 2720, 2731, 2738, 2810, 2820, 2822, 2886, 2888, 2940, 2956, 2958, 2975, 2997, 3028.
Herries, Charles John: 2717.
Herschel, Frederick William: 2234, 2586.
Herschel, Isabella: 2595.
*Herschel, John Frederick William: 2164, 2255, 2586, 2601,
    Faraday writes to: 2361, 2380, 2395, 2406, 2408, 2475, 2587, 2595.
    Writes to Faraday: 2394, 2591, 2693.
Herschel, Margaret Brodie: 2361.
Herschel, William James: 2591.
Hickson, Jane: 2678.
*Hickson, William Edward: Writes to Faraday: 2677, 2678, 2679.
Hillhouse, John Wilson: 2309.
Hillhouse, Ann Hanbury: 2309.
Hjorth, Søren: 2445.
Hodgkin, Sarah Frances: 2626.
Hodgkin, Thomas: Faraday writes to: 2626.
Hodgson, William Ballantyne: 2804.
*Hofmann, August Wilhelm: 2390, 2604, 2634, 2889.
    Writes to Faraday: 2933.
Hogg, Jabez:
    Faraday writes to: 2603.
    Weekly Instructor: 2603.
*Holland, Henry: 2556, 2780.
    Faraday writes to: 2148.
Holmes, Frederick Hale: 2708.

Home, Daniel Douglas: 2971.
Hooker, Joseph Dalton: Writes to Faraday: 2788.
Hooker, William Jackson: 2788.
Household Words: 2291.
Houzeau, Auguste: 2985.
Howard, Peter: 2244.
*Humboldt, Friedrich Wilhelm Heinrich Alexander von: 2255, 2427, 2613, 2706, 2762, 2955.
    Cosmos: 2592, 2648.
    Writes to Faraday: 2313, 2946.
*Hunt, Robert: 2435, 2708.
    Faraday writes to: 2149.
Hunter, John: 2421.
Huygens, Christiaan: 2768.
Hyde, Charles: Faraday writes to: 2271.
Hypnotism: 2724.
Imperial Academy of Sciences, Vienna: 2719, 2790.
India:
    Army medical service: 2555, 2576.
Influenza: 2435.
Inglefield, Edward Augustus: Writes to Faraday: 2665.
Institut Royal des Pays-Bas: 2180, 2186, 2470.
Institution of Civil Engineers: 2657, 2725.
Ipswich Museum: 2362, 2366, 2367, 2424, 2433, 2611, 2754, 2756, 2757, 2758, 2783.
Italy:
    Politics of: 2208.
Jacobi, Moritz Hermann von: 2379.
Jameson, Robert: 2382.
Jerusalem: 2421.
Jocelyn, Robert, Viscount: 2889.
Jones, Edward: 2215.
*Joule, James Prescott: 2193, 2392, 3002.
    Writes to Faraday: 2192.
Journal für Prakticalish Chemie: 2562.
Kane, Robert John: Writes to Faraday: 2553.
Kater, Henry: 2777.
Kerr, Louisa Maria: Faraday writes to: 2155.
Kensington Conversazione: 2852.
Ketley, Joseph: Writes to Faraday: 2827.
Kew:
    Gardens: 2788.
    Observatory: 2331, 2856.
Kidd, John: 2987.

King, Anne Isabella Noel: 2440.
    Writes to Faraday: 2436.
Kirby, William: 2393.
Knoblauch, Karl Hermann: 2328.
Kreil, Karl: 2717.
*Lacey, Henry: 2178, 2889, 2890.
Laffitte, J.: Writes to Faraday: 2812.
Lambert, Johann Heinrich: 2415.
*Lamont, Johann von: 2538, 2554, 2586, 2592.
Lassell, Maria: 2897.
Lassell, William: 2795.
    Writes to Faraday: 2897.
*Latham, Peter Mere: 2175, 2315.
    Writes to Faraday: 2177.
Latham, Robert Gordon: 2804.
Laugier, Auguste Ernest Paul: Writes to Faraday: 2739.
Laugier, Lucie: Writes to Faraday: 2739.
Layard, Austen Henry: 2446.
Le Breton, Anna Letitia: 2171.
Le Breton, Philip Hemery: 2171.
Leeds Philosophical and Literary Society: 2764.
Leigh, Percival: 2295, 2355.
Leighton, George Cargill: Faraday writes to: 2611.
*Leighton, John: 2312.
    Writes to Dundee Glasite Church: 2918.
    Writes to Edinburgh Glasite Church: 2931.
Leighton, John:
    Faraday writes to: 2195.
    Working on art: 2195.
*Leighton, Stephen: 2312, 2785
    Writes to Dundee Glasite Church: 2918.
    Writes to Edinburgh Glasite Church: 2931, 2935.
Lenz, Heinrich Friedrich Emil: 2379.
Lepaute, Augustin-Michel Henry: 2570, 2580.
Leslie, Charles Robert:
    Faraday writes to: 2837.
    Writes to Faraday: 2836.
Leslie, John: 2633.
Létourneau: 2570.
Levées and Balls: 2304, 2306, 2384, 2418, 2453, 2717, 2948.
Leverrier, Urbain Jean Joseph: 2953, 3004.
Lewis, Griffith George: Writes to Faraday: 2502.
L'Hospital, Guillaume-François-Antoine de: 2415.
Liebig, Georg: 2555, 2576.

*Liebig, Justus von: 2511, 2555, 2694, 2735, 2781, 2790, 2859, 2889, 2899, 3025, 3032.
    Annales: 2818.
    Faraday writes to: 2557, 3006.
    Writes to Faraday: 2576.
Light: 2406, 2745.
    Analogy with sound: 2660.
Lighthouses: 2438.
    Catadioptric: 2234, 2809, 3028, 3029.
    Dioptric: 2561, 2565, 2570, 2580, 2593.
    Faraday's ventilating chimney:
        Located at:
            Casket Rocks: 2905.
            Cromer: 2720, 2726, 2771, 2778.
            Dungeness: 2163, 2166, 2172, 2204, 2220.
            Haisbro': 2886, 2907.
            Needles: 2163, 2167, 2168, 2172, 2204, 2220, 2238, 2692.
            Portland: 2228, 2238.
            St Catherine's: 2681, 2692, 2726, 2727.
            South Foreland: 2727.
            Spurn Point: 2738.
    Faraday visits:
        Birmingham (Chance glassworks): 3029.
        Cromer: 2726.
            Has high opinion of keepers: 2726.
        Dungeness: 2204.
        Haisbro': 2907.
        Needles: 2204, 2238, 2692.
        Portland: 2238.
        St Catherine's: 2692.
    Fog signal: 2656, 2660, 2673, 2731, 2733.
    French: 2565, 2590, 2597, 2600, 2643, 2809, 3029.
    Lamps: 2585, 2588, 2593.
    Lenses: 2590, 2593, 2594.
    Lightning conductors: 2574, 2628.
    Specific:
        Beachy Head: 2593, 2997.
        Bishop's Rock: 2628.
        Casket Rocks: 2561, 2565, 2580, 2588, 2594, 2905.
        Cromer: 2720, 2726, 2771, 2778.
        Dungeness: 2163, 2166, 2172, 2204, 2220.
        Eddystone: 2628.
            Model of: 2290.
        Haisbro': 2886, 2907.

        Longships: 2533.
        Nash: 2574.
        Needles: 2163, 2167, 2168, 2172, 2204, 2220, 2234, 2238, 2692, 2825.
        Portland: 2228, 2234, 2238.
        St Agnes: 2593.
        St Catherine's: 2600, 2681, 2692, 2726, 2727, 2825.
        Skerries: 2822, 2825.
        South Foreland: 2600, 2727.
        South Stack: 2656, 2731, 2733.
        Spurn Point: 2738.
        Start Point: 2593, 2600.
        Trevose Head: 2230, 2231.
    Wicks: 2566.
Lightning: 2314, 2315, 2421, 2559, 2567, 2574.
    Conductors: 2300.
        On lighthouses: 2574, 2628.
    Effect on telegraphs: 2847, 2848, 2850, 2938.
    Thunderstorm: 3020.
Lime light: 2159, 2160.
Lind: 2904.
Lindley, John: Faraday writes to: 2244.
Listing, Johann Benedict: 2465.
Literary Gazette: 2279, 2708.
Liverpool Dock Committee: 2732.
Lloyd, Dorothea: 2717.
*Lloyd, Humphrey: 2457, 2489, 2717.
*Logeman, Wilhelm Martin: 2383, 2915.
    Writes to Faraday: 2278, 2893.
London:
    Cab strike: 2710.
    Improvement of: 2179.
    Lord Mayor of: 2676, 2717.
    Medical Gazette: 2401, 2403.
    Plate Glass Company: 2244.
    Orphan Asylum: 2173, 2266, 2582.
    State of Thames: 3003, 3007.
    University of: 2176.
        Examinations: 2478.
        King's College: 2616, 2629, 2645.
        Senate: 2384, 2478.
        University College:
            Professor of Practical Chemistry: 2194, 2196.
Longman: 2627.

Faraday writes to: 3013.
Lorimer, Anthony: Writes to Edinburgh Glasite Church: 2931, 2935.
Lotham, Dr: 2861.
Loveday, Elizabeth: 2548.
Loveday, Thomas: Writes to Edinburgh Glasite Church: 2931, 2935.
*Lovelace, Augusta Ada King, Countess: 2610.
    Children: 2434.
    Faraday writes to: 2434.
    Writes to Faraday: 2440.
Lovelace, William King, Earl: 2434.
Lubbock, John William: 2812.
Lucas, Philip: Writes to Faraday: 2712.
*Lyell, Charles: 2202, 2314, 2757.
    Faraday writes to: 2510.
    Writes to Faraday: 2990, 3010.
Lyell, Mary Elizabeth: 2314, 2757.
Lyttelton, George William, 4th Baron: 2567.
M, Dr: Faraday writes to: 2652.
M, F.W.: Writes to Faraday: 2686, 2688.
MacGregor, Robert: 3007.
Macilwain, George:
    Faraday writes to: 2216.
    Lectures on fever: 2216.
MacKenzie, William: Faraday writes to: 2960, 2970.
Macnaughton, Mrs: 2450.
*Macomie, Alexander:
    Writes to Dundee Glasite Church: 2918.
    Writes to Edinburgh Glasite Church: 2931, 2935.
Macvicar, John Gibson: Faraday writes to: 2763.
*Madan, Frederick: 2163, 2172, 2234,
    Faraday writes to: 2590, 2594.
    Writes to Faraday: 2588, 2593.
Madou, Jean-Baptiste: 2323.
Magnetism: 2214, 2217, 2315, 2333, 2350, 2360, 2379, 2415, 2451, 2457, 2550, 2635, 2786, 2789, 2840, 2858, 2861, 2874, 2899, 2914, 2969, 2985, 3026.
    Alters dimensions: 2391.
    Atmospheric: 2317, 2322, 2326, 2327, 2330, 2331, 2341, 2353, 2407, 2412, 2413, 2537, 2538, 2648, 2649.
        Effect of sun: 2326, 2378, 2409, 2615.
        In *Daily News*: 2349.
    Diamagnetism: 2185, 2211, 2217, 2229, 2237, 2245, 2308, 2327, 2333, 2346, 2350, 2360, 2382, 2430, 2546, 2630, 2632, 2780, 2914, 2942, 2969.

Magnecrystallic action: 2152, 2174, 2183, 2190, 2211, 2239, 2249, 2328, 2354, 2439, 2884, 3027.
    Of gases: 2164, 2884.
    Polarity: 2214, 2250.
Differential torsion balance: 2354, 2364, 2556, 2575, 2962.
Field: 2239, 2317, 2327, 2333, 2378, 2451, 2546.
    Iron filing diagrams: 2475, 2485, 2487, 2494.
    Lines of force: 2326, 2351, 2378, 2468, 2475, 2631, 2633, 2921, 2939, 2942, 2964.
    Mapping: 2482, 2483.
Fluids: 2415, 2942.
Heterodox theories: 2258, 2702, 2709.
Intensity: 2415.
Laws of: 2415, 2702.
New words: 2310, 2311, 2317, 2320, 2322, 2368, 2494, 2496, 2497, 2498, 2633.
Of gases: 2241, 2369, 2439, 2444.
    Oxygen: 2326, 2327, 2332, 2343, 2347, 2368, 2369, 2375, 2378, 2388, 2394, 2396, 2438, 2449, 2526, 2537, 2556, 2667.
    Becquerel: 2363, 2372, 2373, 2378.
Paramagnetism: 2607, 2942.
Physical rather than chemical force: 2388.
Sebastopol as a metaphor for work on: 3027.
Terrestrial: 2190, 2282, 2326, 2327, 2331, 2343, 2344, 2346, 2347, 2379, 2394, 2411, 2423, 2427, 2428, 2535, 2536, 2537, 2586, 2591, 2592, 2615, 2618, 2620, 2651, 2717, 2762.
    Observatories:
        Cape of Good Hope: 2325, 2326, 2343, 2347, 2432, 2618.
        Fort Simpson: 2347.
        Greenwich: 2343, 2347.
            Reduction of observations: 2535.
        Hobarton: 2325, 2343, 2347, 2615.
        Lake Alhabasen: 2347.
        St Helena: 2325, 2326, 2343, 2347, 2618.
        St Petersburg: 2343, 2347.
        Singapore: 2343, 2347.
        Toronto: 2325, 2343, 2347, 2615, 3010.
        Washington: 2343, 2347.
    Periodicity: 2377.
    Vesuvius: 2862.
    Universality of: 2444.
Magnetometer: 2558.
Magnets: 2148, 2149, 2354, 2531, 2934.
    Logeman: 2278, 2354, 2383.

Sheffield: 2382.
*Magnus, Heinrich Gustav: 2218, 2256, 2427, 2451, 2458, 2706, 2858, 2859.
    Writes to Faraday: 2718.
*Magrath, Edward:
    Faraday writes to: 2151, 2187, 2266, 2479, 2480.
    Writes to Faraday: 2188.
Magrath, Henry: 2188.
Maguire, Thomas Herbert: 2433.
Majocchi, Giovanni Alessandro: 2208.
Maltby, Brough George: Writes to Faraday: 2319.
*Manby, Charles: Faraday writes to: 2370, 2654, 2657, 2725, 2829.
Manchester:
    Athenaeum: 2724.
    Free Library: 2530.
Manning, George: Writes to Faraday: 2409.
Mantell, Gideon Algernon:
    Faraday writes to: 2506.
    Writes to Faraday: 2400.
Mantell, Walter Baldock Durrant: 2506.
Mapple, James Lodge: Writes to Faraday: 2855.
Marbling: 2773.
Marburg: 2333.
*Marcet, François: 2571, 2710, 2717, 2786, 2807.
*Marcet, Jane: 2610, 2786, 2999.
Marignac, Jean Charles Galissard de: 2790.
Markland, James Heywood: Faraday writes to: 2671.
*Martin, David W.:
    Writes to Dundee Glasite Church: 2918.
    Writes to Edinburgh Glasite Church: 2931, 2935.
Martin, John: Writes to Faraday: 2179.
*Martin, Thomas Byam:
    Faraday writes to: 2871, 2887, 2910.
    Writes to Faraday: 2882.
*Martineau, Harriet:
    Faraday writes to: 2279.
    History of England: 2277, 2279, 2280.
    Writes to Faraday: 2280.
Martius, Karl Friedrich Philipp von: 2282.
Mason: Faraday writes to: 2821.
Mason, Margaret: 2215.
Mason, William: 2215.
Masson, Antione-Philibert: 2515.
Matches: 3031, 3032.
Mathieu, Charles: Writes to Faraday: 2739.

Mathieu, Claude-Louis: Writes to Faraday: 2739.
Mathieu, Marguerite: Writes to Faraday: 2739.
*Matteucci, Carlo: 2260, 2914, 2921.
    Faraday writes to: 2229, 2647.
    Relations with Du Bois-Reymond: 2640, 2647.
    Writes to Faraday: 2640.
Matteucci, Madam: 2647.
Maury, Matthew Fontaine: 2426.
    Writes to Faraday: 2409.
Maxwell: 2312.
Mayall, John Jabez Edwin: 2416, 2430.
Medico Chirurgical Society: 2388, 2482, 2534.
Medlock, Henry: 2650.
*Melloni, Macedonio: 2817, 2843, 2850, 2955, 2966.
    Death of: 2875, 2881, 2890, 2902.
    Faraday writes to: 2846, 2870.
    Radiant heat: 2255.
    Writes to Faraday: 2255, 2762, 2813, 2834, 2862, 2865.
*Melsens, Louis Henri Fréderic: 2184, 2198, 2200, 2455.
Merian, Peter: 2818.
Merian, Rudolf: 2818.
*Merryweather, George:
    Faraday writes to: 2401, 2404.
    Tempest Prognosticator: 2401.
    Writes to Faraday: 2403.
Meryon, Edward: Faraday writes to: 2745.
Mesmerism: 2330, 2724.
Meteorology: 2730.
*Meyerbeer, Giacomo: 2680, 2991, 3009.
Michall: 2861.
Microscope: 2481.
Middleton, William Fowle: 2393.
Milan, School of Industrial Arts: 2447.
Miles, Mr: 2173.
Miles, Mrs: 2173.
Miles, Miss: Faraday writes to: 2173.
Miller, Rev J.: Faraday writes to: 2175.
Miller, William Allen: 2629.
    Writes to Faraday: 2980.
Milman, Henry Hart:
    Faraday writes to: 2584.
    Writes to Faraday: 2583.
*Milne-Edwards, Henri: 2235, 2240, 2246, 2474.
Milne-Edwards, Laure: 2474.

Mining safety lamp: 2443.
Moigno, François Napoleon Marie: 3031.
    Writes to Faraday: 2869.
Monk, William: 2898.
Monpriut: 2572.
Monticelli, Teodoro: 2255.
*Moore, Harriet Jane:
    Faraday writes to: 2248, 2318, 2740.
    Writes to Faraday: 2314, 2456, 2734.
Moore, Julia: 2314, 2734.
Morris, James: 2741.
    Faraday writes to: 2222.
Morris and Prevost: 3012.
Morse, Samuel Finlay Breese: Writes to Faraday: 2904.
Morson, Thomas Newborn Robert: Faraday writes to: 2664.
Motor: 2812.
Müller, Johannes Peter: 2647.
Murchison, Roderick Impey: Writes to Faraday: 2492.
*Mure, William:
    Faraday writes to: 2670, 2685.
    Writes to Faraday: 2684.
Murray, Miss: Faraday writes to: 2488.
Musschenbroek, Petrus van: 2415.
Mussy Henri Guéneau de: 2913.
Naples Academy: 2255.
Napoleon Bonaparte: 2188.
Nasmyth, James: 2897.
Nassau: 2765.
Nautical Almanac: 2768.
New Quarterly Review: 2674.
New Zealand bird: 2506.
Newcastle fire: 2910, 2977.
*Newman, John: 2208, 2354, 2612, 2691, 2745.
Newsham, Richard: 2298, 2710.
*Newton, Isaac: 2234, 2634, 2678.
    Faraday compared with: 2415, 2861, 3014.
    Gravity: 2631, 2633.
Nicol's prism: 2481.
Niepce de Saint-Victor, Claude M.F.: 2184, 2191.
*Nobert, Friedrich Adolph: 2407.
    Diffraction gratings: 2205, 2269, 2273, 2406.
Nobili, Leopoldo:
    Chromatic scale: 2870.
Noble, Matthew: 2964.

*Northumberland, Algernon Percy, 4th Duke of: 2380, 2492, 2541, 2545, 2624.
Northumberland, Eleanor, Duchess of: Faraday writes to: 2508.
Northumberland House: 2508.
Novello: Faraday writes to: 2417.
Observer: 2215.
Odescalchi, Pietro: Writes to Faraday: 2345.
Odling, William: 2772.
*Oersted, Hans Christian: 2372, 2634.
    Apparatus: 2251.
    Death of: 2396.
    Faraday writes to: 2268.
    Writes to Faraday: 2245.
Orr, Andrew: Faraday writes to: 3015.
O'Shaughnessy, William Brooke: 2651.
Opera: 2835, 2991, 3009.
    Faraday unable to attend: 2680.
Owen, Caroline: 2381.
*Owen, Richard: 2158, 2380, 2571, 2717.
    Faraday writes to: 2253, 2381.
Oxley, John: Writes to Edinburgh Glasite Church: 2931, 2935.
Ozone: 2274, 2287, 2311, 2343, 2353, 2356, 2388, 2413, 2435, 2453, 2534, 2699, 2705, 2735, 2864.
    Liebig's view of: 2735, 2781, 2818, 2859, 2899.
    Ozonometer: 2353, 2356, 2534, 2747, 2749, 2823, 2824, 2830, 2831, 2899.
Page, Charles Grafton: 2321.
Paget, James: 2804.
Painting: 2184, 2836, 2837.
*Pakington, John Somerset: 2568.
    Writes to Faraday: 2559, 2567.
*Palagi, Alessandro: 2799, 2834.
    Faraday writes to: 2967.
    Writes to Faraday: 2945, 2959.
Palmerston, Henry John Temple, 3rd Viscount: 2484, 2977.
*Paradise, William: 2407, 2450, 2703.
    Writes to Edinburgh Glasite Church: 2931, 2935.
Parafin candle: 2725.
Parker, John William: 2409.
    Faraday writes to: 2842.
Parker, Theodore: 2701.
Parker, William: 2871.
Parliament: 2153.
    House of Commons:

            Select Committees:
                        Lighthouses: 2656.
                        National Gallery: 2670, 2684, 2685.
                        New House of Commons: 2304.
Pasteur, Louis: 2376.
Paxton, Joseph: 2473.
Pellatt, Apsley: Faraday writes to: 2165.
Pelletier, Pierre-Joseph: 2401.
*Pelly, John Henry: 2204, 2319, 2600.
Peltier, Ferdinand Athanase: 2263.
*Peltier, Jean Charles Athanase: 2263, 2396, 2401, 2558.
Pepys, John: 2525.
*Percy, Grace: 2219, 2221, 2419.
*Percy, John: 2194, 2196, 2219, 2419, 2706, 2710, 2717.
    Faraday writes to: 2218, 2221.
    Writes to Faraday: 2483.
Perkins, Jacob: 2236, 2252.
Perpetual motion: 2258.
Perry, George: *Conchology*: 2329.
Petřina, Franz Adam: Writes to Faraday: 2213.
*Phillipps, Thomas:
    Faraday writes to: 2270, 2988, 2993.
    Writes to Faraday: 2262, 2267, 2987, 2992.
Phillips: 2727.
Phillips, Ann: 2351, 2352.
*Phillips, John: 2218.
    Faraday writes to: 2352, 2359.
    Writes to Faraday: 2351, 2357.
Philosophical Magazine:
    Baddeley: 2339.
    Breda and Logeman: 2915.
    Draper: 2617.
    Faraday: 2347, 2364, 2375, 2536, 2636, 2715, 2802, 2870, 2902, 2939, 2955, 2964, 2965, 2966.
    Feilitzsch: 2874.
    Francis:
        As editor: 2451, 2636.
    Joule: 2392.
    Melloni: 2813.
    Plateau: 2241.
    Plücker: 2185, 2189.
    Quetelet: 2397, 2412.
    Schoenbein: 2453, 2899.
    Stevenson: 2667.

Tyndall: 2411, 2460.
    As editor: 2915, 2986.
Phinn, Thomas: Faraday writes to: 3019.
Photography: 2184, 2191, 2331, 2416, 2430, 2852.
    Coloured: 2772.
    Flash: 2437.
    Of magnetic observations: 2535, 2536.
Photolysis: 2790, 2864.
Pickersgill, Henry William:
    Daughter: 2477.
    Writes to Faraday: 2477.
Picture conservation: 2684, 2685.
Piddington, Henry: 2339, 2484.
Pincoffs, Peter: 2323.
Pius IX, Pope: 2345.
Plana, Giovanni Antonio Amedeo: Writes to Faraday: 2700.
Plantamour, Emile: 2994.
*Plateau, Joseph Antione Ferdinand:
    Treatise on physics: 2490, 2936.
    Writes to Faraday: 2164, 2241, 2305, 2490, 2936.
*Playfair, Lyon: 2384, 2708.
    Faraday writes to: 2622, 2737, 3008, 3032.
    Writes to Faraday: 2405, 2621, 2755, 2963, 3007, 3031.
*Plücker, Julius: 2217, 2350, 2427, 2451, 2632.
    Faraday writes to: 2185, 2189, 2239, 2250, 2346, 2449, 2780, 2901.
    Writes to Faraday: 2152, 2174, 2183, 2190, 2214, 2237, 2249, 2328, 2439, 2634, 2789, 2884.
*Poggendorff, Johann Christian:
    Annalen: 2813.
        Faraday: 2364.
        Haecker: 2278.
        Plücker: 2174, 2183, 2237, 2250, 2328, 2439, 2634, 2789, 2884.
        Schoenbein: 2893, 2985.
    Diamagnetism: 2350.
    Tyndall: 2451.
*Poisson, Siméon-Denis: 2379, 2415, 2955, 3018.
Polli, Giovanni: 2447.
*Pollock, Juliet: 2663, 2872, 2873.
*Pollock, William Frederick:
    Faraday writes to: 2663, 2873.
    Writes to Faraday: 2872.
Polytechnic Institution: 2531, 2869.
Pooley, Charles: Faraday writes to: 2909.
Portlock, Joseph Ellison: Faraday writes to: 2478, 2495.

Portraits: 2393.
    Of Euler: 2441, 2453.
    Of Faraday: 2396, 2412, 2652.
        By Maguire: 2433.
        By Noble: 2964.
        By Richmond: 2504, 2523.
    Of Henry: 2416.
    Of Quetelet: 2323, 2347, 2412.
Post Office: 2965, 2967.
Potter, John: Faraday writes to: 2530.
Potter, Richard: 2234.
Pouillet, Claude-Servais-Mathias: 2870.
Power, Alexander Bath: Writes to Faraday: 2695.
Prague, Charles University: 2213.
Pratt: 2450.
Presler: 2608.
Prevost, Alexandre Pierre: 2969, 3012.
Prevost, Jean-Louis: 2969.
Prichard, James Henry: 2173.
Priestley, Joseph: 2232.
Proctor, Thomas: Writes to Edinburgh Glasite Church: 2931, 2935.
Proctor, William: Writes to Edinburgh Glasite Church: 2935.
Prout, William: 2435.
Prussia: 2255.
    Order of Merit: 2146, 2706.
    Political unrest: 2152.
Queenwood College: 2451, 2707, 3020.
*Quetelet, Lambert-Adolphe-Jacques: 2397, 2401, 2403, 2558, 2936.
    Faraday writes to: 2263, 2347, 2412.
    Writes to Faraday: 2264, 2323, 2396, 2455.
Railways: 2289, 2430, 2559, 2706, 2818, 3004.
    Holyhead: 2696.
    Waterloo Station: 2845.
Raith, George: Writes to Edinburgh Glasite Church: 2935.
*Ransome, George: 2393.
    Faraday writes to: 2442.
    Writes to Faraday: 2424, 2433.
Rathbone, Hannah Mary: 2867, 2896.
Rathbone, Richard: 2867.
    Faraday writes to: 2896.
Rauscher, Josef Othmar von: 2889.
*Rees, Richard van: 2350, 2921, 2944.
Reeve, Lovell Augustus: Writes to Faraday: 2708, 2713.

*Regnault, Henri Victor: 2199, 2251, 2254, 2398, 2654, 2693, 2955, 2969, 3004.
    Work on animal respiration: 2763.
*Reich, Ferdinand: 2217, 2250, 2268, 2350.
Reid, Andrew: Faraday writes to: 2459.
*Reid, David: 2450.
    Writes to Edinburgh Glasite Church: 2931, 2935.
*Reid, Edward Ker:
    Writes to Dundee Glasite Church: 2918.
    Writes to Edinburgh Glasite Church: 2931, 2935.
*Reid, Elizabeth: 2462, 2703.
    Faraday writes to: 2683.
*Reid, Margery Ann: 2171, 2219, 2503, 2703, 2819, 2964, 3021.
Reid, William Ker: 2703.
Respiration: 2398.
Revolving mirror: 2612, 2616, 2629.
Revue des deux mondes: 2775.
Reynolds, Hannah Mary: 2867.
Reynolds, William: 2896.
    Faraday writes to: 2867.
Richmond, George:
    Faraday writes to: 2523.
    Writes to Faraday: 2504.
Ridout, John: Faraday writes to: 2420.
*Riess, Peter Theophilus:
    Faraday writes to: 2668, 2966.
    Writes to Faraday: 2955, 3014.
Roberton, John David: 2301.
Robinson, Thomas Romney: 2592.
    Faraday writes to: 2289.
Robison, John: 2415.
Roget, Peter Mark: 2631.
Roman Catholics: 2889, 3000.
Ronalds, Francis: 2856.
    Faraday writes to: 2331.
*Rose, Heinrich: 2439, 2706.
    Faraday writes to: 2458.
    Fusible metal: 2254.
Ross, Andrew: 2481.
*Rosse, William Parsons, 3rd Earl of: 2257, 2434, 2600, 2795.
    Faraday writes to: 2302, 2540.
    Writes to Faraday: 2694.
Rossini, Gioacchino Antonio: 2835.
Rowcroft, Charles: Faraday writes to: 2293.

Royal Academy of Sciences, Turin: 2700.
Royal Astronomical Society: 2227, 2257, 2586, 2587, 2591.
Royal College of Chemistry: 2156, 2710.
Royal College of Surgeons: 2158, 2491.
Royal Geographical Society: 2492.
    Dinner: 2680.
Royal Greenwich Observatory: 2535, 2538.
Royal Institution: 2365.
    Annual Meeting: 2410.
    Brodie's views on: 2243.
    Building works: 2542, 2710.
        Painting of: 2215, 2298.
    Faraday: 2750, 2764.
        Appointment to: 2188.
        Lectures at: 2303, 2764
        Support of research: 2398.
    Journal of: 2631.
    Laboratory:
        Apparatus in:
            Great magnet: 2148.
            Hofmann borrows: 2933.
        Experiments in: 2647.
            Talbot: 2437.
        Instruments purchased by Tyndall: 2698, 2704, 2711.
    Lectures at:
        By Buckland: 2253.
        By Davy: 2187.
        By Faraday:
            Domestic chemical philosophy: 2290, 2291, 2292, 2309, 2314.
            Electrical Philosophy: 2420.
            Non-metallic elements: 2534, 2573.
            Static electricity: 2172, 2175, 2176, 2182, 2198, 2208.
                Brande offers to undertake: 2178.
            Static electricity: 2626, 2651, 2664, 2666, 2870.
        By Forbes: 2941.
        By Frankland: 2683.
        By Tyndall: 2683, 2698, 2706.
        By Williamson: 2532.
    Christmas:
        By Faraday:
            Chemical History of a Candle: 2147, 2291, 2292.
            Chemistry: 2602, 2604.
            Chemistry of combustion: 2921, 2924, 2928, 2929.

Voltaic electricity: 2717, 2764.
Friday Evening Discourses: 2175, 2829.
   Admission of lady members: 2154, 2155, 2371, 2380.
   Audience for: 2474, 2596.
   By Airy: 2223, 2227, 2257, 2259, 2410, 2485, 2642, 2917, 2919, 2920, 2924, 2925, 2926.
   By Brockedon: 2756.
   By Carpenter: 2679, 2724.
   By Conolly: 2787.
   By Faraday: 2965.
      Atmospheric electricity: 2263, 2401.
      Atmospheric magnetism: 2318, 2353, 2407, 2413.
      Boussingault et al. on oxygen: 2683, 2685, 2705.
      Electric conduction: 2954, 2986.
      Electric induction: 2776, 2780, 2781, 2782, 2791, 2792, 2793, 2794, 2797, 2799, 2832, 2834, 2936.
      Freezing water: 2832.
      Instead of Buckland: 2253.
      Lines of magnetic force: 2498, 2634.
      Magnetic force: 2596, 2615, 2624, 2625, 2626, 2630, 2632, 2636, 2646.
      Magnetic hypotheses: 2852, 2939.
      Magnetic philosophy: 2924, 2939, 2944, 2964.
      Magnetism of oxygen and nitrogen: 2318, 2353, 2371.
      Ozone: 2274, 2287, 2311, 2343, 2353, 2356, 2388, 2413, 2435, 2436, 2441, 2453.
      Physical lines of magnetic force: 2538.
      Polarity of Bismuth: 2150.
      Preserving plants: 2473.
      Ruhmkorff's apparatus: 2994.
      Trevelyan's experiments on sound: 2704.
   By Grove: 2253, 2376, 2717.
   By Gull: 2394.
   By Hosking: 2425.
   By Lyell: 2510.
   By Russell: 2467.
   By Owen: 2380, 2381, 2717.
   By Sidney: 2182.
   By Stokes: 2601, 2612, 2616, 2637, 2644.
   By Story-Maskelyne: 2376, 2400.
   By Taylor: 2447.
   By Tyndall: 2636, 2638, 2639, 2704, 2707, 2717, 2990, 2994.
   By Ward: 2857.

Expenses: 2638, 2639, 2642, 2644.
Not given:
    By Airy: 2717, 2760, 2766.
    By Buckland: 2253.
    By Faraday: 2474, 2493.
    By Hawkins: 2717.
    By Sabine: 2596.
    By Whewell: 2225, 2226.
Regulations: 2410, 2414.
Tickets for: 2150, 2262, 2267, 2270, 2371, 2394, 2395, 2425, 2626, 2683, 2990.
Laboratory: 2206, 2207, 2209, 2210, 2242, 2247.
On education: 2797, 2804, 2806, 2814.
    By Faraday: 2808, 2819.
        Text of: 2842, 2853, 2909.
    Committee: 2808.
Not given:
    Hooker: 2788.
    Percy: 2710, 2717.
To Prince Albert: 2156, 2157, 2853.
Lecture theatre: 2416, 2430.
    Used by other societies: 2492.
Letter read to: 2648.
Library: 2179, 2202, 2203, 2255, 2329, 2625, 2781, 2864, 2966.
    Committee: 2338.
    Exhibitions in: 2277, 2852.
    Faraday gives notebooks to: 2932.
Managers of: 2240, 2503, 2548, 2612, 2788, 2932.
    Brande's retirement: 2525.
        Replacement for: 2532.
    Brodie: 2206, 2207, 2209, 2210, 2215, 2242, 2243.
    Building work: 2298, 2542.
    Discourse regulations: 2410, 2414.
        Lady members: 2154, 2155.
    Lectures on education: 2797.
    Outside bookings: 2492.
Members:
    Lady:
        Herschel: 2361.
    Melloni: 2255.
    Monticelli: 2255.
    Plücker: 2214.
    Sementini: 2255.
Members Meeting: 2524, 2548, 2790.

Proceedings of the: 2389, 2453.
Prosperity of: 2210.
Season of: 2256, 2354, 2407.
Royal Military Academy: 2235.
    Abel appointed: 2502, 2519.
    Examination: 2478.
    Faraday at: 2158, 2303.
        Retires from: 2495, 2502, 2519.
        Taking up as much time as he can manage: 2478.
Royal Mint: 2391, 2392, 2516, 2525.
Royal Society: 2415, 2441, 2571, 2636, 2723, 2762.
    Bakerian Lecture: 2363, 2364.
    Council: 2301.
    Excise Committee: 2301, 2922, 2923.
    Elections to:
        Brayley: 2798.
        De La Rue: 2273.
        Foreign Members: 2968.
        Frankland: 2621, 2622.
        Hofmann: 2390.
        Tyndall: 2476.
        Wöhler: 2876, 2877, 2892.
    Faraday writes to: 2260, 2457.
    Grant:
        To De La Rue: 2768.
        To Tyndall: 2861.
    Library: 2256.
    Medals of: 2968.
        Copley: 2302.
        Rumford: 2255.
    Membership list: 2146.
    Philosophical Transactions: 2334, 2798, 2821, 2865.
        Papers by:
            Brougham: 2717.
            Christie: 2415.
            Gassiot: 2986.
            Grove: 2512.
            Harris: 2415, 2457, 2489.
            Joule: 2192, 2193, 3002.
            Matteucci: 2260.
            Nobert: 2406.
                Translation of: 2407.
            Sabine: 2377, 2423, 2591, 2613.
            Scoresby: 2391.

        Stevenson: 2649.
        Ward: 2217.
        Wolf: 2586, 2587, 2592.
        Refereeing: 2217, 2260, 2377, 2423, 2457, 2489, 2512, 2717.
    Papers sent to: 2164, 2906.
Royal Society of Arts and Sciences of Mauritius: 2222, 2741.
Royal Society of Copenhagen: 2245.
Royal Society of Edinburgh: 2251, 2813, 2817.
    Transactions: 2415.
Royal Society of Literature: 2503.
Royal Yacht: 2838, 2839.
Royle, John Forbes: Writes to Faraday: 2645.
*Rühmkorff, Heinrich Daniel: 2455, 2777, 2903, 2911.
Russell, Lord John: 2788.
Russell, John Scott: 2467.
    Faraday writes to: 2261.
Russia:
    Anglo-French war against: 2735, 2943, 2965.
        Prefabricated hospital: 2947, 2948.
        Proposed attack on Cronstadt: 2871, 2882, 2910.
        Sebastopol: 3000, 3027.
            As possible cause of gales: 3001.
    Napoleon I's campaign in: 2188.
*Sabine, Edward: 2550.
    Faraday writes to: 2326, 2354, 2596, 2615, 2620, 2922.
    Royal Society: 2331, 2334, 2377, 2423, 2476, 2591, 2595, 2613, 2649, 2861, 2922, 2923.
    Terrestrial magnetism: 2325, 2326, 2354, 2377, 2423, 2538, 2591, 2592, 2613, 2615, 2618, 2620, 2646, 2649, 2659, 2667, 2883.
    Writes to Faraday: 2325, 2592, 2613, 2618, 2923, 2994.
Sabine, Elizabeth Juliana: 2326, 2592.
St Andrews Literary and Philosophical Society: 2648.
Sandby, George: 2724.
Sandemanian Church: 2889, 2890.
    Abstinence from blood: 2918, 2931, 2935.
    Chesterfield: 2931, 2935.
    Dundee: 2918, 2931, 3024.
    Edinburgh: 2931, 2935.
    London: 2819, 2918, 2931, 2935.
        Faraday's possible second exclusion: 2335, 2336, 2337, 2340.
        Wednesday meeting: 2676.
    Newcastle: 2450, 2931, 2935.
    Old Buckenham: 2549, 2931, 2935.
Sarasin: 2441.

Sarasin, Felix: 2274, 2287.
Saturn: 2768.
Saussure, Horace Bénédict de: 2313.
Savage, Sarah: 2890.
Savart, Félix: 2164.
Scheiner, Christoph: 2586.
*Schlagintweit, Adolph: 2313, 2396, 2889.
*Schlagintweit, Hermann Rudolph Alfred: 2313, 2396, 2889.
*Schoenbein, Berta:
    Schoenbein mentions: 2526, 2578, 2607, 2735, 2790, 2832, 2828, 2864, 2943.
*Schoenbein, Christian Friedrich: 2162, 2782, 2859, 2893.
    Faraday writes to: 2287, 2343, 2353, 2356, 2388, 2413, 2453, 2482, 2534, 2604, 2705, 2781, 2832, 2899, 2964.
    Journey down the Danube: 2735, 2943.
    List of letters exchanged: 2832.
    Oxygen: 2526, 2578, 2607, 2790, 2864, 2985.
    Ozone: 2348, 2534, 2607, 2699, 2705, 2781, 2832, 2899.
        Cause of influenza: 2435.
        Faraday's Discourse on: 2274, 2287, 2311, 2343, 2353, 2356, 2388, 2413, 2435, 2441, 2453.
        Ozonometer: 2747, 2749.
    Writes to Faraday: 2274, 2348, 2441, 2526, 2562, 2578, 2607, 2699, 2735, 2790, 2818, 2828, 2864, 2943, 2985.
*Schoenbein, Emilie:
    Faraday mentions: 2781, 2782, 2832.
    Schoenbein mentions: 2526, 2578, 2607, 2735, 2790, 2832, 2828, 2864, 2943.
*Schoenbein, Emilie Wilhelmine Luise:
    Faraday mentions: 2287, 2343, 2413, 2453, 2482, 2534, 2705, 2781, 2832, 2864, 2964.
    Schoenbein mentions: 2526, 2578, 2607, 2735, 2790, 2832, 2828, 2943.
*Schoenbein, Fanny Anna Franziska:
    Schoenbein mentions: 2526, 2578, 2607, 2735, 2790, 2832, 2828, 2864, 2943.
*Schoenbein, Wilhelmine Sophie:
    Schoenbein mentions: 2526, 2578, 2607, 2735, 2790, 2832, 2828, 2864, 2943.
Schönlein Johann Lucas: 2946.
Schrötter, Anton: Writes to Faraday: 2719.
Schumacher, Heinrich Christian: 2396.
*Schwabe, Samuel Heinrich: 2554, 2586, 2591, 2592, 2646.
Schweitzer: 2985.
Scoffern, John: Faraday writes to: 2573.

Scoresby, William:
    Faraday writes to: 2392.
    Writes to Faraday: 2391.
Scorpa, J.: 2879.
Scott, Alexander John: 2769.
Scott, Walter: 3026.
Sea serpent: 2571.
Sementini, Luigi: 2255.
Seymour, Edward Adolphus: Writes to Faraday: 2365.
Shaftesbury, Anthony Ashley Cooper, (from 1851) 7th Earl of:
    Faraday writes to: 2928.
    Writes to Faraday: 2929.
Shaw, John Hope: Faraday writes to: 2764.
Shelley, Jane: 2582.
Shelley, Percy Bysshe: 2582.
*Shepherd, John: 2561, 2565, 2580, 2588, 2590, 2593.
Shuckard, T.L.: Faraday writes to: 2162.
Sidney, Edwin: 2182, 2534.
Silbermann, Johann Theobald: 2716.
Silliman, Benjamin: 2400.
    Faraday writes to: 2402.
Silliman, Benjamin: 2400.
Silliman, Susan: 2400.
Silvering: 2694.
Sims, Robert: 2244.
    Writes to Dundee Glasite Church: 2918.
Smart, Benjamin Humphrey: Writes to Faraday: 3016, 3017.
Smith: Faraday writes to: 2324.
Smith, Albert Richard: Writes to Faraday: 2486.
Smith, George: Writes to Edinburgh Glasite Church: 2931, 2935.
Smith, R.: Writes to Faraday: 2374.
Smithson, James Louis Macie: 2430, 2448.
Smithsonian Institution: 2430, 2448.
    Contributions: 2564.
Smyth, Mrs: Faraday writes to: 2425.
Soap bubbles: 2147.
Society for the Relief of Widows and Orphans of Medical Men: 2581.
Society of Arts: 2159, 2160, 2275, 2517, 2605, 2861, 2989.
Solly, Edward: Faraday writes to: 2605.
Somerville, Mary: Writes to Faraday: 2653.
Sopwith, Thomas: Writes to Faraday: 2795.
Soret, Jacques Louis: 2969.
Sound: 2704, 2707.
Spence, William: Writes to Faraday: 2393, 2419.

Spence, William Blundell: 2393.
Spencer, Frederick, 4th Earl: 2682.
Spottiswood: 2932.
Staite, William Edwards: Writes to Faraday: 2732.
Stark, William: Writes to Edinburgh Glasite Church: 2935.
Stas, Jean-Servais: 2455.
Statham, Samuel: 2746, 2794.
Steam boat: 2430.
Stehlin: 2828, 2899.
Stenhouse, John: 2276.
Stephenson, Robert: Writes to Faraday: 2696.
Stevelly, John: 3026.
Stevens, Henry: Faraday writes to: 2564, 2912.
Stevenson, Alan: 2593.
*Stevenson, William:
    Faraday writes to: 2649.
    Writes to Faraday: 2646, 2648, 2659, 2667.
*Stokes, George Gabriel: 2604, 2607, 2620, 2629, 2832.
    Visits Royal Institution: 2601, 2612.
    Writes to Faraday: 2601, 2612, 2616, 2617, 2637, 2644, 2851, 2927.
Stoppard, William: 2153.
Story-Maskelyne, Mervyn Herbert Nevil: Writes to Faraday: 2376.
Strahan, Paul, Paul and Bates: 2748.
Struve, Otto Wilhelm: 2768.
Sugar extraction: 2184, 2191, 2198.
Sun: 2485.
    Eclipse of: 2485, 2487.
    Meteorological relations: 2586.
    Spectra: 2772.
    Sunspot cycle: 2554.
        Relations with terrestrial magnetism: 2560, 2571, 2586, 2587, 2591, 2620, 2646.
            Herschel on: 2591.
            Sabine on: 2591, 2592, 2596, 2613.
Susan: 2819.
Svanberg, Adolph Ferdinand: Faraday writes to: 2316.
Swedenborg, Emanuel: 2701.
Swiss Naturalists Association: 2311, 2441, 2453, 2864.
Switzerland:
    Political situation: 2571.
*Sykes, William Henry: 2557, 2576.
    Faraday writes to: 2555.
Table moving: 2674, 2679, 2688, 2712, 2724, 2748, 2775, 2816, 2906.
    Cox: 2971, 2972, 2973.

        Detailed descriptions:
            By Allen, H.: 2752.
            By Allen, J.: 2675.
            By Hickson: 2677, 2678.
        Electricity: 2674, 2686, 2691, 2724.
        Electro-biology: 2677.
        Faraday: 2703, 2705, 2797, 2819.
            Experiments: 2691.
            Letter to *Times*: 2691, 2703, 2748.
                Opposed by:
                    Espie: 2701.
                Rumoured to have been withdrawn: 2769, 2770.
                Supported by:
                    Andrews: 2697.
                    Herschel: 2693.
                    Power: 2695.
                    Tyndall: 2698.
        Magnetism: 2674, 2677, 2691, 2701, 2724.
        Odylic influence: 2724.
        Quasi involuntary muscular action: 2691.
        Spirits: 2677.
        Tellurian circuit: 2677.
*Talbot, William Henry Fox: 2759.
    Faraday writes to: 2438.
    Writes to Faraday: 2437.
Tatum, John: 2188.
*Taylor, Alfred Swaine:
    Faraday writes to: 2422.
    Writes to Faraday: 2421, 2977.
Taylor, Brook: 2415.
Taylor, Jeremy: 2497, 2498.
*Taylor, Richard: 2185, 2190, 2241, 2287, 2343, 2387, 2490, 2837.
    Faraday writes to: 2360, 2397.
    Scientific Memoirs: 2305, 2490, 2604, 2936.
Taylor, Tom: Writes to Faraday: 2447.
Tennent, James Emerson:
    Faraday writes to: 2996.
    Writes to Faraday: 2995.
Thenard, Louis Jacques: 2864, 3004.
Thermolysis: 2790, 2864.
Thermoscope: 2834.
Thermometer: 2856.
Thompson, L.: Faraday writes to: 2197.
Thomson: 2450.

Thomson, James: 2169.
Thomson, James: 2251, 2254.
Thomson, Thomas: 2788.
*Thomson, William: 2354, 2489, 2921, 2944, 3026.
    Faraday meets: 2211.
    Faraday writes to: 2252.
    Writes to Faraday: 2169, 2211, 2251, 2254, 2546.
Thorbecke, Johan Rudolf: Writes to Faraday: 2470.
Thynne, John: 2689.
Timber, preservation of: 2181.
Times: 2473, 2571, 2693, 2695, 2697, 2703, 2748, 3001.
    Faraday writes to: 2691, 3003.
Tite, William: 2923.
Titus: 2421.
Todd, Robert Bentley: 2173.
Tomlinson, Charles: Faraday writes to: 2666.
Toronto University: 2451, 2452, 2460, 2463, 2468.
*Towler, George:
    Faraday writes to: 2702, 2709.
    Writes to Faraday: 2258.
Townshend, Chauncey Hare: 2724.
Toynbee, Joseph: Writes to Faraday: 2469, 2833.
Tozer, J.S.: 2500
    Faraday writes to: 2501.
Trajan column: 2300.
Trevelyan, Arthur: 2704, 2707.
Trinity House: 2147.
    Faraday visits: 2163, 2608.
    Faraday writes to: 2878.
Tronchin, Jeanne-Adèle: 2577, 2786.
Tronchin, Louis-Nosky-Rémy: 2786.
Tupper, Arthur Chilver: 2151.
Tupper, Martin Farquhar: 2151.
*Twining, Thomas: 2964, 2965.
    Faraday writes to: 2201, 2224, 2272, 2765, 2845.
Twining, Victorine: 2765.
Tyndale, John William Ware: 2868.
    Writes to Faraday: 2866.
*Tyndall, John: 2489, 2546, 2636, 2884, 2890, 2901, 2914, 2986, 3012.
    British Association:
        1851 Ipswich: 2454, 2463.
        1855 Glasgow: 3026.
    Faraday writes to: 2308, 2344, 2411, 2452, 2468, 2632, 2639, 2672, 2706, 2711, 2859, 2915, 2921, 2932, 3022, 3027.

Galway chair: 2550.
Magnetism: 2328, 2333, 2379, 2451, 2550, 2630, 2632, 2858.
Philosophical Magazine: 2915.
Queenwood College: 2451, 2707, 3020.
Royal Institution:
    Lectures: 2683, 2804.
        Discourses: 2636, 2638, 2639, 2717, 2990, 2994.
    Purchase of apparatus: 2698, 2704, 2711.
Royal Society: 2476, 3002.
Society is the enemy of work: 2704.
Toronto chair: 2451, 2452, 2454, 2460, 2463.
    Testimonials: 2468.
Visits:
    France: 3000, 3004.
    Lake District: 3026.
    Writes to Faraday: 2333, 2379, 2427, 2451, 2454, 2476, 2550, 2630, 2638, 2698, 2704, 2707, 2858, 2861, 2974, 3000, 3004, 3020, 3023, 3026.
Tyrrell, Walter: 3020.
Unidentified correspondents:
    Faraday writes to: 2387, 2682.
    Sarah Faraday writes to: 2623.
    Writes to Faraday: 2816.
Van der Pant, D.F.: Writes to Faraday: 2461.
Van Wetteren: 2383.
Venables, T.E.: Faraday writes to: 2212.
Verdet, Marcel Emile: 2861, 2914.
Vernon, George Venables: Faraday writes to: 2558.
*Victoria, Queen: 2156, 2306, 2517, 2838.
Vilar, Gaston de: Writes to Faraday: 2739.
Vilar, Marie de: Writes to Faraday: 2739.
*Vincent, Benjamin: 2215, 2315, 2710, 2781, 2829, 2852, 2890.
    Faraday writes to: 2309, 2312, 2450, 2548.
    Sarah Faraday writes to: 2549.
    Writes to Dundee Glasite Church: 2918.
    Writes to Edinburgh Glasite Church: 2931, 2935.
Vincent, Janet Young: 2312, 2549.
    Sister: 2312.
Vincent, Thomas: 2548.
*Vincent, William R.:
    Writes to Dundee Glasite Church: 2918.
    Writes to Edinburgh Glasite Church: 2931, 2935.
Vivian, Edward: Faraday writes to: 2730.
*Volta, Alessandro Giuseppe Antonio Anastasio: 2415, 2813, 2834, 2984.
*Volpicelli, Paolo: Writes to Faraday: 2300, 2345, 2949.

Vrolik, Willem:
    Faraday writes to: 2186.
    Writes to Faraday: 2180.
*Vulliamy, Lewis: 2416, 2430, 2448, 2542.
Wagenmann, Paul: 2725.
Walker, Charles Vincent: 2651, 2953.
*Walker, James: 2231, 2574, 2725, 2738, 2778, 2907, 2995.
    Faraday writes to: 2290, 2690.
    Writes to Faraday: 2905.
Walker, Robert: 2745, 2994.
Waller, Augustus Volney: 2634.
Wallich, Nathaniel: Writes to Faraday: 2282, 2445.
Ward, Frederick Oldfield: Faraday writes to: 2815.
Ward, Nathaniel Bagshaw: Writes to Faraday: 2473.
Ward, Stephen Henry: Faraday writes to: 2857.
Ward, William Sykes: 2217.
Warington, Robert: 2265.
    Faraday writes to: 2669.
Warner, Samuel Alfred: 2544.
*Warren, Samuel:
    Faraday writes to: 2625.
    Writes about Faraday: 2624.
    Writes to Faraday: 2624, 2635.
Wartmann, Elie François: 3004.
Water: 2426.
    Bath: 2671.
    Composition of: 2358.
    Freezing point of: 2251, 2252, 2254.
    Physical properties of: 2796.
*Waterhouse, George Robert: 2722, 2723, 2728, 2736.
    Faraday writes to: 2729.
    Writes to Faraday: 2721.
Waterstone: 2785.
Waterworks: 2245.
Watson, Arthur: 2163.
    Faraday writes to: 2166.
*Watson, Joseph John William: 2599, 2606, 2687, 2866, 2878.
    Faraday writes to: 2608, 2619, 2895.
    Writes to Faraday: 2609, 2614, 2868, 2894, 2900.
Way, Albert: Faraday writes to: 2650, 2658.
*Weber, Wilhelm Eduard: 2217, 2239, 2249, 2250, 2268, 2302, 2350.
Webster, John: 2571.
Weekes, William Henry: 2277.
Weichold, Guido: 2934.

*Weld, Charles Richard: Faraday writes to: 2146, 2193, 2217, 2301, 2334, 2877, 3002.
Wellington, Arthur Wellesley, 1st Duke of:
    Funeral of: 2583.
        Faraday declines invitation to: 2584.
Wells, G.: 2673.
Welsh, John: Faraday writes to: 2856.
Westminster Review: 2679.
*Wheatstone, Charles: 2164, 2214, 2264, 2305, 2490, 2616, 2631, 2789, 2836, 2936, 3004.
    Faraday writes to: 2522.
    Writes to Faraday: 2629.
*Whewell, William: 2326, 3018, 3026.
    Delivers letter for Schoenbein: 2562, 2578, 2604.
    Faraday, *Experimental Researches*: 2981, 2982.
    Faraday missed seeing: 2317, 2320.
    Faraday writes to: 2225, 2310, 2317, 2322, 2327, 2496, 2498, 2521, 2537, 2631, 2797, 2804, 2808, 2814, 2942, 2982.
    Friday Evening Discourses: 2225, 2226.
    Lecture on education: 2797, 2804, 2806, 2808, 2814.
    Magnetism: 2327, 2537, 2631, 2633, 2939, 2942.
        New words for:
            Paramagnetic: 2310, 2311, 2317, 2320, 2322:
            Sphondyloid: 2494, 2496, 2496, 2498, 2536.
    Writes to Faraday: 2226, 2311, 2320, 2494, 2497, 2536, 2633, 2806, 2939, 2981.
Whisky: 2394, 2395.
Whiston, William: 2415.
*Whitelaw, George: 2548, 2549.
    Writes to Dundee Glasite Church: 2918, 2932.
    Writes to Edinburgh Glasite Church: 2931, 2935.
Whitworth, Joseph: 2392.
*Wiedemann, Gustav Heinrich: 2790, 2943, 2964, 2985.
Wilcke, Johan Carl: 2865.
Wildy: 2990.
*Wilkins, William Crane: 2172, 2220, 2234, 2238, 2566, 2580, 2590, 2593, 2600, 2720, 2727.
    Writes to Faraday: 2585.
William: 2819.
William II: 3020.
William III: 2180, 2470.
Williams, Penry: 2685.
Williams & Norgate: 2515.
*Williamson, Alexander William: 2194, 2555, 2557.

Writes to Faraday: 2532.
Williamson, William Crawford:
    Faraday writes to: 2770.
    Writes to Faraday: 2769.
Wilson, G.G.: Writes to Faraday: 2748.
*Wilson, George: Writes to Faraday: 2276, 2435, 2941.
Windsor Castle: 2342.
*Wöhler, Friedrich: 2302, 2877, 2992.
    Faraday writes to: 2892.
    Writes to Faraday: 2876.
*Wolf, Johann Rudolf: 2571, 2587, 2591, 2592, 2595, 2613, 2615, 2646.
    Faraday writes to: 2560.
    Writes to Faraday: 2554, 2586.
Wollaston, William Hyde: 2954, 2984.
Woolwich:
    Arsenal: 2354.
    Geyser: 2706.
Woolnough, Charles W.: Faraday writes to: 2773.
Wright: 2710.
*Wrottesley, John:
    Faraday writes to: 2805.
    Writes to Faraday: 2803, 2811, 2968.
Wyatt, Matthew Cotes: Faraday writes to: 2294.
Yachts:
    'America': 2467.
Yarrell, William: 2215.
Yorke, Philip: 2861.
Young, Arthur: Faraday writes to: 2930.
Younghusband, Charles Wright: 2325, 2326.
Zantedeschi, Francesco: 2237.